Solutions Manual

Stephen Hake
John Saxon

SAXON™
PUBLISHERS

Saxon Publishers gratefully acknowledges the contributions of the following individuals in the completion of this project:

Authors: Stephen Hake, John Saxon

Editorial: Chris Braun, Matt Maloney, Brooke Butner, Brian E. Rice

Editorial Support Services: Christopher Davey, Jay Allman, Shelley Turner, Jean Van Vleck, Darlene Terry

Production: Alicia Britt, Karen Hammond, Donna Jarrel, Brenda Lopez, Adriana Maxwell, Cristi D. Whiddon

Project Management: Angela Johnson, Becky Cavnar

Printed in the United States of America

ISBN-13: 978-1-59141-328-8

ISBN-10: 1-59141-328-1

Manufacturing Code: 38 0928 22

4500844482

Solutions for

Lessons and Investigations

Lesson 1, Warm-Up

a. **60**

b. **80**

c. **87**

d. **6**

e. **24**

f. **48**

g. **5**

Problem Solving

+2 +3 +4 +5 +6 +7 +8

1, 3, 6, 10, 15, **21, 28, 36**

Lesson 1, Lesson Practice

a. **\$0.45 per glass; 45¢ per glass**

b. **0**

c. Product of 4 and 4 = 16
Sum of 4 and 4 = 8

$$8\overline{)16} = \mathbf{2};\quad 16,\ 0$$

d.
$$\begin{array}{r} \overset{1}{\$1}.75 \\ \$0.60 \\ +\ \$3.00 \\ \hline \mathbf{\$5.35} \end{array}$$

e.
$$\begin{array}{r} \$\ \overset{1}{\not 2}.\ \overset{9}{\not 0}{}^{1}0 \\ -\ \$\ 0.\ 4\ 7 \\ \hline \mathbf{\$\ 1.\ 5\ 3} \end{array}$$

f.
$$\begin{array}{r} \$0.65 \\ \times\quad 5 \\ \hline \mathbf{\$3.25} \end{array}$$

g.
$$\begin{array}{r} 250 \\ \times\ 24 \\ \hline 1000 \\ 500 \\ \hline \mathbf{6000} \end{array}$$

h. $5\overline{)\$24.00} = \mathbf{\$4.80}$
$$\begin{array}{r} 20 \\ \hline 4\ 0 \\ 4\ 0 \\ \hline 0 \end{array}$$

i. $18\overline{)234} = \mathbf{13}$
$$\begin{array}{r} 18 \\ \hline 54 \\ 54 \\ \hline 0 \end{array}$$

j.
$$\begin{array}{r} 20 \\ +\ \ 4 \\ \hline \mathbf{24} \end{array}$$

k.
$$\begin{array}{r} 20 \\ -\ \ 4 \\ \hline \mathbf{16} \end{array}$$

l.
$$\begin{array}{r} 20 \\ \times\ \ 4 \\ \hline \mathbf{80} \end{array}$$

m. $4\overline{)20} = \mathbf{5}$

Lesson 1, Mixed Practice

1. Product of 5 and 6 = 30
Sum of 5 and 6 = 11
30 − 11 = **19**

2.
$$\begin{array}{r} 8 \\ +\ 9 \\ \hline \mathbf{17} \end{array}$$

3. $4\overline{)\text{dividend}}$ with quotient 8; the dividend is **32.**

4. Product of 6 and 6 = 36
Sum of 6 and 6 = 12

$$12\overline{)36} = \mathbf{3};\quad 36,\ 0$$

5. **Addition, subtraction, multiplication, and division**

6. (a) $12 + 4 = \mathbf{16}$

(b) $12 - 4 = \mathbf{8}$

(c) $12 \cdot 4 = \mathbf{48}$

(d) $\frac{12}{4} = \mathbf{3}$

7.
$$\begin{array}{r} \$\overset{3}{\not4}{}^{1}3.\overset{6}{\not7}{}^{1}4 \\ -\ \$16.59 \\ \hline \mathbf{\$27.15} \end{array}$$

8.
$$\begin{array}{r} 64 \\ \times\ 37 \\ \hline 448 \\ 192 \\ \hline \mathbf{2368} \end{array}$$

9.
$$\begin{array}{r} 7 \\ 8 \\ 4 \\ 6 \\ 9 \\ 3 \\ 5 \\ +\ 7 \\ \hline \mathbf{49} \end{array}$$

10.
$$\begin{array}{r} \overset{1\,2}{364} \\ 52 \\ 867 \\ +\ \ 9 \\ \hline \mathbf{1292} \end{array}$$

11.
$$\begin{array}{r} \overset{3}{\not4}\overset{9}{\not0}\overset{9}{\not0}{}^{1}0 \\ -\ 3625 \\ \hline \mathbf{375} \end{array}$$

12.
$$\begin{array}{r} 316 \\ \times\ 18 \\ \hline 2528 \\ 316 \\ \hline \mathbf{5688} \end{array}$$

13.
$$\begin{array}{r} \mathbf{\$2.18} \\ 20\overline{)\$43.60} \\ 40 \\ \hline 3\ 6 \\ 2\ 0 \\ \hline 160 \\ 160 \\ \hline 0 \end{array}$$

14.
$$\begin{array}{r} 300 \\ \times\ \ 40 \\ \hline \mathbf{12{,}000} \end{array}$$

15.
$$\begin{array}{r} 12 \\ \times\ 8 \\ \hline 96 \end{array} \qquad \begin{array}{r} 96 \\ \times\ 0 \\ \hline \mathbf{0} \end{array}$$

16.
$$\begin{array}{r} \mathbf{309} \\ 12\overline{)3708} \\ 36 \\ \hline 108 \\ 108 \\ \hline 0 \end{array}$$

17.
$$\begin{array}{r} 365 \\ \times\ \ 20 \\ \hline \mathbf{7300} \end{array}$$

18.
$$\begin{array}{r} \mathbf{30\ R\,17} \\ 25\overline{)767} \\ 75 \\ \hline 17 \\ 0 \\ \hline 17 \end{array}$$

19.
$$\begin{array}{r} 30 \\ \times\ 40 \\ \hline \mathbf{1200} \end{array}$$

20.
$$\begin{array}{r} \$\overset{0}{\not1}\overset{9}{\not0}.\overset{9}{\not0}{}^{1}0 \\ -\ \ \$2.34 \\ \hline \mathbf{\$7.66} \end{array}$$

21.
$$\begin{array}{r} \overset{3}{\not4}\overset{9}{\not0}{}^{1}17 \\ -\ 3952 \\ \hline \mathbf{65} \end{array}$$

22.
$$\begin{array}{r} \$2.50 \\ \times\ \ \ 80 \\ \hline \mathbf{\$200.00} \end{array}$$

23.
$$\begin{array}{r} \$2.50 \\ \times\ \ \ 20 \\ \hline \mathbf{\$50.00} \end{array}$$

24.
$$\begin{array}{r} \mathbf{40} \\ 14\overline{)560} \\ 56 \\ \hline 0 \\ 0 \\ \hline 0 \end{array}$$

25. $\begin{array}{r} \mathbf{\$1.25} \\ 8\overline{)\$10.00} \\ \underline{8} \\ 2\,0 \\ \underline{1\,6} \\ 40 \\ \underline{40} \\ 0 \end{array}$

26. **Natural numbers**

27. **\$0.25; 25¢**

28. **All counting numbers are whole numbers.**

29. **Quotient**

30. **Minuend − subtrahend = difference**

Lesson 2, Warm-Up

a. **48**

b. **18**

c. **50**

d. **900**

e. **6000**

f. **600**

g. **0**

Problem Solving

We can make five pairs of addends that each total 11.

$5 \times 11 = \mathbf{55}$

Lesson 2, Lesson Practice

a. **The additive identity is zero. The multiplicative identity is 1.**

b. **Division**

c. $(x + y) + z = x + (y + z)$
Numerical answers may vary.

d. **Commutative property of multiplication**

e. $(5 + 4) + 3 = (9) + 3 = \mathbf{12}$

f. $5 + (4 + 3) = 5 + (7) = \mathbf{12}$

g. $(10 - 5) - 3 = (5) - 3 = \mathbf{2}$

h. $10 - (5 - 3) = 10 - (2) = \mathbf{8}$

i. $(6 \cdot 2) \cdot 5 = (12) \cdot 5 = \mathbf{60}$

j. $6 \cdot (2 \cdot 5) = 6 \cdot (10) = \mathbf{60}$

k. $(12 \div 6) \div 2 = (2) \div 2 = \mathbf{1}$

l. $12 \div (6 \div 2) = 12 \div (3) = \mathbf{4}$

m. $k = n \cdot n$
Eighth term: $k = 8 \cdot 8 = \mathbf{64}$
Ninth term: $k = 9 \cdot 9 = \mathbf{81}$
Tenth term: $k = 10 \cdot 10 = \mathbf{100}$

n. **Each term in the sequence can be found by doubling the preceding term; 16, 32, 64.**

o. $k = (2n) - 1$
$k = (2 \cdot 1) - 1 = (2) - 1 = 1$
$k = (2 \cdot 2) - 1 = (4) - 1 = 3$
$k = (2 \cdot 3) - 1 = (6) - 1 = 5$
$k = (2 \cdot 4) - 1 = (8) - 1 = 7$
First four terms: **1, 3, 5, 7**

Lesson 2, Mixed Practice

1. Product of 2 and 3 $= 6$
Sum of 4 and 5 $= 9$
$9 - 6 = \mathbf{3}$

2. **\$0.04; 4¢**

3. **75¢ per glass; \$0.75 per glass**

4. **Subtraction**

5. $\begin{array}{r} \mathbf{15} \\ 4\overline{)60} \\ \underline{4} \\ 20 \\ \underline{20} \\ 0 \end{array}$

6. $3 \times 5 = 15 \qquad 15 \div 5 = 3$

$5 \times 3 = 15 \qquad 15 \div 3 = 5$

7. **Each term in the sequence can be found by multiplying the preceding term by ten.**

$100(10) = \mathbf{1000}$

$1000(10) = \mathbf{10{,}000}$

8.
$$\begin{array}{r} \$\,\overset{1}{\cancel{2}}\,\overset{9}{\cancel{0}}.\overset{9}{\cancel{0}}{}^{1}0 \\ -\ \$\,1\,4.\,7\,9 \\ \hline \mathbf{\$\,5.\,2\,1} \end{array}$$

9.
$$\begin{array}{r} \$1.54 \\ \times\quad 7 \\ \hline \mathbf{\$10.78} \end{array}$$

10.
$$\begin{array}{r} \mathbf{\$3.75} \\ 8\overline{)\$30.00} \\ \underline{24} \\ 6\,0 \\ \underline{5\,6} \\ 40 \\ \underline{40} \\ 0 \end{array}$$

11.
$$\begin{array}{r} \overset{1\,1}{\$4.36} \\ \$0.75 \\ \$12.00 \\ +\ \$0.06 \\ \hline \mathbf{\$17.17} \end{array}$$

12.
$$\begin{array}{r} \overset{1\,1}{\$4.89} \\ +\ \$0.74 \\ \hline \$5.63 \end{array} \qquad \begin{array}{r} \$\,\overset{0}{\cancel{1}}\,\overset{9}{\cancel{0}}.\overset{9}{\cancel{0}}{}^{1}0 \\ -\ \$\,5.\,6\,3 \\ \hline \mathbf{\$\,4.\,3\,7} \end{array}$$

13.
$$\begin{array}{r} 8 \\ 5 \\ 4 \\ 6 \\ 5 \\ 5 \\ 4 \\ 3 \\ 7 \\ 2 \\ 4 \\ 1 \\ +\ 8 \\ \hline \mathbf{57} \end{array}$$

14.
$$\begin{array}{r} \mathbf{207} \\ 15\overline{)3105} \\ \underline{30} \\ 10 \\ \underline{0} \\ 105 \\ \underline{105} \\ 0 \end{array}$$

15.
$$\begin{array}{r} \mathbf{40\ R\ 30} \\ 40\overline{)1630} \\ \underline{160} \\ 30 \\ \underline{0} \\ 30 \end{array}$$

16. $9 \div 3 = 3$
$$\begin{array}{r} \mathbf{27} \\ 3\overline{)81} \\ \underline{6} \\ 21 \\ \underline{21} \\ 0 \end{array}$$

17.
$$\begin{array}{r} 9 \\ 9\overline{)81} \\ \underline{81} \\ 0 \end{array}$$
$9 \div 3 = \mathbf{3}$

18.
$$\begin{array}{r} \$3.75 \\ \times\quad 10 \\ \hline \mathbf{\$37.50} \end{array}$$

19.
$$\begin{array}{r} \overset{3}{\cancel{4}}\,\overset{14}{\cancel{5}}{}^{1}0 \\ -\quad 7\,8 \\ \hline 3\,7\,2 \end{array} \qquad \begin{array}{r} \overset{2}{\cancel{3}}\,\overset{10}{\cancel{1}}{}^{1}6\,7 \\ -\quad 3\,7\,2 \\ \hline \mathbf{2\,7\,9\,5} \end{array}$$

20.
$$\begin{array}{r} \overset{2}{\cancel{3}}{}^{1}1\,6\,7 \\ -\quad 4\,5\,0 \\ \hline 2\,7\,1\,7 \end{array} \qquad \begin{array}{r} 2\,\overset{6}{\cancel{7}}\,\overset{10}{\cancel{1}}{}^{1}7 \\ -\quad 7\,8 \\ \hline \mathbf{2\,6\,3\,9} \end{array}$$

21.
$$\begin{array}{r} \mathbf{\$1.25} \\ 16\overline{)\$20.00} \\ \underline{16} \\ 4\,0 \\ \underline{3\,2} \\ 80 \\ \underline{80} \\ 0 \end{array}$$

22.
$$\begin{array}{r} 70 \\ \times\ 800 \\ \hline \mathbf{56{,}000} \end{array}$$

23.
$$\begin{array}{r} {\scriptstyle 1\,1\,2} \\ 3714 \\ 268 \\ 47 \\ +\quad 9 \\ \hline \mathbf{4038} \end{array}$$

24.
$$\begin{array}{r} 5 \\ \times\ 4 \\ \hline 20 \end{array} \qquad \begin{array}{r} 20 \\ \times\ \ 3 \\ \hline 60 \end{array} \qquad \begin{array}{r} 60 \\ \times\ \ 2 \\ \hline 120 \end{array} \qquad \begin{array}{r} 120 \\ \times\ \ \ 1 \\ \hline \mathbf{120} \end{array}$$

25.
$$\begin{array}{r} \$1.47 \\ +\ \$8.00 \\ \hline \$9.47 \end{array} \qquad \begin{array}{r} {\scriptstyle 1\ \ 9\ \ \ 9\ \ } \\ \$\,\not{2}\,\not{0}.\,\not{0}\,{}^{1}0 \\ -\quad \$\,9.\,4\,7 \\ \hline \mathbf{\$\,1\,0.\,5\,3} \end{array}$$

26.
$$\begin{array}{r} \$0.45 \\ \times\quad 30 \\ \hline \mathbf{\$13.50} \end{array}$$

27. (a) **Property of zero for multiplication**

(b) **Identity property of multiplication**

28. (a) $18 - 3 = \mathbf{15}$

(b)
$$\begin{array}{r} 18 \\ \times\ \ 3 \\ \hline \mathbf{54} \end{array}$$

(c) $\frac{18}{3} = \mathbf{6}$

(d) $18 + 3 = \mathbf{21}$

29. Zero is called the additive identity because when zero is added to another number, the sum is identical to that number.

30. Dividend ÷ divisor = quotient

Lesson 3, Warm-Up

a. **66**

b. **30**

c. **500**

d. **1500**

e. **250**

f. **900**

g. **7**

Problem Solving

There are $3 \times 1 + 3$, or 6 seats in the first row.

There are $3 \times 2 + 3$, or 9 seats in the second row.

There are $3 \times 3 + 3$, or 12 seats in the third row.

Therefore, there are $3 \times 12 + 3$, or 39 seats in the twelfth row.

Lesson 3, Lesson Practice

a.
$$\begin{array}{r} {\scriptstyle 2\ \ \ } \\ \not{3}\,{}^{1}1 \\ -\ 1\,2 \\ \hline \mathbf{1\,9} \end{array}$$

b.
$$\begin{array}{r} 15 \\ +\ 24 \\ \hline \mathbf{39} \end{array}$$

c.
$$\begin{array}{r} \mathbf{12} \\ 15\overline{)180} \\ \underline{15} \\ 30 \\ \underline{30} \\ 0 \end{array}$$

d.
$$\begin{array}{r} 12 \\ \times\ \ 8 \\ \hline \mathbf{96} \end{array}$$

e.
$$\begin{array}{r} \mathbf{30} \\ 14\overline{)420} \\ \underline{42} \\ 00 \\ \underline{0} \\ 0 \end{array}$$

f.
$$\begin{array}{r} {\scriptstyle 3\ \ \ } \\ \not{4}\,{}^{1}3 \\ -\ 2\,6 \\ \hline \mathbf{1\,7} \end{array}$$

g.
$$\begin{array}{r} 51 \\ -\ 20 \\ \hline \mathbf{31} \end{array}$$

h.
$$\begin{array}{r} \mathbf{52} \\ 7\overline{)364} \\ \underline{35} \\ 14 \\ \underline{14} \\ 0 \end{array}$$

i.
$$\begin{array}{r} \mathbf{2} \\ 12\overline{)24} \\ \underline{24} \\ 0 \end{array}$$

j. $3 + 6 + 12 + 5 = 26$
$30 - 26 = \mathbf{4}$

Lesson 3, Mixed Practice

1. Product of 4 and 4 $= 16$
Sum of 4 and 4 $= 8$
$\frac{16}{8} = \mathbf{2}$

2. Add the subtrahend and the difference to find the minuend.

3. Associative property of addition

4.
$$\begin{array}{r} \overset{1}{\cancel{2}}{}^{1}1 \\ -\quad 7 \\ \hline \mathbf{14} \end{array}$$

5. $\mathbf{3 \cdot 4 = 4 \cdot 3}$

6. (1) $k = 3(1) = 3$
(2) $k = 3(2) = 6$
(3) $k = 3(3) = 9$
(4) $k = 3(4) = 12$
3, 6, 9, 12

7.
$$\begin{array}{r} \overset{0}{\cancel{1}}\,\overset{10}{\cancel{1}}{}^{1}2 \\ -\quad 83 \\ \hline 29 \end{array}$$
$x = \mathbf{29}$

8.
$$\begin{array}{r} \overset{8}{\cancel{9}}{}^{1}6 \\ -\ 27 \\ \hline 69 \end{array}$$
$r = \mathbf{69}$

9.
$$\begin{array}{r} 17 \\ 7\overline{)119} \\ \underline{7} \\ 49 \\ \underline{49} \\ 0 \end{array}$$
$k = \mathbf{17}$

10.
$$\begin{array}{r} \overset{2}{\cancel{3}}\,\overset{9}{\cancel{0}}{}^{1}0 \\ -\ 127 \\ \hline 173 \end{array}$$
$z = \mathbf{173}$

11.
$$\begin{array}{r} 731 \\ +\ 137 \\ \hline 868 \end{array}$$
$m = \mathbf{868}$

12.
$$\begin{array}{r} 16 \\ 25\overline{)400} \\ \underline{25} \\ 150 \\ \underline{150} \\ 0 \end{array}$$
$n = \mathbf{16}$

13.
$$\begin{array}{r} 25 \\ 25\overline{)625} \\ \underline{50} \\ 125 \\ \underline{125} \\ 0 \end{array}$$
$w = \mathbf{25}$

14.
$$\begin{array}{r} 700 \\ \times\quad 60 \\ \hline 42{,}000 \end{array}$$
$x = \mathbf{42{,}000}$

15. (a) $\frac{20}{5} = \mathbf{4}$
(b) $20 - 5 = \mathbf{15}$
(c) $20(5) = \mathbf{100}$
(d) $20 + 5 = \mathbf{25}$

16. $16 \div 2 = 8$
$$\begin{array}{r} \mathbf{12} \\ 8\overline{)96} \\ \underline{8} \\ 16 \\ \underline{16} \\ 0 \end{array}$$

17.
$$\begin{array}{r} 6 \\ 16\overline{)96} \\ \underline{96} \\ 0 \end{array}$$
$6 \div 2 = \mathbf{3}$

18.
```
   1 1 1
  $16.47
  $15.00
+  $0.63
  $32.10
```
$32.10

19.
```
   1 1           4 9 9
  $31.75      $ 5 0. 0 10
+  $6.48    − $ 3 8. 2 3
  $38.23      $ 1 1. 7 7
```
$11.77

20.
```
    47
  × 39
   423
  141
  1833
```
1833

21.
```
   $8.79
 ×    80
 $703.20
```
$703.20

22.
```
      158
  30)4740
     30
     174
     150
      240
      240
        0
```
158

23.
```
   2 16           0 10 9
   3 7 14         1 1 0 10
 −   8 7        −   2 8 7
   2 8 7            8 1 3
```
813

24.
```
   0 10 9          6 11
   1 1 0 10        7 2 16
 −   3 7 4       −   8 7
     7 2 6         6 3 9
```
639

25.
```
   1 2 2
   4736
    271
      9
 +   88
   5104
```
5104

26.
```
   2 9 10 13
   3 0, 1 4 15
 −    4, 2 9 9
   2 5, 8 4 6
```
25,846

27.
```
      60 R 4
  35)2104
     210
      04
       0
       4
```
60 R 4

28.
```
      $1.25
  32)$40.00
     32
      8 0
      6 4
      1 60
      1 60
         0
```
$1.25

29.
```
   $0.48
 ×    40
  $19.20
```
$19.20

30. **One is the multiplicative identity because when any given number is multiplied by 1, the product is identical to the given number.**

Lesson 4, Warm-Up

a. **100**

b. **120**

c. **250**

d. **750**

e. **2500**

f. **40**

g. **100**

Problem Solving

Altogether, there are $1 + 2 + 3 + 4 + 5 + 6$, or 21 dots on a dot cube.

$21 - 7 =$ **14 dots**

Lesson 4, Lesson Practice

a.

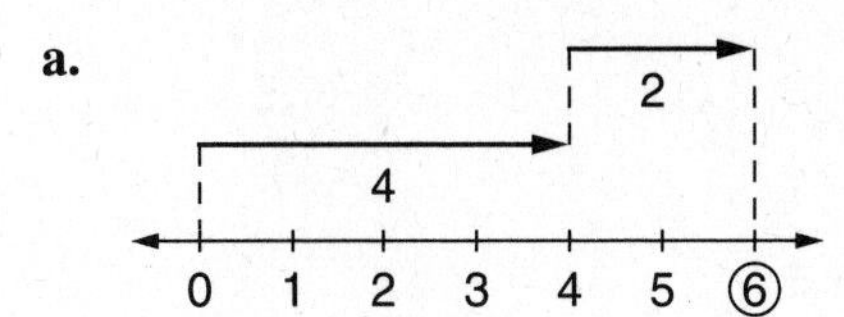

b.

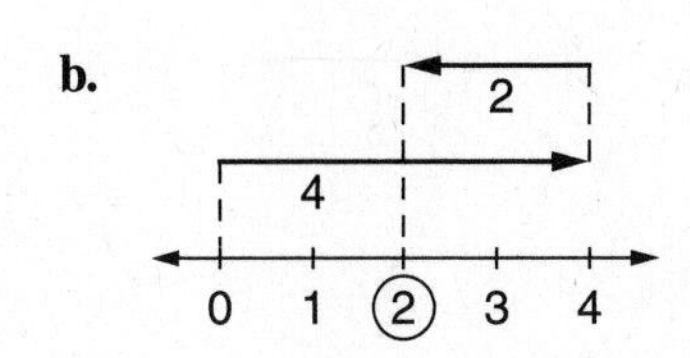

c. 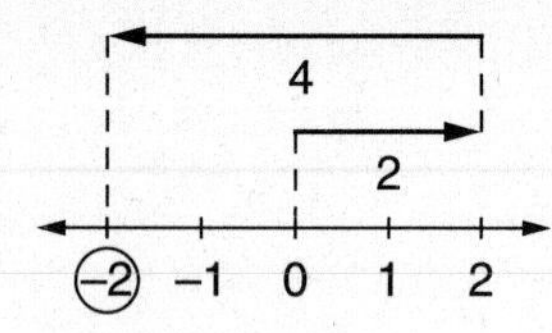

d. **−3, −2, −1, 0**

e. **2 + 3 < 2 × 3**

f. 3 − 4 **<** 4 − 3

g. 2 · 2 **=** 2 + 2

h. **0**

i.
$$\begin{array}{r} \overset{5}{\not 6}\,\overset{12}{\not 3}\,{}^{1}0 \\ -\ 4\ 3\ 6 \\ \hline 1\ 9\ 4 \end{array}$$
−194

j. **−2, −3, −4**

Lesson 4, Mixed Practice

1. Sum of 5 and 4 = 9
Product of 3 and 3 = 9
9 − 9 = **0**

2. 27 − 9 = **18**

3. **Positive numbers**

4. (a) 24 − 6 = **18**

(b) 6 − 24 = **−18**

(c) $\frac{24}{6}$ = **4**

(d)
$$\begin{array}{r} 24 \\ \times\ \ 6 \\ \hline \mathbf{144} \end{array}$$

5. **5 · 2 > 5 + 2**

6. **−2, −1, 0, 1**

7. (a) 3 · 4 **=** 2(6)

(b) −3 **<** −2

(c) 3 − 5 **<** 5 − 3

(d) xy **=** yx

8. **Multiply the divisor by the quotient to find the dividend.**

9. 3
2
−1 0 1 2

10. $\frac{12}{12} = 1$

k = **1**

11.
$$\begin{array}{r} 4 \\ 8 \\ +\ 6 \\ \hline 18 \end{array} \qquad \begin{array}{r} \overset{2}{\not 3}\,{}^{1}0 \\ -\ 1\ 8 \\ \hline 1\ 2 \end{array}$$

n = **12**

12.
$$\begin{array}{r} 654 \\ +\ 123 \\ \hline 777 \end{array}$$

z = **777**

13.
$$\begin{array}{r} \overset{0}{\not 1}\,\overset{9}{\not 0}\,\overset{9}{\not 0}\,{}^{1}0 \\ -\quad 1\ 0\ 1 \\ \hline 8\ 9\ 9 \end{array}$$

m = **899**

14.
$$\begin{array}{r} \$4.95 \\ -\ \$1.45 \\ \hline \$3.50 \end{array}$$

p = **\$3.50**

15.
$$\begin{array}{r} 7 \\ 32\overline{)224} \\ 224 \\ \hline 0 \end{array}$$

k = **7**

16.
$$\begin{array}{r} 24 \\ \times\ \ 8 \\ \hline 192 \end{array}$$

r = **192**

17.
$$\begin{array}{r} \overset{1\,1}{\$3.67} \\ \$0.14 \\ +\ \$52.75 \\ \hline \mathbf{\$56.56} \end{array}$$

18.
$$\begin{array}{r} \$\,\overset{0}{\not 1}\,\overset{9}{\not 0}\,\overset{9}{\not 0}.\overset{9}{\not 0}{}^{1}0 \\ -\quad \$\,3\,6.\,4\,9 \\ \hline \mathbf{\$\,6\,3.\,5\,1} \end{array}$$

19.
$$\begin{array}{r} \$0.36 \\ \times \quad 48 \\ \hline 288 \\ 144 \\ \hline \mathbf{\$17.28} \end{array}$$

20. $5 \cdot 6 = 30$
$$\begin{array}{r} 30 \\ \times \ 7 \\ \hline \mathbf{210} \end{array}$$

21.
$$\begin{array}{r} \mathbf{550} \\ 18\overline{)9900} \\ 90 \\ \hline 90 \\ 90 \\ \hline 00 \\ 0 \\ \hline 0 \end{array}$$

22.
$$\begin{array}{r} 30 \\ \times \ 20 \\ \hline 600 \end{array} \qquad \begin{array}{r} 600 \\ \times \quad 40 \\ \hline \mathbf{24{,}000} \end{array}$$

23.
$$\begin{array}{r} \overset{0}{\not1}\,\overset{12}{\not3}\,{}^{1}0 \\ -\ \ 5\,7 \\ \hline 7\,3 \end{array} \qquad \begin{array}{r} \overset{1}{7}3 \\ +\ 9 \\ \hline \mathbf{82} \end{array}$$

24.
$$\begin{array}{r} \overset{1}{\not2}\,\overset{9}{\not0}\,\overset{10}{\not1}\,{}^{1}4 \\ -\ \ 1\,9\,8\,7 \\ \hline 2\,7 \end{array} \qquad \mathbf{-27}$$

25.
$$\begin{array}{r} \mathbf{\$9.80} \\ 7\overline{)\$68.60} \\ 63 \\ \hline 5\,6 \\ 5\,6 \\ \hline 00 \\ 0 \\ \hline 0 \end{array}$$

26.
$$\begin{array}{r} \overset{1}{\$}\overset{1}{0}.46 \\ +\ \$0.64 \\ \hline \mathbf{\$1.10} \end{array}$$

27.
$$\begin{array}{r} \mathbf{58} \\ 80\overline{)4640} \\ 400 \\ \hline 640 \\ 640 \\ \hline 0 \end{array}$$

28.
$$\begin{array}{r} \$3.75 \\ \times \quad 30 \\ \hline \mathbf{\$112.50} \end{array}$$

29. **Answers may vary. One answer is $(2 \times 3) \times 6 = 2 \times (3 \times 6)$.**

30. **$10 + 20 = 30 \quad 30 - 20 = 10$**
$20 + 10 = 30 \quad 30 - 10 = 20$

Lesson 5, Warm-Up

a. **10**

b. **144**

c. **1000**

d. **275**

e. **500**

f. **500**

g. **15**

Problem Solving

We write a 7 in the ones place of the subtrahend, because $3 + 7 = 10$. We write a 2 in the tens place of the minuend, because $1 + 0 + 1$ (from regrouping) equals 2. We write a 9 in the hundreds column of the difference, because $9 + 6 = 15$. We write a 2 in the thousands place of the subtrahend, because $4 + 2 + 1$ (from regrouping) equals 7.

$$\begin{array}{r} \mathbf{7520} \\ -\ \mathbf{2607} \\ \hline \mathbf{4913} \end{array}$$

Lesson 5, Lesson Practice

a. **3**

b. **Billions**

c. **$(2 \times 1000) + (5 \times 100)$**

d. **Thirty-six million, four hundred twenty-seven thousand, five hundred eighty**

e. **Forty million, three hundred two thousand, ten**

f. **25,206,040**

g. **50,402,100,000**

h. **$15,000,000,000**

LESSON 5, MIXED PRACTICE

1.
$$\begin{array}{r} {}^{1\,1} \\ {}_{1}607 \\ +\ 2393 \\ \hline \mathbf{3000} \end{array}$$

2. **101,000 > 1100**

3. **Fifty million, five hundred seventy-four thousand, six**

4. **2**

5. **250,005,070**

6. $-12 > -15$

Negative twelve is greater than negative fifteen.

7. **−7, −1, 0, 4, 5, 7**

8. **Draw a number line. Start at the origin and draw an arrow 5 units long to the right. From this point draw an arrow 4 units long to the left. The second arrow ends at 1, showing that 5 − 4 = 1. Circle the number 1.**

9. **7 units**

10. $2 \cdot 3 = 6 \qquad 6 \cdot 5 = 30$

$$\begin{array}{r} 32 \\ 30\overline{)960} \\ \underline{90} \\ 60 \\ \underline{60} \\ 0 \end{array}$$

$n =$ **32**

11.
$$\begin{array}{r} 2500 \\ +\ 1367 \\ \hline 3867 \end{array}$$

$a =$ **3867**

12. $17 + 5 = 22$

$$\begin{array}{r} {}^{4} \\ \not{5}{}^{1}0 \\ -\ 2\,2 \\ \hline 2\,8 \end{array}$$

$b =$ **28**

13.
$$\begin{array}{r} {}^{1\ \ 14} \\ \$\,\not{2}\,\not{5}.{}^{1}0\,0 \\ -\ \$\,1\,8.\,7\,0 \\ \hline \$\,6.\,3\,0 \end{array}$$

$k =$ **$6.30**

14.
$$\begin{array}{r} {}^{0\ 9} \\ \not{1}\,\not{0},{}^{1}0\,0\,0 \\ -\ \ 6,\,4\,0\,0 \\ \hline 3,\,6\,0\,0 \end{array}$$

$d =$ **3600**

15.
$$\begin{array}{r} 18 \\ 8\overline{)144} \\ \underline{8} \\ 64 \\ \underline{64} \\ 0 \end{array}$$

$f =$ **18**

16. **(7 × 100,000) + (5 × 10,000)**

17.
$$\begin{array}{r} {}^{1\,1\ 1\,1} \\ 37{,}428 \\ +\ 59{,}775 \\ \hline \mathbf{97{,}203} \end{array}$$

18.
$$\begin{array}{r} {}^{2\ 10\ 9\ 10} \\ \not{3}\,\not{1},\,\not{0}\,\not{1}{}^{1}4 \\ -\ 2\,4,\,7\,6\,7 \\ \hline \mathbf{6,\,2\,4\,7} \end{array}$$

19.
$$\begin{array}{r} {}^{2} \\ {}_{1}45 \\ 362 \\ 7 \\ +\ 4319 \\ \hline \mathbf{4733} \end{array}$$

20.
$$\begin{array}{r} {}^{1\,1\ 1} \\ \$64.59 \\ \$124.00 \\ \$6.30 \\ +\ \ \$0.37 \\ \hline \mathbf{\$195.26} \end{array}$$

21. $12 \div 3 = 4$

$$\begin{array}{r} \mathbf{36} \\ 4\overline{)144} \\ \underline{12} \\ 24 \\ \underline{24} \\ 0 \end{array}$$

22.

$$\begin{array}{r} 12 \\ 12\overline{)144} \\ \underline{12} \\ 24 \\ \underline{24} \\ 0 \end{array}$$

$12 \div 3 = \mathbf{4}$

23.

$$\begin{array}{r} 40 \\ \underline{\times \ 500} \\ \mathbf{20{,}000} \end{array}$$

24.

$$\begin{array}{r} \mathbf{405} \\ 21\overline{)8505} \\ \underline{84} \\ 10 \\ \underline{0} \\ 105 \\ \underline{105} \\ 0 \end{array}$$

25.

$$\begin{array}{r} \$4.\overset{5}{\cancel{6}}{}^{1}0 \\ \underline{-\ \$0.39} \\ \$4.21 \end{array} \qquad \begin{array}{r} \$\overset{0}{\cancel{1}}\overset{9}{\cancel{0}}.\overset{9}{\cancel{0}}{}^{1}0 \\ \underline{-\ \$4.21} \\ \mathbf{\$5.79} \end{array}$$

26.

$$\begin{array}{r} \$0.29 \\ \underline{\times \quad 36} \\ 174 \\ \underline{87\ \ } \\ \mathbf{\$10.44} \end{array}$$

27. (a) **Identity property of multiplication**

(b) **Commutative property of multiplication**

28. **Each term in the sequence can be found by subtracting two from the preceding term.**

$2 - 2 = \mathbf{0}$

$0 - 2 = \mathbf{-2}$

$-2 - 2 = \mathbf{-4}$

29. (a) **Counting numbers or natural numbers**

(b) **Whole numbers**

(c) **Integers**

30. **{. . . , −6, −4, −2}**

Lesson 6, Warm-Up

a. **$7.50**

b. **$15.00**

c. **$0.55**

d. **485**

e. **625**

f. **75**

g. **4**

Problem Solving

12 glubs per lorn × 4 lorns per dort = 48 glubs per dort

48 glubs per dort × $\frac{1}{2}$ dort = **24 glubs**

Lesson 6, Lesson Practice

a. **1, 5, 25**

b. **1, 2, 3, 4, 6, 8, 12, 24**

c. **1, 23**

d. $1 + 2 + 6 + 0 = 9$

1, 2, 3, 4, 5, 6, 7, 9, 10

$$\begin{array}{r} 180 \\ 7\overline{)1260} \\ \underline{7\ \ \ } \\ 56\ \ \\ \underline{56\ \ } \\ 00 \\ \underline{0} \\ 0 \end{array}$$

e. $7 + 3 + 5 + 0 + 0 = 15$

$$\begin{array}{r} 10{,}500 \\ 7\overline{)73{,}500} \\ \underline{7\ \ \ \ \ \ \ } \\ 03\ \ \ \ \ \\ \underline{0\ \ \ \ \ } \\ 35\ \ \ \\ \underline{35\ \ \ } \\ 00\ \\ \underline{0\ } \\ 00 \\ \underline{0} \\ 0 \end{array}$$

1, 2, 3, 4, 5, 6, 7, 10

f. $3 + 6 + 0 + 0 = 9$

1, 2, 3, 4, 5, 6, 8, 9, 10

```
  514 R 2
7)3600
  35
   10
    7
    30
    28
     2
```

g. $1 + 3 + 5 + 6 = 15$

1, 2, 3, 4, 6

```
  193 R 5
7)1356
  7
  65
  63
   26
   21
    5
```

h. **1, 2, 4, 5, 7, 8**

i. Factors of 12: 1, 2, 3, 4, 6, 12
Factors of 20: 1, 2, 4, 5, 10, 20
1, 2, 4

j. Factors of 24: 1, 2, 3, 4, 6, 8, 12, 24
Factors of 40: 1, 2, 4, 5, 8, 10, 20, 40
8

Lesson 6, Mixed Practice

1. Product of 10 and 20 = 200
Sum of 20 and 30 = 50

$$\frac{200}{50} = \mathbf{4}$$

2. (a) Factors of 30: 1, 2, 3, 5, 6, 10, 15, 30
Factors of 40: 1, 2, 4, 5, 8, 10, 20, 40
1, 2, 5, 10

(b) **10**

3. $\{\ldots, -5, -3, -1\}$

4. **407,006,962**

5. $1 + 2 + 3 + 0 + 0 = 6$

1, 2, 3, 4, 5, 6, 10

```
  1,757 R 1
7)12,300
  7
  5 3
  4 9
    40
    35
     50
     49
      1
```

6. $-7 > -11$

Negative seven is greater than negative eleven.

7.

```
  14       57        493 R 5
4)56     8)456     7)3456
  4        40        28
  16        56        65
  16        56        63
   0         0         26
                       21
                        5
```

$3 + 4 + 5 + 6 = 18$

1, 2, 3, 4, 6, 8, 9

8.

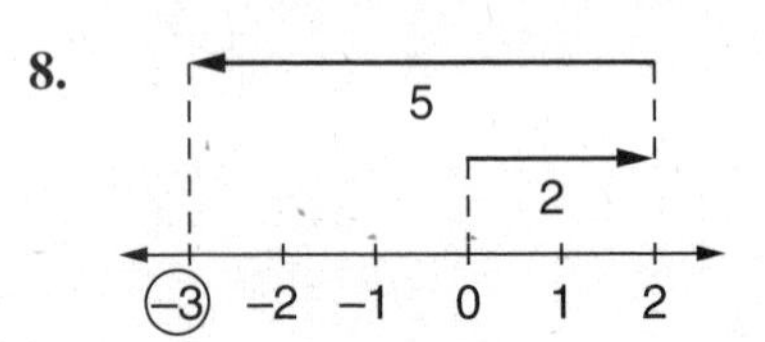

9. $(\mathbf{6 \times 1000}) + (\mathbf{4 \times 100})$

10.

```
  $10.00
−  $4.60
   $5.40
```

$x =$ **\$5.40**

11.

```
   4500
+  3850
   8350
```

$p =$ **8350**

12.

```
  $6.25
8)$50.00
  48
   2 0
   1 6
     40
     40
      0
```

$z =$ **\$6.25**

13.

```
  7        60
  4      − 51
  8         9
  6
  2
  1
  6
  8
+ 9
 51
```

$n =$ **9**

14. $$\begin{array}{r} 1426 \\ -\quad 87 \\ \hline 1339 \end{array}$$

$k = \mathbf{1339}$

15. $$\begin{array}{r} 22 \\ 45\overline{)990} \\ \underline{90} \\ 90 \\ \underline{90} \\ 0 \end{array}$$

$p = \mathbf{22}$

16. $$\begin{array}{r} 32 \\ \times\ 8 \\ \hline 256 \end{array}$$

$z = \mathbf{256}$

17. $$\begin{array}{r} \mathbf{35} \\ 35\overline{)1225} \\ \underline{105} \\ 175 \\ \underline{175} \\ 0 \end{array}$$

18. $$\begin{array}{r} 800 \\ \times\quad 50 \\ \hline \mathbf{40{,}000} \end{array}$$

19. $$\begin{array}{r} \$100.00 \\ -\ \$48.37 \\ \hline \mathbf{\$51.63} \end{array}$$

20. $$\begin{array}{r} 46{,}302 \\ +\ 49{,}998 \\ \hline \mathbf{96{,}300} \end{array}$$

21. $$\begin{array}{r} \mathbf{\$2.25} \\ 20\overline{)\$45.00} \\ \underline{40} \\ 5\,0 \\ \underline{4\,0} \\ 1\,00 \\ \underline{1\,00} \\ 0 \end{array}$$

22. $$\begin{array}{r} 11 \\ \times\ 7 \\ \hline 77 \end{array} \qquad \begin{array}{r} 77 \\ \times\ 13 \\ \hline 231 \\ \underline{77\ } \\ \mathbf{1001} \end{array}$$

23. $$\begin{array}{r} \mathbf{4{,}807}\ \mathbf{R\,8} \\ 9\overline{)43{,}271} \\ \underline{36} \\ 7\,2 \\ \underline{7\,2} \\ 07 \\ \underline{0} \\ 71 \\ \underline{63} \\ 8 \\ \underline{0} \\ 8 \end{array}$$

24. $$\begin{array}{r} 3625 \\ 59 \\ 570 \\ +\quad 8 \\ \hline \mathbf{4262} \end{array}$$

25. $$\begin{array}{r} \$0.48 \\ \$8.49 \\ +\ \$14.00 \\ \hline \mathbf{\$22.97} \end{array}$$

26. $$\begin{array}{r} 430 \\ -\ 58 \\ \hline 372 \end{array} \qquad \begin{array}{r} 1000 \\ -\ 372 \\ \hline \mathbf{628} \end{array}$$

27. $$\begin{array}{r} 140 \\ \times\ 16 \\ \hline 840 \\ \underline{140\ } \\ \mathbf{2240} \end{array}$$

28. $$\begin{array}{r} \$0.25 \\ \times\quad 24 \\ \hline 1\,00 \\ \underline{5\,0\ } \\ \mathbf{\$6.00} \end{array}$$

29. $$\begin{array}{r} \mathbf{\$4.35} \\ 10\overline{)\$43.50} \\ \underline{40} \\ 3\,5 \\ \underline{3\,0} \\ 50 \\ \underline{50} \\ 0 \end{array}$$

30. **Commutative property of multiplication; the order of the factors can be changed without changing the product.**

Lesson 7, Warm-Up

a.  **−5**

b. **$25.00**

c. **65¢**

d. **365**

e. **265**

f. **48**

g. **6**

Problem Solving

We can make ten pairs of addends that each total 21.
$10 \times 21 =$ **210**

Lesson 7, Lesson Practice

a. **Point *A***

b. $XY = XZ - YZ$
$XY = 10 - 6$
$XY =$ **4 cm**

c.

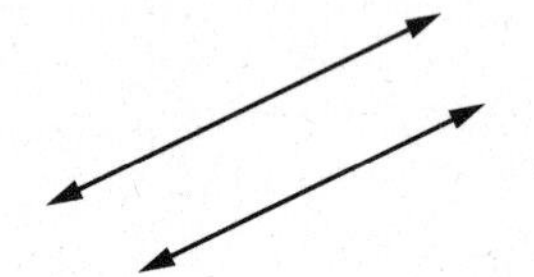

d.

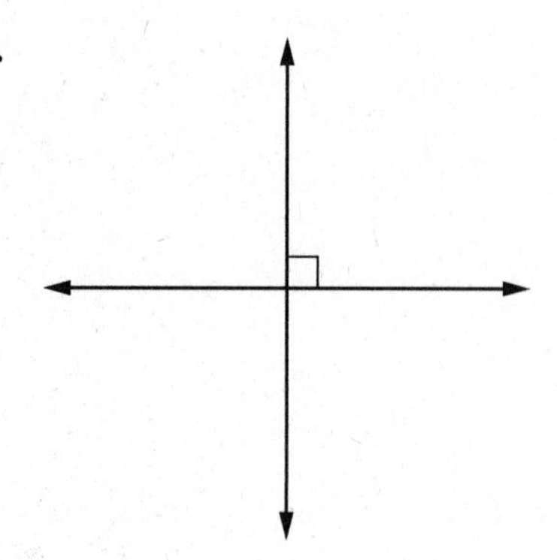

e.

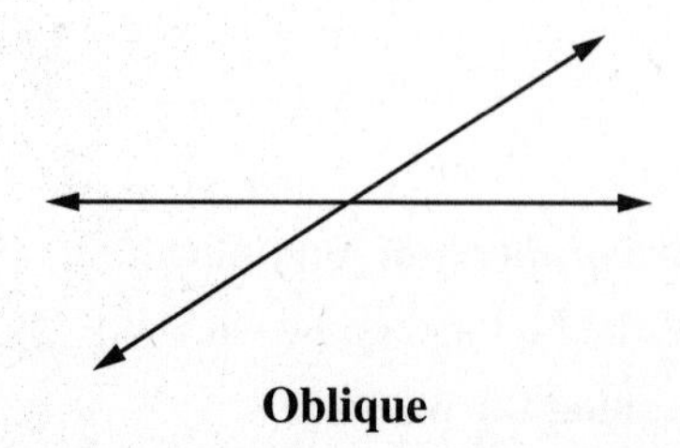

Oblique

f. 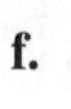

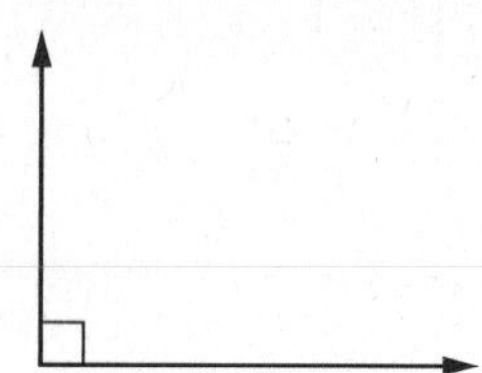

g.

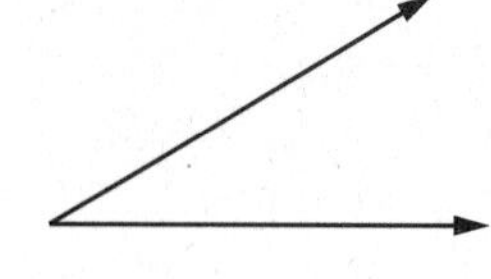

h.

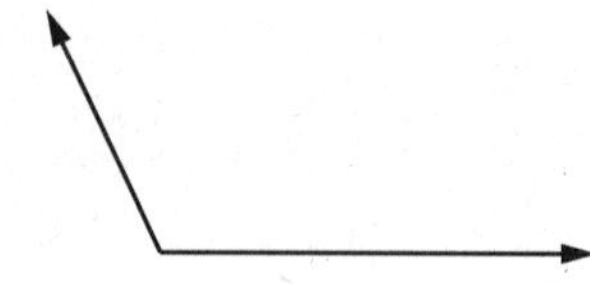

i. **Perpendicular**

Lesson 7, Mixed Practice

1. $7 + 5 =$ **12**

2. **Identity property of multiplication**

3. **1, 2, 5, 10, 25, 50**

4. **2 − 5 = −3**

5. **90,000,000**

6.
```
   115 R 4        132
 8)924          7)924
   8              7
   12             22
    8             21
    44             14
    40             14
     4              0
```
$9 + 2 + 4 = 15$
1, 2, 3, 4, 6, 7

7. **−10, −7, −2, 0, 5, 8**

8. **This is a sequence of perfect squares.**
$11 \cdot 11 =$ **121**
$12 \cdot 12 =$ **144**
$13 \cdot 13 =$ **169**

9. **Posts**

10. (a) Factors of 24: 1, 2, 3, 4, 6, 8, 12, 24
Factors of 32: 1, 2, 4, 8, 16, 32
1, 2, 4, 8

(b) **8**

11. **7 units**

12. $6 \cdot 6 = 36$

$$\begin{array}{r} 34 \\ 36\overline{)1224} \\ \underline{108} \\ 144 \\ \underline{144} \\ 0 \end{array}$$

z = **34**

13.
$$\begin{array}{r} \$100.00 \\ -\ \ \$17.54 \\ \hline \$82.46 \end{array}$$
k = **$82.46**

14.
$$\begin{array}{r} 432 \\ +\ \ 98 \\ \hline 530 \end{array}$$
w = **530**

15.
$$\begin{array}{r} \$1.80 \\ 20\overline{)\$36.00} \\ \underline{20} \\ 16\ 0 \\ \underline{16\ 0} \\ 00 \\ \underline{0} \\ 0 \end{array}$$
x = **$1.80**

16.
$$\begin{array}{r} 200 \\ \times\ \ 20 \\ \hline 4000 \end{array}$$
w = **4000**

17.
$$\begin{array}{r} \mathbf{10} \\ 30\overline{)300} \\ \underline{30} \\ 00 \\ \underline{0} \\ 0 \end{array}$$
x = **10**

18. **The quotient does not have a remainder (the remainder is zero). A number is divisible by 9 if the sum of its digits is divisible by 9. The sum of the digits in 4554 is 18, which is divisible by 9.**

19.
$$\begin{array}{r} 36{,}475 \\ +\ 55{,}984 \\ \hline \mathbf{92{,}459} \end{array}$$

20.
$$\begin{array}{r} 476 \\ \times\ \ 38 \\ \hline 3\ 808 \\ 14\ 28\ \ \\ \hline \mathbf{18{,}088} \end{array}$$

21.
$$\begin{array}{r} \$80.00 \\ -\ \$72.45 \\ \hline \mathbf{\$7.55} \end{array}$$

22.
$$\begin{array}{r} 49 \\ 387 \\ 1579 \\ +\ \ \ 98 \\ \hline \mathbf{2113} \end{array}$$

23.
$$\begin{array}{r} \mathbf{\$1.70} \\ 40\overline{)\$68.00} \\ \underline{40} \\ 28\ 0 \\ \underline{28\ 0} \\ 00 \\ \underline{0} \\ 0 \end{array}$$

24. $8 \cdot 7 = 56$
$$\begin{array}{r} 56 \\ \times\ \ 5 \\ \hline \mathbf{280} \end{array}$$

25. $200 \div 10 = 20$
$$\begin{array}{r} 200 \\ 20\overline{)4000} \\ \underline{40} \\ 00 \\ \underline{0} \\ 00 \\ \underline{0} \\ 0 \end{array}$$

$$\begin{array}{r} 20 \\ 200\overline{)4000} \\ \underline{400} \\ 00 \\ \underline{0} \\ 0 \end{array}$$
$20 \div 10 = 2$

$200 > 2$

26. (a) $200(400)$ = **80,000**

(b) $200 - 400$ = **−200**

(c) $\frac{400}{200}$ = **2**

27. (a) $\angle BMC$ or $\angle CMB$

(b) $\angle AMC$ or $\angle CMA$

28. **Right angle**

29. $\overline{XY}$ (or $\overline{YX}$), $\overline{YZ}$ (or $\overline{ZY}$), $\overline{XZ}$ (or $\overline{ZX}$)

30. **Add m$\overline{XY}$ and m$\overline{YZ}$ to find m$\overline{XZ}$.**

Lesson 8, Warm-Up

a. **−6**

b. **$2.50**

c. **35¢**

d. **375**

e. **317**

f. **2000**

g. **11**

Problem Solving

The permutations of 2, 3, and 5 are 235, 253, 325, 352, 523, and 532. The largest permutation is **532.**

Lesson 8, Lesson Practice

a. $\mathbf{\frac{3}{5}}$

b. $100\% \div 5 = 20\%$
$20\% \times 3 = \mathbf{60\%}$

c. $100\% \div 2 = \mathbf{50\%}$

d.

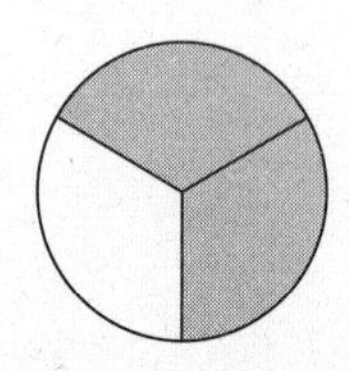

e.

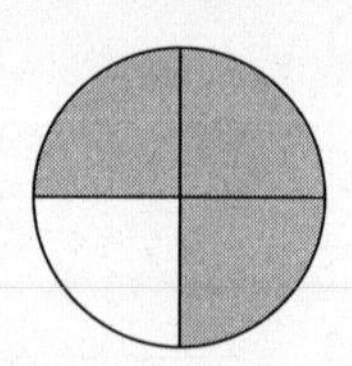

f.

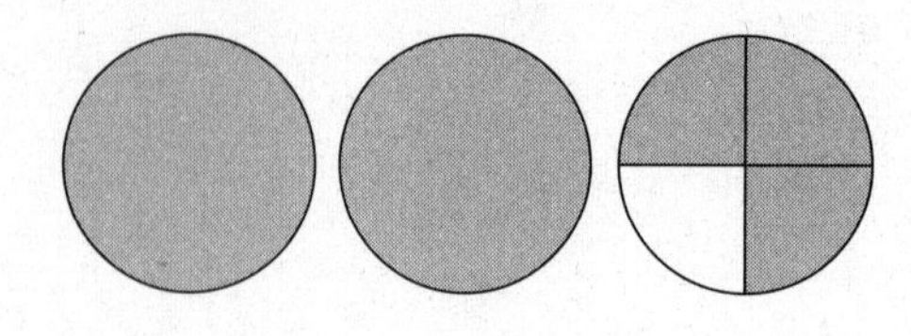

g. $\mathbf{4\frac{2}{3}}$

h. $\mathbf{13\frac{1}{4}}$

i. $\mathbf{3\frac{5}{16}}$ **in.**

j. $\frac{1}{2}$ of $\frac{1}{8} = \mathbf{\frac{1}{16}}$ **inch**

Lesson 8, Mixed Practice

1. $\mathbf{1\frac{3}{4} > 1\frac{3}{5}}$

2. $XY = 2\frac{4}{16} = \mathbf{2\frac{1}{4}}$ **in.**

$YZ = \mathbf{1\frac{1}{16}}$ **in.**

3. Product of 20 and 20 = 400
Sum of 10 and 10 = 20

$$400 \div 20 = \mathbf{20}$$

```
     20
 20)400
    40
     00
      0
      0
```

4.

```
    85        240
 8)680     7)1680
   64        14
    40        28
    40        28
     0         00
                0
                0
```

1 + 6 + 8 + 0 = 15
1, 2, 3, 4, 5, 6, 7, 8

5. $3\frac{4}{5}$

6. (a) $3 + 2 = 2 + 3$

(b) **Commutative property of addition**

7. **Thirty-two billion, five hundred million**

8. (a) $\frac{3}{8}$

(b) $\frac{5}{8}$

9. (a) $100\% \div 5 = 20\%$

$20\% \times 1 = \mathbf{20\%}$

(b) $20\% \times 4 = \mathbf{80\%}$

10. **Denominator**

11.

$$\begin{array}{r} \$2.35 \\ +\ \$4.70 \\ \hline \$7.05 \end{array}$$

$a = \mathbf{\$7.05}$

12.

$$\begin{array}{r} \$60.00 \\ -\ \$25.48 \\ \hline \mathbf{\$34.52} \end{array}$$

$b = \mathbf{\$34.52}$

13.

$$\begin{array}{r} \$7.50 \\ 8\overline{)\$60.00} \\ \underline{56} \\ 4\,0 \\ \underline{4\,0} \\ 00 \\ \underline{0} \\ 0 \end{array}$$

$c = \mathbf{\$7.50}$

14.

$$\begin{array}{r} 10{,}000 \\ -\ \ 5{,}420 \\ \hline 4{,}580 \end{array}$$

$d = \mathbf{4580}$

15.

$$\begin{array}{r} 15 \\ \times\ 15 \\ \hline 75 \\ 15 \\ \hline 225 \end{array}$$

$e = \mathbf{225}$

16.

$$\begin{array}{r} 14 \\ 14\overline{)196} \\ \underline{14} \\ 56 \\ \underline{56} \\ 0 \end{array}$$

$f = \mathbf{14}$

17.

$$\begin{array}{r} 8 \\ 9 \\ 8 \\ 8 \\ 9 \\ +\ 8 \\ \hline 50 \end{array} \qquad \begin{array}{r} 60 \\ -\ 50 \\ \hline 10 \end{array}$$

$n = \mathbf{10}$

18.

$$\begin{array}{r} 400 \\ \times\ \ 500 \\ \hline \mathbf{200{,}000} \end{array}$$

19.

$$\begin{array}{r} \$0.79 \\ \times\ \ \ 30 \\ \hline \mathbf{\$23.70} \end{array}$$

20.

$$\begin{array}{r} {\scriptsize 111} \\ 3625 \\ 431 \\ +\ 687 \\ \hline \mathbf{4743} \end{array}$$

21.

$$\begin{array}{r} \mathbf{120} \\ 50\overline{)6000} \\ \underline{50} \\ 100 \\ \underline{100} \\ 00 \\ \underline{0} \\ 0 \end{array}$$

22.

$$\begin{array}{r} 20 \\ \times\ 10 \\ \hline 200 \end{array} \qquad \begin{array}{r} 200 \\ \times\ \ \ 5 \\ \hline \mathbf{1000} \end{array}$$

23.

$$\begin{array}{r} \mathbf{\$1.50} \\ 18\overline{)\$27.00} \\ \underline{18} \\ 9\,0 \\ \underline{9\,0} \\ 00 \\ \underline{0} \\ 0 \end{array}$$

24. $\begin{array}{r} 576 \\ 6\overline{)3456} \end{array}$
$\begin{array}{r} 30 \\ \hline 45 \\ 42 \\ \hline 36 \\ 36 \\ \hline 0 \end{array}$

25. (a) $\begin{array}{r} 1000 \\ -\ \ \ 11 \\ \hline \mathbf{989} \end{array}$

(b) **−989**

26. $k = 3(10) - 1$
$k = 30 - 1$
$k = \mathbf{29}$

27. $\begin{array}{r} 86 \\ +\ 119 \\ \hline 205 \end{array}$ $\begin{array}{r} 416 \\ -\ 205 \\ \hline 211 \end{array}$

$\begin{array}{r} \overset{3}{\cancel{4}}{}^{1}16 \\ -\ \ 86 \\ \hline 330 \end{array}$ $\begin{array}{r} 330 \\ +\ 119 \\ \hline 449 \end{array}$

211 **<** 449

28. Acute: $\angle \mathbf{CBA}$ (or $\angle \mathbf{ABC}$)
Obtuse: $\angle \mathbf{DAB}$ (or $\angle \mathbf{BAD}$)
Right: $\angle \mathbf{CDA}$ (or $\angle \mathbf{ADC}$)
and $\angle \mathbf{DCB}$ (or $\angle \mathbf{BCD}$)

29. (a) $\overline{\mathbf{CB}}$ (or $\overline{\mathbf{BC}}$)

(b) $\overline{\mathbf{DC}}$ (or $\overline{\mathbf{CD}}$)

30. **$\overline{QR}$ identifies the segment QR, while QR refers to the distance from Q to R. So $\overline{QR}$ is a segment and QR is a length.**

Lesson 9, Warm-Up

a. **−2**

b. **$3.90**

c. **71¢**

d. **542**

e. **540**

f. **10**

g. **15**

Problem Solving

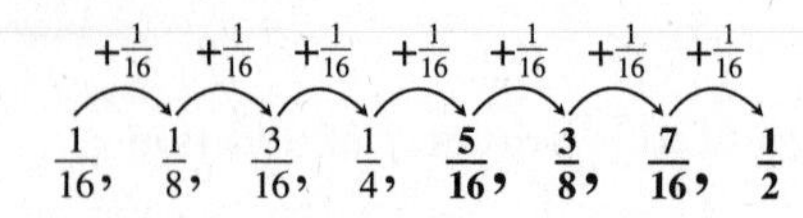

$\frac{1}{16}, \frac{1}{8}, \frac{3}{16}, \frac{1}{4}, \mathbf{\frac{5}{16}}, \mathbf{\frac{3}{8}}, \mathbf{\frac{7}{16}}, \mathbf{\frac{1}{2}}$

Lesson 9, Lesson Practice

a. $\frac{5}{6} + \frac{1}{6} = \frac{6}{6} = \mathbf{1}$

b. $\mathbf{\frac{1}{5}}$

c. $\frac{3}{5} \times \frac{1}{2} = \frac{3}{10}$ $\frac{3}{10} \times \frac{3}{4} = \mathbf{\frac{9}{40}}$

d. $\mathbf{\frac{8}{3}}$

e. $\mathbf{\frac{8}{21}}$

f. $\frac{5}{8} - \frac{5}{8} = \frac{0}{8} = \mathbf{0}$

g. $\mathbf{28\frac{4}{7}\%}$

h. **75%**

i. $\mathbf{\frac{5}{4}}$

j. $\mathbf{\frac{7}{8}}$

k. $\mathbf{\frac{1}{5}}$

l. $\mathbf{\frac{8}{5}}$

m. $\mathbf{\frac{1}{6}}$

n. **$\frac{1}{20}$ of an inch, because $\frac{1}{2}$ of $\frac{1}{10}$ is $\frac{1}{20}$.**

o. $\frac{3}{2}$

p. $\frac{1}{4}$

Lesson 9, Mixed Practice

1. $1 + 2 + 3 = 6$ $\quad 1 \cdot 2 \cdot 3 = 6$ $\quad \frac{6}{6} = \mathbf{1}$

2. **45¢ per pound; \$0.45 per pound**

3. (a) $\frac{1}{2} > \frac{1}{2} \cdot \frac{1}{2}$ **One half is greater than one half times one half.**

   (b) $-2 > -4$ **Negative two is greater than negative four.**

4. **(2 × 10,000) + (6 × 1000)**

5. (a) $\frac{10}{100} = \mathbf{\frac{1}{10}}$

   (b) $100\% \div 10 = \mathbf{10\%}$

6. (a) $\mathbf{\frac{5}{9}}$

   (b) $\mathbf{\frac{4}{9}}$

7. **It is a segment because it has two endpoints.**

8. $LM = \mathbf{1\frac{1}{4}\ in.}$

   $MN = \mathbf{1\frac{1}{4}\ in.}$

   $LN = \mathbf{2\frac{1}{2}\ in.}$

9. (a) **1, 2, 3, 6, 9, 18**

   (b) **1, 2, 3, 4, 6, 8, 12, 24**

   (c) **1, 2, 3, 6**

   (d) **6**

10. (a) $\frac{2}{5} + \frac{2}{5} = \mathbf{\frac{4}{5}}$

    (b) $\frac{2}{5} - \frac{2}{5} = \frac{0}{5} = \mathbf{0}$

11. $200{,}000 - 85{,}000 = 115{,}000$

    $b = \mathbf{115{,}000}$

12. $60\overline{)900}$ = 15; 60, 300, 300, 0

    $c = \mathbf{15}$

13. $\$20.00 - \$5.60 = \$14.40$

    $d = \mathbf{\$14.40}$

14. $12\overline{)\$30.00}$ = \$2.50; 24, 60, 60, 00, 0, 0

    $e = \mathbf{\$2.50}$

15. $\$12.47 + \$98.03 = \$110.50$

    $f = \mathbf{\$110.50}$

16. $5 + 7 + 5 + 7 + 6 + 1 + 2 + 3 + 4 = 40$ $\quad 40 - 40 = 0$

    $n = \mathbf{0}$

17. $\mathbf{2\frac{8}{15}}$

18. $\mathbf{2\frac{7}{8}}$

19. $\mathbf{\frac{3}{16}}$

20. $\begin{array}{r} 106 \\ 17\overline{)1802} \\ \underline{17} \\ 10 \\ \underline{0} \\ 102 \\ \underline{102} \\ 0 \end{array}$

21. $\begin{array}{r} \$8.97 \\ \$110.00 \\ + \quad \$0.53 \\ \hline \mathbf{\$119.50} \end{array}$

22. $\begin{array}{r} \$60.00 \\ - \ \$49.49 \\ \hline \mathbf{\$10.51} \end{array}$

23. $\begin{array}{r} 607 \\ \times \quad 78 \\ \hline 4856 \\ 4249 \\ \hline \mathbf{47{,}346} \end{array}$

24. $\begin{array}{r} \$0.09 \\ \times \quad 56 \\ \hline 54 \\ 45 \\ \hline \mathbf{\$5.04} \end{array}$

25. $\begin{array}{r} 50 \\ \times \ 60 \\ \hline 3000 \end{array} \qquad \begin{array}{r} 3000 \\ \times \quad 70 \\ \hline \mathbf{210{,}000} \end{array}$

26. $\frac{4}{5} \times \frac{2}{3} = \frac{8}{15} \qquad \frac{8}{15} \times \frac{1}{3} = \mathbf{\frac{8}{45}}$

27. $\mathbf{\frac{7}{9}}$

28. (a) **$\angle A$ and $\angle B$**

(b) **$\overline{AC}$** or **$\overline{CA}$**

29. **Each term is half of the preceding term.**

$\frac{1}{8} \times \frac{1}{2} = \mathbf{\frac{1}{16}}$

30. $\mathbf{\frac{5}{2}}$

Lesson 10, Warm-Up

a. **−3**

b. **$12.50**

c. **18¢**

d. **494**

e. **449**

f. **10**

g. **7**

Problem Solving

We write a 2 in the ones place of the minuend because 7 + 5 = 12. We write a 0 in the tens place of the difference because 6 + 1 (from regrouping) equals 7. We write a 4 in the hundreds place of the subtrahend, because 9 + 4 = 13. We write an 8 in the thousands place of the minuend, because 5 + 2 + 1 (from regrouping) equals 8.

$\begin{array}{r} \mathbf{8372} \\ - \ \mathbf{2465} \\ \hline \mathbf{5907} \end{array}$

Lesson 10, Lesson Practice

a. $\begin{array}{r} 8 \\ 4\overline{)35} \\ \underline{32} \\ 3 \end{array}$ $\mathbf{8\frac{3}{4}}$ **inches**

b. $\begin{array}{r} 14 \\ 7\overline{)100\%} \\ \underline{7} \\ 30 \\ \underline{28} \\ 2 \end{array}$ $\mathbf{14\frac{2}{7}\%}$

c. $\begin{array}{r} 2 \\ 5\overline{)12} \\ \underline{10} \\ 2 \end{array}$ $\mathbf{2\frac{2}{5}}$

d. $\begin{array}{r} 2 \\ 6\overline{)12} \end{array}$ **2**

e. $2\frac{12}{7} = \frac{7 \times 2 + 12}{7} = \frac{26}{7}$

$$\begin{array}{r} 3 \\ 7\overline{)26} \\ \underline{21} \\ 5 \end{array} \quad \mathbf{3\frac{5}{7}}$$

f.

g. $\frac{2}{3} + \frac{2}{3} + \frac{2}{3} = \frac{6}{3} = \mathbf{2}$

h. $\frac{7}{3} \times \frac{2}{3} = \frac{14}{9}$

$$\begin{array}{r} 1 \\ 9\overline{)14} \\ \underline{9} \\ 5 \end{array} \longrightarrow \mathbf{1\frac{5}{9}}$$

i. $1\frac{2}{3} + 1\frac{2}{3} = 2\frac{4}{3} = \mathbf{3\frac{1}{3}}$

j. $1\frac{2}{3} = \frac{3 \times 1 + 2}{3} = \mathbf{\frac{5}{3}}$

k. $3\frac{5}{6} = \frac{6 \times 3 + 5}{6} = \mathbf{\frac{23}{6}}$

l. $4\frac{3}{4} = \frac{4 \times 4 + 3}{4} = \mathbf{\frac{19}{4}}$

m. $5\frac{1}{2} = \frac{2 \times 5 + 1}{2} = \mathbf{\frac{11}{2}}$

n. $6\frac{3}{4} = \frac{4 \times 6 + 3}{4} = \mathbf{\frac{27}{4}}$

o. $10\frac{2}{5} = \frac{5 \times 10 + 2}{5} = \mathbf{\frac{52}{5}}$

Lesson 10, Mixed Practice

1. **Answers may vary. One answer is** $\mathbf{(\frac{1}{2} \cdot \frac{1}{3}) \cdot \frac{1}{6} = \frac{1}{2} \cdot (\frac{1}{3} \cdot \frac{1}{6})}$.

2. (a) **Parallel**

   (b) **Perpendicular**

3. $2 + 3 + 4 = 9$

   $2 \times 3 \times 4 = 24$

   $24 - 9 = \mathbf{15}$

4. (a) $100\% \div 10 = 10\%$

   $10\% \times 3 = \mathbf{30\%}$

   (b) **70%**

5. $3\frac{2}{3} = \frac{3 \times 3 + 2}{3} = \mathbf{\frac{11}{3}}$

6. (a) $2 - 2 \mathbf{<} 2 \div 2$

   (b) $\frac{1}{2} + \frac{1}{2} \mathbf{>} \frac{1}{2} \times \frac{1}{2}$

7. $\mathbf{9\frac{5}{6}}$

8.

9. $$\begin{array}{r} 52 \\ 8\overline{)420} \\ \underline{40} \\ 20 \\ \underline{16} \\ 4 \end{array} \qquad \begin{array}{r} 60 \\ 7\overline{)420} \\ \underline{42} \\ 00 \\ \underline{0} \\ 0 \end{array}$$

   $4 + 2 + 0 = 6$

   1, 2, 3, 4, 5, 6, 7

10. $$\begin{array}{r} 36{,}275 \\ -\ 12{,}500 \\ \hline 23{,}775 \end{array}$$

    $x = \mathbf{23{,}775}$

11. $$\begin{array}{r} 22 \\ 18\overline{)396} \\ \underline{36} \\ 36 \\ \underline{36} \\ 0 \end{array}$$

    $y = \mathbf{22}$

12. $$\begin{array}{r} 77{,}000 \\ -\ 39{,}400 \\ \hline 37{,}600 \end{array}$$

    $z = \mathbf{37{,}600}$

13.
$$\begin{array}{r} \$1.25 \\ \times \quad 8 \\ \hline \$10.00 \end{array}$$
$a = \mathbf{\$10.00}$

14.
$$\begin{array}{r} \$8.75 \\ +\ \$16.25 \\ \hline \$25.00 \end{array}$$
$b = \mathbf{\$25.00}$

15.
$$\begin{array}{r} \$75.00 \\ -\ \$37.50 \\ \hline \$37.50 \end{array}$$
$c = \mathbf{\$37.50}$

16.
$$\begin{array}{r} 8 \\ 7 \\ 5 \\ 6 \\ 4 \\ 3 \\ +\ 7 \\ \hline 40 \end{array} \qquad \begin{array}{r} 50 \\ -\ 40 \\ \hline 10 \end{array}$$
$n = \mathbf{10}$

17. $\frac{5}{2} \times \frac{5}{4} = \frac{25}{8}$ $\quad 8\overline{)25}$ = 3 R 1 (24, 1) $\quad \mathbf{3\frac{1}{8}}$

18. $\frac{5}{8} - \frac{5}{8} = \frac{0}{8} = \mathbf{0}$

19. $\frac{11}{20} + \frac{18}{20} = \frac{29}{20} = \frac{20}{20} + \frac{9}{20} = \mathbf{1\frac{9}{20}}$

20.
$$\begin{array}{r} 680 \\ -\ 59 \\ \hline 621 \end{array} \qquad \begin{array}{r} 2000 \\ -\ 621 \\ \hline \mathbf{1379} \end{array}$$

21. $9\overline{)100\%}$ = 11 (9, 10, 9, 1) $\quad \mathbf{11\frac{1}{9}\%}$

22.
$$\begin{array}{r} \$0.89 \\ \$0.57 \\ +\ \$15.74 \\ \hline \mathbf{\$17.20} \end{array}$$

23.
$$\begin{array}{r} 800 \\ \times\ 300 \\ \hline \mathbf{240{,}000} \end{array}$$

24. $2\frac{2}{3} + 2\frac{2}{3} = 4\frac{4}{3} = 4\frac{1}{3} + \frac{3}{3}$

$= 4\frac{1}{3} + 1 = \mathbf{5\frac{1}{3}}$

25. $\mathbf{\frac{8}{27}}$

26. (a) **Ray;** $\overrightarrow{MC}$

(b) **Line;** $\overleftrightarrow{PM}$ or $\overleftrightarrow{MP}$

(c) **Segment;** $\overline{FH}$ or $\overline{HF}$

27. $\mathbf{\frac{9}{5}}$

28. $\frac{1}{2}$ of $2 = \mathbf{1}$

$\frac{1}{2}$ of $1 = \mathbf{\frac{1}{2}}$

$\frac{1}{2}$ of $\frac{1}{2} = \mathbf{\frac{1}{4}}$

29. **C.** $\mathbf{\frac{1}{2}}$

30. (a) $\mathbf{-5}$

(b) $\mathbf{\frac{1}{3}}$

Investigation 1

1. $\mathbf{\frac{1}{4}}$

2. $\mathbf{\frac{1}{8}}$

3. $\mathbf{\frac{1}{6}}$

4. $\mathbf{\frac{1}{12}}$

5. $\mathbf{\frac{1}{6}}$

6. $\mathbf{\frac{1}{12}}$

7. 6 twelfths equals $\frac{1}{2}$.

8.

9.

10.

11.

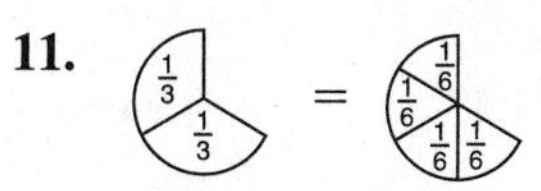

4 sixths equals $\frac{2}{3}$.

12. **9 twelfths** equals $\frac{3}{4}$.

13.

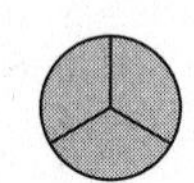

; $\mathbf{\frac{5}{3} = 1\frac{2}{3}}$

14.

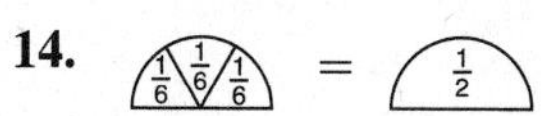

15.

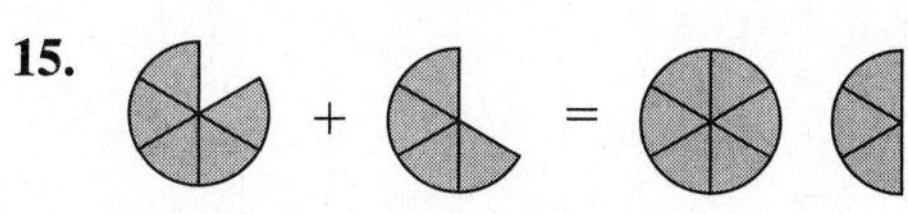

16. Together, $\mathbf{\frac{1}{3}}$ **and** $\mathbf{\frac{1}{6}}$ form a half circle.

17.

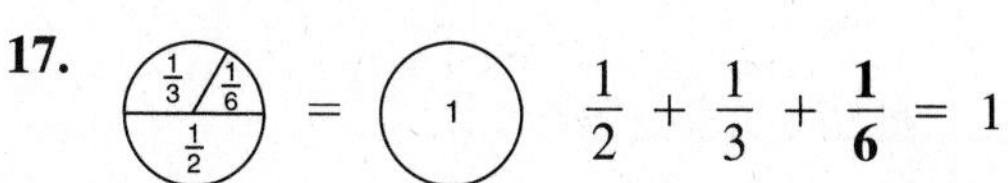

$\frac{1}{2} + \frac{1}{3} + \mathbf{\frac{1}{6}} = 1$

18. $\frac{1}{6} + \mathbf{\frac{1}{12}} = \frac{1}{4}$

19. $\frac{1}{2} + \mathbf{\frac{1}{4}} = \frac{3}{4}$

20. $\frac{1}{4} + \mathbf{\frac{1}{12}} = \frac{1}{3}$

21. $33\frac{1}{3}\% + 33\frac{1}{3}\% = \mathbf{66\frac{2}{3}\%}$

22. $8\frac{1}{3}\% + 8\frac{1}{3}\% + 8\frac{1}{3}\% = \mathbf{25\%}$

23. $12\frac{1}{2}\% + 12\frac{1}{2}\% + 12\frac{1}{2}\% = \mathbf{37\frac{1}{2}\%}$

24. $16\frac{2}{3}\% + 16\frac{2}{3}\% + 16\frac{2}{3}\% = \mathbf{50\%}$

25. $25\% + 8\frac{1}{3}\% = \mathbf{33\frac{1}{3}\%}$

26.

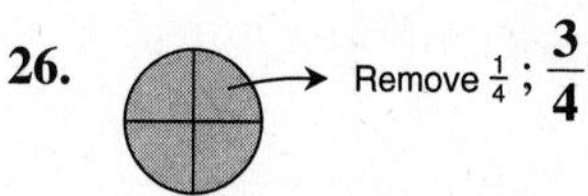

; $\mathbf{\frac{3}{4}}$

27.

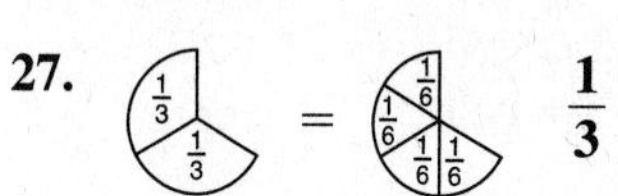

$\mathbf{\frac{1}{3}}$

28. $\mathbf{\frac{1}{4}}$

29. $\mathbf{\frac{1}{4}}$

30. $\mathbf{\frac{2}{4} = \frac{1}{2}; \frac{3}{6} = \frac{1}{2}; \frac{4}{8} = \frac{1}{2}; \frac{6}{12} = \frac{1}{2};}$

$\mathbf{\frac{1}{3} + \frac{1}{6} = \frac{1}{2}; \frac{1}{3} + \frac{2}{12} = \frac{1}{2};}$

$\mathbf{\frac{2}{12} + \frac{2}{6} = \frac{1}{2}; \frac{4}{12} + \frac{1}{6} = \frac{1}{2};}$

$\mathbf{\frac{1}{4} + \frac{2}{8} = \frac{1}{2}; \frac{3}{12} + \frac{2}{8} = \frac{1}{2};}$

$\mathbf{\frac{3}{12} + \frac{1}{4} = \frac{1}{2}}$

Lesson 11, Warm-Up

a. **$8.25**

b. **$4.00**

c. **$4.50**

d. **75**

e. **–750**

f. **14**

g. **7**

Problem Solving

$$\begin{array}{r} 4N = 20¢ \\ +\ 3D = 30¢ \\ \hline 50¢ \end{array}$$

Letha has **4 nickels and 3 dimes.**

Lesson 11, Lesson Practice

a. **$118 + N = 230$**

$$\begin{array}{r} 230 \\ -\ 118 \\ \hline 112 \end{array} \quad \rightarrow \quad \begin{array}{r} 118 \text{ pounds} \\ +\ 112 \text{ pounds} \\ \hline 230 \text{ pounds} \end{array} \quad \text{check}$$

112 pounds

b. **$T + 216 = 400$**

$$\begin{array}{r} 400 \\ -\ 216 \\ \hline 184 \end{array} \quad \rightarrow \quad \begin{array}{r} 216 \text{ turns} \\ +\ 184 \text{ turns} \\ \hline 400 \text{ turns} \end{array} \quad \text{check}$$

184 turns

c. **$254 - H = 126$**

$$\begin{array}{r} 254 \\ -\ 126 \\ \hline 128 \end{array} \quad \rightarrow \quad \begin{array}{r} 126 \text{ horses} \\ +\ 128 \text{ horses} \\ \hline 254 \text{ horses} \end{array} \quad \text{check}$$

128 horses

d. **$P - 36 = 164$**

$$\begin{array}{r} 164 \\ +\ 36 \\ \hline 200 \end{array} \quad \rightarrow \quad \begin{array}{r} 200 \text{ sheets} \\ -\ 36 \text{ sheets} \\ \hline 164 \text{ sheets} \end{array} \quad \text{check}$$

200 sheets

e. **Answers may vary. Sample answer: The price on the tag was $15.00, but after tax the total was $16.13. How much was the tax?**

f. **Answers may vary. Sample answer: There were 32 apples on the tree. After the children picked some apples, 25 apples remained on the tree. How many apples did the children pick?**

Lesson 11, Mixed Practice

1. **$85{,}000 + V = 200{,}000$**

$$\begin{array}{r} 200{,}000 \\ -\ 85{,}000 \\ \hline 115{,}000 \end{array} \quad \rightarrow \quad \begin{array}{r} 85{,}000 \text{ people} \\ +\ 115{,}000 \text{ people} \\ \hline 200{,}000 \text{ people} \end{array} \quad \text{check}$$

115,000 visitors

2. **$M - \$98.03 = \$12.47$**

$$\begin{array}{r} \$98.03 \\ +\ \$12.47 \\ \hline \$110.50 \end{array} \quad \rightarrow \quad \begin{array}{r} \$110.50 \\ -\ \$98.03 \\ \hline \$12.47 \end{array} \quad \text{check}$$

$110.50

3. **$10{,}000 - D = 5420$**

$$\begin{array}{r} 10{,}000 \\ -\ 5{,}420 \\ \hline 4{,}580 \end{array} \quad \rightarrow \quad \begin{array}{r} 5{,}420 \text{ runners} \\ +\ 4{,}580 \text{ runners} \\ \hline 10{,}000 \text{ runners} \end{array} \quad \text{check}$$

4580 runners

4. (a) $\mathbf{\frac{7}{8}}$

(b) $\mathbf{\frac{1}{8}}$

(c) $100\% \div 8 = 12\frac{1}{2}\%$

$12\frac{1}{2}\% \times 1 = \mathbf{12\frac{1}{2}\%}$

5. (a) $\mathbf{-2, 0, \frac{1}{2}, 1}$

(b) $\mathbf{\frac{1}{2}}$

6. $$\begin{array}{r} 4 \\ 8\overline{)35} \text{ inches} \\ \underline{32} \\ 3 \end{array} \qquad \mathbf{4\frac{3}{8} \text{ inches}}$$

7. $\mathbf{1 \cdot 2 < 1 + 2}$

8. $$\begin{array}{r} 100 \text{ million} \\ -\ 89 \text{ million} \\ \hline 11 \text{ million} \end{array}$$

Eleven million

9. (a) **1, 2, 4, 8, 16**

(b) **1, 2, 3, 4, 6, 8, 12, 24**

(c) **1, 2, 4, 8**

(d) **8**

10.
$$\begin{array}{r} 8000 \\ -\ 5340 \\ \hline 2660 \end{array}$$
$k = \mathbf{2660}$

11.
$$\begin{array}{r} 1760 \\ -\ 1320 \\ \hline 440 \end{array}$$
$m = \mathbf{440}$

12. $4 \cdot 9 = 36$
$$\begin{array}{r} 20 \\ 36\overline{)720} \\ 72 \\ \hline 00 \\ 0 \\ \hline 0 \end{array}$$
$n = \mathbf{20}$

13.
$$\begin{array}{r} \$375 \\ -\ \$126 \\ \hline \$249 \end{array}$$
$r = \mathbf{\$249}$

14.
$$\begin{array}{r} 13 \\ 13\overline{)169} \\ 13 \\ \hline 39 \\ 39 \\ \hline 0 \end{array}$$
$s = \mathbf{13}$

15.
$$\begin{array}{r} \$25.00 \\ \times \quad 40 \\ \hline \$1000.00 \end{array}$$
$t = \mathbf{\$1000.00}$

16. $5 \times 20 = 100 \qquad 100 - 100 = 0$

$100 - 5 = 95$
$$\begin{array}{r} 95 \\ \times\ 20 \\ \hline 1900 \end{array}$$
$0 < 1900$

$100 - (5 \times 20) \mathbf{<} (100 - 5) \times 20$

17. $1\frac{5}{9} + 1\frac{5}{9} = 2\frac{10}{9} = 2\frac{1}{9} + \frac{9}{9}$

$= 2\frac{1}{9} + 1 = \mathbf{3\frac{1}{9}}$

18. $\frac{5}{3} \times \frac{2}{3} = \frac{10}{9} = \frac{9}{9} + \frac{1}{9} = \mathbf{1\frac{1}{9}}$

19.
$$\begin{array}{r} 135 \\ \times\ \ 72 \\ \hline 270 \\ 945\ \\ \hline \mathbf{9720} \end{array}$$

20.
$$\begin{array}{r} \mathbf{25} \\ 40\overline{)1000} \\ 80\ \\ \hline 200 \\ 200 \\ \hline 0 \end{array}$$

21.
$$\begin{array}{r} \$1.49 \\ \times \quad 30 \\ \hline \mathbf{\$44.70} \end{array}$$

22.
$$\begin{array}{r} \mathbf{\$4.02} \\ 35\overline{)\$140.70} \\ 140\ \ \ \\ \hline 0\ 7\ \\ 0\ \\ \hline 70 \\ 70 \\ \hline 0 \end{array}$$

23. $\mathbf{\frac{5}{54}}$

24. $\frac{5}{8} + \left(\frac{3}{8} - \frac{1}{8}\right) = \frac{5}{8} + \left(\frac{2}{8}\right) = \mathbf{\frac{7}{8}}$

25. $3\frac{3}{4} = \frac{4 \times 3 + 3}{4} = \mathbf{\frac{15}{4}}$

26. **C. 40%**

27. $\frac{1}{2} + \frac{1}{8} = \frac{4}{8} + \frac{1}{8} = \frac{5}{8}$

$\frac{5}{8} + \frac{1}{8} = \frac{6}{8} = \frac{3}{4}$

$\frac{6}{8} + \frac{1}{8} = \frac{7}{8}$

$\frac{7}{8} + \frac{1}{8} = \frac{8}{8} = 1$

$\mathbf{\frac{5}{8}, \frac{3}{4}, \frac{7}{8}, 1}$

28. (a) $\angle\mathbf{1}, \angle\mathbf{3}$

(b) $\angle\mathbf{2}, \angle\mathbf{4}$

29. $3\frac{1}{2} - 1\frac{7}{8} = 1\frac{5}{8}$

$BC = 1\frac{5}{8}$ **inches**

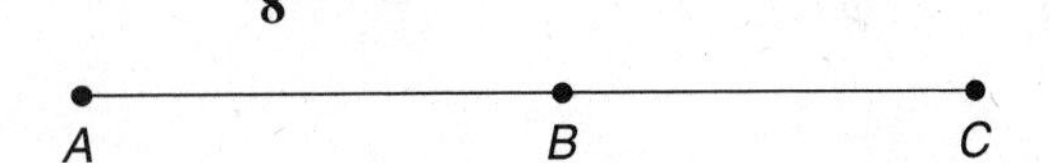

30. $\frac{8}{7}$

Lesson 12, Warm-Up

a. **$7.10**

b. **$12.90**

c. **$7.50**

d. **92**

e. **–1500**

f. **32**

g. **22**

Problem Solving

From the $\frac{1}{2}$-in. mark to the $1\frac{1}{4}$-in. mark is $\frac{3}{4}$ **in.**

Lesson 12, Lesson Practice

a. **1,000,000,000 − 25,000,000 = G**

$$\begin{array}{r} 1{,}000{,}000{,}000 \\ -\quad 25{,}000{,}000 \\ \hline 975{,}000{,}000 \end{array}$$

975,000,000

b. **1791 − 1215 = Y**

$$\begin{array}{r} 1791 \\ -\ 1215 \\ \hline 576 \end{array}$$

576 years

c. **1963 − B = 46**

$$\begin{array}{r} 1963 \\ -\quad 46 \\ \hline 1917 \end{array}$$

John F. Kennedy was born in **1917.**

d. **Answers may vary. Sample answer: Todd is 58 in. tall and Glenda is 55 in. tall. Todd is how many inches taller than Glenda?**

e. **Answers may vary. Sample answer: Rosalie turned 14 in 2003. In what year was she born?**

Lesson 12, Mixed Practice

1. **77,000 − L = 39,400**

$$\begin{array}{r} 77{,}000 \\ -\ 39{,}400 \\ \hline 37{,}600 \end{array} \quad \begin{array}{r} 39{,}400 \text{ fans} \\ +\ 37{,}600 \text{ fans} \\ \hline 77{,}000 \text{ fans} \end{array} \quad \text{check}$$

37,600 fans

2. **B + 18 = 31**

$$\begin{array}{r} 31 \\ -\ 18 \\ \hline 13 \end{array} \quad \begin{array}{r} 18 \text{ bananas} \\ +\ 13 \text{ bananas} \\ \hline 31 \text{ bananas} \end{array} \quad \text{check}$$

13 bananas

3. **1215 − 1066 = Y**

$$\begin{array}{r} 1215 \\ -\ 1066 \\ \hline 149 \end{array}$$

149 years

4. **77,000 − 49,600 = F**

$$\begin{array}{r} 77{,}000 \text{ fans} \\ -\ 49{,}600 \text{ fans} \\ \hline 27{,}400 \text{ fans} \end{array}$$

27,400 fans

5. **Answers may vary. Sample answer: Marla gave the clerk $20.00 to purchase a CD. Marla got back $7.13. How much did the CD cost?**

6. **Identity property of multiplication**

7. **1,000,000 − 23,000 = D**

$$\begin{array}{r} 1{,}000{,}000 \\ -\quad 23{,}000 \\ \hline 977{,}000 \end{array}$$

Nine hundred seventy-seven thousand

8. (a) $2 - 3 = -1$

(b) $\frac{1}{2} > \frac{1}{3}$

9. $\overline{PQ}$ (or $\overline{QP}$), $\overline{QR}$ (or $\overline{RQ}$), $\overline{PR}$ (or $\overline{RP}$)

10.

11. (a) $\mathbf{\frac{3}{4}}$

(b) $100\% \div 4 = 25\%$

$25\% \times 1 = \mathbf{25\%}$

12. **1, 2, 4, 5, 10, 20, 25, 50, 100**

13.
$$\begin{array}{r} 42 \\ 15\overline{)630} \\ \underline{60} \\ 30 \\ \underline{30} \\ 0 \end{array}$$
$x = \mathbf{42}$

14.
$$\begin{array}{r} 3601 \\ +\ 2714 \\ \hline 6315 \end{array}$$
$y = \mathbf{6315}$

15.
$$\begin{array}{r} 2900 \\ -\ \ \ 64 \\ \hline 2836 \end{array}$$
$p = \mathbf{2836}$

16.
$$\begin{array}{r} \$5.00 \\ -\ \$1.53 \\ \hline \$3.47 \end{array}$$
$q = \mathbf{\$3.47}$

17.
$$\begin{array}{r} 60 \\ 20\overline{)1200} \\ \underline{120} \\ 00 \\ \underline{0} \\ 0 \end{array}$$
$r = \mathbf{60}$

18.
$$\begin{array}{r} 16 \\ \times\ 14 \\ \hline 64 \\ \underline{160} \\ 224 \end{array}$$
$m = \mathbf{224}$

19.
$$\begin{array}{r} 72{,}112 \\ -\ 64{,}309 \\ \hline \mathbf{7{,}803} \end{array}$$

20.
$$\begin{array}{r} 453{,}978 \\ +\ 386{,}864 \\ \hline \mathbf{840{,}842} \end{array}$$

21. $\frac{8}{9} - \left(\frac{3}{9} + \frac{5}{9}\right) = \frac{8}{9} - \left(\frac{8}{9}\right) = \mathbf{0}$

22. $\left(\frac{8}{9} - \frac{3}{9}\right) + \frac{5}{9} = \left(\frac{5}{9}\right) + \frac{5}{9} = \frac{10}{9}$

$= \mathbf{1\frac{1}{9}}$

23. $\frac{9}{2} \times \frac{3}{5} = \frac{27}{10} = \mathbf{2\frac{7}{10}}$

24.
$$\begin{array}{r} \mathbf{\$2.48} \\ 15\overline{)\$37.20} \\ \underline{30} \\ 7\ 2 \\ \underline{6\ 0} \\ 1\ 20 \\ \underline{1\ 20} \\ 0 \end{array}$$

25.
$$\begin{array}{r} 4{,}760 \\ 9\overline{)42{,}847} \\ \underline{36} \\ 6\ 8 \\ \underline{6\ 3} \\ 54 \\ \underline{54} \\ 07 \\ \underline{0} \\ 7 \end{array}$$
$\mathbf{4760\frac{7}{9}}$

26.
$$\begin{array}{r} \$4.36 \\ \$15.96 \\ \$0.76 \\ +\ \$35.00 \\ \hline \mathbf{\$56.08} \end{array}$$

27. $\frac{3}{4} + \frac{1}{4} = \frac{4}{4} = 1$

$\frac{4}{4} + \frac{1}{4} = \frac{5}{4} = 1\frac{1}{4}$

$\frac{5}{4} + \frac{1}{4} = \frac{6}{4} = \frac{3}{2}$

1, $\mathbf{1\frac{1}{4}}$**,** $\mathbf{1\frac{1}{2}}$

28. $\mathbf{\frac{3}{2}}$

29. $1\frac{2}{3} = \frac{3 \times 1 + 2}{3} = \frac{5}{3}$

$\frac{5}{3} \times \frac{1}{2} = \mathbf{\frac{5}{6}}$

30. 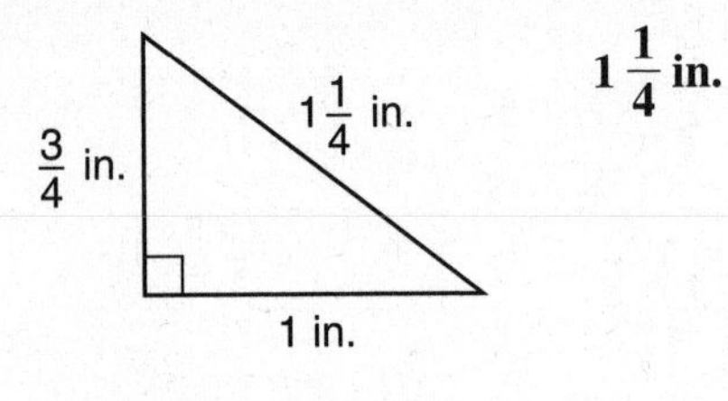

$1\frac{1}{4}$ in.

LESSON 13, WARM-UP

a. **$7.20**

b. **$2.50**

c. **$3.25**

d. **165**

e. **–250**

f. **43**

g. **9**

Problem Solving

J, M, L
J, L, M
M, J, L
M, L, J
L, J, M
L, M, J
There are **6 possible permutations.**

LESSON 13, LESSON PRACTICE

a. **$24 \times 18¢ = M$**

$$\begin{array}{r} \$0.18 \\ \times \quad 24 \\ \hline 72 \\ 36 \\ \hline \$4.32 \end{array}$$

$4.32

b. **$R \times 25 = 375$**

$$\begin{array}{r} 15 \\ 25\overline{)375} \\ 25 \\ \hline 125 \\ 125 \\ \hline 0 \end{array}$$

$15 \times 25 = 375$ check

15 rows

c. **$7P = 1225$**

$$\begin{array}{r} 175 \\ 7\overline{)1225} \\ 7 \\ \hline 52 \\ 49 \\ \hline 35 \\ 35 \\ \hline 0 \end{array}$$

$7 \times 175 = 1225$ check

175 push-ups

d. **Answers may vary. Sample answer: If a dozen doughnuts cost $3.00, what is the cost per doughnut?**

LESSON 13, MIXED PRACTICE

1. **$72,112 - 64,309 = I$**

$$\begin{array}{r} 72,112 \\ -\ 64,309 \\ \hline \mathbf{7,803} \end{array}$$

2. **$60 - N = 17$**

60 night crawlers
− 17 night crawlers
43 night crawlers

3. **$1945 - B = 63$**

$$\begin{array}{r} 1945 \\ -\quad 63 \\ \hline \mathbf{1882} \end{array}$$

4. **$75 \times 12 = B$**

$$\begin{array}{r} 75 \\ \times\ 12 \\ \hline 150 \\ 75 \\ \hline 900 \end{array}$$

900 beach balls

5. **$T \times 8 = 120$**

$$\begin{array}{r} 15 \\ 8\overline{)120} \\ 8 \\ \hline 40 \\ 40 \\ \hline 0 \end{array}$$

$15 \times 8 = 120$ check

15 truckloads

6. Answers may vary. Sample answer: Five tickets for the show cost \$63.75. If all the tickets were the same price, then what was the cost per ticket?

$$5\overline{)\$63.75} = \$12.75$$
$$\begin{array}{r} 5 \\ \hline 13 \\ 10 \\ \hline 37 \\ 35 \\ \hline 25 \\ 25 \\ \hline 0 \end{array}$$

5(\$12.75) = \$63.75 check

\$12.75

7. (5 × 8) − (5 + 8) = 40 − 13 = **27**

8. (a) $\frac{3 \text{ quarters}}{4 \text{ quarters}} = \mathbf{\frac{3}{4}}$

(b) $\frac{75¢}{100¢} =$ **75%**

9. **10 units**

10. (a) **Line;** $\overleftrightarrow{BR}$ (or $\overleftrightarrow{RB}$)

(b) **Segment;** $\overline{TV}$ (or $\overline{VT}$)

(c) **Ray;** $\overrightarrow{MW}$

11. (a) Factors of 24: 1, 2, 3, 4, 6, 8, 12, 24
Factors of 36: 1, 2, 3, 4, 6, 9, 12, 18, 36
1, 2, 3, 4, 6, 12

(b) **12**

12. (a) A: $\mathbf{\frac{6}{7}}$; B: $\mathbf{1\frac{4}{7}}$

(b) $1\frac{4}{7} - \frac{6}{7} = \frac{7 \times 1 + 4}{7} - \frac{6}{7}$
$= \frac{11}{7} - \frac{6}{7} = \mathbf{\frac{5}{7}}$ **units**

13. $36\overline{)1800} = 50$
$$\begin{array}{r} 180 \\ \hline 00 \\ 0 \\ \hline 0 \end{array}$$
c = **50**

14.
$$\begin{array}{r} \$3.77 \\ +\ \$1.64 \\ \hline \$5.41 \end{array}$$
f = **\$5.41**

15.
$$\begin{array}{r} 28 \\ \times\ 7 \\ \hline 196 \end{array}$$
d = **196**

16. $30\overline{)4500} = 150$
$$\begin{array}{r} 30 \\ \hline 150 \\ 150 \\ \hline 00 \\ 0 \\ \hline 0 \end{array}$$
e = **150**

17.
$$\begin{array}{r} 4 \\ 7 \\ 6 \\ 8 \\ 4 \\ 5 \\ 5 \\ 7 \\ 9 \\ 6 \\ +\ 8 \\ \hline 69 \end{array} \qquad \begin{array}{r} 75 \\ -\ 69 \\ \hline 6 \end{array}$$
n = **6**

18.
$$\begin{array}{r} 3674 \\ -\ 2159 \\ \hline 1515 \end{array}$$
a = **1515**

19.
$$\begin{array}{r} 5179 \\ -\ 4610 \\ \hline 569 \end{array}$$
b = **569**

20.
$$\begin{array}{r} 363 \\ 4579 \\ 86 \\ +\ \ 7 \\ \hline \mathbf{5035} \end{array}$$

21. (5 · 4) ÷ (3 + 2) = (20) ÷ (5) = **4**

22. $\frac{5}{3} \cdot \frac{5}{2} = \frac{25}{6} = \mathbf{4\frac{1}{6}}$

23. $3\frac{4}{5} - \left(\frac{2}{5} + 1\frac{1}{5}\right)$

$= \frac{5 \times 3 + 4}{5} - \left(\frac{2}{5} + \frac{5 \times 1 + 1}{5}\right)$

$= \frac{19}{5} - \left(\frac{2}{5} + \frac{6}{5}\right) = \frac{19}{5} - \left(\frac{8}{5}\right)$

$= \frac{11}{5} = \mathbf{2\frac{1}{5}}$

24. $25\overline{)600}$ = **24**

$$\begin{array}{r} 24 \\ 25\overline{)600} \\ \underline{50} \\ 100 \\ \underline{100} \\ 0 \end{array}$$

25.
$$\begin{array}{r} 600 \\ \times\ \ 25 \\ \hline 3000 \\ 1200 \\ \hline \mathbf{15{,}000} \end{array}$$

26.
$$100 \div 10 = 10,\ 1000 \div 10 = 100$$
$$1000 \div 100 = 10,\ 10 \div 10 = 1$$
$$100 > 1$$
$$1000 \div (100 \div 10) \mathbf{>} (1000 \div 100) \div 10$$

27. $2\frac{1}{2} = \frac{2 \times 2 + 1}{2} = \frac{5}{2}$

$\frac{5}{2} \cdot \frac{1}{3} = \mathbf{\frac{5}{6}}$

28. $\frac{11}{12} \cdot \frac{12}{11} = \mathbf{1}$

29. Obtuse: $\angle \boldsymbol{D}$
Acute: $\angle \boldsymbol{A}$
Right: $\angle \boldsymbol{B}$ **and** $\angle \boldsymbol{C}$

30. (a) $\overline{DC}$ (or $\overline{CD}$)

(b) $\overline{CB}$ (or $\overline{BC}$)

Lesson 14, Warm-Up

a. **$6.75**

b. **$6.30**

c. **$1.75**

d. **144**

e. **125**

f. **18**

g. **3**

Problem Solving

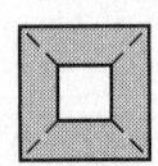

Lesson 14, Lesson Practice

a. **39% + N = 100%**

100% − 39% = 61%

39% + 61% = 100% check

61%

b. $\frac{2}{5} + S = \frac{5}{5}$

$\frac{5}{5} - \frac{2}{5} = \frac{3}{5}$

$\frac{2}{5}$ pioneers did not survive
$+\ \frac{3}{5}$ pioneers survived
$\frac{5}{5}$ total pioneers check

$\mathbf{\frac{3}{5}}$

c. **Answers may vary. Sample answer: If 45% of the candy bars melted, then what percent of the candy bars did not melt?**

Lesson 14, Mixed Practice

1. **65 + C = 142**

142 − 65 = 77

65 grams + 77 grams = 142 grams check

77 grams

2. $\mathbf{\frac{7}{10} + W = \frac{10}{10}}$

$\frac{10}{10} - \frac{7}{10} = \frac{3}{10}$

$\frac{3}{10}$ recruits liked haircut
$+\ \frac{7}{10}$ recruits did not like haircut
$\frac{10}{10}$ total recruits check

$\mathbf{\frac{3}{10}}$

3. **1789 − 1776 = Y**

1789 − 1776 = 13

1776 + 13 = 1789 check

13 years

4. **Answers may vary. Sample answer: If a dozen flavored icicles cost $2.40, then what is the cost per flavored icicle?**

5. $24\% + N_G = 100\%$

$100\% - 24\% = 76\%$

$24\% + 76\% = 100\%$ check

76%

6.

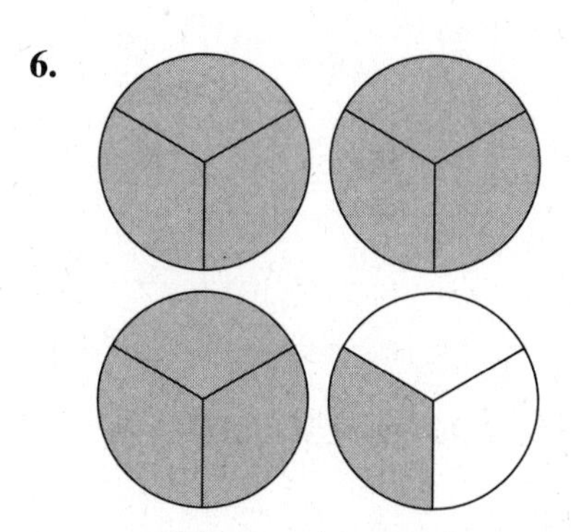

7. **407,042,603**

8. **Property of zero for multiplication**

9. (a) Factors of 40: 1, 2, 4, 5, 8, 10, 20, 40

Factors of 72: 1, 2, 3, 4, 6, 8, 9, 12, 18, 24, 36, 72

1, 2, 4, 8

(b) **8**

10. $\overline{XY}$ (or $\overline{YX}$), $\overline{WX}$ (or $\overline{XW}$), $\overline{WY}$ (or $\overline{YW}$)

11. **Count the number in the group, which is 12. Use this as the denominator. Count the number that are shaded, which is 5. Use this as the numerator; $\frac{5}{12}$.**

12.
```
  623
+ 407
 1030
```
$b =$ **1030**

13.
```
  $20.00
-  $3.47
  $16.53
```
$e =$ **$16.53**

14.
```
     202
35)7070
   70
    07
     0
     70
     70
      0
```
$f =$ **202**

15.
```
  25
× 25
 125
 50
 625
```
$m =$ **625**

16. 5 + 8 + 7 + 6 + 5 + 9 + 4 + 3 + 6 + 4 + 7 + 8 + 5 + 6 = 83

```
  89
- 83
   6
```
$n =$ **6**

17.
```
  1000
-  295
   705
```
$a =$ **705**

18. $3\frac{3}{5} + 2\frac{4}{5} = \frac{18}{5} + \frac{14}{5}$

$= \frac{32}{5} = \mathbf{6\frac{2}{5}}$

19. $\frac{5}{2} \cdot \frac{3}{2} = \frac{15}{4} = \mathbf{3\frac{3}{4}}$

20.
```
  $3.63
  $0.87
+ $0.96
  $5.46
```

21. $5 \cdot 4 \cdot 3 \cdot 2 \cdot 1 = 20 \cdot 3 \cdot 2 \cdot 1$

$= 60 \cdot 2 \cdot 1 = 120 \cdot 1 =$ **120**

22. $\mathbf{\frac{8}{27}}$

23.
```
     45
20)900
   80
   100
   100
     0
```

24.
$$\begin{array}{r} 145 \\ \times\ \ 74 \\ \hline 580 \\ 1015 \\ \hline \mathbf{10{,}730} \end{array}$$

25.
$$\begin{array}{r} \$0.65 \\ \times\ \ \ 30 \\ \hline \mathbf{\$19.50} \end{array}$$

26. $(5)(5 + 5) = 5(10) = \mathbf{50}$

27.
$$\begin{array}{r} 13{,}456 \\ -\ \ 9{,}714 \\ \hline 3{,}742 \end{array} \qquad \mathbf{-3742}$$

28. $1000 - 100 = 900,\ 900 - 10 = 890$
$100 - 10 = 90,\ 1000 - 90 = 910$
$890 < 910$
$(1000 - 100) - 10 \mathbf{<} 1000 - (100 - 10)$

29. (a) **Right angle**

(b) **Straight angle**

(c) **Obtuse angle**

30. $\mathbf{\frac{5}{4}}$

Lesson 15, Warm-Up

a. **$5.25**

b. **$0.40**

c. **$5.02**

d. **224**

e. **75**

f. **26**

g. **5**

Problem Solving

The top partial product has three digits and ends in 0, so the ones digit of the bottom factor must be 5. The bottom partial product has only two digits and ends in 6, so the tens digit of the bottom factor must be 1. To fill in the rest of the digits, we multiply 36 by 15.

$$\begin{array}{r} \mathbf{36} \\ \times\ \mathbf{15} \\ \hline \mathbf{180} \\ \mathbf{360} \\ \hline \mathbf{540} \end{array}$$

Lesson 15, Lesson Practice

a. $\frac{3}{4} \times \frac{5}{5} = \mathbf{\frac{15}{20}}$

$\frac{3}{4} \times \frac{7}{7} = \mathbf{\frac{21}{28}}$

$\frac{3}{4} \times \frac{3}{3} = \mathbf{\frac{9}{12}}$

b. $\frac{3}{4} \times \frac{4}{4} = \mathbf{\frac{12}{16}}$

c. $\frac{4}{5} \times \frac{4}{4} = \frac{\mathbf{16}}{20}$

d. $\frac{3}{8} \times \frac{3}{3} = \frac{9}{\mathbf{24}}$

e. $\frac{3}{5} \times \frac{2}{2} = \mathbf{\frac{6}{10}}$

$\frac{1}{2} \times \frac{5}{5} = \mathbf{\frac{5}{10}}$

$\frac{6}{10} - \frac{5}{10} = \mathbf{\frac{1}{10}}$

f. $\frac{3}{6} = \frac{3 \div 3}{6 \div 3} = \mathbf{\frac{1}{2}}$

g. $\frac{8}{10} = \frac{8 \div 2}{10 \div 2} = \mathbf{\frac{4}{5}}$

h. $\frac{8}{16} = \frac{8 \div 8}{16 \div 8} = \mathbf{\frac{1}{2}}$

i. $\frac{12}{16} = \frac{12 \div 4}{16 \div 4} = \mathbf{\frac{3}{4}}$

j. $\frac{4}{8} = \frac{4 \div 4}{8 \div 4} = \frac{1}{2},\ 4\frac{4}{8} = \mathbf{4\frac{1}{2}}$

k. $\frac{9}{12} = \frac{9 \div 3}{12 \div 3} = \frac{3}{4}, 6\frac{9}{12} = \mathbf{6\frac{3}{4}}$

l. $\mathbf{12\frac{8}{15}}$

m. $\frac{16}{24} = \frac{16 \div 8}{24 \div 8} = \frac{2}{3}, 8\frac{16}{24} = \mathbf{8\frac{2}{3}}$

n. $\frac{5}{12} + \frac{5}{12} = \frac{10}{12} = \frac{10 \div 2}{12 \div 2} = \mathbf{\frac{5}{6}}$

o. $3\frac{7}{10} - 1\frac{1}{10} = 2\frac{6}{10}$

$\frac{6}{10} = \frac{6 \div 2}{10 \div 2} = \frac{3}{5}, 2\frac{6}{10} = \mathbf{2\frac{3}{5}}$

p. $\frac{5}{8} \cdot \frac{2}{3} = \frac{10}{24} = \frac{10 \div 2}{24 \div 2} = \mathbf{\frac{5}{12}}$

q. $90\% = \frac{90}{100}$

$\frac{90 \div 10}{100 \div 10} = \mathbf{\frac{9}{10}}$

r. $75\% = \frac{75}{100}$

$\frac{75 \div 25}{100 \div 25} = \mathbf{\frac{3}{4}}$

s. $5\% = \frac{5}{100}$

$\frac{5 \div 5}{100 \div 5} = \mathbf{\frac{1}{20}}$

t. $\frac{2}{3} \cdot \frac{2}{2} = \frac{4}{6}$

$\frac{4}{6} - \frac{1}{6} = \frac{3}{6} = \frac{3 \div 3}{6 \div 3} = \mathbf{\frac{1}{2}}$

Lesson 15, Mixed Practice

1. $\mathbf{1998 - B = 75}$

$$\begin{array}{r} 1998 \\ -\ \ 75 \\ \hline \mathbf{1923} \end{array}$$

2. $\mathbf{27 + 38 + 56 = T}$

$$\begin{array}{r} 27 \text{ geese} \\ 38 \text{ geese} \\ +\ 56 \text{ geese} \\ \hline \mathbf{121 \text{ geese}} \end{array}$$

3. $40\% = \frac{40}{100}$

$\frac{40}{100} \div \frac{20}{20} = \mathbf{\frac{2}{5}}$

4. $\mathbf{60C = 9000}$

$$\begin{array}{r} 150 \\ 60\overline{)9000} \\ 60 \\ \hline 300 \\ 300 \\ \hline 00 \\ 0 \\ \hline 0 \end{array}$$

150 bushels

5. ______________________________

____

$2\frac{1}{2}\text{ in.} - 1\frac{7}{8}\text{ in.} = \frac{5}{2}\text{ in.} - \frac{15}{8}\text{ in.}$

$= \frac{20}{8}\text{ in.} - \frac{15}{8}\text{ in.} = \mathbf{\frac{5}{8}\text{ in.}}$

6. $\mathbf{3 \cdot 5 > 3 + 5}$

7. The one-digit factors of 2100 are:
1 (All numbers are divisible by 1.)
2 (List digit is even.)
3 (Sum of digits is divisible by 3.)
4 (Number ends in two zeros.)
5 (Last digit is zero.)
6 (Number is divisible by both 2 and 3.)
7 (Check by using division.)

8. (a) $\frac{6}{8} = \frac{6 \div 2}{8 \div 2} = \mathbf{\frac{3}{4}}$

(b) $\frac{6}{10} = \frac{6 \div 2}{10 \div 2} = \frac{3}{5}, 2\frac{6}{10} = \mathbf{2\frac{3}{5}}$

9. $\frac{2}{3} \cdot \frac{3}{3} = \mathbf{\frac{6}{9}}$

$\frac{2}{3} \cdot \frac{5}{5} = \mathbf{\frac{10}{15}}$

$\frac{2}{3} \cdot \frac{6}{6} = \mathbf{\frac{12}{18}}$

10. (a) $\frac{3}{5} \cdot \frac{4}{4} = \mathbf{\frac{12}{20}}$

(b) $\frac{1}{2} \cdot \frac{10}{10} = \mathbf{\frac{10}{20}}$

(c) $\frac{3}{4} \cdot \frac{5}{5} = \mathbf{\frac{15}{20}}$

11. (a) $\overleftrightarrow{QS}$ (or $\overleftrightarrow{SQ}$) or $\overleftrightarrow{QR}$ (or $\overleftrightarrow{RQ}$) or $\overleftrightarrow{RS}$ (or $\overleftrightarrow{SR}$)

(b) $\overrightarrow{RT}, \overrightarrow{RQ}, \overrightarrow{RS}$

(c) $\angle TRS$ or $\angle SRT$

12. (a) $\mathbf{3\frac{2}{3}}$

(b) $\frac{12}{3} = \frac{12 \div 3}{3 \div 3} = \frac{4}{1} = \mathbf{4}$

(c) $\mathbf{4\frac{1}{3}}$

13.
$$11(6 + 7) = 11(13) = 143$$
$$66 + 77 = 143$$
$$143 = 143$$
$$(11)(6 + 7) \mathbf{=} 66 + 77$$

14.
$$\begin{array}{r} 50 \\ -\ 39 \\ \hline 11 \end{array}$$
$b = \mathbf{11}$

15.
$$\begin{array}{r} 50 \\ 6\overline{)300} \\ \underline{30} \\ 00 \\ \underline{0} \\ 0 \end{array}$$
$a = \mathbf{50}$

16.
$$\begin{array}{r} \$5.00 \\ +\ \$0.05 \\ \hline \$5.05 \end{array}$$
$c = \mathbf{\$5.05}$

17.
$$\begin{array}{r} 35 \\ \times\ 35 \\ \hline 175 \\ \underline{105} \\ 1225 \end{array}$$
$w = \mathbf{1225}$

18.
$$80\% = \frac{80}{100}$$
$$\frac{80}{100} \div \frac{20}{20} = \mathbf{\frac{4}{5}}$$

19.
$$35\% = \frac{35}{100}$$
$$\frac{35}{100} \div \frac{5}{5} = \mathbf{\frac{7}{20}}$$

20. **8**

21. $\frac{2}{5} + \frac{3}{5} + \frac{4}{5} = \frac{9}{5} = \mathbf{1\frac{4}{5}}$

22. $3\frac{5}{8} - 1\frac{3}{8} = 2\frac{2}{8} = \mathbf{2\frac{1}{4}}$

23. $\frac{4}{3} \cdot \frac{3}{4} = \frac{12}{12} = \mathbf{1}$

24. $\frac{3}{4} + \frac{3}{4} = \frac{6}{4} = 1\frac{2}{4} = \mathbf{1\frac{1}{2}}$

25. $\frac{7}{5} + \frac{8}{5} = \frac{15}{5} = \mathbf{3}$

26.
$$\frac{11}{12} - \frac{1}{12} = \frac{10}{12}$$
$$\frac{10 \div 2}{12 \div 2} = \mathbf{\frac{5}{6}}$$

27.
$$\frac{5}{6} \cdot \frac{2}{3} = \frac{10}{18}$$
$$\frac{10 \div 2}{18 \div 2} = \mathbf{\frac{5}{9}}$$

28. (a) $\frac{4}{8} + \frac{4}{8} = \frac{8}{8} = \mathbf{1}$

(b) $\frac{4}{8} - \frac{4}{8} = \frac{0}{8} = \mathbf{0}$

29.
$$\frac{1}{3} \cdot \frac{2}{2} = \frac{2}{6}$$
$$\frac{2}{6} + \frac{1}{6} = \frac{3}{6} = \mathbf{\frac{1}{2}}$$

30.
$$2\frac{2}{3} = \frac{8}{3}$$
$$\frac{8}{3} \cdot \frac{1}{4} = \frac{8}{12}$$
$$\frac{8 \div 4}{12 \div 4} = \mathbf{\frac{2}{3}}$$

Lesson 16, Warm-Up

a. **–10**

b. **$1.50**

c. **82¢**

d. **92**

e. **125**

f. $\mathbf{\frac{1}{6}}$

g. **2**

Problem Solving

$$\begin{array}{rl} 5P = & 5¢ \\ 2D = & 20¢ \\ + \ 1Q = & 25¢ \\ \hline & 50¢ \end{array} \qquad \begin{array}{rl} 6N = & 30¢ \\ + \ 2D = & 20¢ \\ \hline & 50¢ \end{array}$$

Fiona could have either **5 pennies, 2 dimes, and 1 quarter** or **6 nickels and 2 dimes.**

Lesson 16, Lesson Practice

a. **2 yards**

b. **2 quarts**

c. **8 lb**

d. $\frac{3}{8}$ in. + $\frac{5}{8}$ in. = $\frac{8}{8}$ in. = **1 in.**

e.

$$\begin{array}{r} 32° \text{ F} \\ + \ 180° \text{ F} \\ \hline \mathbf{212° \text{ F}} \end{array}$$

f. 2(3 ft + 4 ft) = 2(7 ft) = **14 ft**

g. 1 ton = 2000 pounds

$$\begin{array}{r} 2000 \text{ pounds} \\ - \ 1000 \text{ pounds} \\ \hline \mathbf{1000 \text{ pounds}} \end{array}$$

h. $\mathbf{\frac{1}{2}}$ **ounce**

Lesson 16, Mixed Practice

1. **35 +** $\mathbf{N_C}$ **= 118**

$$\begin{array}{r} 118 \text{ sports cars} \\ - \ 35 \text{ sports cars} \\ \hline \mathbf{83 \text{ sports cars}} \end{array}$$

2. **18*C* = 4500**

$$\begin{array}{r} 250 \\ 18\overline{)4500} \\ \underline{36} \\ 90 \\ \underline{90} \\ 00 \\ \underline{0} \\ 0 \end{array}$$

250 cartons

3. **324 − *F* = 27**

$$\begin{array}{r} 324 \text{ ducks} \\ - \ 27 \text{ ducks} \\ \hline \mathbf{297 \text{ ducks}} \end{array}$$

4. **(2 × 100) + (5 × 10)**

5. (a) $\frac{8}{10} = \frac{4}{5}$

(b) $\frac{8}{5} = 1\frac{3}{5}$

$\frac{8}{5} > 1\frac{2}{5}$

6. ***AB* is 1$\frac{3}{8}$ in.; *CB* is 1$\frac{3}{8}$ in.; *CA* is 2$\frac{3}{4}$ in.**

7. (a) $\frac{8}{12} = \frac{8 \div 4}{12 \div 4} = \mathbf{\frac{2}{3}}$

(b) $40\% = \frac{40}{100}$

$\frac{40}{100} \div \frac{20}{20} = \mathbf{\frac{2}{5}}$

(c) $\frac{10}{12} = \frac{10 \div 2}{12 \div 2} = \frac{5}{6}$, $6\frac{10}{12} = \mathbf{6\frac{5}{6}}$

8.

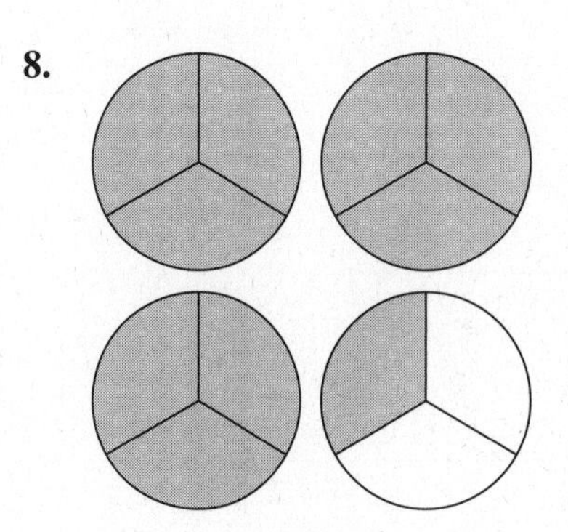

9. (a) $\frac{5}{6} \cdot \frac{4}{4} = \mathbf{\frac{20}{24}}$

(b) $\frac{3}{8} \cdot \frac{3}{3} = \mathbf{\frac{9}{24}}$

(c) $\frac{1}{4} \cdot \frac{6}{6} = \mathbf{\frac{6}{24}}$

10. (a) $100\% \div 3 = \mathbf{33\frac{1}{3}\%}$

(b) $\frac{1 \text{ quart}}{4 \text{ quarts}} = \mathbf{\frac{1}{4}}$

11. $7\overline{)630}$ = 90

63
00
0
0

1, 2, 3, 5, 6, 7, 9

12. (a) $\mathbf{2\frac{2}{7}}$

(b) $3\frac{16}{8} = 3\frac{2}{1} = \frac{5}{1} = \mathbf{5}$

(c) $2\frac{16}{9} = \frac{34}{9} = \mathbf{3\frac{7}{9}}$

13. **Identity property of multiplication**

14.

$$\begin{array}{r} 1776 \\ +\ \ 87 \\ \hline 1863 \end{array}$$

$m = \mathbf{1863}$

15.

$$\begin{array}{r} \$16.25 \\ -\ \$10.15 \\ \hline \$6.10 \end{array}$$

$b = \mathbf{\$6.10}$

16. $13\overline{)1001}$ = 77

91
91
91
0

$n = \mathbf{77}$

17. $42\overline{)1764}$ = 42

168
84
84
0

$d = \mathbf{42}$

18. $3\frac{3}{4} - 1\frac{1}{4} = 2\frac{2}{4} = \mathbf{2\frac{1}{2}}$

19. $\frac{3}{10}$ in. $+ \frac{8}{10}$ in. $= \frac{11}{10}$ in. $= \mathbf{1\frac{1}{10}}$ **in.**

20. $\frac{3}{4} \times \frac{1}{3} = \frac{3}{12} = \frac{3 \div 3}{12 \div 3} = \mathbf{\frac{1}{4}}$

21. $\frac{4}{3} \cdot \frac{3}{2} = \frac{12}{6} = \mathbf{2}$

22. $16\overline{)10{,}000}$ = **625**

96
40
32
80
80
0

23. $\frac{100\%}{8} = \frac{100\% \div 4}{8 \div 4} = \frac{25\%}{2}$

$= \mathbf{12\frac{1}{2}\%}$

24. $9\overline{)70{,}000}$ = 7,777

63
70
63
70
63
70
63
7

$\mathbf{7{,}777\frac{7}{9}}$

25.

$$\begin{array}{r} 45 \\ \times\ 45 \\ \hline 225 \\ 180\ \ \\ \hline \mathbf{2025} \end{array}$$

26. **Each term can be found by adding $\frac{1}{16}$ to the preceding term $\left(\text{or } K = \frac{1}{16}n\right)$.**

$\frac{3}{16} + \frac{1}{16} = \frac{4}{16} = \frac{1}{4}$

$\frac{4}{16} + \frac{1}{16} = \frac{5}{16}$

$\frac{5}{16} + \frac{1}{16} = \frac{6}{16}, \frac{6 \div 2}{16 \div 2} = \frac{3}{8}$

$\mathbf{\frac{1}{4}, \frac{5}{16}, \frac{3}{8}}$

27. **Acute angle, obtuse angle**

28. $2\frac{1}{2} = \frac{5}{2}, 1\frac{2}{3} = \frac{5}{3}$

$\frac{5}{2} \times \frac{5}{3} = \frac{25}{6}$

$= \mathbf{4\frac{1}{6}}$

29. $\frac{2}{3} \cdot \frac{2}{2} = \frac{4}{6}$

$\frac{4}{6} + \frac{1}{6} = \mathbf{\frac{5}{6}}$

30. 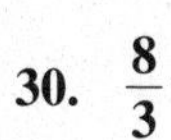 $\mathbf{\frac{8}{3}}$

Lesson 17, Warm-Up

a. **$5.00**

b. **$0.36**

c. **$3.60**

d. **165**

e. **125**

f. **16**

g. **5**

Problem Solving

Nelson's fingers covered two opposite faces of the dot cube. The dots on opposite sides of a dot cube total 7. Opposite 3 dots are 4 dots, and opposite 5 dots are 2 dots. By elimination, Nelson's fingers must have covered the faces with **1 dot** and **6 dots.**

Lesson 17, Lesson Practice

a.  **90°**

b. **50°**

c. **115°**

d. **65°**

e. **130°**

f. **23°**

g.

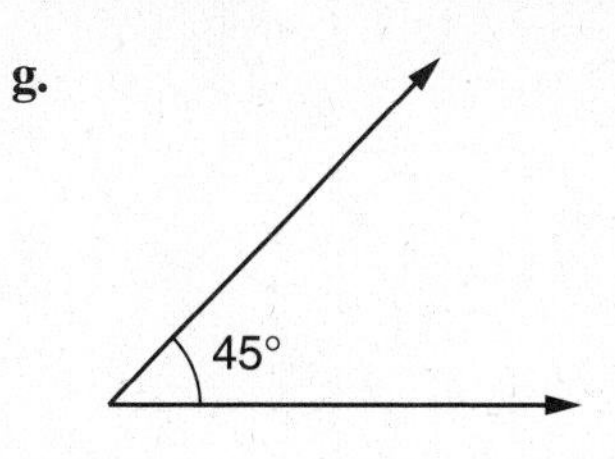

h.

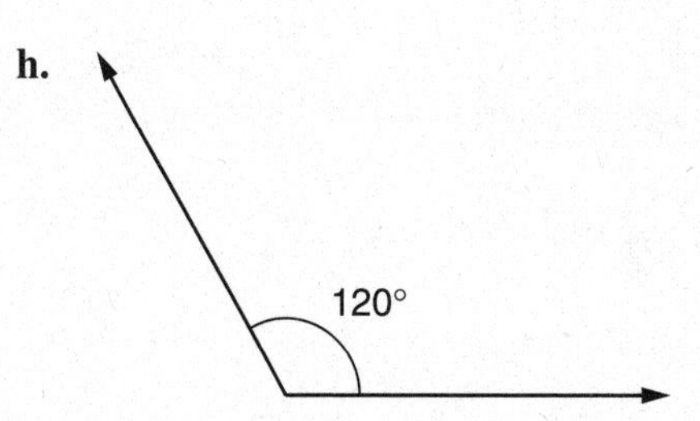

i.

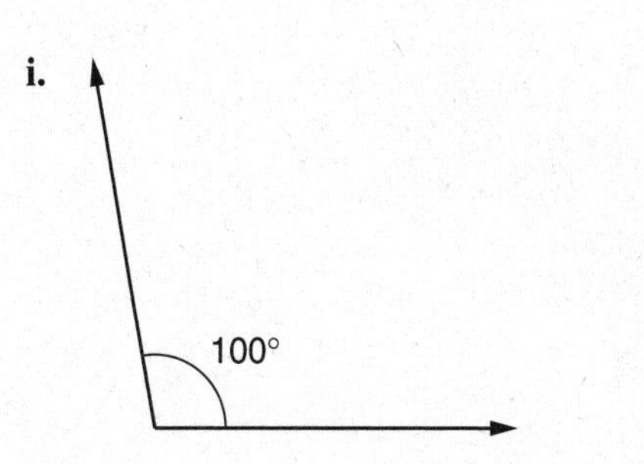

j.

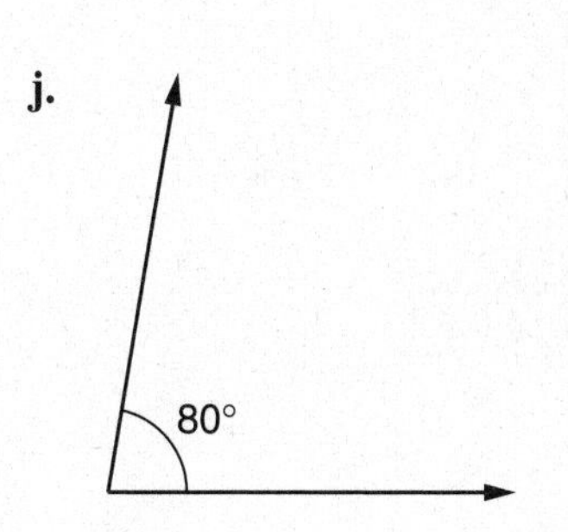

k. $\mathbf{\frac{1}{2}}$ **degree**

Lesson 17, Mixed Practice

1. **2420 + 5090 = *T***

$$\begin{array}{r} 2420 \text{ soldiers} \\ +\ 5090 \text{ soldiers} \\ \hline \mathbf{7510 \text{ soldiers}} \end{array}$$

2. $\mathbf{\frac{3}{20} + F = \frac{20}{20}}$

$$\begin{array}{r} \frac{20}{20} \\ -\ \frac{3}{20} \\ \hline \mathbf{\frac{17}{20}} \end{array}$$

3. $\mathbf{15S = 210}$

$$\begin{array}{r} 14 \\ 15\overline{)210} \\ \underline{15} \\ 60 \\ \underline{60} \\ 0 \end{array}$$

14 children

4. $\mathbf{1620 - 1492 = D}$

$$\begin{array}{r} 1620 \\ -\ 1492 \\ \hline 128 \end{array}$$

128 years

5. C. $\mathbf{\frac{5}{3}}$

6. (a) $\overleftrightarrow{QR}$ (or $\overleftrightarrow{RQ}$)

(b) $\overleftrightarrow{RT}$ (or $\overleftrightarrow{TR}$)

(c) **90°**

7. (a) $\frac{12 \div 4}{16 \div 4} = \mathbf{\frac{3}{4}}$

(b) $\frac{12 \div 6}{18 \div 6} = \frac{2}{3}, 3\frac{12}{18} = \mathbf{3\frac{2}{3}}$

(c) $\frac{25}{100} \div \frac{25}{25} = \mathbf{\frac{1}{4}}$

8. 2 lb = 2 (16 oz) = 32 oz

32 oz + 8 oz = **40 oz**

9. (a) $\frac{2}{9} \cdot \frac{2}{2} = \mathbf{\frac{4}{18}}$

(b) $\frac{1}{3} \cdot \frac{6}{6} = \mathbf{\frac{6}{18}}$

(c) $\frac{5}{6} \cdot \frac{3}{3} = \mathbf{\frac{15}{18}}$

10.

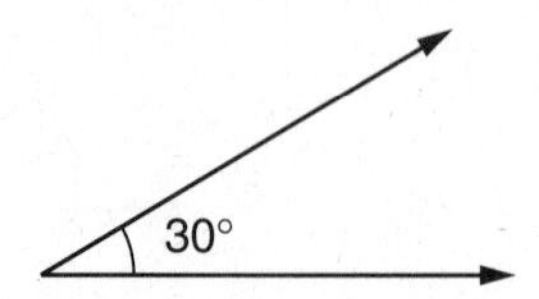

11. (a) Factors of 20: 1, 2, 4, 5, 10, 20

Factors of 50: 1, 2, 5, 10, 25, 50

1, 2, 5, 10

(b) **10**

12.

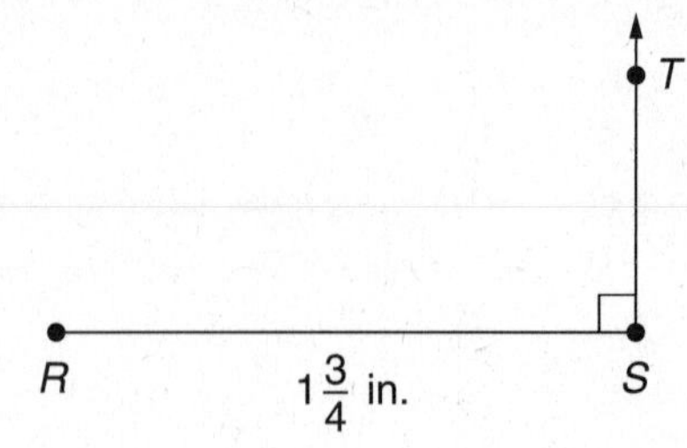

13. (a) $\frac{8}{4} - \frac{4}{8} = \frac{8}{4} - \frac{4 \div 2}{8 \div 2}$

$= \frac{8}{4} - \frac{2}{4} = \frac{6}{4} = \mathbf{\frac{3}{2}}$ or $\mathbf{1\frac{1}{2}}$

(b) $4 - \frac{4}{8} = \frac{4}{1} - \frac{4}{8}$

$= \frac{4}{1} \cdot \frac{8}{8} - \frac{4}{8} = \frac{32}{8} - \frac{4}{8}$

$= \frac{28}{8} = \frac{28 \div 4}{8 \div 4} = \mathbf{\frac{7}{2}}$ or $\mathbf{3\frac{1}{2}}$

14.

$$\begin{array}{r} 141 \\ +\ 231 \\ \hline 372 \end{array}$$

x = **372**

15.

$$\begin{array}{r} \$25.00 \\ -\ \$6.30 \\ \hline \$18.70 \end{array}$$

y = **$18.70**

16.

$$\begin{array}{r} \$3.75 \\ 8\overline{)\$30.00} \\ \underline{24} \\ 60 \\ \underline{56} \\ 40 \\ \underline{40} \\ 0 \end{array}$$

w = **$3.75**

17. 100% ÷ 20% = 5

m = **5**

18. $3\frac{5}{6} - 1\frac{1}{6} = 2\frac{4}{6}$

$\frac{4 \div 2}{6 \div 2} = \frac{2}{3}$

$2\frac{4}{6} = \mathbf{2\frac{2}{3}}$

19. $\frac{1}{2} \cdot \frac{2}{3} = \frac{2}{6} = \mathbf{\frac{1}{3}}$

20.
$$\begin{array}{r} \mathbf{\$2.50} \\ 40\overline{)\$100.00} \\ \underline{80} \\ 200 \\ \underline{200} \\ 00 \\ \underline{0} \\ 0 \end{array}$$

21.
$$\begin{array}{r} 55 \\ \underline{\times\ 55} \\ 275 \\ \underline{275\ } \\ \mathbf{3025} \end{array}$$

22. $2(8\text{ in.} + 6\text{ in.}) = 2(14\text{ in.})$
$= \mathbf{28\text{ in.}}$

23. $\frac{3}{4}\text{ in.} + \frac{3}{4}\text{ in.} = \frac{6}{4}\text{ in.}$
$= \frac{3}{2}\text{ in.} = \mathbf{1\frac{1}{2}\text{ in.}}$

24. $\frac{15}{16}\text{ in.} - \frac{3}{16}\text{ in.} = \frac{12 \div 4}{16 \div 4}\text{ in.}$
$= \mathbf{\frac{3}{4}\text{ in.}}$

25. $\frac{1}{2} \cdot \frac{4}{3} \cdot \frac{9}{2} = \frac{4}{6} \cdot \frac{9}{2}$
$= \frac{36}{12} = \mathbf{3}$

26.
$$\begin{array}{r} \$20.25 \\ \underline{-\ \$15.17} \\ \$5.08 \end{array}$$
$5 bill, 1 nickel, 3 pennies

27. (a) $\left(\frac{1}{2} \cdot \frac{3}{4}\right) \cdot \frac{2}{3} = \left(\frac{3}{8}\right) \cdot \frac{2}{3}$
$= \frac{6 \div 6}{24 \div 6}$
$= \frac{1}{4}$

$\frac{1}{2} \cdot \left(\frac{3}{4} \cdot \frac{2}{3}\right) = \frac{1}{2} \cdot \left(\frac{6}{12}\right)$
$= \frac{6}{24} = \frac{1}{4}, \frac{1}{4} = \frac{1}{4}$

$\left(\frac{1}{2} \cdot \frac{3}{4}\right) \cdot \frac{2}{3} = \frac{1}{2}\left(\frac{3}{4} \cdot \frac{2}{3}\right)$

(b) **Associative property of multiplication**

28. **Answers may vary. Sample answer: If 85% of Shyla's answers were correct, then what percent were not correct?**

29. $3\frac{3}{4} = \frac{4 \times 3 + 3}{4} = \mathbf{\frac{15}{4}}$
reciprocal $= \mathbf{\frac{4}{15}}$

30. $\frac{3}{4} \cdot \frac{2}{2} = \frac{6}{8}$
$\frac{6}{8} + \frac{5}{8} = \frac{11}{8} = \mathbf{1\frac{3}{8}}$

Lesson 18, Warm-Up

a. **$5.50**

b. **$16.50**

c. **$7.50**

d. **144**

e. **125**

f. $\mathbf{\frac{1}{8}}$

g. **7**

Problem Solving
8, 17, 32
8, 32, 17
17, 8, 32
17, 32, 8
32, 8, 17
32, 17, 8

Lesson 18, Lesson Practice

a. **Octagon**

b. **Square**

c. **Acute angle**

d. **Yes**

e. **No**

f. $\angle B$

g. **equal in measure** (Each angle is a right angle with a measure of 90°.)

h.

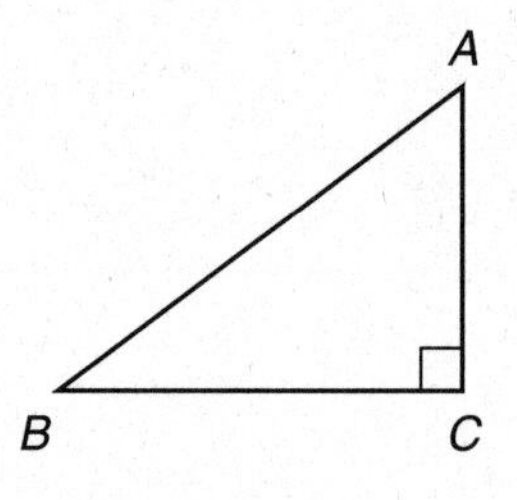

LESSON 18, MIXED PRACTICE

1. $\mathbf{6D = 3300}$

$$\begin{array}{r} 550 \\ 6\overline{)3300} \\ \underline{30} \\ 30 \\ \underline{30} \\ 00 \\ \underline{0} \\ 0 \end{array}$$

550 miles

2. $\mathbf{456 + 517 = T}$

$$\begin{array}{r} 456 \text{ miles} \\ + \ 517 \text{ miles} \\ \hline \mathbf{973 \text{ miles}} \end{array}$$

3. $\mathbf{3977 + W = 5000}$

$$\begin{array}{r} 5000 \text{ meters} \\ - \ 3977 \text{ meters} \\ \hline \mathbf{1023 \text{ meters}} \end{array}$$

4. $\mathbf{1{,}000{,}000{,}000 - 10{,}000{,}000 = D}$

$$\begin{array}{r} 1{,}000{,}000{,}000 \\ - \quad 10{,}000{,}000 \\ \hline 990{,}000{,}000 \end{array}$$

Nine hundred ninety million

5. (a) $\mathbf{-1, 0, \frac{3}{4}, 1, \frac{5}{3}}$

(b) $\mathbf{-1, 0}$

6. **Side** AD (or **side** DA)

7. (a) $\mathbf{-2}$

(b) **4**

8. (a) $2\% = \frac{2}{100} = \frac{2}{100} \div \frac{2}{2} = \mathbf{\frac{1}{50}}$

(b) $\frac{12 \div 4}{20 \div 4} = \mathbf{\frac{3}{5}}$

(c) $\frac{15 \div 5}{20 \div 5} = \frac{3}{4}, \ 6\frac{15}{20} = \mathbf{6\frac{3}{4}}$

9. (a) $\frac{4}{5} \cdot \frac{6}{6} = \mathbf{\frac{24}{30}}$

(b) $\frac{2}{3} \cdot \frac{10}{10} = \mathbf{\frac{20}{30}}$

(c) $\frac{1}{6} \cdot \frac{5}{5} = \mathbf{\frac{5}{30}}$

10. $8 - 5 = 3$

3 sides

11. (a)

(b) **Acute angles**

12. (a) **25%**

(b) $\mathbf{\frac{6}{8}}$ or $\mathbf{\frac{3}{4}}$

13. **Identity property of multiplication**

14. $\frac{5}{8} + \frac{3}{8} = \frac{8}{8}$

$x = \mathbf{\frac{8}{8}}$ or **1**

15. $\frac{7}{10} - \frac{3}{10} = \frac{4}{10}$

$y = \mathbf{\frac{4}{10}}$ or $\mathbf{\frac{2}{5}}$

16. $\frac{5}{6} - \frac{1}{6} = \frac{4}{6}$

$m = \mathbf{\frac{4}{6}}$ or $\mathbf{\frac{2}{3}}$

17. $\mathbf{\frac{4}{3}}$

18. $5\frac{7}{10} - \frac{3}{10} = 5\frac{4}{10} = \mathbf{5\frac{2}{5}}$

19. $\frac{3}{2} \cdot \frac{2}{4} = \frac{6}{8} = \mathbf{\frac{3}{4}}$

20.
$$\begin{array}{r} \mathbf{45} \\ 45\overline{)2025} \\ \underline{180} \\ 225 \\ \underline{225} \\ 0 \end{array}$$

21.
$$\begin{array}{r} 750 \\ \times \quad 80 \\ \hline \mathbf{60{,}000} \end{array}$$

22.
$$\begin{array}{r} 21 \\ \times\ 21 \\ \hline 21 \\ 42\ \ \\ \hline \mathbf{441} \end{array}$$

23. $2(50 \text{ in.} + 40 \text{ in.}) = 2(90 \text{ in.})$
$= \mathbf{180 \text{ in.}}$

24. $\frac{8}{16} = \frac{1}{2} = \frac{50}{100} = \mathbf{50\%}$

25. (a) $360° \div 4 = \mathbf{90°}$

(b) $360° \div 6 = \mathbf{60°}$

26. (a)

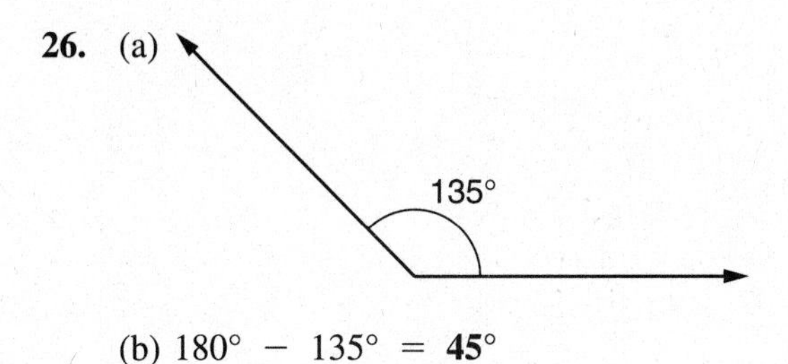

(b) $180° - 135° = \mathbf{45°}$

27. (a) $\mathbf{\Delta \mathit{SQR}}$

(b) $\mathbf{\Delta \mathit{XYZ}}$

(c) $\mathbf{\angle \mathit{F}}$

28. $\frac{1}{2} \cdot \frac{3}{3} = \frac{3}{6}$

$\frac{1}{3} \cdot \frac{2}{2} = \frac{2}{6}$

$\frac{3}{6} + \frac{2}{6} = \mathbf{\frac{5}{6}}$

29. $2\frac{1}{4} = \frac{9}{4}$

$\frac{9}{4} \cdot \frac{4}{3} = \frac{36}{12} = \mathbf{3}$

30.

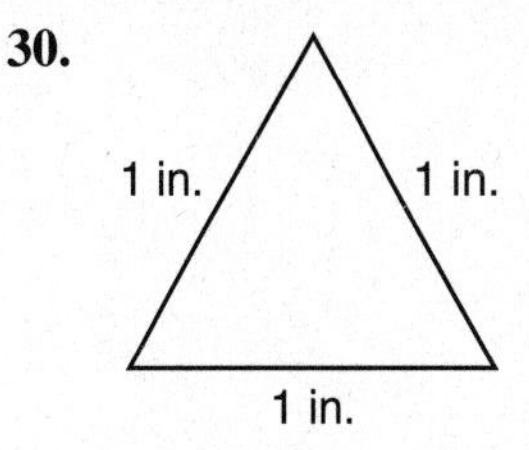

Regular

LESSON 19, WARM-UP

a. **$10.00**

b. **$1.20**

c. **24¢**

d. **224**

e. **375**

f. **60**

g. **12**

Problem Solving

$25 + 12 + 8 = 45$ tickets
$45 \div 3 = 15$ tickets
Yin should give Bobby 3 tickets and Mary 7 tickets. (Then they will each have 15.)

LESSON 19, LESSON PRACTICE

a. Perimeter $= 3 \text{ in.} + 3 \text{ in.} + 2 \text{ in.} + 5 \text{ in.}$
$= \mathbf{13 \text{ in.}}$

b. Perimeter $= 5(5 \text{ cm}) = \mathbf{25 \text{ cm}}$

c. Perimeter $= 8(12 \text{ in.}) = \mathbf{96 \text{ in.}}$

d. Missing length $= 10 \text{ in.} - 4 \text{ in.} = 6 \text{ in.}$
Missing height $= 5 \text{ in.} - 2 \text{ in.} = 3 \text{ in.}$
Perimeter $= 10 \text{ in.} + 2 \text{ in.} + 6 \text{ in.}$
$+ 3 \text{ in.} + 4 \text{ in.} + 5 \text{ in.}$
$= \mathbf{30 \text{ in.}}$

e. $100 \text{ feet} \div 4 = \mathbf{25 \text{ feet}}$

f. $\frac{3}{4}$ in.

$\frac{3}{4}$ in.

$$\text{Perimeter} = 4\left(\frac{3}{4}\text{ in.}\right)$$
$$= \frac{4}{1}\cdot\frac{3}{4}\text{ in.} = \frac{12}{4}\text{ in.}$$
$$= \mathbf{3\text{ in.}}$$

LESSON 19, MIXED PRACTICE

1. $\mathbf{\frac{1}{8} + N_R = \frac{8}{8}}$

$$\begin{array}{r} \frac{8}{8} \\ -\ \frac{1}{8} \\ \hline \mathbf{\frac{7}{8}} \end{array}$$

2. $\mathbf{F - 76 = 124}$

$$\begin{array}{r} 124\text{ people} \\ +\ \ 76\text{ people} \\ \hline \mathbf{200\text{ people}} \end{array}$$

3. $\mathbf{84 \times 6 = T}$

$$\begin{array}{r} 84 \\ \times\ \ 6 \\ \hline 504 \end{array}$$

504 slices

4. 4(20 years) + 7 years = **87 years**

5. (a) **Eighteen million, seven hundred thousand**

   (b) $\mathbf{(8 \times 100) + (7 \times 10) + (4 \times 1)}$

6. $\mathbf{3 - 7 = -4}$

7. **Water freezes at 32°F. Water boils at 212°F.**

8. Perimeter = 6 cm + 6 cm + 8 cm + 8 cm
   = **28 cm**

9. (a) $3\frac{16}{24} = 3\frac{16 \div 8}{24 \div 8} = \mathbf{3\frac{2}{3}}$

   (b) $\frac{15 \div 3}{24 \div 3} = \mathbf{\frac{5}{8}}$

   (c) $4\% = \frac{4}{100} = \frac{4 \div 4}{100 \div 4} = \mathbf{\frac{1}{25}}$

10. (a) $\frac{3}{4}\cdot\frac{9}{9} = \mathbf{\frac{27}{36}}$

    (b) $\frac{4}{9}\cdot\frac{4}{4} = \mathbf{\frac{16}{36}}$

11.

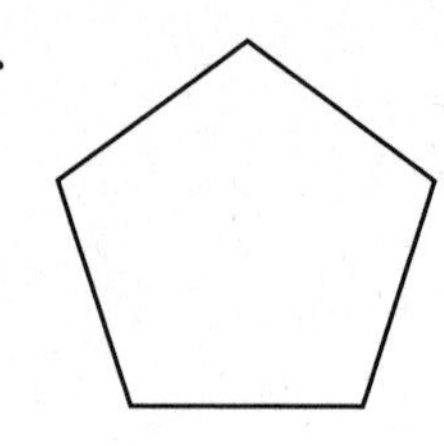

12. **Octagon**

13. (a) **90°**

    (b) 4(90°) = **360°**

14. $k = \frac{1}{8}\cdot 8 = \frac{1}{8}\cdot\frac{8}{1} = \frac{8}{8} = \mathbf{1}$

15.
$$\begin{array}{r} 8998 \\ -\ 1547 \\ \hline 7451 \end{array}$$

$a = \mathbf{7451}$

16.
$$\begin{array}{r} \$1.37 \\ 30\overline{)\$41.10} \\ 30 \\ \hline 111 \\ 90 \\ \hline 210 \\ 210 \\ \hline 0 \end{array}$$

$b = \mathbf{\$1.37}$

17.
$$\begin{array}{r} 23 \\ \$0.32\overline{)\$7.36} \\ 64 \\ \hline 96 \\ 96 \\ \hline 0 \end{array}$$

$c = \mathbf{23}$

18.
$$\begin{array}{r} \$30.10 \\ -\ \$26.57 \\ \hline \$3.53 \end{array}$$

$d = \mathbf{\$3.53}$

19. $\frac{2}{3} + \frac{2}{3} + \frac{2}{3} = \frac{6}{3} = \mathbf{2}$

20. $3\frac{7}{8} - \frac{5}{8} = 3\frac{2}{8} = \mathbf{3\frac{1}{4}}$

21. $\frac{2}{3} \cdot \frac{3}{7} = \frac{6}{21} = \mathbf{\frac{2}{7}}$

22. $3\frac{7}{8} + \frac{5}{8} = 3\frac{12}{8} = 4\frac{4}{8} = \mathbf{4\frac{1}{2}}$

23.
$$\begin{array}{r} 50 \\ \times\ 50 \\ \hline \mathbf{2500} \end{array}$$

24.
$$\begin{array}{r} \mathbf{9{,}100} \\ 11\overline{)100{,}100} \\ \underline{99} \\ 1\,1 \\ \underline{1\,1} \\ 00 \\ \underline{0} \\ 00 \\ \underline{0} \\ 0 \end{array}$$

25. (a) **2**

(b) $2 \cdot 5 = \mathbf{10}$

26.

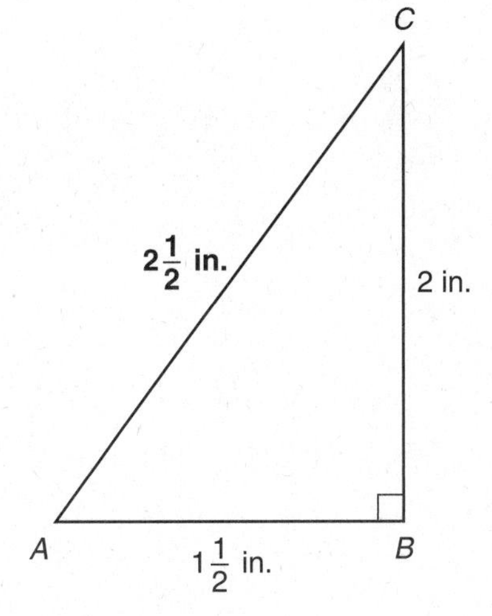

27. $3\frac{1}{3} = \frac{10}{3}$

$\frac{10}{3} \cdot \frac{3}{2} = \frac{30}{6} = \mathbf{5}$

28. $\frac{1}{2} \cdot \frac{5}{5} = \frac{5}{10}$

$\frac{9}{10} - \frac{5}{10} = \frac{4}{10} = \mathbf{\frac{2}{5}}$

29. $100\% \div 3 = \mathbf{33\frac{1}{3}\%}$

30. Missing length = 10 in. − 6 in. = 4 in.
Missing height = 7 in. − 4 in. = 3 in.
Perimeter = 10 in. + 4 in. + 6 in. + 3 in. + 4 in. + 7 in.
= **34 in.**

Lesson 20, Warm-Up

a. **$7.25**

b. **$3.60**

c. **$0.68**

d. **215**

e. **500**

f. $\mathbf{\frac{3}{8}}$

g. **1**

Problem Solving

We write 1's in the hundreds and tens places of the quotient, because the subtrahends of the first two subtraction stages have only one digit each. We know that the minuend and subtrahend of the final subtraction stages are both 48 (because the difference is 0), so we write a 6 in the ones digit of the quotient (because $6 \times 8 = 48$). To find the dividend, we multiply 116 by 8.

$$\begin{array}{r} \mathbf{116} \\ \mathbf{8\overline{)928}} \\ \underline{\mathbf{8}} \\ \mathbf{12} \\ \underline{\mathbf{8}} \\ \mathbf{48} \\ \underline{\mathbf{48}} \\ \mathbf{0} \end{array}$$

Lesson 20, Lesson Practice

a. **Four cubed**
$4 \cdot 4 \cdot 4 = \mathbf{64}$

b. **One half squared**
$\frac{1}{2} \cdot \frac{1}{2} = \mathbf{\frac{1}{4}}$

c. **Ten to the sixth power**
$10 \cdot 10 \cdot 10 \cdot 10 \cdot 10 \cdot 10 = \mathbf{1{,}000{,}000}$

d. **Base is 10; exponent is 3.**

e. $2^3 \cdot 2^2 = 2 \cdot 2 \cdot 2 \cdot 2 \cdot 2$
$= \mathbf{2^5}$

f. $\frac{2^6}{2^2} = \frac{2 \cdot 2 \cdot 2 \cdot 2 \cdot 2 \cdot 2}{2 \cdot 2}$
$= \frac{64}{4} = 16 = 2 \cdot 2 \cdot 2 \cdot 2 = \mathbf{2^4}$

g. **10**

h. **20**

i. **15**

j. Area $= 15 \text{ m} \times 10 \text{ m}$
$= \mathbf{150 \text{ m}^2}$

k. Area $= 2 \text{ in.} \times 5 \text{ in.}$
$= \mathbf{10 \text{ in.}^2}$

l. Area $= 4 \text{ cm} \times 4 \text{ cm}$
$= \mathbf{16 \text{ cm}^2}$

m. $20 \text{ cm} \div 4 = 5 \text{ cm}$
Area $= 5 \text{ cm} \times 5 \text{ cm} = \mathbf{25 \text{ cm}^2}$

n. Area $= 100$ yards $\times$ 100 yards
$=$ **10,000 square yards**

Lesson 20, Mixed Practice

1. $\mathbf{4A = 628}$

$$\begin{array}{r} 157 \\ 4\overline{)628} \\ \underline{4} \\ 22 \\ \underline{20} \\ 28 \\ \underline{28} \\ 0 \end{array}$$

157 tenants

2. $\mathbf{P - 36 = 46}$

$$\begin{array}{r} 46 \text{ parrots} \\ + \ 36 \text{ parrots} \\ \hline \mathbf{82 \text{ parrots}} \end{array}$$

3. $\mathbf{225 + N_T = 600}$

$$\begin{array}{r} 600 \text{ fish} \\ - \ 225 \text{ fish} \\ \hline \mathbf{375 \text{ fish}} \end{array}$$

4. $\mathbf{21{,}050 + 48{,}972 = T}$

$$\begin{array}{r} 21{,}050 \\ + \ 48{,}972 \\ \hline \mathbf{70{,}022} \end{array}$$

5. $k = 2^6 = 2 \cdot 2 \cdot 2 \cdot 2 \cdot 2 \cdot 2$
$= \mathbf{64}$

6. (a) $\mathbf{-2, -\frac{1}{2}, 0, \frac{1}{3}, 1}$

(b) $\mathbf{\frac{1}{3}, -\frac{1}{2}}$

7. **B.** $\mathbf{33\frac{1}{3}\%}$

8. **Side** ***DC*** (or **side** ***CD***) and **side** ***AB*** (or **side** ***BA***)

9. (a) $\left(\frac{1}{3}\right)^3 = \frac{1}{3} \cdot \frac{1}{3} \cdot \frac{1}{3} = \mathbf{\frac{1}{27}}$

(b) $10^4 = 10 \cdot 10 \cdot 10 \cdot 10$
$= \mathbf{10{,}000}$

(c) $\sqrt{12^2} = \sqrt{144} = \mathbf{12}$

10. (a) $\frac{2}{9} \cdot \frac{4}{4} = \mathbf{\frac{8}{36}}$

(b) $\frac{3}{4} \cdot \frac{9}{9} = \mathbf{\frac{27}{36}}$

11. (a) **1, 2, 5, 10**

(b) **1, 7**

(c) **1**

12. 2 feet $=$ 24 inches
24 inches $\div$ 4 $=$ **6 inches**

13.

$$\begin{array}{r} 54 \\ - \ 36 \\ \hline 18 \end{array}$$

$a = \mathbf{18}$

14.

$$\begin{array}{r} 46 \\ - \ 20 \\ \hline 26 \end{array}$$

$w = \mathbf{26}$

15.

$$\begin{array}{r} 12 \\ 5\overline{)60} \\ \underline{5} \\ 10 \\ \underline{10} \\ 0 \end{array}$$

$x = \mathbf{12}$

16.
$$\begin{array}{r} 100 \\ -\ 64 \\ \hline 36 \end{array}$$
$m = \mathbf{36}$

17. $5^4 \cdot 5^2 = 5 \cdot 5 \cdot 5 \cdot 5 \cdot 5 \cdot 5 = 5^6$
$n = \mathbf{6}$

18.
$$\begin{array}{r} 15 \\ 4\overline{)60} \\ \underline{4} \\ 20 \\ \underline{20} \\ 0 \end{array}$$
$y = \mathbf{15}$

19. $1\frac{8}{9} + 1\frac{7}{9} = 2\frac{15}{9} = 3\frac{6}{9} = \mathbf{3\frac{2}{3}}$

20. $\frac{5}{2} \cdot \frac{5}{6} = \frac{25}{12} = \mathbf{2\frac{1}{12}}$

21.
$$\begin{array}{r} \mathbf{705} \\ 9\overline{)6345} \\ \underline{63} \\ 04 \\ \underline{0} \\ 45 \\ \underline{45} \\ 0 \end{array}$$

22.
$$\begin{array}{r} 360 \\ \times\ 25 \\ \hline 1800 \\ \underline{720} \\ \mathbf{9000} \end{array}$$

23. $\frac{3}{4} - \left(\frac{1}{4} + \frac{2}{4}\right) = \frac{3}{4} - \left(\frac{3}{4}\right) = \mathbf{0}$

24. $\left(\frac{3}{4} - \frac{1}{4}\right) + \frac{2}{4} = \frac{2}{4} + \frac{2}{4}$
$= \frac{4}{4} = \mathbf{1}$

25. (a) $\frac{3}{10} + \frac{3}{10} = \frac{6}{10} = \mathbf{\frac{3}{5}}$

(b) $\frac{3}{10} \cdot \frac{3}{10} = \mathbf{\frac{9}{100}}$

26. $\frac{1}{2} \cdot \frac{5}{5} = \frac{5}{10}$
$\frac{5}{10} + \frac{3}{10} = \frac{8}{10} = \mathbf{\frac{4}{5}}$

27. $1\frac{4}{5} = \frac{9}{5}$
$\frac{9}{5} \cdot \frac{1}{3} = \frac{9}{15} = \mathbf{\frac{3}{5}}$

28. **Identity property of multiplication**

29. (a) Perimeter = 12 in. + 12 in. + 12 in. + 12 in. = **48 in.** or **4 ft**

(b) Area = 12 in. × 12 in. = **144 in.^{2}** or **1 ft^2**

30. Missing length = 10 in. − 5 in. = 5 in.
Missing height = 8 in. − 4 in. = 4 in.
Perimeter = 10 in. + 4 in. + 5 in. + 4 in. + 5 in. + 8 in.
= **36 in.**

INVESTIGATION 2

1. **See student work.**

2. **See student work.**

3. **See student work.**

4. **60°**

5. 60° + 60° + 60° = **180°**

6. **a six-point star with a regular hexagon inside**

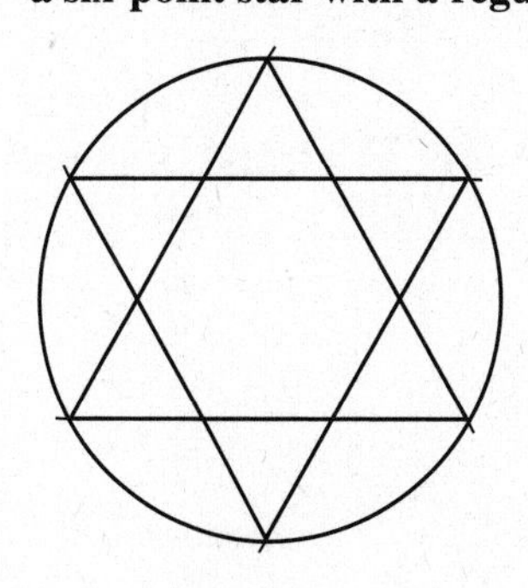

7. **See student work.**

8. **120°**

9. $\frac{1}{3} = \mathbf{33\frac{1}{3}\%}$

10. **60°**

11. $\frac{1}{6} = 16\frac{2}{3}\%$

12. **circumference**

13. **diameter**

14. **radius**

15. **arc**

16. **sector**

17. **concentric circles**

18. **chord**

19. **inscribed polygon**

20. **semicircle**

21. **central angle**

22. **radius**

23. **center**

24. **inscribed angle**

LESSON 21, WARM-UP

a. **$2.24**

b. **$0.65**

c. **$4.25**

d. **204**

e. **4**

f. **12**

g. **1**

Problem Solving

We know Sam can read 20 pages in 30 minutes. Two hundred pages is 10 times as many pages as 20 pages. So we multiply 30 minutes by 10 and then convert to hours:

30 minutes × 10 = 300 minutes

300 minutes ÷ 60 minutes per hour = **5 hours**

LESSON 21, LESSON PRACTICE

a. **2, 3, 5, 7, 11, 13, 17, 19, 23, 29**

b. **Composite number**

c.

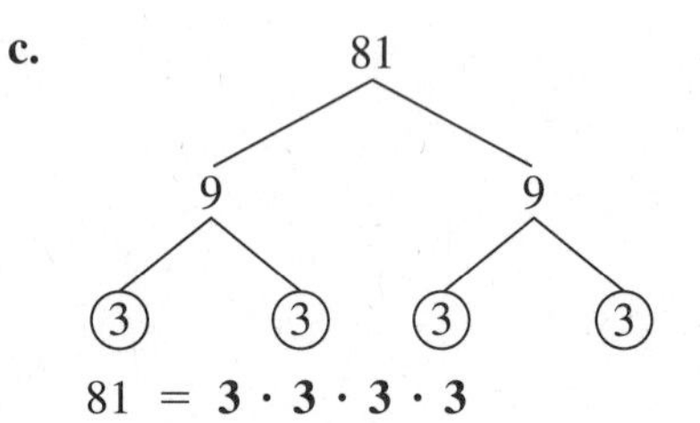

81 = **3 · 3 · 3 · 3**

d. 360 = **2 · 2 · 2 · 3 · 3 · 5**

$$\begin{array}{r} 1 \\ 5\overline{)5} \\ 3\overline{)15} \\ 3\overline{)45} \\ 2\overline{)90} \\ 2\overline{)180} \\ 2\overline{)360} \end{array}$$

e.

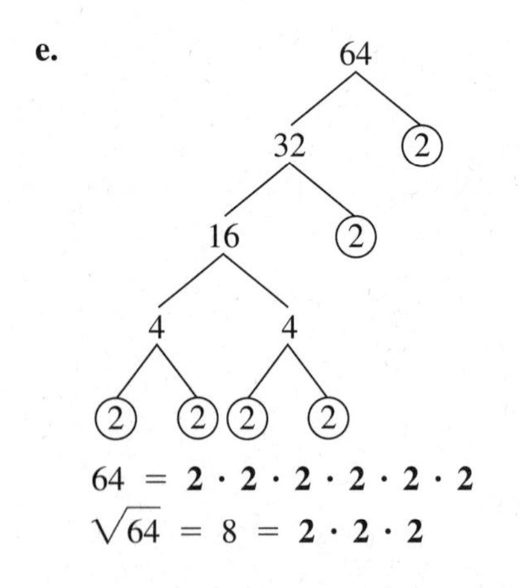

64 = **2 · 2 · 2 · 2 · 2 · 2**

$\sqrt{64} = 8 = \mathbf{2 \cdot 2 \cdot 2}$

LESSON 21, MIXED PRACTICE

1. $\mathbf{\frac{2}{3}} + N_G = \mathbf{\frac{3}{3}}$

$\frac{3}{3} - \frac{2}{3} = \mathbf{\frac{1}{3}}$

2. $\mathbf{7Q = 343}$

$$\begin{array}{r} 49 \\ 7\overline{)343} \\ \underline{28} \\ 63 \\ \underline{63} \\ 0 \end{array}$$

49 quills

3. $\mathbf{2{,}000{,}000{,}000 - 21{,}000{,}000 = D}$

$$\begin{array}{r} 2{,}000{,}000{,}000 \\ -\quad 21{,}000{,}000 \\ \hline 1{,}979{,}000{,}000 \end{array}$$

One billion, nine hundred seventy-nine million

4. $\mathbf{\$14{,}289 + \$824 = N}$

$$\begin{array}{r} \$14{,}289 \\ +\quad \$824 \\ \hline \mathbf{\$15{,}113} \end{array}$$

5. (a) $3\,\dfrac{12 \div 3}{21 \div 3} = \mathbf{3\dfrac{4}{7}}$

(b) $\dfrac{12 \div 12}{48 \div 12} = \mathbf{\dfrac{1}{4}}$

(c) $12\% = \dfrac{12}{100} \div \dfrac{4}{4} = \mathbf{\dfrac{3}{25}}$

6. **53, 59**

7. (a) $50 = \mathbf{2 \cdot 5 \cdot 5}$

50
25 (2)
(5) (5)

$50 = 2 \cdot 5 \cdot 5$

(b) $60 = \mathbf{2 \cdot 2 \cdot 3 \cdot 5}$

60
30 (2)
15 (2)
(3) (5)

$60 = 2 \cdot 2 \cdot 3 \cdot 5$

(c) $300 = \mathbf{2 \cdot 2 \cdot 3 \cdot 5 \cdot 5}$

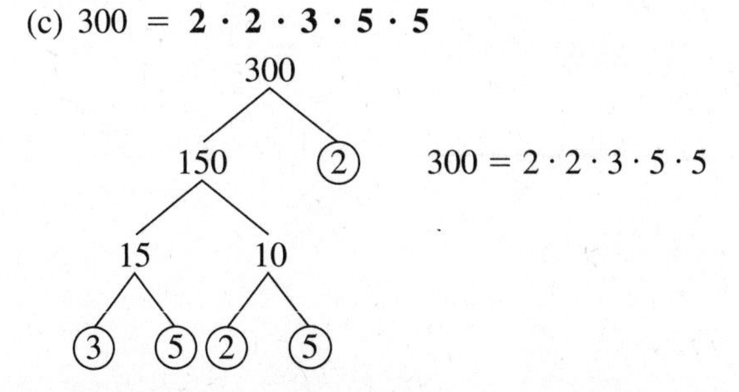

$300 = 2 \cdot 2 \cdot 3 \cdot 5 \cdot 5$

8. **Point *C*; The tick mark between points *B* and *C* is halfway between 1000 and 2000, which is 1500, so points *A* and *B* are eliminated. Point *C* is closer to 1500 than 2000, so *C* is the best choice. Point *D* is too close to 2000 to represent 1610.**

9. (a) $\dfrac{2}{3} \cdot \dfrac{5}{5} = \mathbf{\dfrac{10}{15}}$

(b) $\dfrac{3}{5} \cdot \dfrac{3}{3} = \mathbf{\dfrac{9}{15}}$

(c) $\dfrac{8 \div 4}{12 \div 4} = \mathbf{\dfrac{2}{3}}$

10. (a) **3**

(b) **9**

11. 12 inches $\div$ 4 = 3 inches

Area = 3 inches $\times$ 3 inches

= **9 square inches**

12.

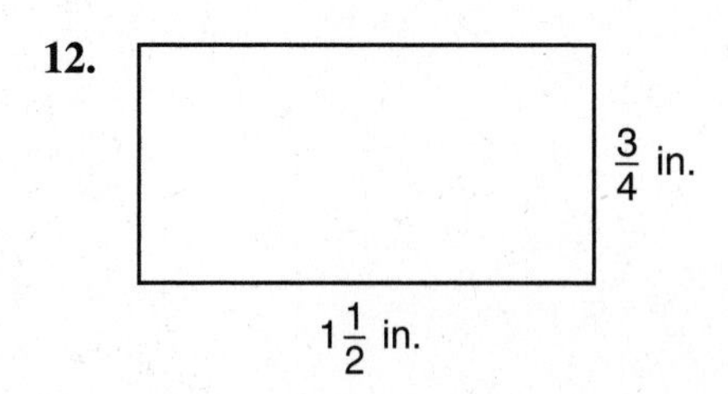

(a) $2 \times \dfrac{3}{4}\text{ in.} = \dfrac{6}{4}\text{ in.} = \mathbf{1\dfrac{1}{2}\text{ in.}}$

(b) $\text{Perimeter} = \dfrac{3}{4}\text{ in.} + \dfrac{3}{4}\text{ in.} + 1\dfrac{1}{2}\text{ in.}$

$+ 1\dfrac{1}{2}\text{ in.} = \dfrac{6}{4}\text{ in.} + 2\dfrac{2}{2}\text{ in.}$

$= \dfrac{6}{4}\text{ in.} + \dfrac{6}{2}\text{ in.} = \dfrac{6}{4}\text{ in.}$

$+ \dfrac{12}{4}\text{ in.} = \dfrac{18}{4}\text{ in.} = \mathbf{4\dfrac{1}{2}\text{ in.}}$

13. Missing length = 3 in. + 12 in. = 15 in.

Missing height = 8 in. − 5 in. = 3 in.

Perimeter = 8 in. + 15 in. + 5 in.

+ 12 in. + 3 in. + 3 in.

= **46 in.**

14. **Check polygon for five sides; one possibility:**

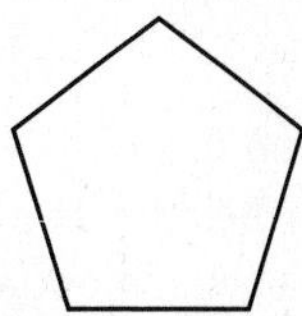

15. $1 - \frac{3}{5} = \frac{5}{5} - \frac{3}{5} = \frac{2}{5}$

$p = \mathbf{\frac{2}{5}}$

16. $1 \cdot \frac{5}{3}$

$q = \mathbf{\frac{5}{3}}$

17.
$$\begin{array}{r} 25 \\ \times\ 50 \\ \hline 1250 \end{array}$$

$w = \mathbf{1250}$

18. $\frac{5}{6} - \frac{1}{6} = \frac{4}{6}$

$f = \mathbf{\frac{4}{6}}$ or $\mathbf{\frac{2}{3}}$

19. $1\frac{2}{3} + 3\frac{2}{3} = 4\frac{4}{3} = 5\frac{1}{3}$

$m = \mathbf{5\frac{1}{3}}$

20.
$$\begin{array}{r} \mathbf{17} \\ 3\overline{)51} \\ 3 \\ \hline 21 \\ 21 \\ \hline 0 \end{array}$$

$c = \mathbf{17}$

21. $\frac{2}{3} + \frac{2}{3} + \frac{2}{3} = \frac{6}{3} = \mathbf{2}$

22. $\left(\frac{2}{3}\right)^3 = \frac{2}{3} \cdot \frac{2}{3} \cdot \frac{2}{3} = \mathbf{\frac{8}{27}}$

23. (a) $225 = \mathbf{3 \cdot 3 \cdot 5 \cdot 5}$

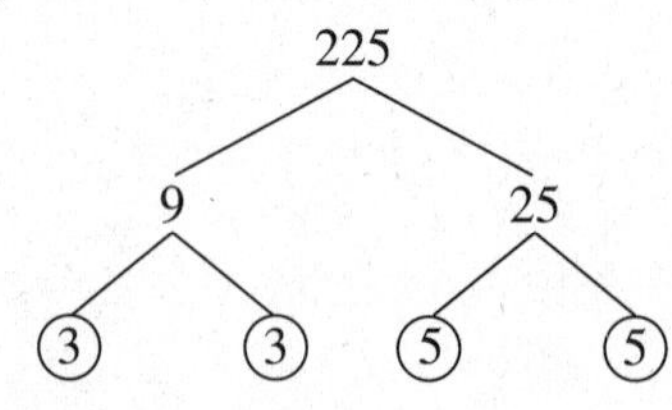

(b) $\sqrt{225} = 15$

$15 = \mathbf{3 \cdot 5}$

24. If we divide the numerator and the denominator of a fraction by their GCF, we reduce the fraction to lowest terms in one step.

25.

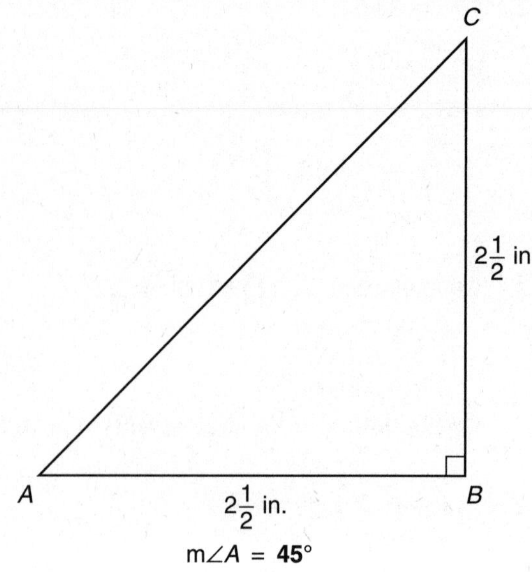

m∠A = **45°**

26. $1\frac{3}{4} = \frac{7}{4}$

$\frac{7}{4} \times \frac{3}{2} = \frac{21}{8} = \mathbf{2\frac{5}{8}}$

27. (a) $\overline{\mathbf{CB}}$ (or $\overline{\mathbf{BC}}$)

(b) $\overline{\mathbf{AB}}$ (or $\overline{\mathbf{BA}}$)

(c) $\overline{\mathbf{MC}}$ **and** $\overline{\mathbf{MB}}$

(d) **∠*ABC*** (or **∠*CBA***)

28. $\frac{1 \text{ quart}}{4 \text{ quarts}} = \frac{1}{4}, \frac{1}{4} \cdot \frac{25}{25} = \frac{25}{100} = \mathbf{25\%}$

29. (a) $\mathbf{a + b = b + a}$

(b) **Commutative property of addition**

30. (a) **Δ*KLJ***

(b) **Δ*DEF***

(c) **∠*S***

Lesson 22, Warm-Up

a. **$2.53**

b. **$8.00**

c. **$2.11**

d. **371**

e. **5**

f. **6**

g. **octagon**

Problem Solving

$2\frac{2}{3}$ ft × 12 in. = 32 in.

$2\frac{2}{3}$ yd × 36 in. = 96 in.

96 in. − 32 in. = **64 in.**

Lesson 22, Lesson Practice

a.–b.

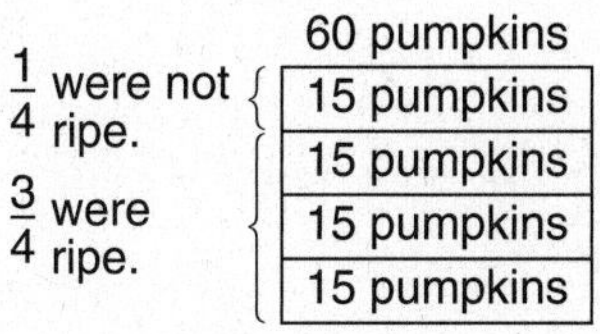

60 pumpkins

$\frac{1}{4}$ were not ripe. — 15 pumpkins

$\frac{3}{4}$ were ripe. — 15 pumpkins | 15 pumpkins | 15 pumpkins

a. 3 × 15 pumpkins = **45 pumpkins**

b. **15 pumpkins**

c.–d.

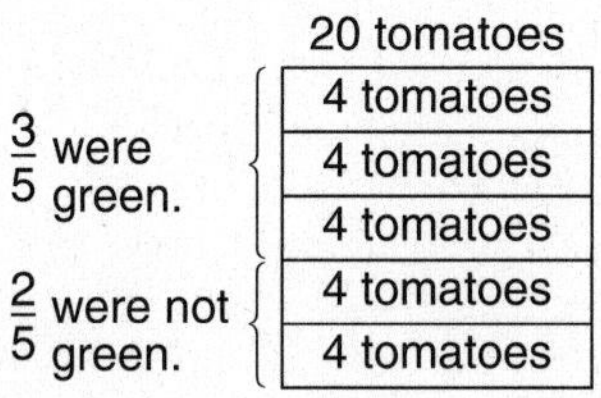

20 tomatoes

$\frac{3}{5}$ were green. — 4 tomatoes | 4 tomatoes | 4 tomatoes

$\frac{2}{5}$ were not green. — 4 tomatoes | 4 tomatoes

c. 100% − 60% = 40%

$$40\% = \frac{40}{100} \div \frac{20}{20} = \mathbf{\frac{2}{5}}$$

d. 3 × 4 tomatoes = **12 tomatoes**

e. **See student work.**

Lesson 22, Mixed Practice

1. **28 + 30 + 23 = *T***

28 books
30 books
+ 23 books
81 books

2. **3*R* = 81**

$3\overline{)81}$ = 27; 6, 21, 21, 0

27 books

3. **126,000 − *L* = 79,000**

126,000
− 79,000
47,000

4. **10,313 − 2700 = *D***

10,313
− 2,700
7,613

Seven thousand, six hundred thirteen

5.

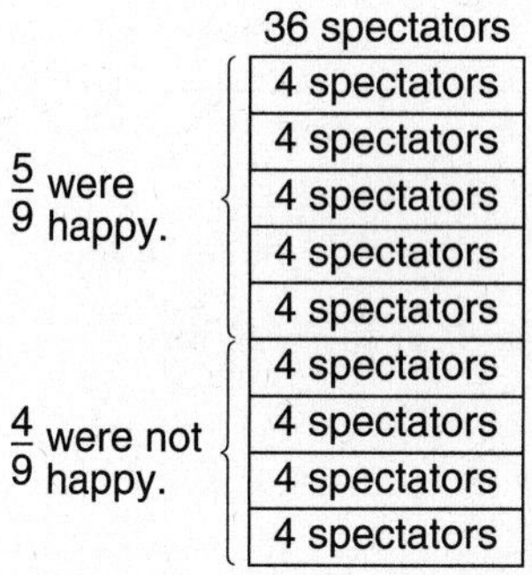

36 spectators

$\frac{5}{9}$ were happy. — 4 spectators | 4 spectators | 4 spectators | 4 spectators | 4 spectators

$\frac{4}{9}$ were not happy. — 4 spectators | 4 spectators | 4 spectators | 4 spectators

(a) 5 × 4 spectators = **20 spectators**

(b) 4 × 4 spectators = **16 spectators**

6.

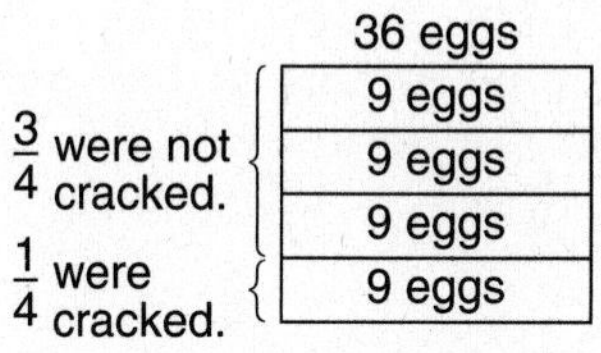

36 eggs

$\frac{3}{4}$ were not cracked. — 9 eggs | 9 eggs | 9 eggs

$\frac{1}{4}$ were cracked. — 9 eggs

$$25\% = \frac{25}{100} \div \frac{25}{25} = \frac{1}{4}$$

(a) $\frac{4}{4} - \frac{1}{4} = \mathbf{\frac{3}{4}}$

(b) 3 × 9 eggs = **27 eggs**

7. (a) $\mathbf{\frac{2}{5}}$

(b) $\frac{6}{10} = \frac{6}{10} \times \frac{10}{10} = \frac{60}{100} =$ **60%**

8. (a) **4**

(b) 4 × 3 = **12**

9. (a) **0**

(b) **Property of zero for multiplication**

10. $\frac{3}{3} - \left(\frac{1}{3} \cdot \frac{3}{1}\right) = \frac{3}{3} - \left(\frac{3}{3}\right) = 0$

$\left(\frac{3}{3} - \frac{1}{3}\right) \cdot \frac{3}{1} = \left(\frac{2}{3}\right) \cdot \frac{3}{1} = \frac{6}{3} = 2$

$\mathbf{0 < 2}$

11.

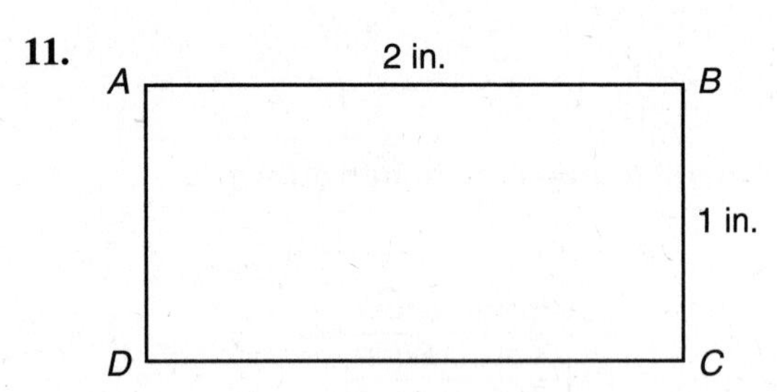

(a) Perimeter = 2 in. + 2 in. + 1 in. + 1 in. = **6 in.**

(b) Area = 2 in. × 1 in. = $\mathbf{2\ in.^2}$

(c)

$$\begin{array}{r} 90° \\ 90° \\ 90° \\ +\ 90° \\ \hline \mathbf{360°} \end{array}$$

12. (a) 32

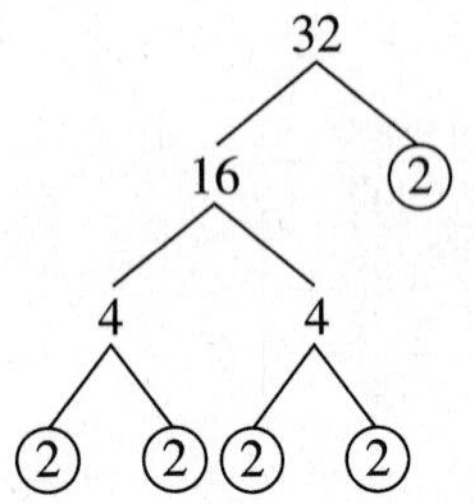

$32 = \mathbf{2 \cdot 2 \cdot 2 \cdot 2 \cdot 2}$

(b)

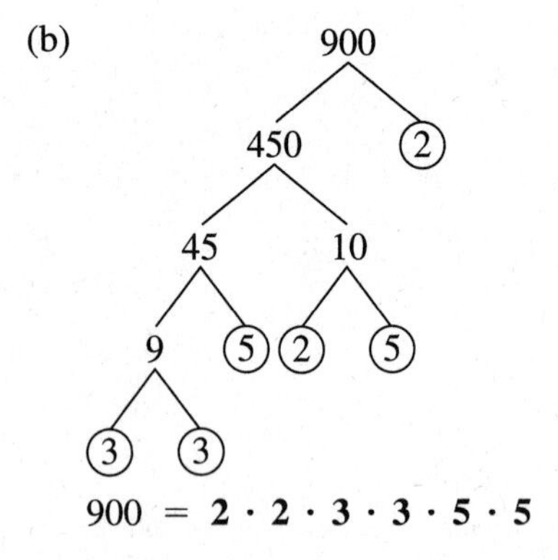

$900 = \mathbf{2 \cdot 2 \cdot 3 \cdot 3 \cdot 5 \cdot 5}$

(c) $\sqrt{900} = 30$

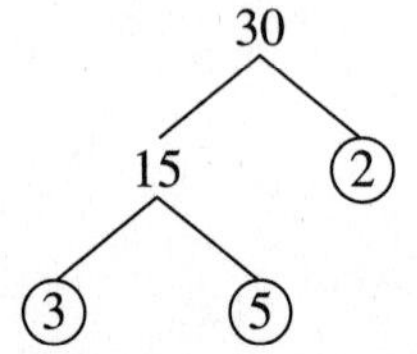

30

$30 = \mathbf{2 \cdot 3 \cdot 5}$

13. (a) $\frac{5}{6} \cdot \frac{10}{10} = \mathbf{\frac{50}{60}}$

(b) $\frac{3}{5} \cdot \frac{12}{12} = \mathbf{\frac{36}{60}}$

(c) $\frac{7}{12} \cdot \frac{5}{5} = \mathbf{\frac{35}{60}}$

14. $\frac{50}{60} + \frac{36}{60} + \frac{35}{60} = \frac{121}{60} = \mathbf{2\frac{1}{60}}$

15. (a) $\mathbf{-2, -\frac{2}{3}, 0, 1, \frac{3}{2}}$

(b) $\mathbf{1, \frac{3}{2}}$

16. $\frac{11}{12} - \frac{5}{12} = \frac{6}{12}$

$a = \mathbf{\frac{6}{12}}$ or $\mathbf{\frac{1}{2}}$

17.

$$\begin{array}{r} 10 \\ 90\overline{)900} \\ \underline{90} \\ 00 \\ \underline{0} \\ 0 \end{array}$$

$c = \mathbf{10}$

18.

$$\begin{array}{r} 11 \\ 11\overline{)121} \\ \underline{11} \\ 11 \\ \underline{11} \\ 0 \end{array}$$

$x = \mathbf{11}$

19. $2\frac{2}{3} + 1\frac{1}{3} = 3\frac{3}{3} = 4$

$y = \mathbf{4}$

20. $10^2 \cdot 10^5 =$

$10 \cdot 10 \cdot 10 \cdot 10 \cdot 10 \cdot 10 \cdot 10 = 10^7$

$n = \mathbf{7}$

21. $\frac{5}{6} + \frac{5}{6} + \frac{5}{6} = \frac{15}{6} = \mathbf{2\frac{1}{2}}$

22. $\frac{15}{2} \cdot \frac{10}{3} = \frac{150}{6} = \mathbf{25}$

23. $\left(\frac{5}{6}\right)^2 = \frac{5}{6} \cdot \frac{5}{6} = \mathbf{\frac{25}{36}}$

24. **30**

25. $\mathbf{\frac{9}{5}}$

26. $1\frac{1}{2} = \frac{3}{2}, 1\frac{2}{3} = \frac{5}{3}$

$\frac{3}{2} \times \frac{5}{3} = \frac{15}{6} = \mathbf{2\frac{1}{2}}$

27. 1 lb = 16 oz

16 oz + 5 oz = **21 oz**

28.

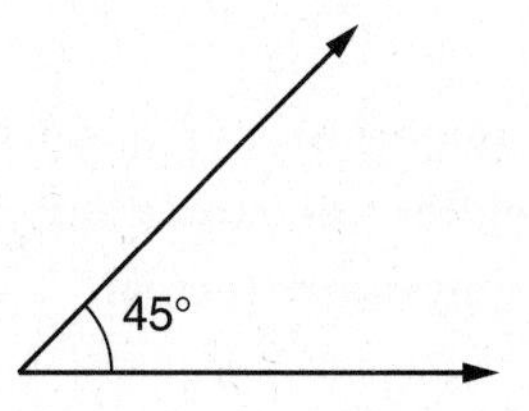

29. $\frac{1}{10} \times \frac{1}{10} = \mathbf{\frac{1}{100}}$

30. **−1**

Lesson 23, Warm-Up

a. **$4.63**

b. **$0.25**

c. **–51**

d. **496**

e. **4**

f. **38**

g. **5**

Problem Solving

Imani can take 2 × 3, or **6 different routes.**

Possible diagram:

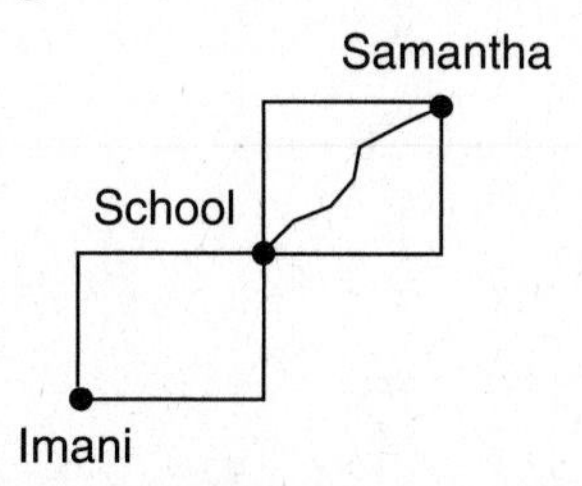

Lesson 23, Lesson Practice

a. $\begin{array}{r} 7 \\ -\ 2\frac{1}{3} \\ \hline \end{array} \longrightarrow \begin{array}{r} 6\frac{3}{3} \\ -\ 2\frac{1}{3} \\ \hline \mathbf{4\frac{2}{3}} \end{array}$

b. $\begin{array}{r} 6\frac{2}{5} \\ -\ 1\frac{4}{5} \\ \hline \end{array} \xrightarrow{5 + \frac{5}{5} + \frac{2}{5}} \begin{array}{r} 5\frac{7}{5} \\ -\ 1\frac{4}{5} \\ \hline \mathbf{4\frac{3}{5}} \end{array}$

c. $\begin{array}{r} 5\frac{1}{6} \\ -\ 1\frac{5}{6} \\ \hline \end{array} \xrightarrow{4 + \frac{6}{6} + \frac{1}{6}} \begin{array}{r} 4\frac{7}{6} \\ -\ 1\frac{5}{6} \\ \hline 3\frac{2}{6} \end{array}$

$3\frac{2}{6} = \mathbf{3\frac{1}{3}}$

d. $\begin{array}{r} 100\% \\ -\ 12\frac{1}{2}\% \\ \hline \end{array} \longrightarrow \begin{array}{r} 99\frac{2}{2}\% \\ -\ 12\frac{1}{2}\% \\ \hline \mathbf{87\frac{1}{2}}\% \end{array}$

e. $\begin{array}{r} 83\frac{1}{3}\% \\ -\ 16\frac{2}{3}\% \\ \hline \end{array} \xrightarrow{\left(82 + \frac{3}{3} + \frac{1}{3}\right)\%} \begin{array}{r} 82\frac{4}{3}\% \\ -\ 16\frac{2}{3}\% \\ \hline \mathbf{66\frac{2}{3}}\% \end{array}$

Lesson 23, Mixed Practice

1. **18 × 36 = *E***

$\begin{array}{r} 18 \\ \times\ 36 \\ \hline 108 \\ 54\ \\ \hline 648 \end{array}$ **648 exposures**

2. **50,000,000 − 250,000 = *D***

```
  50,000,000
−    250,000
  49,750,000
```

Forty-nine million, seven hundred fifty thousand

3. **259 + 269 + 307 = *T***

```
  259 people
  269 people
+ 307 people
  835 people
```

835 people

4. **16*P* = $14.24**

```
     $0.89
16)$14.24
    12 8
     1 44
     1 44
        0
```

89¢ per pound

5.

56 restaurants

$\frac{3}{8}$ were closed.	7 restaurants
	7 restaurants
	7 restaurants
$\frac{5}{8}$ were open.	7 restaurants
	7 restaurants
	7 restaurants
	7 restaurants
	7 restaurants

(a) 3 × 7 restaurants = **21 restaurants**

(b) 5 × 7 restaurants = **35 restaurants**

6. $40\% = \frac{40}{100} \div \frac{20}{20} = \mathbf{\frac{2}{5}}$

30 children

$\frac{3}{5}$ were girls.	6 children
	6 children
	6 children
$\frac{2}{5}$ were boys.	6 children
	6 children

(a) 2 × 6 children = 12 children; **12 boys**

(b) $\frac{5}{5} - \frac{2}{5} = \frac{3}{5}$

3 × 6 children = 18 children; **18 girls**

7. 1 yard = 36 inches

```
     115
36)4140
    36
     54
     36
     180
     180
       0
```

115 yards

8. (a) **5**

(b) 5 × 3 = **15**

9. **Express the mixed number as an improper fraction. Then switch the numerator and the denominator of the improper fraction.**

10. (a) $\frac{2}{3} \cdot \frac{3}{2} = \frac{5}{5}$

(b) $\frac{12}{36} < \frac{12}{24}$

11. $2\frac{1}{4} = \frac{9}{4},\ 3\frac{1}{3} = \frac{10}{3}$

$\frac{9}{4} \times \frac{10}{3} = \frac{90}{12} = \frac{15}{2} = \mathbf{7\frac{1}{2}}$

12. (a) $\frac{3}{4} \cdot \frac{10}{10} = \mathbf{\frac{30}{40}}$

(b) $\frac{2}{5} \cdot \frac{8}{8} = \mathbf{\frac{16}{40}}$

(c) $\frac{15 \div 5}{40 \div 5} = \mathbf{\frac{3}{8}}$

13. (a)

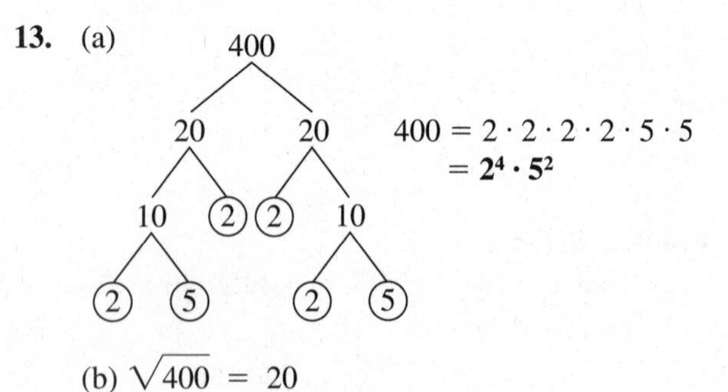

(b) $\sqrt{400} = 20$

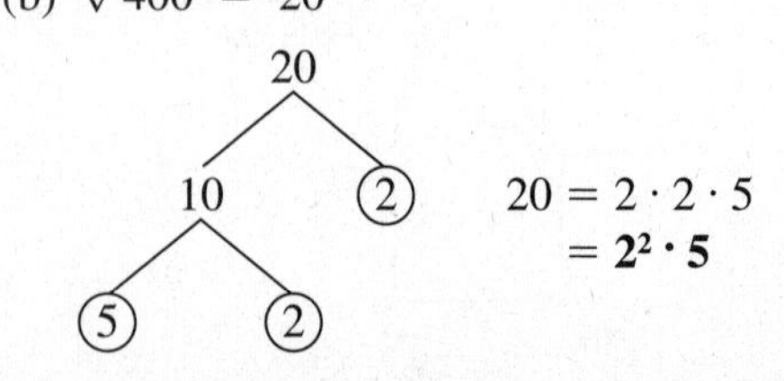

14. (a) **Acute angle**

(b) **Right angle**

(c) **Obtuse angle**

(d) $\overrightarrow{DC}$

15. **Check polygon for eight sides; one possibility:**

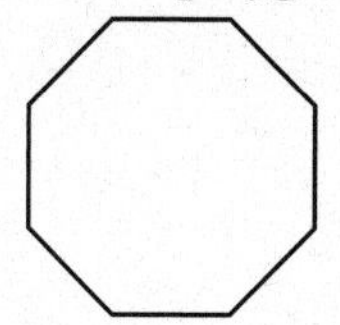

16. $7\overline{)105}$ = 15
7
35
35
0

$w = \mathbf{15}$

17. $2x = 100,\ 100 \div 2 = 50$

$x = \mathbf{50}$

18. $6\frac{3}{4} - 1\frac{1}{4} = 5\frac{2}{4}$

$x = \mathbf{5\frac{2}{4}}$ or $\mathbf{5\frac{1}{2}}$

19. $1\frac{5}{8} + 4\frac{1}{8} = 5\frac{6}{8}$

$x = \mathbf{5\frac{6}{8}}$ or $\mathbf{5\frac{3}{4}}$

20. $5 - 3\frac{1}{3} \longrightarrow 4\frac{3}{3} - 3\frac{1}{3} = \mathbf{1\frac{2}{3}}$

21. $83\frac{1}{3}\% - 66\frac{2}{3}\% \xrightarrow{\left(82 + \frac{3}{3} + \frac{1}{3}\right)\%} 82\frac{4}{3}\% - 66\frac{2}{3}\% = \mathbf{16\frac{2}{3}\%}$

22. $\frac{7}{12} + \left(\frac{1}{4} \cdot \frac{1}{3}\right) = \frac{7}{12} + \left(\frac{1}{12}\right)$

$= \frac{8}{12} = \mathbf{\frac{2}{3}}$

23. $\frac{7}{8} - \left(\frac{3}{4} \cdot \frac{1}{2}\right) = \frac{7}{8} - \left(\frac{3}{8}\right) = \frac{4}{8} = \mathbf{\frac{1}{2}}$

24.

$\overline{AC}$ is about 2 inches long.

Perimeter $= 2 \text{ in.} + 1 \text{ in.} + 1\frac{3}{4} \text{ in.}$

$= 4\frac{3}{4} \text{ in.}$

The perimeter is about $4\frac{3}{4}$ inches.

25. **About 30°**

26. Perimeter $= 14 \text{ ft} + 14 \text{ ft} + 12 \text{ ft} + 12 \text{ ft}$

$= \mathbf{52 \text{ ft}}$

27. $\frac{3}{4} \times \frac{1}{3} = \frac{3}{12} = \mathbf{\frac{1}{4}}$

28. $\frac{3}{4} \cdot \frac{3}{3} = \frac{9}{12}$

$\frac{2}{3} \cdot \frac{4}{4} = \frac{8}{12}$

$\frac{9}{12} - \frac{8}{12} = \mathbf{\frac{1}{12}}$

29. (a) $\mathbf{4^3, 5^3}$

(b) $4^3 = 4 \cdot 4 \cdot 4 = 64$

$5^3 = 5 \cdot 5 \cdot 5 = 125$

64, 125

30. (a) $\overline{AB}$ (or $\overline{BA}$)

(b) $\angle CMB$ (or $\angle BMC$)

(c) $\angle ACB$ (or $\angle BCA$)

Lesson 24, Warm-Up

a. **$6.72**

b. **$15.00**

c. **64¢**

d. **260**

e. **5**

f. **8**

g. **15**

Problem Solving

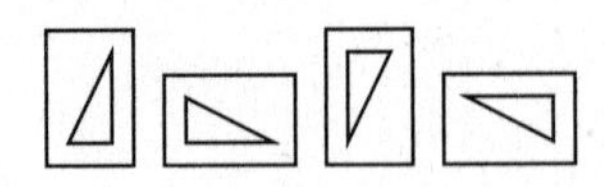

Lesson 24, Lesson Practice

a. $\frac{48}{144} = \frac{\overset{1}{\cancel{2}} \cdot \overset{1}{\cancel{2}} \cdot \overset{1}{\cancel{2}} \cdot \overset{1}{\cancel{2}} \cdot \overset{1}{\cancel{3}}}{\underset{1}{\cancel{2}} \cdot \underset{1}{\cancel{2}} \cdot \underset{1}{\cancel{2}} \cdot \underset{1}{\cancel{2}} \cdot \underset{1}{\cancel{3}} \cdot 3} = \mathbf{\frac{1}{3}}$

b. $\frac{90}{324} = \frac{\overset{1}{\cancel{2}} \cdot \overset{1}{\cancel{3}} \cdot \overset{1}{\cancel{3}} \cdot 5}{\underset{1}{\cancel{2}} \cdot 2 \cdot \underset{1}{\cancel{3}} \cdot \underset{1}{\cancel{3}} \cdot 3 \cdot 3} = \mathbf{\frac{5}{18}}$

c. $90 = 2 \cdot 3 \cdot 3 \cdot 5$

$324 = 2 \cdot 2 \cdot 3 \cdot 3 \cdot 3 \cdot 3$

$\text{GCF} = 2 \cdot 3 \cdot 3 = \mathbf{18}$

d. $\frac{\overset{1}{\cancel{5}}}{8} \cdot \frac{3}{\underset{2}{\cancel{10}}} = \mathbf{\frac{3}{16}}$

e. $\frac{\overset{\overset{1}{\cancel{2}}}{\cancel{8}}}{\underset{\underset{1}{\cancel{3}}}{\cancel{15}}} \cdot \frac{\overset{1}{\cancel{5}}}{\underset{\underset{1}{\cancel{3}}}{\cancel{12}}} \cdot \frac{\overset{\overset{1}{\cancel{3}}}{\cancel{9}}}{\underset{5}{\cancel{10}}} = \mathbf{\frac{1}{5}}$

f. $\frac{\overset{1}{\cancel{8}}}{\underset{1}{\cancel{3}}} \cdot \frac{\overset{\overset{1}{\cancel{2}}}{\cancel{6}}}{7} \cdot \frac{5}{\underset{\underset{1}{\cancel{2}}}{\cancel{16}}} = \mathbf{\frac{5}{7}}$

g. $\frac{\overset{1}{\cancel{2}} \cdot \overset{1}{\cancel{2}} \cdot \overset{1}{\cancel{3}} \cdot \overset{1}{\cancel{3}}}{\underset{1}{\cancel{3}} \cdot \underset{1}{\cancel{3}} \cdot \underset{1}{\cancel{3}}} \cdot \frac{\overset{1}{\cancel{3}} \cdot 5}{\underset{1}{\cancel{2}} \cdot \underset{1}{\cancel{2}} \cdot 2 \cdot 3} = \mathbf{\frac{5}{6}}$

Lesson 24, Mixed Practice

1. $3026 - 2895 = D$

$$\begin{array}{r} 3026 \text{ miles} \\ -\ 2895 \text{ miles} \\ \hline \mathbf{131 \text{ miles}} \end{array}$$

2. $15 \times 24 = M$

$$\begin{array}{r} 15 \\ \times\ 24 \\ \hline 60 \\ 30 \\ \hline 360 \end{array}$$

360 microprocessors

3. $75\% = \frac{75}{100} \div \frac{25}{25} = \frac{3}{4}$

	$30.00
$\frac{1}{4}$ not spent	$ 7.50
$\frac{3}{4}$ spent	$ 7.50
	$ 7.50
	$ 7.50

(a) $\mathbf{\frac{3}{4}}$

(b)
$$\begin{array}{r} \$7.50 \\ \times\quad 3 \\ \hline \mathbf{\$22.50} \end{array}$$

4. Diameter $= 2 \times$ radius
1 yard $=$ 36 inches
36 inches $= 2 \times$ radius
Radius $=$ 36 inches $\div\ 2 =$ **18 inches**

5. 30 steps $\div\ 3 =$ 10 steps
10 steps $\times\ 2 =$ **20 steps**

6. (a) **8**

(b) $8 \times 3 = \mathbf{24}$

7. (a) $\mathbf{\frac{1}{3}}$

(b) $\mathbf{\frac{1}{3}}$

8. (a) $\frac{540}{600} = \frac{\overset{1}{\cancel{2}} \cdot \overset{1}{\cancel{2}} \cdot \overset{1}{\cancel{3}} \cdot 3 \cdot 3 \cdot \overset{1}{\cancel{5}}}{\underset{1}{\cancel{2}} \cdot \underset{1}{\cancel{2}} \cdot 2 \cdot \underset{1}{\cancel{3}} \cdot \underset{1}{\cancel{5}} \cdot 5} = \mathbf{\frac{9}{10}}$

(b) $2 \cdot 2 \cdot 3 \cdot 5 = \mathbf{60}$

9. (a) **Acute angle**

(b) **Right angle**

(c) **Obtuse angle**

10. Equivalent fractions are formed by multiplying or dividing a fraction by a fraction equal to 1. To change from fifths to thirtieths, multiply $\frac{3}{5}$ by $\frac{6}{6}$.

11. (a)
$$\begin{aligned} 10{,}000 &= 1000 \cdot 10 \\ 1000 &= 2 \cdot 2 \cdot 2 \cdot 5 \cdot 5 \cdot 5 \\ 10 &= 2 \cdot 5 \\ 1000 \cdot 10 &= 2 \cdot 2 \cdot 2 \cdot 5 \cdot 5 \cdot 5 \cdot 2 \cdot 5 \\ &= \mathbf{2^4 \cdot 5^4} \end{aligned}$$

(b)
$$\begin{aligned} \sqrt{10{,}000} &= 100 \\ 100 &= 2 \cdot 2 \cdot 5 \cdot 5 = \mathbf{2^2 \cdot 5^2} \end{aligned}$$

12. (a)

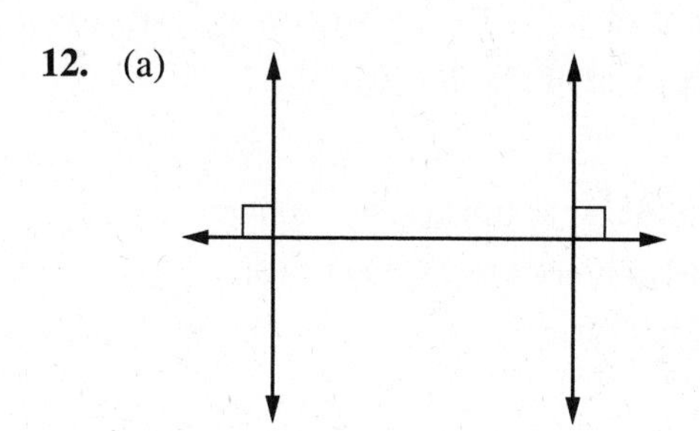

(b) **Right angles**

13. (a)
$$\begin{aligned} 1 \text{ yard} &= 36 \text{ inches} \\ 36 \text{ inches} \div 4 &= \mathbf{9 \text{ inches}} \end{aligned}$$

(b)
$$\begin{aligned} \text{Area} &= 9 \text{ inches} \times 9 \text{ inches} \\ &= \mathbf{81 \text{ square inches}} \end{aligned}$$

14. Commutative property

15. $4\frac{7}{12} - 1\frac{1}{12} = 3\frac{6}{12} = 3\frac{1}{2}$

$x = \mathbf{3\frac{1}{2}}$

16. $2\frac{3}{4} + 3\frac{3}{4} = 5\frac{6}{4} = 6\frac{1}{2}$

$w = \mathbf{6\frac{1}{2}}$

17.
$$\begin{array}{r} 12 \\ 8\overline{)100} \\ \underline{8} \\ 20 \\ \underline{16} \\ 4 \\ \underline{0} \\ 4 \end{array}$$

$12\frac{4}{8} = 12\frac{1}{2}$

$m = \mathbf{12\frac{1}{2}}$

18.
$$\begin{array}{r} \$0.28 \\ \times \quad 12 \\ \hline 56 \\ 28 \\ \hline \$3.36 \end{array}$$

$n = \mathbf{\$3.36}$

19. $\dfrac{10^5}{10^2} = \dfrac{\overset{1}{\cancel{10}} \cdot \overset{1}{\cancel{10}} \cdot 10 \cdot 10 \cdot 10}{\underset{1}{\cancel{10}} \cdot \underset{1}{\cancel{10}}}$

$= 10 \cdot 10 \cdot 10 = \mathbf{10^3}$ or $\mathbf{1000}$

20. $\sqrt{9} - \sqrt{4^2} = 3 - 4 = \mathbf{-1}$

21.
$$\begin{array}{r} 100\% \\ -\ 66\frac{2}{3}\% \\ \hline \end{array} \longrightarrow \begin{array}{r} 99\frac{3}{3}\% \\ -\ 66\frac{2}{3}\% \\ \hline \mathbf{33\frac{1}{3}\%} \end{array}$$

22.
$$\begin{array}{r} 5\frac{1}{8} \\ -\ 1\frac{7}{8} \\ \hline \end{array} \xrightarrow{4 + \frac{8}{8} + \frac{1}{8}} \begin{array}{r} 4\frac{9}{8} \\ -\ 1\frac{7}{8} \\ \hline 3\frac{2}{8} \end{array}$$

$3\frac{2}{8} = \mathbf{3\frac{1}{4}}$

23. $\left(\frac{5}{6}\right)^2 = \frac{5}{6} \cdot \frac{5}{6} = \mathbf{\frac{25}{36}}$

24. $\dfrac{\overset{1}{\cancel{3}}}{\underset{1}{\cancel{4}}} \cdot \dfrac{1}{\underset{1}{\cancel{2}}} \cdot \dfrac{\overset{\overset{1}{\cancel{2}}}{\cancel{8}}}{\underset{3}{\cancel{9}}} = \mathbf{\dfrac{1}{3}}$

25. Heptagon

26. (a) $10 \cdot 100 = \mathbf{1000}$

(b) $10 - 100 = \mathbf{-90}$

(c) $\dfrac{10}{100} = \mathbf{\dfrac{1}{10}}$

27.
$$\begin{aligned} \text{Missing length} &= 10 \text{ yards} + 10 \text{ yards} \\ &= 20 \text{ yards} \\ \text{Missing height} &= 25 \text{ yards} - 20 \text{ yards} \\ &= 5 \text{ yards} \\ \text{Perimeter} &= 25 \text{ yards} + 10 \text{ yards} + 5 \text{ yards} \\ &\quad + 10 \text{ yards} + 20 \text{ yards} + 20 \text{ yards} \\ &= \mathbf{90 \text{ yards}} \end{aligned}$$

28. $\frac{1}{4} \cdot \frac{3}{3} = \frac{3}{12}, \frac{1}{6} \cdot \frac{2}{2} = \frac{2}{12}$

$\frac{3}{12} + \frac{2}{12} = \mathbf{\frac{5}{12}}$

29. **$\angle DAC$** and **$\angle BCA$** (or **$\angle CAD$** and **$\angle ACB$**); **$\angle DCA$** and **$\angle BAC$** (or **$\angle ACD$** and **$\angle CAB$**)

30. (a) $\mathbf{-1, -\frac{1}{2}, 0, \frac{1}{2}, 1}$

(b) $1 + \frac{1}{2} = 1\frac{1}{2}$

$1\frac{1}{2} + \frac{1}{2} = 1\frac{2}{2} = 2$

$2 + \frac{1}{2} = 2\frac{1}{2}$

$\mathbf{1\frac{1}{2}, 2, 2\frac{1}{2}}$

Lesson 25, Warm-Up

a. **$4.64**

b. **$6.00**

c. **$0.76**

d. **252**

e. **6**

f. **9**

g. **7**

Problem Solving

We know that the ones digit of the bottom factor must be either 1 or 6 (because the ones digit of the product is 6). The tens digit of each factor must be either 1 or 2, because the hundreds digit of the product is 2. Two combinations of factors complete the problem (accept either combination).

$$\begin{array}{r} \mathbf{16} \\ \times\ \mathbf{16} \\ \hline \mathbf{96} \\ \mathbf{16} \\ \hline \mathbf{256} \end{array} \quad \text{or} \quad \begin{array}{r} \mathbf{26} \\ \times\ \mathbf{11} \\ \hline \mathbf{26} \\ \mathbf{26} \\ \hline \mathbf{286} \end{array}$$

Lesson 25, Lesson Practice

a. $1 \div \frac{2}{3} = \mathbf{\frac{3}{2}}$

$\frac{3}{4} \div \frac{2}{3} = \frac{3}{4} \times \frac{3}{2} = \frac{9}{8} = \mathbf{1\frac{1}{8}}$

b. $1 \div \frac{3}{4} = \frac{4}{3}$

$3 \div \frac{3}{4} = 3 \times \frac{4}{3} = \frac{12}{3} = \mathbf{4}$

c. **Instead of dividing by the divisor, multiply by the reciprocal of the divisor.**

d. **Pressing this key changes the number previously entered to its reciprocal (in decimal form).**

e. $1 \div \frac{2}{3} = \frac{3}{2}$

$\frac{3}{5} \div \frac{2}{3} = \frac{3}{5} \times \frac{3}{2} = \mathbf{\frac{9}{10}}$

f. $1 \div \frac{1}{4} = \frac{4}{1}$

$\frac{7}{8} \div \frac{1}{4} = \frac{7}{\cancel{8}_2} \times \frac{\cancel{4}^1}{1} = \frac{7}{2} = \mathbf{3\frac{1}{2}}$

g. $1 \div \frac{2}{3} = \frac{3}{2}$

$\frac{5}{6} \div \frac{2}{3} = \frac{5}{\cancel{6}_2} \times \frac{\cancel{3}^1}{2} = \frac{5}{4} = \mathbf{1\frac{1}{4}}$

Lesson 25, Mixed Practice

1. **$6B = 324$**

$$\begin{array}{r} 54 \\ 6\overline{)324} \\ 30 \\ \hline 24 \\ 24 \\ \hline 0 \end{array}$$

54 boxes

2.

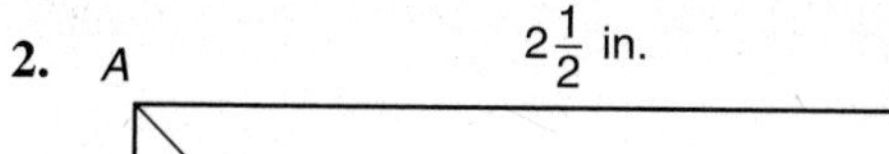
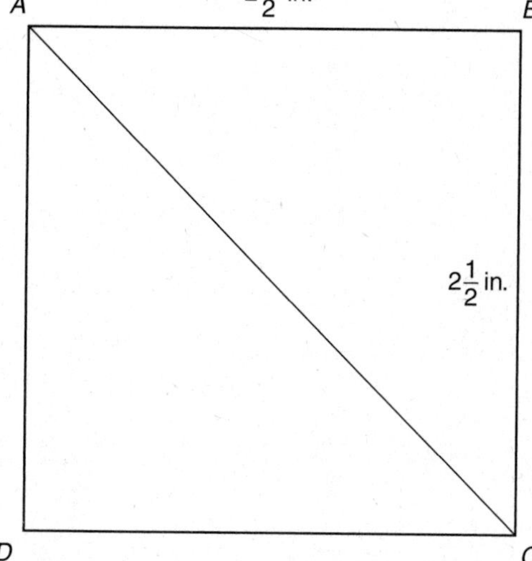

(a) Perimeter $= 2\frac{1}{2}$ in. $+ 2\frac{1}{2}$ in. $+ 2\frac{1}{2}$ in. $+ 2\frac{1}{2}$ in. $= 8\frac{4}{2}$ in. $=$ **10 in.**

(b) **90°**

(c) $180° - 90° = 90°$
$90° \div 2 =$ **45°**

(d) $90° + 45° + 45° =$ **180°**

3. 56 relatives ÷ 2 = **28 relatives**

4.
28 players
− 7 players
21 players

5. 28 players ÷ 2 = **14 players**

6. $70\% = \frac{70}{100} \div \frac{10}{10} = \frac{7}{10}$

310 pages

$\frac{7}{10}$ have been read.
31 pages
31 pages
31 pages
31 pages
31 pages
31 pages
31 pages

$\frac{3}{10}$ have not been read.
31 pages
31 pages
31 pages

(a) 7 × 31 pages = **217 pages**

(b) 3 × 31 pages = **93 pages**

7. (a) $1 \div \frac{3}{4} = \mathbf{\frac{4}{3}}$

(b) $\frac{7}{8} \div \frac{3}{4} = \frac{7}{\cancel{8}_2} \times \frac{\cancel{4}^1}{3} = \frac{7}{6} = \mathbf{1\frac{1}{6}}$

8. **C.** $\mathbf{\frac{2}{5}}$

A little less than half is shaded. We eliminate $\frac{2}{3}$, which is more than $\frac{1}{2}$. Since $\frac{2}{4}$ equals $\frac{1}{2}$, and $\frac{2}{5}$ is a little less than $\frac{1}{2}$, we choose $\frac{2}{5}$.

9. $84 = 2 \cdot 2 \cdot 3 \cdot 7$
$210 = 2 \cdot 3 \cdot 5 \cdot 7$

$$\frac{\cancel{2}^1 \cdot 2 \cdot \cancel{3}^1 \cdot \cancel{7}^1}{\cancel{2}_1 \cdot \cancel{3}_1 \cdot 5 \cdot \cancel{7}_1} = \mathbf{\frac{2}{5}}$$

10. (a) $\mathbf{\frac{10}{9}}$

(b) $\mathbf{\frac{1}{8}}$

(c) $2\frac{3}{8} = \frac{19}{8}, \mathbf{\frac{8}{19}}$

11. $\frac{3}{4} \cdot \frac{5}{5} = \frac{15}{20}, \frac{4}{5} \cdot \frac{4}{4} = \frac{16}{20}$

$\frac{15}{20} + \frac{16}{20} = \frac{31}{20} = \mathbf{1\frac{11}{20}}$

12. $640 = 40 \cdot 16 = 2^3 \cdot 5 \cdot 16$
$= 2 \cdot 2 \cdot 2 \cdot 5 \cdot 2 \cdot 2 \cdot 2 \cdot 2$
$= \mathbf{2^7 \cdot 5}$

13. $2\frac{2}{3} = \frac{8}{3}, 2\frac{1}{4} = \frac{9}{4}$

$\frac{\cancel{8}^2}{\cancel{3}_1} \times \frac{\cancel{9}^3}{\cancel{4}_1} = \mathbf{6}$

14. (a) A: $4\frac{4}{6} = \mathbf{4\frac{2}{3}}$

B: $5\frac{3}{6} = \mathbf{5\frac{1}{2}}$

(b) $5\frac{3}{6} - 4\frac{4}{6}$ $\xrightarrow{4 + \frac{6}{6} + \frac{3}{6}}$ $4\frac{9}{6} - 4\frac{4}{6} = \mathbf{\frac{5}{6}}$

15. (a)

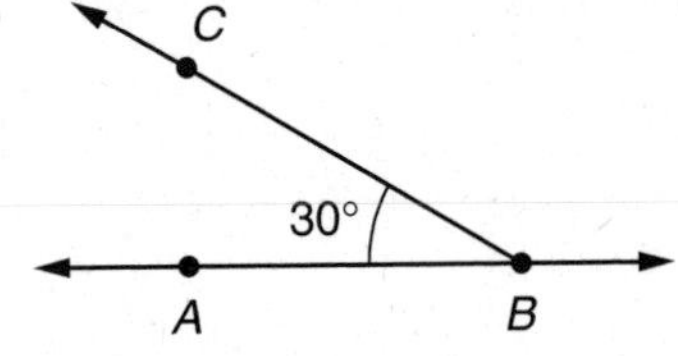

(b) **Acute angle**

16. $$\begin{array}{r} 3 \\ -\ 1\frac{7}{12} \\ \hline \end{array} \longrightarrow \begin{array}{r} 2\frac{12}{12} \\ -\ 1\frac{7}{12} \\ \hline y = \mathbf{1\frac{5}{12}} \end{array}$$

17. $5\frac{7}{8} + 4\frac{5}{8} = 9\frac{12}{8} = 10\frac{4}{8} = 10\frac{1}{2}$

$x = \mathbf{10\frac{1}{2}}$

18. $$\begin{array}{r} 45° \\ 8\overline{)360°} \\ \underline{32} \\ 40 \\ \underline{40} \\ 0 \end{array}$$

$n = \mathbf{45°}$

19. $1^3 = 1 \cdot 1 \cdot 1 = 1, \frac{3}{4}$

$m = \mathbf{\frac{3}{4}}$

20. $6\frac{1}{6} + 1\frac{5}{6} = 7\frac{6}{6} = \mathbf{8}$

21. $\frac{\cancel{3}^{1}}{\cancel{4}_{1}} \cdot \frac{\cancel{3}^{1}}{\cancel{9}_{3}} \cdot \frac{\cancel{8}^{2}}{\cancel{15}_{3}} = \mathbf{\frac{2}{9}}$

22. $1 \div \frac{2}{1} = \frac{1}{2}$

$\frac{4}{5} \div \frac{2}{1} = \frac{\cancel{4}^{2}}{5} \times \frac{1}{\cancel{2}_{1}} = \mathbf{\frac{2}{5}}$

23. $1 \div \frac{6}{5} = \frac{5}{6}$

$\frac{8}{5} \div \frac{6}{5} = \frac{\cancel{8}^{4}}{\cancel{5}_{1}} \times \frac{\cancel{5}^{1}}{\cancel{6}_{3}} = \frac{4}{3} = \mathbf{1\frac{1}{3}}$

24. $1 \div \frac{5}{6} = \frac{6}{5}$

$\frac{3}{7} \div \frac{5}{6} = \frac{3}{7} \times \frac{6}{5} = \mathbf{\frac{18}{35}}$

25. $$\begin{array}{r} 12 \\ 8\overline{)100} \\ \underline{8} \\ 20 \\ \underline{16} \\ 4 \end{array}$$

$\frac{0}{4}$

$12\frac{4}{8}\% = \mathbf{12\frac{1}{2}\%}$

26. $\mathbf{\frac{5}{3}}$

27. (a) $2^2 \cdot 2^3 = 4 \cdot 8$
$\mathbf{4 \cdot 8 = 8 \cdot 4}$ or
$\mathbf{32 = 32}$

(b) **2**

28. Perimeter = 6 inches + 6 inches
+ 6 inches + 6 inches
+ 6 inches + 6 inches
= 36 inches

36 inches ÷ 12 = **3 feet**

29. Perimeter = 4 in. + 2 in. + 2 in.
+ 2 in. + 2 in. + 4 in.
= **16 in.**

30. Third prime number = 5, **−5**

Lesson 26, Warm-Up

a. **$9.54**

b. **$30.00**

c. **93¢**

d. **222**

e. $\mathbf{1\frac{1}{3}}$

f. **16**

g. **0**

Problem Solving

$\$20.00 \div 5 = \$4.00$

$\$20.00 + \$4.00 = \mathbf{\$24.00}$

Lesson 26, Lesson Practice

a.

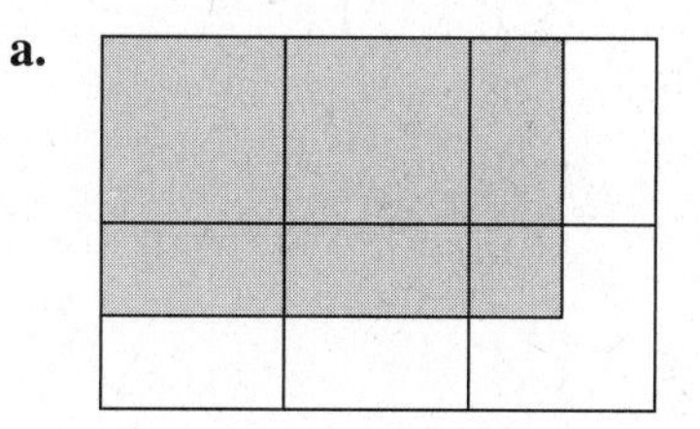

$$\text{Area} = 2\text{ in.}^2 + \frac{3}{2}\text{ in.}^2 + \frac{1}{4}\text{ in.}^2$$

$$= 2\text{ in.}^2 + \frac{6}{4}\text{ in.}^2 + \frac{1}{4}\text{ in.}^2$$

$$= 2\frac{7}{4}\text{ in.}^2 = \mathbf{3\frac{3}{4}\text{ in.}^2}$$

check: $1\frac{1}{2}\text{ in.} \times 2\frac{1}{2}\text{ in.}$

$$= \frac{3}{2}\text{ in.} \times \frac{5}{2}\text{ in.} = \frac{15}{4}\text{ in.}^2$$

$$= 3\frac{3}{4}\text{ in.}^2$$

b. $6\frac{2}{3} \times \frac{3}{5} = \frac{\overset{4}{\cancel{20}}}{\underset{1}{\cancel{3}}} \times \frac{\overset{1}{\cancel{3}}}{\underset{1}{\cancel{5}}} = \mathbf{4}$

c. $2\frac{1}{3} \times 3\frac{1}{2} = \frac{7}{3} \times \frac{7}{2} = \frac{49}{6} = \mathbf{8\frac{1}{6}}$

d. $3 \times 3\frac{3}{4} = \frac{3}{1} \times \frac{15}{4} = \frac{45}{4} = \mathbf{11\frac{1}{4}}$

e. $1\frac{2}{3} \div 3 = 1\frac{2}{3} \times \frac{1}{3} = \frac{5}{3} \times \frac{1}{3} = \mathbf{\frac{5}{9}}$

f. $2\frac{1}{2} \div 3\frac{1}{3} = \frac{5}{2} \div \frac{10}{3}$

$$= \frac{\overset{1}{\cancel{5}}}{2} \times \frac{3}{\underset{2}{\cancel{10}}} = \mathbf{\frac{3}{4}}$$

g. $5 \div \frac{2}{3} = \frac{5}{1} \times \frac{3}{2} = \frac{15}{2} = \mathbf{7\frac{1}{2}}$

h. $2\frac{2}{3} \div 1\frac{1}{3} = \frac{8}{3} \div \frac{4}{3}$

$$= \frac{\overset{2}{\cancel{8}}}{\underset{1}{\cancel{3}}} \times \frac{\overset{1}{\cancel{3}}}{\underset{1}{\cancel{4}}} = \mathbf{2}$$

i. $1\frac{1}{3} \div 2\frac{2}{3} = \frac{4}{3} \div \frac{8}{3} = \frac{\overset{1}{\cancel{4}}}{\underset{1}{\cancel{3}}} \times \frac{\overset{1}{\cancel{3}}}{\underset{2}{\cancel{8}}} = \mathbf{\frac{1}{2}}$

j. $4\frac{1}{2} \times 1\frac{2}{3} = \frac{\overset{3}{\cancel{9}}}{2} \times \frac{5}{\underset{1}{\cancel{3}}} = \frac{15}{2} = \mathbf{7\frac{1}{2}}$

Lesson 26, Mixed Practice

1. $\mathbf{23 + M = 61}$

$$\begin{array}{r} 61 \text{ millimeters} \\ -\ 23 \text{ millimeters} \\ \hline \mathbf{38\ millimeters} \end{array}$$

2. $\mathbf{26 \times 85¢ = T}$

$$\begin{array}{r} \$0.85 \\ \times\ \ 26 \\ \hline 510 \\ 170\ \\ \hline \mathbf{\$22.10} \end{array}$$

3. $\mathbf{1453 - 330 = B}$

$$\begin{array}{r} 1453 \\ -\ \ 330 \\ \hline 1123 \end{array}$$ **1123 years**

4. $\mathbf{\$20.00 - S = \$11.25}$

$$\begin{array}{r} \$20.00 \\ -\ \$11.25 \\ \hline \mathbf{\$8.75} \end{array}$$

5. $\mathbf{12 \times 12 = P}$

$$\begin{array}{r} 12 \\ \times\ 12 \\ \hline 24 \\ 12\ \\ \hline 144 \end{array}$$

144 pencils

6.

60 marbles

$\frac{2}{5}$ were blue.	12 marbles
	12 marbles
$\frac{3}{5}$ were not blue.	12 marbles
	12 marbles
	12 marbles

$$40\% = \frac{40}{100} \div \frac{20}{20} = \mathbf{\frac{2}{5}}$$

(a) 2×12 marbles $=$ **24 marbles**

(b) 3×12 marbles $=$ **36 marbles**

7. 1 ton $=$ 2000 pounds

$$\frac{\overset{500}{\cancel{2000}} \text{ pounds}}{\underset{1}{\cancel{4}}} = \textbf{500 pounds}$$

8. (a) $\mathbf{\frac{3}{100}}$

(b) $\frac{97}{100} = \mathbf{97\%}$

9. (a) $210 = 2 \cdot 3 \cdot 5 \cdot 7$

$252 = 2 \cdot 2 \cdot 3 \cdot 3 \cdot 7$

$$\frac{\overset{1}{\cancel{2}} \cdot \overset{1}{\cancel{3}} \cdot 5 \cdot \overset{1}{\cancel{7}}}{\underset{1}{\cancel{2}} \cdot 2 \cdot \underset{1}{\cancel{3}} \cdot 3 \cdot \underset{1}{\cancel{7}}} = \mathbf{\frac{5}{6}}$$

(b) GCF $= 2 \cdot 3 \cdot 7 =$ **42**

10. (a) $\mathbf{\frac{9}{5}}$

(b) $\mathbf{\frac{4}{23}}$

(c) $\mathbf{\frac{1}{7}}$

11. (a) $\frac{5}{8} \cdot \frac{3}{3} = \mathbf{\frac{15}{24}}$

(b) $\frac{5}{12} \cdot \frac{2}{2} = \mathbf{\frac{10}{24}}$

(c) $\frac{15}{24} + \frac{10}{24} = \frac{25}{24} = \mathbf{1\frac{1}{24}}$

12. Check polygon for 7 sides; one possibility:

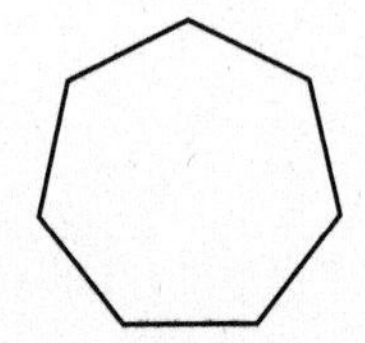

13.

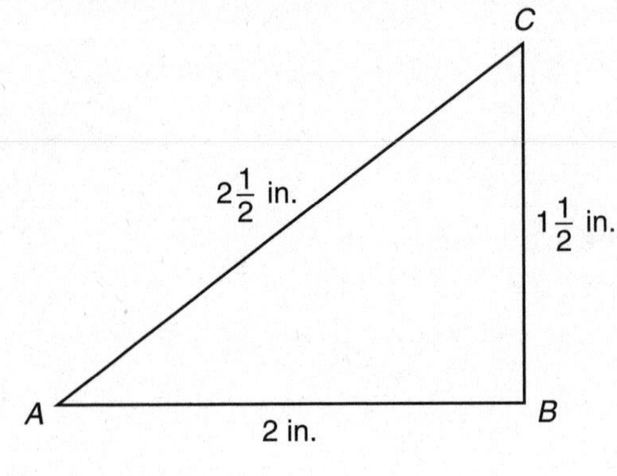

$AC = \mathbf{2\frac{1}{2}}$ **in.**

14. (a) $\mathbf{-3, 0, \frac{5}{6}, 1, \frac{4}{3}}$

(b) **0, 1**

15. $6\frac{5}{12} + 8\frac{11}{12} = 14\frac{16}{12} = 15\frac{4}{12}$

$x = \mathbf{15\frac{1}{3}}$

16.

$$\begin{array}{r} 180 \\ -\ 75 \\ \hline 105 \end{array}$$

$y =$ **105**

17.

$$\begin{array}{r} 30° \\ 12\overline{)360°} \\ 36 \\ \hline 00 \\ 0 \\ \hline 0 \end{array}$$

$w = \mathbf{30°}$

18.

$$\begin{array}{r} 100 \\ -\ 58\frac{1}{3} \\ \hline \end{array} \longrightarrow \begin{array}{r} 99\frac{3}{3} \\ -\ 58\frac{1}{3} \\ \hline w = \mathbf{41\frac{2}{3}} \end{array}$$

19. (a) Area $=$ 10 in. $\times$ 10 in.

$=$ **100 in.²**

(b) $\frac{100 \text{ in.}^2}{2} = \mathbf{50 \text{ in.}^2}$

20. $9\frac{1}{9} \xrightarrow{8 + \frac{9}{9} + \frac{1}{9}} 8\frac{10}{9}$

$$\begin{array}{r} 9\frac{1}{9} \\ -\ 4\frac{4}{9} \\ \hline \end{array} \qquad \begin{array}{r} 8\frac{10}{9} \\ -\ 4\frac{4}{9} \\ \hline 4\frac{6}{9} \end{array}$$

$4\frac{6}{9} = \mathbf{4\frac{2}{3}}$

21. $\frac{\cancel{5}^{1}}{8} \cdot \frac{\cancel{3}^{1}}{\cancel{10}_{2}} \cdot \frac{1}{\cancel{6}_{2}} = \mathbf{\frac{1}{32}}$

22. $\left(2\frac{1}{2}\right)^2 = 2\frac{1}{2} \times 2\frac{1}{2}$

$= \frac{5}{2} \times \frac{5}{2} = \frac{25}{4} = \mathbf{6\frac{1}{4}}$

23. $1\frac{3}{5} \div 2\frac{2}{3}$

$= \frac{8}{5} \div \frac{8}{3} = \frac{\cancel{8}^{1}}{5} \times \frac{3}{\cancel{8}_{1}} = \mathbf{\frac{3}{5}}$

24. $3\frac{1}{3} \div 4 = \frac{10}{3} \div \frac{4}{1}$

$= \frac{\cancel{10}^{5}}{3} \times \frac{1}{\cancel{4}_{2}} = \mathbf{\frac{5}{6}}$

25. $5 \cdot 1\frac{3}{4} = \frac{5}{1} \times \frac{7}{4} = \frac{35}{4}$

$= \mathbf{8\frac{3}{4}}$

26. $\sqrt{10^2 \cdot 10^4}$

$= \sqrt{10 \cdot 10 \cdot 10 \cdot 10 \cdot 10 \cdot 10}$

$= \sqrt{1{,}000{,}000} =$ **1000** or $\mathbf{10^3}$

27.
```
      459
36)16,524
   14 4
    2 12
    1 80
      324
      324
        0
```

28. (a) $3 - \frac{6}{3} = 3 - 2 = \mathbf{1}$

(b) $\frac{3 \cdot \cancel{6}^{1}}{\cancel{6}_{1}} = \mathbf{3}$

(c) $\frac{\cancel{3}^{1}}{\cancel{6}_{1}} \cdot \frac{\cancel{6}^{1}}{\cancel{3}_{1}} = \mathbf{1}$

29. $k = 3(9) - 2 = 27 - 2 = \mathbf{25}$

30. $\frac{1}{2} \times 90° = \frac{90°}{2} = \mathbf{45°}$

Lesson 27, Warm-Up

a. **$5.73**

b. **$12.50**

c. **420**

d. **210**

e. $\mathbf{1\frac{1}{2}}$

f. **18**

g. **0**

Problem Solving

No, Simon could not have been telling the truth. The two possible combinations of eight dots are 5, 2, and 1; and 4, 3, and 1. However, 5 and 2 are on opposite faces, and 4 and 3 are on opposite faces.

Lesson 27, Lesson Practice

a. 8, 16, 24, 32, (40), 48, . . .
10, 20, 30, (40), 50, . . .
LCM (8, 10) = **40**

b.
$$\begin{aligned} 4 &= 2 \cdot 2 \\ 6 &= 2 \cdot 3 \\ 10 &= 2 \cdot 5 \\ \text{LCM}(4, 6, 10) &= 2 \cdot 2 \cdot 3 \cdot 5 = 4 \cdot 15 \\ &= \mathbf{60} \end{aligned}$$

c.
$$24 = 2 \cdot 2 \cdot 2 \cdot 3$$
$$40 = 2 \cdot 2 \cdot 2 \cdot 5$$
$$\text{LCM}(24, 40) = 2 \cdot 2 \cdot 2 \cdot 3 \cdot 5 = 24 \cdot 5 = \mathbf{120}$$

d.
$$30 = 2 \cdot 3 \cdot 5$$
$$75 = 3 \cdot 5 \cdot 5$$
$$\text{LCM}(30, 75) = 2 \cdot 3 \cdot 5 \cdot 5 = \mathbf{150}$$

e. $\left(7\frac{1}{2}\right)2 = 15; \left(1\frac{1}{2}\right)2 = 3$

$15 \div 3 = \mathbf{5}$

f. Sample answer: **240 ÷ 4 = 60**

g. Sample answer: $\dfrac{\mathbf{\$6.00 \div 6}}{\mathbf{12 \div 6}} = \dfrac{\mathbf{\$1.00}}{\mathbf{2}} = \mathbf{50¢}$

h. Sample answer: **280 ÷ 10 = 28**

Lesson 27, Mixed Practice

1. 11,460 + 9420 + 8916 = *P*

$$\begin{array}{r} 11{,}460 \\ 9{,}420 \\ +\ 8{,}916 \\ \hline \mathbf{29{,}796} \end{array}$$

2. 6 · 12 = *I*

$$\begin{array}{r} 12 \\ \times\ 6 \\ \hline 72 \end{array}$$

72 inches

3. $0.15 per egg; Some equivalent division problems:

$0.90 ÷ 6 = $0.15

$0.60 ÷ 4 = $0.15

$0.45 ÷ 3 = $0.15

$0.30 ÷ 2 = $0.15

4. C. 10^9

5.

712 fruit flies

$\frac{3}{8}$ had red eyes.	89 fruit flies
	89 fruit flies
	89 fruit flies
$\frac{5}{8}$ did not have red eyes.	89 fruit flies
	89 fruit flies
	89 fruit flies
	89 fruit flies
	89 fruit flies

(a) 3 × 89 fruit flies = **267 fruit flies**

(b) 5 × 89 fruit flies = **445 fruit flies**

6. (a) 30 in. − 6 in. − 6 in. = 18 in.

18 in. ÷ 2 = **9 in.**

(b) Area = 6 in. × 9 in. = **54 in.^{2}**

7.
$$25 = 5 \cdot 5$$
$$45 = 3 \cdot 3 \cdot 5$$
$$\text{LCM}(25, 45) = 3 \cdot 3 \cdot 5 \cdot 5 = \mathbf{225}$$

8. 3500

9. (a) $\dfrac{24}{100} \div \dfrac{4}{4} = \dfrac{\mathbf{6}}{\mathbf{25}}$

(b) $36 = 2 \cdot 2 \cdot 3 \cdot 3$

$180 = 2 \cdot 2 \cdot 3 \cdot 3 \cdot 5$

$$\frac{\overset{1}{\cancel{2}} \cdot \overset{1}{\cancel{2}} \cdot \overset{1}{\cancel{3}} \cdot \overset{1}{\cancel{3}}}{\underset{1}{\cancel{2}} \cdot \underset{1}{\cancel{2}} \cdot \underset{1}{\cancel{3}} \cdot \underset{1}{\cancel{3}} \cdot 5} = \frac{\mathbf{1}}{\mathbf{5}}$$

10. (a)
$$\begin{array}{r} 102° \text{ F} \\ -\ 32° \text{ F} \\ \hline \mathbf{70° \text{ F}} \end{array}$$

(b)
$$\begin{array}{r} 212° \text{ F} \\ -\ 102° \text{ F} \\ \hline \mathbf{110° \text{ F}} \end{array}$$

11. (a) $\dfrac{5}{12} \cdot \dfrac{3}{3} = \dfrac{\mathbf{15}}{\mathbf{36}}$

(b) $\dfrac{1}{6} \cdot \dfrac{6}{6} = \dfrac{\mathbf{6}}{\mathbf{36}}$

(c) $\dfrac{7}{9} \cdot \dfrac{4}{4} = \dfrac{\mathbf{28}}{\mathbf{36}}$

(d) **Identity property of multiplication**

12. (a)

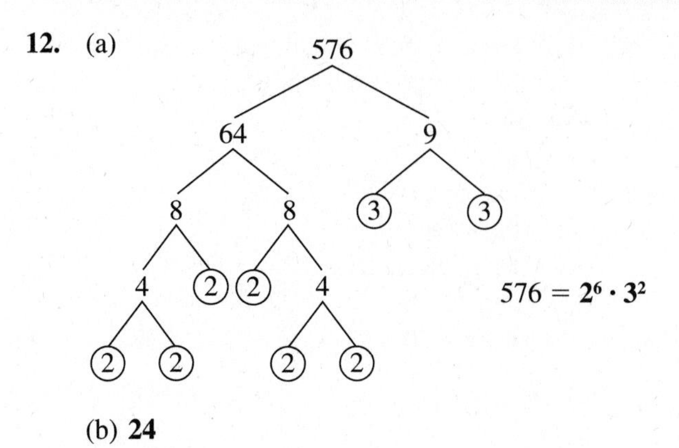

(b) **24**

13. $5\frac{5}{6} \times 6\frac{6}{7} = \frac{\overset{5}{\cancel{35}}}{\underset{1}{\cancel{6}}} \times \frac{\overset{8}{\cancel{48}}}{\underset{1}{\cancel{7}}} = \mathbf{40}$

14. (a) **Obtuse angle**

(b) $\overline{AB}$ (or $\overline{BA}$) and $\overline{ED}$ (or $\overline{DE}$)

15. (a) $\mathbf{\frac{1}{2}}$

(b) $\mathbf{\frac{1}{2}}$

(c) $\mathbf{\frac{1}{2}}$

16. (a) Perimeter $= 3\text{ ft} + 3\text{ ft} + 6\text{ ft} + 6\text{ ft}$
$= \mathbf{18\text{ ft}}$

(b) Area $= 3\text{ ft} \times 6\text{ ft} = \mathbf{18\text{ ft}^2}$

17.
```
      36°
  10)360°
     30
     ---
      60
      60
      ---
       0
```
$y = \mathbf{36°}$

18. $12^2 - 2^4 = 144 - 16 = 128$

$p = \mathbf{128}$

19. $5\frac{1}{8} - 1\frac{3}{8} \xrightarrow{4 + \frac{8}{8} + \frac{1}{8}} 4\frac{9}{8} - 1\frac{3}{8} = 3\frac{6}{8}$

$n = 3\frac{6}{8} = \mathbf{3\frac{3}{4}}$

20. $4\frac{1}{3} + 6\frac{2}{3} = 10\frac{3}{3} = 11$

$m = \mathbf{11}$

21. $10 - 1\frac{3}{5} \longrightarrow 9\frac{5}{5} - 1\frac{3}{5} = \mathbf{8\frac{2}{5}}$

22. $5\frac{1}{3} \cdot 1\frac{1}{2} = \frac{\overset{8}{\cancel{16}}}{\underset{1}{\cancel{3}}} \cdot \frac{\overset{1}{\cancel{3}}}{\underset{1}{\cancel{2}}} = \mathbf{8}$

23. $3\frac{1}{3} \div \frac{5}{6} = \frac{10}{3} \div \frac{5}{6}$

$= \frac{\overset{2}{\cancel{10}}}{\underset{1}{\cancel{3}}} \times \frac{\overset{2}{\cancel{6}}}{\underset{1}{\cancel{5}}} = \mathbf{4}$

24. $5\frac{1}{4} \div 3 = \frac{21}{4} \div \frac{3}{1}$

$= \frac{\overset{7}{\cancel{21}}}{4} \times \frac{1}{\underset{1}{\cancel{3}}} = \frac{7}{4} = \mathbf{1\frac{3}{4}}$

25. $\frac{\overset{1}{\cancel{3}}}{\underset{2}{\cancel{6}}} \cdot \frac{\overset{1}{\cancel{\overset{3}{\cancel{9}}}}}{\underset{2}{\cancel{8}}} \cdot \frac{\overset{1}{\cancel{4}}}{\underset{1}{\cancel{\underset{3}{\cancel{15}}}}} = \mathbf{\frac{1}{4}}$

26. $\frac{8}{9} - \left(\frac{7}{9} - \frac{5}{9}\right) = \frac{8}{9} - \left(\frac{2}{9}\right) = \frac{6}{9} = \mathbf{\frac{2}{3}}$

27. 1 yard = 36 inches

$\frac{36\text{ inches}}{2} = 18\text{ inches}$

radius $= \frac{18\text{ inches}}{2} = \mathbf{9\text{ inches}}$

28. Example: $1.50 ÷ 2 = **75¢**

29. (a) Missing length $= 5\text{ in.} - 3\text{ in.}$
$= 2\text{ in.}$

Missing height $= 5\text{ in.} - 3\text{ in.}$
$= 2\text{ in.}$

Perimeter $= 5\text{ in.} + 5\text{ in.}$
$+ 2\text{ in.}$
$+ 3\text{ in.} + 3\text{ in.} + 2\text{ in.}$
$= \mathbf{20\text{ in.}}$

(b) Area $= 25\text{ in.}^2 - 9\text{ in.}^2 = \mathbf{16\text{ in.}^2}$

30. (a) $\overline{CB}$ (or $\overline{BC}$)

(b) $\overline{AB}$ (or $\overline{BA}$)

(c) $\angle AMC$ (or $\angle CMA$)

(d) $\angle ABC$ (or $\angle CBA$ or $\angle ABM$ or $\angle MBA$) and $\angle BAM$ (or $\angle MAB$)

(e) $\overline{MA}$ (or $\overline{AM}$) and $\overline{MB}$ (or $\overline{BM}$)

Lesson 28, Warm-Up

a. **$9.22**

b. **$175.00**

c. **$3.71**

d. **424**

e. $\mathbf{1\frac{1}{4}}$

f. **10**

g. **$4.00**

Problem Solving

(B, P, J, G); (B, P, G, J); (B, J, P, G); (B, J, G, P); (B, G, P, J); (B, G, J, P); (P, B, J, G); (P, B, G, J); (P, J, B, G); (P, J, G, B); (P, G, B, J); (P, G, J, B); (J, B, P, G); (J, B, G, P); (J, P, B, G); (J, P, G, B); (J, G, B, P); (J, G, P, B); (G, B, P, J); (G, B, J, P); (G, P, B, J); (G, P, J, B); (G, J, B, P); (G, J, P, B)

24 permutations

Lesson 28, Lesson Practice

a.

$$\begin{array}{r} \$20.00 \\ -\ \$5.36 \\ \hline \$14.64 \end{array}$$

$$\begin{array}{r} \$4.88 \\ 3\overline{)\$14.64} \\ 12 \\ \hline 2\ 6 \\ 2\ 4 \\ \hline 24 \\ 24 \\ \hline 0 \end{array}$$

$4.88

b. 32 children ÷ 8 = 4 children

5 × 4 children = 20 children

20 boys

c.

$$\begin{array}{r} 28 \text{ passengers} \\ 29 \text{ passengers} \\ 30 \text{ passengers} \\ +\ 25 \text{ passengers} \\ \hline 112 \text{ passengers} \end{array}$$

$$\begin{array}{r} 28 \\ 4\overline{)112} \\ 8 \\ \hline 32 \\ 32 \\ \hline 0 \end{array}$$

28 passengers

d.

$$\begin{array}{r} 46 \\ 37 \\ 34 \\ 31 \\ 29 \\ +\ 24 \\ \hline 201 \end{array} \qquad \begin{array}{r} 33 \\ 6\overline{)201} \\ 18 \\ \hline 21 \\ 18 \\ \hline 3 \end{array}$$

$33\frac{3}{6} = \mathbf{33\frac{1}{2}}$

e.

$$\begin{array}{r} 40 \\ +\ 70 \\ \hline 110 \end{array} \qquad \begin{array}{r} 55 \\ 2\overline{)110} \\ 10 \\ \hline 10 \\ 10 \\ \hline 0 \end{array}$$

55; 55

f. **B. 84; The average score must fall between the highest and lowest scores.**

Lesson 28, Mixed Practice

1.

$$\begin{array}{r} 242 \text{ pounds} \\ 236 \text{ pounds} \\ 248 \text{ pounds} \\ 268 \text{ pounds} \\ +\ 226 \text{ pounds} \\ \hline 1220 \text{ pounds} \end{array} \qquad \begin{array}{r} 244 \\ 5\overline{)1220} \\ 10 \\ \hline 22 \\ 20 \\ \hline 20 \\ 20 \\ \hline 0 \end{array}$$

244 pounds

2. 5 minutes = 5 × 60 seconds

= 300 seconds

$$\begin{array}{r} 300 \text{ seconds} \\ +\ \ 14 \text{ seconds} \\ \hline \textbf{314 seconds} \end{array}$$

3.

$$\begin{array}{r} \$15.99 \\ \times \quad 3 \\ \hline \$47.97 \end{array}$$

$$\begin{array}{r} \$47.97 \\ + \ \$24.95 \\ \hline \mathbf{\$72.92} \end{array}$$

4.

$$\begin{array}{r} 1492 \\ - \quad 41 \\ \hline \mathbf{1451} \end{array}$$

5. $75\% = \frac{75}{100} = \frac{3}{4}$

5000 meters

Salma led $\frac{3}{4}$.	1250 meters
	1250 meters
	1250 meters
Salma did not lead $\frac{1}{4}$.	1250 meters

(a)

$$\begin{array}{r} 1250 \\ \times \quad 3 \\ \hline 3750 \end{array}$$

3750 meters

(b) **1250 meters**

6. (a) Perimeter = 4 in. + 4 in. + 8 in. + 8 in. = **24 in.**

(b) Area = 4 in. × 8 in. = **32 in.^{2}**

7. (a) **3, 6, 9, 12, 15, 18**

(b) **4, 8, 12, 16, 20, 24**

(c) **12**

(d)

$$\begin{aligned} 27 &= 3 \cdot 3 \cdot 3 \\ 36 &= 2 \cdot 2 \cdot 3 \cdot 3 \\ \text{LCM}(27, 36) &= 2 \cdot 2 \cdot 3 \cdot 3 \cdot 3 \\ &= \mathbf{108} \end{aligned}$$

8. (a) **280**

(b) **300**

9.

$$56 = 2 \cdot 2 \cdot 2 \cdot 7$$

$$240 = 2 \cdot 2 \cdot 2 \cdot 2 \cdot 3 \cdot 5$$

$$\frac{\overset{1}{\cancel{2}} \cdot \overset{1}{\cancel{2}} \cdot \overset{1}{\cancel{2}} \cdot 7}{\underset{1}{\cancel{2}} \cdot \underset{1}{\cancel{2}} \cdot \underset{1}{\cancel{2}} \cdot 2 \cdot 3 \cdot 5} = \mathbf{\frac{7}{30}}$$

10.

$$\begin{array}{r} 1760 \\ 3\overline{)5280} \\ \underline{3} \\ 22 \\ \underline{21} \\ 18 \\ \underline{18} \\ 00 \\ \underline{0} \\ 0 \end{array}$$

1760 yards

11. (a) $\frac{7}{8} \cdot \frac{3}{3} = \mathbf{\frac{21}{24}}$

(b) $\frac{11}{12} \cdot \frac{2}{2} = \mathbf{\frac{22}{24}}$

(c) $\frac{21}{24} + \frac{22}{24} = \frac{43}{24} = \mathbf{1\frac{19}{24}}$

12. (a)

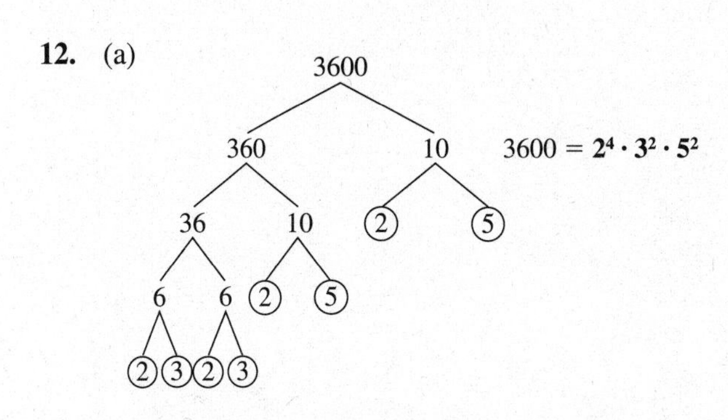

(b) $\sqrt{3600} = \sqrt{60^2} = \mathbf{60}$

13. **Add the six numbers. Then divide the sum by 6.**

14. (a)

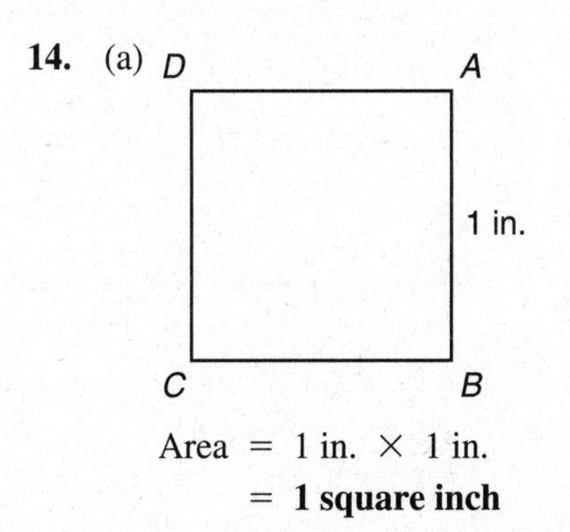

Area = 1 in. × 1 in.
= **1 square inch**

(b)–(c)

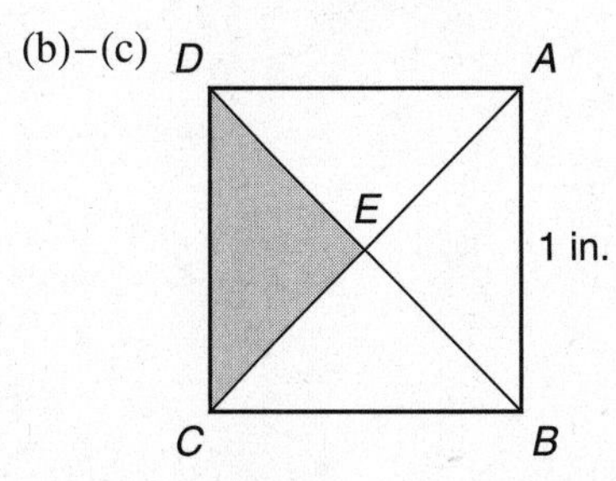

(d) **25%**

15. (a) $\mathbf{-1, 0, \frac{1}{10}, 1, \frac{11}{10}}$

(b) $\mathbf{-1, 1}$

16. $12\overline{)360°}$ = 30°
36
00
0
0

$y = \mathbf{30°}$

17. $10^2 - 8^2 = 100 - 64 = 36$
$m = \mathbf{36}$

18. $60\overline{)180}$ = 3
180
0

$w = \mathbf{3}$

19. $4\frac{5}{12} - 1\frac{1}{12} = 3\frac{4}{12} = \mathbf{3\frac{1}{3}}$

20. $8\frac{7}{8} + 3\frac{3}{8} = 11\frac{10}{8} = 12\frac{2}{8} = \mathbf{12\frac{1}{4}}$

21. $12 - 8\frac{1}{8} \longrightarrow 11\frac{8}{8} - 8\frac{1}{8} = \mathbf{3\frac{7}{8}}$

22. $6\frac{2}{3} \cdot 1\frac{1}{5} = \frac{\cancel{20}^{4}}{\cancel{3}_{1}} \cdot \frac{\cancel{6}^{2}}{\cancel{5}_{1}} = \mathbf{8}$

23. $\left(1\frac{1}{2}\right)^2 \div 7\frac{1}{2} = \left(\frac{3}{2}\right)^2 \div \frac{15}{2}$

$= \left(\frac{3}{2} \cdot \frac{3}{2}\right) \div \frac{15}{2} = \frac{9}{4} \div \frac{15}{2}$

$= \frac{\cancel{9}^{3}}{\cancel{4}_{2}} \times \frac{\cancel{2}^{1}}{\cancel{15}_{5}} = \mathbf{\frac{3}{10}}$

24. $8 \div 2\frac{2}{3} = \frac{8}{1} \div \frac{8}{3} = \frac{\cancel{8}^{1}}{1} \times \frac{3}{\cancel{8}_{1}} = \mathbf{3}$

25. $80\overline{)10{,}000}$ = 125
8 0
2 00
1 60
400
400
0

26. $\frac{3}{4} - \left(\frac{1}{2} \div \frac{2}{3}\right)$

$= \frac{3}{4} - \left(\frac{1}{2} \times \frac{3}{2}\right) = \frac{3}{4} - \left(\frac{3}{4}\right) = \mathbf{0}$

27. (a) $3^4 = 3 \cdot 3 \cdot 3 \cdot 3 = 9 \cdot 9 = \mathbf{81}$

(b) $3^2 + 4^2 = 9 + 16 = \mathbf{25}$

28. **Check polygon for ten sides; one possibility:**

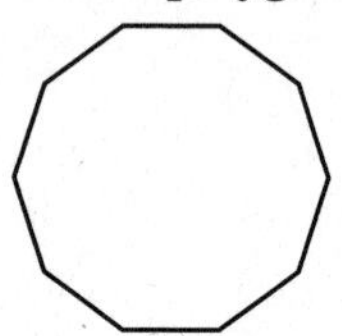

29. (a) $\angle ACD$

(b) $\overline{CB}$

(c) Area $= 2\left(7\frac{1}{2} \text{ in.}^2\right)$

$= 2 \times \frac{15}{2} \text{ in.}^2 = \mathbf{15 \text{ in.}^2}$

30.

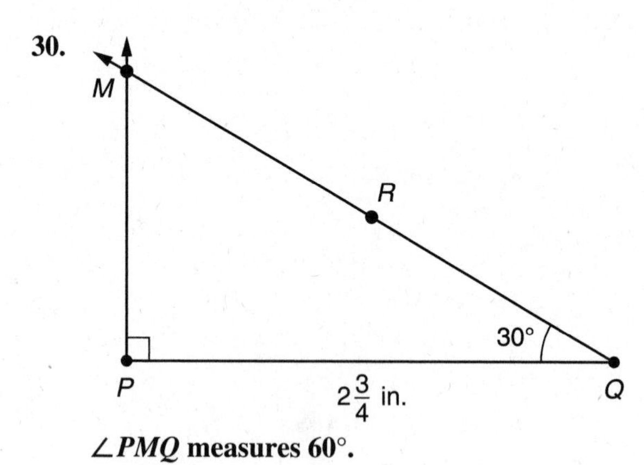

$\angle PMQ$ **measures 60°.**

LESSON 29, WARM-UP

a. **$7.30**

b. **$1.25**

c. **$1.02**

d. **198**

e. $\mathbf{1\frac{2}{3}}$

f. **12**

g. **10**

Problem Solving

Huck was facing **south.** He was **3 paces west of the tree.**

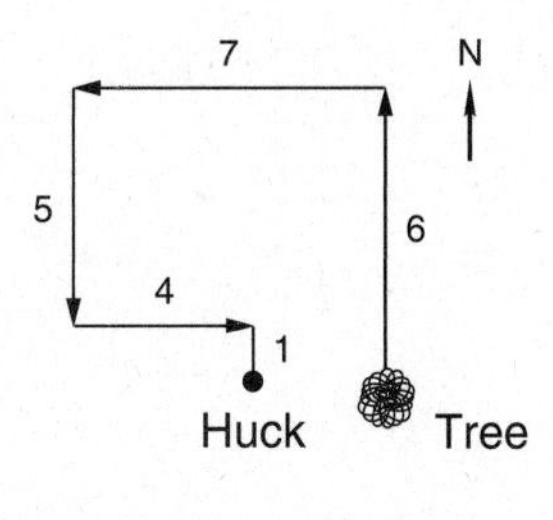

Lesson 29, Lesson Practice

a. 17⑥0 ⟶ **1800**

b. 5④89 ⟶ **5000**

c. 186,②82 ⟶ **186,000**

d. $7986 - 3074$
$\downarrow \quad \downarrow$
$8000 - 3000 = \mathbf{5000}$

e. $300 \times 30 = \mathbf{9000}$

f. $\frac{5860}{19} \longrightarrow \frac{6000}{20} = \mathbf{300}$

g. $12\frac{1}{4} \div 3\frac{7}{8}$
$\downarrow \quad \downarrow$
$12 \div 4 = \mathbf{3}$

h. Area $= 1\frac{7}{8}$ in. $\times 1\frac{1}{8}$ in.
$= \frac{15}{8}$ in. $\times \frac{9}{8}$ in. $= \frac{135}{64}$ in.^{2}
$= \mathbf{2\frac{7}{64}}$ **in.^{2}**

The answer is reasonable because the estimated area is 2 in. × 1 in. = 2 in.^{2}.

Lesson 29, Mixed Practice

1. 16 feet = 16 (12 inches)

16×12: 32, 16, 192 inches

192 inches + 8 inches = **200 inches**

2. **The cost per pound means the cost for each pound. We divide \$3.68, which is the cost for 8 pounds, by 8.**

3. $75 + 70 + 80 + 80 + 85 + 90 = 480$

$480 \div 6 = \mathbf{80}$ (6)480: 48, 00, 0, 0)

4. 1,000,000,000,000
− 219,800,000,000
780,200,000,000

Seven hundred eighty billion, two hundred million

5. $40\% = \frac{40}{100} \div \frac{20}{20} = \frac{2}{5}$

80 chips

$\frac{2}{5}$ were blue.	16 chips
	16 chips
$\frac{3}{5}$ were not blue.	16 chips
	16 chips
	16 chips

(a) 2 × 16 chips = **32 chips**

(b) 3 × 16 chips = **48 chips**

6. (a)
$4 = 2 \cdot 2$
$6 = 2 \cdot 3$
$8 = 2 \cdot 2 \cdot 2$
LCM (4, 6, 8) $= 2 \cdot 2 \cdot 2 \cdot 3 = \mathbf{24}$

(b)
$16 = 2 \cdot 2 \cdot 2 \cdot 2$
$36 = 2 \cdot 2 \cdot 3 \cdot 3$
LCM (16, 36) $= 2 \cdot 2 \cdot 2 \cdot 2 \cdot 3 \cdot 3$
$= \mathbf{144}$

7. (a) Perimeter $= \frac{3}{4}$ in. $+ \frac{3}{4}$ in. $+ \frac{3}{4}$ in. $+ \frac{3}{4}$ in. $= \frac{12}{4}$ in.
$=$ **3 in.**

(b) Area $= \frac{3}{4}$ in. $\times \frac{3}{4}$ in. $= \mathbf{\frac{9}{16}}$ **in.^{2}**

8. (a) 3⑥6 ⟶ **400**

(b) 36⑥ ⟶ **370**

9. $6143 + 4952$

$\downarrow \quad \downarrow$

$6000 + 5000 = \mathbf{11{,}000}$

10. (a) $\frac{3}{4} \cdot 5\frac{1}{3} \cdot 1\frac{1}{8}$

$\downarrow \quad \downarrow \quad \downarrow$

$1 \cdot 5 \cdot 1 = \mathbf{5}$

(b) $\frac{3}{4} \cdot 5\frac{1}{3} \cdot 1\frac{1}{8}$

$= \frac{\cancel{3}^{1}}{\cancel{4}_{2}} \cdot \frac{\cancel{16}^{\cancel{2}^{1}}}{\cancel{3}_{1}} \cdot \frac{9}{\cancel{8}_{1}} = \frac{9}{2} = \mathbf{4\frac{1}{2}}$

11. (a) $\frac{2}{3} \cdot \frac{10}{10} = \mathbf{\frac{20}{30}}$

(b) $\frac{25 \div 5}{30 \div 5} = \mathbf{\frac{5}{6}}$

12. $1{,}000{,}000{,}000 =$
$1{,}000{,}000 \cdot 1000 =$
$1000 \cdot 1000 \cdot 1000 =$
$2 \cdot 2 \cdot 2 \cdot 5 \cdot 5 \cdot 5 \cdot 2 \cdot 2 \cdot 2 \cdot 5 \cdot 5 \cdot 5$
$\cdot 2 \cdot 2 \cdot 2 \cdot 5 \cdot 5 \cdot 5$
$= \mathbf{2^9 \cdot 5^9}$

13. (a) **50%**

(b) **50%**

(c) **50%**

14. $BC = 2 \text{ in.} + 2 \text{ in.} = 4 \text{ in.}$
$AC = 2 \text{ in.} + 4 \text{ in.} = 6 \text{ in.}$
$AF + CD = AC = 6 \text{ in.}$
$AF = 3 \text{ in.}$
$CD = 3 \text{ in.}$

(a) Perimeter $= 3 \text{ in.} + 3 \text{ in.} + 2 \text{ in.} + 2 \text{ in.} = \mathbf{10 \text{ in.}}$

(b) Area $= 4 \text{ in.} \times 3 \text{ in.} = \mathbf{12 \text{ in.}^2}$

15. (a) $\angle$***AFB***

(b) **90°**

16. $8^2 = 64$

$$\begin{array}{r} 16 \\ 4\overline{)64} \\ \underline{4} \\ 24 \\ \underline{24} \\ 0 \end{array}$$

$m = \mathbf{16}$

17.
$$\begin{array}{rcr} 15 & \longrightarrow & 14\frac{9}{9} \\ -\ 4\frac{4}{9} & & -\ 4\frac{4}{9} \\ \hline & & x = \mathbf{10\frac{5}{9}} \end{array}$$

18. $3\frac{5}{9} + 4\frac{7}{9} = 7\frac{12}{9} = 8\frac{3}{9} = 8\frac{1}{3}$

$n = \mathbf{8\frac{1}{3}}$

19.
$$\begin{array}{rcr} 6\frac{1}{3} & \xrightarrow{5 + \frac{3}{3} + \frac{1}{3}} & 5\frac{4}{3} \\ -\ 5\frac{2}{3} & & -\ 5\frac{2}{3} \\ \hline & & \mathbf{\frac{2}{3}} \end{array}$$

20. $\frac{\cancel{20}^{4}}{3} \times \frac{1}{\cancel{5}_{1}} = \frac{4}{3} = \mathbf{1\frac{1}{3}}$

21. $1\frac{2}{3} \div 3\frac{1}{2} = \frac{5}{3} \div \frac{7}{2} =$

$\frac{5}{3} \times \frac{2}{7} = \mathbf{\frac{10}{21}}$

22.
$$\begin{array}{r} \$7.49 \\ \times \quad 24 \\ \hline 2996 \\ 1498 \\ \hline \mathbf{\$179.76} \end{array}$$

23. **Round $5\frac{1}{3}$ to 5 and round $4\frac{7}{8}$ to 5. Then multiply the rounded numbers. The product of the mixed numbers is about 25.**

24. (a) $10^3 \cdot 10^3 = 10 \cdot 10 \cdot 10 \cdot 10 \cdot 10 \cdot 10$
$= 10^6$

$m = \mathbf{6}$

(b) $\frac{10^6}{10^3} = \frac{\cancel{10}^{1} \cdot \cancel{10}^{1} \cdot \cancel{10}^{1} \cdot 10 \cdot 10 \cdot 10}{\cancel{10}_{1} \cdot \cancel{10}_{1} \cdot \cancel{10}_{1}}$

$= 10 \cdot 10 \cdot 10 = 10^3$

$n = \mathbf{3}$

25. $k = 2^5 + 1 = 2 \cdot 2 \cdot 2 \cdot 2 \cdot 2 + 1$
$= 32 + 1 = \mathbf{33}$

26. (a) Diameter $= 2\ (1 \text{ inch})$
$= \textbf{2 inches}$

(b) Perimeter $= 6\ (1 \text{ inch})$
$= \textbf{6 inches}$

27. $\frac{2}{3} \cdot \frac{2}{2} = \frac{4}{6}; \frac{1}{2} \cdot \frac{3}{3} = \frac{3}{6}$

$\frac{4}{6} - \frac{3}{6} = \mathbf{\frac{1}{6}}$

28. (a) **Acute angle**

(b) **Obtuse angle**

(c) **Straight angle**

29. 1 quart = 4 cups = 32 ounces
2 cups = 16 ounces
32 ounces − 16 ounces = **16 ounces**

30. Missing height = 6 in. − 4 in.
= 2 in.
Missing length = 8 in. + 3 in.
= 11 in.
Perimeter = 4 in. + 8 in. + 2 in.
+ 3 in. + 6 in. + 11 in.
= **34 in.**

Lesson 30, Warm-Up

a. **$3.98**

b. **$150.00**

c. **9**

d. **420**

e. $\mathbf{4\frac{1}{3}}$

f. **15**

g. **6**

Problem Solving

We know that the ones digit of the first subtrahend is 9, because $3 \times 3 = 9$. This means that the tens digit of the dividend is also 9, because their difference is 0. So the tens digit of the divisor is either 4 or 5 (6 is too large), and the first subtrahend is either 129 or 159. By trying each one, we find that the divisor is 53 and the first subtrahend is 159. From this information we find that the thousands digit of the dividend is 1. We know that the ones digit of the second minuend is 6, because 6 is brought down from the ones digit of the dividend. Therefore, the ones digit of the quotient must be 2. Since the difference of the second subtraction stage is 0, both the minuend and subtrahend are 106.

$$\begin{array}{r} \mathbf{32} \\ \mathbf{53)\overline{1696}} \\ \underline{\mathbf{159}} \\ \mathbf{106} \\ \underline{\mathbf{106}} \\ \mathbf{0} \end{array}$$

Lesson 30, Lesson Practice

a. $\frac{3}{5} \cdot \frac{2}{2} = \frac{6}{10}$

$\mathbf{\frac{6}{10} < \frac{7}{10}}$

b. $\frac{5}{12} \cdot \frac{5}{5} = \frac{25}{60}; \frac{7}{15} \cdot \frac{4}{4} = \frac{28}{60}$

$\mathbf{\frac{25}{60} < \frac{28}{60}}$

c.
$$\begin{array}{rl} & \frac{3}{4} \cdot \frac{6}{6} = \frac{18}{24} \\ & \frac{5}{6} \cdot \frac{4}{4} = \frac{20}{24} \\ + & \frac{3}{8} \cdot \frac{3}{3} = \frac{9}{24} \\ \hline & \frac{47}{24} \end{array}$$

$\frac{47}{24} = \mathbf{1\frac{23}{24}}$

d.
$$\begin{array}{rl} & 7\frac{5}{6} = 7\frac{5}{6} \\ - & 2\frac{1}{2} = 2\frac{3}{6} \\ \hline & 5\frac{2}{6} = \mathbf{5\frac{1}{3}} \end{array}$$

e.
$$\begin{array}{rl} 4\frac{3}{4} = & 4\frac{6}{8} \\ +\ 5\frac{5}{8} = & 5\frac{5}{8} \\ \hline & 9\frac{11}{8} = \mathbf{10\frac{3}{8}} \end{array}$$

f.
$$4\frac{1}{6}\cdot\frac{3}{3} = 4\frac{3}{18}$$
$$2\frac{5}{9}\cdot\frac{2}{2} = 2\frac{10}{18}$$

$$\begin{array}{r} 4\frac{3}{18} \\ -\ 2\frac{10}{18} \\ \hline \end{array} \xrightarrow{3 + \frac{18}{18} + \frac{3}{18}} \begin{array}{r} 3\frac{21}{18} \\ -\ 2\frac{10}{18} \\ \hline \mathbf{1\frac{11}{18}} \end{array}$$

g.
$$\frac{3}{25} = \frac{3}{5\cdot 5},\ \frac{2}{45} = \frac{2}{3\cdot 3\cdot 5}$$
$$3\cdot 3\cdot 5\cdot 5 = 225$$
$$\begin{array}{rl} \frac{3}{25}\cdot\frac{9}{9} = & \frac{27}{225} \\ -\ \frac{2}{45}\cdot\frac{5}{5} = & \frac{10}{225} \\ \hline & \mathbf{\frac{17}{225}} \end{array}$$

Lesson 30, Mixed Practice

1.
$$\begin{array}{r} 76 \text{ inches} \\ 77 \text{ inches} \\ 77 \text{ inches} \\ 78 \text{ inches} \\ +\ 82 \text{ inches} \\ \hline 390 \text{ inches} \end{array}$$

$$\begin{array}{r} 78 \\ 5\overline{)390} \\ \underline{35} \\ 40 \\ \underline{40} \\ 0 \end{array}$$

78 inches

2.
$$\begin{array}{r} \$0.87 \\ \times\ \ 6 \\ \hline \$5.22 \end{array} \qquad \begin{array}{r} \$10.00 \\ -\ \$5.22 \\ \hline \mathbf{\$4.78} \end{array}$$

3. Barney is correct. By estimating, we know the total is closer to 12,000 pounds than 120,000 pounds.

4.
$$\begin{array}{r} 260 \\ -\ 140 \\ \hline 120 \end{array} \qquad \frac{120 \div 10}{260 \div 10} = \frac{12}{26} = \mathbf{\frac{6}{13}}$$

5.
$$30\% = \frac{30}{100} \div \frac{10}{10} = \frac{3}{10}$$

	2140 miles
$\frac{3}{10}$ completed	214 miles
	214 miles
	214 miles
$\frac{7}{10}$ not completed	214 miles
	214 miles
	214 miles
	214 miles
	214 miles
	214 miles
	214 miles

(a) 3×214 miles = **642 miles**

(b) 7×214 miles = **1498 miles**

6. 5 feet = 5(12 inches) = 60 inches
60 inches ÷ 4 = **15 inches**

7.
$$18 = 2\cdot 3\cdot 3$$
$$30 = 2\cdot 3\cdot 5$$
$$2\cdot 3\cdot 3\cdot 5 = 90$$
$$\begin{array}{rl} \frac{1}{18}\cdot\frac{5}{5} = & \frac{5}{90} \\ -\ \frac{1}{30}\cdot\frac{3}{3} = & \frac{3}{90} \\ \hline & \frac{2}{90} = \mathbf{\frac{1}{45}} \end{array}$$

8. (a) 36,④67 ⟶ **36,000**

(b) 36,4⑥7 ⟶ **36,500**

9.
$$\begin{array}{r} \mathbf{600} \\ 50\overline{)30{,}000} \\ \underline{30\,0} \\ 00 \\ \underline{0} \\ 00 \\ \underline{0} \\ 0 \end{array}$$

10. (a) $\frac{32}{100} \div \frac{4}{4} = \mathbf{\frac{8}{25}}$

(b) $\frac{48}{72} = \frac{\overset{1}{\cancel{2}}\cdot\overset{1}{\cancel{2}}\cdot\overset{1}{\cancel{2}}\cdot 2\cdot\overset{1}{\cancel{3}}}{\underset{1}{\cancel{2}}\cdot\underset{1}{\cancel{2}}\cdot\underset{1}{\cancel{2}}\cdot\underset{1}{\cancel{3}}\cdot 3}$
$= \mathbf{\frac{2}{3}}$

11. $\frac{5}{6}\cdot\frac{4}{4} = \frac{20}{24},\ \frac{7}{8}\cdot\frac{3}{3} = \frac{21}{24}$

$\mathbf{\frac{20}{24} < \frac{21}{24}}$

12. (a) Area = 3 in. × 3 in. = **9 in.²**

(b) Area = 4 in. × 4 in. = **16 in.²**

(c) 16 in.² + 9 in.² = **25 in.²**

13. (a) Perimeter = 3 in. + 3 in. + 1 in. + 4 in. + 4 in. + 4 in. + 3 in. = **22 in.**

(b) **The perimeter of the hexagon is 6 in. less than the combined perimeter of the squares because a 3 in. side of the smaller square and the adjoining 3 in. portion of a side of the larger square are not part of the perimeter of the hexagon.**

14. (a)

5184 → 9, 576
9 → (3), (3)
576 → 9, 64
9 → (3), (3)
64 → 8, 8
8 → 4, (2)
8 → (2), 4
4 → (2), (2)
4 → (2), (2)

5184 = $\mathbf{2^6 \cdot 3^4}$

(b) $\sqrt{5184} = \sqrt{2^6 \cdot 3^4}$
$= \mathbf{2^3 \cdot 3^2 = 72}$

15.

```
  5
  7
  9
 11
 12
 13
 24
 25
 26
+ 28
 160
```

```
    16
10)160
   10
    60
    60
     0
```

16. **1, 2, 3, 5, 6, 7, 9**

17. $6^3 = 6 \cdot 6 \cdot 6 = 36 \cdot 6 = 216$

```
   36
6)216
  18
   36
   36
    0
```

w = **36**

18.
$90° + 30° = 120°$
$180° - 120° = 60°$
$a = \mathbf{60°}$

19.

```
   $1.25
36)$45.00
    36
     9 0
     7 2
     1 80
     1 80
        0
```

p = **$1.25**

20.

```
    $3.75
  ×    32
      750
     1125
  $120.00
```

t = **$120.00**

21.
$$\frac{1}{2} \cdot \frac{3}{3} = \frac{3}{6}$$
$$+\ \frac{1}{3} \cdot \frac{2}{2} = \frac{2}{6}$$
$$\mathbf{\frac{5}{6}}$$

22.
$$\frac{3}{4} \cdot \frac{3}{3} = \frac{9}{12}$$
$$-\ \frac{1}{3} \cdot \frac{4}{4} = \frac{4}{12}$$
$$\mathbf{\frac{5}{12}}$$

23.
$$2\frac{5}{6} = 2\frac{5}{6}$$
$$-\ 1\frac{1}{2} = 1\frac{3}{6}$$
$$1\frac{2}{6}$$
$$1\frac{2}{6} = \mathbf{1\frac{1}{3}}$$

24. $\frac{4}{5} \cdot 1\frac{2}{3} \cdot 1\frac{1}{8}$

$$= \frac{\overset{1}{\cancel{4}}}{\underset{1}{\cancel{5}}} \cdot \frac{\overset{1}{\cancel{5}}}{\underset{1}{\cancel{3}}} \cdot \frac{\overset{3}{\cancel{9}}}{\underset{2}{\cancel{8}}} = \frac{3}{2} = \mathbf{1\frac{1}{2}}$$

25. $1\frac{3}{4} \div 2\frac{2}{3} = \frac{7}{4} \div \frac{8}{3}$

$$= \frac{7}{4} \times \frac{3}{8} = \mathbf{\frac{21}{32}}$$

26. $3 \div 1\frac{7}{8} = \frac{3}{1} \div \frac{15}{8}$

$= \frac{\overset{1}{\cancel{3}}}{1} \times \frac{8}{\underset{5}{\cancel{15}}} = \frac{8}{5} = \mathbf{1\frac{3}{5}}$

27. $3\frac{2}{3} + 1\frac{5}{6}$

$\downarrow \quad \downarrow$

$4 + 2 = \mathbf{6}$

$$\begin{array}{r} 3\frac{2}{3} = 3\frac{4}{6} \\ + 1\frac{5}{6} = 1\frac{5}{6} \\ \hline 4\frac{9}{6} = \mathbf{5\frac{1}{2}} \end{array}$$

28. $5\frac{1}{8} - 1\frac{3}{4}$

$\downarrow \quad \downarrow$

$5 - 2 = \mathbf{3}$

$1\frac{3}{4} \cdot \frac{2}{2} = 1\frac{6}{8}$

$$\begin{array}{r} 5\frac{1}{8} \\ - 1\frac{6}{8} \\ \hline \end{array} \xrightarrow{4 + \frac{8}{8} + \frac{1}{8}} \begin{array}{r} 4\frac{9}{8} \\ - 1\frac{6}{8} \\ \hline \mathbf{3\frac{3}{8}} \end{array}$$

29. One possibility:

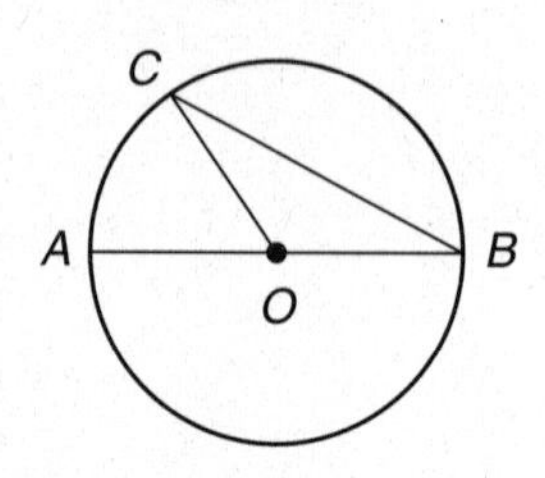

30. (a) $\overline{AB}$ (or $\overline{BA}$)

(b) $\overline{OA}, \overline{OB}, \overline{OC}$

(c) $\angle BOC$ (or $\angle COB$)

INVESTIGATION 3

1.

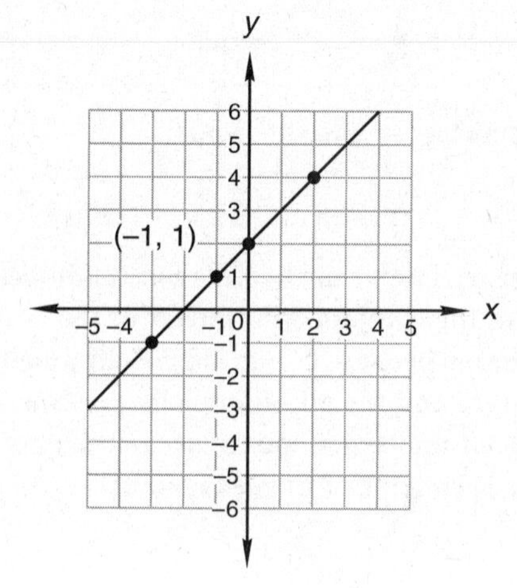

The point **(–1, 1)** is located on the line in the second quadrant.

2. (–2, –2)

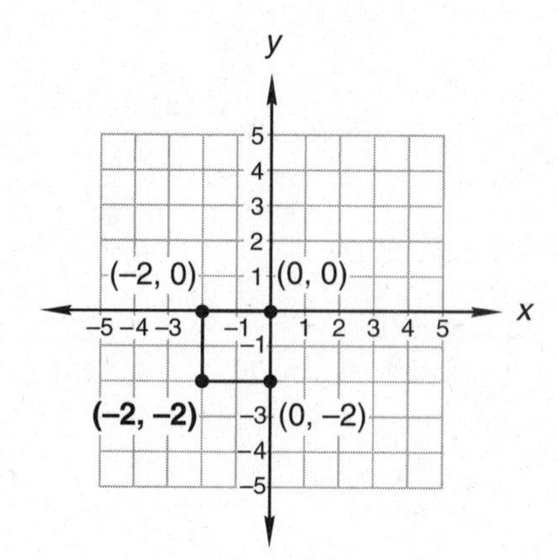

3.

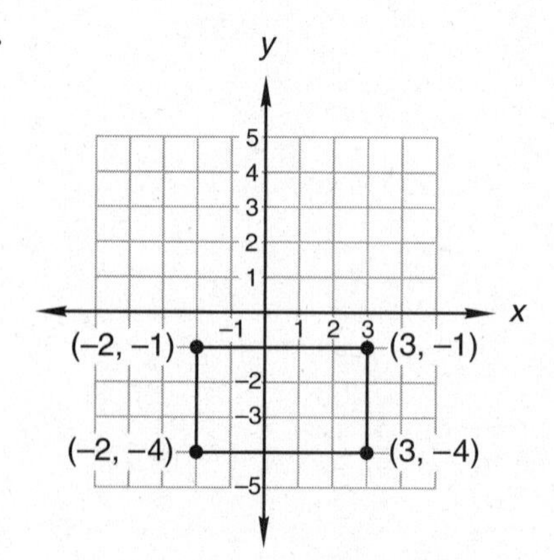

The length of the rectangle is 5 units. The width of the rectangle is 3 units.

Perimeter: $2(5) + 2(3) =$ **16 units**

Area: $5 \times 3 =$ **15 sq. units**

4.

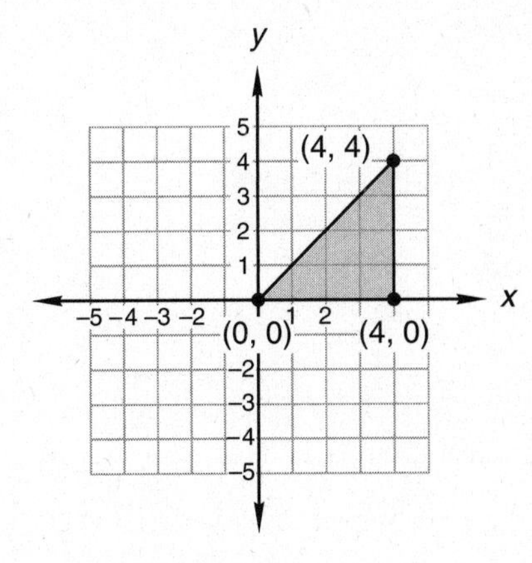

Six whole squares plus 4 half squares totals
8 square units.

5.

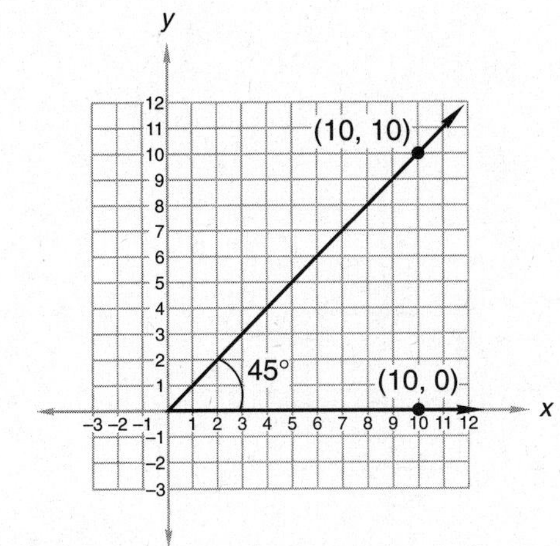

45°

6. (a) **3rd**

(b) **1st**

(c) **4th**

(d) **2nd**

7.

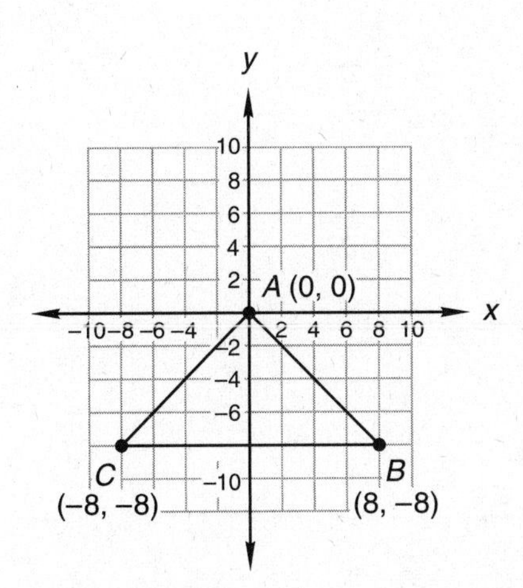

m∠A = 90°;
m∠B = 45°;
m∠C = 45°

8.

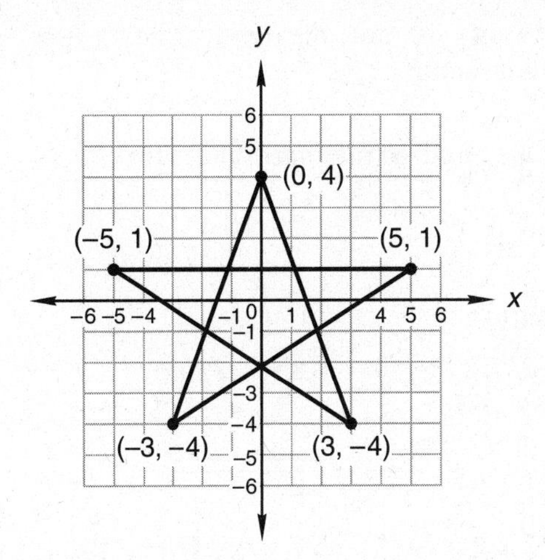

9. **See student work.**

Lesson 31, Warm-Up

a. **\$3.01**

b. **\$2.45**

c. **8**

d. $\mathbf{\frac{3}{4}}$

e. **19**

f. **45**

g. **3**

Problem Solving

This is a sequence of perfect squares, beginning with 10^2.
…, 100, 121, 144, **169, 196, 225,** …

Lesson 31, Lesson Practice

a. $\mathbf{\frac{3}{100}}$**; 0.03**

b. $\mathbf{\frac{3}{10}}$**; 0.3**

c. **3**

d. **4**

e. **Twenty-five and one hundred thirty-four thousandths**

f. **One hundred and one hundredth**

g. **102.3**

h. **0.0125**

i. **300.075**

Lesson 31, Mixed Practice

1.
$$\begin{array}{r} \$26.47 \\ +\ \$32.54 \\ \hline \$59.01 \end{array} \qquad \begin{array}{r} \$89.89 \\ -\ \$59.01 \\ \hline \mathbf{\$30.88} \end{array}$$

2.
$$\begin{array}{r} 326 \text{ pages} \\ 288 \text{ pages} \\ 349 \text{ pages} \\ +\ 401 \text{ pages} \\ \hline 1364 \text{ pages} \end{array}$$

$$4\overline{)1364} = \mathbf{341 \text{ pages}}$$

3.
$$\begin{array}{r} \mathbf{\$1.33} \\ 12\overline{)\$15.96} \\ 12 \\ \hline 3\,9 \\ 3\,6 \\ \hline 36 \\ 36 \\ \hline 0 \end{array}$$

4.
$$\begin{array}{r} 1607 \\ -\ 1492 \\ \hline \mathbf{115 \text{ years}} \end{array}$$

5. **Divide the perimeter of the square by 4 to find the length of a side. Then multiply the length of a side by 6 to find the perimeter of the hexagon.**

6. $80\% = \frac{80}{100} = \mathbf{\frac{4}{5}}$

20 questions

$\frac{4}{5}$ were correct.	4 questions
	4 questions
	4 questions
	4 questions
$\frac{1}{5}$ were incorrect.	4 questions

(a) 4×4 questions = **16 questions**

(b) **4 questions**

7. (a) **500,000**

(b) **481,000**

8. $50{,}000 - 20{,}000 = \mathbf{30{,}000}$

9. (a) $\mathbf{\frac{7}{100}}$

(b) **0.07**

(c) $\frac{7}{100} = \mathbf{7\%}$

10. **7**

11. (a) $\frac{3}{10} \mathbf{=} 0.3$

(b) $\frac{3}{100} \mathbf{<} 0.3$

12.

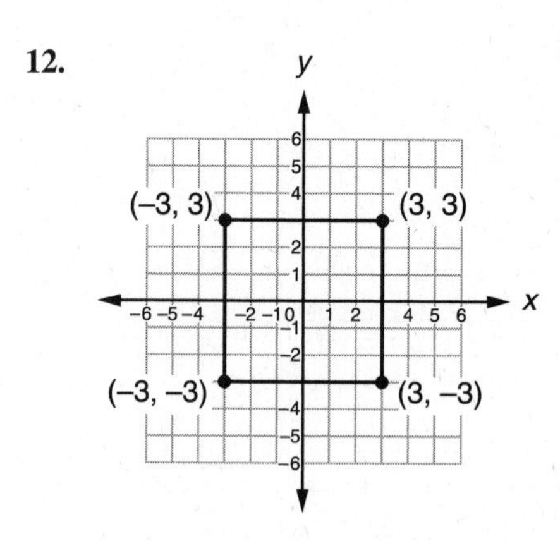

(a) Perimeter = 4(6 units) = **24 units**

(b) Area = (6 units)(6 units)
= **36 units^2**

13. (a) $\frac{15 \div 3}{24 \div 3} = \mathbf{\frac{5}{8}}$

(b) $\frac{7}{12} \cdot \frac{2}{2} = \mathbf{\frac{14}{24}}$

(c) $\frac{4 \div 4}{24 \div 4} = \mathbf{\frac{1}{6}}$

14. (a)

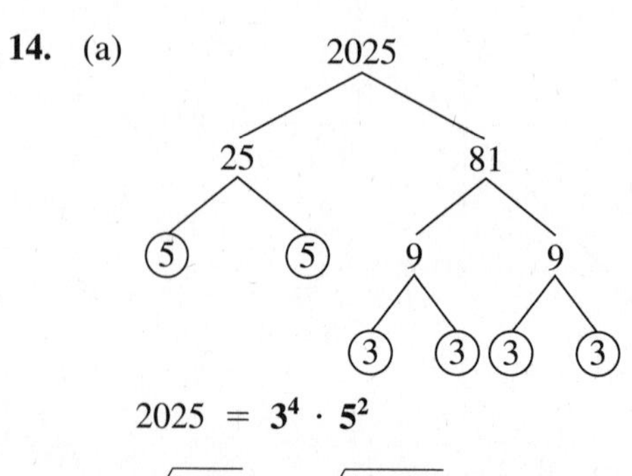

$2025 = \mathbf{3^4 \cdot 5^2}$

(b) $\sqrt{2025} = \sqrt{3^4 \cdot 5^2}$
$= 3^2 \cdot 5 = \mathbf{45}$

15. **One possibility:**

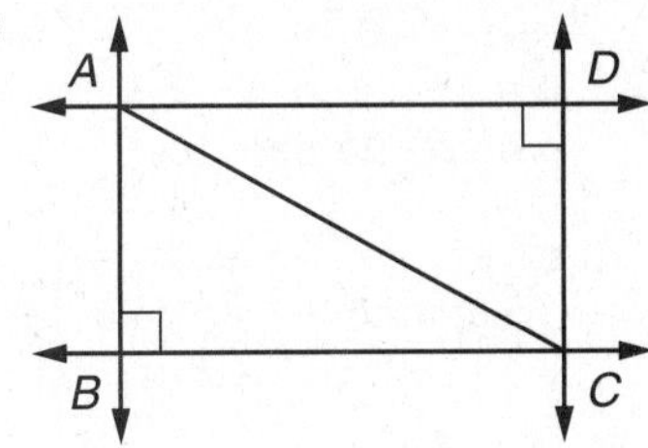

(a) **Rectangle**

(b) $\angle BCA$ (or $\angle ACB$)

16. $\dfrac{\overset{2}{\cancel{6}} \cdot \overset{4}{\cancel{12}}}{\underset{1}{\underset{\cancel{3}}{\cancel{9}}}} = 8 \qquad n = \mathbf{8}$

17.

$$\begin{array}{r} 90^\circ \\ +\ 45^\circ \\ \hline 135^\circ \end{array} \qquad \begin{array}{r} 180^\circ \\ -\ 135^\circ \\ \hline 45^\circ \end{array}$$

$b = \mathbf{45^\circ}$

18.

$$\begin{array}{r} \$220.15 \\ -\ \ \$98.75 \\ \hline \$121.40 \end{array}$$

$w = \mathbf{\$121.40}$

19.

$$\begin{array}{r} \$4.65 \\ \times\ \ \ 48 \\ \hline 37\ 20 \\ 186\ 0\ \\ \hline \$223.20 \end{array}$$

$m = \mathbf{\$223.20}$

20.

$$\begin{array}{r} \frac{1}{2} \cdot \frac{3}{3} = \frac{3}{6} \\ +\ \frac{2}{3} \cdot \frac{2}{2} = \frac{4}{6} \\ \hline \frac{7}{6} = \mathbf{1\frac{1}{6}} \end{array}$$

21. $\dfrac{\overset{1}{\cancel{3}}}{\underset{2}{\cancel{4}}} \cdot \dfrac{\overset{1}{\cancel{2}}}{\underset{1}{\cancel{3}}} = \dfrac{1}{2}$

$\dfrac{1}{2} - \dfrac{1}{2} = \mathbf{0}$

22.

$$\begin{array}{r} 3\frac{5}{6} = 3\frac{5}{6} \\ -\ \frac{1}{3} \cdot \frac{2}{2} = \ \ \frac{2}{6} \\ \hline 3\frac{3}{6} = \mathbf{3\frac{1}{2}} \end{array}$$

23. $\dfrac{5}{8} \cdot 2\dfrac{2}{5} \cdot \dfrac{4}{9}$

$= \dfrac{\overset{1}{\cancel{5}}}{\underset{2}{\cancel{8}}} \cdot \dfrac{\overset{1}{\overset{\cancel{3}}{\cancel{12}}}}{\underset{1}{\cancel{5}}} \cdot \dfrac{4}{\underset{3}{\cancel{9}}} = \dfrac{4}{6} = \mathbf{\dfrac{2}{3}}$

24. $2\dfrac{2}{3} \div 1\dfrac{3}{4}$

$= \dfrac{8}{3} \div \dfrac{7}{4} = \dfrac{8}{3} \times \dfrac{4}{7}$

$= \dfrac{32}{21} = \mathbf{1\dfrac{11}{21}}$

25. $1\dfrac{7}{8} \div 3 = \dfrac{15}{8} \div \dfrac{3}{1}$

$= \dfrac{\overset{5}{\cancel{15}}}{8} \times \dfrac{1}{\underset{1}{\cancel{3}}} = \mathbf{\dfrac{5}{8}}$

26.

$$\begin{array}{r} 3\frac{1}{2} = 3\frac{3}{6} \\ +\ 1\frac{5}{6} = 1\frac{5}{6} \\ \hline 4\frac{8}{6} = 5\frac{2}{6} = \mathbf{5\frac{1}{3}} \end{array}$$

27.

$$\begin{array}{r} 5\frac{1}{4} = 5\frac{2}{8} \\ -\ 1\frac{5}{8} = 1\frac{5}{8} \\ \hline \end{array} \longrightarrow \begin{array}{r} 4\frac{10}{8} \\ -\ 1\frac{5}{8} \\ \hline \mathbf{3\frac{5}{8}} \end{array}$$

28.

$$\begin{array}{r} \frac{4}{3} = \frac{16}{12} \\ +\ \frac{3}{4} = \frac{9}{12} \\ \hline \frac{25}{12} = \mathbf{2\frac{1}{12}} \end{array}$$

29. $k = 10^6 = 10 \cdot 10 \cdot 10 \cdot 10 \cdot 10 \cdot 10$

$= 1{,}000{,}000$

One million

30. (a) **180°**

(b) **90°**

(c) **45°**

Lesson 32, Warm-Up

a. **\$2.77**

b. **\$2.00**

c. **20**

d. $\frac{3}{5}$

e. **25**

f. **12**

g. **2**

Problem Solving

12 in. ÷ $\frac{3}{4}$ in. per penny = **16 pennies**

Lesson 32, Lesson Practice

a. 2 meters = 2(100 centimeters)
= **200 centimeters**

b. **A 1-gallon jug can hold a little less than four liters. (Have the student check the label on a gallon bottle; 3.78 liters.)**

c. **1000 × 2.2 pounds is about 2200 pounds.**

d. A 100°C difference is equivalent to a difference of 180°F. Divide both by 10°. A 10°C difference is equivalent to a difference of 18°F.
18°F

e. 3 kilometers = 3000 meters

$$\begin{array}{r} 3000 \text{ meters} \\ -\ 800 \text{ meters} \\ \hline \textbf{2200 meters} \end{array}$$

f. 30 cm = 300 mm

$$\begin{array}{r} 300 \text{ mm} \\ -\ 120 \text{ mm} \\ \hline \textbf{180 mm} \end{array}$$

Lesson 32, Mixed Practice

1.
$$\begin{array}{r} 4248 \\ 3584 \\ +\ 9418 \\ \hline 17{,}250 \end{array}$$

$$\begin{array}{r} \textbf{5,750} \\ 3\overline{)17{,}250} \\ 15 \\ \hline 2\,2 \\ 2\,1 \\ \hline 15 \\ 15 \\ \hline 00 \\ 00 \\ \hline 0 \end{array}$$

2. $60\overline{)206}$ = 3 R 26
3 hours 26 minutes

3. $\frac{440}{1760}$ mile = $\frac{22}{88}$ mile = $\frac{\mathbf{1}}{\mathbf{4}}$ **mile**

4. 20 cm ÷ 4 = 5 cm
Perimeter = (5)(5 cm) = **25 cm**

5. (a) **3,000,000**

(b) **3,200,000**

6.
$$\begin{array}{r} 300 \\ \times\ 500 \\ \hline \textbf{150,000} \end{array}$$

7.

200 songs

$\frac{5}{8}$ were about love and chivalry.	25 songs
	25 songs
	25 songs
	25 songs
	25 songs
$\frac{3}{8}$ were not about love and chivalry.	25 songs
	25 songs
	25 songs

(a) 5(25 songs) = **125 songs**

(b) 3(25 songs) = **75 songs**

8. (a) $\frac{\mathbf{9}}{\mathbf{10}}$

(b) **0.9**

(c) $\frac{9}{10} = \frac{90}{100} =$ **90%**

9. **Three and twenty-five thousandths**

10. **76.05**

11. **\$30.00 ÷ 5 = \$6.00**

12. (a) **(2 × 1000) + (5 × 100)**

(b)

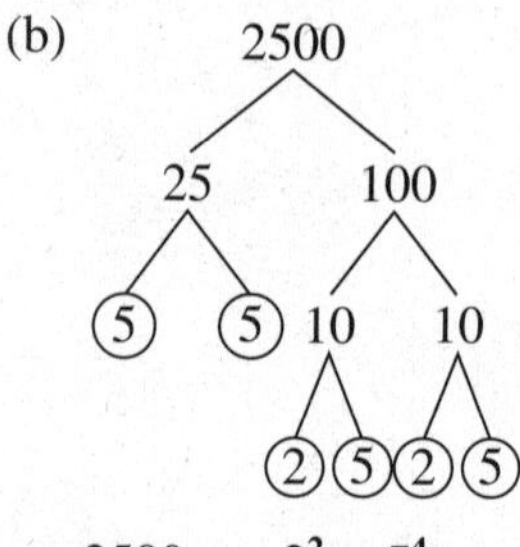

2500 = $\mathbf{2^2 \cdot 5^4}$

(c) $\sqrt{2500} = \sqrt{2^2 \cdot 5^4}$
$= 2 \cdot 5^2 =$ **50**

13. $35\overline{)\$21.00}$ — **\$0.60 per liter**

14.

15. (a) Area = (6 cm)(6 cm) = **36 cm²**

(b) Area = (8 cm)(8 cm) = **64 cm²**

(c) Area = 36 cm² + 64 cm² = **100 cm²**

16. Perimeter = 6 cm + 6 cm + 6 cm + 2 cm + 8 cm + 8 cm + 8 cm = **44 cm**

17. $\frac{\overset{5}{\cancel{10}} \cdot \overset{3}{\cancel{6}}}{\underset{1}{\underset{\cancel{2}}{\cancel{4}}}} = 15 \qquad w = \mathbf{15}$

18. $180° - 65° = 115°$

$s = \mathbf{115°}$

19. $\frac{1}{4} = \frac{2}{8}$, $\frac{3}{8} = \frac{3}{8}$, $+\frac{1}{2} = \frac{4}{8}$; $\frac{9}{8} = \mathbf{1\frac{1}{8}}$

20. $\frac{5}{6} = \frac{10}{12}$, $-\frac{3}{4} = \frac{9}{12}$; $\mathbf{\frac{1}{12}}$

21. $\frac{5}{16} = \frac{25}{80}$, $-\frac{3}{20} = \frac{12}{80}$; $\mathbf{\frac{13}{80}}$

22. $\frac{8}{9} \cdot 1\frac{1}{5} \cdot 10 = \frac{8}{\underset{3}{\cancel{9}}} \cdot \frac{\overset{2}{\cancel{6}}}{\underset{1}{\cancel{5}}} \cdot \frac{\overset{2}{\cancel{10}}}{1}$

$= \frac{32}{3} = \mathbf{10\frac{2}{3}}$

23. $6\frac{1}{6} = 6\frac{1}{6} \longrightarrow 5\frac{7}{6}$

$-2\frac{1}{2} = 2\frac{3}{6} \qquad -2\frac{3}{6}$

$3\frac{4}{6}$

$3\frac{4}{6} = \mathbf{3\frac{2}{3}}$

24. $4\frac{5}{8} = 4\frac{5}{8}$, $+1\frac{1}{2} = 1\frac{4}{8}$; $5\frac{9}{8} = \mathbf{6\frac{1}{8}}$

25. $\frac{2}{3} \div \frac{1}{2} = \frac{2}{3} \cdot \frac{2}{1} = \frac{4}{3}$

$\frac{2}{3} + \frac{4}{3} = \frac{6}{3} = \mathbf{2}$

26. $\frac{\overset{1}{\overset{\cancel{5}}{\cancel{25}}}}{\underset{1}{\underset{\cancel{4}}{\cancel{36}}}} \cdot \frac{\overset{1}{\cancel{9}}}{\underset{1}{\underset{\cancel{2}}{\cancel{10}}}} \cdot \frac{\overset{1}{\overset{\cancel{2}}{\cancel{8}}}}{\underset{3}{\cancel{15}}} = \mathbf{\frac{1}{3}}$

27. $5\frac{2}{5} \div \frac{9}{10}$

↓ ↓

$5 \div 1 = \mathbf{5}$

$5\frac{2}{5} \div \frac{9}{10} = \frac{\overset{3}{\cancel{27}}}{\underset{1}{\cancel{5}}} \times \frac{\overset{2}{\cancel{10}}}{\underset{1}{\cancel{9}}} = \mathbf{6}$

28. $7\frac{3}{4} + 1\frac{7}{8}$

↓ ↓

$8 + 2 = \mathbf{10}$

$7\frac{3}{4} = 7\frac{6}{8}$, $+1\frac{7}{8} = 1\frac{7}{8}$; $8\frac{13}{8} = \mathbf{9\frac{5}{8}}$

29.

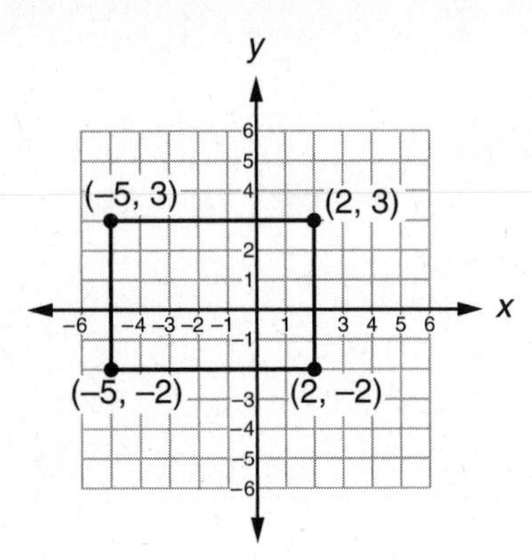

(a) **(2, 3)**

(b) Area = (5 units)(7 units) = **35 units²**

30. (a) $\overline{BC}$ (or $\overline{CB}$)

(b) ∠*AOC* (or ∠*COA*) or ∠*BOC* (or ∠*COB*)

(c) ∠*ABC* (or ∠*CBA*) or ∠*BCO* (or ∠*OCB*)

LESSON 33, WARM-UP

a. **$1.85**

b. **$3.30**

c. **9**

d. $\mathbf{\frac{4}{5}}$

e. **20**

f. **25**

g. **12**

Problem Solving

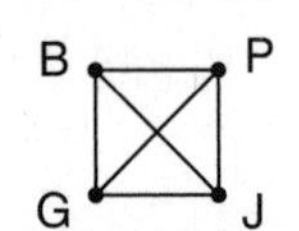

6 handshakes

There were **6 handshakes** in all.

LESSON 33, LESSON PRACTICE

a. 10.30 = 10.3

b. 5.06 < 5.60

c. 1.1 > 1.099

d. 3.1415⑨ → **3.1416**

e. 3⑥5.2418 ⟶ 400.~~0000~~ ⟶ **400**

f. 57.④32 ⟶ 57.~~000~~ ⟶ **57**

g. 10.2~~000~~ ⟶ **10.2**

h. 8.⑥5 ⟶ 9
21.⑦ ⟶ 22
11.⓪38 ⟶ 11

$$\begin{array}{r} 9 \\ 22 \\ +\ 11 \\ \hline \mathbf{42} \end{array}$$

LESSON 33, MIXED PRACTICE

1. **Multiply 12 inches by 5 to find the number of inches in 5 feet. Then add 8 inches to find the total number of inches in 5 feet 8 inches.**

2.
$$\begin{array}{r} 42°F \\ 43°F \\ 38°F \\ 47°F \\ 51°F \\ 52°F \\ +\ 49°F \\ \hline 322°F \end{array} \qquad 7\overline{)322°F} = \mathbf{46°F}$$

3.
$$\begin{array}{r} 120{,}310 \text{ people} \\ -\ 87{,}196 \text{ people} \\ \hline \mathbf{33{,}114 \text{ people}} \end{array}$$

4. 75 − 15 = 60; 60 − 15 = 45
60, 45

5. 24 cm ÷ 6 = 4 cm
Perimeter = (4 cm) 8 = **32 cm**

6.

	60 questions
$\frac{2}{3}$ were not T–F.	20 questions
	20 questions
$\frac{1}{3}$ were T–F.	20 questions

(a) **20 questions**

(b) 2 × 20 questions
= **40 questions**

(c) $\frac{100\%}{3} = \mathbf{33\frac{1}{3}\%}$

7.

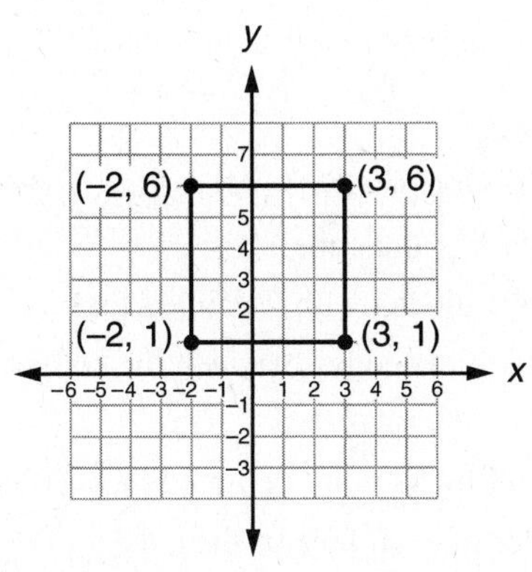

Area = (5 units)(5 units) = **25 units²**

8. (a) $15.7\underline{3}$⑤91 ⟶ **15.74**

(b) $1\underline{5}$.⑦3591 ⟶ 16
$\underline{3}$.①4 ⟶ 3
16 × 3 = **48**

9. (a) **One hundred fifty and thirty-five thousandths**

(b) **Fifteen ten-thousandths**

10. (a) **0.125**

(b) **100.025**

11. (a) 0.128 **<** 0.14

(b) 0.03 **>** 0.0015

12. (a) **4 cm**

(b) **40 mm**

13.

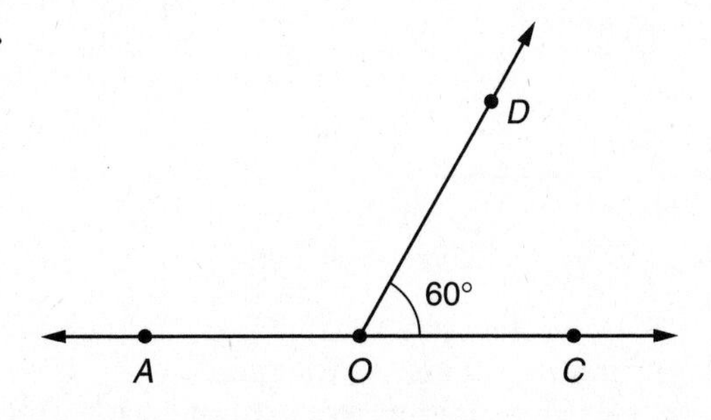

14. $n \cdot 0 = \mathbf{0}$

15.
$63 = 3 \cdot 3 \cdot 7$
$49 = 7 \cdot 7$
$3 \cdot 3 \cdot 7 \cdot 7 = \mathbf{441}$

16. $\frac{\overset{1}{\cancel{4}} \cdot \overset{9}{\cancel{18}}}{\underset{1}{\underset{\cancel{2}}{\cancel{8}}}} = 9$

$m = \mathbf{9}$

17.
$$\begin{array}{r} 180^\circ \\ -\ 135^\circ \\ \hline 45^\circ \end{array}$$

$a = \mathbf{45^\circ}$

18.
$$\begin{array}{rcl} \frac{3}{4} & = & \frac{6}{8} \\ \frac{5}{8} & = & \frac{5}{8} \\ +\ \frac{1}{2} & = & \frac{4}{8} \\ \hline & & \frac{15}{8} = \mathbf{1\frac{7}{8}} \end{array}$$

19.
$$\begin{array}{rcl} \frac{3}{4} & = & \frac{9}{12} \\ -\ \frac{1}{6} & = & \frac{2}{12} \\ \hline & & \mathbf{\frac{7}{12}} \end{array}$$

20.
$$\begin{array}{rcl} 4\frac{1}{2} & = & 4\frac{4}{8} \\ -\ \frac{3}{8} & = & \frac{3}{8} \\ \hline & & \mathbf{4\frac{1}{8}} \end{array}$$

21. $\frac{3}{8} \cdot 2\frac{2}{5} \cdot 3\frac{1}{3}$

$= \frac{\overset{1}{\cancel{3}}}{\underset{1}{\underset{\cancel{2}}{\cancel{8}}}} \cdot \frac{\overset{3}{\cancel{12}}}{\underset{1}{\cancel{5}}} \cdot \frac{\overset{1}{\overset{\cancel{2}}{\cancel{10}}}}{\underset{1}{\cancel{3}}} = \mathbf{3}$

22. $2\frac{7}{10} \div 5\frac{2}{5} = \frac{27}{10} \div \frac{27}{5}$

$= \frac{\overset{1}{\cancel{27}}}{\underset{2}{\cancel{10}}} \times \frac{\overset{1}{\cancel{5}}}{\underset{1}{\cancel{27}}} = \mathbf{\frac{1}{2}}$

23. $5 \div 4\frac{1}{6} = \frac{5}{1} \div \frac{25}{6}$

$= \frac{\overset{1}{\cancel{5}}}{1} \times \frac{6}{\underset{5}{\cancel{25}}} = \frac{6}{5} = \mathbf{1\frac{1}{5}}$

24. $6\frac{1}{2} = 6\frac{3}{6} \longrightarrow 5\frac{9}{6}$

$- 2\frac{5}{6} = 2\frac{5}{6} \qquad - 2\frac{5}{6}$

$3\frac{4}{6}$

$3\frac{4}{6} = \mathbf{3\frac{2}{3}}$

25. $\frac{1}{2} \div \frac{2}{3} = \frac{1}{2} \times \frac{3}{2} = \frac{3}{4}$

$\frac{3}{4} + \frac{3}{4} = \frac{6}{4} = \mathbf{1\frac{1}{2}}$

26.
```
     $2.50
16)$40.00
    32
     8 0
     8 0
       00
       00
        0
```

27. (a) $54 - 54 = 0$

$y = \mathbf{0}$

(b) **Identity property of addition**

28. **The quotient will be greater than 1 because a larger number is divided by a smaller number.**

29. **The mixed numbers are greater than 8 and 5, so the sum is greater than 13. The mixed numbers are less than 9 and 6, so the sum is less than 15.**

30.

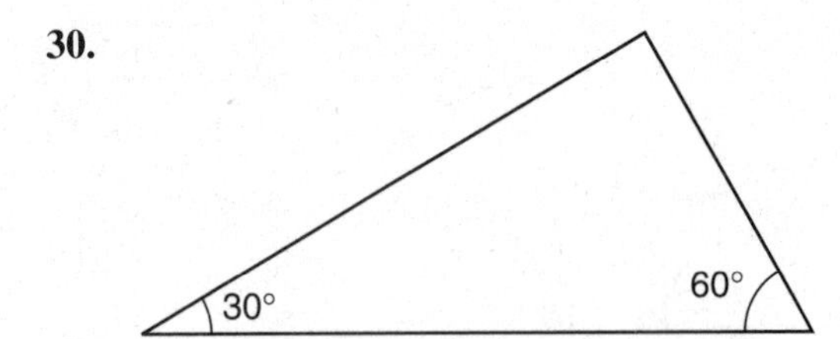

Lesson 34, Warm-Up

a. **\$5.50**

b. **\$2.40**

c. **18**

d. $\mathbf{\frac{2}{3}}$

e. **42**

f. **24**

g. **3**

Problem Solving

There are **8 blocks** with 3 painted faces. (Blocks at the corners of the cube.)

There are **12 blocks** with 2 painted faces. (Blocks along the edges between the corner blocks.)

There are **6 blocks** with 1 painted face. (Blocks in the center of each face of the cube.)

There is **1 block** with no painted faces. (Block at the center of the cube.)

Lesson 34, Lesson Practice

a. **1.6 cm**

b. **16 mm**

c. $\mathbf{\frac{1}{2}}$ **mm** or **0.5 mm**

d. **0.75 meter**

e. **157 centimeters**

f. **2.65**

g. **10.01**

h. **5 cm**

i. 3.5 cm + 1.2 cm = **4.7 cm**

j. 40 mm − 12 mm = **28 mm**

Lesson 34, Mixed Practice

1.

188 raisins
212 raisins
+ 203 raisins
603 raisins

$3\overline{)603}$ = **201 raisins**

2.

1032 parts per million
− 497 parts per million
535 parts per million

3.

$12.55
+ $3.95
$16.50

4.

1903
+ 66
1969

5. Perimeter = 6(6 inches)
= 36 inches

$\frac{36 \text{ inches}}{4}$ = **9 inches**

6. $40\% = \frac{40}{100} = \mathbf{\frac{2}{5}}$

	$4.00
$\frac{2}{5}$ is saved.	$0.80
	$0.80
$\frac{3}{5}$ is not saved.	$0.80
	$0.80
	$0.80

(a) $0.80 × 2 = **$1.60**

(b) $0.80 × 3 = **$2.40**

7. **First round 396 to 400 and 71 to 70. Then multiply 400 by 70.**

8. 7.493⑥2 ⟶ **7.494**

9. (a) **Two hundred and two hundredths**

(b) **One thousand, six hundred twenty-five millionths**

10. (a) **0.000175**

(b) **3030.03**

11. (a) 6.174 < 6.17401

(b) 14.276 > 1.4276

12. (a) **2.7 cm**

(b) **27 mm**

13. **8.25**

14.

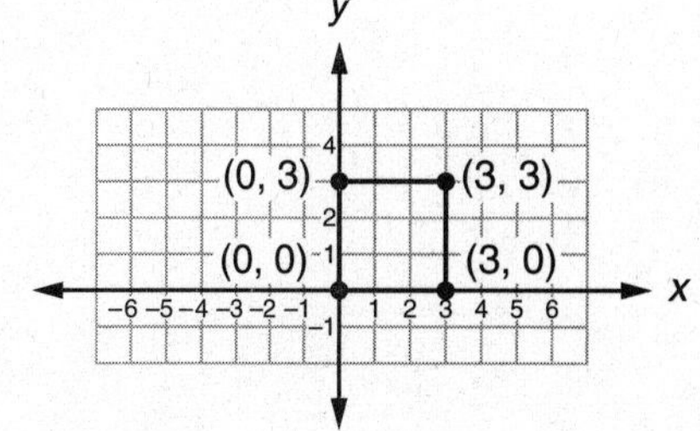

(a) **(3, 0)**

(b) Area = (3 units)(3 units)
= **9 units²**

15. (a) **7.5**

(b) **0.75**

16. $\frac{\cancel{15}^{5} \cdot \cancel{20}^{\cancel{10}^{5}}}{\cancel{12}_{\cancel{6}_{\cancel{2}_{1}}}} = 25$

y = **25**

17.

180°
− 74°
106°

c = **106°**

18.

$\frac{5}{6} = \frac{5}{6}$

$\frac{2}{3} = \frac{4}{6}$

$+\ \frac{1}{2} = \frac{3}{6}$

$\frac{12}{6} = \mathbf{2}$

19.

$\frac{5}{36} = \frac{10}{72}$

$-\ \frac{1}{24} = \frac{3}{72}$

$\mathbf{\frac{7}{72}}$

20. $$\begin{array}{r} 5\frac{1}{6} = 5\frac{1}{6} \longrightarrow \\ -\ 1\frac{2}{3} = 1\frac{4}{6} \\ \hline \end{array} \qquad \begin{array}{r} 4\frac{7}{6} \\ -\ 1\frac{4}{6} \\ \hline 3\frac{3}{6} \end{array}$$

$3\frac{3}{6} = \mathbf{3\frac{1}{2}}$

21. $\frac{1}{10} \cdot 2\frac{2}{3} \cdot 3\frac{3}{4}$

$= \frac{1}{\cancel{10}_{\cancel{5}_1}} \cdot \frac{\cancel{8}^{\cancel{2}^1}}{\cancel{3}_1} \cdot \frac{\cancel{15}^{\cancel{5}^1}}{\cancel{4}_1} = \mathbf{1}$

22. $5\frac{1}{4} \div 1\frac{2}{3} = \frac{21}{4} \div \frac{5}{3}$

$= \frac{21}{4} \times \frac{3}{5} = \frac{63}{20} = \mathbf{3\frac{3}{20}}$

23. $3\frac{1}{5} \div 4 = \frac{16}{5} \div \frac{4}{1}$

$= \frac{\cancel{16}^4}{5} \times \frac{1}{\cancel{4}_1} = \mathbf{\frac{4}{5}}$

24. $$\begin{array}{r} 6\frac{7}{8} = 6\frac{7}{8} \\ +\ 4\frac{1}{4} = 4\frac{2}{8} \\ \hline 10\frac{9}{8} = \mathbf{11\frac{1}{8}} \end{array}$$

25. $\frac{5}{\cancel{6}_2} \cdot \frac{\cancel{3}^1}{4} = \frac{5}{8}$

$\frac{1}{8} + \frac{5}{8} = \frac{6}{8} = \mathbf{\frac{3}{4}}$

26. (a) 3.6 cm − 2.4 cm = **1.2 cm**

(b) 36 mm − 24 mm = **12 mm**

27. **A. 2^5**

28. **0.3575, 0.36, 0.365**

29. $\frac{10}{5} - 5 = 2 - 5 = \mathbf{-3}$

30. *Note:* Not actual size. See student work.

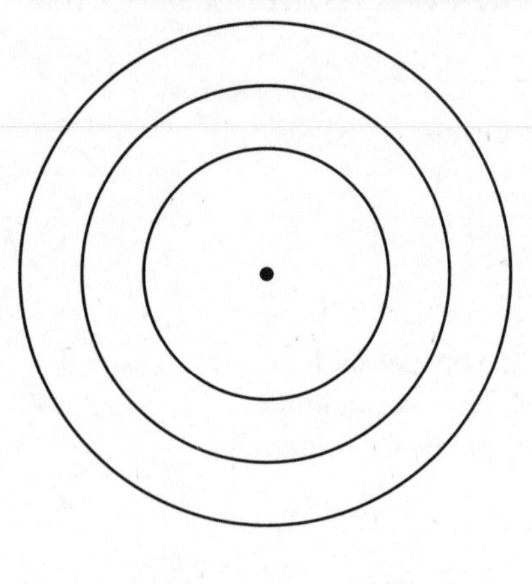

Lesson 35, Warm-Up

a. **\$5.51**

b. **\$3.20**

c. **27**

d. $\mathbf{\frac{5}{8}}$

e. **14**

f. **10**

g. **decade**

Problem Solving

The subtrahends of the subtraction stages are all multiples of 2, 4, and 8. Since the remainder is 5, the divisor must be 8. Therefore, the thousands digit of the quotient is 1, the hundreds digit of the quotient is 2, and the tens digit of the quotient is 3. Because the fourth subtrahend has only one digit, that subtrahend is 8 and the ones digit of the quotient is 1. To find the dividend, we multiply 1231 by 8 and add the remainder.

$$\begin{array}{r} \mathbf{1231\ R\ 5} \\ \mathbf{8\overline{)9853}} \\ \mathbf{8} \\ \hline \mathbf{18} \\ \mathbf{16} \\ \hline \mathbf{25} \\ \mathbf{24} \\ \hline \mathbf{13} \\ \mathbf{8} \\ \hline \mathbf{5} \end{array}$$

LESSON 35, LESSON PRACTICE

a.
$$\begin{array}{r} \overset{1}{1}.2 \\ 3.45 \\ +\ 23.6 \\ \hline \mathbf{28.25} \end{array}$$

b.
$$\begin{array}{r} \overset{1}{4}.5 \\ 0.51 \\ {}_{1}6 \\ +\ 12.4 \\ \hline \mathbf{23.41} \end{array}$$

c.
$$\begin{array}{r} \overset{2}{0}.2 \\ 0.4 \\ 0.6 \\ +\ 0.8 \\ \hline 2.0 \end{array} = \mathbf{2}$$

d.
$$\begin{array}{r} 3\ \overset{5}{\not 6}.\ \overset{11}{\not 2}\,{}^{1}\!\not 7\ 4 \\ -\quad 5.\ 3\ 9 \\ \hline \mathbf{3\ 0.\ 8\ 8\ 4} \end{array}$$

e.
$$\begin{array}{r} 1\ \overset{5}{\not 6}.\ \overset{16}{\not 7}\ \overset{9}{\not 0}\,{}^{1}0 \\ -\quad 1.\ 9\ 3\ 6 \\ \hline \mathbf{1\ 4.\ 7\ 6\ 4} \end{array}$$

f.
$$\begin{array}{r} 1\ \overset{1}{\not 2}.\ \overset{9}{\not 0}\ \overset{9}{\not 0}\,{}^{1}0 \\ -\quad 0.\ 8\ 7\ 5 \\ \hline \mathbf{1\ 1.\ 1\ 2\ 5} \end{array}$$

g.
$$\begin{array}{r} 4.20 \\ \times\ 0.24 \\ \hline 1680 \\ 840 \\ \hline 1.0080 \end{array} = \mathbf{1.008}$$

h.
$$\begin{array}{r} 0.12 \\ \times\ 0.06 \\ \hline \mathbf{0.0072} \end{array}$$

i.
$$\begin{array}{r} 5.4 \\ \times\quad 7 \\ \hline \mathbf{37.8} \end{array}$$

j. $3 \times 2 \times 1 = 6$

$0.\_\_6 \longrightarrow$ **0.006**

k.
$$\begin{array}{r} 0.04 \\ \times\quad 10 \\ \hline 00 \\ 04 \\ \hline 0.40 \end{array} \longrightarrow \mathbf{0.4}$$

l.
$$\begin{array}{r} 0.045 \\ \times\quad 0.6 \\ \hline 0270 \\ 000 \\ \hline 0.0270 \end{array} \longrightarrow \mathbf{0.027}$$

m.
$$\begin{array}{r} \mathbf{2.4} \\ 6\overline{)14.4} \\ 12 \\ \hline 2\,4 \\ 2\,4 \\ \hline 0 \end{array}$$

n.
$$\begin{array}{r} \mathbf{0.006} \\ 8\overline{)0.048} \\ 48 \\ \hline 0 \end{array}$$

o.
$$\begin{array}{r} \mathbf{0.68} \\ 5\overline{)3.40} \\ 3\,0 \\ \hline 40 \\ 40 \\ \hline 0 \end{array}$$

p.
$$\begin{array}{r} \mathbf{0.05} \\ 6\overline{)0.30} \\ 30 \\ \hline 0 \end{array}$$

LESSON 35, MIXED PRACTICE

1. **Add all the bills together and divide by 6.**

2.
$$\begin{array}{r} 2\frac{1}{2} = 2\frac{2}{4} \\ -\ 1\frac{3}{4} = 1\frac{3}{4} \\ \hline \end{array} \longrightarrow \begin{array}{r} 1\frac{6}{4} \\ -\ 1\frac{3}{4} \\ \hline \mathbf{\frac{3}{4}} \end{array}$$ **gallon**

3.
$$\begin{array}{r} \$1.30 \\ 12\overline{)\$15.60} \\ 12 \\ \hline 3\,6 \\ 3\,6 \\ \hline 00 \\ 00 \\ \hline 0 \end{array} \qquad \begin{array}{r} \$1.75 \\ -\ \$1.30 \\ \hline \mathbf{\$0.45} \end{array}$$

4. 1 minute = 60 seconds
60 seconds + 3 seconds = 63 seconds
63 seconds − 5 seconds = **58 seconds**

5. Perimeter = 5(16 cm) = 80 cm

$\frac{80 \text{ cm}}{4}$ = **20 cm**

6.

54 fish

6 fish
6 fish
6 fish
6 fish
6 fish
6 fish
6 fish
6 fish
6 fish

$\frac{2}{9}$ were guppies. (first 2 rows)
$\frac{7}{9}$ were not guppies. (remaining 7 rows)

(a) 2(6 fish) = **12 fish**

(b) 7(6 fish) = **42 fish**

7. (a) Area = (10 cm)(10 cm) = **100 cm²**

(b) Area = (6 cm)(6 cm) = **36 cm²**

(c) Area = 100 cm² − 36 cm²
= **64 cm²**

8. (a) $\mathbf{\frac{99}{100}}$

(b) **0.99**

(c) $\frac{99}{100}$ = **99%**

9.

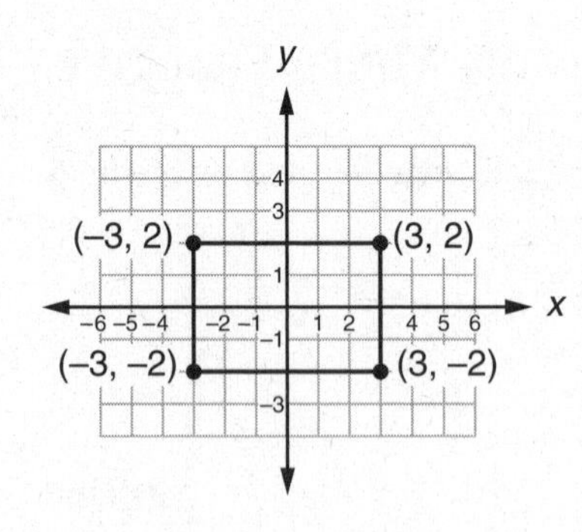

(a) **(3, 2)**

(b) Area = (4 units)(6 units)
= **24 units²**

10. (a) **One hundred and seventy-five thousandths**

(b) **0.00025**

11. (a) **3.5 centimeters**

(b) **35 millimeters**

12.
```
   $1.50
 ×    12
    3 00
   15 0
  $18.00
```

13. **3.37**

14. **0.5 × 0.7 = 0.35**

15. **1.25**

16. $\frac{\overset{3}{\cancel{9}} \cdot \overset{2}{\cancel{10}}}{\underset{1}{\cancel{\underset{\cancel{3}}{\cancel{15}}}}} = 6 \qquad x = \mathbf{6}$

17.
```
   5.83
 − 4.6
   1.23
```
f = **1.23**

18.
```
    5.8
 8)46.4
   40
    6 4
    6 4
      0
```
y = **5.8**

19.
```
    12
 +  3.4
   15.4
```
w = **15.4**

20.
```
     1
    3.65
    0.9
    1 8
 + 15.23
   27.78
```

21.
$$1\frac{1}{2} = 1\frac{6}{12}$$
$$2\frac{2}{3} = 2\frac{8}{12}$$
$$+\ 3\frac{3}{4} = 3\frac{9}{12}$$
$$6\frac{23}{12} = \mathbf{7\frac{11}{12}}$$

22. $1\frac{1}{2} \cdot 2\frac{2}{3} \cdot 3\frac{3}{4} = \frac{\overset{1}{\cancel{3}}}{\underset{1}{\cancel{2}}} \cdot \frac{\overset{\overset{1}{\cancel{2}}}{\cancel{8}}}{\underset{1}{\cancel{3}}} \cdot \frac{15}{\underset{1}{\cancel{4}}} = \mathbf{15}$

23.
$$\begin{array}{r} \frac{1}{2} = \frac{3}{6} \\ + \frac{1}{3} = \frac{2}{6} \\ \hline \frac{5}{6} \end{array}$$

$$1\frac{1}{6} \longrightarrow \frac{7}{6}$$
$$-\frac{5}{6} \qquad -\frac{5}{6}$$
$$\frac{2}{6} = \mathbf{\frac{1}{3}}$$

24.
$$\begin{array}{r} 3\frac{1}{12} = 3\frac{1}{12} \\ - 1\frac{3}{4} = 1\frac{9}{12} \\ \hline \end{array} \longrightarrow \begin{array}{r} 2\frac{13}{12} \\ - 1\frac{9}{12} \\ \hline 1\frac{4}{12} \end{array}$$

$$1\frac{4}{12} = \mathbf{1\frac{1}{3}}$$

25.
$$\begin{array}{r} \mathbf{0.12} \\ 10\overline{)1.20} \\ \underline{1\,0} \\ 20 \\ \underline{20} \\ 0 \end{array}$$

26. $3 \times 4 \times 5 = 60$

$0.\underset{\smile\smile}{60} \longrightarrow \mathbf{0.06}$

27.
$$\begin{array}{r} 3\frac{1}{2} = 3\frac{2}{4} \\ + 1\frac{3}{4} = 1\frac{3}{4} \\ \hline 4\frac{5}{4} = 5\frac{1}{4} \end{array}$$

$$\begin{array}{r} 4 \\ - 3\frac{1}{8} \\ \hline \end{array} \longrightarrow \begin{array}{r} 3\frac{8}{8} \\ - 3\frac{1}{8} \\ \hline \frac{7}{8} \end{array}$$

$$5\frac{1}{4} \div \frac{7}{8} = \frac{21}{4} \div \frac{7}{8} = \frac{\overset{3}{\cancel{21}}}{\underset{1}{\cancel{4}}} \times \frac{\overset{2}{\cancel{8}}}{\underset{1}{\cancel{7}}} = \mathbf{6}$$

28. $36.45 - 4.912$

$\downarrow \qquad \downarrow$

$36 - 5 = \mathbf{31}$

$$\begin{array}{r} 3\overset{5}{\cancel{6}}.{}^{1}4\,\overset{4}{\cancel{5}}{}^{1}0 \\ - \quad 4.\,9\,1\,2 \\ \hline \mathbf{31.538} \end{array}$$

29. 4.2×0.9

$\downarrow \qquad \downarrow$

$4 \times 1 = \mathbf{4}$

$$\begin{array}{r} 4.2 \\ \times\ 0.9 \\ \hline \mathbf{3.78} \end{array}$$

30.

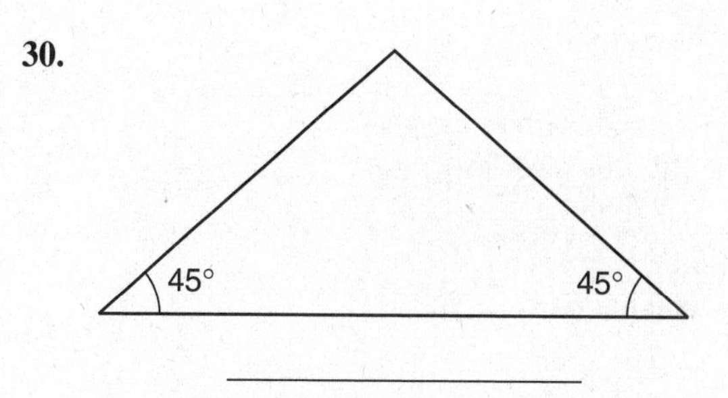

LESSON 36, WARM-UP

a. **$14.50**

b. **$6.00**

c. **16**

d. $\mathbf{\frac{4}{5}}$

e. **3**

f. **36**

g. **9**

Problem Solving

math: m
English: e
science: s
history: h

(m, e, s, h); (m, e, h, s); (m, s, e, h); (m, s, h, e); (m, h, e, s); (m, h, s, e); (e, m, s, h); (e, m ,h, s); (e, s, m, h); (e, s, h, m); (e, h, m, s); (e, h, s, m); (s, m, e, h); (s, m, h, e); (s, e, m, h); (s, e, h, m); (s, h, m, e); (s, h, e, m); (h, m, e, s); (h, m, s, e); (h, e, m, s); (h, e, s, m); (h, s, m, e); (h, s, e, m)
There are **24 possible permutations** of the four subjects.

Lesson 36, Lesson Practice

a. $\dfrac{\text{big fish}}{\text{little fish}} = \dfrac{90}{240} = \mathbf{\dfrac{3}{8}}$

b. $\begin{array}{r} 14 \text{ girls} \\ + \ ? \text{ boys} \\ \hline 30 \text{ total} \end{array} \longrightarrow \begin{array}{r} 14 \text{ girls} \\ + \ 16 \text{ boys} \\ \hline 30 \text{ total} \end{array}$

$\dfrac{\text{boys}}{\text{girls}} = \dfrac{16}{14} = \mathbf{\dfrac{8}{7}}$

c. $\begin{array}{r} 3 \text{ won} \\ + \ ? \text{ lost} \\ \hline 8 \text{ total} \end{array} \longrightarrow \begin{array}{r} 3 \text{ won} \\ + \ 5 \text{ lost} \\ \hline 8 \text{ total} \end{array}$

$\dfrac{\text{won}}{\text{lost}} = \mathbf{\dfrac{3}{5}}$

d. $\begin{array}{r} 5 \text{ red marbles} \\ + \ 3 \text{ blue marbles} \\ \hline 8 \text{ total} \end{array}$

$\dfrac{\text{blue marbles}}{\text{total marbles}} = \mathbf{\dfrac{3}{8}}$

e. $\dfrac{3}{6} = \mathbf{\dfrac{1}{2}}$

f. $\mathbf{\dfrac{1}{4}}$

g. $\dfrac{0}{4} = \mathbf{0}$

h. $\dfrac{4}{4} = \mathbf{1}$

i. $\dfrac{2}{4} = \mathbf{\dfrac{1}{2}}$

j. $\mathbf{\dfrac{3}{4}}$

Lesson 36, Mixed Practice

1. $\begin{array}{r} 14 \text{ blue} \\ + \ ? \text{ red} \\ \hline 32 \text{ total} \end{array} \longrightarrow \begin{array}{r} 14 \text{ blue} \\ + \ 18 \text{ red} \\ \hline 32 \text{ total} \end{array}$

$\dfrac{\text{red}}{\text{blue}} = \dfrac{18}{14} = \mathbf{\dfrac{9}{7}}$

2. $\begin{array}{r} 23 \text{ inches} \\ 21 \text{ inches} \\ + \ 16 \text{ inches} \\ \hline 60 \text{ inches} \end{array}$

$3\overline{)60 \text{ inches}}$ = **20 inches**

3. $\begin{array}{r} 35 \text{ pages} \\ \times \ 7 \\ \hline \mathbf{245 \text{ pages}} \end{array}$

4. $\begin{array}{r} 59.48 \text{ seconds} \\ - \ 56.24 \text{ seconds} \\ \hline \mathbf{3.24 \text{ seconds}} \end{array}$

5. $40\% = \dfrac{40}{100} = \dfrac{2}{5}$

30 players

$\frac{2}{5}$ had never played rugby. { 6 players / 6 players

$\frac{3}{5}$ had played rugby. { 6 players / 6 players / 6 players

(a) 2×6 players $=$ **12 players**

(b) $\dfrac{\text{had played}}{\text{had not played}} = \mathbf{\dfrac{3}{2}}$

6. **One way to find *BC* in millimeters is to first convert *AB* to 40 mm and *AC* to 95 mm. Then subtract 40 mm from 95 mm.**

7. (a) Area $= (8 \text{ cm})(13 \text{ cm}) = \mathbf{104 \text{ cm}^2}$

(b) Perimeter $= 8 \text{ cm} + 8 \text{ cm} + 13 \text{ cm} + 13 \text{ cm} = \mathbf{42 \text{ cm}}$

8. $\begin{array}{r} 3600 \\ 2900 \\ + \ 900 \\ \hline \mathbf{7400} \end{array}$

9. (a) 6.857①42 ⟶ **6.857**

(b) 6.8571420 ⟶ 7
1.9870 ⟶ 2
$7 \times 2 = \mathbf{14}$

10. (a) **12,000,000**

(b) **0.000012**

11. $\begin{array}{r} 3 \text{ red marbles} \\ 4 \text{ white marbles} \\ + \ 5 \text{ blue marbles} \\ \hline 12 \text{ total} \end{array}$

(a) $\dfrac{\text{red marbles}}{\text{total marbles}} = \dfrac{3}{12} = \mathbf{\dfrac{1}{4}}$

(b) $\dfrac{\text{white marbles}}{\text{total marbles}} = \dfrac{4}{12} = \mathbf{\dfrac{1}{3}}$

(c) $\dfrac{\text{blue marbles}}{\text{total marbles}} = \mathbf{\dfrac{5}{12}}$

(d) $\dfrac{\text{green marbles}}{\text{total marbles}} = \dfrac{0}{12} = \mathbf{0}$

12. (a) **4.2 cm**

(b) **42 mm**

13. **13.56**

14. (a) $85\% = \frac{85}{100} = \mathbf{\frac{17}{20}}$

(b) $\frac{144}{600} = \frac{\cancel{2}^1 \cdot \cancel{2}^1 \cdot \cancel{2}^1 \cdot 2 \cdot \cancel{3}^1 \cdot 3}{\cancel{2}_1 \cdot \cancel{2}_1 \cdot \cancel{2}_1 \cdot \cancel{3}_1 \cdot 5 \cdot 5} = \mathbf{\frac{6}{25}}$

15. Estimate: $\mathbf{6\frac{3}{4}}$ **hr** or **7 hr**

$8 per hour

$8 × 7 hr = $56

She earned a little less than $56.

16. (a) $\angle MPN$ (or $\angle NPM$)

(b) $\angle LPM$ (or $\angle MPL$)

(c) $\angle LPN$ (or $\angle NPL$)

17. $8y = 12^2$

$8y = 144$

$$\begin{array}{r} 18 \\ 8\overline{)144} \\ 8 \\ \hline 64 \\ 64 \\ \hline 0 \end{array}$$

$y = \mathbf{18}$

18.

$$\begin{array}{r} 1.2 \\ \times \quad 4 \\ \hline 4.8 \end{array}$$

$w = \mathbf{4.8}$

19. $4.27 + 16.3 + 10$

↓ ↓ ↓

$4 + 16 + 10 = \mathbf{30}$

$$\begin{array}{r} {}_1 4.27 \\ 16.3 \\ +\ 10. \\ \hline \mathbf{30.57} \end{array}$$

20. $4.2 - 0.42$

↓ ↓

$4 - 0 = \mathbf{4}$

$$\begin{array}{r} \overset{3}{\cancel{4}}.\overset{1}{\cancel{2}}{}^1 0 \\ -\ 0.\,4\,2 \\ \hline \mathbf{3.\,7\,8} \end{array}$$

21.

$$\begin{array}{rcl} 3\frac{1}{2} & = & 3\frac{6}{12} \\ 1\frac{1}{3} & = & 1\frac{4}{12} \\ +\ 2\frac{1}{4} & = & 2\frac{3}{12} \\ \hline & & 6\frac{13}{12} = \mathbf{7\frac{1}{12}} \end{array}$$

22. $3\frac{1}{2} \cdot 1\frac{1}{3} \cdot 2\frac{1}{4}$

$= \frac{7}{2} \cdot \frac{\overset{1}{\cancel{4}}}{\underset{1}{\cancel{3}}} \cdot \frac{\overset{3}{\cancel{9}}}{\underset{1}{\cancel{4}}} = \frac{21}{2} = \mathbf{10\frac{1}{2}}$

23.

$$\begin{array}{rcl} \frac{2}{3} & = & \frac{4}{6} \\ -\ \frac{1}{2} & = & \frac{3}{6} \\ \hline & & \frac{1}{6} \end{array}$$

$3\frac{5}{6} - \frac{1}{6} = 3\frac{4}{6} = \mathbf{3\frac{2}{3}}$

24.

$$\begin{array}{rcl} 8\frac{5}{12} & = & 8\frac{5}{12} \\ -\ 3\frac{2}{3} & = & 3\frac{8}{12} \\ \hline \end{array} \longrightarrow \begin{array}{r} 7\frac{17}{12} \\ -\ 3\frac{8}{12} \\ \hline 4\frac{9}{12} = \mathbf{4\frac{3}{4}} \end{array}$$

25. $2\frac{3}{4} \div 4\frac{1}{2} = \frac{11}{4} \div \frac{9}{2}$

$= \frac{11}{\underset{2}{\cancel{4}}} \times \frac{\overset{1}{\cancel{2}}}{9} = \mathbf{\frac{11}{18}}$

26. $\frac{2}{3} \div \frac{1}{2} = \frac{2}{3} \times \frac{2}{1} = \frac{4}{3} = 1\frac{1}{3}$

$$\begin{array}{r} 5 \\ -\ 1\frac{1}{3} \\ \hline \end{array} \longrightarrow \begin{array}{r} 4\frac{3}{3} \\ -\ 1\frac{1}{3} \\ \hline \mathbf{3\frac{2}{3}} \end{array}$$

27.

$$\begin{array}{r} \mathbf{0.175} \\ 8\overline{)1.400} \\ 8 \\ \hline 60 \\ 56 \\ \hline 40 \\ 40 \\ \hline 0 \end{array}$$

28. $2 \times 3 \times 4 = 24$

$0.\underset{\smile\smile}{24} \longrightarrow$ **0.024**

29. (a)

$$\begin{array}{r} 12.25 \\ \times \quad 10 \\ \hline 122.50 \end{array} = \mathbf{122.5}$$

(b)

$$\begin{array}{r} \mathbf{1.225} \\ 10\overline{)12.250} \\ \underline{10} \\ 2\,2 \\ \underline{2\,0} \\ 25 \\ \underline{20} \\ 50 \\ \underline{50} \\ 0 \end{array}$$

30. Answers may vary. Sample answer:

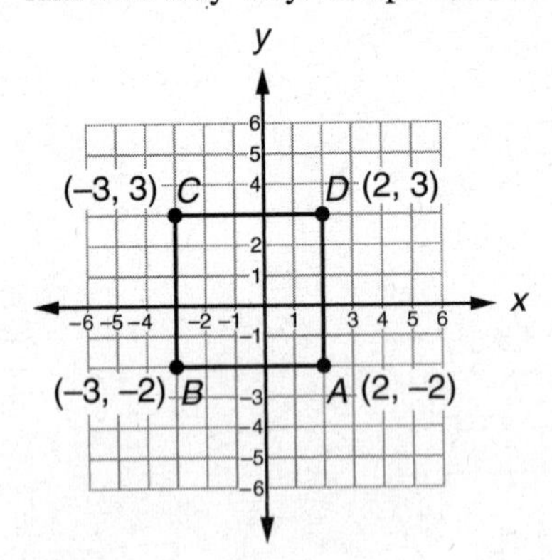

Coordinates: ***A* (2, −2), *B* (−3, −2), *C* (−3, 3), *D* (2, 3)**

Lesson 37, Warm-Up

a. **\$4.65**

b. **\$6.25**

c. **21**

d. $\mathbf{\frac{3}{5}}$

e. **2**

f. **18**

g. **30**

Problem Solving

One sheet of newspaper contains a maximum of four pages (left front, left back, right front, right back). Therefore, the fewest number of sheets of paper in the section is $36 \div 4$, or **9 sheets.**

Lesson 37, Lesson Practice

a. $\text{Area} = \dfrac{5\text{ cm} \cdot 12\text{ cm}}{2} = \dfrac{60\text{ cm}^2}{2}$

$= \mathbf{30\ cm^2}$

b. $\text{Area} = \dfrac{12\text{ cm} \cdot 8\text{ cm}}{2} = \dfrac{96\text{ cm}^2}{2}$

$= \mathbf{48\ cm^2}$

c. $\text{Area} = \dfrac{6\text{ cm} \cdot 6\text{ cm}}{2} = \dfrac{36\text{ cm}^2}{2}$

$= \mathbf{18\ cm^2}$

d.

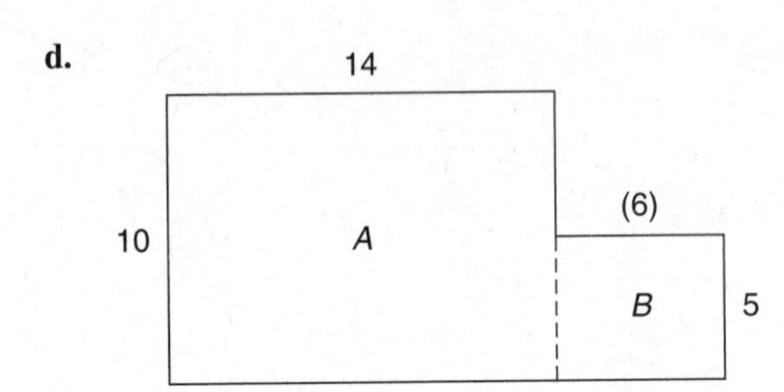

$$\begin{array}{rcl} \text{Area } A &=& 14\text{ m} \times 10\text{ m} = 140\text{ m}^2 \\ +\ \text{Area } B &=& 6\text{ m} \times 5\text{ m} = 30\text{ m}^2 \\ \hline && \text{Total} = \mathbf{170\ m^2} \end{array}$$

or

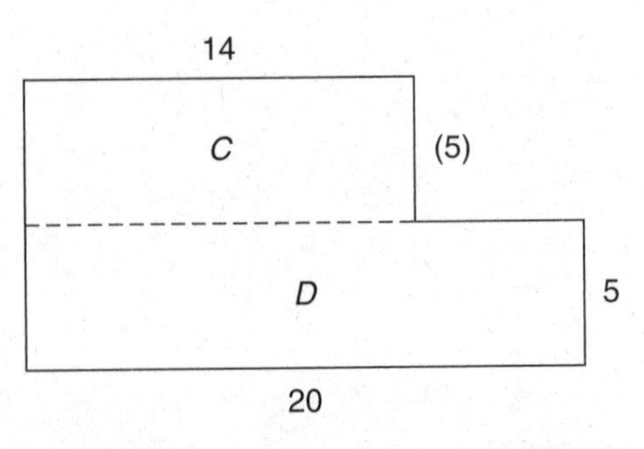

$$\begin{array}{rcl} \text{Area } C &=& 14\text{ m} \times 5\text{ m} = 70\text{ m}^2 \\ +\ \text{Area } D &=& 20\text{ m} \times 5\text{ m} = 100\text{ m}^2 \\ \hline && \text{Total} = \mathbf{170\ m^2} \end{array}$$

e.

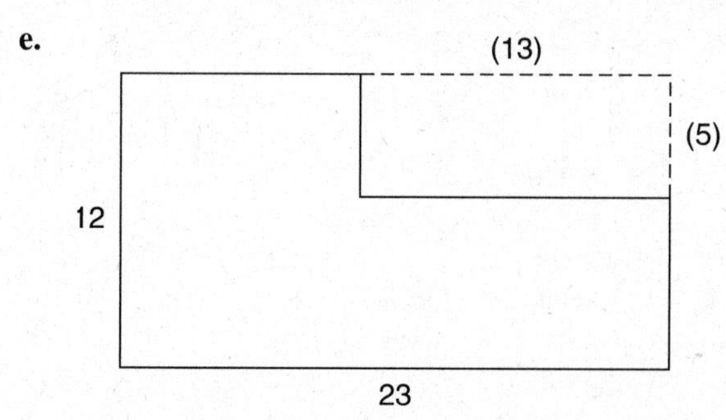

$$\begin{array}{rl} \text{Area of large} = 12\text{ cm} \times 23\text{ cm} = & 276\text{ cm}^2 \\ -\ \text{Area of small} = 13\text{ cm} \times 5\text{ cm} = & 65\text{ cm}^2 \\ \hline \text{Total} = & \mathbf{211\ cm^2} \end{array}$$

f.

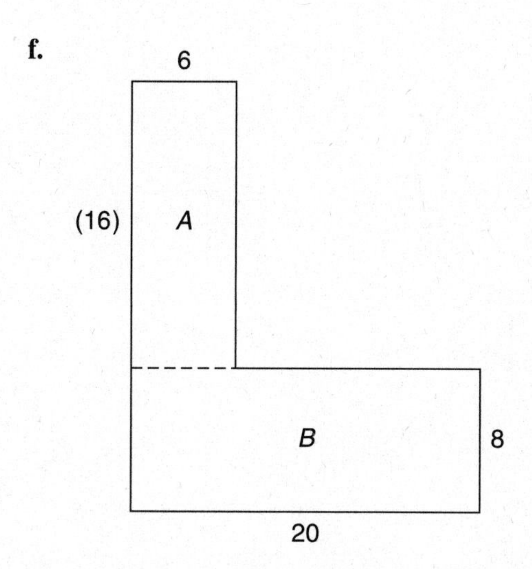

$$\begin{array}{rl} \text{Area } A = 16\text{ in.} \times 6\text{ in.} = & 96\text{ in.}^2 \\ +\ \text{Area } B = 20\text{ in.} \times 8\text{ in.} = & 160\text{ in.}^2 \\ \hline \text{Total} = & \mathbf{256\ in.^2} \end{array}$$

or

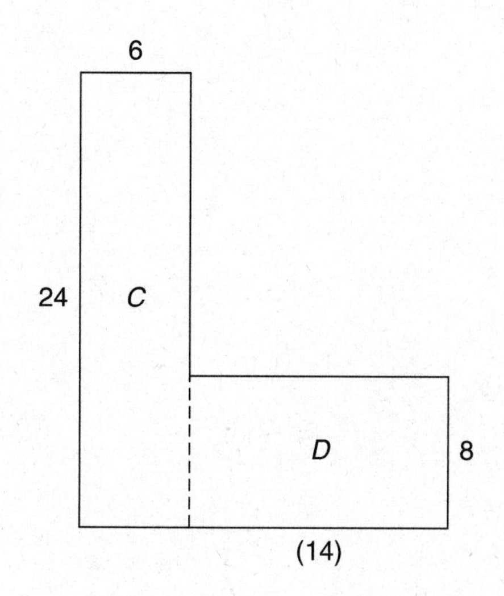

$$\begin{array}{rl} \text{Area } C = 24\text{ in.} \times 6\text{ in.} = & 144\text{ in.}^2 \\ +\ \text{Area } D = 14\text{ in.} \times 8\text{ in.} = & 112\text{ in.}^2 \\ \hline \text{Total} = & \mathbf{256\ in.^2} \end{array}$$

g. $A = \frac{1}{2}bh;\ A = \frac{bh}{2}$

Lesson 37, Mixed Practice

1.
$$\begin{array}{r} 2\text{ won} \\ +\ ?\text{ lost} \\ \hline 3\text{ total} \end{array} \longrightarrow \begin{array}{r} 2\text{ won} \\ +\ 1\text{ lost} \\ \hline 3\text{ total} \end{array}$$

$$\frac{\text{won}}{\text{lost}} = \mathbf{\frac{2}{1}}$$

2.
$$\begin{array}{r} 47 \\ 53 \\ 62 \\ 56 \\ 46 \\ +\ 48 \\ \hline 312\text{ cars} \end{array} \qquad \begin{array}{r} \mathbf{52\ cars} \\ 6\overline{)312}\text{ cars} \\ \underline{30} \\ 12 \\ \underline{12} \\ 0 \end{array}$$

3.
$$\begin{array}{r} \overset{2}{1}1.6\text{ seconds} \\ 11.3\text{ seconds} \\ 11.2\text{ seconds} \\ +\ 10.9\text{ seconds} \\ \hline 45.0\text{ seconds} \end{array} = \mathbf{45\ seconds}$$

4. **Subtract $1.30 from $10 to find how much the 3 gallons of milk cost. Then divide that number by 3 to find how much each gallon cost.**

5.

	18 holes
$\frac{2}{3}$ were par holes.	6 holes
	6 holes
$\frac{1}{3}$ were not par holes.	6 holes

(a) 2 × 6 holes = **12 holes**

(b) **6 holes**

6.

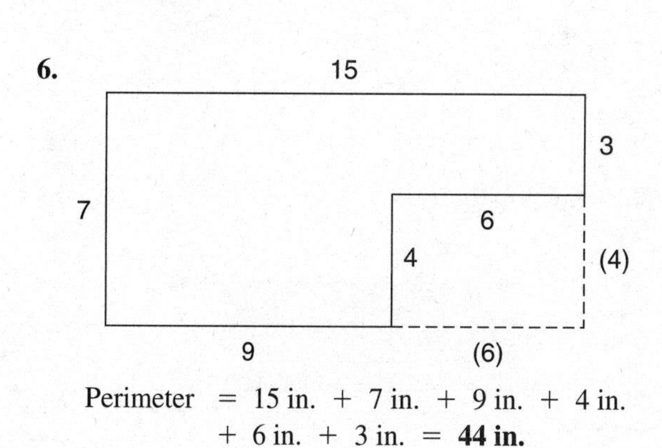

Perimeter = 15 in. + 7 in. + 9 in. + 4 in. + 6 in. + 3 in. = **44 in.**

7.
$$\begin{array}{rl} \text{Area of large rectangle} = & 7\text{ in.} \times 15\text{ in.} \\ = & 105\text{ in.}^2 \\ -\ \text{Area of small rectangle} = & 6\text{ in.} \times 4\text{ in.} \\ = & 24\text{ in.}^2 \\ \hline \text{Area of figure} = & \mathbf{81\ in.^2} \end{array}$$

8. (a) $\frac{5}{6} \cdot \frac{3}{3} = \mathbf{\frac{15}{18}}$

(b) $\frac{9 \div 3}{24 \div 3} = \mathbf{\frac{3}{8}}$

(c) $\frac{3}{4} \cdot \frac{5}{5} = \frac{15}{\mathbf{20}}$

9. (a) **0.49**

(b) **0.51**

(c) $\frac{51}{100} = \mathbf{51\%}$

10. (a) 3184.56④1 ⟶ **3184.56**

(b) 31⑧4.5641 ⟶ 3200.~~0000~~ ⟶ **3200**

11. (a) **Twenty-five hundred-thousandths**

(b) **60.07**

12. (a) $2\% = \frac{2}{100} = \mathbf{\frac{1}{50}}$

(b) $\frac{720}{1080} = \frac{\cancel{2} \cdot \cancel{2} \cdot \cancel{2} \cdot 2 \cdot \cancel{3} \cdot \cancel{3} \cdot \cancel{5}}{\cancel{2} \cdot \cancel{2} \cdot \cancel{2} \cdot \cancel{3} \cdot \cancel{3} \cdot 3 \cdot \cancel{5}} = \mathbf{\frac{2}{3}}$

13. $\mathbf{1\frac{1}{8}}$ **in.**

14. **One possibility:**

15. (a) Perimeter = 15 cm + 15 cm + 18 cm = **48 cm**

(b) Area $= \frac{18 \text{ cm} \cdot 12 \text{ cm}}{2} = \mathbf{108 \text{ cm}^2}$

16.
$$\begin{array}{r} 0.2 \\ +\ 0.3 \\ \hline 0.5 \end{array} \qquad \begin{array}{r} 0.2 \\ \times\ 0.3 \\ \hline 0.06 \end{array}$$

0.5 > 0.06

17.
$$\begin{array}{r} 1 \text{ heads} \\ +\ 1 \text{ tails} \\ \hline 2 \text{ total} \end{array} \qquad \frac{\text{heads}}{\text{total}} = \mathbf{\frac{1}{2}}$$

18. $\frac{7 \cdot \overset{2}{\cancel{8}}}{\underset{1}{\cancel{4}}} = 14 \qquad x = \mathbf{14}$

19.
$$\begin{array}{r} \overset{3}{\cancel{4}}.^{1}2 \\ -\ 1.\ 7 \\ \hline \mathbf{2.\ 5} \end{array}$$

20.
$$\begin{array}{r} 0.45 \\ +\ 3.6 \\ \hline \mathbf{4.05} \end{array}$$

21.
$$\begin{array}{r} \mathbf{1.5} \\ 3\overline{)4.5} \\ 3 \\ \hline 1\,5 \\ 1\,5 \\ \hline 0 \end{array}$$

22. $\frac{3}{5} \cdot 12 \cdot 4\frac{1}{6} = \frac{3}{\underset{1}{\cancel{5}}} \cdot \frac{\overset{2}{\cancel{12}}}{1} \cdot \frac{\overset{5}{\cancel{25}}}{\underset{1}{\cancel{6}}}$

$= \mathbf{30}$

23.
$$\begin{array}{rcl} \frac{5}{6} & = & \frac{10}{12} \\ 1\frac{3}{4} & = & 1\frac{9}{12} \\ +\ 2\frac{1}{2} & = & 2\frac{6}{12} \\ \hline & & 3\frac{25}{12} = \mathbf{5\frac{1}{12}} \end{array}$$

24.
$$\begin{array}{rcl} \frac{5}{8} & = & \frac{5}{8} \\ \frac{1}{2} & = & \frac{4}{8} \\ +\ \frac{3}{8} & = & \frac{3}{8} \\ \hline & & \frac{12}{8} = \mathbf{1\frac{1}{2}} \end{array}$$

25.
$$\begin{array}{rcl} 3\frac{9}{20} & = & 3\frac{27}{60} \\ -\ 1\frac{5}{12} & = & 1\frac{25}{60} \\ \hline & & 2\frac{2}{60} = \mathbf{2\frac{1}{30}} \end{array}$$

26. $\frac{a}{b} = a \div b = 3\frac{1}{3} \div 5$

$= \frac{10}{3} \div \frac{5}{1} = \frac{\overset{2}{\cancel{10}}}{3} \times \frac{1}{\underset{1}{\cancel{5}}} = \mathbf{\frac{2}{3}}$

27. $2 \cdot 2 \cdot 2 \cdot 2 \cdot 2 \cdot 2 = 2^6$

28. (a)
$$\begin{array}{r} 0.25 \\ \times\ 10 \\ \hline 2.50 \end{array} = \mathbf{2.5}$$

(b)
$$\begin{array}{r} \mathbf{0.025} \\ 10\overline{)0.250} \\ 20 \\ \hline 50 \\ 50 \\ \hline 0 \end{array}$$

29.

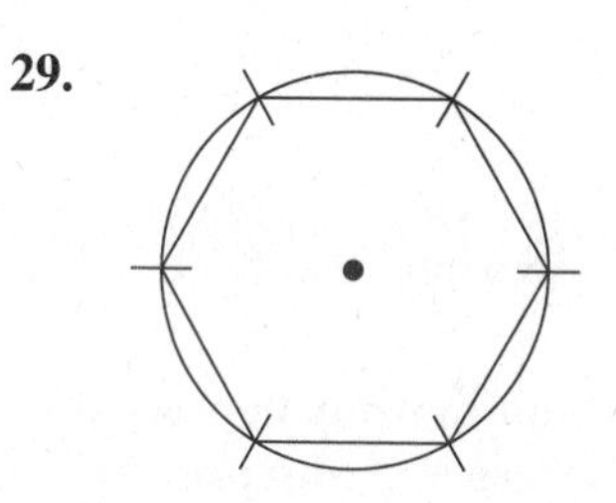

30. **Fourth quadrant**

Lesson 38, Warm-Up

a. **$6.44**

b. **$7.50**

c. **25**

d. $\mathbf{\frac{1}{2}}$

e. **60**

f. **35**

g. **6**

Problem Solving

The perimeter of the rectangle is the same as the length of the string (2 yards, or 6 feet):

$2l + 2w = 6$ ft

Since the rectangle is twice as long as it is wide, we can substitute $2w$ for l in the perimeter equation. Then we solve for w:

$$\begin{aligned} 2(2w) + 2w &= 6 \text{ ft} \\ 4w + 2w &= 6 \text{ ft} \\ 6w &= 6 \text{ ft} \\ w &= 1 \text{ ft} \end{aligned}$$

The width of the rectangle is 1 ft, and the length is 2 ft. We find the area in square feet by multiplying:

$A = l \times w$

$A = 2 \text{ ft} \times 1 \text{ ft}$

$A =$ **2 sq. ft**

Lesson 38, Lesson Practice

a. January: $4 \times 10{,}000$ doughnuts
$= 40{,}000$ doughnuts
February: $6 \times 10{,}000$ doughnuts
$= 60{,}000$ doughnuts

$$\begin{array}{r} 60{,}000 \text{ doughnuts} \\ -\ 40{,}000 \text{ doughnuts} \\ \hline \mathbf{20{,}000 \text{ doughnuts}} \end{array}$$

b.
$$\begin{array}{r} 500 \text{ cans} \\ 800 \text{ cans} \\ 900 \text{ cans} \\ +\ 400 \text{ cans} \\ \hline \mathbf{2600 \text{ cans}} \end{array}$$

c. **Test 4**

d. $\frac{4}{24} = \mathbf{\frac{1}{6}}$

Lesson 38, Mixed Practice

1.
$$\begin{array}{r} 7 \text{ civilians} \\ +\ 3 \text{ soldiers} \\ \hline 10 \text{ total} \end{array}$$

$\frac{\text{soldiers}}{\text{total}} = \mathbf{\frac{3}{10}}$

2.
$$\begin{array}{r} \mathbf{115 \text{ pages}} \\ 3\overline{)345} \text{ pages} \\ 3 \\ \hline 04 \\ 3 \\ \hline 15 \\ 15 \\ \hline 0 \end{array}$$

3. 5 minutes = 5(60 seconds) = 300 seconds

$$\begin{array}{r} 300 \text{ seconds} \\ +\ 52 \text{ seconds} \\ \hline \mathbf{352 \text{ seconds}} \end{array}$$

4.

$$\begin{array}{r} 900 \text{ cans} \\ -\ 400 \text{ cans} \\ \hline \mathbf{500 \text{ cans}} \end{array}$$

5.

$$\begin{array}{r} 60 \\ 70 \\ 75 \\ 70 \\ 80 \\ 85 \\ +\ 85 \\ \hline 525 \end{array} \qquad 7\overline{)525} = \mathbf{75}$$

6. **See student work.**

7.

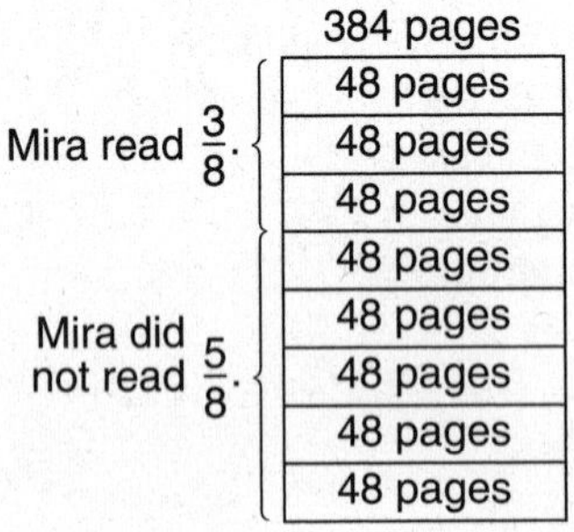

(a) 3 × 48 pages = **144 pages**

(b) $\frac{3}{8} + \frac{1}{8} = \frac{4}{8} = \frac{1}{2}$ **48 pages**

8.

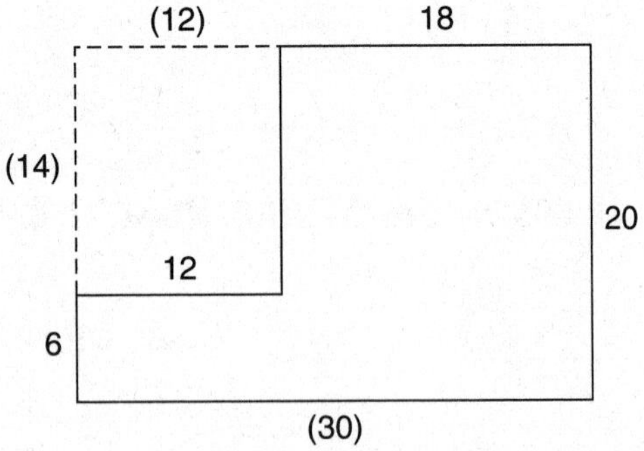

(a) Area of large rectangle = 30 in. × 20 in. = 600 in.^{2}

Area of small rectangle = 12 in. × 14 in. = 168 in.^{2}

Area of figure = 600 in.^{2} − 168 in.^{2} = **432 in.^{2}**

(b) Perimeter = 18 in. + 20 in. + 30 in. + 6 in. + 12 in. + 14 in. = **100 in.**

9. (a) $\frac{7}{9} \cdot \frac{2}{2} = \mathbf{\frac{14}{18}}$

(b) $\frac{20}{36} \div \frac{4}{4} = \mathbf{\frac{5}{9}}$

(c) $\frac{4}{5} \cdot \frac{6}{6} = \frac{24}{\mathbf{30}}$

10. (a) 2⑨86.34157 ⟶ 3000.~~00000~~ ⟶ **3000**

(b) 2986.341⑤7 ⟶ **2986.342**

11. Probability of stopping on

$1 = \frac{3}{8}$

$2 = \frac{2}{8} = \frac{1}{4}$

$3 = \frac{2}{8} = \frac{1}{4}$

$4 = \frac{1}{8}$

(a) **1**

(b) **4**

12. (a) **1.2 cm**

(b) **12 mm**

13. Perimeter = 1.2 cm + 1 cm + 1.2 cm + 1 cm = **4.4 cm**

14. **The number 3.4 is about halfway between 3 and 4. Point *B* is too close to 3 to represent 3.4. So the best choice is point *C*.**

15. (a) $\overline{AC}$ (or $\overline{CA}$)

(b) $\overline{BC}$ (or $\overline{CB}$)

16. (a) Area = $\frac{6 \text{ cm} \cdot 6 \text{ cm}}{2}$ = **18 cm^2**

(b) Area = $\frac{6 \text{ cm} \cdot 6 \text{ cm}}{2}$ = **18 cm^2**

(c) Area = 18 cm^2 + 18 cm^2 = **36 cm^2**

17.

$$\begin{array}{r} 6.7 \\ -\ 4.3 \\ \hline 2.4 \end{array}$$

a = **2.4**

18.

$$\begin{array}{r} \overset{1}{4}.7 \\ +\ 3.6 \\ \hline 8.3 \end{array}$$

m = **8.3**

19. $\begin{array}{r} 0.45 \\ 10\overline{)4.50} \\ \underline{4\,0} \\ 50 \\ \underline{50} \\ 0 \end{array}$

$w = \mathbf{0.45}$

20. $\begin{array}{r} 2.5 \\ \times\ 2.5 \\ \hline 125 \\ \underline{50\ } \\ 6.25 \end{array}$

$x = \mathbf{6.25}$

21. $\begin{array}{r} {}_1\overset{1}{5}.37 \\ 27.7 \\ +\ \ 4. \\ \hline \mathbf{37.07} \end{array}$

22. $\begin{array}{r} \mathbf{0.25} \\ 5\overline{)1.25} \\ \underline{1\,0} \\ 25 \\ \underline{25} \\ 0 \end{array}$

23. $\frac{5}{9} \cdot 6 \cdot 2\frac{1}{10}$

$\frac{\overset{1}{\cancel{5}}}{\underset{1}{\underset{\cancel{3}}{\cancel{9}}}} \cdot \frac{\overset{1}{\overset{\cancel{2}}{\cancel{6}}}}{1} \cdot \frac{\overset{7}{\cancel{21}}}{\underset{1}{\underset{\cancel{2}}{\cancel{10}}}} = \mathbf{7}$

24.

$\frac{5}{8} = \frac{5}{8}$

$\frac{3}{4} = \frac{6}{8}$

$+\ \frac{1}{2} = \frac{4}{8}$

$\frac{15}{8} = \mathbf{1\frac{7}{8}}$

25. $5 \div 3\frac{1}{3} = \frac{5}{1} \div \frac{10}{3} = \frac{\overset{1}{\cancel{5}}}{1} \times \frac{3}{\underset{2}{\cancel{10}}}$

$= \frac{3}{2} = \mathbf{1\frac{1}{2}}$

26.

$\frac{1}{2} = \frac{5}{10}$

$-\ \frac{1}{5} = \frac{2}{10}$

$\frac{3}{10}$

$\frac{3}{10} - \frac{3}{10} = \mathbf{0}$

27. **A.** $\mathbf{4 \cdot 4^2}$

28. (a) **125 mL**

(b) $\begin{array}{r} 1000\text{ mL} \\ -\ \ 125\text{ mL} \\ \hline \mathbf{875\text{ mL}} \end{array}$

29. **Switch 916.42 and 916.37**

30.

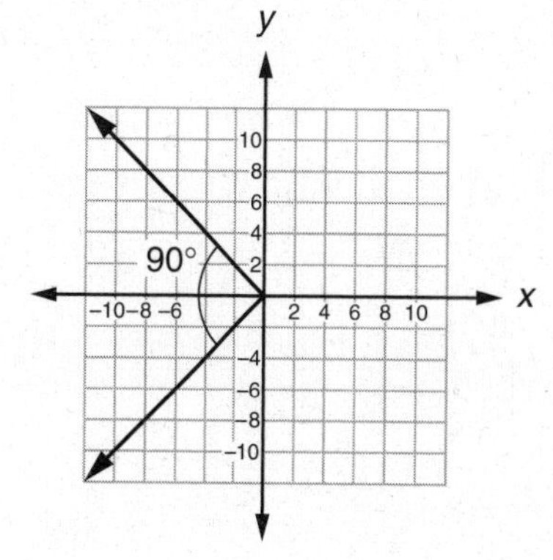

LESSON 39, WARM-UP

a. **$2.45**

b. **$7.20**

c. **35**

d. $\mathbf{\frac{3}{4}}$

e. **16**

f. **32**

g. **0**

Problem Solving

Even, 2–4–6, yes

Odd, 1–3–5, yes

LESSON 39, LESSON PRACTICE

a.

$\frac{a}{12} = \frac{6}{8}$

$8 \cdot a = 12 \cdot 6$

$8a = 72$

$a = \frac{72}{8}$

$a = \mathbf{9}$

$\frac{9}{12} = \frac{6}{8}$

b.
$$\frac{30}{b} = \frac{20}{16}$$
$$20 \cdot b = 30 \cdot 16$$
$$20b = 480$$
$$b = \frac{480}{20}$$
$$b = \mathbf{24}$$
$$\frac{30}{24} = \frac{20}{16}$$

c.
$$\frac{14}{21} = \frac{c}{15}$$
$$21 \cdot c = 15 \cdot 14$$
$$21c = 210$$
$$c = \frac{210}{21}$$
$$c = \mathbf{10}$$
$$\frac{14}{21} = \frac{10}{15}$$

d.
$$\frac{30}{25} = \frac{24}{d}$$
$$30 \cdot d = 24 \cdot 25$$
$$30d = 600$$
$$d = \mathbf{20}$$
$$\frac{30}{25} = \frac{24}{20}$$

e.
$$\frac{30}{100} = \frac{n}{40}$$
$$100 \cdot n = 40 \cdot 30$$
$$100n = 1200$$
$$n = \frac{1200}{100}$$
$$n = \mathbf{12}$$
$$\frac{30}{100} = \frac{12}{40}$$

f.
$$\frac{m}{100} = \frac{9}{12}$$
$$12 \cdot m = 100 \cdot 9$$
$$12m = 900$$
$$m = \frac{900}{12}$$
$$m = \mathbf{75}$$
$$\frac{75}{100} = \frac{9}{12}$$

LESSON 39, MIXED PRACTICE

1.
```
  64 inches
− 61 inches
   3 inches
```
3 inches

2. **Between his thirteenth and fourteenth birthdays**

3. $\dfrac{\text{princes}}{\text{princesses}} = \dfrac{12}{16} = \mathbf{\dfrac{3}{4}}$

4.
```
   497
   513
   436
 + 410
  1856 miles
```
```
   464 miles
4)1856
  16
   25
   24
    16
    16
     0
```
464 miles

5.
```
    $4.50
  ×    52
      900
    2250
  $234.00
```
$234.00

6.

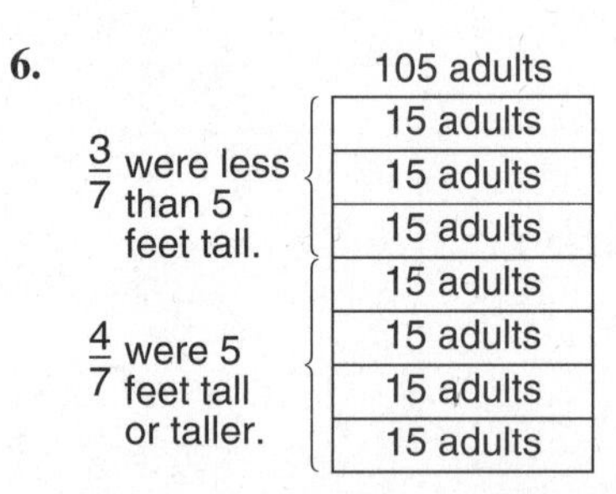

(a) 3×15 adults $=$ **45 adults**

(b) 4×15 adults $=$ **60 adults**

7.

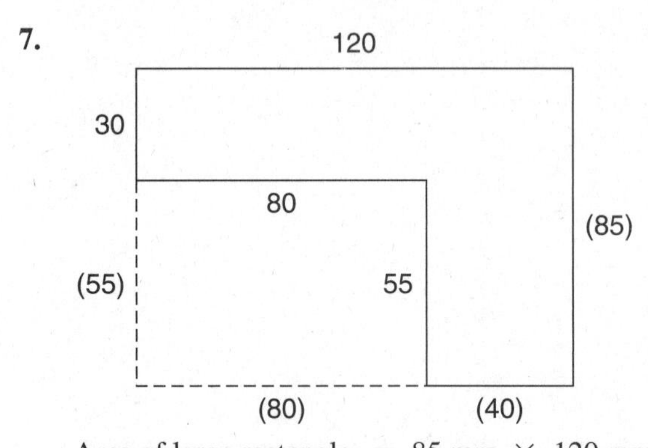

Area of large rectangle $= 85 \text{ mm} \times 120 \text{ mm}$
$= 10{,}200 \text{ mm}^2$

Area of small rectangle $= 55 \text{ mm} \times 80 \text{ mm}$
$= 4400 \text{ mm}^2$

Area of figure $= 10{,}200 \text{ mm}^2 - 4400 \text{ mm}^2$
$= \mathbf{5800 \text{ mm}^2}$

8. Perimeter $= 120 \text{ mm} + 85 \text{ mm} + 40 \text{ mm} + 55 \text{ mm} + 80 \text{ mm} + 30 \text{ mm}$
$= \mathbf{410 \text{ mm}}$

9. (a) **2.5**

(b) $2\frac{5}{10} = \mathbf{2\frac{1}{2}}$

10. (a) $0.91\underline{6}ⓖ6666 \longrightarrow$ **0.92**

(b) $0.9166\underline{6}⑥6 \longrightarrow$ **0.91667**

11. $9.16 \longrightarrow 9$

$\$1.39\frac{9}{10} \longrightarrow \$1.40$

$$\begin{array}{r} \$1.40 \\ \times \quad 9 \\ \hline \mathbf{\$12.60} \end{array}$$

12. (a) **100.075**

(b) **0.175**

13. (a) $\angle RPS$ (or $\angle SPR$)

(b) $\angle QPR$ (or $\angle RPQ$)

(c) $\angle QPS$ (or $\angle SPQ$)

14. **Each term can be found by dividing the previous term by 10.**

$0.1 \div 10 = 0.01$

$0.01 \div 10 = 0.001$

$0.001 \div 10 = 0.0001$

0.01, 0.001, 0.0001

15.
$$\frac{8}{12} = \frac{6}{x}$$
$$8 \cdot x = 6 \cdot 12$$
$$8x = 72$$
$$x = \frac{72}{8}$$
$$x = \mathbf{9}$$

16.
$$\frac{16}{y} = \frac{2}{3}$$
$$2 \cdot y = 16 \cdot 3$$
$$2y = 48$$
$$y = \frac{48}{2}$$
$$y = \mathbf{24}$$

17.
$$\frac{21}{14} = \frac{n}{4}$$
$$14 \cdot n = 4 \cdot 21$$
$$14n = 84$$
$$n = \frac{84}{14}$$
$$n = \mathbf{6}$$

18.
$$\begin{array}{r} 0.\overset{6}{\cancel{7}}{}^{1}5 \\ -\ 0.\ 3\ 6 \\ \hline 0.\ 3\ 9 \end{array}$$
$m = \mathbf{0.39}$

19.
$$\begin{array}{r} \overset{0}{\cancel{1}}.{}^{1}4 \\ -\ 0.\ 8 \\ \hline 0.\ 6 \end{array}$$
$w = \mathbf{0.6}$

20.
$$\begin{array}{r} 0.9 \\ 8\overline{)7.2} \\ 7\,2 \\ \hline 0 \end{array}$$
$x = \mathbf{0.9}$

21.
$$\begin{array}{r} 1.2 \\ \times\ 0.4 \\ \hline 0.48 \end{array}$$
$y = \mathbf{0.48}$

22. $9.6 + 12 + 8.59$

$\downarrow \quad \downarrow \quad \downarrow$

$10 + 12 + 9 = \mathbf{31}$

$$\begin{array}{r} \overset{1}{{}_{2}9}.6 \\ 12. \\ +\ 8.59 \\ \hline \mathbf{30.19} \end{array}$$

23. $3.15 - (2.1 - 0.06)$

$\downarrow \quad \downarrow \quad \downarrow$

$3 - (2 - 0) = \mathbf{1}$

$$\begin{array}{r} 2.{}^{0}\cancel{1}{}^{1}0 \\ -\ 0.\ 0\ 6 \\ \hline 2.\ 0\ 4 \end{array} \qquad \begin{array}{r} 3.15 \\ -\ 2.04 \\ \hline \mathbf{1.11} \end{array}$$

24.
$$\begin{array}{rcl} 4\frac{5}{12} & = & 4\frac{10}{24} \\ +\ 6\frac{5}{8} & = & 6\frac{15}{24} \\ \hline & & 10\frac{25}{24} = \mathbf{11\frac{1}{24}} \end{array}$$

25.
$$\begin{array}{rcl} 4\frac{1}{4} & = & 4\frac{5}{20} \longrightarrow \\ -\ 1\frac{3}{5} & = & 1\frac{12}{20} \\ \hline \end{array} \qquad \begin{array}{r} 3\frac{25}{20} \\ -\ 1\frac{12}{20} \\ \hline \mathbf{2\frac{13}{20}} \end{array}$$

26. $8\frac{1}{3} \cdot 1\frac{4}{5} = \frac{\overset{5}{\cancel{25}}}{\underset{1}{\cancel{3}}} \cdot \frac{\overset{3}{\cancel{9}}}{\underset{1}{\cancel{5}}} = \mathbf{15}$

27. $5\frac{5}{6} \div 7 = \frac{35}{6} \div \frac{7}{1}$

$= \frac{\overset{5}{\cancel{35}}}{6} \times \frac{1}{\underset{1}{\cancel{7}}} = \mathbf{\frac{5}{6}}$

28. (a) Perimeter = 15 mm + 15 mm + 18 mm
= **48 mm**

(b) Area $= \frac{18 \text{ mm} \cdot 12 \text{ mm}}{2} = \mathbf{108\ mm^2}$

29. Odd primes: 3, 5

$\frac{2}{6} = \mathbf{\frac{1}{3}}$

30. $\frac{2}{3} = \frac{8}{12}$

$\frac{1}{2} = \frac{6}{12}$

$\frac{5}{6} = \frac{10}{12}$

$\mathbf{\frac{1}{2}, \frac{7}{12}, \frac{2}{3}, \frac{5}{6}}$

LESSON 40, WARM-UP

a. **$18.00**

b. **$5.00**

c. **18**

d. $\mathbf{\frac{3}{8}}$

e. **23**

f. **40**

g. **$4.30**

Problem Solving

Since the ones digit of the second minuend is 8, the ones digit of the dividend must be 8 and the ones digit of the quotient must be 4. Since the second minuend has only two digits, the ones digit of the divisor must be 1. We can multiply 24 by 17 to find the other digits.

```
    24
17)408
   34
    68
    68
     0
```

LESSON 40, LESSON PRACTICE

a. **Each angle measures 60° because the angles equally share 180°.**

$\mathbf{\frac{180°}{3} = 60°}$

b. **20°. Angle *ACB* and ∠*ACD* are complementary:**

90° − 70° = 20°

c. **Angle *CAB* measures 70° because it is the third angle of a triangle whose other angles measure 90° and 20°:**

180° − (90° + 20°) = 180° − 110° = 70°

d. **They are not vertical angles. Their angles are equal in measure, but they are not nonadjacent angles formed by two intersecting lines.**

e. Angle x is the third angle of a triangle:

$m\angle x = 180° - (40° + 80°)$

$= 180° - 120° = \mathbf{60°}$

Angle x and $\angle y$ are supplementary angles:

$m\angle y = 180° - 60° = \mathbf{120°}$

Angles z and x are vertical angles:

$m\angle z = \mathbf{60°}$

LESSON 40, MIXED PRACTICE

1. (a) 3 red marbles
+ 2 white marbles
5 total marbles

$\frac{\text{white marbles}}{\text{total marbles}} = \mathbf{\frac{2}{5}}$

(b) $\mathbf{\frac{2}{5}}$

2. (a) 6 minutes = 6(60 seconds)
= 360 seconds

6 minutes + 20 seconds
= **380 seconds**

(b) $\frac{380 \text{ seconds}}{4} = \mathbf{95\ seconds}$

3.

```
   18
× 24 miles
   72
  36
 432 miles
```

432 miles

4.
$$\begin{array}{r} 103.4°\text{F} \\ -\ 98.6°\text{F} \\ \hline \mathbf{4.8°F} \end{array}$$

5. (a) Perimeter = 35 mm + 35 mm + 70 mm + 70 mm = **210 mm**

(b) Area = 35 mm × 70 mm = **2450 mm²**

6.

200 sheep	
$\frac{5}{8}$ grazed.	25 sheep
	25 sheep
	25 sheep
	25 sheep
	25 sheep
$\frac{3}{8}$ drank.	25 sheep
	25 sheep
	25 sheep

(a) $5 \cdot 25$ sheep = **125 sheep**

(b) $3 \cdot 25$ sheep = **75 sheep**

7. BC = 100 mm − (30 mm + 45 mm)
= 100 mm − 75 mm = 25 mm
25 mm = **2.5 cm**

8. (a) 0.083(3)33 ⟶ **0.083**

(b) 0.0(8)3333 ⟶ **0.1**

9. (a) **Twelve and fifty-four thousandths**

(b) **Ten and eleven hundredths**

10.

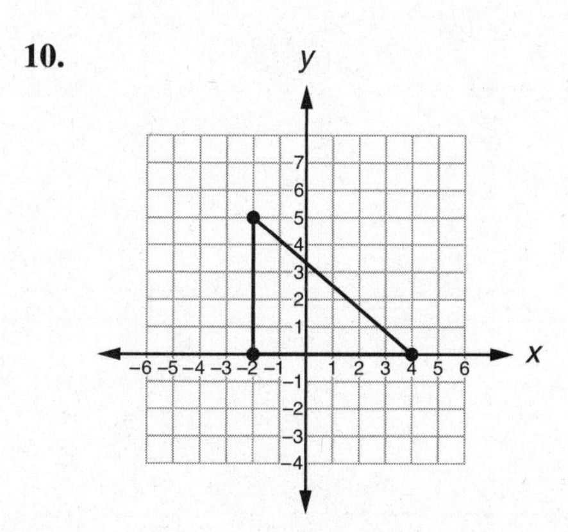

Area = $\frac{6 \text{ units} \cdot 5 \text{ units}}{2}$ = **15 units²**

11. **0.76**

12. (a) $\angle ACB$ = 180° − (70° + 75°)
= 180° − 145° = **35°**

(b) $\angle ACD$ = 180° − 35° = **145°**

(c) $\angle DCE$ = **35°**

13. $\angle BCE$ (or $\angle ECB$)

14. (a) **Identity property of multiplication**

(b) $5\overline{)120}$ = **24**

15. $10w = 25 \cdot 8$

$w = \frac{25 \cdot 8}{10}$

$w = \mathbf{20}$

16. $9n = 1.5 \cdot 6$

$n = \frac{(1.5)(6)}{9}$

$n = \mathbf{1}$

17. $9m = 12 \cdot 15$

$m = \frac{12 \cdot 15}{9}$

$m = \mathbf{20}$

18.
$$\begin{array}{r} \overset{3}{\cancel{4}}.{}^{1}0 \\ -\ 1.8 \\ \hline 2.2 \end{array}$$

$a = \mathbf{2.2}$

19.
$$\begin{array}{r} \overset{1}{3}.9 \\ +\ 0.39 \\ \hline 4.29 \end{array}$$

$t = \mathbf{4.29}$

20. 12 cm = 0.12 m

$$\begin{array}{r} 1.{}^{1}\cancel{2}{}^{1}0 \text{ m} \\ -\ 0.12 \text{ m} \\ \hline \mathbf{1.08 \text{ m}} \end{array}$$

21.
$$\begin{array}{r} 0.15 \\ \times\ 0.05 \\ \hline \mathbf{0.0075} \end{array}$$

22.
$$\begin{array}{r} 15 \\ \times\ 1.5 \\ \hline 75 \\ 15\ \\ \hline \mathbf{22.5} \end{array}$$

23.
$$\begin{array}{r} \mathbf{1.2} \\ 12\overline{)14.4} \\ 12\ \ \\ \hline 2\,4 \\ 2\,4 \\ \hline 0 \end{array}$$

24.
$$\begin{array}{r} \overset{3}{\cancel{4}}.\overset{9}{\cancel{0}}{}^{1}0 \\ -\ 1.25 \\ \hline 2.75 \end{array} \qquad \begin{array}{r} \overset{4}{\cancel{5}}.\overset{15}{\cancel{6}}{}^{1}0 \\ -\ 2.75 \\ \hline \mathbf{2.85} \end{array}$$

25.
$$\begin{array}{r} 3.14 \\ +\ 1.20 \\ \hline 4.34 \end{array} \qquad \begin{array}{r} \overset{4}{\cancel{5}}.\overset{9}{\cancel{0}}{}^{1}0 \\ -\ 4.34 \\ \hline \mathbf{0.66} \end{array}$$

26. $6\frac{1}{4} \cdot 1\frac{3}{5} = \frac{\overset{5}{\cancel{25}}}{\underset{1}{\cancel{4}}} \cdot \frac{\overset{2}{\cancel{8}}}{\underset{1}{\cancel{5}}} = \frac{10}{1} = \mathbf{10}$

27. $7 \div 5\frac{5}{6} = \frac{7}{1} \div \frac{35}{6} = \frac{\overset{1}{\cancel{7}}}{1} \cdot \frac{6}{\underset{5}{\cancel{35}}}$

$= \frac{6}{5} = \mathbf{1\frac{1}{5}}$

28.
$$\begin{array}{rcl} \frac{8}{15} & = & \frac{40}{75} \\ +\ \frac{12}{25} & = & \frac{36}{75} \\ \hline & & \frac{76}{75} = \mathbf{1\frac{1}{75}} \end{array}$$

29.
$$\begin{array}{rcl} 4\frac{2}{5} & = & 4\frac{8}{20} \\ -\ 1\frac{3}{4} & = & -1\frac{15}{20} \\ \hline \end{array} \longrightarrow \begin{array}{r} 3\frac{28}{20} \\ -\ 1\frac{15}{20} \\ \hline \mathbf{2\frac{13}{20}} \end{array}$$

30. $3\underline{1}.⑨75 \longrightarrow 32$

$\frac{1}{4} \times 32 = \frac{1}{\underset{1}{\cancel{4}}} \times \frac{\overset{8}{\cancel{32}}}{1} = \mathbf{8}$

INVESTIGATION 4

1. **41**

2. $58 - 21 = \mathbf{37}$

3. There are 35 scores in the list. Half of 35 is $17\frac{1}{2}$. This means there are 17 scores below the median and 17 scores above the median. We count 17 scores up from the lowest score and find that **40** is the median.

4. There are 17 scores in each quartile. Half of 17 is $8\frac{1}{2}$. We count 8 scores up from the lowest score and find that **32** is the first quartile. We count 8 scores up from the median and find that **47** is the third quartile.

5.

Stem	Leaf
1	5
2	6 6 7 8 9 9
3	0|2 3 5 6 8 8 8 8|
4	0 1 2 3 5 5 6 7|7 8
5	0 2 4 5 7 8

2 | 9 represents a score of 29

6. There are 32 scores in the list. Since there are an even number of scores, the median is the mean of the two middle scores, which are 38 and 40. So the median is $(38 + 40) \div 2$, or **39.** We find the quartiles using the same method. The lower quartile is the mean of the two middle scores in the lower half, which are 30 and 32. So the lower quartile is $(30 + 32) \div 2$, or **31.** The upper quartile is the mean of the two middle scores in the upper half, which are 47 and 47. So the upper quartile is $(47 + 47) \div 2$, or **47.**

7. **38**

8. **58 and 15**

9. $58 - 15 = \mathbf{43}$

10. $47 - 31 = \mathbf{16}$

11.

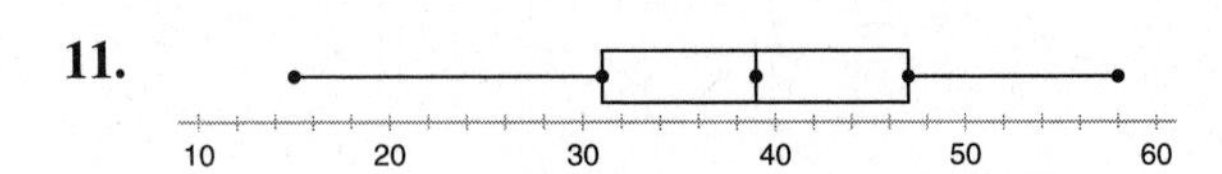

12. **15**

LESSON 41, WARM-UP

a. **700**

b. **15.4**

c. **25**

d. **9**

e. **1200**

f. **21**

g. $\mathbf{1\frac{1}{6}, \frac{1}{6}, \frac{1}{3}, 1\frac{1}{3}}$

Problem Solving

1 + 2 + 3 + 4 + 5 + 6 + 7 + 8 + 9
= **45 blocks**

Lesson 41, Lesson Practice

a. $A = (15 \text{ in.})(8 \text{ in.}) = \mathbf{120 \text{ in.}^2}$

b. $\frac{(6 \text{ ft})(8 \text{ ft})}{2} = \mathbf{24 \text{ ft}^2}$

c. **One possibility:** $x(y + z) = xy + xz$

d. **6(15) = 90**
(6 · 20) − (6 · 5) = 120 − 30 = 90

e. $p = 2(l + w)$
$p = 2l + 2w$

f. **One way is to add 6 and 4. Then multiply the sum by 2. Another way is to multiply 6 by 2 and 4 by 2. Then add the products.**

Lesson 41, Mixed Practice

1. $\frac{\text{gazelles}}{\text{wildebeests}} = \frac{150}{200} = \mathbf{\frac{3}{4}}$

2. 105 points
112 points
98 points
113 points
+ 107 points
535 points

$5\overline{)535 \text{ points}}$ = **107 points**

3. 19 feet 6 inches = 19 (12 inches) + 6 inches = **234 inches**

4. (a) **Associative property of addition**
(b) **Associative property of multiplication**
(c) **Distributive property**

5. 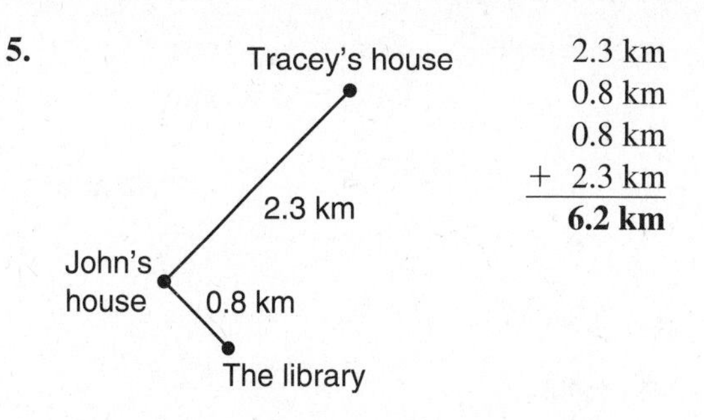

2.3 km
0.8 km
0.8 km
+ 2.3 km
6.2 km

6. (a) 100% − 70% = 30%
$30\% = \frac{30}{100} = \mathbf{\frac{3}{10}}$

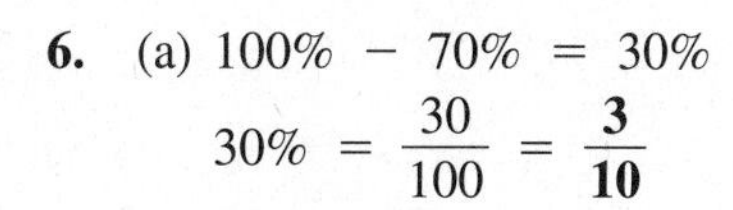

(b) $\frac{\text{water area}}{\text{land area}} = \mathbf{\frac{7}{3}}$

7. (a) **30 correct answers**
(b) **26 correct answers**
(c) **34 correct answers**
(d) **11 correct answers**

8. 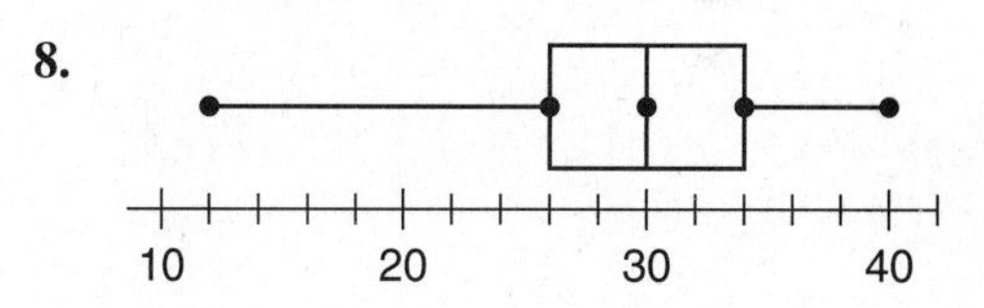

9. 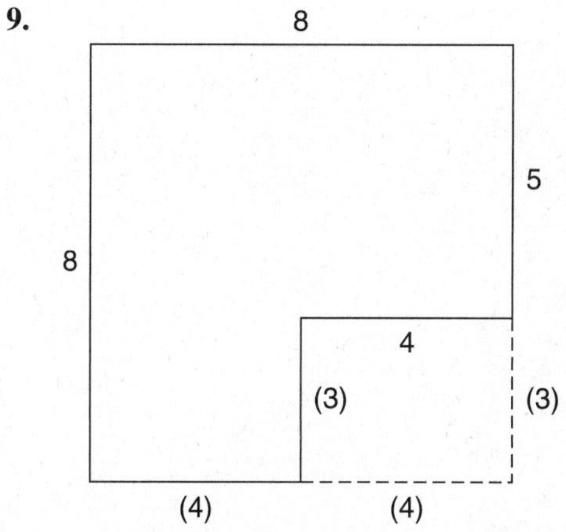

(a) Area of large rectangle = (8 ft)(8 ft) = 64 ft²
Area of small rectangle = (3 ft)(4 ft) = 12 ft²
Area of figure = 64 ft² − 12 ft² = **52 ft²**

(b) Perimeter = 8 ft + 5 ft + 4 ft + 3 ft + 4 ft + 8 ft = **32 ft**

10. (a) **3.6**
(b) $3\frac{6}{10} = \mathbf{3\frac{3}{5}}$

11. 2 + 3 + 5 + 7 = **17**

12. (a) Perimeter $= 13 \text{ mm} + 15 \text{ mm} + 14 \text{ mm} = \mathbf{42 \text{ mm}}$

(b) Area $= \dfrac{(14 \text{ mm})(12 \text{ mm})}{2} = \mathbf{84 \text{ mm}^2}$

13. (a) **0.00067**

(b) **100.023**

14. $2\pi r = 2(3.14)(10)$
$= 2(31.4) = \mathbf{62.8}$

15. $\dfrac{3}{5} \cdot \dfrac{14}{14} = \dfrac{42}{70}$

$\dfrac{1}{2} \cdot \dfrac{35}{35} = \dfrac{35}{70}$

$\dfrac{5}{7} \cdot \dfrac{10}{10} = \dfrac{50}{70}$

$\mathbf{\dfrac{35}{70}, \dfrac{42}{70}, \dfrac{50}{70}}$

16. Area $= (5.6 \text{ cm})(3.4 \text{ cm}) = \mathbf{19.04 \text{ cm}^2}$

17. $\dfrac{x}{2.4} = \dfrac{10}{16}$

$16x = 2.4(10)$

$x = \dfrac{24}{16}$

$x = 1\dfrac{8}{16} = 1\dfrac{1}{2} = \mathbf{1.5}$

18. $\dfrac{18}{8} = \dfrac{m}{20}$

$8m = 18 \cdot 20$

$m = \dfrac{360}{8}$

$m = \mathbf{45}$

19. $7.\overset{5}{\cancel{6}}{}^{1}0 - 3.45 = 4.15$

$a = \mathbf{4.15}$

20. $3\overline{)0.144} = \mathbf{0.048}$

12
24
24
0

$y = \mathbf{0.048}$

21. $8\overline{)7.400} = \mathbf{0.925}$

7 2
20
16
40
40
0

22. $0.4 \times 0.6 = 0.24$ $\qquad 0.24 \times 0.02 = \mathbf{0.0048}$

23. $5\overline{)4.315} = \mathbf{0.863}$

4 0
31
30
15
15
0

24. $100\overline{)6.500} = \mathbf{0.065}$

6 00
500
500
0

25. $3\dfrac{1}{3} = 3\dfrac{4}{12}$

$1\dfrac{5}{6} = 1\dfrac{10}{12}$

$+ \dfrac{7}{12} = \dfrac{7}{12}$

$4\dfrac{21}{12} = 5\dfrac{9}{12} = \mathbf{5\dfrac{3}{4}}$

26. $4 \longrightarrow 3\dfrac{4}{4}$

$-1\dfrac{1}{4} \qquad -1\dfrac{1}{4}$

$2\dfrac{3}{4}$

$4\dfrac{1}{6} = 4\dfrac{2}{12} \longrightarrow 3\dfrac{14}{12}$

$-2\dfrac{3}{4} = 2\dfrac{9}{12} \qquad -2\dfrac{9}{12}$

$\mathbf{1\dfrac{5}{12}}$

27. $3\dfrac{1}{5} \cdot 2\dfrac{5}{8} \cdot 1\dfrac{3}{7}$

$= \dfrac{\overset{2}{\cancel{16}}}{\underset{1}{\cancel{5}}} \cdot \dfrac{\overset{3}{\cancel{21}}}{\underset{1}{\cancel{8}}} \cdot \dfrac{\overset{2}{\cancel{10}}}{\underset{1}{\cancel{7}}} = \mathbf{12}$

28. $4\frac{1}{2} \div 6 = \frac{9}{2} \div \frac{6}{1} = \frac{\overset{3}{\cancel{9}}}{2} \times \frac{1}{\underset{2}{\cancel{6}}} = \mathbf{\frac{3}{4}}$

29. (a) $(12 \cdot 7) + (12 \cdot 13)$
$= 84 + 156 = 240$
or
$12(7 + 13) = 12 \cdot 20 = 240$
$(12 \cdot 7) + (12 \cdot 13) \mathbf{=} 12(7 + 13)$

(b) **Distributive property**

30. $m\angle x = 180° - (90° + 42°)$
$= 180° - 132° = \mathbf{48°}$
$m\angle y = 180° - 48° = \mathbf{132°}$
$m\angle z = m\angle x = \mathbf{48°}$

Lesson 42, Warm-Up

a. **$2.34**

b. **40**

c. **5**

d. **2**

e. **6**

f. **28**

g. **25**

Problem Solving

There are 16 cups in a gallon.
$16 - 8 = 8$ cups
$8 - 4 = 4$ cups
$4 - 2 = 2$ cups
$2 - 1 =$ **1 cup will be left in the gallon container.**

Lesson 42, Lesson Practice

a. $\mathbf{2.\overline{72}}$

b. $\mathbf{0.8\overline{16}}$

c. $0.\overline{6} = 0.6666\ldots$
$0.66\underline{6}⑥\ldots \longrightarrow$ **0.667**

d. $5.3\overline{81} = 5.38181\ldots$
$5.38\underline{1}⑧1\ldots \longrightarrow$ **5.382**

e.
$$\begin{array}{r} 0.141666\ldots \\ 12\overline{)1.7000000\ldots} \\ \underline{12} \\ 50 \\ \underline{48} \\ 20 \\ \underline{12} \\ 80 \\ \underline{72} \\ 80 \\ \underline{72} \\ 80 \end{array}$$
$\mathbf{0.141\overline{6}}$

f. $0.141\underline{6}⑥6\ldots \longrightarrow$ **0.1417**

Lesson 42, Mixed Practice

1.
$$\begin{array}{r} 2 \text{ boys} \\ \underline{+\ ?\text{ girls}} \\ 5 \text{ total} \end{array} \longrightarrow \begin{array}{r} 2 \text{ boys} \\ \underline{+\ 3\text{ girls}} \\ 5 \text{ total} \end{array}$$
$\frac{\text{boys}}{\text{girls}} = \mathbf{\frac{2}{3}}$

2.
$$\begin{array}{r} \textbf{27 magazines} \\ 16\overline{)432} \\ \underline{32} \\ 112 \\ \underline{112} \\ 0 \end{array}$$

3.
$$\begin{array}{r} 23 \text{ miles} \\ \underline{\times\ 7}\phantom{\text{ miles}} \\ \textbf{161 miles} \end{array}$$

4.

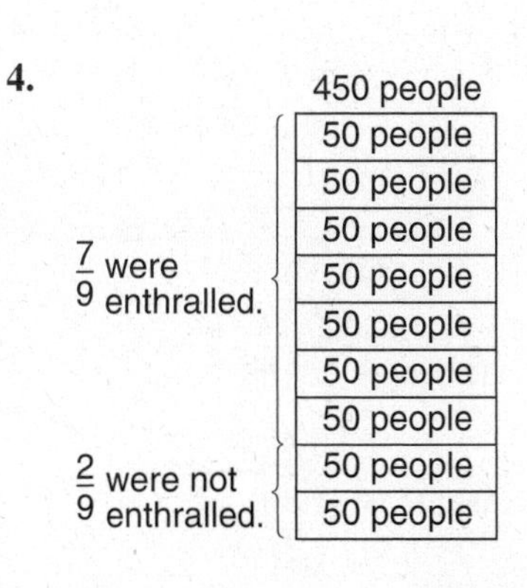

(a) 7(50 people) = **350 people**

(b) 2(50 people) = **100 people**

5. (a) $5.1\overline{6} = 5.16666\ldots$

$5.166\underline{6}⑥\ldots \longrightarrow$ **5.1667**

(b) $5.\overline{27} = 5.272727\ldots$

$5.272\underline{7}②7\ldots \longrightarrow$ **5.2727**

6. (a)

$$\begin{array}{r} 19 \text{ tests} \\ -\ 11 \text{ tests} \\ \hline \mathbf{8 \text{ tests}} \end{array}$$

(b) $\frac{9}{30} = \mathbf{\frac{3}{10}}$

7. Area $= \frac{(6 \text{ units})(6 \text{ units})}{2} =$ **18 units²**

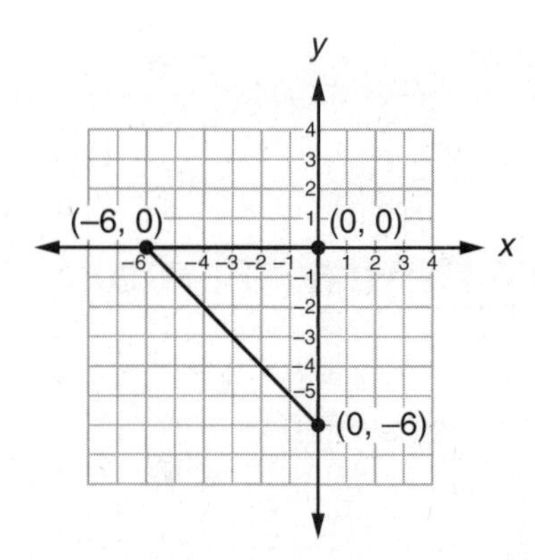

8. (a) Perimeter $= 10 \text{ in.} + 8 \text{ in.} + 10 \text{ in.} + 20 \text{ in.} + 20 \text{ in.} + 12 \text{ in.}$

$=$ **80 in.**

(b)

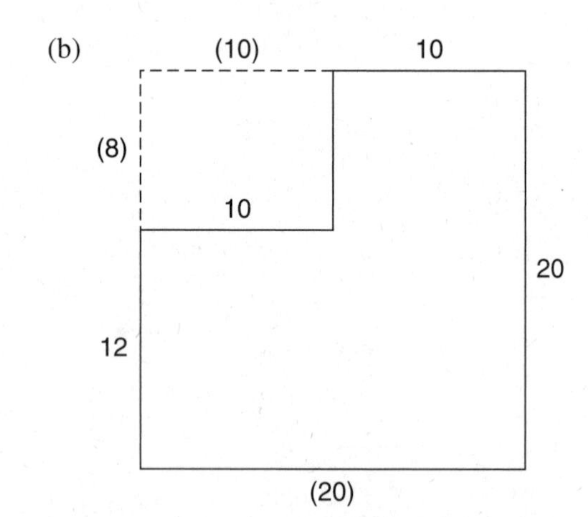

Area of large rectangle $= 20 \text{ in.} \times 20 \text{ in.} = 400 \text{ in.}^2$

Area of small rectangle $= 10 \text{ in.} \times 8 \text{ in.} = 80 \text{ in.}^2$

Area of figure $= 400 \text{ in.}^2 - 80 \text{ in.}^2$

$=$ **320 in.²**

9.

$$\begin{array}{r} 0.15454\ldots \\ 11\overline{)1.700000\ldots} \\ \underline{1\,1} \\ 60 \\ \underline{55} \\ 50 \\ \underline{44} \\ 60 \\ \underline{55} \\ 50 \\ \underline{44} \\ 60 \end{array}$$

(a) **$0.1\overline{54}$**

(b) $0.15\underline{4}⑤4\ldots \longrightarrow$ **0.155**

10.

$$\begin{array}{r} \overset{1}{0.027} \\ +\ 0.58 \\ \hline \mathbf{0.607} \end{array}$$

11. $\frac{2}{6} = \mathbf{\frac{1}{3}}$

12. (a) **One possibility:**

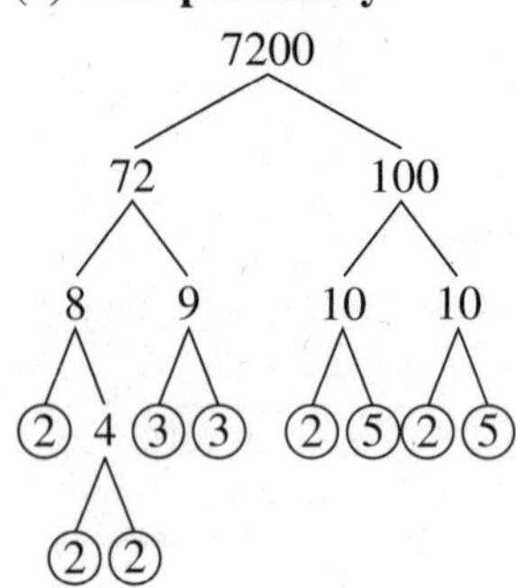

(b) $\mathbf{2^5 \cdot 3^2 \cdot 5^2}$

13.

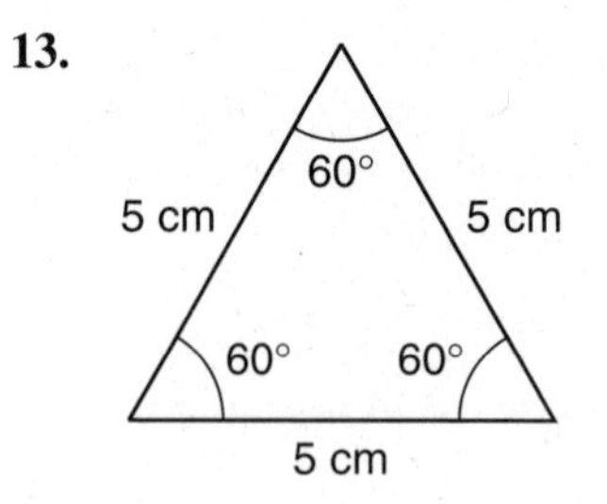

14. $12 = 2 \cdot 2 \cdot 3$

$15 = 3 \cdot 5$

$\text{LCM}(12, 15) = 2 \cdot 2 \cdot 3 \cdot 5 =$ **60**

15. $\frac{21}{24} = \frac{w}{40}$

$24w = 21 \cdot 40$

$w = \frac{840}{24}$

$w =$ **35**

16. $\dfrac{1.2}{x} = \dfrac{9}{6}$

$9x = (1.2)6$

$x = \dfrac{7.2}{9}$

$x = \mathbf{0.8}$

17. $\begin{array}{r} \overset{0}{\cancel{1}}\,\overset{1}{\cancel{4}}\overset{3}{.}{}^{1}0 \\ -\ \ 9.\ 6 \\ \hline 4.\ 4 \end{array}$

$m = \mathbf{4.4}$

18. $\begin{array}{r} 1.63 \\ +\ 4.2 \\ \hline 5.83 \end{array}$

$n = \mathbf{5.83}$

19. $\frac{1}{2}(12)(10) = \frac{1}{2}(120) = \mathbf{60}$

20. $\mathbf{4(11) = 44}$

or

$\mathbf{(4 \cdot 5) + (4 \cdot 6) = 20 + 24 = 44}$

21. **343**

22. $\begin{array}{r} \$4.56 \\ \times\ 0.08 \\ \hline \$0.3648 \end{array} \longrightarrow \mathbf{\$0.36}$

23. $\mathbf{24 \div 6 = 4}$

24. (a) **Diameter**

(b) **Radius**

25. $\begin{array}{r} \mathbf{1.775} \\ 4\overline{)7.100} \\ \underline{4} \\ 3\,1 \\ \underline{2\,8} \\ 30 \\ \underline{28} \\ 20 \\ \underline{20} \\ 0 \end{array}$

26. $6\frac{1}{4} = 6\frac{3}{12}$

$5\frac{5}{12} = 5\frac{5}{12}$

$+\ \frac{2}{3} = \frac{8}{12}$

$11\frac{16}{12} = 12\frac{4}{12} = \mathbf{12\frac{1}{3}}$

27. $4\frac{1}{6} = 4\frac{2}{12} \longrightarrow 3\frac{14}{12}$

$-\ 1\frac{1}{4} = 1\frac{3}{12} \qquad -\ 1\frac{3}{12}$

$2\frac{11}{12}$

$4 \longrightarrow 3\frac{12}{12}$

$-\ 2\frac{11}{12} \qquad -\ 2\frac{11}{12}$

$\mathbf{1\frac{1}{12}}$

28. $6\frac{2}{5} \cdot 2\frac{5}{8} \cdot 2\frac{6}{7} =$

$\dfrac{\overset{4}{\cancel{32}}}{\underset{1}{\cancel{5}}} \cdot \dfrac{\overset{3}{\cancel{21}}}{\underset{1}{\cancel{8}}} \cdot \dfrac{\overset{4}{\cancel{20}}}{\underset{1}{\cancel{7}}} = \mathbf{48}$

29. **The quotient is greater than 1 because the dividend is greater than the divisor.**

$6 \div 4\frac{1}{2} = \frac{6}{1} \div \frac{9}{2} = \frac{\overset{2}{\cancel{6}}}{1} \times \frac{2}{\underset{3}{\cancel{9}}} = \frac{4}{3}$

$= \mathbf{1\frac{1}{3}}$

30. $m\angle a = 180° - 40° = \mathbf{140°}$

$m\angle b = 180° - (90° + 40°)$

$= 180° - 130° = \mathbf{50°}$

$m\angle c = 180° - 50° = \mathbf{130°}$

Lesson 43, Warm-Up

a. **$2.88**

b. **0.035**

c. **7**

d. **39**

e. **$60.00**

f. **60**

g. $\mathbf{\frac{11}{12}, \frac{5}{12}, \frac{1}{6}, 2\frac{2}{3}}$

Problem Solving

The total can be any whole number from 4 (all ones) to 24 (all sixes), which is **21 different numbers.**

Lesson 43, Lesson Practice

a. $0.24 = \frac{24}{100} = \mathbf{\frac{6}{25}}$

b. $45.6 = 45\frac{6}{10} = \mathbf{45\frac{3}{5}}$

c. $2.375 = 2\frac{375}{1000} = \mathbf{2\frac{3}{8}}$

d.
$$\begin{array}{r} \mathbf{5.75} \\ 4\overline{)23.00} \\ \underline{20} \\ 3\,0 \\ \underline{2\,8} \\ 20 \\ \underline{20} \\ 0 \end{array}$$

e. $4\frac{3}{5} = \frac{23}{5}$

$$\begin{array}{r} \mathbf{4.6} \\ 5\overline{)23.0} \\ \underline{20} \\ 3\,0 \\ \underline{3\,0} \\ 0 \end{array}$$

f.
$$\begin{array}{r} \mathbf{0.625} \\ 8\overline{)5.000} \\ \underline{4\,8} \\ 20 \\ \underline{16} \\ 40 \\ \underline{40} \\ 0 \end{array}$$

g.
$$\begin{array}{r} 0.8333\ldots \\ 6\overline{)5.0000\ldots} \\ \underline{4\,8} \\ 20 \\ \underline{18} \\ 20 \\ \underline{18} \\ 20 \end{array}$$

$\mathbf{0.8\overline{3}}$

h. $8\% = \frac{8}{100} = \mathbf{0.08}$

i. $12.5\% = \frac{12.5}{100} = \mathbf{0.125}$

j. $150\% = \frac{150}{100} = \mathbf{1.50}$

k. $6\frac{1}{2}\% = 6.5\% = \frac{6.5}{100} = \mathbf{0.065}$

Lesson 43, Mixed Practice

1. $\frac{\text{Celtic soldiers}}{\text{total soldiers}} = \mathbf{\frac{2}{7}}$

2. (a) 11 minutes 44 seconds =
11(60 seconds) + 44 seconds =
704 seconds

(b)
$$\begin{array}{r} \textbf{88 seconds} \\ 8\overline{)704 \text{ seconds}} \\ \underline{64}\phantom{4 \text{ seconds}} \\ 64\phantom{\text{ seconds}} \\ \underline{64}\phantom{\text{ seconds}} \\ 0\phantom{\text{ seconds}} \end{array}$$

3.
$$\begin{array}{r} \overset{1}{\cancel{2}}\,\overset{1}{\cancel{1}}.{}^{1}0 \text{ gallons} \\ -\ 1\,3.\,3 \text{ gallons} \\ \hline \textbf{7. 7 gallons} \end{array}$$

4.
$$\begin{array}{r} 1{,}000{,}200{,}000 \text{ people} \\ -\ \ 725{,}000{,}000 \text{ people} \\ \hline \textbf{275,200,000 people} \end{array}$$

5.

15 games

$\frac{2}{3}$ won.	5 games
	5 games
$\frac{1}{3}$ lost.	5 games

(a) 2(5 games) = **10 games**

(b) $\frac{\text{won}}{\text{lost}} = \mathbf{\frac{2}{1}}$

6. (a) **13.5 seconds**

(b) **12.8 seconds**

(c) **14.7 seconds**

7.

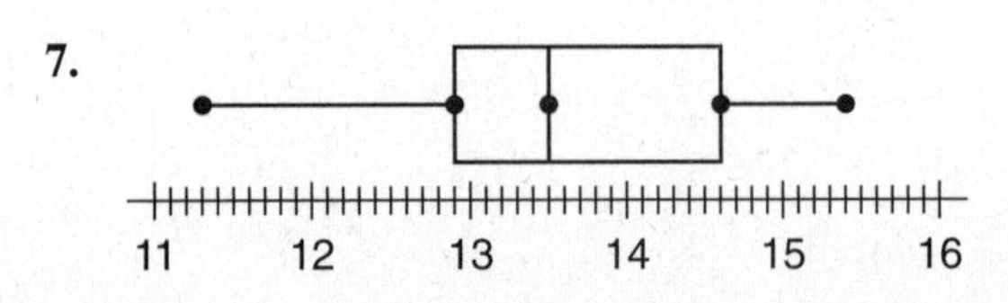

8. $\frac{120 \text{ mm}}{6} = 20 \text{ mm}$

Perimeter $= (4)(20 \text{ mm}) = \mathbf{80 \text{ mm}}$

9. (a) $0.375 = \frac{375}{1000} = \mathbf{\frac{3}{8}}$

(b) $5.55 = 5\frac{55}{100} = \mathbf{5\frac{11}{20}}$

10. (a) $2\frac{2}{5} = \frac{12}{5}$

$$\begin{array}{r} \mathbf{2.4} \\ 5\overline{)12.0} \\ \underline{10} \\ 2\,0 \\ \underline{2\,0} \\ 0 \end{array}$$

(b)
$$\begin{array}{r} \mathbf{0.125} \\ 8\overline{)1.000} \\ \underline{8} \\ 20 \\ \underline{16} \\ 40 \\ \underline{40} \\ 0 \end{array}$$

11. (a) $0.\overline{45} = 0.4545\ldots$

$0.454⑤\ldots \longrightarrow \mathbf{0.455}$

(b) $3.\overline{142857} =$

$3.142857142857\ldots$

$3.142⑧57142857 \longrightarrow \mathbf{3.143}$

12.
$$\begin{array}{r} 0.15833\ldots \\ 12\overline{)1.90000\ldots} \\ \underline{1\,2} \\ 70 \\ \underline{60} \\ 100 \\ \underline{96} \\ 40 \\ \underline{36} \\ 40 \\ \underline{36} \\ 40 \end{array}$$

(a) $\mathbf{0.158\overline{3}}$

(b) $0.158③\ldots \longrightarrow \mathbf{0.158}$

13.
$$\begin{array}{r} \overset{3}{\cancel{4}}.\overset{9}{\cancel{0}}\,\overset{14}{\cancel{5}}\,{}^{1}0 \\ -\ 0.\,1\,6\,7 \\ \hline \mathbf{3.\,8\,8\,3} \end{array}$$

14.

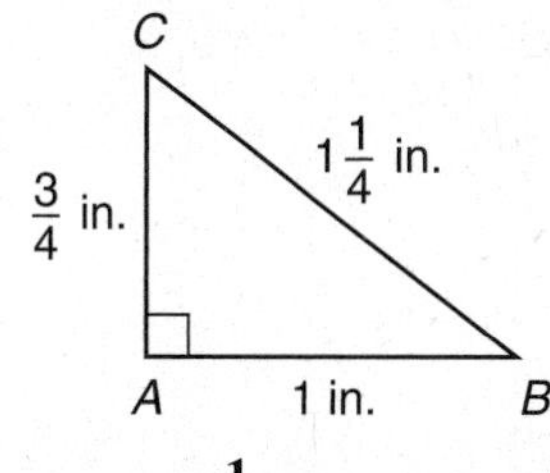

$BC = \mathbf{1\frac{1}{4} \text{ inches}}$

15. $\frac{26}{52} = \mathbf{\frac{1}{2}}$

16. (a) **One possibility:**

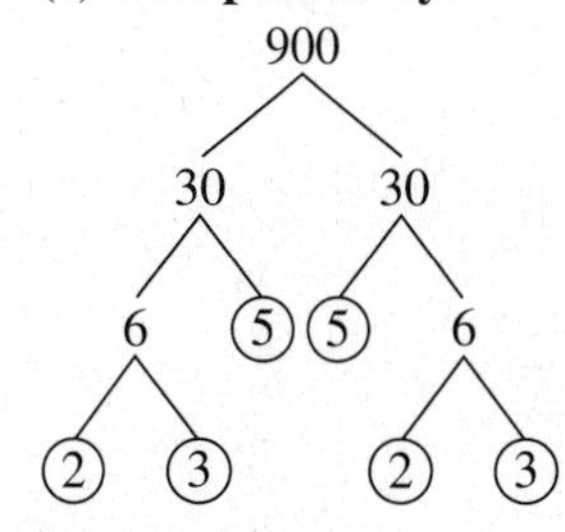

(b) $\mathbf{2^2 \cdot 3^2 \cdot 5^2}$

(c) $\sqrt{900} = 30 = \mathbf{2 \cdot 3 \cdot 5}$

17. 1 liter = 1000 milliliters

$\frac{1000 \text{ milliliters}}{2 \text{ milliliters}} = 500$

500 eyedroppers

18. (a) $8\% = \frac{8}{100} = \mathbf{0.08}$

(b)
$$\begin{array}{r} \$8.90 \\ \times\ 0.08 \\ \hline \$0.7120 \end{array} \longrightarrow \mathbf{\$0.71}$$

19. (a) Perimeter $= 0.6 \text{ m} + 1 \text{ m} + 0.8 \text{ m}$
$= \mathbf{2.4 \text{ m}}$

(b) Area $= \frac{(0.8 \text{ m})(0.6 \text{ m})}{2} = \mathbf{0.24 \text{ m}^2}$

20. $\frac{32}{2} = \frac{320}{20}$

The division problems are equivalent problems because the quotients are equal.

21. $2(3 + 4) = 2(7) = \mathbf{14}$

22. $\frac{10}{18} = \frac{c}{4.5}$

$18c = 10 \cdot 4.5$

$c = \frac{45}{18}$

$c = 2\frac{1}{2}$ or **2.5**

23.
$$\begin{array}{r} 1.\overset{8}{\cancel{9}}{}^{1}0 \\ -\ 0.42 \\ \hline 1.48 \end{array}$$
$w = \mathbf{1.48}$

24.
$$\begin{array}{r} \mathbf{1.625} \\ 4\overline{)6.500} \\ \underline{4} \\ 2\,5 \\ \underline{2\,4} \\ 10 \\ \underline{8} \\ 20 \\ \underline{20} \\ 0 \end{array}$$

25.
$$\begin{array}{rcl} 3\frac{3}{10} & = & 3\frac{9}{30} \\ -\ 1\frac{11}{15} & = & 1\frac{22}{30} \\ \hline \end{array} \longrightarrow \begin{array}{r} 2\frac{39}{30} \\ -\ 1\frac{22}{30} \\ \hline \mathbf{1\frac{17}{30}} \end{array}$$

26.
$$\begin{array}{rcl} 5\frac{1}{2} & = & 5\frac{5}{10} \\ 6\frac{3}{10} & = & 6\frac{3}{10} \\ +\ \frac{4}{5} & = & \frac{8}{10} \\ \hline \end{array}$$
$$11\frac{16}{10} = 12\frac{6}{10} = \mathbf{12\frac{3}{5}}$$

27. $7\frac{1}{2} \cdot 3\frac{1}{3} \cdot \frac{4}{5} = \frac{\overset{5}{\cancel{15}}}{\underset{1}{\cancel{2}}} \cdot \frac{\overset{\overset{1}{\cancel{2}}}{\cancel{10}}}{\underset{1}{\cancel{3}}} \cdot \frac{4}{\underset{1}{\cancel{5}}} = 20$

$20 \div 5 = \mathbf{4}$

28. **(5, 10)**

29. $m\angle a = 180° - 110° = \mathbf{70°}$
$m\angle b = 180° - (70° + 50°)$
$= 180° - 120° = \mathbf{60°}$
$m\angle c = 180° - 60° = \mathbf{120°}$

30. (a) **180°**

(b) **90°**

(c) **45°**

LESSON 44, WARM-UP

a. **$3.20**

b. **0.05**

c. **6**

d. **60**

e. **1**

f. **150**

g. **6**

Problem Solving

3 and 11

(Do not accept 1 and 13, because 1 is not prime.)

LESSON 44, LESSON PRACTICE

a.
$$\begin{array}{r} \mathbf{13\ R\ 3} \\ 4\overline{)55} \\ \underline{4} \\ 15 \\ \underline{12} \\ 3 \end{array}$$

b.
$$\begin{array}{r} \mathbf{13\frac{3}{4}} \\ 4\overline{)55} \\ \underline{4} \\ 15 \\ \underline{12} \\ 3 \end{array}$$

c.
$$\begin{array}{r} \mathbf{13.75} \\ 4\overline{)55.00} \\ \underline{4} \\ 15 \\ \underline{12} \\ 3\,0 \\ \underline{2\,8} \\ 20 \\ \underline{20} \\ 0 \end{array}$$

d.
```
    1.8333 . . .
 3)5.5000 . . .
   3
   2 5
   2 4
     10
      9
     10
      9
     10
```
1.833③ . . . ⟶ **1.833**

e.
```
   23 R 1
 4)93
   8
   13
   12
    1
```
23 horses are in three stables, and 24 horses are in the fourth stable.
23, 23, 23, and 24 horses

Lesson 44, Mixed Practice

1. $\frac{\text{length}}{\text{width}} = \frac{24}{18} = \mathbf{\frac{4}{3}}$

2.
```
    90         89
    95     10)890
    90
    85
    80
    85
    90
    80
    95
 + 100
   890
```

3. $\frac{\text{able to find a job}}{\text{total}} = \mathbf{\frac{3}{5}}$

4.
```
   $0.34        $20.00
 ×    50      − $17.00
  $17.00         $3.00
```
(**$3.00**)

5.
```
   2. 9 8 0
 − 0. 0 9 7
   2. 8 8 3
```
Two and eight hundred eighty-three thousandths

6.

300 runners

50 runners
50 runners
50 runners
50 runners
50 runners
50 runners

$\frac{5}{6}$ finished. (first five rows)
$\frac{1}{6}$ did not finish. (last row)

(a) **50 runners**

(b) $\frac{\text{finished}}{\text{did not finish}} = \frac{250}{50} = \mathbf{\frac{5}{1}}$

7.

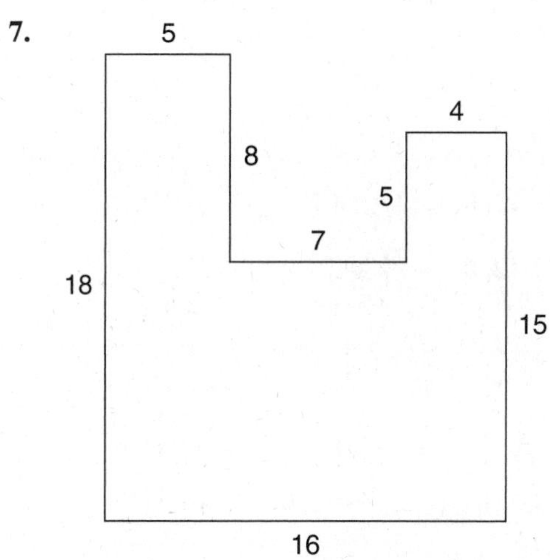

Perimeter = 18m + 5m + 8m + 7m + 5m + 4m + 15m + 16m
= **78 m**

8. (a) $0.75 = \frac{75}{100} = \mathbf{\frac{3}{4}}$

(b)
```
   0.625
 8)5.000
   4 8
     20
     16
      40
      40
       0
```
(**0.625**)

(c) $125\% = \frac{125}{100} = \mathbf{1.25}$

9. $\frac{\text{hearts}}{\text{total cards}} = \frac{13}{52} = \mathbf{\frac{1}{4}}$

10. **B. (2 · 3) + (2 · 4)**

11. 10 + 5 = 15
15 + 6 = 21
21 + 7 = 28
15, 21, 28

12.
$$\begin{array}{r} 0.49090\ldots \\ 11\overline{)5.40000\ldots} \\ \underline{4\,4} \\ 1\,00 \\ \underline{99} \\ 10 \\ \underline{0} \\ 100 \\ \underline{99} \\ 10 \\ \underline{0} \\ 10 \end{array}$$

(a) $\mathbf{0.4\overline{90}}$

(b) $0.490\underline{9}0\ldots \longrightarrow$ **0.491**

13. $2 \times 3 \times 5 \times 7 =$ **210**

14. (a) $\mathbf{-12, 0, 0.12, \frac{1}{2}, 1.2}$

(b) $\mathbf{-12, 0}$

15. (a) $12\left(1\frac{1}{2}\text{ inches}\right)$

$= 12\left(\frac{3}{2}\text{ inches}\right) = 6 \times 3\text{ inches}$

$=$ **18 inches**

(b) 1 yard $=$ 36 inches

$36\text{ inches} \div 1\frac{1}{2} = \frac{36}{1} \div \frac{3}{2}$

$= \frac{\overset{12}{\cancel{36}}}{1} \times \frac{2}{\underset{1}{\cancel{3}}} =$ **24 books**

16.
$$\begin{array}{r} \overset{1\,1}{2.46} \\ +\ 2.54 \\ \hline 5.00 \end{array} = \mathbf{5}$$

17. 3×10 meters $=$ **30 meters**

18. (a) Area $= (2.5\text{ cm})(2.5\text{ cm}) =$ $\mathbf{6.25\ cm^2}$

(b) Perimeter $= 4(2.5\text{ cm}) =$ **10 cm**

19.

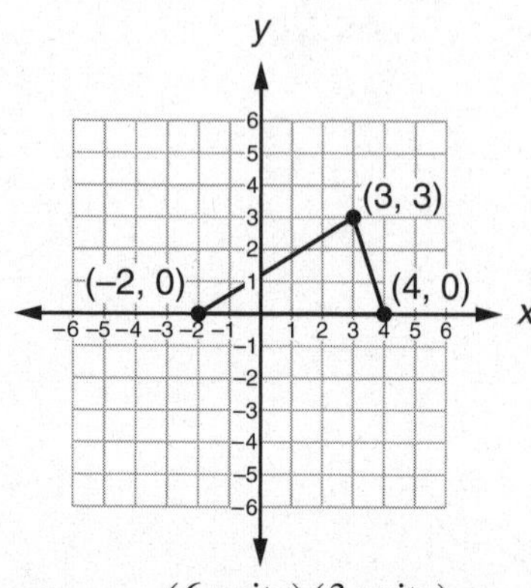

$\text{Area} = \frac{(6\text{ units})(3\text{ units})}{2} = \mathbf{9\ units^2}$

20.
$$\frac{25}{15} = \frac{n}{1.2}$$
$$15n = 25 \cdot 1.2$$
$$n = \frac{30}{15}$$
$$n = \mathbf{2}$$

21.
$$\frac{p}{90} = \frac{4}{18}$$
$$18p = 90 \cdot 4$$
$$p = \frac{360}{18}$$
$$p = \mathbf{20}$$

22.
$$\begin{array}{r} \overset{3}{\cancel{4}}.\overset{9}{\cancel{1}}\cancel{0}{}^{1}0 \\ -\ 3.\,1\,4 \\ \hline 0.\,8\,6 \end{array}$$

$x =$ **0.86**

23.
$$\begin{array}{r} \overset{0}{\cancel{1}}.{}^{1}0 \\ -\ 0.\,1 \\ \hline 0.\,9 \end{array}$$

$z =$ **0.9**

24.
$$\begin{array}{r} \mathbf{2.0525} \\ 8\overline{)16.4200} \\ \underline{16} \\ 0\,4 \\ \underline{0} \\ 42 \\ \underline{40} \\ 20 \\ \underline{16} \\ 40 \\ \underline{40} \\ 0 \end{array}$$

25.
$$\begin{array}{r} \mathbf{0.017} \\ 9\overline{)0.153} \\ \underline{9} \\ 63 \\ \underline{63} \\ 0 \end{array}$$

26.
$$\begin{array}{rcl} 5\frac{3}{4} & = & 5\frac{9}{12} \\ \frac{5}{6} & = & \frac{10}{12} \\ +\ 2\frac{1}{2} & = & 2\frac{6}{12} \\ \hline & & 7\frac{25}{12} = \mathbf{9\frac{1}{12}} \end{array}$$

27. $5 - 1\frac{5}{6} \longrightarrow 4\frac{6}{6} - 1\frac{5}{6} = 3\frac{1}{6}$ $\qquad 3\frac{1}{3} - 3\frac{1}{6} = 3\frac{2}{6} - 3\frac{1}{6} = \mathbf{\frac{1}{6}}$

28. $3\frac{3}{4} \cdot 3\frac{1}{3} \cdot 8 = \frac{\overset{5}{\cancel{15}}}{\underset{1}{\cancel{4}}} \cdot \frac{10}{\underset{1}{\cancel{3}}} \cdot \frac{\overset{2}{\cancel{8}}}{1} = \mathbf{100}$

29. $7 \div 10\frac{1}{2} = \frac{7}{1} \div \frac{21}{2} = \frac{\overset{1}{\cancel{7}}}{1} \times \frac{2}{\underset{3}{\cancel{21}}} = \mathbf{\frac{2}{3}}$

30. (a) $m\angle ABD = 180° - (35° + 90°)$
$= 180° - 125° = \mathbf{55°}$
The other two angles of ΔABD measure 35° and 90°. For the sum to be 180°, m∠ABD must be 55°.

(b) $m\angle CBD = 90° - 55° = \mathbf{35°}$
Since the figure is a rectangle, m∠ABC is 90°. We found that m∠ABD is 55°; ∠CBD is the complement of ∠ABD, so m∠CBD is 35°.

(c) $m\angle BDC = 90° - 35° = \mathbf{55°}$
Angle BDC is the complement of a 35° angle. Also, ∠BDC is the third angle of a triangle whose other two angles measure 35° and 90°.

Lesson 45, Warm-Up

a. **$10.50**

b. **125**

c. **15**

d. **100**

e. **12**

f. **50**

g. $\mathbf{1\frac{5}{12}, \frac{1}{12}, \frac{1}{2}, 1\frac{1}{8}}$

Problem Solving

Since there are three digits and one digit in the factors, respectively, and four digits in the product (and because the first digit of the product is 8), the bottom factor must be very close to 10. We write 9 as the bottom factor. We write a 2 in the ones place of the top factor, because $2 \times 9 = 18$. We write a 3 in the tens place of the factor because 8×9 plus the 1 from regrouping equals 73. The hundreds digit of the top factor must be 9, because the thousands digit of the product is 8. We write an 8 in the hundreds place of the product, because 9×9 plus the 7 from regrouping equals 88.

982
$\times$ **9**
8838

Lesson 45, Lesson Practice

a. $06.\overline{)51.6}$ = **8.6**
48
36
36
0

b. $009.\overline{)14.4}$ = **1.6**
9
54
54
0

c. $007.\overline{)2380.}$ = **340.**
21
28
28
00
0
0

d. $008.\overline{)2400.}$ = **300.**
24
00
0
00
0
0

e. $\$0.75\overline{)\$12.00}$ = **16 pens**

$$\begin{array}{r} 16 \\ 0.75\overline{)12.00} \\ \underline{75} \\ 450 \\ \underline{450} \\ 0 \end{array}$$

f. **If we multiply $\frac{0.25}{0.5}$ by $\frac{10}{10}$, the result is $\frac{2.5}{5}$. Since $\frac{10}{10}$ equals 1, we have not changed the value by multiplying—we have only changed the form.**

Lesson 45, Mixed Practice

1. $\dfrac{\text{raisins}}{\text{nuts}} = \dfrac{\mathbf{3}}{\mathbf{5}}$

2. $\$1 + \$0.40(8)$
$= \$1 + \$3.20 = \mathbf{\$4.20}$

3. $$\begin{array}{r} 54.05 \\ -\ 50.04 \\ \hline 4.01 \end{array}$$
Four and one hundredth

4. (a) $$\begin{array}{r} 22 \text{ votes} \\ -\ 18 \text{ votes} \\ \hline \mathbf{4 \text{ votes}} \end{array}$$

(b) $\dfrac{\text{Carlos's votes}}{\text{total votes}} = \dfrac{14}{70} = \dfrac{\mathbf{1}}{\mathbf{5}}$

5. Riders on the Giant Gyro

$\frac{4}{7}$ were euphoric.	$\frac{1}{7}$ of riders
	$\frac{1}{7}$ of riders
	$\frac{1}{7}$ of riders
	$\frac{1}{7}$ of riders
$\frac{3}{7}$ were vertiginous.	$\frac{1}{7}$ of riders
	$\frac{1}{7}$ of riders
	$\frac{1}{7}$ of riders

(a) $\dfrac{\mathbf{3}}{\mathbf{7}}$

(b) $\dfrac{\text{euphoric riders}}{\text{vertiginous riders}} = \dfrac{\mathbf{4}}{\mathbf{3}}$

6. $10 = 2 \cdot 5$
$16 = 2 \cdot 2 \cdot 2 \cdot 2$
$\text{LCM}(10, 16) = 2 \cdot 2 \cdot 2 \cdot 2 \cdot 5 = \mathbf{80}$

7. $5^2 \times 10^2 = 5 \cdot 5 \cdot 10 \cdot 10$
$= 25 \cdot 100 = \mathbf{2500}$

8. (a) $56 \text{ cm} - 20 \text{ cm} = 36 \text{ cm}$
$\dfrac{36 \text{ cm}}{2} = \mathbf{18 \text{ cm}}$

(b) $\text{Area} = (10 \text{ cm})(18 \text{ cm})$
$= \mathbf{180 \text{ cm}^2}$

9. (a) $\mathbf{62\frac{1}{2}}$

(b) **0.09**

(c) $7.5\% = \dfrac{7.5}{100} = \mathbf{0.075}$

10. (a) 23.54545④5 . . . ⟶ **23.54545**

(b) 0.91666⑥ . . . ⟶ **0.91667**

11. 2 kilograms = 2(1000 grams)
= **2000 grams**

12. $$\begin{array}{r} 0.065 \\ \times\ \$5.00 \\ \hline \$0.32500 \end{array} \longrightarrow \mathbf{\$0.33}$$

13. $$\begin{array}{r} 0.566\ldots \\ 9\overline{)5.100\ldots} \\ \underline{4\,5} \\ 60 \\ \underline{54} \\ 60 \\ \underline{54} \\ 6 \end{array}$$

(a) 0.566⑥ . . . ⟶ **0.567**

(b) $\mathbf{0.5\overline{6}}$

14. $\dfrac{\text{aces}}{\text{total cards}} = \dfrac{4}{52} = \dfrac{\mathbf{1}}{\mathbf{13}}$

15. 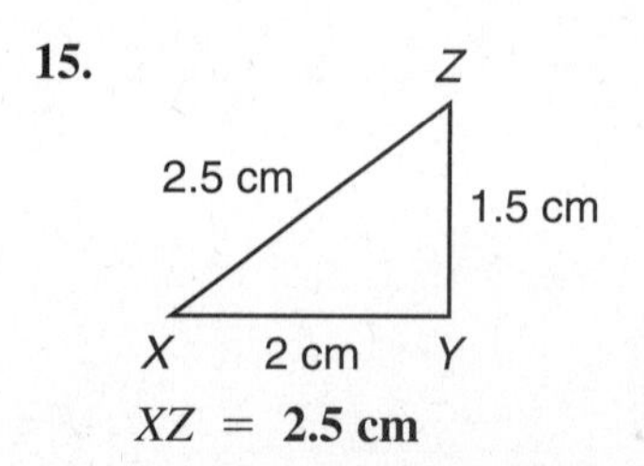

$XZ = \mathbf{2.5 \text{ cm}}$

16. (a) Perimeter $= 2.5\text{ cm} + 1.5\text{ cm} + 2\text{ cm}$
$= \mathbf{6\text{ cm}}$

(b) Area $= \dfrac{(1.5\text{ cm})(2\text{ cm})}{2}$
$= \mathbf{1.5\text{ cm}^2}$

17. $\dfrac{3}{w} = \dfrac{25}{100}$
$25w = 3 \cdot 100$
$w = \dfrac{300}{25}$
$w = \mathbf{12}$

18. $\dfrac{1.2}{4.4} = \dfrac{3}{a}$
$1.2a = 3 \cdot 4.4$
$a = \dfrac{13.2}{1.2}$
$a = \mathbf{11}$

19.
$$\begin{array}{r} \overset{0}{\cancel{1}}.\overset{1}{\cancel{2}}{}^{1}0 \\ -\ 0.\,2\,3 \\ \hline 0.\,9\,7 \end{array}$$
$m = \mathbf{0.97}$

20.
$$\begin{array}{r} \overset{1\,1}{0.65} \\ +\ 1.97 \\ \hline 2.62 \end{array}$$
$r = \mathbf{2.62}$

21.
$$\begin{array}{r} 0.15 \\ \times\ 0.15 \\ \hline 75 \\ 15 \\ \hline \mathbf{0.0225} \end{array}$$

22.
$$\begin{array}{r} 1.2 \\ \times\ 2.5 \\ \hline 6\,0 \\ 2\,4 \\ \hline 3.00 \end{array}$$
$3.00 = 3$
$3 \times 4 = \mathbf{12}$

23.
$$\begin{array}{r} \mathbf{2.828} \\ 5\overline{)14.140} \\ \underline{10} \\ 4\,1 \\ \underline{4\,0} \\ 14 \\ \underline{10} \\ 40 \\ \underline{40} \\ 0 \end{array}$$

24.
$$\begin{array}{r} \mathbf{0.8} \\ 012.\overline{)009.6} \\ \underline{9\,6} \\ 0 \end{array}$$

25.
$$\begin{array}{rcl} \dfrac{5}{8} & = & \dfrac{15}{24} \\ \dfrac{5}{6} & = & \dfrac{20}{24} \\ +\ \dfrac{5}{12} & = & \dfrac{10}{24} \\ \hline & & \dfrac{45}{24} = 1\dfrac{21}{24} = \mathbf{1\dfrac{7}{8}} \end{array}$$

26.
$$\begin{array}{rcl} 2\dfrac{1}{3} & = & 2\dfrac{4}{12} \\ -\ 1\dfrac{1}{4} & = & 1\dfrac{3}{12} \\ \hline & & 1\dfrac{1}{12} \end{array}$$

$$\begin{array}{rcl} 4\dfrac{1}{2} & = & 4\dfrac{6}{12} \\ -\ 1\dfrac{1}{12} & = & 1\dfrac{1}{12} \\ \hline & & \mathbf{3\dfrac{5}{12}} \end{array}$$

27. $\dfrac{7}{15} \cdot 10 \cdot 2\dfrac{1}{7} = \dfrac{\overset{1}{\cancel{7}}}{\underset{1}{\cancel{15}}} \cdot \dfrac{10}{1} \cdot \dfrac{\overset{1}{\cancel{15}}}{\underset{1}{\cancel{7}}} = \mathbf{10}$

28. $6\dfrac{3}{5} \div 1\dfrac{1}{10} = \dfrac{33}{5} \div \dfrac{11}{10}$

$= \dfrac{\overset{3}{\cancel{33}}}{\underset{1}{\cancel{5}}} \times \dfrac{\overset{2}{\cancel{10}}}{\underset{1}{\cancel{11}}} = \mathbf{6}$

29.
$$\begin{array}{r} 33\frac{1}{3} \\ \$021.\overline{)\$700.} \\ \underline{63} \\ 70 \\ \underline{63} \\ 7 \end{array}$$

33 pencils

30. (a) **See student work.**

(b) **The sum of the angle measures of a triangle is 180°.**

Lesson 46, Warm-Up

a. **$7.38**

b. **0.036**

c. **10**

d. **10**

e. **12**

f. **72**

g. **24**

Problem Solving

Yes

Example: 123 → 321
→ 321 − 123 = 198
→ 198 ÷ 9 = 22

Lesson 46, Lesson Practice

a. $\frac{\$1.12}{28 \text{ ounces}} = \frac{\$0.04}{1 \text{ ounce}}$

4¢ per ounce

b. $\frac{\$0.55}{11 \text{ ounces}} = \frac{\$0.05}{1 \text{ ounce}}$

5¢ per ounce

c. $\frac{\$1.98}{18 \text{ ounces}} = \frac{\$0.11}{1 \text{ ounce}}$

$\frac{\$2.28}{24 \text{ ounces}} = \frac{\$0.095}{1 \text{ ounce}}$

The 24-ounce jar is the better buy because 9.5¢ per ounce is less than 11¢ per ounce.

d. $\frac{416 \text{ mi}}{8 \text{ hr}} = 52 \frac{\text{mi}}{\text{hr}}$

52 mi/hr or **52 mph**

e. $\frac{322 \text{ mi}}{14 \text{ gal}} = 23 \frac{\text{mi}}{\text{gal}}$

23 mi/gal or **23 mpg**

f. (a) $\frac{44 \text{ euros}}{40 \text{ dollars}} = \frac{11 \text{ euros}}{10 \text{ dollars}}$

11 euros per 10 dollars

(b) $\frac{40 \text{ dollars}}{44 \text{ euros}} = \frac{10 \text{ dollars}}{11 \text{ euros}}$

10 dollars per 11 euros

g.
$$\begin{array}{r} \$36.89 \\ \times \quad .07 \\ \hline \$2.5823 \end{array} \longrightarrow \mathbf{\$2.58}$$

h.
$$\begin{array}{r} \$36.89 \\ + \ \$2.58 \\ \hline \mathbf{\$39.47} \end{array}$$

i.
$$\begin{array}{r} \$6.95 \\ \$0.95 \\ + \ \$2.45 \\ \hline \$10.35 \end{array}$$

$$\begin{array}{r} \$10.35 \\ \times \quad .06 \\ \hline \$0.6210 \end{array} \longrightarrow \$0.62$$

$$\begin{array}{r} \$10.35 \\ + \ \$0.62 \\ \hline \mathbf{\$10.97} \end{array}$$

j.
$$\begin{array}{r} \$15 \\ \times \ 0.15 \\ \hline 75 \\ 15\ \ \\ \hline \$2.25 \end{array}$$

About $2.25

k.
$$\begin{array}{r} 0.15 \\ \times \quad 12 \\ \hline 30 \\ 15\ \ \\ \hline \mathbf{\$1.80} \end{array}$$

l. **One way is to think of 10% of the bill and then double that number.**

Lesson 46, Mixed Practice

1. $\frac{\$2.40}{16 \text{ ounces}} = \frac{\$0.15}{1 \text{ ounce}}$

$\frac{\$1.92}{12 \text{ ounces}} = \frac{\$0.16}{1 \text{ ounce}}$

Brand X = $0.15/ounce
Brand Y = $0.16/ounce
Brand X is the better buy.

2. $\dfrac{702 \text{ kilometers}}{6 \text{ hours}} = 117 \dfrac{\text{kilometers}}{\text{hour}}$

117 kilometers per hour

3. $\dfrac{\text{sheep}}{\text{cows}} = \dfrac{48}{36} = \mathbf{\dfrac{4}{3}}$

4.

$$\begin{array}{r} \$2.86 \\ \$2.83 \\ \$2.98 \\ +\ \$3.09 \\ \hline \$11.76 \end{array}$$

$$\begin{array}{r} \mathbf{\$2.94} \\ 4\overline{)\$11.76} \\ 8 \\ \hline 3\,7 \\ 3\,6 \\ \hline 16 \\ 16 \\ \hline 0 \end{array}$$

5.

$$\begin{array}{r} 3.\,\overset{1}{\cancel{2}}{}^{1}0 \\ -\ 2.\,0\,3 \\ \hline 1.\,1\,7 \end{array}$$

One and seventeen hundredths

6. 2 feet = 24 inches

$24 \text{ inches} \div 1\dfrac{1}{2} = \dfrac{24 \text{ inches}}{1} \div \dfrac{3}{2}$

$= \dfrac{\overset{8}{\cancel{24}}}{1} \times \dfrac{2}{\underset{1}{\cancel{3}}}$

$=$ **16 books**

7.

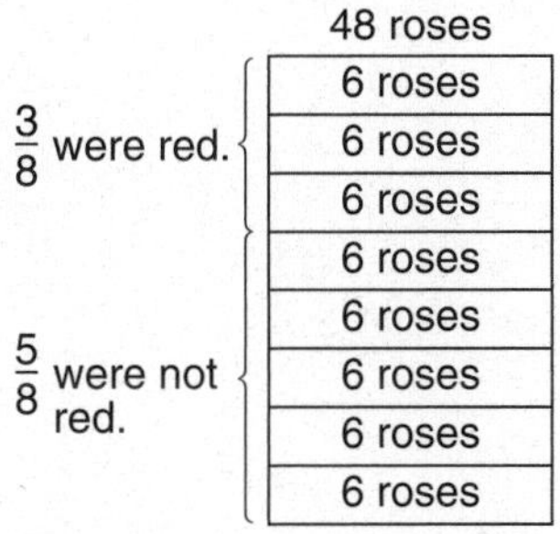

48 roses	
$\frac{3}{8}$ were red.	6 roses
	6 roses
	6 roses
$\frac{5}{8}$ were not red.	6 roses
	6 roses
	6 roses
	6 roses
	6 roses

(a) 3(6 roses) = **18 roses**

(b) 5(6 roses) = **30 roses**

(c) $\dfrac{\text{not red roses}}{\text{total roses}} = \dfrac{30}{48} = \mathbf{\dfrac{5}{8}}$

8. (a) 3.0303 **<** 3.303

(b) 0.6 **=** 0.600

9. 100 yards = 100(3 feet)

= **300 feet**

10. (a) $0.080 = \dfrac{8}{100} = \mathbf{\dfrac{2}{25}}$

(b) $37\dfrac{1}{2}\% = 37.5\%$

$= \dfrac{37.5}{100} = \mathbf{0.375}$

(c)

$$\begin{array}{r} 0.0909\ldots \\ 11\overline{)1.0000\ldots} \\ 99 \\ \hline 10 \\ 0 \\ \hline 100 \\ 99 \\ \hline 1 \end{array}$$

$\mathbf{0.\overline{09}}$

11. (a)

$$\begin{array}{r} \$14.95 \\ \times\ \ 0.07 \\ \hline \$1.0465 \end{array} \longrightarrow \mathbf{\$1.05}$$

(b)

$$\begin{array}{r} \$14.95 \\ +\ \ \$1.05 \\ \hline \mathbf{\$16.00} \end{array}$$

12.

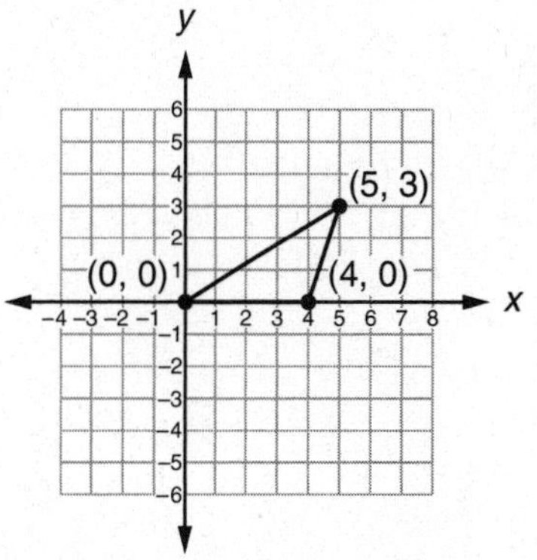

(0, 0) (5, 3) (4, 0)

Area $= \dfrac{(4 \text{ units})(3 \text{ units})}{2}$

$=$ **6 units²**

13. $\dfrac{\text{face cards}}{\text{total cards}} = \dfrac{12}{52} = \mathbf{\dfrac{3}{13}}$

14.

$$\begin{array}{r} 2 \\ 3 \\ 5 \\ 7 \\ +\ 11 \\ \hline 28 \end{array}$$

$$\begin{array}{r} \mathbf{5.6} \\ 5\overline{)28.0} \\ 25 \\ \hline 3\,0 \\ 3\,0 \\ \hline 0 \end{array}$$

15. 0.3(0.4 + 0.5) = 0.3(0.9) = 0.27

or

0.3(0.4 + 0.5) = 0.12 + 0.15 = 0.27

16. (a) Perimeter = 3 in. + 5 in. + 4 in. + 3 in. + 11 in. + 3 in. + 4 in. + 5 in. = **38 in.**

(b)

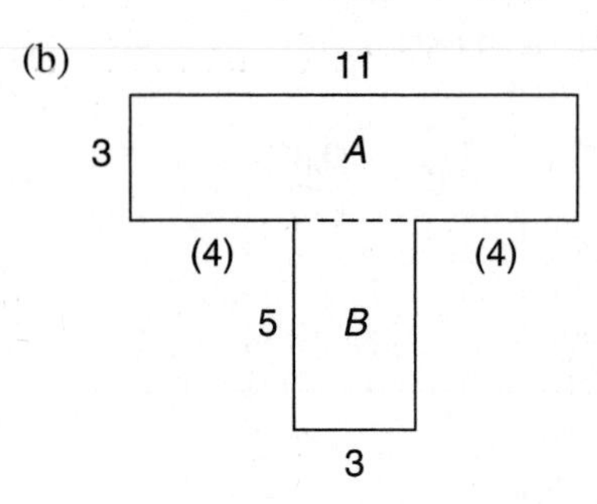

$$\begin{array}{rcl} \text{Area } A &=& (11 \text{ in.})(3 \text{ in.}) = 33 \text{ in.}^2 \\ + \text{ Area } B &=& (5 \text{ in.})(3 \text{ in.}) = 15 \text{ in.}^2 \\ \hline & & \text{Area of figure} = \mathbf{48 \text{ in.}^2} \end{array}$$

17. (a) **180°**

(b) $\frac{360°}{3} = \mathbf{120°}$

(c) $\frac{360°}{6} = \mathbf{60°}$

18. $\frac{10}{12} = \frac{2.5}{a}$

$10a = 2.5 \cdot 12$

$a = \frac{30}{10}$

$a = \mathbf{3}$

19. $\frac{6}{8} = \frac{b}{100}$

$8b = 100 \cdot 6$

$b = \frac{600}{8}$

$b = \mathbf{75}$

20. $\begin{array}{r} 4.7 \\ -\ 1.2 \\ \hline 3.5 \end{array}$

$w = \mathbf{3.5}$

21. $\frac{10^2}{10} = \frac{100}{10} = 10$

$x = \mathbf{10}$

22. $1\frac{11}{18} + 2\frac{11}{24}$

$\downarrow \quad \downarrow$

$2 + 2 = \mathbf{4}$

$$\begin{array}{rcl} 1\frac{11}{18} &=& 1\frac{44}{72} \\ + 2\frac{11}{24} &=& 2\frac{33}{72} \\ \hline & & 3\frac{77}{72} = \mathbf{4\frac{5}{72}} \end{array}$$

23. $5\frac{5}{6} - \left(3 - 1\frac{1}{3}\right)$

$\downarrow \quad \downarrow \quad \downarrow$

$6 - (3 - 1) = 6 - 2 = \mathbf{4}$

$$\begin{array}{rcl} 3 &=& 2\frac{3}{3} \\ -\ 1\frac{1}{3} &=& 1\frac{1}{3} \\ \hline & & 1\frac{2}{3} \end{array}$$

$$\begin{array}{rcl} 5\frac{5}{6} &=& 5\frac{5}{6} \\ -\ 1\frac{2}{3} &=& 1\frac{4}{6} \\ \hline & = & \mathbf{4\frac{1}{6}} \end{array}$$

24. $\frac{2}{3} \times 4 \times 1\frac{1}{8} = \frac{\overset{1}{\cancel{2}}}{\underset{1}{\cancel{3}}} \times \frac{\overset{1}{\cancel{4}}}{1} \times \frac{\overset{3}{\cancel{9}}}{\underset{\underset{1}{\cancel{2}}}{\cancel{8}}} = \mathbf{3}$

25. $6\frac{2}{3} \div 4 = \frac{20}{3} \div \frac{4}{1} = \frac{\overset{5}{\cancel{20}}}{3} \times \frac{1}{\underset{1}{\cancel{4}}}$

$= \frac{5}{3} = \mathbf{1\frac{2}{3}}$

26. $\begin{array}{r} \overset{4}{\cancel{5}}.\overset{1}{\cancel{2}}{}^{1}0 \\ -\ 0.57 \\ \hline 4.63 \end{array} \qquad \begin{array}{r} \overset{1}{3}.45 \\ 6 \\ +\ 4.63 \\ \hline \mathbf{14.08} \end{array}$

27. $0016.\overline{)2400.}$ = **150.**

$$\begin{array}{r} 16 \\ \hline 80 \\ 80 \\ \hline 00 \\ 0 \\ \hline 0 \end{array}$$

28. **Round $6\frac{7}{8}$ to 7 and round $5\frac{1}{16}$ to 5. Then multiply 7 by 5.**

29. (a) $\angle$***CAB***

(b) $\overline{\mathbf{BA}}$

30. (a) $m\angle B = m\angle ADC = \mathbf{60°}$

(b) $m\angle CAB = 180° - (60° + 45°) = 180° - 105° = \mathbf{75°}$

(c) $m\angle CAD = m\angle ACB = \mathbf{45°}$

LESSON 47, WARM-UP

a. **$41.00**

b. **15**

c. **6**

d. **$100.00**

e. **32**

f. **50**

g. $\frac{9}{10}, \frac{1}{10}, \frac{1}{5}, 1\frac{1}{4}$

Problem Solving

Each chicken lays 1 egg in 2 days, so in 4 days 4 chickens can lay **8 eggs.**

LESSON 47, LESSON PRACTICE

a. $400 + 50 + 6$
$(4 \times 10^2) + (5 \times 10^1) + (6 \times 10^0)$

b. $1000 + 700 + 60$
$(1 \times 10^3) + (7 \times 10^2) + (6 \times 10^1)$

c. $100{,}000 + 80{,}000 + 6000$
$(1 \times 10^5) + (8 \times 10^4) + (6 \times 10^3)$

d. 24.25×10^3
$= 24.25 \times 1000$
$= \mathbf{24{,}250}$

e. $25 \times 10^6 = 25 \times 1{,}000{,}000$
$= \mathbf{25{,}000{,}000}$

f. $12.5 \div 10^3$
$= 12.5 \div 1000$
$= \mathbf{0.0125}$

g. $4.8 \div 10^4 = 4.8 \div 10{,}000$
$= \mathbf{0.00048}$

h. $10^3 \cdot 10^4 = 10^7$

i. $10^8 \div 10^2 = 10^6$

j. **2,500,000**

k. **15,000,000,000**

l. **1,600,000,000,000**

LESSON 47, MIXED PRACTICE

1. (a) **True**

   (b) **True**

2. $\frac{\text{walked}}{\text{rode in car}} = \frac{10}{12} = \mathbf{\frac{5}{6}}$

3. $\frac{\text{rode in car}}{\text{total}} = \frac{12}{33} = \mathbf{\frac{4}{11}}$

4.
```
  1
  1.2         1.56
  1.4      5)7.80
  1.5        5
  1.7        2 8
+ 2          2 5
  7.8          30
               30
                0
```
1.56

5. (a) **134,800,000 viewers**

   (b) $5000 + 200 + 80$
   $(5 \times 10^3) + (2 \times 10^2) + (8 \times 10^1)$

6. 40 contestants

$\frac{1}{8}$ answered correctly: 5 contestants

$\frac{7}{8}$ did not answer correctly: 5 contestants, 5 contestants, 5 contestants, 5 contestants, 5 contestants, 5 contestants, 5 contestants

   (a) **5 contestants**

   (b) 7(5 contestants) = **35 contestants**

7. (a)
```
     10 2/3
12)128
   12
    08
     0
     8
```
10 glasses

   (b) **11 glasses**

8. (a) **Answers may vary.**

(b) **Answers may vary.**

9. (a) $0.375 = \frac{375}{1000} = \mathbf{\frac{3}{8}}$

(b) $62\frac{1}{2}\% = 62.5\% = \frac{62.5}{100} = \mathbf{0.625}$

10.
$$\begin{array}{r} \$56.40 \\ \times \quad 0.08 \\ \hline \$4.5120 \end{array} \longrightarrow \mathbf{\$4.51}$$

11. (a) 53,714.545④ . . .
$\longrightarrow$ **53,714.545**

(b) 53,⑦14.5454 . . .
$\longrightarrow$ **54,000**

12. (a) $10^5 \cdot 10^2 = 10^7$

(b) $10^8 \div 10^4 = 10^4$

13. **3.03**

14. $BC = 6$ cm, $AF = 6$ cm
$AB = 6$ cm, $FE = 9$ cm
$ED = 12$ cm
Perimeter = 6 cm + 6 cm + 6 cm
+ 3 cm + 12 cm + 9 cm
= **42 cm**

15.

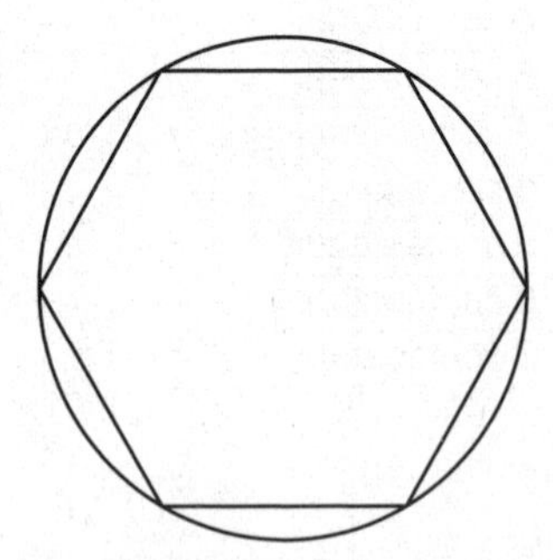

(a) Diameter = 2(1 inch) = **2 inches**

(b) Perimeter = 6(1 inch) = **6 inches**

16.
$$\frac{6}{10} = \frac{w}{100}$$
$$10w = 6 \cdot 100$$
$$w = \frac{600}{10}$$
$$w = \mathbf{60}$$

17.
$$\frac{3.6}{x} = \frac{16}{24}$$
$$16x = (3.6)(24)$$
$$x = \frac{86.4}{16}$$
$$x = \mathbf{5.4}$$

18.
$$\begin{array}{r} 1.5 \\ \times \ 1.5 \\ \hline 75 \\ 15\ \\ \hline 2.25 \end{array}$$
$a = \mathbf{2.25}$

19.
$$\begin{array}{r} \overset{8}{\cancel{9}}.{}^{1}8 \\ -\ 8.9 \\ \hline 0.9 \end{array}$$
$x = \mathbf{0.9}$

20. $4\frac{1}{5} + 5\frac{1}{3} + \frac{1}{2}$
↓ ↓ ↓
$4 + 5 + 1 = \mathbf{10}$

$$\begin{array}{rcl} 4\frac{1}{5} & = & 4\frac{6}{30} \\ 5\frac{1}{3} & = & 5\frac{10}{30} \\ +\ \frac{1}{2} & = & \frac{15}{30} \\ \hline & & 9\frac{31}{30} = \mathbf{10\frac{1}{30}} \end{array}$$

21. $6\frac{1}{8} - \left(5 - 1\frac{2}{3}\right)$
↓ ↓ ↓
$6 - (5 - 2) = 6 - 3 = \mathbf{3}$

$$\begin{array}{r} 5 \\ -\ 1\frac{2}{3} \\ \hline \end{array} \longrightarrow \begin{array}{r} 4\frac{3}{3} \\ -\ 1\frac{2}{3} \\ \hline 3\frac{1}{3} \end{array}$$

$$\begin{array}{rcl} 6\frac{1}{8} & = & 6\frac{3}{24} \\ -\ 3\frac{1}{3} & = & 3\frac{8}{24} \\ \hline \end{array} \longrightarrow \begin{array}{r} 5\frac{27}{24} \\ -\ 3\frac{8}{24} \\ \hline \mathbf{2\frac{19}{24}} \end{array}$$

22. $\sqrt{16 \cdot 25} = \sqrt{400} = \mathbf{20}$

23. 3.6×10^3
$= 3.6 \times 1000$
$= \mathbf{3600}$

24. $8\frac{1}{3} \times 3\frac{3}{5} \times \frac{1}{3}$

$= \frac{\overset{5}{\cancel{25}}}{\underset{1}{\cancel{3}}} \times \frac{\overset{\overset{2}{\cancel{6}}}{\cancel{18}}}{\underset{1}{\cancel{5}}} \times \frac{1}{\underset{1}{\cancel{3}}}$

$= \mathbf{10}$

25. $3\frac{1}{8} \div 6\frac{1}{4} = \frac{25}{8} \div \frac{25}{4}$

$= \frac{\overset{1}{\cancel{25}}}{\underset{2}{\cancel{8}}} \times \frac{\overset{1}{\cancel{4}}}{\underset{1}{\cancel{25}}} = \mathbf{\frac{1}{2}}$

26.

$$\begin{array}{r} \overset{2\,1}{26.7} \\ 3.45 \\ 0.036 \\ 12 \\ +\ 8.7 \\ \hline \mathbf{50.886} \end{array}$$

27. (a) Perimeter = 15 in. + 13 in. + 14 in.
= **42 in.**

(b) Area $= \frac{(14 \text{ in.})(12 \text{ in.})}{2}$
$= \mathbf{84\ in.^2}$

28. $125 \div 10^2 = 125 \div 100 = 1.25$
$0.125 \times 10^2 = 0.125 \times 100 = 12.5$
$\mathbf{1.25 < 12.5}$

29. $\frac{2}{3} = \frac{8}{12}$

$\frac{1}{2} = \frac{6}{12}$ $\mathbf{\frac{1}{2}, \frac{7}{12}, \frac{2}{3}, \frac{5}{6}}$

$\frac{5}{6} = \frac{10}{12}$

30. (a) $\angle a = 180° - 130° = \mathbf{50°}$

(b) $\angle b = 180° - (65° + 50°)$
$= 180° - 115° = \mathbf{65°}$

(c) **Together, $\angle b$ and $\angle c$ form a straight angle that measures 180°. To find the measure of $\angle c$, we subtract the measure of $\angle b$ from 180°.**

Lesson 48, Warm-Up

a. **$245.00**

b. **1.275**

c. **6**

d. **4**

e. **24**

f. **30**

g. **1**

Problem Solving

The sum of the counting numbers 1 through 9 is 45, which means the total of all three columns is 45. So the sum of the numbers in each column is 45 ÷ 3, which equals **15.**

Lesson 48, Lesson Practice

a.

$$\begin{array}{r} 0.66\ldots \\ 3\overline{)2.00\ldots} \\ \underline{1\,8} \\ 20 \\ \underline{18} \\ 2 \end{array} \qquad \mathbf{0.\overline{6}}$$

b. $\frac{2}{3} \times 100\% = \frac{200\%}{3} = \mathbf{66\frac{2}{3}\%}$

c. $1.1 = \mathbf{1\frac{1}{10}}$

d. $1.1 \times 100\% = \mathbf{110\%}$

e. $4\% = \frac{4}{100} = \mathbf{\frac{1}{25}}$

f. $4\% = \frac{4}{100} = \mathbf{0.04}$

Lesson 48, Mixed Practice

1. $\frac{80 \text{ kilometers}}{2.5 \text{ hours}} = \mathbf{32\ \frac{kilometers}{hour}}$

2. $\frac{1008}{1323} = \frac{2 \cdot 2 \cdot 2 \cdot 2 \cdot \overset{1}{\cancel{3}} \cdot \overset{1}{\cancel{3}} \cdot \overset{1}{\cancel{7}}}{\underset{1}{\cancel{3}} \cdot \underset{1}{\cancel{3}} \cdot 3 \cdot \underset{1}{\cancel{7}} \cdot 7}$

$= \mathbf{\frac{16}{21}}$

3.
$$\begin{array}{r} 1867 \\ -\ 1803 \\ \hline \textbf{64 years} \end{array}$$

4. (a) $\dfrac{\text{blue marbles}}{\text{total marbles}} = \mathbf{\dfrac{7}{12}}$

(b) $\dfrac{\text{red marbles}}{\text{blue marbles}} = \mathbf{\dfrac{5}{7}}$

5. $\dfrac{\$0.90}{6 \text{ ounces}} = \dfrac{\$0.15}{1 \text{ ounce}}$

$\dfrac{\$1.26}{9 \text{ ounces}} = \dfrac{\$0.14}{1 \text{ ounce}}$

6-ounce can is \$0.15 per ounce; 9-ounce can is \$0.14 per ounce; 9-ounce can is the better buy.

6.
$$\begin{array}{r} 2550 \\ +\ 2900 \\ \hline 5450 \end{array}$$

$$\begin{array}{r} \mathbf{2725} \\ 2\overline{)5450} \\ \underline{4} \\ 14 \\ \underline{14} \\ 5 \\ \underline{4} \\ 10 \\ \underline{10} \\ 0 \end{array}$$

7.

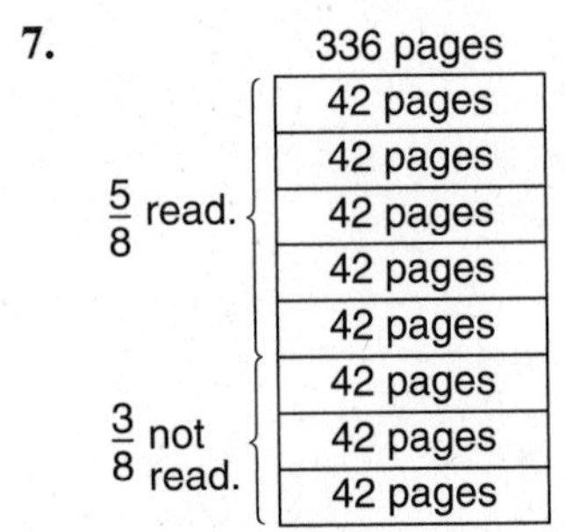

(a) $5(42 \text{ pages}) =$ **210 pages**

(b) $3(42 \text{ pages}) =$ **126 pages**

8. (a) $2\overline{)1.0}$ → **0.5**

(b) $\dfrac{1}{2} \times 100\% = \dfrac{100\%}{2} = \mathbf{50\%}$

(c) $0.1 = \mathbf{\dfrac{1}{10}}$

(d) $0.1 \times 100\% = \mathbf{10\%}$

(e) $25\% = \dfrac{25}{100} = \mathbf{\dfrac{1}{4}}$

(f) $25\% = \dfrac{25}{100} = \mathbf{0.25}$

9. (a) $100\% - (10\% + 12\% + 20\% + 25\% + 20\%) = 100\% - 87\% = \mathbf{13\%}$

(b) $20\% = \dfrac{20}{100} = \mathbf{\dfrac{1}{5}}$

(c) $2(\$3200) = \mathbf{\$6400}$

10. $0.545④ \ldots \longrightarrow$ **0.545**

11. (a) **See student answer.**

(b) **5 centimeters**

12. (a) **The exponent is 3 and the base is 5.**

(b) $10^4 \cdot 10^4 = 10^8$

13. $1 \text{ foot} = 12 \text{ inches}$

$\dfrac{12 \text{ inches}}{6} =$ **2 inches**

14.

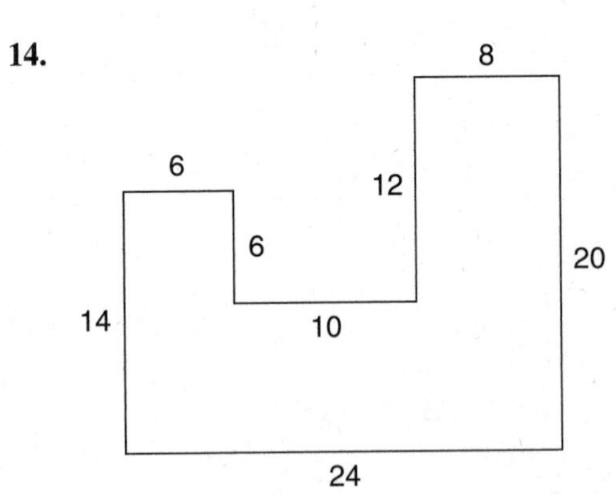

Perimeter $= 24 \text{ cm} + 14 \text{ cm} + 6 \text{ cm} + 6 \text{ cm} + 10 \text{ cm} + 12 \text{ cm} + 8 \text{ cm} + 20 \text{ cm}$
$=$ **100 cm**

15. $\dfrac{78 \text{ miles}}{1.2 \text{ gallons}} = 65 \dfrac{\text{miles}}{\text{gallon}}$

65 mpg

16. $\dfrac{6}{100} = \dfrac{15}{w}$

$6w = 100 \cdot 15$

$w = \dfrac{1500}{6}$

$w = \mathbf{250}$

17. $\dfrac{20}{x} = \dfrac{15}{12}$

$15x = 20 \cdot 12$

$x = \dfrac{240}{15}$

$x = \mathbf{16}$

18. $6\overline{)1.44}$ = 0.24

$$\begin{array}{r} 0.24 \\ 6\overline{)1.44} \\ 1\,2 \\ \hline 24 \\ 24 \\ \hline 0 \end{array}$$

$m = \mathbf{0.24}$

19.
$$\begin{array}{rcl} \frac{1}{2} & = & \frac{3}{6} \\ -\ \frac{1}{3} & = & \frac{2}{6} \\ \hline f & = & \mathbf{\frac{1}{6}} \end{array}$$

20. $2^5 + 1^4 + 3^3$
$= 2 \cdot 2 \cdot 2 \cdot 2 \cdot 2 + 1 + 3 \cdot 3 \cdot 3$
$= 32 + 1 + 27 = \mathbf{60}$

21. $\sqrt{10^2 \cdot 6^2} = \sqrt{100 \cdot 36} = \sqrt{3600}$
$= \mathbf{60}$

22.
$$\begin{array}{rcl} 1\frac{1}{4} & = & 1\frac{3}{12} \\ +\ 1\frac{1}{6} & = & 1\frac{2}{12} \\ \hline & & 2\frac{5}{12} \end{array}$$

$$\begin{array}{rcl} 3\frac{5}{6} & = & 3\frac{10}{12} \\ -\ 2\frac{5}{12} & = & 2\frac{5}{12} \\ \hline & & \mathbf{1\frac{5}{12}} \end{array}$$

23.
$$\begin{array}{rcl} 4 & \longrightarrow & 3\frac{3}{3} \\ -\ \frac{2}{3} & & -\ \frac{2}{3} \\ \hline & & 3\frac{1}{3} \end{array}$$

$$\begin{array}{rcl} 8\frac{3}{4} & = & 8\frac{9}{12} \\ +\ 3\frac{1}{3} & = & 3\frac{4}{12} \\ \hline & & 11\frac{13}{12} = \mathbf{12\frac{1}{12}} \end{array}$$

24. $\frac{15}{16} \cdot \frac{24}{25} \cdot 1\frac{1}{9}$

$= \frac{\overset{1}{\cancel{\overset{\cancel{3}}{\cancel{15}}}}}{\underset{1}{\cancel{\underset{\cancel{2}}{\cancel{16}}}}} \cdot \frac{\overset{1}{\cancel{\overset{\cancel{3}}{\cancel{24}}}}}{\underset{1}{\cancel{\underset{\cancel{5}}{\cancel{25}}}}} \cdot \frac{\overset{1}{\cancel{\overset{\cancel{2}}{\cancel{10}}}}}{\underset{1}{\cancel{\underset{\cancel{3}}{\cancel{9}}}}} = \mathbf{1}$

25. $2\frac{2}{3} \div 4 = \frac{8}{3} \div \frac{4}{1}$
$= \frac{\overset{2}{\cancel{8}}}{3} \times \frac{1}{\underset{1}{\cancel{4}}} = \frac{2}{3}$

$1\frac{1}{3} \div \frac{2}{3} = \frac{4}{3} \div \frac{2}{3} = \frac{\overset{2}{\cancel{4}}}{\underset{1}{\cancel{3}}} \times \frac{\overset{1}{\cancel{3}}}{\underset{1}{\cancel{2}}}$
$= \mathbf{2}$

26. $\frac{a}{b} = \frac{\$13.93}{0.07}$

$$\begin{array}{r} \mathbf{\$199.} \\ 007.\overline{)\$1393.} \\ 7 \\ \hline 69 \\ 63 \\ \hline 63 \\ 63 \\ \hline 0 \end{array}$$

$199.00

27.

y, x, (−1, −1), (5, −1), (5, −4)

Area $= \frac{(6 \text{ units})(3 \text{ units})}{2} = \mathbf{9\ sq.\ units}$

28. $\frac{\text{friends with more than one sibling}}{\text{total friends}}$
$= \frac{8}{20} = \mathbf{\frac{2}{5}}$

29.
$$\begin{array}{r} \$50.00 \\ \times\ \ .075 \\ \hline 25000 \\ 35000 \\ \hline \$3.75000 \end{array} \longrightarrow \$3.75$$

$$\begin{array}{r} \$50.00 \\ +\ \ \$3.75 \\ \hline \mathbf{\$53.75} \end{array}$$

30. $m\angle a = 180° - (50° + 90°)$
$= 180° - 140° = \mathbf{40°}$
$m\angle b = \mathbf{50°}$
$m\angle c = 180° - 50° = \mathbf{130°}$

Lesson 49, Warm-Up

a. **$52.00**

b. **257.5**

c. **10**

d. **6**

e. **20**

f. **60**

g. $\frac{14}{15}$, $\frac{4}{15}$, $\frac{1}{5}$, $1\frac{4}{5}$

Problem Solving

Adam: A
Blanca: B
Chad: C

1. A and B
2. A and C
3. B and C

Note: Order does not matter in selecting *combinations.* "A and B" is the same combination as "B and A," even though they are different *permutations.*

Lesson 49, Lesson Practice

a. $\begin{array}{r} 5 \\ 12\overline{)70} \\ \underline{60} \\ 10 \end{array}$ **5 feet 10 in.**

b. 6 feet = 6(12 inches)
= 72 inches
72 inches + 3 inches
= **75 inches**

c. 20 in. = 1 ft 8 in.

$\begin{array}{r} 1 \text{ ft } 8 \text{ in.} \\ \underline{+\ 5 \text{ ft}} \\ \mathbf{6 \text{ ft } 8 \text{ in.}} \end{array}$

d. $\begin{array}{rrr} 2 \text{ yd} & 1 \text{ ft} & 8 \text{ in.} \\ +\ 1 \text{ yd} & 2 \text{ ft} & 9 \text{ in.} \\ \hline 3 \text{ yd} & 3 \text{ ft} & 17 \text{ in.} \end{array}$

17 in. = 1 ft 5 in.

$\begin{array}{r} 1 \text{ ft } 5 \text{ in.} \\ +\ 3 \text{ ft} \\ \hline 4 \text{ ft } 5 \text{ in.} \end{array} \longrightarrow$ 3 yd 4 ft 5 in.

4 ft = 1 yd 1 ft

$\begin{array}{r} 1 \text{ yd } 1 \text{ ft} \\ +\ 3 \text{ yd} \\ \hline 4 \text{ yd } 1 \text{ ft} \end{array} \longrightarrow$ **4 yd 1 ft 5 in.**

e. $\begin{array}{rrr} 5 \text{ hr} & 42 \text{ min} & 53 \text{ s} \\ +\ 6 \text{ hr} & 17 \text{ min} & 27 \text{ s} \\ \hline 11 \text{ hr} & 59 \text{ min} & 80 \text{ s} \end{array}$

80 s = 1 min 20 s

$\begin{array}{rr} 1 \text{ min} & 20 \text{ s} \\ +\ 59 \text{ min} & \\ \hline 60 \text{ min} & 20 \text{ s} \end{array} \longrightarrow$ 11 hr 60 min 20 s

60 min = 1 hr
11 hr + 1 hr = 12 hr
12 hr 20 s

Lesson 49, Mixed Practice

1. $\begin{array}{r} 0.2 \\ +\ 0.05 \\ \hline 0.25 \end{array}$ $\begin{array}{r} 0.05 \\ \times\ 0.2 \\ \hline 0.010 \end{array} \longrightarrow 0.01$ $\begin{array}{r} \mathbf{25} \\ 001.\overline{)025.} \\ \underline{2} \\ 05 \\ \underline{5} \\ 0 \end{array}$

2. $\begin{array}{r} \mathbf{9.2 \text{ yards}} \\ 20\overline{)184.0} \\ \underline{180} \\ 4\,0 \\ \underline{4\,0} \\ 0 \end{array}$

3. $\begin{array}{r} \$0.25 \\ 24\overline{)\$6.00} \\ \underline{4\,8} \\ 1\,20 \\ \underline{1\,20} \\ 0 \end{array}$

25¢ per arrow

4. 3(8 sides) + 2(6 sides) + 5 sides + 2(4 sides)
= 24 sides + 12 sides + 5 sides + 8 sides
= **49 sides**

5.

$$\begin{array}{r} \overset{2\ 1}{6.21} \\ 4.38 \\ 7.5 \\ 6.3 \\ 5.91 \\ +\ 8.04 \\ \hline 38.34 \end{array} \qquad \begin{array}{r} \mathbf{6.39} \\ 6\overline{)38.34} \\ \underline{36} \\ 2\,3 \\ \underline{1\,8} \\ 54 \\ \underline{54} \\ 0 \end{array}$$

6.

72 billy goats	
$\frac{2}{9}$ were gruff.	8 billy goats
	8 billy goats
$\frac{7}{9}$ were cordial.	8 billy goats
	8 billy goats
	8 billy goats
	8 billy goats
	8 billy goats
	8 billy goats
	8 billy goats

(a) 7(8 billy goats) = **56 billy goats**

(b) $\dfrac{\text{gruff billy goats}}{\text{cordial billy goats}} = \mathbf{\dfrac{2}{7}}$

7. **0.5, $0.\overline{54}$, $0.\overline{5}$**

8. (a) **Answers may vary.**

(b) $\mathbf{2\frac{5}{8}}$ **inches**

9. (a) $0.9 \times 100\% = \mathbf{90\%}$

(b) $1\frac{3}{5} \times 100\%$

$= \frac{8}{5} \times 100\% = \frac{800\%}{5}$

$= \mathbf{160\%}$

(c) $\frac{5}{6} \times 100\% = \frac{500\%}{6}$

$\frac{250\%}{3} = \mathbf{83\frac{1}{3}\%}$

10. (a) $75\% = \frac{75}{100} = \mathbf{\frac{3}{4}}$

(b) $75\% = \frac{75}{100} = \mathbf{0.75}$

(c) $5\% = \frac{5}{100} = \mathbf{\frac{1}{20}}$

(d) $5\% = \frac{5}{100} = \mathbf{0.05}$

11.

$$\begin{array}{r} 62 \\ \times\ 60 \\ \hline \end{array}$$
3720 times

12. even primes: 2

$\dfrac{\text{even primes}}{\text{total}} = \mathbf{\dfrac{1}{6}}$

13. (a) Area = (1 in.)(1 in.) = $\mathbf{1\ in.^2}$

(b) Area = $\left(\frac{1}{2}\text{ in.}\right)\left(\frac{1}{2}\text{ in.}\right) = \mathbf{\frac{1}{4}\ in.^2}$

(c) Area = $1\text{ in.}^2 - \frac{1}{4}\text{ in.}^2$

$= \frac{4}{4}\text{ in.}^2 - \frac{1}{4}\text{ in.}^2 = \mathbf{\frac{3}{4}\ in.^2}$

14. Perimeter = $\frac{1}{2}$ in. + 1 in. + 1 in.

$+ \frac{1}{2}$ in. $+ \frac{1}{2}$ in. $+ \frac{1}{2}$ in.

$= 2$ in. $+ \frac{4}{2}$ in. $= 2$ in. $+ 2$ in.

$=$ **4 in.**

15. (a) **8 cm**

(b) **6 cm**

(c) **4.8 cm**

16. $\frac{y}{100} = \frac{18}{45}$

$45y = 100 \cdot 18$

$y = \frac{1800}{45}$

$y = \mathbf{40}$

17. $\frac{35}{40} = \frac{1.4}{m}$

$35m = (1.4)(40)$

$m = \frac{56}{35}$

$m = 1\frac{21}{35} = 1\frac{3}{5} = \mathbf{1.6}$

18.

$$\begin{array}{rcl} \frac{1}{2} & = & \frac{3}{6} \\ -\ \frac{1}{6} & = & \frac{1}{6} \\ \hline & & \frac{2}{6} = \frac{1}{3} \end{array}$$

$n = \mathbf{\frac{1}{3}}$

19.

$$\begin{array}{r} 0.29 \\ 9\overline{)2.61} \\ \underline{1\,8} \\ 81 \\ \underline{81} \\ 0 \end{array}$$

$d = \mathbf{0.29}$

20. $\sqrt{100} + 4^3 = 10 + 4 \cdot 4 \cdot 4$
$= 10 + 64 = \mathbf{74}$

21. $3.14 \times 10^4 = 3.14 \times 10{,}000 = \mathbf{31{,}400}$

22.
$$\begin{array}{r} 4\frac{1}{6} = 4\frac{1}{6} \\ -\ 2\frac{1}{2} = 2\frac{3}{6} \\ \hline \end{array} \longrightarrow \begin{array}{r} 3\frac{7}{6} \\ -\ 2\frac{3}{6} \\ \hline 1\frac{4}{6} = 1\frac{2}{3} \end{array}$$

$$\begin{array}{r} 3\frac{3}{4} = 3\frac{9}{12} \\ +\ 1\frac{2}{3} = 1\frac{8}{12} \\ \hline 4\frac{17}{12} = \mathbf{5\frac{5}{12}} \end{array}$$

23. $3\frac{3}{4} \div 1\frac{1}{2} = \frac{15}{4} \div \frac{3}{2} = \frac{\overset{5}{\cancel{15}}}{\underset{2}{\cancel{4}}} \times \frac{\overset{1}{\cancel{2}}}{\underset{1}{\cancel{3}}} = \frac{5}{2}$

$6\frac{2}{3} \cdot \frac{5}{2} = \frac{\overset{10}{\cancel{20}}}{3} \cdot \frac{5}{\underset{1}{\cancel{2}}} = \frac{50}{3} = \mathbf{16\frac{2}{3}}$

24.
$$\begin{array}{r} 3 \text{ days } \ 8 \text{ hr } \ 15 \text{ min} \\ +\ 2 \text{ days } \ 15 \text{ hr } \ 45 \text{ min} \\ \hline 5 \text{ days } \ 23 \text{ hr } \ 60 \text{ min} \end{array}$$

60 min = 1 hr; 23 hr + 1 hr = 24 hr
24 hr = 1 day
5 days + 1 day = **6 days**

25.
$$\begin{array}{r} 1 \text{ yd } 2 \text{ ft } \ 6 \text{ in.} \\ +\ 2 \text{ yd } 1 \text{ ft } \ 9 \text{ in.} \\ \hline 3 \text{ yd } 3 \text{ ft } 15 \text{ in.} \end{array}$$

15 in. = 1 ft 3 in.

$$\begin{array}{r} 1 \text{ ft } 3 \text{ in.} \\ +\ 3 \text{ ft} \quad\ \ \\ \hline 4 \text{ ft } 3 \text{ in.} \end{array} \longrightarrow 3 \text{ yd } 4 \text{ ft } 3 \text{ in.}$$

4 ft = 1 yd 1 ft

$$\begin{array}{r} 1 \text{ yd } 1 \text{ ft} \\ +\ 3 \text{ yd} \quad\ \ \\ \hline 4 \text{ yd } 1 \text{ ft} \end{array} \longrightarrow \textbf{4 yd 1 ft 3 in.}$$

26.
$$\begin{array}{r} \mathbf{\$300.} \\ 0.06\overline{)\$18.00} \\ 18 \\ \hline 00 \\ 0 \\ \hline 00 \\ 0 \\ \hline 0 \end{array}$$

$300.00

27. **Round 35.675 to 36. Round $2\frac{7}{8}$ to 3. Then divide 36 by 3.**

28.
$$\begin{array}{r} \$18.50 \\ +\ \ \$3.50 \\ \hline \$22.00 \end{array}$$

$$\begin{array}{r} \$22.00 \\ \times\ \ \ 0.06 \\ \hline \$1.3200 \end{array} \longrightarrow \$1.32$$

$$\begin{array}{r} \$22.00 \\ \$1.32 \\ \hline \mathbf{\$23.32} \end{array}$$

29. $LWH = (0.5)(0.2)(0.1)$

$$\begin{array}{r} 0.5 \\ \times\ 0.2 \\ \hline 0.10 \end{array} \longrightarrow 0.1 \qquad \begin{array}{r} 0.1 \\ \times\ 0.1 \\ \hline \mathbf{0.01} \end{array}$$

30. $m\angle a = \mathbf{32°}$
$m\angle b = 180° - (90° + 32°)$
$= 180° - 122° = \mathbf{58°}$
$m\angle c = 180° - 58° = \mathbf{122°}$

Lesson 50, Warm-Up

a. **$240**

b. **1.25**

c. **8**

d. **15**

e. **10**

f. **16**

g. **–5**

Problem Solving

The only 1-digit number that is a factor of 1001 is 7. To find the top factor, we divide 1001 by 7.

$$\begin{array}{r} \mathbf{143} \\ \times\ \ \ \mathbf{7} \\ \hline \mathbf{1001} \end{array}$$

Lesson 50, Lesson Practice

a. $\frac{\textbf{1 yd}}{\textbf{36 in.}}$ and $\frac{\textbf{36 in.}}{\textbf{1 yd}}$

b. $\frac{\textbf{100 cm}}{\textbf{1 m}}$ and $\frac{\textbf{1 m}}{\textbf{100 cm}}$

c. $\frac{\textbf{16 oz}}{\textbf{1 lb}}$ and $\frac{\textbf{1 lb}}{\textbf{16 oz}}$

d. $10 \cancel{\text{yards}} \cdot \frac{36 \text{ inches}}{1 \cancel{\text{yard}}}$
$= \textbf{360 inches}$

e. $24 \cancel{\text{ft}} \cdot \frac{1 \text{ yd}}{3 \cancel{\text{ft}}}$
$= \textbf{8 yd}$

f. $24 \cancel{\text{shillings}} \cdot \frac{12 \text{ pence}}{1 \cancel{\text{shilling}}}$
$= \textbf{288 pence}$

Lesson 50, Mixed Practice

1.
$$\begin{array}{r} 3.5 \\ \times\ 0.4 \\ \hline 1.40 \end{array} \longrightarrow 1.4 \qquad \begin{array}{r} 3.5 \\ +\ 0.4 \\ \hline 3.9 \end{array}$$
$$\begin{array}{r} 3.9 \\ -\ 1.4 \\ \hline \textbf{2.5} \end{array}$$

2. (a) $\frac{\text{parts with 1}}{\text{total parts}} = \frac{4}{10} = \mathbf{\frac{2}{5}}$

(b) $\frac{3}{5} \times 100\% = \frac{300\%}{5} = \mathbf{60\%}$

(c) $\frac{\text{numbers greater than 2}}{\text{total}} = \mathbf{\frac{3}{10}}$

3. $\frac{\$1.17}{13 \text{ ounces}} = \frac{\$0.09}{1 \text{ ounce}}$

$\frac{\$1.44}{18 \text{ ounces}} = \frac{\$0.08}{1 \text{ ounce}}$

13-ounce box = 9¢ per ounce
18-ounce box = 8¢ per ounce
18-ounce box is the better buy.

4. $\frac{20 \text{ miles}}{2.5 \text{ hours}} = \mathbf{8 \frac{miles}{hour}}$

5. $\$2 + \$0.50(5)$
$= \$2 + \$2.50 = \mathbf{\$4.50}$

6. $60\overline{)420}$ — quotient: **7 hours**

7.

30 football players

$\frac{2}{5}$ were endomorphic.	6 players
	6 players
$\frac{3}{5}$ were not endomorphic.	6 players
	6 players
	6 players

(a) 2(6 football players)
$=$ **12 football players**

(b) $\frac{3}{5} \times 100\% = \frac{300\%}{5} = \mathbf{60\%}$

8. **B. 40%**

9.
$$\begin{array}{r} 0.8333\ldots \\ 6\overline{)5.0000\ldots} \\ 4\,8 \\ \hline 20 \\ 18 \\ \hline 20 \\ 18 \\ \hline 20 \\ 18 \\ \hline 2 \end{array}$$
3.8333

10. 7,500,000
$= 7{,}000{,}000 + 500{,}000$
$= \mathbf{(7 \times 10^6) + (5 \times 10^5)}$

11. (a) $0.6 \times 100\% = \mathbf{60\%}$

(b) $\frac{1}{6} \times 100\% = \frac{100\%}{6}$
$= \mathbf{16\frac{2}{3}\%}$

(c) $1\frac{1}{2} \times 100\% = \frac{3}{2} \times 100\%$
$= \frac{300\%}{2} = \mathbf{150\%}$

12. (a) $30\% = \frac{30}{100} = \mathbf{\frac{3}{10}}$

(b) $10\overline{)3.0}$ quotient 0.3; $3\,0$; 0 **0.3**

(c) $250\% = \frac{250}{100} = \frac{25}{10} = \frac{5}{2} = \mathbf{2\frac{1}{2}}$

(d) $250\% = 2\frac{1}{2} = \mathbf{2.5}$

13. **97**

14. (a) Area $= (8 \text{ cm})(12 \text{ cm})$
$= \mathbf{96 \text{ cm}^2}$

(b) Area $= \frac{(6 \text{ cm})(8 \text{ cm})}{2} = \mathbf{24 \text{ cm}^2}$

(c) Area $= 96 \text{ cm}^2 + 24 \text{ cm}^2$
$= \mathbf{120 \text{ cm}^2}$

15.

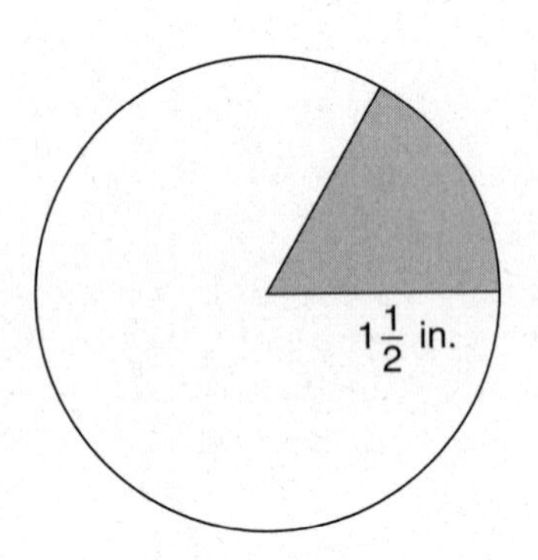

16. $\frac{10}{x} = \frac{7}{42}$

$7x = 10 \cdot 42$

$x = \frac{420}{7}$

$x = \mathbf{60}$

17. $\frac{1.5}{1} = \frac{w}{4}$

$1w = (1.5)4$

$w = \frac{6}{1}$

$w = \mathbf{6}$

18.
$$\begin{array}{r} 5.\overset{5}{\cancel{6}}{}^{1}0 \\ -\ 3.56 \\ \hline 2.04 \end{array}$$

$y = \mathbf{2.04}$

19.
$$\begin{array}{rcl} \frac{3}{20} & = & \frac{9}{60} \\ -\ \frac{1}{15} & = & \frac{4}{60} \\ \hline & & \frac{5}{60} = \frac{1}{12} \end{array}$$

$w = \mathbf{\frac{1}{12}}$

20. (a) **Distributive property**

(b) **Commutative property of addition**

(c) **Identity property of multiplication**

21. **B. 10^4**

22.

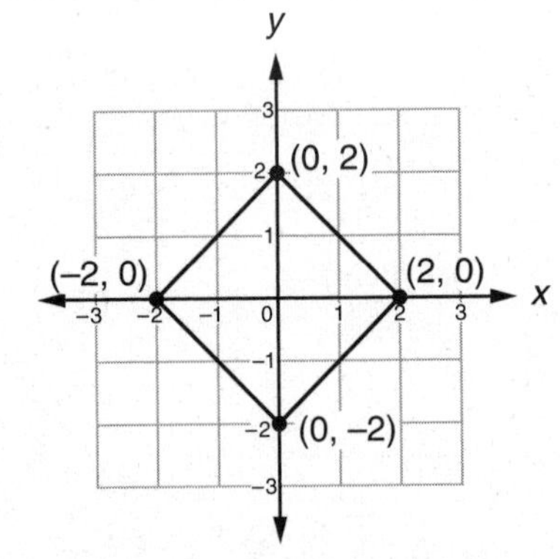

(a) **(0, 2)**

(b) Area $= 4(1 \text{ sq. units}) + 8\left(\frac{1}{2} \text{ sq. units}\right)$
$= 4 \text{ sq. units} + 4 \text{ sq. units}$
$= \mathbf{8 \text{ sq. units}}$

23. $4\overline{)10}$ quotient $\mathbf{2\frac{1}{2}}$ **cookies**; 8; 2

24. (a) **14**

(b) **15**

(c) **See student work.**

25. $\mathbf{\frac{10 \text{ mm}}{1 \text{ cm}}, \frac{1 \text{ cm}}{10 \text{ mm}}}$

$\overset{16}{\cancel{160}}\ \cancel{\text{mm}} \cdot \frac{1 \text{ cm}}{\underset{1}{\cancel{10}}\ \cancel{\text{mm}}} = \mathbf{16 \text{ cm}}$

26.
$$\begin{array}{rrr} 4 \text{ yd} & 2 \text{ ft} & 7 \text{ in.} \\ +\ 3 \text{ yd} & & 5 \text{ in.} \\ \hline 7 \text{ yd} & 2 \text{ ft} & 12 \text{ in.} \end{array}$$

12 in. = 1 ft, 1 ft + 2 ft = 3 ft
3 ft = 1 yd, 7 yd + 1 yd = **8 yd**

27. $1\frac{3}{4} \div 2\frac{1}{3} = \frac{7}{4} \div \frac{7}{3}$

$= \frac{\overset{1}{\cancel{7}}}{4} \times \frac{3}{\underset{1}{\cancel{7}}} = \frac{3}{4}$

$$\begin{array}{rcl} 5\frac{1}{6} = 5\frac{2}{12} & \longrightarrow & 4\frac{14}{12} \\ -\ \frac{3}{4} = \frac{9}{12} & & -\ \frac{9}{12} \\ \hline & & \mathbf{4\frac{5}{12}} \end{array}$$

28. $3\frac{1}{8} \cdot 2\frac{2}{5} = \frac{\overset{5}{\cancel{25}}}{\underset{2}{\cancel{8}}} \cdot \frac{\overset{3}{\cancel{12}}}{\underset{1}{\cancel{5}}} = \frac{15}{2}$

$$\begin{array}{rcl} 3\frac{5}{7} & = & 3\frac{10}{14} \\ +\ \frac{15}{2} & = & \frac{105}{14} \\ \hline & & 3\frac{115}{14} = \mathbf{11\frac{3}{14}} \end{array}$$

29. (a) $m\angle BAC = m\angle CDB =$ **60°**

(b) $m\angle BCA = 180° - (70° + 60°)$
$= 180° - 130° =$ **50°**

(c) $m\angle CBD = m\angle BCA =$ **50°**

30. **4(5 − 3)** or **4(5 − 3)**
4(2) **20 − 12**
8 **8**

INVESTIGATION 5

1.

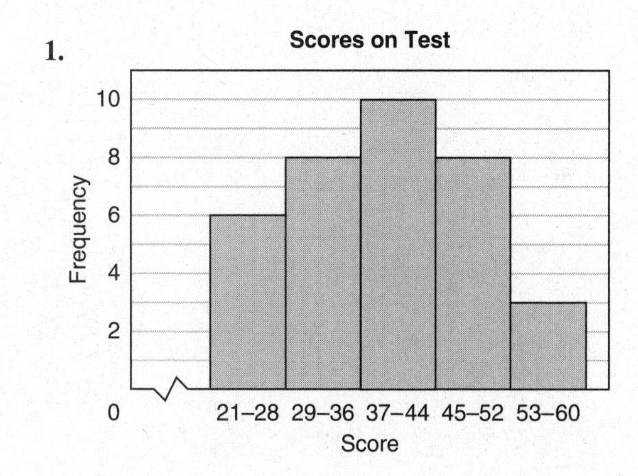

2. **The graph on the left creates the visual impression that sales doubled, because the vertical scale starts at 400 units instead of being equally divided from 0 units to 600 units.**

3.

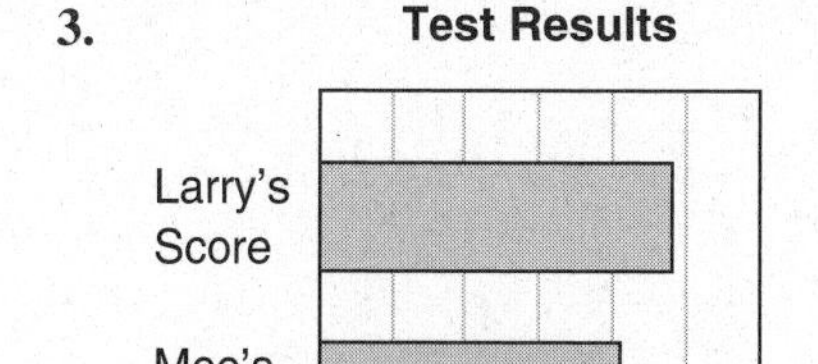
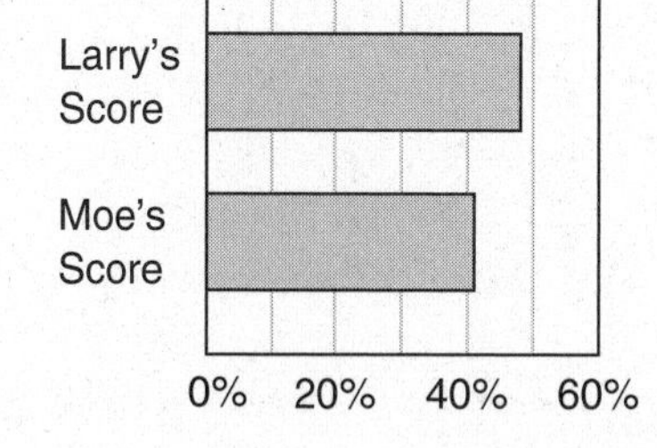

4.

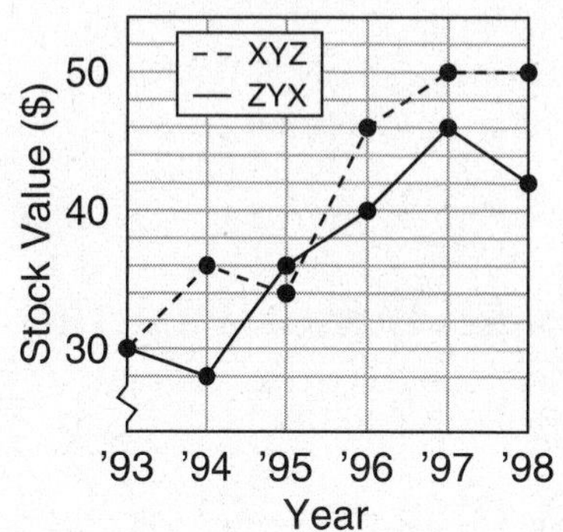

5.

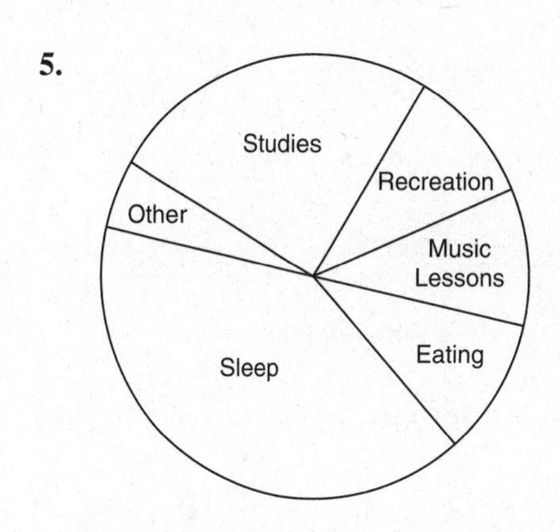

Central angles:
Studies: **90°**
Recreation: **36°**
Music Lessons: **36°**
Eating: **36°**
Sleeping: **144°**
Other: **18°**

Extensions

a. **See student work.**

b. **See student work.**

LESSON 51, WARM-UP

a. **$14.00**

b. **450**

c. **12**

d. **5000 m**

e. **200**

f. **25**

g. $\mathbf{1\frac{3}{8}, \frac{3}{8}, \frac{7}{16}, 1\frac{3}{4}}$

Problem Solving

The 1st perfect square is $1^2 = 1$, the 2nd is $2^2 = 4$, and so on.
The 1000th perfect square is 1000^2.
$1000 \times 1000 =$ **1,000,000**

Lesson 51, Lesson Practice

a. $1.5000000 \longrightarrow \mathbf{1.5 \times 10^7}$

b. $4.00000000000 \longrightarrow \mathbf{4 \times 10^{11}}$

c. $5.090000 \longrightarrow \mathbf{5.09 \times 10^6}$

d. $2.50000000000 \longrightarrow \mathbf{2.5 \times 10^{11}}$

e. $3400000. \longrightarrow$ **3,400,000**

f. $500000000. \longrightarrow$ **500,000,000**

g. $100000. \longrightarrow$ **100,000**

h. $150000. \longrightarrow 150,000$
$1500000. \longrightarrow 1,500,000$
$1.5 \times 10^5 \mathbf{<} 1.5 \times 10^6$

i. $1000000. \longrightarrow 1,000,000$
$1,000,000 = 1,000,000$
one million $\mathbf{=}$ 1×10^6

Lesson 51, Mixed Practice

1. **3 tests**

2. $70 + 80 + 75 + 85 + 90 = 400$

$400 \div 5 = \mathbf{80}$

3. $\frac{9 \text{ in.}}{6} = \frac{3}{2} \text{ in.}$

$5\left(\frac{3}{2} \text{ in.}\right) = \frac{15}{2} \text{ in.} = \mathbf{7\frac{1}{2} \text{ in.}}$

4. $\frac{\$1.98}{6 \text{ cans}} = \frac{\$0.33}{1 \text{ can}}$

$\$0.40 - \$0.33 = \$0.07$

7¢ per can

5. (a) $\frac{\text{unconvinced people}}{\text{total people}} = \mathbf{\frac{2}{7}}$

(b) $\frac{\text{convinced people}}{\text{unconvinced people}} = \mathbf{\frac{5}{2}}$

6. (a) $1.2000000 \longrightarrow \mathbf{1.2 \times 10^7}$

(b) $1.7600 \longrightarrow \mathbf{1.76 \times 10^4}$

7. (a) $12000. \longrightarrow$ **12,000**

(b) $5000000. \longrightarrow$ **5,000,000**

8. (a) $1.000 \div 8 = \mathbf{0.125}$

```
   0.125
8)1.000
  8
  20
  16
   40
   40
    0
```

(b) $87\frac{1}{2}\% = \frac{87.5}{100} = \mathbf{0.875}$

9. (a) **30,000**

(b) **5000**

10. (a) $40\% = \frac{40}{100} = \mathbf{\frac{2}{5}}$

(b) $40\% = \frac{40}{100} = \frac{4}{10} = \mathbf{0.4}$

(c) $4\% = \frac{4}{100} = \mathbf{\frac{1}{25}}$

(d) $4\% = \frac{4}{100} = \mathbf{0.04}$

11. (a) $0.5 \times 360° = \mathbf{180°}$

(b) $0.25 \times 360° = \mathbf{90°}$

(c) $0.125 \times 360° = \mathbf{45°}$

(d) $0.125 \times 360° = \mathbf{45°}$

12.
$$\begin{array}{r} \$15.80 \\ \times \quad 0.05 \\ \hline \$0.7900 \end{array} \longrightarrow \$0.79$$
$$\begin{array}{r} \$15.80 \\ + \quad \$0.79 \\ \hline \mathbf{\$16.59} \end{array}$$

13. **0**

14. (a) $\angle \boldsymbol{Z}$

(b) $\overline{\boldsymbol{DC}}$

15. Perimeter $= 2\text{ m} + 5\text{ m} + 4\text{ m} + 6\text{ m} + 8\text{ m} + 7\text{ m} + 14\text{ m} + 18\text{ m} = \mathbf{64\text{ m}}$

16.

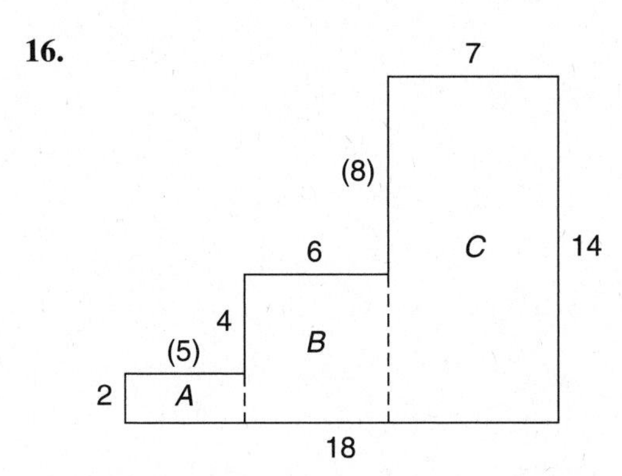

$$\begin{array}{rcl} \text{Area } A = & (2\text{ m})(5\text{ m}) = & 10\text{ m}^2 \\ \text{Area } B = & (6\text{ m})(6\text{ m}) = & 36\text{ m}^2 \\ + \text{ Area } C = & (14\text{ m})(7\text{ m}) = & 98\text{ m}^2 \\ \hline & \text{Area of figure} = & \mathbf{144\text{ m}^2} \end{array}$$

17.
$$\frac{24}{x} = \frac{60}{40}$$
$$60x = 24 \cdot 40$$
$$x = \frac{960}{60}$$
$$x = \mathbf{16}$$

18.
$$\frac{6}{4.2} = \frac{n}{7}$$
$$4.2n = 6 \cdot 7$$
$$n = \frac{42}{4.2}$$
$$n = \mathbf{10}$$

19.
$$\begin{array}{r} 1.68 \\ 5\overline{)8.40} \\ \underline{5} \\ 3\,4 \\ \underline{3\,0} \\ 40 \\ \underline{40} \\ 0 \end{array}$$
$m = \mathbf{1.68}$

20.
$$\begin{array}{r} 6.50 \\ -\ 5.06 \\ \hline 1.44 \end{array}$$
$y = \mathbf{1.44}$

21. $5^2 + 3^3 + \sqrt{64}$
$= 25 + 27 + 8 = \mathbf{60}$

22. $16\text{ cm} \cdot \dfrac{10\text{ mm}}{1\text{ cm}} = \mathbf{160\text{ mm}}$

23.
$$\begin{array}{rrr} 5\text{ days} & 18\text{ hr} & 50\text{ min} \\ +\ 2\text{ days} & 8\text{ hr} & 25\text{ min} \\ \hline 7\text{ days} & 26\text{ hr} & 75\text{ min} \end{array}$$
$75\text{ min} = 1\text{ hr}\quad 15\text{ min}$
$$\begin{array}{rr} 1\text{ hr} & 15\text{ min} \\ +\ 26\text{ hr} & \\ \hline 27\text{ hr} & 15\text{ min} \end{array}$$
$27\text{ hr} = 1\text{ day}\quad 3\text{ hr}$
$$\begin{array}{rrr} 1\text{ day} & 3\text{ hr} & 15\text{ min} \\ +\ 7\text{ days} & & \\ \hline \mathbf{8\text{ days}} & \mathbf{3\text{ hr}} & \mathbf{15\text{ min}} \end{array}$$

24.
$$\begin{array}{rrr} 3\text{ yd} & 2\text{ ft} & 5\text{ in.} \\ +\ 1\text{ yd} & & 9\text{ in.} \\ \hline 4\text{ yd} & 2\text{ ft} & 14\text{ in.} \end{array}$$
$14\text{ in.} = 1\text{ ft}\quad 2\text{ in.}$
$$\begin{array}{rr} 1\text{ ft} & 2\text{ in.} \\ +\ 2\text{ ft} & \\ \hline 3\text{ ft} & 2\text{ in.} \end{array}$$
$3\text{ ft} = 1\text{ yd};\ 1\text{ yd} + 4\text{ yd} = 5\text{ yd}$

5 yd 2 in.

25.
$$\begin{array}{rcl} 5\frac{1}{4} & = & 5\frac{2}{8} \\ -\ 3\frac{7}{8} & = & 3\frac{7}{8} \\ \hline \end{array} \longrightarrow \begin{array}{r} 4\frac{10}{8} \\ -\ 3\frac{7}{8} \\ \hline 1\frac{3}{8} \end{array}$$
$$\begin{array}{rcl} 6\frac{2}{3} & = & 6\frac{16}{24} \\ +\ 1\frac{3}{8} & = & 1\frac{9}{24} \\ \hline & & 7\frac{25}{24} = \mathbf{8\frac{1}{24}} \end{array}$$

26.
$$2\frac{2}{3} \div 1\frac{1}{2} = \frac{8}{3} \div \frac{3}{2}$$
$$= \frac{8}{3} \times \frac{2}{3} = \frac{16}{9}$$
$$3\frac{1}{3} \times \frac{16}{9} = \frac{10}{3} \times \frac{16}{9} = \frac{160}{27}$$
$$= \mathbf{5\frac{25}{27}}$$

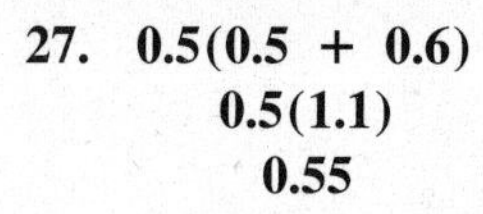

27. **$0.5(0.5 + 0.6)$**
$0.5(1.1)$
0.55

or

$0.5(0.5 + 0.6)$
$0.25 + 0.3$
0.55

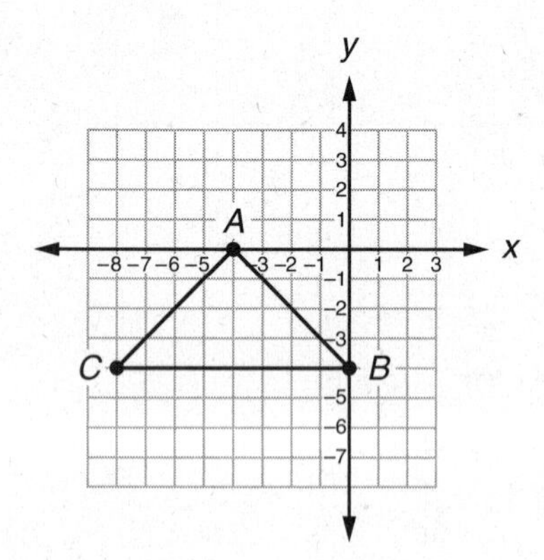

28. **$m\angle A = 90°$; $m\angle B = 45°$;**
$m\angle C = 45°$

29. Area $= \dfrac{(8 \text{ units})(4 \text{ units})}{2}$
$=$ **16 sq. units**

30. **180°F**

Lesson 52, Warm-Up

a. **\$4.50**

b. **0.045**

c. **18**

d. **2.5 m**

e. **600**

f. **180**

g. **200 km**

Problem Solving

If $3x + 5 = 80$, then $3x = 75$ because $75 + 5 = 80$. If $3x = 75$, then $\boldsymbol{x = 25}$ because $3 \cdot 25 = 75$. We test the solution by multiplying 25 by 3 and then adding 5 to get 80.

Lesson 52, Lesson Practice

a. $5 + 5 \cdot 5 - 5 \div 5$
$5 + 25 - 1$
$30 - 1$
29

b. $50 - 8 \cdot 5 + 6 \div 3$
$50 - 40 + 2$
$10 + 2$
12

c. $24 - 8 - 6 \cdot 2 \div 4$
$24 - 8 - 12 \div 4$
$24 - 8 - 3$
$16 - 3$
13

d. $\dfrac{2^3 + 3^2 + 2 \cdot 5}{3}$
$\dfrac{8 + 9 + 2 \cdot 5}{3}$
$\dfrac{8 + 9 + 10}{3}$
$\dfrac{17 + 10}{3}$
$\dfrac{27}{3}$
9

e. $(5)(3) - (3)(4)$
$15 - 12$
3

f. $(6)(4) + \dfrac{(6)}{(2)}$
$24 + 3$
27

g. $\left(\dfrac{2}{3}\right) - \left(\dfrac{2}{3}\right)\left(\dfrac{3}{4}\right)$
$\dfrac{2}{3} - \dfrac{1}{2}$
$\dfrac{4}{6} - \dfrac{3}{6}$
$\boldsymbol{\dfrac{1}{6}}$

Lesson 52, Mixed Practice

1. $(2 \cdot 3 \cdot 5) \div (2 + 3 + 5)$
$= 30 \div 10 =$ **3**

2. $100 - 4(7) = 100 - 28 = 72$
$72 \div 9 = 8$
8 nonagons

3. $\begin{array}{r} 202.020 \\ -\ \ 25.217 \\ \hline \mathbf{176.803} \end{array}$

4. (a) $3\overline{)\$5.9500\ldots}$ = \$1.9833 . . .

$\begin{array}{l} 3 \\ 29 \\ 27 \\ \ 25 \\ \ 24 \\ \ \ 10 \\ \ \ \ 9 \\ \ \ 10 \\ \ \ \ 9 \\ \ \ \ 1 \end{array}$

\$1.98 per tape

(b) $\begin{array}{r} \$5.95 \\ \times\ \ 0.07 \\ \hline \$0.4165 \end{array} \longrightarrow \$0.42$

$\begin{array}{r} \$5.95 \\ +\ \$0.42 \\ \hline \mathbf{\$6.37} \end{array}$

5. (a) $\begin{array}{r} 35 \\ \times\ \ 4 \\ \hline \mathbf{140} \end{array}$ **140 pages**

(b) $\begin{array}{r} 330 \text{ pages} \\ -\ 140 \text{ pages} \\ \hline \mathbf{190 \text{ pages}} \end{array}$

6. $75\% = \frac{75}{100} = \mathbf{\frac{3}{4}}$

60 passengers	
$\frac{3}{4}$ disembarked.	15 passengers
	15 passengers
	15 passengers
$\frac{1}{4}$ did not disembark.	15 passengers

(a) **45 passengers**

(b) **25%**

7. (a) $3.750000 \longrightarrow \mathbf{3.75 \times 10^6}$

(b) $8.0000000 \longrightarrow \mathbf{8 \times 10^7}$

8. (a) $2050000. \longrightarrow \mathbf{2{,}050{,}000}$

(b) $40. \longrightarrow \mathbf{40}$

9. (a) $8\overline{)3.000}$ = **0.375**

$\begin{array}{l} 2\,4 \\ \ \ 60 \\ \ \ 56 \\ \ \ \ 40 \\ \ \ \ 40 \\ \ \ \ \ 0 \end{array}$

(b) $6.5\% = \frac{6.5}{100} = \mathbf{0.065}$

10. $3.\overline{27} = 3.2727\ldots$
3.273

11. (a) $250\% = \frac{250}{100} = \frac{5}{2} = \mathbf{2\frac{1}{2}}$

(b) $250\% = 2\frac{1}{2} = \mathbf{2.5}$

(c) $25\% = \frac{25}{100} = \mathbf{\frac{1}{4}}$

(d) $25\% = \frac{25}{100} = \mathbf{0.25}$

12. (a) $9\overline{)70}$ = $\mathbf{7\frac{7}{9}}$

$\begin{array}{l} 63 \\ \ 7 \end{array}$

(b) $9\overline{)70.00\ldots}$ = 7.77 . . . $\mathbf{7.\overline{7}}$

$\begin{array}{l} 63 \\ \ 70 \\ \ 63 \\ \ \ 70 \\ \ \ 63 \\ \ \ \ 7 \end{array}$

13. **0.99**

14.

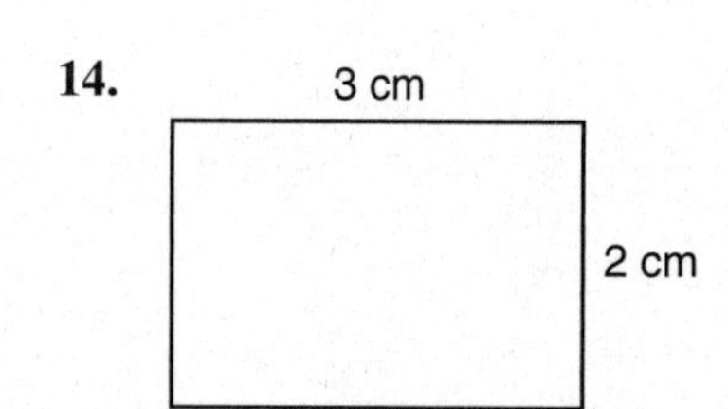

(a) Perimeter = 2(30 mm) + 2(20 mm)
= 60 mm + 40 mm = **100 mm**

(b) Area = (2 cm)(3 cm) = $\mathbf{6\ cm^2}$

15. (a) Area $= \frac{(12 \text{ cm})(6 \text{ cm})}{2} = \mathbf{36\ cm^2}$

(b) Area $= \frac{(8 \text{ cm})(6 \text{ cm})}{2} = \mathbf{24\ cm^2}$

(c) Area $= 36\text{ cm}^2 + 24\text{ cm}^2 = \mathbf{60\ cm^2}$

16. $\frac{8}{f} = \frac{56}{105}$

$56f = 105 \cdot 8$

$f = \frac{840}{56}$

$f = \mathbf{15}$

17. $\frac{12}{15} = \frac{w}{2.5}$

$15w = 12 \cdot 2.5$

$w = \frac{30}{15}$

$w = \mathbf{2}$

18.
$$\begin{array}{r} 20.0 \\ -\ \ 6.8 \\ \hline 13.2 \end{array}$$

$p = \mathbf{13.2}$

19.
$$\begin{array}{r} 6.4 \\ +\ 3.6 \\ \hline 10.0 \end{array} = 10$$

$q = \mathbf{10}$

20. $5^3 - 10^2 - \sqrt{25}$

$= 125 - 100 - 5 = \mathbf{20}$

21. $4 + 4 \cdot 4 - 4 \div 4$

$4 + 16 - 1$

$20 - 1$

$\mathbf{19}$

22. $\frac{4.8 - 0.24}{(0.2)(0.6)}$

$= \frac{4.8 - 0.24}{0.12}$

$= \frac{4.56}{0.12} = \mathbf{38}$

23.
$$\begin{array}{rrr} 5\text{ hr} & 45\text{ min} & 30\text{ s} \\ +\ 2\text{ hr} & 53\text{ min} & 55\text{ s} \\ \hline 7\text{ hr} & 98\text{ min} & 85\text{ s} \end{array}$$

85 s = 1 min 25 s

$$\begin{array}{rr} 1\text{ min} & 25\text{ s} \\ +\ 98\text{ min} & \\ \hline 99\text{ min} & 25\text{ s} \end{array}$$

99 min = 1 hr 39 min

$$\begin{array}{rrr} 1\text{ hr} & 39\text{ min} & 25\text{ s} \\ +\ 7\text{ hr} & & \\ \hline \mathbf{8\text{ hr}} & \mathbf{39\text{ min}} & \mathbf{25\text{ s}} \end{array}$$

24. $5\frac{1}{3} \cdot 2\frac{1}{2} = \frac{\overset{8}{\cancel{16}}}{3} \cdot \frac{5}{\underset{1}{\cancel{2}}}$

$= \frac{40}{3} = 13\frac{1}{3}$

$$\begin{array}{rcr} 6\frac{3}{4} & = & 6\frac{9}{12} \\ +\ 13\frac{1}{3} & = & 13\frac{4}{12} \\ \hline & & 19\frac{13}{12} \end{array} = \mathbf{20\frac{1}{12}}$$

25. $3\frac{3}{4} \div 2 = \frac{15}{4} \div \frac{2}{1}$

$= \frac{15}{4} \times \frac{1}{2} = \frac{15}{8} = 1\frac{7}{8}$

$$\begin{array}{rcr} 5\frac{1}{2} & = & 5\frac{4}{8} \longrightarrow \\ -\ 1\frac{7}{8} & = & 1\frac{7}{8} \\ \hline \end{array} \qquad \begin{array}{r} 4\frac{12}{8} \\ -\ 1\frac{7}{8} \\ \hline \mathbf{3\frac{5}{8}} \end{array}$$

26. $9 + 13 + 8 + 70 = \mathbf{100}$

$$\begin{array}{r} 8.575 \\ 12.625 \\ 8.4 \\ +\ 70.4 \\ \hline 100.000 \end{array} = \mathbf{100}$$

27.
$$\begin{array}{r} 1.25 \\ \times\ \ 0.8 \\ \hline 1.000 \end{array} = 1$$

$1 \times 10^6 = \mathbf{1{,}000{,}000}$

28. $(4)(0.5) + \frac{(4)}{(0.5)}$

$2 + 8$

$\mathbf{10}$

29. $1.4\ \cancel{\text{meters}} \cdot \frac{100\text{ centimeters}}{1\ \cancel{\text{meter}}}$

$= \mathbf{140\ centimeters}$

30. $\frac{\text{Favorite sport is basketball.}}{\text{total}} = \frac{10}{30}$

$= \mathbf{\frac{1}{3}}$

Lesson 53, Warm-Up

a. **$10.00**

b. **127.5**

c. **35**

d. **350 mm**

e. $\mathbf{\frac{1}{4}}$

f. **27**

g. **121**

Problem Solving

Zero. Allen did not form a triangle, since the longest side of a triangle must be shorter than the sum of the two other sides.

Lesson 53, Lesson Practice

a. $\mathbf{\frac{15 \text{ chairs}}{1 \text{ row}}; \frac{1 \text{ row}}{15 \text{ chairs}}}$

b. $18 \cancel{\text{rows}} \cdot \frac{15 \text{ chairs}}{1 \cancel{\text{row}}} = \mathbf{270 \text{ chairs}}$

c. $\mathbf{\frac{24 \text{ miles}}{1 \text{ gallon}}; \frac{1 \text{ gallon}}{24 \text{ miles}}}$

d. $\overset{20}{\cancel{160}} \cancel{\text{miles}} \cdot \frac{1 \text{ gallon}}{\underset{3}{\cancel{24}} \cancel{\text{miles}}}$

$= \frac{20}{3} \text{ gallons} = \mathbf{6\frac{2}{3} \text{ gallons}}$

Lesson 53, Mixed Practice

1. (a) **16 boys**

(b) **16 girls**

2. $\frac{\text{January through June birthdays}}{\text{total birthdays}}$

$= \frac{16}{32} = \frac{1}{2}$

$\frac{1}{2} \times 100\% = \frac{100\%}{2} = \mathbf{50\%}$

3. $\frac{\text{April through June boys' birthdays}}{\text{total boys' birthdays}} = \mathbf{\frac{5}{16}}$

4. (a)

$$\begin{array}{r} \$3.95 \\ \$4.47 \\ \$4.95 \\ \underline{\$4.95} \\ \$18.32 \end{array}$$

$$\begin{array}{r} \mathbf{\$4.58 \text{ per book}} \\ 4\overline{)\$18.32} \\ \underline{16} \\ 2\,3 \\ \underline{2\,0} \\ 32 \\ \underline{32} \\ 0 \end{array}$$

(b)

$$\begin{array}{r} \$18.32 \\ \times \quad 0.08 \\ \hline \$1.4656 \end{array} \longrightarrow \$1.47 \text{ tax}$$

$$\begin{array}{r} \$18.32 \\ \underline{\$1.47} \\ \mathbf{\$19.79 \text{ total}} \end{array}$$

5. 840 gerbils

840 gerbils	
$\frac{7}{12}$ were hiding.	70 gerbils
	70 gerbils
	70 gerbils
	70 gerbils
	70 gerbils
	70 gerbils
	70 gerbils
$\frac{5}{12}$ were not hiding.	70 gerbils
	70 gerbils
	70 gerbils
	70 gerbils
	70 gerbils

(a) $\mathbf{\frac{5}{12}}$

(b) **350 gerbils**

6. (a) $1{,}000000000000 \longrightarrow \mathbf{1 \times 10^{12}}$

(b) $4{,}75000 \longrightarrow \mathbf{4.75 \times 10^5}$

7. (a) $700. \longrightarrow$ **700**

(b) $2500000. \longrightarrow 2{,}500{,}000$

$250000. \longrightarrow 250{,}000$

$2.5 \times 10^6 \mathbf{>} 2.5 \times 10^5$

8. (a) $35 \cancel{\text{yd}} \cdot \frac{3 \text{ ft}}{1 \cancel{\text{yd}}} = \mathbf{105 \text{ ft}}$

(b) $\overset{20}{\cancel{2000}} \cancel{\text{cm}} \cdot \frac{1 \text{ m}}{\underset{1}{\cancel{100}} \cancel{\text{cm}}} = \mathbf{20 \text{ m}}$

9.

$$\begin{aligned} 54 &= 2 \cdot 3 \cdot 3 \cdot 3 \\ 36 &= 2 \cdot 2 \cdot 3 \cdot 3 \\ \text{LCM } (54, 36) &= 2 \cdot 2 \cdot 3 \cdot 3 \cdot 3 \\ &= \mathbf{108} \end{aligned}$$

10.

$$\begin{array}{r} 20{,}000 \\ -\ 12{,}000 \\ \hline \mathbf{8{,}000} \end{array}$$

11. (a) $150\% = \frac{150}{100} = \frac{3}{2} = \mathbf{1\frac{1}{2}}$

(b) $150\% = \frac{3}{2} = 1\frac{1}{2} = \mathbf{1.5}$

(c) $15\% = \frac{15}{100} = \mathbf{\frac{3}{20}}$

d) $15\% = \frac{15}{100} = \mathbf{0.15}$

12. (a) $\frac{4}{5} \times 100\% = \frac{400\%}{5} = \mathbf{80\%}$

(b) $0.06 = \frac{6}{100} = \mathbf{6\%}$

13. 2 m = 200 cm

$$\begin{array}{r} 200 \text{ cm} \\ -\ 165 \text{ cm} \\ \hline \mathbf{35 \text{ cm}} \end{array}$$

14.

8

7

(11) A

(4)

B 4

12

(a)

$$\begin{array}{rcrcr} \text{Area } A & = & (11 \text{ ft})(8 \text{ ft}) & = & 88 \text{ ft}^2 \\ +\ \text{Area } B & = & (4 \text{ ft})(4 \text{ ft}) & = & 16 \text{ ft}^2 \\ \hline & & \text{Area of figure} & = & \mathbf{104 \text{ ft}^2} \end{array}$$

(b) Perimeter = 8 ft + 7 ft + 4 ft + 4 ft + 12 ft + 11 ft = **46 ft**

15. (a) $\mathbf{\frac{1.6 \text{ C\$}}{1 \text{ US\$}}; \frac{1 \text{ US\$}}{1.6 \text{ C\$}}}$

(b) $160 \cancel{\text{US\$}} \cdot \frac{1.6 \text{ C\$}}{1 \cancel{\text{US\$}}} = \mathbf{256 \text{ C\$}}$

16.

$$\frac{18}{100} = \frac{90}{p}$$

$$18p = 100 \cdot 90$$

$$p = \frac{9000}{18}$$

$$p = \mathbf{500}$$

17.

$$\frac{6}{9} = \frac{t}{1.5}$$

$$9t = 6 \cdot 1.5$$

$$t = \frac{9}{9}$$

$$t = \mathbf{1}$$

18.

$$\begin{array}{r} 8.00 \\ -\ 7.25 \\ \hline 0.75 \end{array}$$

$m = \mathbf{0.75}$

19. $\frac{1.5}{10} = 0.15$

$n = \mathbf{0.15}$

20. $\sqrt{81} + 9^2 - 2^5$

$= 9 + 81 - 32 = \mathbf{58}$

21. $16 \div 4 \div 2 + 3 \times 4$

$4 \div 2 + 12$

$2 + 12$

14

22.

$$\begin{array}{rrr} 3 \text{ yd} & 1 \text{ ft} & 7\frac{1}{2} \text{ in.} \\ + \quad & 2 \text{ ft} & 6\frac{1}{2} \text{ in.} \\ \hline 3 \text{ yd} & 3 \text{ ft} & 14 \text{ in.} \end{array}$$

14 in. = 1 ft 2 in.

$$\begin{array}{rr} 1 \text{ ft} & 2 \text{ in.} \\ +\ 3 \text{ ft} & \\ \hline 4 \text{ ft} & 2 \text{ in.} \end{array}$$

4 ft = 1 yd 1 ft

$$\begin{array}{rrr} 1 \text{ yd} & 1 \text{ ft} & 2 \text{ in.} \\ +\ 3 \text{ yd} & & \\ \hline \mathbf{4 \text{ yd}} & \mathbf{1 \text{ ft}} & \mathbf{2 \text{ in.}} \end{array}$$

23. $5\frac{5}{6} \div 2\frac{1}{3} = \frac{35}{6} \div \frac{7}{3}$

$= \frac{\overset{5}{\cancel{35}}}{\underset{2}{\cancel{6}}} \times \frac{\overset{1}{\cancel{3}}}{\underset{1}{\cancel{7}}} = \frac{5}{2} = 2\frac{1}{2}$

$$\begin{array}{rcr} 12\frac{2}{3} & = & 12\frac{4}{6} \\ +\ 2\frac{1}{2} & = & 2\frac{3}{6} \\ \hline & & 14\frac{7}{6} = \mathbf{15\frac{1}{6}} \end{array}$$

24. $1\frac{1}{2} \cdot 3\frac{1}{5} = \frac{3}{\cancel{2}_1} \cdot \frac{\cancel{16}^8}{5} = \frac{24}{5} = 4\frac{4}{5}$

$$\begin{array}{rcr} 8\frac{3}{5} & \longrightarrow & 7\frac{8}{5} \\ -\ 4\frac{4}{5} & & -\ 4\frac{4}{5} \\ \hline & & \mathbf{3\frac{4}{5}} \end{array}$$

25. $3.875 \times 10^1 \longrightarrow 38.75$

$$\begin{array}{r} 10.6 \\ 4.2 \\ 16.4 \\ +\ 38.75 \\ \hline \mathbf{69.95} \end{array}$$

26. $7 \times 4 =$ **28**

27. $\frac{(6)(0.9)}{(0.9)(5)}$

$\frac{5.4}{4.5}$

1.2

28. $30\overline{)1000}$ with quotient $33\frac{1}{3}$

33 flats

29. $\frac{\text{guessing wrong answer}}{\text{all answers}} = \mathbf{\frac{4}{5}}$

30.
$$\begin{aligned} m\angle a &= 180° - 130° = \mathbf{50°} \\ m\angle b &= 180° - (90° + 50°) \\ &= 180° - 140° = \mathbf{40°} \\ m\angle c &= 180° - (40° + 60°) \\ &= 180° - 100° = \mathbf{80°} \end{aligned}$$

Lesson 54, Warm-Up

a. **$18.00**

b. **1.275**

c. **8**

d. **150 cm**

e. **3**

f. **27**

g. **$1\frac{3}{20}, \frac{7}{20}, \frac{3}{10}, 1\frac{7}{8}$**

Problem Solving

At first, there were twice as many boys as girls ($\frac{1}{3}$ were girls, so $\frac{2}{3}$ were boys). When 3 boys left, there were the same number of boys and girls. Therefore, there were **3 girls** in the room.

Lesson 54, Lesson Practice

a.

	Ratio	Actual Count
Girls	9	63
Boys	7	*B*

$$\frac{9}{7} = \frac{63}{B}$$
$$9B = 441$$
$$B = \mathbf{49\ boys}$$

b.

	Ratio	Actual Count
Sparrows	5	*S*
Blue jays	3	15

$$\frac{5}{3} = \frac{S}{15}$$
$$3S = 75$$
$$S = \mathbf{25\ sparrows}$$

c.

	Ratio	Actual Count
Tagged fish	2	90
Untagged fish	9	*U*

$$\frac{2}{9} = \frac{90}{U}$$
$$2U = 810$$
$$U = \mathbf{405\ untagged\ fish}$$

d. **See student work.**

Lesson 54, Mixed Practice

1. $1776 + 50 = 1826$

$$\begin{array}{r} 1826 \\ -\ 1743 \\ \hline \mathbf{83\ years} \end{array}$$

2.

$$\begin{array}{r} 190 \text{ cm} \\ 195 \text{ cm} \\ 197 \text{ cm} \\ 201 \text{ cm} \\ +\ 203 \text{ cm} \\ \hline 986 \text{ cm} \end{array} \qquad 5\overline{)986}\ = 197.2$$

197 cm

3.

	Ratio	Actual Count
Winners	5	1200
Losers	4	L

$$\frac{5}{4} = \frac{1200}{L}$$

$5L = 4800$

$L =$ **960 losers**

4. $2.6 \cancel{\text{pounds}} \cdot \dfrac{\$1.75}{1 \cancel{\text{pound}}}$

$=$ **$4.55**

5. $4 = 2 \cdot 2$

$6 = 2 \cdot 3$

$\text{LCM}(4, 6) = 2 \cdot 2 \cdot 3 = 12$

$\text{GCF}(4, 6) = 2$

$\dfrac{12}{2} = \mathbf{6}$

6.

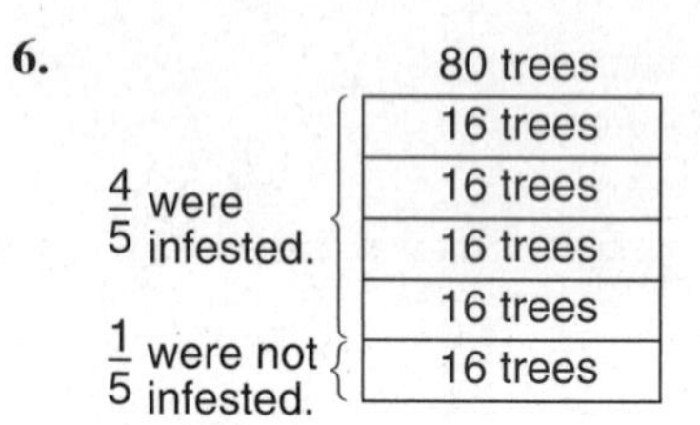

(a) **64 trees**

(b) **16 trees**

7. (a) $4.05000 \longrightarrow$ **4.05×10^5**

(b) $004000. \longrightarrow$ **4000**

8. (a) 10^8

(b) 10^4

9. (a) $\overset{1760}{\cancel{5280}}\ \cancel{\text{ft}} \cdot \dfrac{1 \text{ yd}}{\underset{1}{\cancel{3}}\ \cancel{\text{ft}}} =$ **1760 yd**

(b) $300\ \cancel{\text{cm}} \cdot \dfrac{10 \text{ mm}}{1\ \cancel{\text{cm}}} =$ **3000 mm**

10. **3.1416**

11. (a) $0.4 \times 360° =$ **144°**

(b) $0.3 \times 360° =$ **108°**

(c) $0.2 \times 360° =$ **72°**

(d) $0.1 \times 360° =$ **36°**

12. (a) $4\ \cancel{\text{hours}} \cdot \dfrac{60 \text{ miles}}{1\ \cancel{\text{hour}}} =$ **240 miles**

(b) $\overset{5}{\cancel{300}}\ \cancel{\text{miles}} \cdot \dfrac{1 \text{ hour}}{\underset{1}{\cancel{60}}\ \cancel{\text{miles}}} =$ **5 hours**

13. **B.** $\mathbf{2^4}$

14. Perimeter $= 5 \text{ cm} + 8 \text{ cm} + 3 \text{ cm}$
$+ 5 \text{ cm} + 6 \text{ cm} + 9 \text{ cm}$
$+ 2 \text{ cm} + 4 \text{ cm} =$ **42 cm**

15.

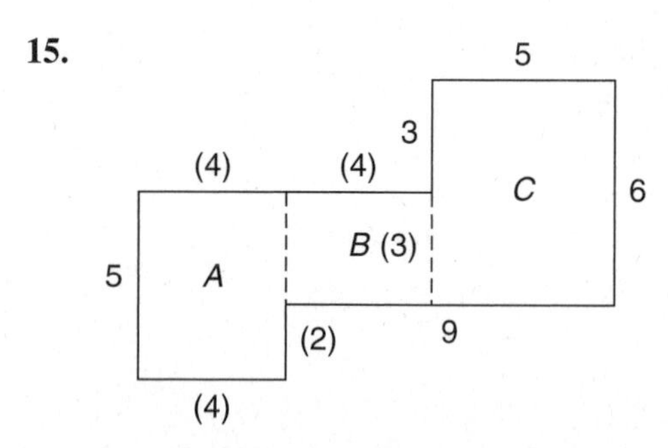

$$\begin{array}{rcl} \text{Area } A &=& (4 \text{ cm})(5 \text{ cm}) = 20 \text{ cm}^2 \\ \text{Area } B &=& (4 \text{ cm})(3 \text{ cm}) = 12 \text{ cm}^2 \\ +\ \text{Area } C &=& (5 \text{ cm})(6 \text{ cm}) = 30 \text{ cm}^2 \\ \hline \text{Area of figure} &=& \mathbf{62 \text{ cm}^2} \end{array}$$

16. (a) **Identity property of addition**

(b) **Distributive property**

(c) **Associative property of addition**

17. 0.5 in.

0.5 in.

(a) Perimeter $= 4(0.5 \text{ in.}) =$ **2 inches**

(b) Area $= (0.5 \text{ in.})(0.5 \text{ in.})$
$=$ **0.25 square inch**

18. **The average score is likely to be below the median score. The mean "balances" low scores with high scores. The scores above the median are not far enough above the median to shift the mean to be at or above the median.**

19.
$$\begin{array}{r} 6.2 \\ -\ 4.1 \\ \hline 2.1 \end{array}$$
$x = \mathbf{2.1}$

20.
$$\begin{array}{r} 1.2 \\ +\ 0.21 \\ \hline 1.41 \end{array}$$
$y = \mathbf{1.41}$

21. $\frac{24}{r} = \frac{36}{27}$
$36r = 24 \cdot 27$
$r = \frac{648}{36}$
$r = \mathbf{18}$

22.
$$\begin{array}{r} 6.25 \\ \times\ 0.16 \\ \hline 3750 \\ 625 \\ \hline 1.0000 \end{array} = 1$$
$w = \mathbf{1}$

23. $11^2 + 1^3 - \sqrt{121}$
$= 121 + 1 - 11 = \mathbf{111}$

24. $24 - 4 \times 5 \div 2 + 5$
$24 - 20 \div 2 + 5$
$24 - 10 + 5$
$14 + 5$
$\mathbf{19}$

25. $\frac{(2.5)^2}{2(2.5)} = \frac{6.25}{5} = \mathbf{1.25}$

26.

	1 week	5 days	14 hr
+	2 weeks	6 days	10 hr
	3 weeks	11 days	24 hr

24 hr = 1 day
11 days + 1 day = 12 days
12 days = 1 week 5 days

	1 week	5 days
+	3 weeks	
	4 weeks	**5 days**

27.
$$\begin{array}{rcl} 9\frac{1}{2} = 9\frac{3}{6} & \longrightarrow & 8\frac{9}{6} \\ -\ 6\frac{2}{3} = 6\frac{4}{6} & & -\ 6\frac{4}{6} \\ \hline & & 2\frac{5}{6} \end{array}$$

$$\begin{array}{r} 3\frac{5}{10} = 3\frac{15}{30} \\ +\ 2\frac{5}{6} = 2\frac{25}{30} \\ \hline 5\frac{40}{30} = 6\frac{10}{30} = \mathbf{6\frac{1}{3}} \end{array}$$

28. $6 \div 3\frac{2}{3} = \frac{6}{1} \div \frac{11}{3}$
$= \frac{6}{1} \times \frac{3}{11} = \frac{18}{11}$
$7\frac{1}{3} \cdot \frac{18}{11} = \frac{\overset{2}{\cancel{22}}}{\underset{1}{\cancel{3}}} \cdot \frac{\overset{6}{\cancel{18}}}{\underset{1}{\cancel{11}}}$
$= \mathbf{12}$

29.

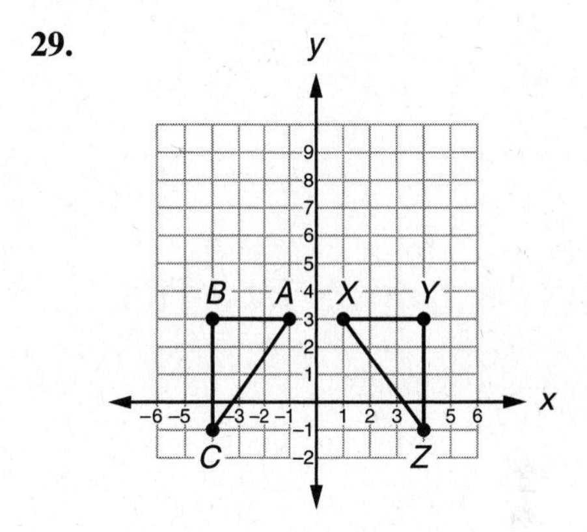

30. (a) **Yes**

(b) **Yes**

(c) $\angle C$

Lesson 55, Warm-Up

a. **$5.00**

b. **37.5**

c. **40**

d. **3 km**

e. $\mathbf{\frac{4}{9}}$

f. **75**

g. **75 pages**

Problem Solving

We know that the ones digit of the top partial product is 9 because the ones digit of the product is 9. Therefore, the ones digits of the factors are either 3 and 3 or 7 and 7. By checking both possibilities, we find that the ones digits of the factors are 7 and 7. Because the bottom partial product has only 3 digits and its tens digit is 3, we find that the tens digit of the bottom factor is 1. We know that the hundreds digit of the top factor is either 4 or 5 because the thousands digit of the top partial product is 3. Because the thousands digit of the product is 9, the hundreds digit of the top factor must be 5.

$$\begin{array}{r} \mathbf{537} \\ \times\ \mathbf{17} \\ \hline \mathbf{3759} \\ \mathbf{537} \\ \hline \mathbf{9129} \end{array}$$

Lesson 55, Lesson Practice

a. $5(18 \text{ points}) = \mathbf{90\ points}$

b. $4(45) = 180$
$24 + 36 + 52 + n = 180$
$n = \mathbf{68}$

c. $5(91) = 455$
$6(89) = 534$

$$\begin{array}{r} 534 \\ -\ 455 \\ \hline \mathbf{79} \end{array}$$

Lesson 55, Mixed Practice

1.

	Ratio	Actual Count
Sailboats	7	56
Rowboats	4	R

$\frac{7}{4} = \frac{56}{R}$

$7R = 56 \cdot 4$

$R = \mathbf{32\ rowboats}$

2. $4(85) = 340$
$76 + 78 + 81 + n = 340$
$n = \mathbf{105}$

3. $12\overline{)\$8.64}$ = $0.72

$$\begin{array}{r} \$0.72 \\ 12\overline{)\$8.64} \\ 8\,4 \\ \hline 24 \\ 24 \\ \hline 0 \end{array} \qquad \begin{array}{r} \$0.89 \\ -\ \$0.72 \\ \hline \$0.17 \end{array}$$

$0.17 per container

4. $BC - AB = 2\frac{2}{8} \text{ in.} - 1\frac{6}{8} \text{ in.}$

$$\begin{array}{r} 2\frac{2}{8} \text{ in.} \\ -\ 1\frac{6}{8} \text{ in.} \\ \hline \end{array} \longrightarrow \begin{array}{r} 1\frac{10}{8} \text{ in.} \\ -\ 1\frac{6}{8} \text{ in.} \\ \hline \frac{4}{8} \text{ in.} \end{array} = \mathbf{\frac{1}{2}\ in.}$$

5.

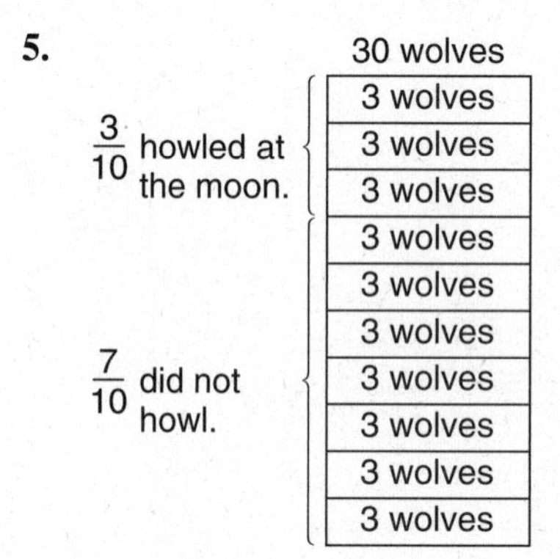

(a) **9 wolves**

(b) $\frac{3}{10} \times 100\% = \mathbf{30\%}$

6. (a) $6.75000000 \longrightarrow \mathbf{6.75 \times 10^8}$

(b) $186000. \longrightarrow \mathbf{186{,}000}$

7. (a) 10^{10}

(b) 10^6

8. (a) $24 \text{ feet} \cdot \frac{12 \text{ inches}}{1 \text{ foot}} = \mathbf{288\ inches}$

(b) $\overset{50}{\cancel{500}} \text{ millimeters} \cdot \frac{1 \text{ centimeter}}{\underset{1}{\cancel{10}} \text{ millimeters}}$

$= \mathbf{50\ centimeters}$

9. $0.02 \cdot 0.025 = \mathbf{0.0005}$

10.
$$\begin{array}{r} \$3.25 \\ +\ \$1.10 \\ \hline \$4.35 \end{array} \qquad \begin{array}{r} \$4.35 \\ \times\ 0.07 \\ \hline \$0.3045 \end{array} \longrightarrow \$0.30$$

$$\begin{array}{r} \$4.35 \\ +\ \$0.30 \\ \hline \mathbf{\$4.65} \end{array}$$

11. (a) $5\overline{)1.0}$ with quotient $\mathbf{0.2}$

(b) $\frac{1}{5} \times 100\% = \frac{100\%}{5} = \mathbf{20\%}$

(c) $0.1 = \mathbf{\frac{1}{10}}$

(d) $0.1 \times 100\% = \mathbf{10\%}$

(e) $75\% = \frac{75}{100} = \mathbf{\frac{3}{4}}$

(f) $75\% = \frac{75}{100} = \mathbf{0.75}$

12. (a) $\overline{AD}$ (or $\overline{DA}$)

(b) $\overline{DC}$ (or $\overline{CD}$) **and** $\overline{AH}$ (or $\overline{HA}$)

(c) $\angle DAB$ (or $\angle BAD$)

13. (a) Area $= (6\text{ cm})(8\text{ cm}) = \mathbf{48\ cm^2}$

(b) Area $= \frac{(4\text{ cm})(8\text{ cm})}{2} = \mathbf{16\ cm^2}$

(c) Area $= 48\text{ cm}^2 + 16\text{ cm}^2 = \mathbf{64\ cm^2}$

14. 6 feet 2 inches = 6(12 inches) + 2 inches
= 74 inches

$$\begin{array}{r} 74\text{ inches} \\ -\ 68\text{ inches} \\ \hline \mathbf{6\ inches} \end{array}$$

15. (a) $\mathbf{\frac{5\ laps}{4\ min}; \frac{4\ min}{5\ laps}}$

(b) $\overset{5}{\cancel{20}}\ \cancel{\text{min}} \cdot \frac{5\text{ laps}}{\underset{1}{\cancel{4}}\ \cancel{\text{min}}} = \mathbf{25\ laps}$

(c) $\overset{4}{\cancel{20}}\ \cancel{\text{laps}} \cdot \frac{4\text{ min}}{\underset{1}{\cancel{5}}\ \cancel{\text{laps}}} = \mathbf{16\ minutes}$

16. $\frac{1}{2}\left(\frac{1}{4} + \frac{1}{2}\right)$ or $\frac{1}{2}\left(\frac{1}{4} + \frac{1}{2}\right)$

$\frac{1}{2}\left(\frac{3}{4}\right) \qquad \frac{1}{8} + \frac{1}{4}$

$\mathbf{\frac{3}{8}} \qquad \mathbf{\frac{3}{8}}$

17.
$$\frac{30}{70} = \frac{21}{x}$$
$$30x = 21 \cdot 70$$
$$x = \frac{1470}{30}$$
$$x = \mathbf{49}$$

18. $25\overline{)10000}$ with quotient 400

$w = \mathbf{400}$

19. $2 + 7 + 5 = \mathbf{14}$

$$\begin{array}{rcl} 2\frac{5}{12} & = & 2\frac{10}{24} \\ 6\frac{5}{6} & = & 6\frac{20}{24} \\ +\ 4\frac{7}{8} & = & 4\frac{21}{24} \\ \hline & & 12\frac{51}{24} = 14\frac{3}{24} = \mathbf{14\frac{1}{8}} \end{array}$$

20. $6 - (7 - 5) = 6 - 2 = \mathbf{4}$

$$\begin{array}{rcl} 7\frac{1}{3} & = & 7\frac{5}{15} \\ -\ 4\frac{4}{5} & = & 4\frac{12}{15} \\ \hline \end{array} \longrightarrow \begin{array}{r} 6\frac{20}{15} \\ -\ 4\frac{12}{15} \\ \hline 2\frac{8}{15} \end{array}$$

$$\begin{array}{r} 6 \\ -\ 2\frac{8}{15} \\ \hline \end{array} \longrightarrow \begin{array}{r} 5\frac{15}{15} \\ -\ 2\frac{8}{15} \\ \hline \mathbf{3\frac{7}{15}} \end{array}$$

21. $10\ \cancel{\text{yd}} \cdot \frac{36\text{ in.}}{1\ \cancel{\text{yd}}} = \mathbf{360\ in.}$

22.
$$\begin{array}{rrr} 8\text{ yd} & 2\text{ ft} & 7\text{ in.} \\ + & & 5\text{ in.} \\ \hline 8\text{ yd} & 2\text{ ft} & 12\text{ in.} \end{array}$$
12 in. = 1 ft
2 ft + 1 ft = 3 ft
3 ft = 1 yd
8 yd + 1 yd = 9 yd
9 yd

23. $12^2 - 4^3 - 2^4 - \sqrt{144}$
$= 144 - 64 - 16 - 12 = \mathbf{52}$

24. $50 + 30 \div 5 \cdot 2 - 6$
$50 + 6 \cdot 2 - 6$
$50 + 12 - 6$
$62 - 6$
$\mathbf{56}$

25. $6\frac{2}{3} \cdot 5\frac{1}{4} \cdot 2\frac{1}{10}$

$$= \frac{\overset{1}{\overset{\cancel{2}}{\cancel{20}}}}{\underset{1}{\cancel{3}}} \cdot \frac{\overset{7}{\cancel{21}}}{\underset{2}{\cancel{4}}} \cdot \frac{21}{\underset{1}{\cancel{10}}} = \frac{147}{2} = \mathbf{73\frac{1}{2}}$$

26. $3\frac{1}{3} \div 3 \div 2\frac{1}{2}$

$$= \frac{10}{3} \div \frac{3}{1} \div \frac{5}{2}$$

$$= \left(\frac{10}{3} \times \frac{1}{3}\right) \div \frac{5}{2} = \frac{10}{9} \div \frac{5}{2}$$

$$= \frac{\overset{2}{\cancel{10}}}{9} \times \frac{2}{\underset{1}{\cancel{5}}} = \mathbf{\frac{4}{9}}$$

27.

$$\begin{array}{r} 6.000 \\ -\ 1.359 \\ \hline 4.641 \end{array} \qquad \begin{array}{r} 3.47 \\ +\ 4.641 \\ \hline \mathbf{8.111} \end{array}$$

28.

$$\begin{array}{r} 0.28 \\ \times\ 0.6 \\ \hline 0.168 \end{array} \qquad \begin{array}{r} 0.168 \\ \times\ 0.01 \\ \hline \mathbf{0.00168} \end{array}$$

29. $75\overline{)\$1500.00}$ = **$20.00**

30.
$$\begin{aligned} m\angle a &= 180° - (90° + 52°) \\ &= 180° - 142° = \mathbf{38°} \\ m\angle b &= 90° - 38° = \mathbf{52°} \\ m\angle c &= 90° - 52° = \mathbf{38°} \end{aligned}$$

Lesson 56, Warm-Up

a. **75**

b. **0.025**

c. **12**

d. **50 cm**

e. **400**

f. **$35.00**

g. **−3**

Problem Solving

For each possible hundreds digit (1−9), there are 10 palindromes. (E.g., 101, 111, …, 181, 191 for the hundreds digit 1.)

$9 \times 10 = \mathbf{90}$

Lesson 56, Lesson Practice

a. (60 min)

$$\begin{array}{rrr} \overset{2}{\cancel{3}}\text{ hr} & & 3\text{ s} \\ -\ 1\text{ hr} & 15\text{ min} & 55\text{ s} \\ \hline \end{array} \longrightarrow$$

$$\begin{array}{rrr} \overset{2}{\cancel{3}}\text{ hr} & \overset{59}{\cancel{60}}\text{ min} & \overset{63}{\cancel{3}}\text{ s} \\ -\ 1\text{ hr} & 15\text{ min} & 55\text{ s} \\ \hline \mathbf{1\text{ hr}} & \mathbf{44\text{ min}} & \mathbf{8\text{ s}} \end{array}$$

b. (12 in.)

$$\begin{array}{rrr} 8\text{ yd} & \overset{0}{\cancel{1}}\text{ ft} & 5\text{ in.} \\ -\ 3\text{ yd} & 2\text{ ft} & 7\text{ in.} \\ \hline \end{array} \longrightarrow$$

(3 ft)

$$\begin{array}{rrr} \overset{7}{\cancel{8}}\text{ yd} & \overset{0}{\cancel{1}}\text{ ft} & \overset{17}{\cancel{5}}\text{ in.} \\ -\ 3\text{ yd} & 2\text{ ft} & 7\text{ in.} \\ \hline & & 10\text{ in.} \end{array} \longrightarrow$$

$$\begin{array}{rrr} \overset{7}{\cancel{8}}\text{ yd} & \overset{3}{\overset{\cancel{0}}{\cancel{1}}}\text{ ft} & \overset{17}{\cancel{5}}\text{ in.} \\ -\ 3\text{ yd} & 2\text{ ft} & 7\text{ in.} \\ \hline \mathbf{4\text{ yd}} & \mathbf{1\text{ ft}} & \mathbf{10\text{ in.}} \end{array}$$

c. (60 min)

$$\begin{array}{rrr} 2\text{ days} & \overset{2}{\cancel{3}}\text{ hr} & 30\text{ min} \\ -\ 1\text{ day} & 8\text{ hr} & 45\text{ min} \\ \hline \end{array} \longrightarrow$$

(24 hr)

$$\begin{array}{rrr} \overset{1}{\cancel{2}}\text{ days} & \overset{2}{\cancel{3}}\text{ hr} & \overset{90}{\cancel{30}}\text{ min} \\ -\ 1\text{ day} & 8\text{ hr} & 45\text{ min} \\ \hline & & 45\text{ min} \end{array} \longrightarrow$$

$$\begin{array}{rrr} \overset{1}{\cancel{2}}\text{ days} & \overset{26}{\overset{\cancel{2}}{\cancel{3}}}\text{ hr} & \overset{90}{\cancel{30}}\text{ min} \\ -\ 1\text{ day} & 8\text{ hr} & 45\text{ min} \\ \hline & \mathbf{18\text{ hr}} & \mathbf{45\text{ min}} \end{array}$$

Lesson 56, Mixed Practice

1.

$$\begin{array}{r} 0.0329 \\ -\ 0.0320 \\ \hline 0.0009 \end{array}$$

Nine ten-thousandths

2.

	Ratio	Actual Count
Length	4	12 feet
Width	3	*W*

(a)
$$\begin{aligned} \frac{4}{3} &= \frac{12\text{ feet}}{W} \\ 4W &= 3(12\text{ feet}) \\ W &= \mathbf{9\text{ feet}} \end{aligned}$$

(b)
$$\begin{aligned} \text{Perimeter} &= 2(12\text{ feet}) + 2(9\text{ feet}) \\ &= 24\text{ feet} + 18\text{ feet} = \mathbf{42\text{ feet}} \end{aligned}$$

3. $\$2 + \$0.50(4) = \$2 + \$2 = \mathbf{\$4}$

4.
$$4(85) = 340$$
$$340 + 90 = 430$$
$$\frac{430}{5} = \mathbf{86}$$

5.
$$\frac{\$1.50}{12 \text{ ounces}} = \frac{\$0.125}{1 \text{ ounce}}$$
$$\frac{\$1.92}{16 \text{ ounces}} = \frac{\$0.12}{1 \text{ ounce}}$$

Brand X = 12.5¢ per ounce
Brand Y = 12¢ per ounce
Brand Y is the better buy.

6. (a) $\frac{\text{igneous rocks}}{\text{total rocks}} = \mathbf{\frac{3}{8}}$

(b) $\frac{\text{igneous rocks}}{\text{metamorphic rocks}} = \mathbf{\frac{3}{5}}$

(c) $\frac{5}{8} \times 100\% = \frac{500\%}{8} = \mathbf{62\frac{1}{2}\%}$

7. (a) **∠*QPR*** (or **∠*RPQ***) **and ∠*TPS*** (or **∠*SPT***)

∠*RPS* (or **∠*SPR***) **and ∠*QPT*** (or **∠*TPQ***)

(b) **∠*RPQ*** (or **∠*QPR***) **and ∠*SPT*** (or **∠*TPS***)

8. (a) $6{,}10000 \longrightarrow \mathbf{6.1 \times 10^5}$

(b) $15000. \longrightarrow \mathbf{15{,}000}$

9. (a) $\overset{9}{\cancel{216}}\ \cancel{\text{hours}} \cdot \frac{1 \text{ day}}{\underset{1}{\cancel{24}}\ \cancel{\text{hours}}} = \mathbf{9 \text{ days}}$

(b) $5\ \cancel{\text{minutes}} \cdot \frac{60 \text{ seconds}}{1\ \cancel{\text{minute}}} = \mathbf{300 \text{ seconds}}$

10. (a)
$$\begin{array}{r} 0.166\ldots \\ 6\overline{)1.000\ldots} \\ \underline{6} \\ 40 \\ \underline{36} \\ 40 \\ \underline{36} \\ 4 \end{array}$$

0.17

(b) $\frac{1}{6} \times 100\% = \frac{100\%}{6} = \mathbf{16\frac{2}{3}\%}$

11. $1{,}000{,}000\ \cancel{\text{dollars}} \cdot \frac{100 \text{ pennies}}{1\ \cancel{\text{dollar}}}$
$= 100{,}000{,}000 \text{ pennies} = \mathbf{1 \times 10^8 \text{ pennies}}$

12. $1{,}1000000 \longrightarrow 1.1 \times 10^7$
11 million **>** 1.1×10^6

13. **70**

14.
$$\frac{100°}{180°} = \frac{10°}{\mathbf{F}}$$
$$\mathbf{F}(100°) = (180°)(10°)$$
$$= \mathbf{18° \ F}$$

15. Perimeter = 20 mm + 30 mm + 10 mm + 15 mm + 15 mm + 8 mm + 5 mm + 7 mm = **110 mm**

16.

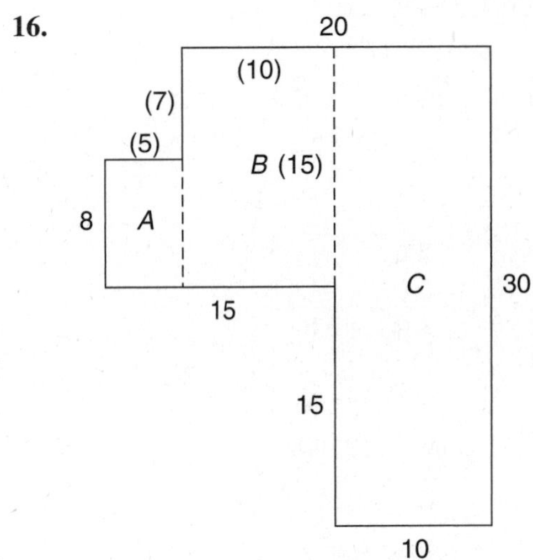

$$\begin{aligned} \text{Area } A &= (5 \text{ mm})(8 \text{ mm}) = 40 \text{ mm}^2 \\ \text{Area } B &= (10 \text{ mm})(15 \text{ mm}) = 150 \text{ mm}^2 \\ + \text{ Area } C &= (30 \text{ mm})(10 \text{ mm}) = 300 \text{ mm}^2 \\ \hline \text{Area of figure} &= \mathbf{490 \text{ mm}^2} \end{aligned}$$

17.
$$\frac{3}{2.5} = \frac{48}{c}$$
$$3c = 48(2.5)$$
$$c = \frac{120}{3}$$
$$c = \mathbf{40}$$

18.
$$\begin{array}{r} 0.75 \\ +\ 0.75 \\ \hline 1.50 \end{array} = 1.5$$
$k = \mathbf{1.5}$

19. $15^2 - 5^3 - \sqrt{100} = 225 - 125 - 10$
$= 100 - 10 = \mathbf{90}$

20. $6 + 12 \div 3 \cdot 2 - 3 \cdot 4$
$6 + 4 \cdot 2 - 12$
$6 + 8 - 12$
$14 - 12$
$\mathbf{2}$

21.

5 yd	2 ft	3 in.
+ 2 yd	2 ft	9 in.
7 yd	4 ft	12 in.

12 in. = 1 ft
4 ft + 1 ft = 5 ft
5 ft = 1 yd 2 ft

1 yd	2 ft
+ 7 yd	
8 yd	**2 ft**

22.

5 yd	$\overset{1}{\cancel{2}}$ ft → (12 in.)	3 in.
− 2 yd	2 ft	9 in.

→ (3 ft)

$\overset{4}{\cancel{5}}$ yd	$\overset{1}{\cancel{2}}$ ft	$\overset{15}{\cancel{3}}$ in.
− 2 yd	2 ft	9 in.
		6 in.

→

$\overset{4}{\cancel{5}}$ yd	$\overset{4}{\cancel{1}}$ $\cancel{2}$ ft	$\overset{15}{\cancel{3}}$ in.
− 2 yd	2 ft	9 in.
2 yd	**2 ft**	**6 in.**

23. $\frac{88 \text{ km}}{1 \cancel{\text{hr}}} \cdot 4 \cancel{\text{hr}} = \mathbf{352\ km}$

24. $5\frac{1}{6} = 5\frac{2}{12} \longrightarrow 4\frac{14}{12}$

$-1\frac{1}{4} = 1\frac{3}{12} \qquad -1\frac{3}{12}$

$3\frac{11}{12}$

$2\frac{3}{4} = 2\frac{9}{12}$

$+3\frac{11}{12} = 3\frac{11}{12}$

$5\frac{20}{12} = 6\frac{8}{12} = \mathbf{6\frac{2}{3}}$

25. $3\frac{3}{4} \cdot 2\frac{1}{2} \div 3\frac{1}{8}$

$= \left(\frac{15}{4} \cdot \frac{5}{2}\right) \div \frac{25}{8}$

$= \frac{75}{8} \div \frac{25}{8} = \frac{\overset{3}{\cancel{75}}}{\underset{1}{\cancel{8}}} \times \frac{\overset{1}{\cancel{8}}}{\underset{1}{\cancel{25}}} = \mathbf{3}$

26. $3\frac{3}{4} \div 2\frac{1}{2} \cdot 3\frac{1}{8}$

$= \left(\frac{15}{4} \div \frac{5}{2}\right) \cdot \frac{25}{8} = \left(\frac{\overset{3}{\cancel{15}}}{\underset{2}{\cancel{4}}} \times \frac{\overset{1}{\cancel{2}}}{\underset{1}{\cancel{5}}}\right) \cdot \frac{25}{8}$

$= \frac{3}{2} \cdot \frac{25}{8} = \frac{75}{16} = \mathbf{4\frac{11}{16}}$

27. **The first five numbers in the sequence are the squares of the first five counting numbers. So the 99th number in the sequence is 99^2.**

28. **See student work. If the triangle is drawn and measured accurately, the longest side is twice the length of the shortest side.**

29. 0.5 meter = 50 centimeters

radius = $\frac{50 \text{ centimeters}}{2}$

= **25 centimeters**

30.

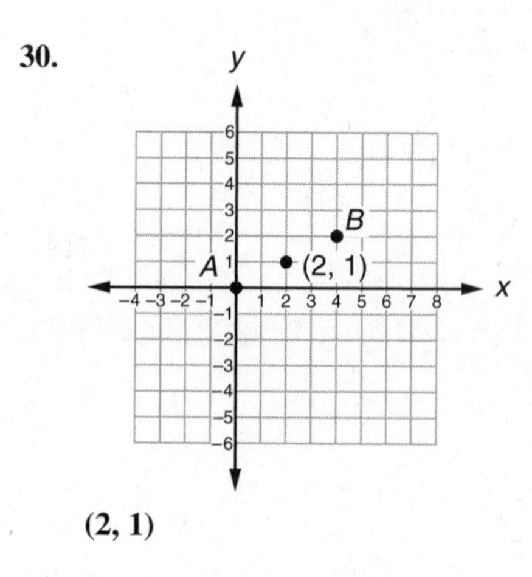

(2, 1)

Lesson 57, Warm-Up

a. **128**

b. **4200**

c. **28**

d. **0.5 L**

e. **100**

f. **$10.00**

g. $\mathbf{2\frac{1}{2}}$

Problem Solving

|←—600 ft—→|
● ● ● ● ● ● ● ●
1 2 3 4 5 6 7 8

Lesson 57, Lesson Practice

a. $5^{-2} = \frac{1}{5^2} = \mathbf{\frac{1}{25}}$

b. $3^0 = \mathbf{1}$

c. $10^{-4} = \frac{1}{10^4} = \mathbf{\frac{1}{10,000}}$ **or 0.0001**

d. 0000002.5 ⟶ $\mathbf{2.5 \times 10^{-7}}$

e. 000000001. ⟶ $\mathbf{1 \times 10^{-9}}$

f. 0001.05 ⟶ $\mathbf{1.05 \times 10^{-4}}$

g. .00000045 ⟶ **0.00000045**

h. .001 ⟶ **0.001**

i. .0000125 ⟶ **0.0000125**

j. $1 \times 10^{-3} = 0.001$
$1 \times 10^{2} = 100$
$1 \times 10^{-3} \mathbf{<} 1 \times 10^{2}$

k. $2.5 \times 10^{-2} = 0.025$
$2.5 \times 10^{-3} = 0.0025$
$2.5 \times 10^{-2} \mathbf{>} 2.5 \times 10^{-3}$

Lesson 57, Mixed Practice

1.

	Ratio	Actual Count
Walkers	5	315
Riders	3	R

$\frac{5}{3} = \frac{315}{R}$
$5R = 3(315)$
$R = \mathbf{189\ riders}$

2. $5(88) = 440$
$6(90) = 540$
$540 - 440 = \mathbf{100}$

3. $2(\$34.95) + \$0.18(300)$
$= \$69.90 + \$54.00$
$= \mathbf{\$123.90}$

4. 1 quart = 2 pints
$2\overline{)\$0.52}$ = **\$0.26 per pint**
4
12
12
0

5. 1 hour: 12 minutes, 12 minutes, 12 minutes, 12 minutes, 12 minutes
finished in $\frac{4}{5}$ of an hour (four of the 12-minute parts)

(a) **24 minutes**

(b) $\frac{2}{5} \times 100\% = \frac{200\%}{5} = \mathbf{40\%}$

6. (a) 1.86000 ⟶ $\mathbf{1.86 \times 10^5}$
(b) 000004. ⟶ $\mathbf{4 \times 10^{-5}}$

7. (a) 32.5 ⟶ **32.5**
(b) 000001.5 ⟶ **0.0000015**

8. (a) $2^{-3} = \frac{1}{2^3} = \mathbf{\frac{1}{8}}$
(b) $5^0 = \mathbf{1}$
(c) $10^{-2} = \frac{1}{10^2} = \mathbf{\frac{1}{100}}$ **or 0.01**

9. $\overset{2}{\cancel{2000}}\ \cancel{\text{milliliters}} \cdot \frac{1 \text{ liter}}{\underset{1}{\cancel{1000}}\ \cancel{\text{milliliters}}}$
$= \mathbf{2\ liters}$

10. $\frac{2}{6} = \mathbf{\frac{1}{3}}$

11. $24\overline{)\$330.00}$ = **\$13.75**
24
90
72
18 0
16 0
1 20
1 20
0

12.

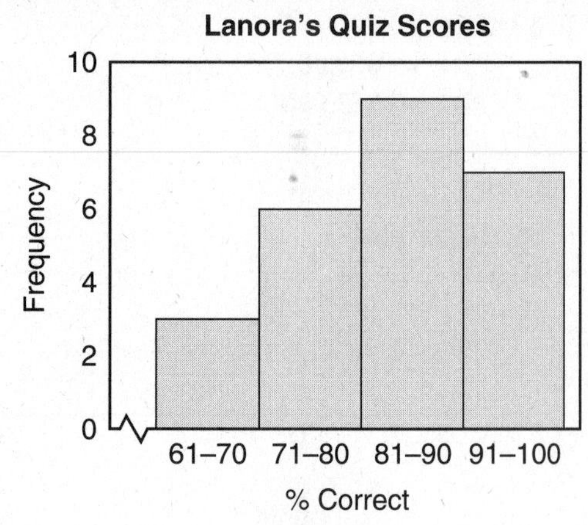

13. (a) $2.5 \times 10^{-2} = 0.025$

$2.5 \div 10^2 = \frac{2.5}{100} = 0.025$

$2.5 \times 10^{-2} \mathbf{=} 2.5 \div 10^2$

(b) $1 \times 10^{-6} = 0.000001$

one millionth $\mathbf{=} 1 \times 10^{-6}$

(c) $3^0 = 1, 2^0 = 1$

$3^0 \mathbf{=} 2^0$

14. Perimeter $= 4\text{ yd} + 3\text{ yd} + 1\text{ yd} + 1\text{ yd} + 1.5\text{ yd} + 2\text{ yd} + 1.5\text{ yd} + 4\text{ yd} = \mathbf{18\ yd}$

15.

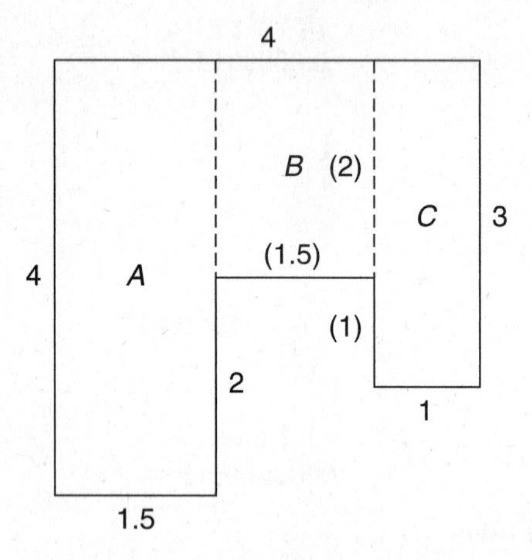

$$\begin{aligned} \text{Area } A &= (1.5\text{ yd})(4\text{ yd}) = 6\text{ yd}^2 \\ \text{Area } B &= (1.5\text{ yd})(2\text{ yd}) = 3\text{ yd}^2 \\ + \text{ Area } C &= (1\text{ yd})(3\text{ yd}) = 3\text{ yd}^2 \\ \hline \text{Area of figure} &= \mathbf{12\ yd^2} \end{aligned}$$

16. $4(5)(0.5) = 20(0.5) = \mathbf{10}$

17. $\$20.00 \div 4 = \mathbf{\$5.00}$

18. $y = 3(12) + 5 = 36 + 5 = \mathbf{41}$

19. $20^2 + 10^3 - \sqrt{36}$

$= 400 + 1000 - 6 = \mathbf{1394}$

20. $48 \div 12 \div 2 + 2(3)$

$4 \div 2 + 6$

$2 + 6$

$\mathbf{8}$

21.

$$\begin{array}{rrr} 3\text{ yd} & \overset{1}{\not{2}}\text{ ft} & 1\text{ in.} \\ -\ 1\text{ yd} & 2\text{ ft} & 3\text{ in.} \\ \hline \end{array} \quad (12\text{ in.}) \longrightarrow$$

$$\begin{array}{rrr} \overset{2}{\not{3}}\text{ yd} & \overset{1}{\not{2}}\text{ ft} & \overset{13}{\not{1}}\text{ in.} \\ -\ 1\text{ yd} & 2\text{ ft} & 3\text{ in.} \\ \hline & & 10\text{ in.} \end{array} \quad (3\text{ ft}) \longrightarrow$$

$$\begin{array}{rrr} \overset{2}{\not{3}}\text{ yd} & \overset{4}{\not{1}}\not{2}\text{ft} & \overset{13}{\not{1}}\text{ in.} \\ -\ 1\text{ yd} & 2\text{ ft} & 3\text{ in.} \\ \hline \mathbf{1\ yd} & \mathbf{2\ ft} & \mathbf{10\ in.} \end{array}$$

22.

$$\begin{array}{rrrr} 4\text{ gal} & 3\text{ qt} & 1\text{ pt} & 6\text{ oz} \\ +\ 1\text{ gal} & 2\text{ qt} & 1\text{ pt} & 5\text{ oz} \\ \hline 5\text{ gal} & 5\text{ qt} & 2\text{ pt} & 11\text{ oz} \end{array}$$

$2\text{ pt} = 1\text{ qt}$

$$\begin{array}{rr} 1\text{ qt} & 11\text{ oz} \\ +\ 5\text{ qt} & \\ \hline 6\text{ qt} & 11\text{ oz} \end{array}$$

$6\text{ qt} = 1\text{ gal } 2\text{ qt}$

$$\begin{array}{rrr} 1\text{ gal} & 2\text{ qt} & 11\text{ oz} \\ +\ 5\text{ gal} & & \\ \hline \mathbf{6\ gal} & \mathbf{2\ qt} & \mathbf{11\ oz} \end{array}$$

23. $\overset{3}{\not{48}}\ \not{oz} \cdot \frac{1\text{ pt}}{\underset{1}{\not{16}}\ \not{oz}} = \mathbf{3\ pt}$

24. $7 \div 1\frac{3}{4} = \frac{7}{1} \div \frac{7}{4} =$

$\frac{\overset{1}{\not{7}}}{1} \times \frac{4}{\underset{1}{\not{7}}} = 4$

$5\frac{1}{3} \cdot 4 = \frac{16}{3} \cdot \frac{4}{1} = \frac{64}{3} = \mathbf{21\frac{1}{3}}$

25.

$$\begin{array}{rcl} 5\frac{1}{6} &=& 5\frac{4}{24} \\ 3\frac{5}{8} &=& 3\frac{15}{24} \\ +\ 2\frac{7}{12} &=& 2\frac{14}{24} \\ \hline && 10\frac{33}{24} = 11\frac{9}{24} = \mathbf{11\frac{3}{8}} \end{array}$$

26.

$$\begin{array}{rcl} \frac{1}{20} &=& \frac{9}{180} \\ -\ \frac{1}{36} &=& \frac{5}{180} \\ \hline && \frac{4}{180} = \mathbf{\frac{1}{45}} \end{array}$$

27. $4.6 \times 10^{-2} = 0.046$

$$\begin{array}{r} 0.46 \\ +\ 0.046 \\ \hline \mathbf{0.506} \end{array}$$

28.
$$\begin{array}{r} 2.300 \\ -\ 0.575 \\ \hline 1.725 \end{array} \qquad \begin{array}{r} 10.000 \\ -\ 1.725 \\ \hline \mathbf{8.275} \end{array}$$

29.
$$\begin{array}{r} 0.24 \\ \times\ 0.15 \\ \hline 1\,20 \\ 2\,4 \\ \hline 0.03\,60 \end{array} = 0.036$$

$$\begin{array}{r} 0.036 \\ \times\ \ 0.05 \\ \hline 0.00180 \end{array} = \mathbf{0.0018}$$

30. $70\overline{)0.140}$ = 0.002 $\qquad 2\overline{)10{,}000}$ = **5000**

Lesson 58, Warm-Up

a. **215**

b. **0.0042**

c. **15**

d. **1500 g**

e. **10**

f. **$22.00**

g. **1.8, 0.6, 0.72, 2**

Problem Solving

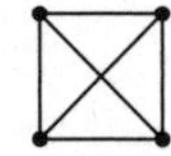

6 handshakes

Lesson 58, Activity

1. **There will be at least one line of symmetry, the line along the fold.**

2. **at least two lines of symmetry (one per fold)**

Lesson 58, Lesson Practice

a.

b.

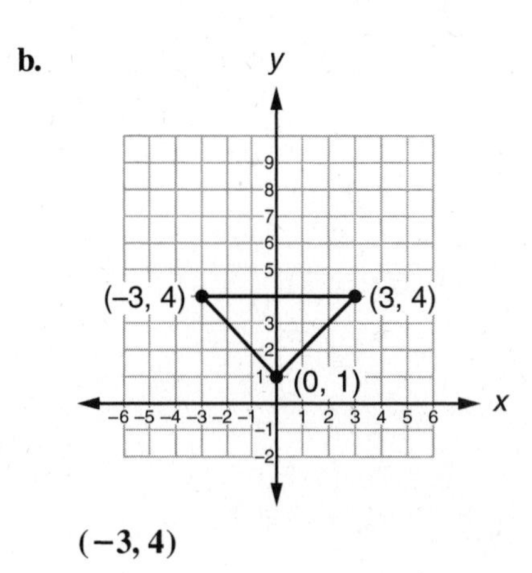

(−3, 4)

c. $7(5) = \mathbf{35}$

d. $1 + (4) = \mathbf{5}$

Lesson 58, Mixed Practice

1. 10(1.4 kilometers) = **14 kilometers**

2. 5($0.75) = **$3.75**

3. $17n = 340$

$n = \frac{340}{17}$

$n = 20$

$$\begin{array}{r} 17 \\ +\ 20 \\ \hline \mathbf{37} \end{array}$$

4. (a) $\frac{\text{won}}{\text{lost}} = \frac{3}{9} = \mathbf{\frac{1}{3}}$

(b) $\frac{\text{lost}}{\text{total games}} = \frac{9}{12} = \mathbf{\frac{3}{4}}$

(c) $\frac{1}{4} \times 100\% = \frac{100\%}{4} = \mathbf{25\%}$

5. $5(120) = 600$

$$\begin{array}{r} 600 \\ 118 \\ 124 \\ +\ 142 \\ \hline 984 \end{array} = \text{total score for 8 games}$$

Average score = **123**

$$\begin{array}{r} \mathbf{123} \\ 8\overline{)984} \\ 8 \\ \hline 18 \\ 16 \\ \hline 24 \\ 24 \\ \hline 0 \end{array}$$

6.

	60 questions
$\frac{3}{5}$ were multiple-choice.	12 questions
	12 questions
	12 questions
$\frac{2}{5}$ were not multiple-choice.	12 questions
	12 questions

(a) **36 questions**

(b) $\frac{2}{5} \times 100\% = \frac{200\%}{5} = \mathbf{40\%}$

7. (a) $\overline{OA}, \overline{OB}, \overline{OC}$

(b) $\overline{AC}$ (or $\overline{CA}$), $\overline{BC}$ (or $\overline{CB}$)

(c) **60°**

(d) **30°**

8. (a) 15000000. ⟶ **15,000,000**

(b) .00025 ⟶ **0.00025**

(c) $10^{-1} = \frac{1}{10}$ **or 0.1**

9. 20 qt = 5 gal

10.

```
     19.16 . . .
18)345.00 . . .
   18
   165
   162
     3 0
     1 8
     1 20
     1 08
       12
```

19

11. **5, 0, −5**

12. (a) $6\overline{)1.000\ldots}$ = $0.1\overline{6}6\ldots$ $\mathbf{0.1\overline{6}}$

(b) $\frac{1}{6} \times 100\% = \frac{100\%}{6} = \mathbf{16\frac{2}{3}\%}$

(c) $16\% = \frac{16}{100} = \mathbf{\frac{4}{25}}$

(d) $16\% = \frac{16}{100} = \mathbf{0.16}$

13. $2(4) = \mathbf{8}$

14. (a) $m\angle ACB = 180° - (90° + 35°)$
$= 180° - 125° = \mathbf{55°}$

(b) $\angle ACD = 180° - 55° = \mathbf{125°}$

(c) $\angle CAD = 180° - (35° + 125°)$
$= 180° - 160° = \mathbf{20°}$

15.

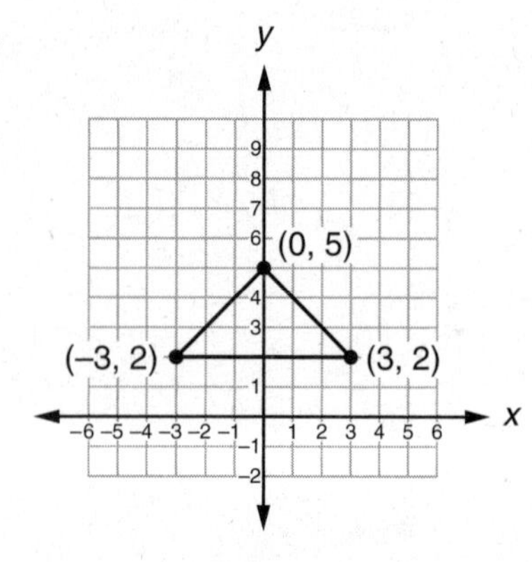

(a) **(3, 2)**

(b) Area $= \frac{(6 \text{ units})(3 \text{ units})}{2}$
$= \mathbf{9 \text{ units}^2}$

16. **5 lines**

17. (a) $\overset{7}{\cancel{210}}\ \cancel{\text{miles}} \cdot \frac{1 \text{ hr}}{\underset{2}{\cancel{60}}\ \cancel{\text{miles}}}$
$= \frac{7}{2} \text{ hr} = \mathbf{3\frac{1}{2} \text{ hr}}$

(b) $\overset{3}{\cancel{210}}\ \cancel{\text{miles}} \cdot \frac{1 \text{ hr}}{\underset{1}{\cancel{70}}\ \cancel{\text{miles}}}$
$= \mathbf{3 \text{ hr}}$

18. $\frac{1.5}{2} = \frac{7.5}{w}$
$1.5w = 2(7.5)$
$w = \frac{15}{1.5}$
$w = \mathbf{10}$

19.

```
  1.70
– 0.17
  1.53
```

$y = \mathbf{1.53}$

20. $10^3 - 10^2 + 10^1 - 10^0$
$1000 - 100 + 10 - 1$
$900 + 10 - 1$
$910 - 1$
909

21. $6 + 3(2) - 4 - (5 + 3)$
$6 + 6 - 4 - 8$
$12 - 4 - 8$
$8 - 8$
$\mathbf{0}$

22.

	1 gal	2 qt	1 pt
+	1 gal	2 qt	1 pt
	2 gal	4 qt	2 pt

2 pt = 1 qt
4 qt + 1 qt = 5 qt
5 qt = 1 gal 1 qt

	1 gal	1 qt
+	2 gal	
	3 gal	**1 qt**

23.

		2 (60 min)	
	1 day	~~3~~ hr	15 min
−		8 hr	30 min

→

	0 (24 hr)	2	75
	~~1~~ day	~~3~~ hr	~~15~~ min
−		8 hr	30 min
			45 min

→

	0	26 ~~2~~	75
	~~1~~ day	~~3~~ hr	~~15~~ min
−		8 hr	30 min
		18 hr	**45 min**

24. $2 \text{ mi} \cdot \frac{5280 \text{ ft}}{1 \text{ mi}} = \mathbf{10{,}560 \text{ ft}}$

25.
$5\frac{3}{4} = 5\frac{9}{12} \longrightarrow 4\frac{21}{12}$
$-1\frac{5}{6} = 1\frac{10}{12} \qquad -1\frac{10}{12}$
$\qquad 3\frac{11}{12}$

$10 \longrightarrow 9\frac{12}{12}$
$-3\frac{11}{12} \qquad -3\frac{11}{12}$
$\qquad \mathbf{6\frac{1}{12}}$

26.
$2\frac{1}{5} = 2\frac{2}{10}$
$+5\frac{1}{2} = 5\frac{5}{10}$
$\qquad 7\frac{7}{10}$

$7\frac{7}{10} \div 2\frac{1}{5} = \frac{77}{10} \div \frac{11}{5}$

$= \frac{\overset{7}{\cancel{77}}}{\underset{2}{\cancel{10}}} \times \frac{\overset{1}{\cancel{5}}}{\underset{1}{\cancel{11}}} = \frac{7}{2} = \mathbf{3\frac{1}{2}}$

27. $6 \div 4\frac{1}{2} = \frac{6}{1} \div \frac{9}{2} =$

$\frac{\overset{2}{\cancel{6}}}{1} \times \frac{2}{\underset{3}{\cancel{9}}} = \frac{4}{3}$

$3\frac{3}{4} \cdot \frac{4}{3} = \frac{\overset{5}{\cancel{15}}}{\underset{1}{\cancel{4}}} \cdot \frac{\overset{1}{\cancel{4}}}{\underset{1}{\cancel{3}}} = \mathbf{5}$

28. $(6)^2 - 4(3.6)(2.5)$
$= 36 - 36 = \mathbf{0}$

29. (a) $\mathbf{2.5, 2, \frac{3}{2}, 0, -\frac{1}{2}, -1}$

(b) **−1, 0, 2**

30. **Lindsey could double both numbers before dividing, forming the equivalent division problem 70 ÷ 5. She could also double both of these numbers to form 140 ÷ 10.**

Lesson 59, Warm-Up

a. **324**

b. **0.5**

c. **2**

d. **1.85 m**

e. **16**

f. **$35.00**

g. **$60.00**

Problem Solving

The side length of both shapes is 100 cm ÷ 5, or 20 cm.
20 cm × 20 cm = **400 sq. cm**

Lesson 59, Lesson Practice

a.

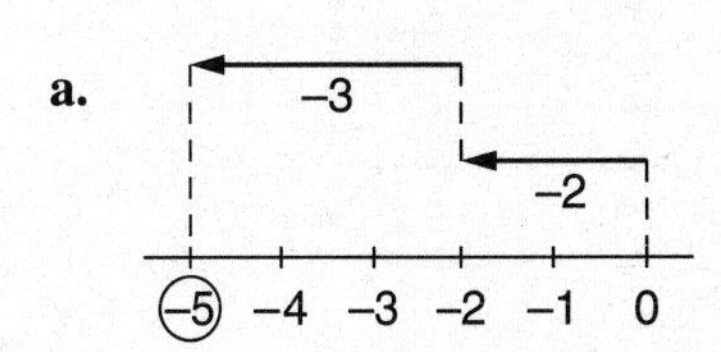

b.

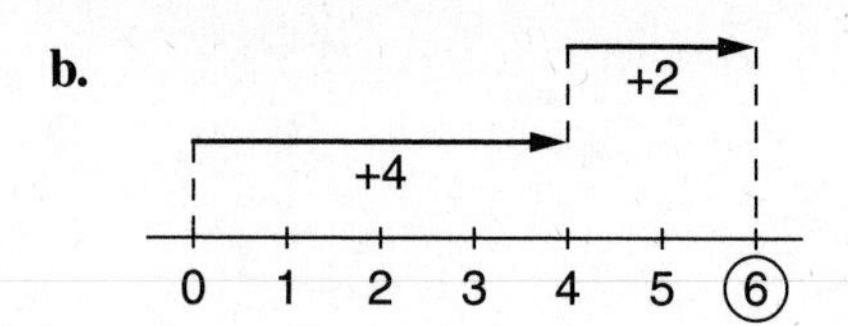

c.

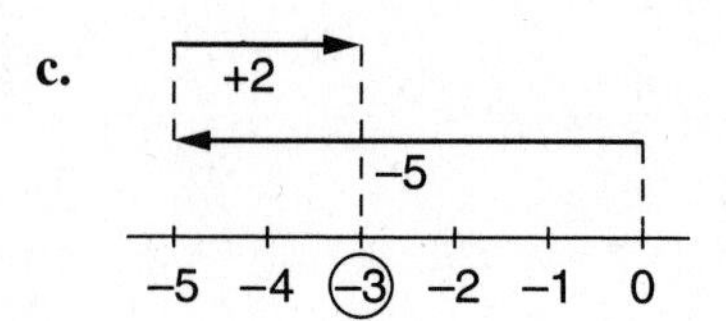

d.

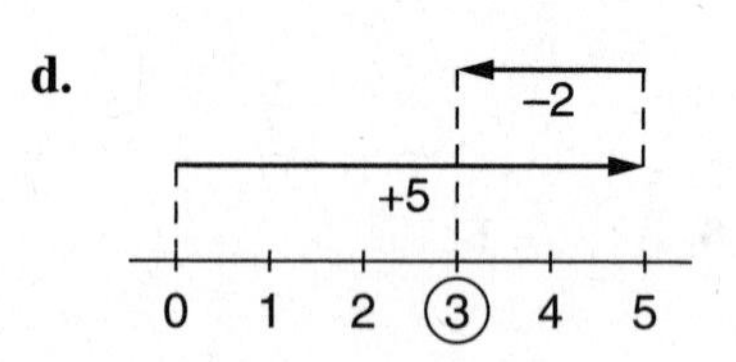

e.

+4

−4

−4 −3 −2 −1 0

f.

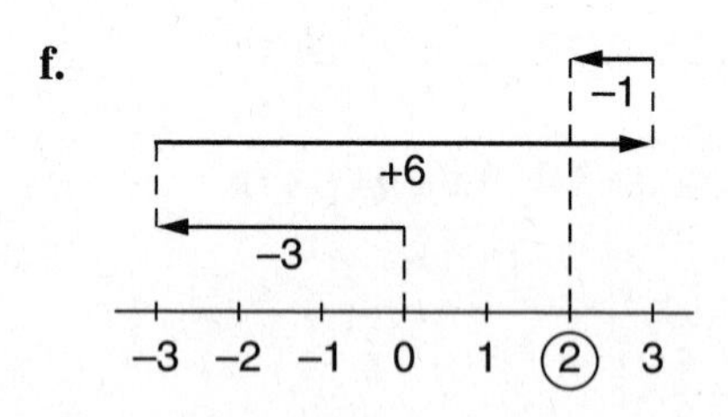

g. $|-3| + |3| = 3 + 3 = \mathbf{6}$

h. $|3 - 3| = |0| = \mathbf{0}$

i. $|5 - 3| = |2| = \mathbf{2}$

j.

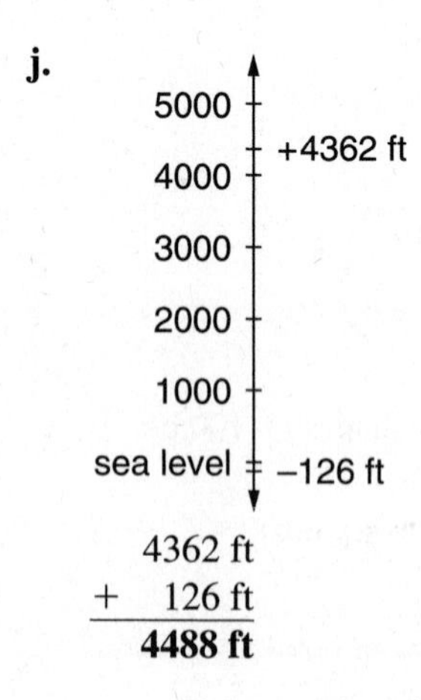

$$\begin{array}{r} 4362 \text{ ft} \\ +\ 126 \text{ ft} \\ \hline \mathbf{4488 \text{ ft}} \end{array}$$

k.

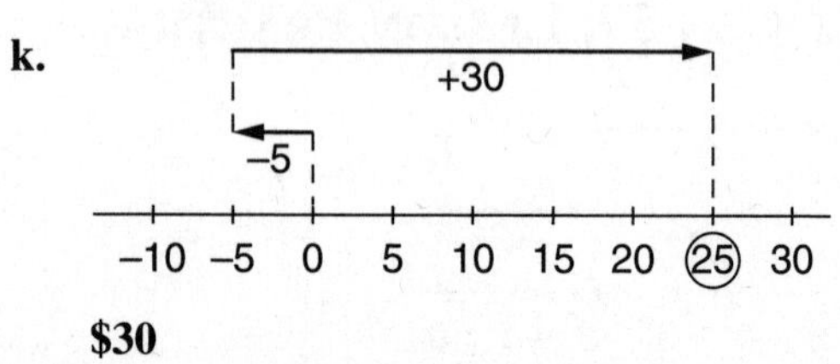

$30

Lesson 59, Mixed Practice

1. $2(\$2.35) + \$0.60(6)$
$= \$4.70 + \$3.60$
$= \mathbf{\$8.30}$

2.
$$\begin{array}{r} 90°\text{F} \\ -\ 85°\text{F} \\ \hline \mathbf{5°F} \end{array}$$

3.
$$\begin{array}{r} 82°\text{F} \\ 84°\text{F} \\ 86°\text{F} \\ 88°\text{F} \\ 84°\text{F} \\ 90°\text{F} \\ +\ 88°\text{F} \\ \hline 602°\text{F} \end{array} \qquad 7\overline{)602°\text{F}} = \mathbf{86°F}$$

4.

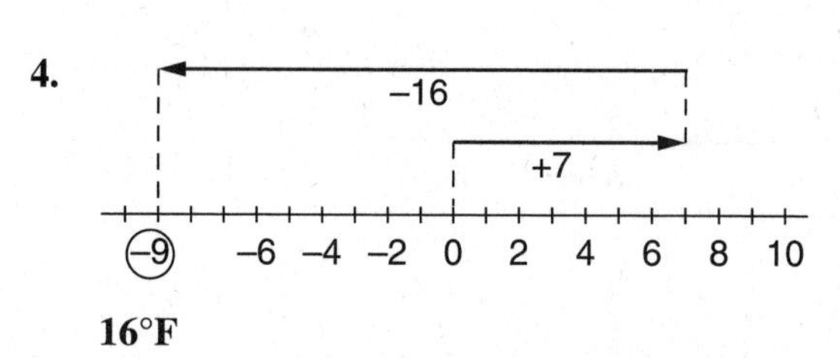

16°F

5.

	Ratio	Actual Count
Sonorous voices	7	S
Discordant voices	4	56

$$\frac{7}{4} = \frac{S}{56}$$
$4S = 7(56)$
$S = \mathbf{98 \text{ voices}}$

6.

20 games

$\frac{3}{4}$ won. — 5 games / 5 games / 5 games

$\frac{1}{4}$ failed to win. — 5 games

(a) **15 games**

(b) $\frac{1}{4} \times 100\% = \frac{100\%}{4} = \mathbf{25\%}$

7. $|-3| = 3, |3| = 3$
$3 = 3$
$|-3| \mathbf{=} |3|$

8. (a) 4,000000000000 ⟶ $\mathbf{4 \times 10^{12}}$

(b) 3670000000. miles ⟶ **3,670,000,000 miles**

9. (a) .000001 meter ⟶ **0.000001 meter**

(b)
$$1 \text{ millimeter} = 0.001 \text{ meter}$$
$$1 \times 10^{-3} \text{ meter} = 0.001 \text{ meter}$$
$$1 \text{ millimeter} = 1 \times 10^{-3} \text{ meter}$$

10. $\overset{3}{\cancel{300}}\ \cancel{\text{mm}} \cdot \dfrac{1 \text{ m}}{\underset{10}{\cancel{1000}}\ \cancel{\text{mm}}} = \dfrac{3}{10} \text{ m} = \mathbf{0.3 \text{ m}}$

11. (a) $12\% = \dfrac{12}{100} = \mathbf{\dfrac{3}{25}}$

(b) $12\% = \dfrac{12}{100} = \mathbf{0.12}$

(c) $3\overline{)1.00\ldots}$ = 0.33 . . . $\mathbf{0.\overline{3}}$

(d) $\dfrac{1}{3} \times 100\% = \dfrac{100\%}{3} = \mathbf{33\dfrac{1}{3}\%}$

12. (a) Number line from −3 to 2: arrow +2 from 0 to 2, arrow −5 from 2 to −3; answer (−3) circled.

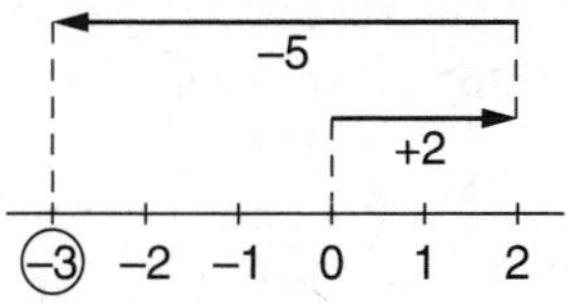

(b) Number line from −2 to 3: arrow −2 from 0 to −2, arrow +5 from −2 to 3; answer (3) circled.

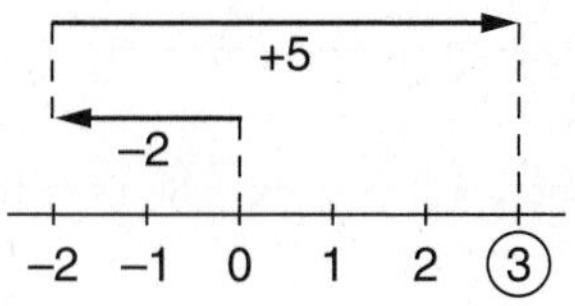

13. $8 + (12) = \mathbf{20}$

14.
$$\text{Perimeter} = 50 \text{ mm} + 60 \text{ mm} + 35 \text{ mm} + 15 \text{ mm} + 15 \text{ mm} + 25 \text{ mm} + 30 \text{ mm} + 20 \text{ mm} = \mathbf{250 \text{ mm}}$$

15. Figure: region A (50 by 20), region B ((20) by (25)), region C (35 by 15); labels 50, 20, 30, (20), 60, (25), (15), 15, 35.

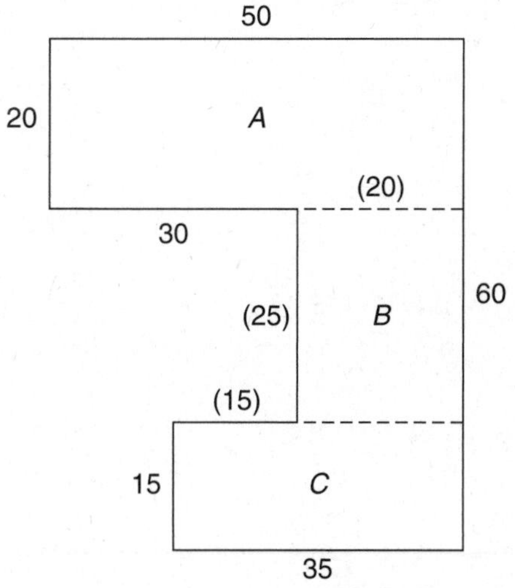

$$\text{Area } A = (50 \text{ mm})(20 \text{ mm}) = 1000 \text{ mm}^2$$
$$\text{Area } B = (20 \text{ mm})(25 \text{ mm}) = 500 \text{ mm}^2$$
$$+ \text{ Area } C = (35 \text{ mm})(15 \text{ mm}) = 525 \text{ mm}^2$$
$$\text{Area of figure} = \mathbf{2025 \text{ mm}^2}$$

16. $8\overline{)4.40}$ = 0.55

$w = \mathbf{0.55}$

17. $\dfrac{0.8}{1} = \dfrac{x}{1.5}$

$1x = (0.8)(1.5)$

$x = \mathbf{1.2}$

18.
$$\frac{17}{30} = \frac{34}{60}$$
$$-\ \frac{11}{20} = \frac{33}{60}$$
$$n = \mathbf{\frac{1}{60}}$$

19. $7\overline{)0.364}$ = 0.052
35
14
14
0

$m = \mathbf{0.052}$

20. $2^{-1} + 2^{-1} = \dfrac{1}{2} + \dfrac{1}{2} = \dfrac{2}{2} = \mathbf{1}$

21. $\sqrt{64} - 2^3 + 4^0 = 8 - 8 + 1 = \mathbf{1}$

22.
3 yd 2 ft $7\frac{1}{2}$ in.
+ 1 yd $5\frac{1}{2}$ in.
4 yd 2 ft 13 in.

13 in. = 1 ft 1 in.

1 ft 1 in.
2 ft
3 ft 1 in.

3 ft = 1 yd, 4 yd + 1 yd = 5 yd

5 yd 1 in.

23.
$\overset{0}{\cancel{1}}$ qt ⟶ (2 pt) 1 pt 6 oz
− 1 pt 12 oz ⟶

$\overset{0}{\cancel{1}}$ qt $\overset{2}{\cancel{3}}$ ⟶ (16 oz) $\cancel{1}$ pt 6 oz
− 1 pt 12 oz ⟶

$\overset{0}{\cancel{1}}$ qt $\overset{2}{\cancel{3}}$ $\cancel{1}$ pt $\overset{22}{\cancel{6}}$ oz
− 1 pt 12 oz
1 pt 10 oz

24. $2\frac{1}{2}\ \cancel{\text{hr}} \cdot \frac{50\text{ mi}}{1\ \cancel{\text{hr}}} = \mathbf{125\ mi}$

25. $\frac{5}{9} \cdot 12 = \frac{5}{\cancel{9}_3} \cdot \frac{\cancel{12}^4}{1} = \frac{20}{3}$

$\frac{20}{3} \div 6\frac{2}{3} = \frac{20}{3} \div \frac{20}{3}$

$= \frac{\cancel{20}^1}{\cancel{3}_1} \times \frac{\cancel{3}^1}{\cancel{20}_1} = \mathbf{1}$

26. $4 - (4 - 1) = 4 - 3 = 1$

$$\begin{array}{r} 4 \\ -\ 1\frac{1}{9} \\ \hline \end{array} \longrightarrow \begin{array}{r} 3\frac{9}{9} \\ -\ 1\frac{1}{9} \\ \hline 2\frac{8}{9} \end{array}$$

$$\begin{array}{r} 3\frac{5}{6} = 3\frac{15}{18} \\ -\ 2\frac{8}{9} = 2\frac{16}{18} \\ \hline \end{array} \longrightarrow \begin{array}{r} 2\frac{33}{18} \\ -\ 2\frac{16}{18} \\ \hline \mathbf{\frac{17}{18}} \end{array}$$

27. $(6 + 6) \div 6 = 12 \div 6 = 2$

$$\begin{array}{r} 5\frac{5}{8} = 5\frac{5}{8} \\ +\ 6\frac{1}{4} = 6\frac{2}{8} \\ \hline 11\frac{7}{8} \end{array}$$

$11\frac{7}{8} \div 6\frac{1}{4} = \frac{95}{8} \div \frac{25}{4}$

$= \frac{\cancel{95}^{19}}{\cancel{8}_2} \times \frac{\cancel{4}^1}{\cancel{25}_5} = \frac{19}{10} = \mathbf{1\frac{9}{10}}$

28. $(0.1) - (0.2)(0.3)$
$= 0.1 - 0.06 = \mathbf{0.04}$

29.
$$\begin{array}{r} \$18.00 \\ \times\ 0.065 \\ \hline 9000 \\ 10800 \\ \hline \$1.17000 \end{array} = \mathbf{\$1.17}$$

30. $\frac{\text{candidate with most votes}}{\text{total votes}} = \frac{10}{25}$
$= \mathbf{\frac{2}{5}}$

Lesson 60, Warm-Up

a. **161**

b. **4.35**

c. **10**

d. **7.5 cm**

e. **7**

f. **80¢**

g. **30¢**

Problem Solving

The only 1-digit number that is a factor of 1101 is 3. To find the top factor, we divide 1101 by 3.

$$\begin{array}{r} \mathbf{367} \\ \times\ \mathbf{3} \\ \hline \mathbf{1101} \end{array}$$

Lesson 60, Lesson Practice

a. $W_N = \frac{4}{5} \times 71$

$W_N = \frac{284}{5}$

$W_N = \mathbf{56\frac{4}{5}}$

b. $\mathbf{\frac{3}{8} \times 3\frac{3}{7} = W_N}$

$\frac{3}{\cancel{8}_1} \times \frac{\cancel{24}^3}{7} = W_N$

$\frac{9}{7} = \mathbf{1\frac{2}{7}} = W_N$

c. $\mathbf{W_N = 0.6 \times 145}$
$W_N = \mathbf{87}$

d. $\mathbf{0.75 \times 14.4 = W_N}$
$\mathbf{10.8} = W_N$

e. $\mathbf{W_N = 0.5 \times 150}$
$W_N = \mathbf{75}$

f. $\mathbf{0.03 \times \$39 = M}$
$\mathbf{\$1.17} = M$

g. $W_N = 0.25 \times 64$

$W_N = \mathbf{16}$

h. $0.12 \times \$250{,}000 = C$

$\mathbf{\$30{,}000} = C$

Lesson 60, Mixed Practice

1.
$$\begin{array}{r} 7.0021 \\ -\ 5.7840 \\ \hline \mathbf{1.2181} \end{array}$$

2. $\overset{100}{\cancel{\$20}} \cdot \dfrac{1 \text{ board}}{\underset{1}{\cancel{\$0.20}}} = \mathbf{100\ boards}$

3. $\dfrac{72}{n} = 12$

$n = 6$

$72 \times 6 = \mathbf{432}$

4. (a) $\dfrac{1}{5} \times 100\% = \mathbf{20\%}$

(b) $\dfrac{\text{trumpeters who hit the note}}{\text{trumpeters who did not hit the note}} = \mathbf{\dfrac{4}{1}}$

5. $5(77 \text{ inches}) = 385 \text{ inches}$

$$\begin{array}{r} 71 \text{ inches} \\ 74 \text{ inches} \\ 78 \text{ inches} \\ +\ 78 \text{ inches} \\ \hline 301 \text{ inches} \end{array} \qquad \begin{array}{r} 385 \text{ inches} \\ -\ 301 \text{ inches} \\ \hline \mathbf{84 \text{ inches}} \end{array}$$

6. (a) $\mathbf{8 \times 10^{-8}}$

(b) $\mathbf{6.75 \times 10^{10}}$

7.

	96 members
$\frac{2}{3}$ approved.	32 members
	32 members
$\frac{1}{3}$ did not approve.	32 members

(a) **64 members**

(b) $\dfrac{1}{3} \times 100\% = \mathbf{33\frac{1}{3}\%}$

8.

30,000
20,000 — 23,000 feet
10,000
sea level
−10,000 — −9000 feet

$$\begin{array}{r} 23{,}000 \text{ feet} \\ +\ 9{,}000 \text{ feet} \\ \hline \mathbf{32{,}000 \text{ feet}} \end{array}$$

9. $W_N = \dfrac{3}{4} \times 17$

$W_N = \dfrac{51}{4}$

$W_N = \mathbf{12\frac{3}{4}}$

10. $0.4 \times \$65 = P$

$\mathbf{\$26} = P$

11. (a) $3\overline{)1.00\ldots}$ = $0.33\ldots$ $\qquad \dfrac{1}{3} \mathbf{>} 0.33$

(b) $|5 - 3| = |2| = 2$

$|3 - 5| = |-2| = 2$

$|5 - 3| \mathbf{=} |3 - 5|$

12. (a) $8\overline{)1.000}$ = $\mathbf{0.125}$

(b) $\dfrac{1}{8} \times 100\% = \mathbf{12\frac{1}{2}\%}$

(c) $125\% = \dfrac{125}{100} = \mathbf{1\frac{1}{4}}$

(d) $125\% = \dfrac{125}{100} = \mathbf{1.25}$

13. (a)

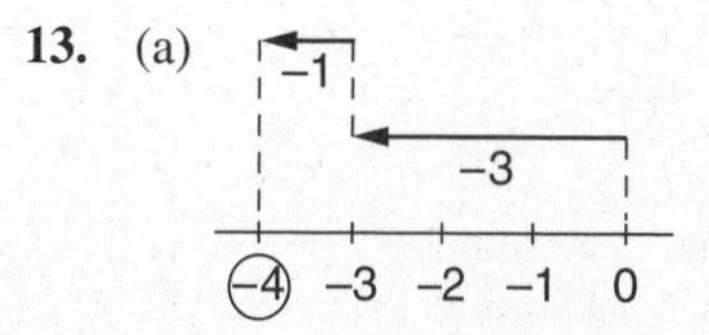

(b)

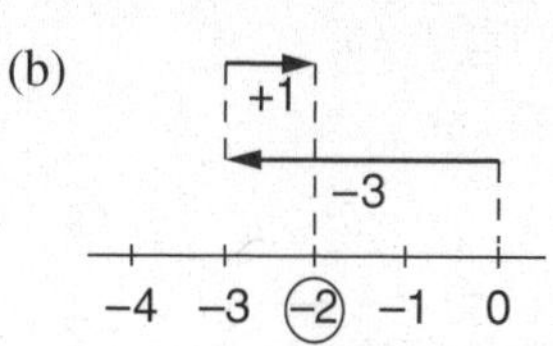

14. (a) $\mathbf{2^4 \cdot 3^2 \cdot 5^2}$

(b) $\sqrt{3600} = 60 = \mathbf{2^2 \cdot 3 \cdot 5}$

15. (a) $360° \times \frac{1}{2} = \frac{360°}{2} = \mathbf{180°}$

(b) $360° \times \frac{1}{3} = \frac{360°}{3} = \mathbf{120°}$

(c) $360° \times \frac{1}{6} = \frac{360°}{6} = \mathbf{60°}$

16. (a) $\boldsymbol{\Delta CDB}$

(b) $\boldsymbol{\Delta CEA}$

17. (a) $\text{Area} = \frac{(6 \text{ ft})(8 \text{ ft})}{2} = \mathbf{24\ ft^2}$

(b) $\text{Area} = \frac{(12 \text{ ft})(16 \text{ ft})}{2} = \mathbf{96\ ft^2}$

18. $\text{Area}(\Delta DEF) = \frac{(6 \text{ ft})(8 \text{ ft})}{2} = 24 \text{ ft}^2$

$96 \text{ ft}^2 - 24 \text{ ft}^2 - 24 \text{ ft}^2 = \mathbf{48\ ft^2}$

19.

$\frac{1}{20} = \frac{3}{60}$

$+ \frac{1}{30} = \frac{2}{60}$

$\frac{5}{60} = \frac{1}{12}$

$p = \mathbf{\frac{1}{12}}$

20. $9\overline{)0.117}$ = 0.013

9
27
27
0

$m = \mathbf{0.013}$

21. $3^2 + 4(3 + 2) - 2^3 \cdot 2^{-2} + \sqrt{36}$

$9 + 4(3 + 2) - 8 \cdot \frac{1}{4} + 6$

$9 + 4(5) - 2 + 6$

$9 + 20 - 2 + 6$

$29 - 2 + 6$

$27 + 6$

33

22.

3 days	16 hr	48 min
+ 1 day	15 hr	54 min
4 days	31 hr	102 min

102 min = 1 hr 42 min

1 hr	42 min
+ 31 hr	
32 hr	42 min

32 hr = 1 day 8 hr

1 day	8 hr	42 min
+ 4 days		
5 days	**8 hr**	**42 min**

23.

$19\frac{3}{4} = 19\frac{18}{24}$

$27\frac{7}{8} = 27\frac{21}{24}$

$+ 24\frac{5}{6} = 24\frac{20}{24}$

$70\frac{59}{24} = \mathbf{72\frac{11}{24}}$

24. $\frac{5}{6} \cdot 4 = \frac{5}{\cancel{6}_3} \cdot \frac{\cancel{4}^2}{1} = \frac{10}{3} = 3\frac{1}{3}$

$3\frac{3}{5} = 3\frac{9}{15}$

$- 3\frac{1}{3} = 3\frac{5}{15}$

$\mathbf{\frac{4}{15}}$

25. $1\frac{1}{4} \div \frac{5}{12} = \frac{5}{4} \div \frac{5}{12}$

$= \frac{\cancel{5}^1}{\cancel{4}_1} \times \frac{\cancel{12}^3}{\cancel{5}_1} = 3$

$3 \div 24 = \frac{3}{24} = \mathbf{\frac{1}{8}}$ or **0.125**

26.

$0.650 - 0.065 = 0.585$

$6.500 - 0.585 = \mathbf{5.915}$

27. $3 \div 0.03 = 100$

$0.3 \div 100 = \mathbf{0.003}$

28. $3.5 \text{ ~~centimeters~~} \cdot \frac{1 \text{ meter}}{100 \text{ ~~centimeters~~}}$

$= \frac{3.5}{100} \text{ meter} = \mathbf{0.035\ meter}$

29. **The first division problem can be multiplied by $\frac{100}{100}$ to form the second division problem. Since $\frac{100}{100}$ equals 1, the quotients are the same.**
One possibility: $\frac{\$1.50}{\$0.25} = \frac{150¢}{25¢}$

30.

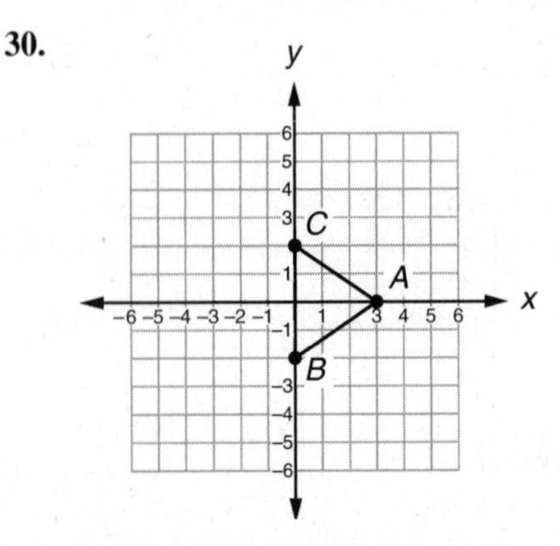

(0, 2)

Investigation 6

1. ***A, C***

2. ***C, D***

3. ***A, B, C, D***

4. ***E***

5. ***F, G***

6. ***A, B, C, D, G***

7. ***A, B, C, D***

8. ***E***

9. ***F, G***

10. ***C, D***

11. ***A, C***

12. 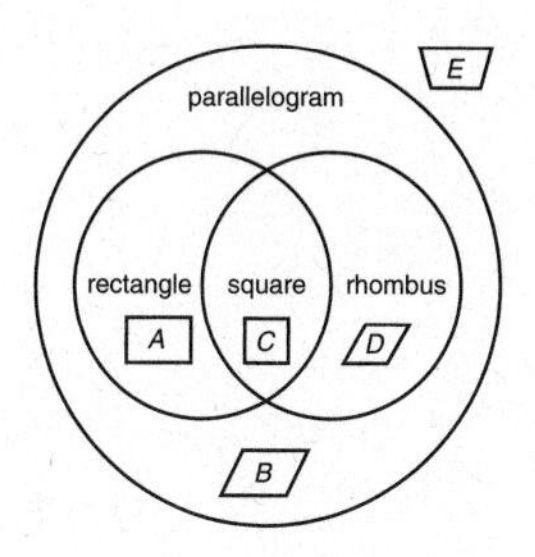

13. **No.** Figure K does not have four right angles, so it is not a rectangle.

14. **Yes.** Figure K has two pairs of parallel sides, so it is a parallelogram.

15. **Yes.** The lengths of the sides were not changed, so the perimeters of the figures are equal.

16. **The area of Figure K is less than the area of Figure J. The area becomes less and less the more the sides are shifted.**

17. ***G***

18. 2(2 ft) + 2(3 ft) = **10 ft**

19. 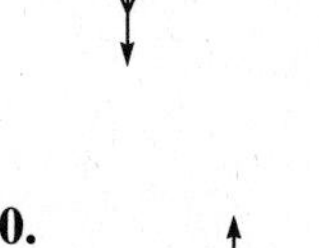

20.

21. 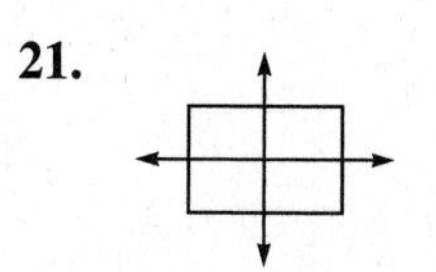

22. 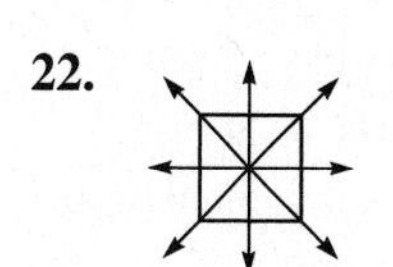

23. 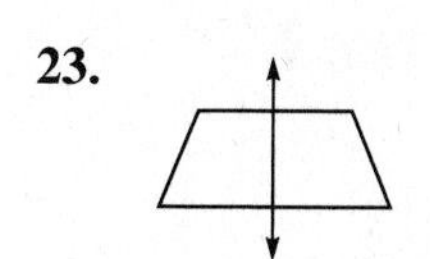

24. **See student work; Yes.**

25. ***A, B, C, D***

26. **True. A square is a parallelogram with four right angles.**

27. **True. All rectangles have two pairs of parallel sides.**

28. **False. All squares have two pairs of parallel sides, and trapezoids have only one pair of parallel sides.**

29. **True. Some parallelograms have four right angles.**

30. Quadrilaterals (Parallelograms) (Trapezoids)

LESSON 61, WARM-UP

a. 230

b. 0.24

c. 30

d. 1500 m

e. 1

f. $2.10

g. $21.20

Problem Solving

Since 31^2 equals 961 and 32^2 equals 1024, we know there are **31 perfect squares** less than 1000.

LESSON 61, ACTIVITY 1

See student work.

LESSON 61, ACTIVITY 2

1. **Answers may vary; however, the measures of the two angles should be equal.**

2. **Answers may vary; however, the measures of the two angles should be equal.**

3. **180°**

LESSON 61, LESSON PRACTICE

a. Perimeter $= 10 \text{ cm} + 12 \text{ cm} + 10 \text{ cm} + 12 \text{ cm} = \mathbf{44 \text{ cm}}$

Area $= (12 \text{ cm})(8 \text{ cm}) = \mathbf{96 \text{ cm}^2}$

b. Perimeter $= 10 \text{ cm} + 13 \text{ cm} + 10 \text{ cm} + 13 \text{ cm} = \mathbf{46 \text{ cm}}$

Area $= (10 \text{ cm})(12 \text{ cm}) = \mathbf{120 \text{ cm}^2}$

c. Perimeter $= 10 \text{ cm} + 10 \text{ cm} + 10 \text{ cm} + 10 \text{ cm} = \mathbf{40 \text{ cm}}$

Area $= (10 \text{ cm})(9 \text{ cm}) = \mathbf{90 \text{ cm}^2}$

d. $m\angle d = 180° - 75° = \mathbf{105°}$

e. $m\angle e = 180° - 105° = \mathbf{75°}$

f. $m\angle f = m\angle d = \mathbf{105°}$

g. $m\angle g = m\angle e = \mathbf{75°}$

h. $m\angle A = m\angle C = \mathbf{60°}$

i. $m\angle ADB = 180° - (90° + 60°) = 180° - 150° = \mathbf{30°}$

j. $m\angle ABC = 180° - 60° = \mathbf{120°}$

LESSON 61, MIXED PRACTICE

1. $\frac{1}{2}$ gallon $=$ 2 quarts $=$ 4 pints

$4\overline{)\$1.12}$ = **$0.28 per pint**

2.

	Ratio	Actual Count
Oatmeal	2	3 cups
Brown sugar	1	B

$$\frac{2}{1} = \frac{3 \text{ cups}}{B}$$

$$2B = 1(3 \text{ cups})$$

$$B = \frac{3}{2} \text{ cups} = \mathbf{1\frac{1}{2} \text{ cups}}$$

3. $3(55.0 \text{ seconds}) = 165 \text{ seconds}$

$54.3 \text{ seconds} + 56.1 \text{ seconds} + n = 165 \text{ seconds}$

$n = 165 \text{ seconds} - 110.4 \text{ seconds}$

$n = \mathbf{54.6 \text{ seconds}}$

4. $\frac{9 \text{ miles}}{60 \text{ minutes}} = \frac{9 \text{ miles}}{1 \text{ hour}} = \mathbf{9 \text{ miles per hour}}$

5.
$$\begin{array}{r} 63{,}100{,}000 \\ -\ 7{,}060{,}000 \\ \hline 56{,}040{,}000 \end{array}$$

$56{,}040{,}000 = \mathbf{5.604 \times 10^7}$

6. (a) $\frac{7}{10} \times 100\% = \frac{700\%}{10} = \mathbf{70\%}$

(b) $\frac{\text{news area}}{\text{advertisement area}} = \mathbf{\frac{3}{7}}$

(c) $\frac{\text{advertisement area}}{\text{total area}} = \mathbf{\frac{7}{10}}$

7. (a) $\mathbf{1.05 \times 10^{-3}}$

(b) **302,000**

8. $\dfrac{128}{192} = \dfrac{\overset{1}{\cancel{2}} \cdot \overset{1}{\cancel{2}} \cdot \overset{1}{\cancel{2}} \cdot \overset{1}{\cancel{2}} \cdot \overset{1}{\cancel{2}} \cdot \overset{1}{\cancel{2}} \cdot 2}{\underset{1}{\cancel{2}} \cdot \underset{1}{\cancel{2}} \cdot \underset{1}{\cancel{2}} \cdot \underset{1}{\cancel{2}} \cdot \underset{1}{\cancel{2}} \cdot \underset{1}{\cancel{2}} \cdot 3} = \mathbf{\dfrac{2}{3}}$

9. $\mathbf{1760\ \cancel{yards} \cdot \dfrac{3\ feet}{1\ \cancel{yard}} = 5280\ feet}$

10. (a) **Parallelogram**

(b) **Trapezoid**

11. (a) Area $= (4\text{ m})(6\text{ m}) = \mathbf{24\ m^2}$

(b) Area $= \dfrac{(2\text{ m})(4\text{ m})}{2} = \mathbf{4\ m^2}$

(c) Area $= 24\text{ m}^2 + 4\text{ m}^2 = \mathbf{28\ m^2}$

12. (a) **Obtuse angle**

(b) **Right angle**

(c) **Acute angle**

13. (a) **8**

(b) **6**

(c) **9**

(d) **2**

14. (a) Perimeter $= 12\text{ cm} + 16\text{ cm} + 12\text{ cm} + 16\text{ cm} = \mathbf{56\ cm}$

(b) Area $= (16\text{ cm})(10\text{ cm}) = \mathbf{160\ cm^2}$

(c)

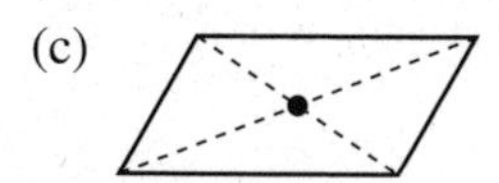

15. (a) $m\angle a = 180° - (59° + 61°)$
$= 180° - 120° = \mathbf{60°}$

(b) $m\angle b = \mathbf{61°}$

(c) $m\angle c = \mathbf{59°}$

(d) $m\angle d = m\angle a = \mathbf{60°}$

16. $2\ \cancel{\text{centimeters}} \cdot \dfrac{1\text{ meter}}{100\ \cancel{\text{centimeters}}}$

$= \dfrac{2}{100}\text{ meter} = \mathbf{0.02\ meter}$

17.

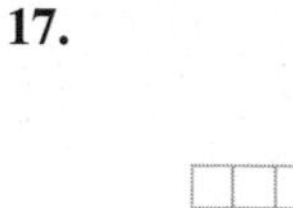

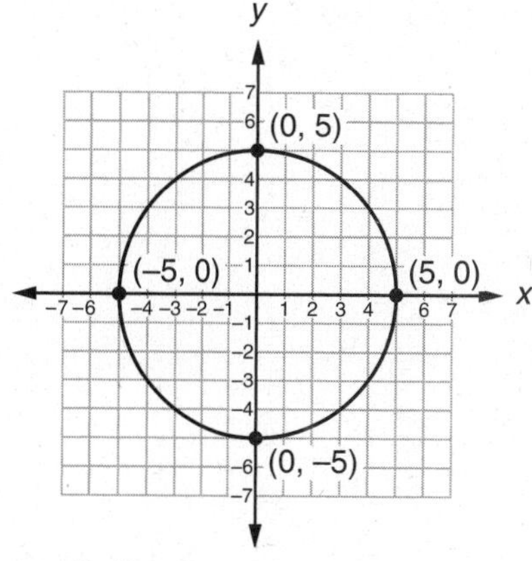

(a) **(0, 5), (0, −5)**

(b) **10 units**

18. **The scale is balanced so the 3 items on the left have a total mass of 50 g. The labeled masses total 15 g, so the cube must be 35 g because 35 g + 15 g = 50 g.**

19.

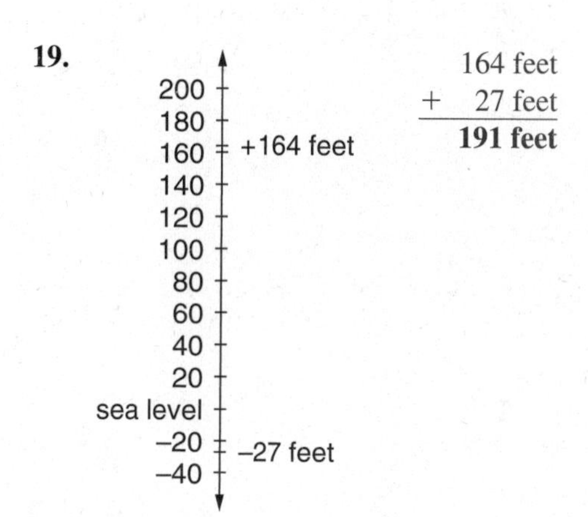

$$\begin{array}{r} 164\text{ feet} \\ +\ \ 27\text{ feet} \\ \hline \mathbf{191\ feet} \end{array}$$

20. $10 + 10 \times 10 - 10 \div 10$
$10 + 100 - 1$
$110 - 1$
109

21. $2^0 - 2^{-3} = 1 - \dfrac{1}{2^3}$

$= 1 - \dfrac{1}{8} = \dfrac{8}{8} - \dfrac{1}{8} = \mathbf{\dfrac{7}{8}}$

22. $70\text{ cm} = 0.7\text{ m}$

$$\begin{array}{r} 4.5\text{ m} \\ +\ 0.7\text{ m} \\ \hline \mathbf{5.2\ m} \end{array}$$

23. $2.75\ \cancel{L} \cdot \dfrac{1000\text{ mL}}{1\ \cancel{L}} = \mathbf{2750\ mL}$

24. $$\begin{array}{r} 3\frac{1}{3} = 3\frac{2}{6} \\ -\ 1\frac{1}{2} = 1\frac{3}{6} \\ \hline \end{array} \longrightarrow \begin{array}{r} 2\frac{8}{6} \\ -\ 1\frac{3}{6} \\ \hline 1\frac{5}{6} \end{array}$$

$$\begin{array}{r} 5\frac{7}{8} = 5\frac{21}{24} \\ +\ 1\frac{5}{6} = 1\frac{20}{24} \\ \hline 6\frac{41}{24} = \mathbf{7\frac{17}{24}} \end{array}$$

25. $4\frac{4}{5} \cdot 1\frac{1}{9} \cdot 1\frac{7}{8}$

$$= \frac{\overset{1}{\cancel{\overset{\cancel{3}}{24}}}}{\underset{1}{\cancel{5}}} \cdot \frac{10}{\underset{\underset{1}{\cancel{3}}}{\cancel{9}}} \cdot \frac{\overset{1}{\cancel{\overset{\cancel{3}}{15}}}}{\underset{1}{\cancel{8}}} = \mathbf{10}$$

26. $3\frac{1}{5} \div 8 = \frac{16}{5} \div \frac{8}{1} = \frac{\overset{2}{\cancel{16}}}{5} \times \frac{1}{\underset{1}{\cancel{8}}} = \mathbf{\frac{2}{5}}$

$6\frac{2}{3} \div \frac{2}{5} = \frac{20}{3} \div \frac{2}{5} = \frac{\overset{10}{\cancel{20}}}{3} \times \frac{5}{\underset{1}{\cancel{2}}} = \frac{50}{3}$

$= \mathbf{16\frac{2}{3}}$

27. $$\begin{array}{r} 0.8 \\ +\ 0.97 \\ \hline 1.77 \end{array} \qquad \begin{array}{r} 12.00 \\ -\ 1.77 \\ \hline \mathbf{10.23} \end{array}$$

28. $$\begin{array}{r} 0.05 \\ \times\ 2.4 \\ \hline 0.12 \end{array} \qquad \begin{array}{r} 0.005 \\ \times\ 0.12 \\ \hline \mathbf{0.0006} \end{array}$$

29. $4 \times 10^2 = 4 \times 100 = 400$
$0.2 \div 400 = \mathbf{0.0005}$

30. $4 \div 0.25 = 16$
$0.36 \div 16 = \mathbf{0.0225}$

LESSON 62, WARM-UP

a. **43**

b. **0.025**

c. **3**

d. **2.5 kg**

e. **1**

f. **$16.00**

g. **0**

Problem Solving

1024 − 512 = 512 m
512 − 256 = 256 m
256 − 128 = 128 m
After Alice, 128 m remains. If the pattern continues, the baton will get closer to, but will not cross, the finish line.

LESSON 62, LESSON PRACTICE

a. **Right triangle**

b. **Obtuse triangle**

c. **Acute triangle**

d. **Scalene triangle**

e. **Equilateral triangle**

f. **Isosceles triangle**

g. Perimeter = 3 cm + 4 cm + 4 cm = **11 cm**

h. $\angle L, \angle N, \angle M$

LESSON 62, MIXED PRACTICE

1. 4(15 minutes) = 60 minutes
60 minutes + 10 minutes = 70 minutes
= 1 hour 10 minutes
2:40 p.m.

2. $$\begin{array}{r} 40{,}060 \text{ miles} \\ -\ 39{,}872 \text{ miles} \\ \hline \mathbf{188 \text{ miles}} \end{array}$$

3. $\frac{188 \text{ miles}}{8 \text{ gallons}} = \mathbf{23.5 \text{ miles per gallon}}$

4.
$$24 \cdot w = 288$$
$$w = 12$$
$$24 \div 12 = \mathbf{2}$$

5.

	Ratio	Actual Count
Bolsheviks	9	144
Czarists	8	C

$$\frac{9}{8} = \frac{144}{C}$$
$$9C = 1152$$
$$C = \textbf{128 czarists}$$

6.

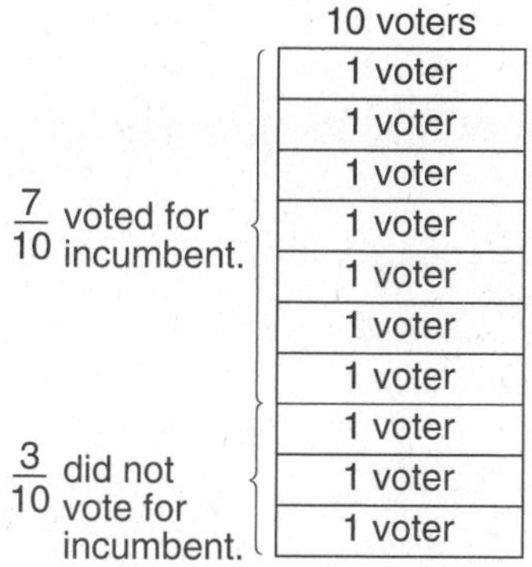

(a) $\frac{7}{10} \times 100\% = \frac{700\%}{10} = \mathbf{70\%}$

(b) $\mathbf{\frac{3}{10}}$

7. $\mathbf{W_N = \frac{5}{6} \times 3\frac{1}{3}}$

$$W_N = \frac{5}{\cancel{6}_3} \times \frac{\cancel{10}^5}{3} = \frac{25}{9} = \mathbf{2\frac{7}{9}}$$

8.
$$\begin{array}{r} \$10,000 \\ \times \quad 0.085 \\ \hline 50\ 000 \\ 800\ 00 \\ \hline \$850.000 \end{array} = \$850$$
$$\$10,000 + \$850 = \mathbf{\$10,850}$$

9. **186,000; one hundred eighty-six thousand**

10. 1 quart ⓒ< 1 liter

11.

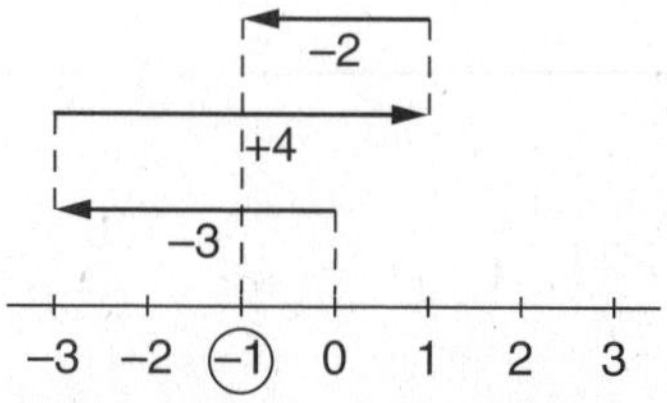

12. (a) $8\overline{)5.000}$ = **0.625**

(b) $\frac{5}{8} \times 100\% = \frac{500\%}{8} = \mathbf{62\frac{1}{2}\%}$

(c) $275\% = \frac{275}{100} = \frac{11}{4} = \mathbf{2\frac{3}{4}}$

(d) $275\% = \frac{275}{100} = \mathbf{2.75}$

13. $(12) + \frac{(12)}{(3)} - 3 =$

$12 + 4 - 3 = 16 - 3 = \mathbf{13}$

14. (a) 2^8

(b) 2^2

(c) 2^0

(d) 2^{-2}

15. (a) **△ZWY**

(b) **△WYX**

(c) **△ZWX**

16.

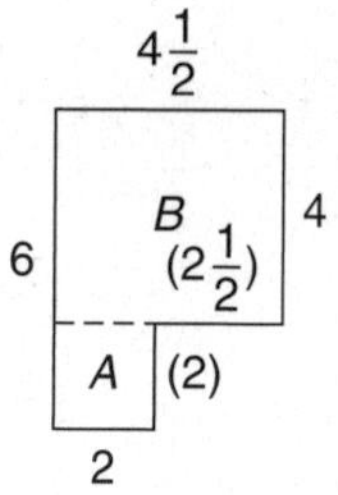

(a) Perimeter $= 4\frac{1}{2}$ in. + 4 in. + $2\frac{1}{2}$ in.
+ 2 in. + 2 in. + 6 in.
= **21 in.**

(b)

	Area A =	(2 in.) (2 in.)	=	4 in.^{2}
+	Area B =	($4\frac{1}{2}$ in.)(4 in.)	=	18 in.^{2}
		Area of figure	=	**22 in.^{2}**

17. (a) **Isosceles triangle**

(b) $\frac{180° - 90°}{2} = \frac{90°}{2} = \mathbf{45°}$

(c) Area $= \frac{(6 \text{ cm})(6 \text{ cm})}{2} = \mathbf{18 \text{ cm}^2}$

(d) **∠C**

18. $7\overline{)1.428}$ = **0.204**

19. $\frac{30}{70} = \frac{w}{\$2.10}$

$70w = 30(\$2.10)$

$w = \frac{\$63.00}{70}$

$w = \mathbf{\$0.90}$

20. $5^2 + 2^5 - \sqrt{49} = 25 + 32 - 7 = \mathbf{50}$

21. $3(8) - (5)(2) + 10 \div 2$

$24 - 10 + 5$

$14 + 5$

19

22.
$$\begin{array}{rrrr} & 1\text{ yd} & 2\text{ ft} & 3\frac{3}{4}\text{ in.} \\ + & & 2\text{ ft} & 6\frac{1}{2}\text{ in.} \\ \hline & 1\text{ yd} & 4\text{ ft} & 10\frac{1}{4}\text{ in.} \end{array}$$

4 ft = 1 yd 1 ft

$$\begin{array}{rrrr} & 1\text{ yd} & 1\text{ ft} & 10\frac{1}{4}\text{ in.} \\ + & 1\text{ yd} & & \\ \hline & \mathbf{2\text{ yd}} & \mathbf{1\text{ ft}} & \mathbf{10\frac{1}{4}\text{ in.}} \end{array}$$

23. 1 L = 1000 mL,
1000 mL − 50 mL = **950 mL**

24. $\frac{\overset{1}{\cancel{60}}\text{ mi}}{\cancel{1\text{ hr}}} \cdot \frac{\cancel{1\text{ hr}}}{\underset{1}{\cancel{60}}\text{ min}} = \mathbf{1\ \frac{mi}{min}}$

25.
$$\begin{array}{rrl} & 2\frac{7}{24} & = 2\frac{28}{96} \\ + & 3\frac{9}{32} & = 3\frac{27}{96} \\ \hline & & \mathbf{5\frac{55}{96}} \end{array}$$

26. $4\frac{1}{5} \div 1\frac{3}{4} = \frac{21}{5} \div \frac{7}{4} = \frac{\overset{3}{\cancel{21}}}{5} \times \frac{4}{\underset{1}{\cancel{7}}} = \frac{12}{5}$

$2\frac{2}{5} \div \frac{12}{5} = \frac{12}{5} \div \frac{12}{5} = \frac{\overset{1}{\cancel{12}}}{\underset{1}{\cancel{5}}} \times \frac{\overset{1}{\cancel{5}}}{\underset{1}{\cancel{12}}} = \mathbf{1}$

27. $7\frac{1}{2} \div \frac{2}{3} = \frac{15}{2} \div \frac{2}{3}$

$= \frac{15}{2} \times \frac{3}{2} = \frac{45}{4} = 11\frac{1}{4}$

$$\begin{array}{rcr} 20 & \longrightarrow & 19\frac{4}{4} \\ -\ 11\frac{1}{4} & & -\ 11\frac{1}{4} \\ \hline & & \mathbf{8\frac{3}{4}} \end{array}$$

28.

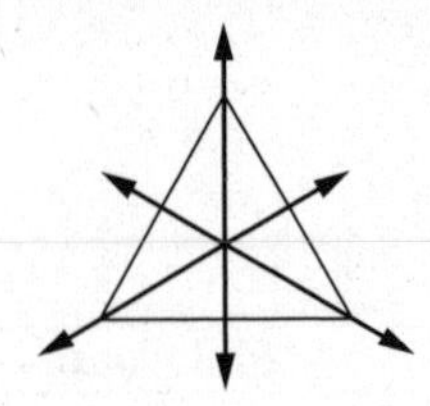

29. $|3 - 4| = |-1| = \mathbf{1}$

30.
$$\begin{array}{rr} & 1000\text{ g} \\ - & 250\text{ g} \\ \hline & \mathbf{750\text{ g}} \end{array}$$

Lesson 63, Warm-Up

a. **1230**

b. **0.004**

c. **6**

d. **500 mL**

e. **14**

f. **$9.00**

g. **1400**

Problem Solving

The trip to town took one hour. The return trip took 2 hours. The round trip was 120 miles. So we have

$$\frac{\mathbf{120\text{ mi}}}{\mathbf{1\text{ hr} + 2\text{ hr}}} = \mathbf{40\text{ mph}}$$

Lesson 63, Lesson Practice

a. $30 - [40 - (10 - 2)]$

$30 - [40 - (8)]$

$30 - [32]$

$\mathbf{-2}$

b. $100 - 3[2(6 - 2)]$

$100 - 3[2(4)]$

$100 - 3[8]$

$100 - 24$

76

c. $\dfrac{10 + 9 \cdot 8 - 7}{6 \cdot 5 - 4 - 3 + 2}$

$\dfrac{10 + 72 - 7}{30 - 4 - 3 + 2}$

$\dfrac{75}{25}$

3

d. $\dfrac{1 + 2(3 + 4) - 5}{10 - 9(8 - 7)}$

$\dfrac{1 + 14 - 5}{10 - 9}$

$\dfrac{10}{1}$

10

e. $12 + 3(8 - |-2|)$

$12 + 3(8 - 2)$

$12 + 3(6)$

$12 + 18$

30

Lesson 63, Mixed Practice

1. $\$6(3) + \$6(2\frac{1}{2}) = \$18 + \$15 =$ **$33**

2. $30\ \cancel{\text{min}} \cdot \dfrac{70 \text{ times}}{1\ \cancel{\text{min}}} = 2100 \text{ times}$

$30\ \cancel{\text{min}} \cdot \dfrac{150 \text{ times}}{1\ \cancel{\text{min}}} = 4500 \text{ times}$

4500 times
− 2100 times
2400 times

3.

	Ratio	Actual Count
Brachiopods	2	B
Trilobites	9	720

$\dfrac{2}{9} = \dfrac{B}{720}$

$9B = 2 \cdot 720$

$B =$ **160 brachiopods**

4. $5\ \cancel{\text{days}} \cdot \dfrac{18 \text{ miles}}{1\ \cancel{\text{day}}} = 90 \text{ miles}$

90 miles
16 miles
+ 21 miles
127 miles

1017 miles
− 127 miles
890 miles

5.

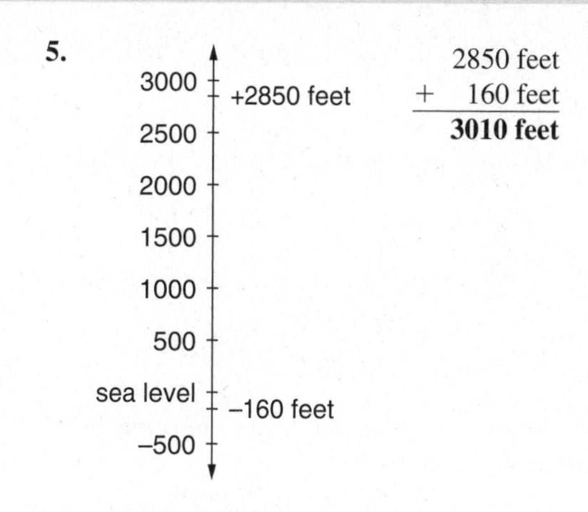

6. **One hundred forty-nine million, six hundred thousand kilometers**

7.

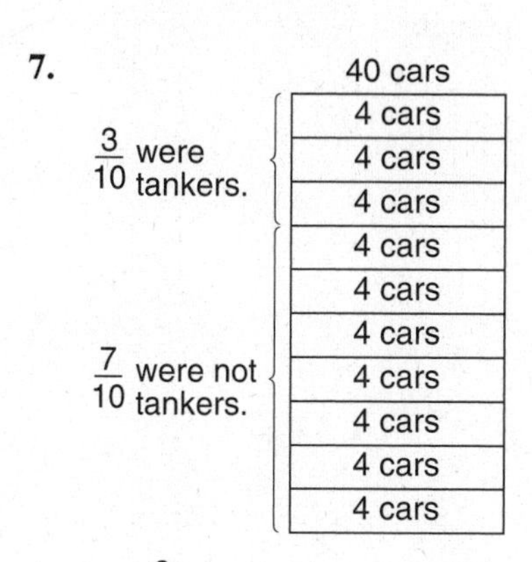

(a) $\mathbf{\dfrac{3}{10}}$

(b) $\dfrac{7}{10} \times 100\% = \dfrac{700\%}{10} =$ **70%**

8. **Two thousandths mile per hour**

9. **$1.5\ \cancel{\text{km}} \cdot \dfrac{1000 \text{ m}}{1\ \cancel{\text{km}}} = 1500 \text{ m}$**

10. $12\overline{)4360.0\ldots}$ = 363.3 . . .

36
76
72
40
36
40
36
4

$363.\overline{3}$

11.

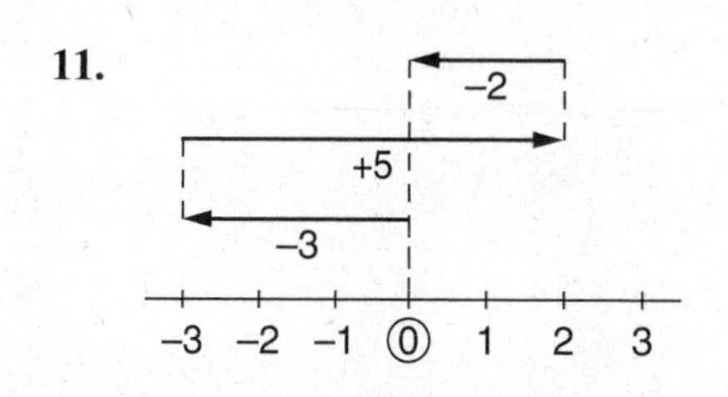

12. (a) $33\% = \mathbf{\frac{33}{100}}$

(b) $33\% = \frac{33}{100} = \mathbf{0.33}$

(c) $3\overline{)1.0}$ with quotient $\mathbf{0.\overline{3}}$

(d) $\frac{1}{3} \times 100\% = \frac{100\%}{3} = \mathbf{33\frac{1}{3}\%}$

13. **Divide the "in" number by 3 to find the "out" number.**

14. $\frac{\text{red face card}}{\text{total cards}} = \frac{6}{52} = \mathbf{\frac{3}{26}}$

15. (a) **Isosceles triangle**

(b) Perimeter = 5 cm + 5 cm + 5 cm
= **15 cm**

(c) $\mathbf{\Delta ABC}$

16. (a) $m\angle BAC = \frac{180°}{3} = \mathbf{60°}$

(b) $m\angle ADB = \frac{180°}{3} = \mathbf{60°}$

(c) $m\angle BDC = 180° - 60° = \mathbf{120°}$

(d) $m\angle DBA = \frac{180°}{3} = \mathbf{60°}$

(e) $m\angle DBC = \frac{180° - 120°}{2} = \frac{60°}{2} = \mathbf{30°}$

(f) $m\angle DCB = m\angle DBC = \mathbf{30°}$

17. $\frac{\text{length of shortest side}}{\text{length of longest side}} = \frac{5}{10} = \mathbf{\frac{1}{2}}$

18.
$$\frac{5}{18} = \frac{10}{36}$$
$$-\ \frac{1}{12} = \frac{3}{36}$$
$$\mathbf{\frac{7}{36}}$$

19. $2 \div 0.4 = \mathbf{5}$

20. $3[24 - (8 + 3 \cdot 2)] - \frac{6 + 4}{|-2|}$

$3[24 - (8 + 6)] - \frac{6 + 4}{2}$

$3[24 - (14)] - \frac{10}{2}$

$3[10] - 5$

$30 - 5$

25

21. $3^3 - \sqrt{3^2 + 4^2}$

$27 - \sqrt{9 + 16}$

$27 - \sqrt{25}$

$27 - 5$

22

22.
$\overset{0}{\not 1}$ week 2 days 7 hr → (7 days)
− 5 days 9 hr

$\overset{0}{\not 1}$ week $\overset{8}{\not 9}\!\!\!\not 2$ days 7 hr → (24 hr)
− 5 days 9 hr

$\overset{0}{\not 1}$ week $\overset{8}{\not 9}\!\!\!\not 2$ days $\overset{31}{\not 7}$ hr
− 5 days 9 hr
3 days 22 hr

23. $\frac{20\text{ mi}}{1\text{ gal}} \cdot \frac{1\text{ gal}}{4\text{ qt}} = \mathbf{5\ \frac{mi}{qt}}$

24.
$$4\frac{2}{3} = 4\frac{12}{18}$$
$$3\frac{5}{6} = 3\frac{15}{18}$$
$$+\ 2\frac{5}{9} = 2\frac{10}{18}$$
$$9\frac{37}{18} = \mathbf{11\frac{1}{18}}$$

25. $12\frac{1}{2} \cdot 4\frac{4}{5} \cdot 3\frac{1}{3}$

$= \frac{\overset{5}{\cancel{25}}}{\underset{1}{\cancel{2}}} \cdot \frac{\overset{8}{\cancel{24}}}{\underset{1}{\cancel{5}}} \cdot \frac{\overset{5}{\cancel{10}}}{\underset{1}{\cancel{3}}} = \mathbf{200}$

26. $1\frac{2}{3} \div 3 = \frac{5}{3} \div \frac{3}{1}$

$= \frac{5}{3} \times \frac{1}{3} = \frac{5}{9}$

$$6\frac{1}{3} = 6\frac{3}{9} \longrightarrow 5\frac{12}{9}$$
$$-\ \frac{5}{9} = \frac{5}{9} \qquad -\ \frac{5}{9}$$
$$\mathbf{5\frac{7}{9}}$$

27. $(3)^2 + 2(3)(4) + (4)^2$
$= 9 + 24 + 16 = \mathbf{49}$

28.

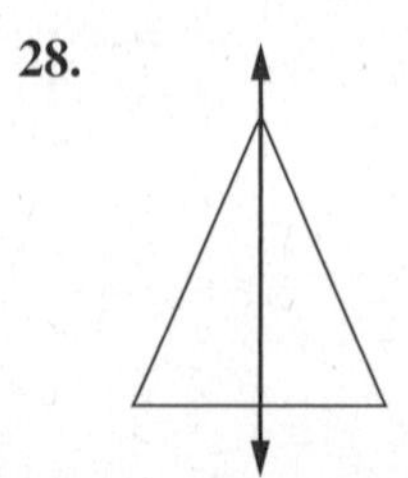

29. (a)

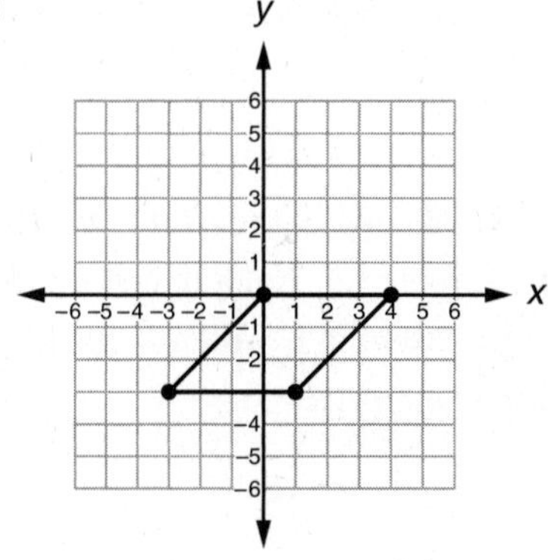

(b) Area = (4 units)(3 units)
= **12 sq. units**

(c) $\frac{180° - 90°}{2} = \frac{90°}{2} = \mathbf{45°}$

30. $3\overline{)750\text{ g}}$ = **250 g**

Lesson 64, Warm-Up

a. **180**

b. **750**

c. **10**

d. **200 mm**

e. **5**

f. **$1.00**

g. **0.5**

Problem Solving

The LCM of 2, 3, 4, 5, and 6 is 60, so Dean's deck has **60 cards.**

Lesson 64, Lesson Practice

a. $(-56) + (+96) = \mathbf{+40}$

b. $(-28) + (-145) = \mathbf{-173}$

c. $(-5) + (+7) + (+9) + (-3)$
$(+2) + (+9) + (-3)$
$(+11) + (-3)$
$\mathbf{+8}$

d. $(-3) + (-8) + (+15)$
$(-11) + (+15)$
$\mathbf{+4}$

e. $(-12) + (-9) + (+16)$
$(-21) + (+16)$
$\mathbf{-5}$

f. $(+12) + (-18) + (+6)$
$(-6) + (+6)$
$\mathbf{0}$

g. $\left(-3\frac{5}{6}\right) + \left(+5\frac{1}{3}\right)$
$\left(-3\frac{5}{6}\right) + \left(+5\frac{2}{6}\right)$
$\left(-3\frac{5}{6}\right) + \left(+4\frac{8}{6}\right)$
$+1\frac{3}{6} = \mathbf{+1\frac{1}{2}}$

h. $(-1.6) + (-11.47)$
$\mathbf{-13.07}$

i. **(+$250) + (−$300) + (+$525); The net result was a gain of $475.**

Lesson 64, Mixed Practice

1.
$$\begin{array}{r} 2{,}000{,}000{,}000{,}000 \\ -\ \ 750{,}000{,}000{,}000 \\ \hline 1{,}250{,}000{,}000{,}000 \end{array}$$
$\mathbf{1.25 \times 10^{12}}$

2. $2.25 + $0.15(42)
= $2.25 + $6.30 = $8.55
$10 − $8.55 = **$1.45**

3. 5($0.25) + 3($0.10) + 2($0.05)
= $1.25 + $0.30 + $0.10
= $1.65

$35\overline{)165.00}$ = 4.71

4 packages

4. $53 + 59 = \mathbf{112}$

5.

$$\begin{array}{r} 1.74 \\ 2.8 \\ 3.4 \\ 0.96 \\ 2 \\ +\ 1.22 \\ \hline 12.12 \end{array}$$

$12.12 \div 6 = \mathbf{2.02}$

6.

	1200 serfs
$\frac{2}{5}$ were conscripted.	240 serfs
	240 serfs
$\frac{3}{5}$ were not conscripted.	240 serfs
	240 serfs
	240 serfs

(a) **480 serfs**

(b) $\frac{3}{5} \times 100\% = \frac{300\%}{5} = \mathbf{60\%}$

7. $\mathbf{W_N = \frac{5}{9} \times 100}$

$W_N = \frac{500}{9}$

$W_N = \mathbf{55\frac{5}{9}}$

8. (a) **Sixteen million degrees Celsius**

(b) **Seven millionths meter**

9. (a) 1.6×10^{7} (>) 7×10^{-6}

(b) 7×10^{-6} (>) 0

(c) 2^{-3} (<) 2^{-2}

10.

$$\begin{array}{r} 16.285\ldots \\ 28\overline{)456.000\ldots} \\ \underline{28} \\ 176 \\ \underline{168} \\ 8\,0 \\ \underline{5\,6} \\ 2\,40 \\ \underline{2\,24} \\ 160 \\ \underline{140} \\ 20 \end{array}$$

(a) $\mathbf{16\frac{2}{7}}$

(b) **16.29**

(c) **16**

11. (a) $(-63) + (-14) = \mathbf{-77}$

(b) $(-16) + (+20) + (-32) = \mathbf{-28}$

12. $(-\$327) + (+\$280) = -\$47$; **a loss of $47**

13. (a) **60°**

(b) **The chords are $\overline{AB}$ (or $\overline{BA}$), $\overline{BC}$ (or $\overline{CB}$), and $\overline{CA}$ (or $\overline{AC}$). Each chord is shorter than the diameter, which is the longest chord of a circle.**

14. $\left(\frac{2}{3}\right) + \left(\frac{2}{3}\right)\left(\frac{3}{4}\right)$

$= \frac{2}{3} + \frac{1}{2} = \frac{4}{6} + \frac{3}{6} = \frac{7}{6} = \mathbf{1\frac{1}{6}}$

15.

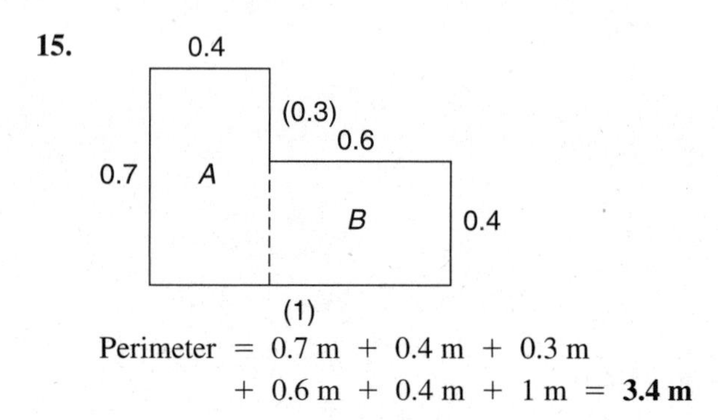

Perimeter $= 0.7\text{ m} + 0.4\text{ m} + 0.3\text{ m} + 0.6\text{ m} + 0.4\text{ m} + 1\text{ m} = \mathbf{3.4\text{ m}}$

16.

$$\begin{array}{lcl} \text{Area } A = (0.7\text{ m})(0.4\text{ m}) & = & 0.28\text{ m}^2 \\ +\ \text{Area } B = (0.6\text{ m})(0.4\text{ m}) & = & 0.24\text{ m}^2 \\ \hline \text{Area of figure} & = & \mathbf{0.52\text{ m}^2} \end{array}$$

17. $12x = 84$ $\quad 12y = 48$

$x = 7$ $\quad y = 4$

$7 \cdot 4 = \mathbf{28}$

18.

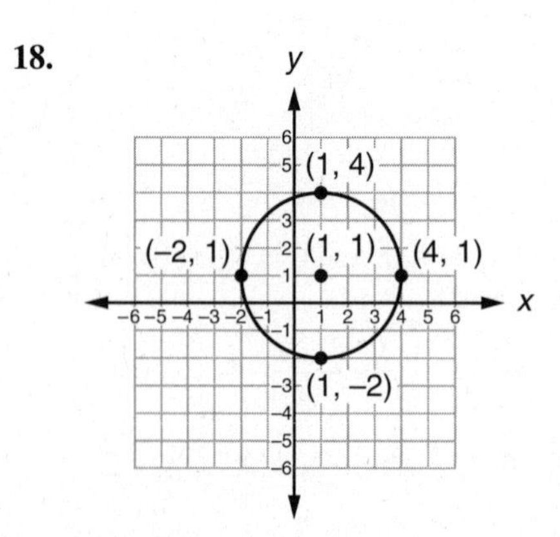

B. (−2, 1)

19. $\frac{4}{9} + \frac{2}{9} = \frac{6}{9} = \mathbf{\frac{2}{3}}$

20. $\frac{10}{25} = \mathbf{\frac{2}{5}}$ or **0.4**

21. $\dfrac{3^2 + 4^2}{\sqrt{3^2 + 4^2}} = \dfrac{9 + 16}{\sqrt{9 + 16}} = \dfrac{25}{\sqrt{25}} = \dfrac{25}{5}$
$= \mathbf{5}$

22. $6 \div 2\frac{1}{2} = \frac{6}{1} \div \frac{5}{2}$
$= \frac{6}{1} \times \frac{2}{5} = \frac{12}{5} = 2\frac{2}{5}$

$2\frac{4}{5} \div 2\frac{2}{5} = \frac{14}{5} \div \frac{12}{5}$
$= \frac{\overset{7}{\cancel{14}}}{\underset{1}{\cancel{5}}} \times \frac{\overset{1}{\cancel{5}}}{\underset{6}{\cancel{12}}} = \frac{7}{6} = \mathbf{1\frac{1}{6}}$

23. $100 - [20 + 5(4) + 3(2 + 4^0)]$
$100 - [20 + 20 + 3(2 + 1)]$
$100 - [20 + 20 + 9]$
$100 - [49]$
$\mathbf{51}$

24.

	5 gal	2 qt	1 pt	7 oz
+	1 gal	1 qt	1 pt	9 oz
	6 gal	3 qt	2 pt	16 oz

16 oz = 1 pt, 2 pt + 1 pt = 3 pt
3 pt = 1 qt 1 pt

	1 qt	1 pt
+	3 qt	
	4 qt	1 pt

4 qt = 1 gal, 6 gal + 1 gal = 7 gal, **7 gal 1 pt**

25. $\left(1\frac{1}{2}\right)^2 - \left(4 - 2\frac{1}{3}\right)$
$= \frac{9}{4} - \left(4 - 2\frac{1}{3}\right)$
$= 2\frac{1}{4} - \left(4 - 2\frac{1}{3}\right)$

$4 \longrightarrow 3\frac{3}{3}$
$-\ 2\frac{1}{3} \qquad -\ 2\frac{1}{3}$
$1\frac{2}{3}$

$2\frac{1}{4} = 2\frac{3}{12} \longrightarrow 1\frac{15}{12}$
$-\ 1\frac{2}{3} = 1\frac{8}{12} \qquad -\ 1\frac{8}{12}$
$\mathbf{\frac{7}{12}}$

26.
$0.010 - 0.001 = 0.009$
$0.100 - 0.009 = \mathbf{0.091}$

27. $5.1 \div 1.5 = 3.4, 5.1 \div 3.4 = \mathbf{1.5}$

28. $1.0 \div 5 = 0.2$
$4.375 - 3.200 = \mathbf{1.175}$

29. $\dfrac{\text{even primes}}{\text{total}} = \mathbf{\dfrac{1}{6}}$

30. (a) $m\angle B = 180° - 58° = \mathbf{122°}$

(b) $m\angle BCD = m\angle A = \mathbf{58°}$

(c) $m\angle BCM = 180° - 58° = \mathbf{122°}$

Lesson 65, Warm-Up

a. **21**

b. **0.125**

c. **6**

d. **750 mm**

e. **25**

f. **$3.60**

g. **$21.40**

Problem Solving

Since the subtrahend in the first subtraction stage has only two digits, we know that the divisor is a two-digit number less than 12 (because 9 × 12 results in a three-digit product). We also know the divisor is even, because it is double the remainder. The divisor is 10. To find the dividend, we multiply 91 by 10 and add 5 (one half of the divisor).

$915 \div 10 = 91\frac{1}{2}$
90
15
10
5

Lesson 65, Lesson Practice

a.

	Ratio	Actual Count
Acrobats	3	A
Clowns	5	C
Total	8	72

$$\frac{5}{8} = \frac{C}{72}$$
$$8C = 360$$
$$C = \textbf{45 clowns}$$

b.

	Ratio	Actual Count
Young men	8	240
Young women	9	W
Total	17	T

$$\frac{8}{17} = \frac{240}{T}$$
$$8T = 4080$$
$$T = \textbf{510 young people}$$

Lesson 65, Mixed Practice

1. (a) $\dfrac{\$2.40}{5 \text{ pounds}} = \dfrac{\$0.48}{1 \text{ pound}}$

$0.48 per pound

(b) $\$0.48 \times 8 = \mathbf{\$3.84}$

2. (a) $(0.3)(0.4) + (0.3)(0.5) = 0.12 + 0.15 = 0.27$

$0.3(0.4 + 0.5) = 0.3(0.9) = 0.27$

$\mathbf{0.27 = 0.27}$

(b) **Distributive property**

3.

	Ratio	Actual Count
Big fish	4	B
Little fish	11	L
Total	15	1320

$$\frac{4}{15} = \frac{B}{1320}$$
$$15B = 5280$$
$$B = \textbf{352 big fish}$$

4. $\dfrac{350 \text{ miles}}{15 \text{ gallons}} = 23.\overline{3}\,\dfrac{\text{miles}}{\text{gallon}}$

$\mathbf{23.\overline{3}\,\dfrac{miles}{gallon}}$

5. $\dfrac{1}{2} + \dfrac{1}{4} = \dfrac{2}{4} + \dfrac{1}{4} = \dfrac{3}{4}$

$\dfrac{3}{4} \div 2 = \dfrac{3}{4} \times \dfrac{1}{2} = \mathbf{\dfrac{3}{8}}$

6. $\mathbf{1.2 \times 10^{10}}$

7.

60 eggs

$\frac{1}{6}$ were cracked. { 10 eggs

$\frac{5}{6}$ were not cracked. { 10 eggs, 10 eggs, 10 eggs, 10 eggs, 10 eggs

(a) **50 eggs**

(b) $\dfrac{\text{cracked eggs}}{\text{uncracked eggs}} = \mathbf{\dfrac{1}{5}}$

(c) $\dfrac{1}{6} \times 100\% = \mathbf{16\dfrac{2}{3}\%}$

8. (a) One possibility:

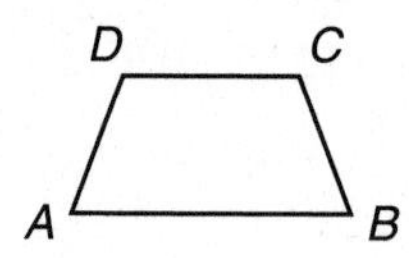

(b) **Trapezoid**

9. (a) Area $= \dfrac{(6 \text{ cm})(4 \text{ cm})}{2} = \mathbf{12\ cm^2}$

(b) Area $= \dfrac{(6 \text{ cm})(4 \text{ cm})}{2} = \mathbf{12\ cm^2}$

(c) Area $= \dfrac{(6 \text{ cm})(4 \text{ cm})}{2} = \mathbf{12\ cm^2}$

10. $0.76 + 0.88 = 1.64$

$1.64 \div 2 = \mathbf{0.82}$

11. $\mathbf{W_N = 0.75 \times 64}$

$W_N = \mathbf{48}$

12. $\mathbf{t = 0.08 \times \$7.40}$

$t = \mathbf{\$0.59}$

13. (a) $(-3) + (-8) = \mathbf{-11}$

(b) $(+3) + (-8) = \mathbf{-5}$

(c) $(-3) + (+8) + (-5) = \mathbf{0}$

14.

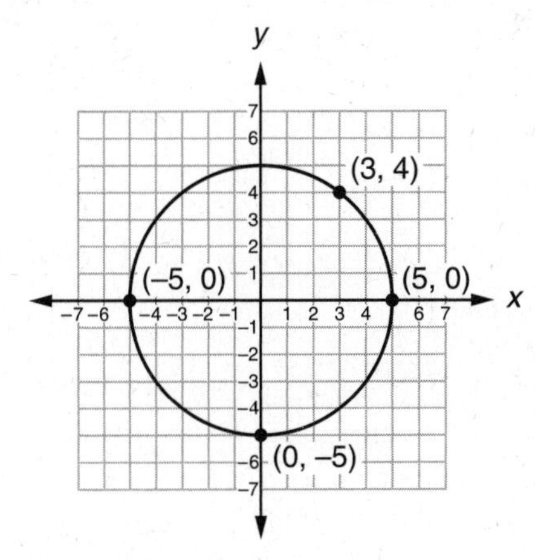

(a) **(5, 0), (−5, 0)**

(b) **10 units**

15. **0.95 ~~liters~~ · $\frac{\textbf{1000 milliliters}}{\textbf{1 \sout{liter}}}$**
= **950 milliliters**

16. $(5)(0.2) + 5 + \frac{5}{0.2}$
$= 1 + 5 + 25 = \mathbf{31}$

17. **27 blocks**

18. (a) $\angle \boldsymbol{COD}$ or $\angle \boldsymbol{DOC}$

(b) $\angle \boldsymbol{AOB}$ or $\angle \boldsymbol{BOA}$

19. $20 \times 5 = \mathbf{100}$

20. (a) $m\angle A = 180° - (62° + 59°)$
$= 180° - 121° = \mathbf{59°}$

(b) $\overline{\boldsymbol{AB}}$ or $\overline{\boldsymbol{BA}}$

(c) **Isosceles triangle**

(d) ***C***

21. (a) **Arrange the numbers in order, and look for the middle number. Since there is an even number of scores, there are two middle numbers. So the median is the mean of the two middle numbers.**

(b) $16 + 17 = 33$
$33 \div 2 = \mathbf{16.5}$

22. (a) **False**

(b) **True**

23.
$$\begin{array}{r} 2.20 \text{ meters} \\ -\ 2.15 \text{ meters} \\ \hline 0.05 \text{ meters} \end{array} = \textbf{5 centimeters}$$

24. $\frac{10^3 \cdot 10^3}{10^2} = \frac{10^6}{10^2} = \mathbf{10^4 \text{ or } 10{,}000}$

25.
(24 hr)
(60 min)
$$\begin{array}{r} \overset{3}{\cancel{4}} \text{ days } \overset{4}{\cancel{5}} \text{ hr } 15 \text{ min} \\ -\ \ 1 \text{ day } 7 \text{ hr } 50 \text{ min} \\ \hline \end{array} \longrightarrow$$

$$\begin{array}{rrr} \overset{3}{\cancel{4}} \text{ days} & \overset{28}{\cancel{\overset{4}{\cancel{5}}}} \text{ hr} & \overset{75}{\cancel{15}} \text{ min} \\ -\ 1 \text{ day} & 7 \text{ hr} & 50 \text{ min} \\ \hline \textbf{2 days} & \textbf{21 hr} & \textbf{25 min} \end{array}$$

26. $4.5 \div (0.4 + 0.5) = 4.5 \div 0.9 = \mathbf{5}$

27. $\frac{3 + 0.6}{3 - 0.6} = \frac{3.6}{2.4} = \mathbf{1.5}$

28. $1\frac{1}{6} \cdot 3 = \frac{7}{\underset{2}{\cancel{6}}} \cdot \frac{\overset{1}{\cancel{3}}}{1} = \frac{7}{2} = 3\frac{1}{2}$

$4\frac{1}{5} \div 3\frac{1}{2} = \frac{21}{5} \div \frac{7}{2}$

$= \frac{\overset{3}{\cancel{21}}}{5} \times \frac{2}{\underset{1}{\cancel{7}}} = \frac{6}{5} = \mathbf{1\frac{1}{5}}$

29. $3^2 + \sqrt{4 \cdot 7 - 3}$
$= 9 + \sqrt{28 - 3}$
$= 9 + \sqrt{25} = 9 + 5 = \mathbf{14}$

30. $|-3| + 4[(5 - 2)(3 + 1)]$
$3 + 4[(3)(4)]$
$3 + 4[12]$
$3 + 48$
$\mathbf{51}$

Lesson 66, Warm-Up

a. **6.85**

b. **0.0012**

c. **8**

d. **2 meters**

e. **20**

f. **6**

g. 75

Problem Solving

$1001 \div 7 = 143$
$143 \div 11 = 13$
The three prime factors of 1001 are **7, 11, and 13.**

LESSON 66, LESSON PRACTICE

a. $C = \pi d$
$C \approx 3.14(8\text{ in.})$
$C \approx \mathbf{25.12\text{ in.}}$

b. $C = \pi d$
$C \approx \frac{22}{7}(42\text{ mm})$
$C \approx \mathbf{132\text{ mm}}$

c. $C = \pi d$
$C = \pi(4\text{ ft})$
$C = \mathbf{4\pi\text{ ft}}$

d. $C = \pi d$
$C \approx 3.14(6\text{ inches})$
$C \approx \mathbf{18.84\text{ inches}}$

LESSON 66, MIXED PRACTICE

1. $100\% - (42\% + 25\%) = \mathbf{33\%}$

$$\begin{array}{r} \$25{,}000 \\ \times \quad 0.25 \\ \hline \mathbf{\$6250} \end{array}$$

2. $10\left(1\frac{1}{4}\text{ miles}\right) = \frac{25}{2}\text{ miles} = \mathbf{12\frac{1}{2}\text{ miles}}$

3. $(1.9)(2.2) - (1.9 + 2.2) = 4.18 - 4.1$
$= \mathbf{0.08}$

4.

	Ratio	Actual Count
Dimes	5	D
Quarters	8	Q
Total	13	520

$\frac{5}{13} = \frac{D}{520}$
$13D = 2600$
$D = \mathbf{200\text{ dimes}}$

5. $\mathbf{9 \times 10^8\text{ miles}}$

6.

400 acres

40 acres
40 acres
40 acres
40 acres
40 acres
40 acres
40 acres
40 acres
40 acres
40 acres

$\frac{3}{10}$ were planted with alfalfa. (first 3 rows)

$\frac{7}{10}$ were not planted with alfalfa. (remaining 7 rows)

(a) $\frac{3}{10} \times 100\% = \frac{300\%}{10} = \mathbf{30\%}$

(b) **280 acres**

7. (a) $\frac{12}{30} = \mathbf{\frac{2}{5}}$

(b) $\frac{2}{5} \times 100\% = \frac{200\%}{5} = \mathbf{40\%}$

(c) **It is more likely that a randomly selected watermelon is not ripe, because fewer than half the watermelons are ripe.**

8. (a) $C = \pi d$
$C \approx 3.14(21\text{ in.})$
$C \approx \mathbf{65.94\text{ in.}}$

(b) $C = \pi d$
$C \approx \frac{22}{7}(21\text{ in.})$
$C \approx \mathbf{66\text{ in.}}$

9. (a) $\text{Area} = (14\text{ cm})(24\text{ cm}) = \mathbf{336\text{ cm}^2}$

(b) $\text{Area} = \frac{(14\text{ cm})(24\text{ cm})}{2} = \mathbf{168\text{ cm}^2}$

(c) $\text{Perimeter} = 25\text{ cm} + 14\text{ cm} + 25\text{ cm}$
$= \mathbf{64\text{ cm}}$

10. $\mathbf{3.25 \times 10^{10}}$

11. $W_N = \mathbf{0.9 \times 3500}$
$W_N = \mathbf{3150}$

12. $W_N = \mathbf{\frac{5}{6} \times 2\frac{2}{5}}$

$W_N = \frac{\overset{1}{\cancel{5}}}{\underset{1}{\cancel{6}}} \times \frac{\overset{2}{\cancel{12}}}{\underset{1}{\cancel{5}}}$

$W_N = \mathbf{2}$

13. (a) $0.45 = \frac{45}{100} = \mathbf{\frac{9}{20}}$

(b) $0.45 = \frac{45}{100} = \mathbf{45\%}$

(c) $7.5\% = \frac{7.5}{100} = \mathbf{\frac{3}{40}}$

(d) $7.5\% = \frac{7.5}{100} = \mathbf{0.075}$

14. (a) $(5) + (-4) + (6) + (-1)$
$= 1 + (6) + (-1)$
$= 7 + (-1) = \mathbf{6}$

(b) $3 + (-5) + (+4) + (-2)$
$= -2 + (+4) + (-2)$
$= 2 + (-2) = \mathbf{0}$

15. $\mathbf{1.4\ \cancel{\text{kilograms}} \cdot \frac{1000\ \text{grams}}{1\ \cancel{\text{kilogram}}} = 1400\ \text{grams}}$

16. (a) $m\angle a = 180° - (90° + 35°)$
$= 180° - 125° = \mathbf{55°}$

(b) $m\angle b = m\angle a = \mathbf{55°}$

(c) $m\angle c = 180° - 55° = \mathbf{125°}$

(d) $m\angle d = m\angle b = \mathbf{55°}$

(e) $m\angle e = m\angle c = \mathbf{125°}$

17. $(3000)(500)(20) = \mathbf{30{,}000{,}000}$

18.

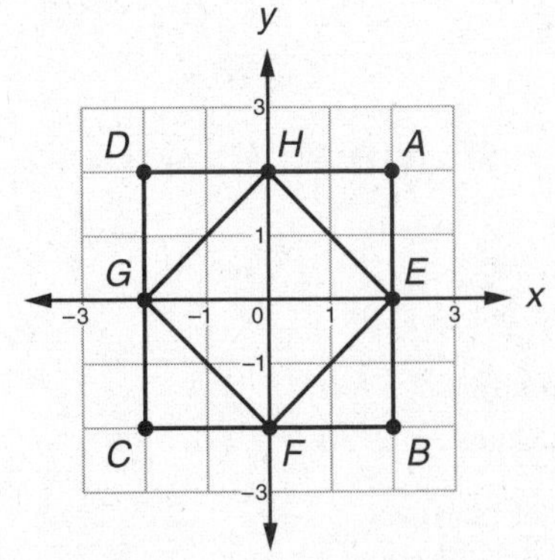

(a) Area $= (4\text{ units})(4\text{ units}) = \mathbf{16\ units^2}$

(b) **4 units**

(c) Area $= 4(1\text{ unit}^2) + 8\left(\frac{1}{2}\text{ unit}^2\right)$

$= 4\text{ units}^2 + 4\text{ units}^2 = \mathbf{8\ units^2}$

(d) $\mathbf{\sqrt{8}\ units}$

19. $\frac{0.9}{1.5} = \frac{12}{n}$
$0.9n = 1.5(12)$
$n = \frac{18}{0.9}$
$n = \mathbf{20}$

20.
$$\begin{array}{rcr} \frac{11}{12} & = & \frac{22}{24} \\ -\ \frac{11}{24} & = & \frac{11}{24} \\ \hline & & \mathbf{\frac{11}{24}} \end{array}$$

21. $2^1 - 2^0 - 2^{-1} = 2 - 1 - \frac{1}{2}$
$= 1 - \frac{1}{2} = \frac{2}{2} - \frac{1}{2} = \mathbf{\frac{1}{2}}$

22.
$$\begin{array}{rrr} & 4\text{ lb} & 12\text{ oz} \\ + & 1\text{ lb} & 7\text{ oz} \\ \hline & 5\text{ lb} & 19\text{ oz} \end{array}$$
19 oz = 1 lb 3 oz
$$\begin{array}{rrr} & 1\text{ lb} & 3\text{ oz} \\ + & 5\text{ lb} & \\ \hline & \mathbf{6\ lb} & \mathbf{3\ oz} \end{array}$$

23. $\frac{3\ \cancel{\text{ft}}}{1\text{ yd}} \cdot \frac{12\text{ in.}}{1\ \cancel{\text{ft}}} = \mathbf{36\ \frac{in.}{yd}}$

24. $16 \div (0.8 \div 0.04) = 16 \div 20 = \mathbf{0.8}$

25. $0.4[0.5 - (0.6)(0.7)]$
$0.4[0.5 - 0.42]$
$0.4[0.08]$
$\mathbf{0.032}$

26. $\frac{3}{8} \cdot 1\frac{2}{3} \cdot 4 \div 1\frac{2}{3}$

$= \left(\frac{\cancel{3}^{1}}{8} \cdot \frac{5}{\cancel{3}_{1}}\right) \cdot \frac{4}{1} \div \frac{5}{3}$

$= \frac{5}{\cancel{8}_{2}} \cdot \frac{\cancel{4}^{1}}{1} \div \frac{5}{3} = \frac{5}{2} \div \frac{5}{3}$

$= \frac{\cancel{5}^{1}}{2} \times \frac{3}{\cancel{5}_{1}} = \frac{3}{2} = \mathbf{1\frac{1}{2}}$

27. $30 - 5[4 + (3)(2) - 5]$
$30 - 5[4 + 6 - 5]$
$30 - 5[10 - 5]$
$30 - 5[5]$
$30 - 25$
$\mathbf{5}$

28. **One possibility: If a dozen flavored icicles cost \$2.88, what is the price of each flavored icicle?**

29. $\frac{9 \text{ ounces}}{2} = \mathbf{4\frac{1}{2}}$ **ounces**

30. (a) $\overline{AB}$ **or** $\overline{BA}$**,** $\overline{BC}$ **or** $\overline{CB}$

(b) **Isosceles triangle**

(c) **90°**

LESSON 67, WARM-UP

a. **33.6**

b. **3850**

c. **5**

d. **200 dm**

e. **100**

f. **18**

g. **720**

Problem Solving

Tom can read 5 pages in 4 minutes. Two hundred pages is 40 times as many pages as 5 pages. To find the number of minutes it will take Tom to read 200 pages, we multiply 4 minutes by 40:
40 minutes × 40 = 160 minutes

Jerry can read 4 pages in 5 minutes. Two hundred pages is 50 times as many pages as 4 pages. To find the number of minutes it will take Jerry to read 200 pages, we multiply 5 minutes by 50:
5 minutes × 50 = 250 minutes

Tom will finish his book 250 − 160, or **90 minutes** before Jerry.

LESSON 67, LESSON PRACTICE

a. **Triangular prism**

b. **Cone**

c. **Rectangular prism**

d. **5 faces**

e. **9 edges**

f. **6 vertices**

g.

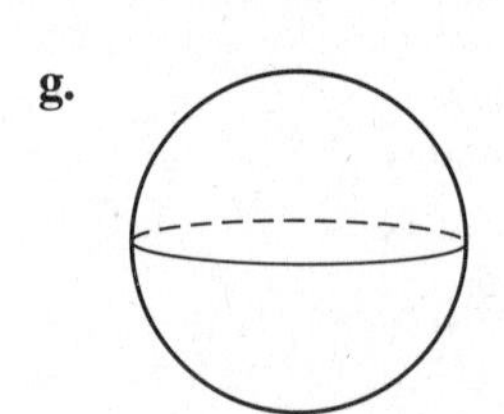

h.

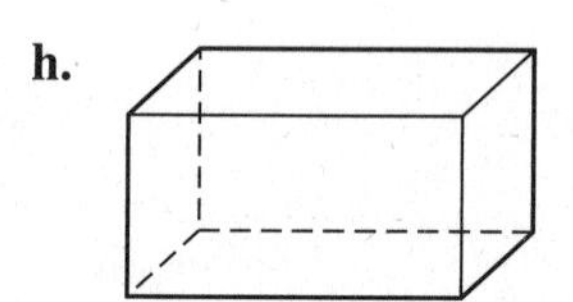

i.

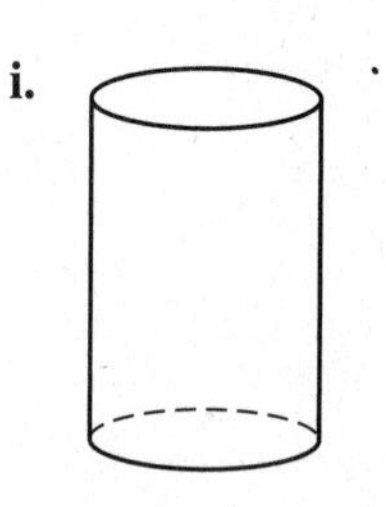

j. **Triangular prism**

k. $3 \text{ cm} \times 3 \text{ cm} = 9 \text{ cm}^2$

$6 \times 9 \text{ cm}^2 = \mathbf{54 \text{ cm}^2}$

LESSON 67, MIXED PRACTICE

1. (a) $\frac{\text{red marbles}}{\text{blue marbles}} = \frac{20}{40} = \mathbf{\frac{1}{2}}$

(b) $\frac{\text{white marbles}}{\text{red marbles}} = \frac{30}{20} = \mathbf{\frac{3}{2}}$

(c) $\frac{\text{not white marbles}}{\text{total marbles}} = \frac{60}{90} = \mathbf{\frac{2}{3}}$

2. $\left(\frac{1}{3} + \frac{1}{2}\right) - \left(\frac{1}{3} \times \frac{1}{2}\right)$

$= \left(\frac{2}{6} + \frac{3}{6}\right) - \left(\frac{1}{6}\right)$

$= \frac{5}{6} - \frac{1}{6} = \frac{4}{6} = \mathbf{\frac{2}{3}}$

3. 180 pounds − $165\frac{1}{2}$ pounds = **$14\frac{1}{2}$ pounds**

4. (a) 94 points + 85 points + 85 points = 264 points

$3\overline{)264}$ = **88 points**

(b) 5(92 points) = 460 points
460 points + 264 points = 724 points

$8\overline{)724.0}$ = **90.5 points**

5.

	Ratio	Actual Count
Diamonds	5	*D*
Rubies	2	*R*
Total	7	210

$\frac{5}{7} = \frac{D}{210}$

$7D = 1050$

$D =$ **150 diamonds**

6. 360 dolls

$\frac{4}{5}$ were sold.	72 dolls
	72 dolls
	72 dolls
	72 dolls
$\frac{1}{5}$ were not sold.	72 dolls

(a) **288 dolls**

(b) $\frac{1}{5} \times 100\% = \frac{100\%}{5} =$ **20%**

7. (a) **12 edges**

(b) **6 faces**

(c) **8 vertices**

8. (a) Area = $\frac{(12 \text{ m})(9 \text{ m})}{2}$ = **54 m²**

(b) Perimeter = 5 m + 5 m + 6 m
= **16 m**

(c) 90° − 37° = **53°**

(d) **The right triangle is not symmetrical.**

9. 7.65 + 7.83 = 15.48

$2\overline{)15.48}$ = **7.74**

10. **2.5×10^{-3}**

11. $W_N = 0.24 \times 75$
$W_N =$ **18**

12. $W_N = 1.2 \times 12$
$W_N =$ **14.4**

13. (a) (−2) + (−3) + (−4) = **−9**

(b) (+2) + (−3) + (+4) = **3**

14. (a) $4\% = \frac{4}{100} =$ **$\frac{1}{25}$**

(b) $4\% = \frac{4}{100} =$ **0.04**

(c) $8\overline{)7.000}$ = **0.875**

(d) $\frac{7}{8} \times 100\% = \frac{700\%}{8}$
= **87.5%** or **$87\frac{1}{2}\%$**

15. $\cancel{700}^{70} \cancel{\text{mm}} \cdot \frac{1 \text{ cm}}{\cancel{10}_{1} \cancel{\text{mm}}}$ = **70 cm**

16. (−\$560) + (+\$850) + (−\$280) = +\$10
A gain of \$10

17. **Multiply the "in" number by 7 to find the "out" number.** 1 × 7 = **7**

18. (a) **7856.43**

(b) **7900**

19. $C = \pi d$
$C \approx 3.14(24 \text{ inches})$
$C \approx 75.36$ inches ≈ **75 inches**

20. (a) **∠A and ∠B**

(b) **∠B and ∠D**

21. (a) **2(5 ft + 3 ft)**
2(8 ft)
16 ft

or

2(5 ft + 3 ft)
10 ft + 6 ft
16 ft

(b) **Distributive property**

22. $$\begin{aligned}\frac{2.5}{w} &= \frac{15}{12}\\ 15w &= 2.5(12)\\ w &= \frac{30}{15}\\ w &= \mathbf{2}\end{aligned}$$

23. $$\begin{aligned}&9 + 8\{7 \cdot 6 - 5[4 + (3 - 2 \cdot 1)]\}\\ &9 + 8\{7 \cdot 6 - 5[4 + (1)]\}\\ &9 + 8\{7 \cdot 6 - 5[5]\}\\ &9 + 8\{7 \cdot 6 - 25\}\\ &9 + 8\{42 - 25\}\\ &9 + 8\{17\}\\ &9 + 136\\ &\mathbf{145}\end{aligned}$$

24.

$\overset{0}{\cancel{1}}$ yd (3 ft)		
−	1 ft	3 in.

$\overset{0}{\cancel{1}}$ yd	$\overset{2}{\cancel{3}}$ ft	(12 in.)
−	1 ft	3 in.

$\overset{0}{\cancel{1}}$ yd	$\overset{2}{\cancel{3}}$ ft	12 in.
−	1 ft	3 in.
	1 ft	**9 in.**

25. $$\begin{aligned}&6.4 - (0.6 - 0.04)\\ &= 6.4 - 0.56 = \mathbf{5.84}\end{aligned}$$

26. $$\frac{3 + 0.6}{3(0.6)} = \frac{3.6}{1.8} = \mathbf{2}$$

27. $$\begin{aligned}1\frac{2}{3} &= 1\frac{8}{12}\\ + 3\frac{1}{4} &= 3\frac{3}{12}\\ \hline &\ 4\frac{11}{12}\end{aligned}$$

$$\begin{aligned}4\frac{11}{12} &= 4\frac{11}{12}\\ - 1\frac{5}{6} &= 1\frac{10}{12}\\ \hline &\ \mathbf{3\frac{1}{12}}\end{aligned}$$

28. $$\begin{aligned}&\frac{3}{5} \div 3\frac{1}{5} \cdot 5\frac{1}{3} \cdot |-1|\\ &= \left(\frac{3}{5} \div \frac{16}{5}\right) \cdot \frac{16}{3} \cdot 1\\ &= \left(\frac{3}{\cancel{5}_1} \times \frac{\cancel{5}^1}{16}\right) \cdot \frac{16}{3} \cdot 1\\ &= \frac{\cancel{3}^1}{\cancel{16}_1} \cdot \frac{\cancel{16}^1}{\cancel{3}_1} \cdot 1 = \mathbf{1}\end{aligned}$$

29. $$\begin{aligned}3 \div 1\frac{2}{3} &= \frac{3}{1} \div \frac{5}{3}\\ &= \frac{3}{1} \times \frac{3}{5} = \frac{9}{5} = 1\frac{4}{5}\end{aligned}$$

$$\begin{aligned}3\frac{3}{4} \div 1\frac{4}{5} &= \frac{15}{4} \div \frac{9}{5}\\ &= \frac{\cancel{15}^5}{4} \times \frac{5}{\cancel{9}_3} = \frac{25}{12} = \mathbf{2\frac{1}{12}}\end{aligned}$$

30. $$\begin{aligned}&5^2 - \sqrt{4^2} + 2^{-2}\\ &= 25 - 4 + \frac{1}{4} = \mathbf{21\frac{1}{4}}\end{aligned}$$

Lesson 68, Warm-Up

a. **1.25**

b. $\mathbf{\frac{3}{4}}$

c. **9**

d. **30**

e. **200 cm**

f. **8**

g. **7 m; 3 m²**

Problem Solving

6 triangles

(three small triangles, one whole figure, and two triangles composed of pairs of the small triangles)

Lesson 68, Lesson Practice

a. $$\begin{aligned}&(-3) - (+2)\\ &(-3) + [-(+2)]\\ &(-3) + [-2] = \mathbf{-5}\end{aligned}$$

b. $$\begin{aligned}&(-3) - (-2)\\ &(-3) + [-(-2)]\\ &(-3) + [2] = \mathbf{-1}\end{aligned}$$

c. $$\begin{aligned}&(+3) - (2)\\ &(+3) + [-(+2)]\\ &(+3) + [-2] = \mathbf{1}\end{aligned}$$

d. $(-3) - (+2) - (-4)$
$(-3) + [-(+2)] + [-(-4)]$
$(-3) + [-2] + [4] = \mathbf{-1}$

e. $(-8) + (-3) - (+2)$
$(-8) + (-3) + [-(+2)]$
$(-8) + (-3) + [-2] = \mathbf{-13}$

f. $(-8) - (+3) + (-2)$
$(-8) + [-(+3)] + (-2)$
$(-8) + [-3] + (-2) = \mathbf{-13}$

Lesson 68, Mixed Practice

1.
$$\begin{array}{r} 1037\text{ g} \\ -\ 350\text{ g} \\ \hline \mathbf{687\text{ g}} \end{array}$$

2.

	Ratio	Actual Count
Pentagons	3	12
Hexagons	5	H

$\frac{3}{5} = \frac{12}{H}$
$3H = 60$
$H = \mathbf{20\ hexagons}$

3. $\left(\frac{1}{4} + \frac{1}{2}\right) \div \left(\frac{1}{4} \times \frac{1}{2}\right)$
$= \left(\frac{2}{8} + \frac{4}{8}\right) \div \left(\frac{1}{8}\right)$
$= \frac{6}{8} \div \frac{1}{8} = \frac{6}{\cancel{8}_1} \times \frac{\cancel{8}^1}{1} = \mathbf{6}$

4. (a) $4\overline{)\$1.24}$ = **\$0.31 per pen**

(b) $100(\$0.31) = \mathbf{\$31.00}$

5. (a) $\frac{60\text{ miles}}{5\text{ hours}} = 12\frac{\text{miles}}{\text{hr}}$, **12 miles per hour**

(b) $\frac{5\text{ hours}}{60\text{ miles}} = \frac{300\text{ minutes}}{60\text{ miles}}$
$= \frac{5\text{ minutes}}{1\text{ mile}}$, **5 minutes per mile**

6. $1000\ \cancel{\text{meters}} \cdot \frac{1\text{ second}}{331\ \cancel{\text{meters}}} = \frac{1000}{331}$ seconds
About 3 seconds

7. (a) **88**

(b) **84**

(c)
$$\begin{array}{r} 72 \\ 80 \\ 84 \\ 88 \\ 100 \\ 88 \\ +\ 76 \\ \hline 588 \end{array}$$
$7\overline{)588}$ = **84**

8.
$$\begin{array}{r} 8.4 \\ +\ 9.8 \\ \hline 18.2 \end{array}$$
$2\overline{)18.2}$ = **9.1**

9. (a) **12 cubes**

(b) **Rectangular prism**

10. (a) $C = \pi d$
$C \approx 3.14(40\text{ cm})$
$C \approx \mathbf{125.6\ cm}$

(b) $C = \pi d$
$C = \pi(40\text{ cm})$
$C = \mathbf{40\pi\ cm}$

11.

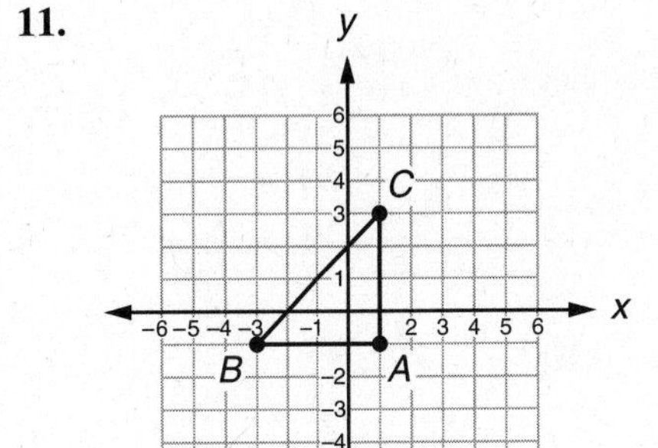

(a) **Right triangle**

(b) **Isosceles triangle**

(c) ***A***

(d) $m\angle B = \frac{90°}{2} = \mathbf{45°}$

(e) Area $= \frac{(4\text{ units})(4\text{ units})}{2} = \mathbf{8\ sq.\ units}$

12. $20{,}000 \times 30{,}000 = 600{,}000{,}000$
$= \mathbf{6 \times 10^8}$

13. $W_N = 0.75 \times 400$
$W_N = \mathbf{300}$

14. $W_N = 1.5 \times 1.5$ or $W_N = 1\frac{1}{2} \times 1\frac{1}{2}$
$W_N = \mathbf{2.25}$

$$W_N = \frac{3}{2} \times \frac{3}{2}$$
$$W_N = \frac{9}{4}$$
$$W_N = \mathbf{2\frac{1}{4}}$$

15. (a) $(-4) - (-6)$
$(-4) + [-(-6)]$
$(-4) + [6] = \mathbf{2}$

(b) $(-4) - (+6)$
$(-4) + [-(+6)]$
$(-4) + [-6] = \mathbf{-10}$

(c) $(-6) - (-4)$
$(-6) + [-(-4)]$
$(-6) + [4] = \mathbf{-2}$

(d) $(+6) - (-4)$
$(+6) + [-(-4)]$
$(+6) + [4] = \mathbf{10}$

16. $(4 \text{ in.})(4 \text{ in.}) = 16 \text{ in.}^2$
$6(16 \text{ in.}^2) = \mathbf{96\ in.^2}$

17. (a) $25\overline{)3.00}$ with quotient $\mathbf{0.12}$

(b) $\frac{3}{25} \times 100\% = \frac{300\%}{25} = \mathbf{12\%}$

(c) $120\% = \frac{120}{100} = \frac{6}{5} = \mathbf{1\frac{1}{5}}$

(d) $120\% = \frac{120}{100} = \mathbf{1.2}$

18. $(4)^2 + 2(4)(5) + (5)^2 = 16 + 40 + 25$
$= \mathbf{81}$

19. (a) **Rectangular prism**

(b) **Cone**

(c) **Cylinder**

20. (a) $m\angle DCA = \frac{90°}{2} = \mathbf{45°}$

(b) $m\angle DAC = 120° - 45° = \mathbf{75°}$

(c) $m\angle CAB = m\angle DCA = \mathbf{45°}$

(d) $m\angle ABC = m\angle CDA = \mathbf{60°}$

(e) $m\angle BCA = 120° - 45° = \mathbf{75°}$

(f) $m\angle BCD = 180° - 60° = \mathbf{120°}$

21. **One possibility: How many $0.25 pens can you buy with $3.00?**

22. $$\frac{4}{c} = \frac{3}{7\frac{1}{2}}$$
$$3c = 4\left(7\frac{1}{2}\right)$$
$$3c = 4\left(\frac{15}{2}\right)$$
$$3c = \frac{30}{3}$$
$$c = \mathbf{10}$$

23. $\frac{(1.5)^2}{15} = \mathbf{0.15}$

24.
```
  0
  1 gal (4 qt)
−        1 qt  1 pt  1 oz  →

  0    3
  1 gal 4 qt (2 pt)
−        1 qt  1 pt  1 oz  →

  0     3     1
  1 gal 4 qt  2 pt (16 oz)
−        1 qt  1 pt   1 oz  →

  0     3     1
  1 gal 4 qt  2 pt  16 oz
−        1 qt  1 pt   1 oz
         2 qt        15 oz
```
(**2 qt 15 oz**)

25. $16 \div (0.04 \div 0.8) = 16 \div 0.05 = \mathbf{320}$

26. $10 - [0.1 - (0.01)(0.1)]$
$10 - [0.1 - 0.001]$
$10 - 0.099$
$\mathbf{9.901}$

27. $$\frac{5}{8} + \frac{2}{3} \cdot \frac{3}{4} - \frac{3}{4}$$
$$= \frac{5}{8} + \frac{1}{2} - \frac{3}{4} = \frac{5}{8} + \frac{4}{8} - \frac{6}{8}$$
$$= \frac{9}{8} - \frac{6}{8} = \mathbf{\frac{3}{8}}$$

28. $$4\frac{1}{2} \cdot 3\frac{3}{4} \div 1\frac{2}{3}$$
$$= \left(\frac{9}{2} \cdot \frac{15}{4}\right) \div \frac{5}{3} = \frac{135}{8} \div \frac{5}{3}$$
$$= \frac{\overset{27}{\cancel{135}}}{8} \times \frac{3}{\underset{1}{\cancel{5}}} = \frac{81}{8} = \mathbf{10\frac{1}{8}}$$

29. $\sqrt{5^2 - 2^4} = \sqrt{25 - 16} = \sqrt{9} = \mathbf{3}$

30. $3 + 6[10 - (3 \cdot 4 - 5)]$
$3 + 6[10 - (7)]$
$3 + 6[3]$
$3 + 18$
$\mathbf{21}$

Lesson 69, Warm-Up

a. **2.5**

b. **0.075**

c. **2**

d. **630**

e. **2 dm**

f. **16**

g. $\mathbf{3\frac{1}{2}}$

Problem Solving

$3w = 12 + w$
$3w - w = 12 + w - w$
$2w \div 2 = 12 \div 2$
$w = \mathbf{6\ oz}$

Lesson 69, Lesson Practice

a. $0.16 \times 10^6 = 1.6 \times 10^{-1} \times 10^6$
$= \mathbf{1.6 \times 10^5}$

b. $24 \times 10^{-7} = 2.4 \times 10^1 \times 10^{-7}$
$= \mathbf{2.4 \times 10^{-6}}$

c. $30 \times 10^5 = 3 \times 10^1 \times 10^5$
$= \mathbf{3 \times 10^6}$

d. $0.75 \times 10^{-8} = 7.5 \times 10^{-1} \times 10^{-8}$
$= \mathbf{7.5 \times 10^{-9}}$

e. $14.4 \times 10^8 = 1.44 \times 10^1 \times 10^8$
$= \mathbf{1.44 \times 10^9}$

f. $12.4 \times 10^{-5} = 1.24 \times 10^1 \times 10^{-5}$
$= \mathbf{1.24 \times 10^{-4}}$

Lesson 69, Mixed Practice

1. (a) **6.5**

(b) **6.5**

(c) 7.0
6.5
6.5
7.4
7.0
6.5
+ 6.0
46.9

$$7\overline{)46.9} = \mathbf{6.7}$$

(d) $7.4 - 6.0 = \mathbf{1.4}$

2.

	Ratio	Actual Count
Won	5	15
Lost	3	L
Total	8	T

$\frac{5}{8} = \frac{15}{T}$
$5T = 120$
$T = \mathbf{24\ games}$

3. $\overset{5}{\cancel{10}}\ \text{laps} \cdot \frac{\overset{3}{\cancel{6}}\ \text{minutes}}{\underset{\underset{1}{\cancel{2}}}{\cancel{4}}\ \text{laps}} = \mathbf{15\ minutes}$

4. (a) $15 \times 10^5 = 1.5 \times 10^1 \times 10^5$
$= \mathbf{1.5 \times 10^6}$

(b) $0.15 \times 10^5 = 1.5 \times 10^{-1} \times 10^5$
$= \mathbf{1.5 \times 10^4}$

5. (a) $\frac{\text{do not believe in giants}}{\text{total Lilliputians}} = \mathbf{\frac{3}{5}}$

(b) $\frac{2}{5} = \frac{L}{60}$
$5L = 120$
$L = \mathbf{24\ Lilliputians}$

(c) $\mathbf{\frac{2}{5}}$

6. $C = \pi d$
$C \approx 3.14(40\text{ cm})$
$C \approx 125.6\text{ cm} \approx \mathbf{126\ cm}$

7. (a) **Sphere**

(b) **Cylinder**

(c) **Cone**

8. (a) Perimeter $= \frac{10}{16}\text{ in.} + \frac{10}{16}\text{ in.} + \frac{10}{16}\text{ in.}$

$= \frac{30}{16}\text{ in.} = \frac{15}{8}\text{ in.} = \mathbf{1\frac{7}{8}\text{ in.}}$

(b) **60°**

(c)

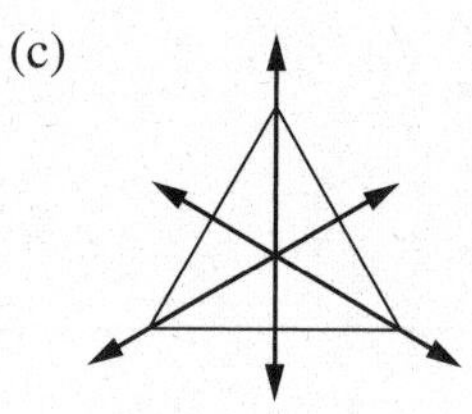

9. (a)
$(-4) + (-5) - (-6)$
$(-4) + (-5) + [-(-6)]$
$(-4) + (-5) + [6]$
$\mathbf{-3}$

(b)
$(-2) + (-3) - (-4) - (+5)$
$(-2) + (-3) + [-(-4)] + [-(+5)]$
$(-2) + (-3) + [4] + [-5]$
$\mathbf{-6}$

10. (a) $C = \pi d$
$C \approx 3.14(7\text{ cm})$
$C \approx \mathbf{21.98\text{ cm}}$

(b) $C = \pi d$
$C \approx \frac{22}{7}(7\text{ cm})$
$C \approx \mathbf{22\text{ cm}}$

11.

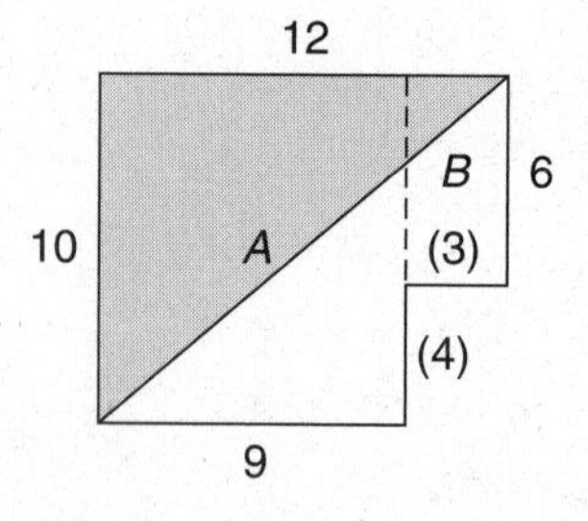

(a)

Area A	= (9 mm)(10 mm)	=	90 mm²
+ Area B	= (3 mm)(6 mm)	=	18 mm²
	Area of figure	=	**108 mm²**

(b) Area $= \frac{(12\text{ mm})(10\text{ mm})}{2} = \mathbf{60\text{ mm}^2}$

(c) $\frac{60}{108} = \mathbf{\frac{5}{9}}$

12. $W_N = \mathbf{\frac{1}{2} \times 200}$
$W_N = \mathbf{100}$

13. $W_N = \mathbf{2.5 \times 4.2}$
$W_N = \mathbf{10.5}$

14. (a) $20\overline{)3.00}$ = **0.15**

(b) $\frac{3}{20} \times 100\% = \frac{300\%}{20} = \mathbf{15\%}$

(c) $150\% = \frac{150}{100} = \mathbf{1\frac{1}{2}}$

(d) $150\% = \frac{150}{100} = \mathbf{1.5}$

15. (a) **∠*TPQ*** or **∠*QPT***

(b) **∠*SPR*** or **∠*RPS***

(c) $m\angle QPT = 360° - (125° + 90°)$
$= 360° - 215° = \mathbf{145°}$

16. $(4)^2 - \sqrt{(4)} + (4)(0.5) - (4)^0$
$= 16 - 2 + 2 - 1 = \mathbf{15}$

17. **Multiply the "in" number by 2, then subtract 1 to find the "out" number.**
$4 \times 2 - 1 = 8 - 1 = \mathbf{7}$

18.
```
      13.0909 . . .
11)144.0000 . . .
   11
    34
    33
     1 0
       0
     1 00
       99
        10
         0
        100
```

(a) $\mathbf{13.\overline{09}}$

(b) **13**

19. $1.8(20°\text{C}) + 32 = \mathbf{68°F}$

20. $(19)(2) = 38$
$7 + 31 = 38$
7 and 31

21.
$\frac{15}{16} = \frac{15}{16}$
$-\ \frac{5}{8} = \frac{10}{16}$
$\mathbf{\frac{5}{16}}$

22. $\frac{a}{8} = \frac{3\frac{1}{2}}{2}$

$2a = 3\frac{1}{2}(8)$

$2a = \frac{7}{2}(8)$

$2a = 28$

$a = \mathbf{14}$

23. $(4 \div 2) \cdot 3 = 2 \cdot 3 = \mathbf{6}$

$3\frac{3}{4} \div 1\frac{2}{3} = \frac{15}{4} \div \frac{5}{3}$

$= \frac{\overset{3}{\cancel{15}}}{4} \times \frac{3}{\underset{1}{\cancel{5}}} = \frac{9}{4}$

$3 \cdot \frac{9}{4} = \frac{27}{4} = \mathbf{6\frac{3}{4}}$

24. $4 + (5 \div 1) = 4 + 5 = \mathbf{9}$

or

$5 + (5 \div 1) = 5 + 5 = \mathbf{10}$

$5\frac{1}{6} \div 1\frac{1}{3} = \frac{31}{6} \div \frac{4}{3}$

$= \frac{31}{\underset{2}{\cancel{6}}} \times \frac{\overset{1}{\cancel{3}}}{4} = \frac{31}{8} = 3\frac{7}{8}$

$$\begin{array}{rcl} 4\frac{1}{2} & = & 4\frac{4}{8} \\ + \ 3\frac{7}{8} & = & 3\frac{7}{8} \\ \hline & & 7\frac{11}{8} = \mathbf{8\frac{3}{8}} \end{array}$$

25.

$$\begin{array}{rr} & 5 \text{ ft} \quad 7 \text{ in.} \\ + & 6 \text{ ft} \quad 8 \text{ in.} \\ \hline & 11 \text{ ft} \quad 15 \text{ in.} \end{array}$$

15 in. = 1 ft 3 in.

$$\begin{array}{rr} & 1 \text{ ft} \quad 3 \text{ in.} \\ + & 11 \text{ ft} \\ \hline & \mathbf{12 \text{ ft} \quad 3 \text{ in.}} \end{array}$$

26. $\frac{350 \cancel{m}}{1 \cancel{s}} \cdot \frac{60 \cancel{s}}{1 \text{ min}} \cdot \frac{1 \text{ km}}{1000 \cancel{m}}$

$= \frac{21{,}000 \text{ km}}{1000 \text{ min}} = \mathbf{21 \frac{km}{min}}$

27. $6 - (0.5 \div 4) = 6 - 0.125$

$= \mathbf{5.875}$

28. $\begin{array}{r} \mathbf{\$100.00} \\ 75\overline{)\$7500} \end{array}$

29. $\frac{432}{675} = \frac{2 \cdot 2 \cdot 2 \cdot 2 \cdot \overset{1}{\cancel{3}} \cdot \overset{1}{\cancel{3}} \cdot \overset{1}{\cancel{3}}}{\underset{1}{\cancel{3}} \cdot \underset{1}{\cancel{3}} \cdot \underset{1}{\cancel{3}} \cdot 5 \cdot 5}$

$= \mathbf{\frac{16}{25}}$

30. (a) $2\frac{1}{4} = 2.25, \quad 2.25 + 0.15 = \mathbf{2.4}$

(b) $6.5 = 6\frac{1}{2}$

$$\begin{array}{rcl} 6\frac{1}{2} & = & 6\frac{3}{6} \\ + \ \frac{5}{6} & = & \frac{5}{6} \\ \hline & & 6\frac{8}{6} = 7\frac{2}{6} = \mathbf{7\frac{1}{3}} \end{array}$$

Lesson 70, Warm-Up

a. **8.1**

b. **625**

c. **6**

d. **$240.00**

e. **2000 mm**

f. **15**

g. **2 m; 0.25 m^2**

Problem Solving

The only 1-digit factor of 679 is 7. To find the top factor, we divide 679 by 7.

$$\begin{array}{r} \mathbf{97} \\ \times \ \mathbf{7} \\ \hline \mathbf{679} \end{array}$$

Lesson 70, Lesson Practice

a. (6 sugar cubes) (3 sugar cubes)
= 18 sugar cubes

$\frac{18 \text{ sugar cubes}}{\cancel{\text{layer}}} \times 4 \cancel{\text{layers}} = \mathbf{72 \text{ sugar cubes}}$

b. (10 1-cm cubes) (10 1-cm cubes)
= 100 1-cm cubes,

$\frac{100 \text{ 1-cm cubes}}{1 \cancel{\text{layer}}} \times 10 \cancel{\text{layers}}$

= **1000 1-cm cubes**

c. (10 cubes)(4 cubes) = 40 cubes

$\frac{40 \text{ cubes}}{1 \text{ layer}} \times 6 \text{ layers} = 240 \text{ cubes}$

240 ft³

d. **See student work.**

Lesson 70, Mixed Practice

1. $\frac{2(38 \text{ kilometers})}{4 \text{ hours}} =$ **19 kilometers per hour**

2.

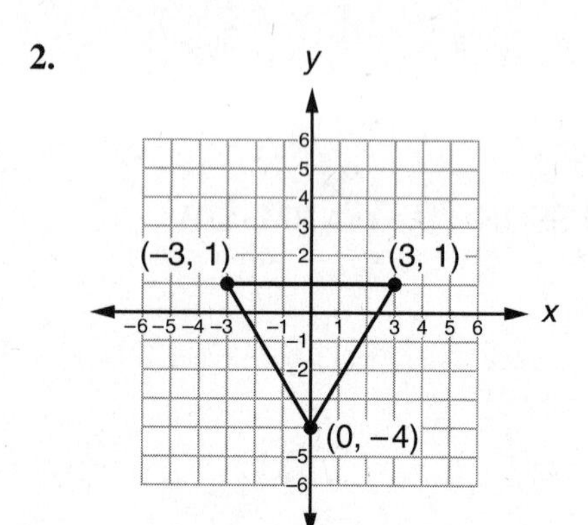

(a) **(−3, 1)**

(b) Area $= \frac{(6 \text{ units})(5 \text{ units})}{2} =$ **15 sq. units**

3. **A little too large. The actual value of π is greater than 3. Dividing the circumference, 600 cm, by π results in a measurement less than 200 cm.**

4. (a) $\frac{\$1.29}{3 \text{ pounds}} = \frac{\$0.43}{1 \text{ pound}}$

\$0.43 per pound

(b) 10(\$0.43) = **\$4.30**

5. (0.7 + 0.6) − (0.9 × 0.8)
= 1.3 − 0.72 = **0.58**

6. (a) $\frac{3}{4}$(188 hits) = **141 hits**

(b) $\frac{1}{4} \times 100\% = \frac{100\%}{4} =$ **25%**

7. **$2\frac{1}{4}$-inch mark**

8. (5 1-cm cubes)(5 1-cm cubes)
= 25 1-cm cubes,

$\frac{25 \text{ 1-cm cubes}}{1 \text{ layer}} \times 3 \text{ layers}$

= **75 1-cm cubes**

9. (a) $C = \pi d$
$C = \pi(2 \text{ in.})$
$C =$ **2π in.**

(b) $C = \pi d$
$C \approx 3.14(1 \text{ in.})$
$C \approx$ **3.14 in.**

10. (a) $12 \times 10^{-6} = 1.2 \times 10^{1} \times 10^{-6}$
$=$ **1.2×10^{-5}**

(b) $0.12 \times 10^{-6} = 1.2 \times 10^{-1} \times 10^{-6}$
$=$ **1.2×10^{-7}**

11.
```
  0.74
  0.83
+ 0.98
  2.55
```
$2.55 \div 3 =$ **0.85**

12. **1.25 kilograms** $\cdot$ **$\frac{1000 \text{ grams}}{1 \text{ kilogram}}$**

= **1250 grams**

13. (a) 2^9

(b) 2^3

(c) 2^{-3}

(d) 2^0

14. $W_N = \frac{1}{6} \times 100$

$W_N = \frac{100}{6}$

$W_N =$ **$16\frac{2}{3}$**

15. (a) $14\% = \frac{14}{100} =$ **$\frac{7}{50}$**

(b) $14\% = \frac{14}{100} =$ **0.14**

(c) $5.00\ldots \div 6 =$ **$0.8\overline{3}$**

(d) $\frac{5}{6} \times 100\% = \frac{500\%}{6} =$ **$83\frac{1}{3}\%$**

16. (a) $(-6) - (-4) + (+2)$
$(-6) + [-(-4)] + (+2)$
$(-6) + [4] + (+2) = \mathbf{0}$

(b) $(-5) + (-2) - (-7) - (+9)$
$(-5) + (-2) + [-(-7)] + [-(+9)]$
$(-5) + (-2) + [7] + [-9]$
$\mathbf{-9}$

17. $(0.4)(0.3) - (0.4 - 0.3)$
$= 0.12 - 0.1 = \mathbf{0.02}$

18. **29,375**

19. $6 \times 8 = \mathbf{48}$

20. **Pyramid;**

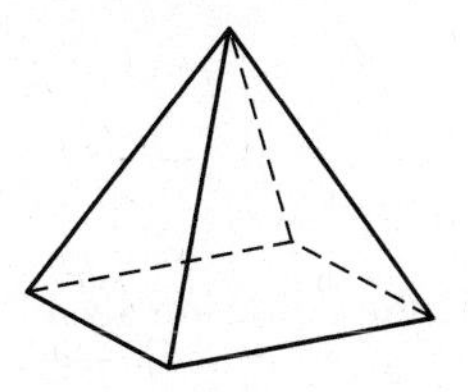

21. $2\text{ ft} \times 2\text{ ft} = 4\text{ ft}^2$,
$6(4\text{ ft}^2) = \mathbf{24\ ft^2}$

22.
$$\begin{array}{r} 4.3 \\ +\ 0.8 \\ \hline \mathbf{5.1} \end{array}$$

23.
$\frac{2}{d} = \frac{1.2}{1.5}$
$1.2d = 2(1.5)$
$d = \frac{3}{1.2}$
$d = \mathbf{2.5}$

24.
$$\begin{array}{rr} \overset{9}{\cancel{10}}\text{ lb} & (16\text{ oz}) \\ -\ 6\text{ lb} & 7\text{ oz} \\ \hline \end{array} \longrightarrow$$
$$\begin{array}{rr} \overset{9}{\cancel{10}}\text{ lb} & 16\text{ oz} \\ -\ 6\text{ lb} & 7\text{ oz} \\ \hline \mathbf{3\ lb} & \mathbf{9\ oz} \end{array}$$

25. $\frac{\$5.25}{1\ \cancel{\text{hr}}} \cdot \frac{8\ \cancel{\text{hr}}}{1\ \cancel{\text{day}}} \cdot \frac{5\ \cancel{\text{days}}}{1\text{ week}}$
$= \mathbf{\frac{\$210.00}{week}}$

26. $1\frac{2}{3} \cdot 3 = \frac{5}{\cancel{3}_1} \cdot \frac{\cancel{3}^1}{1} = 5$

$3\frac{3}{4} \div 5 = \frac{15}{4} \div \frac{5}{1}$
$= \frac{\cancel{15}^3}{4} \times \frac{1}{\cancel{5}_1} = \mathbf{\frac{3}{4}}$

27. $4\frac{1}{2} + 5\frac{1}{6} - 1\frac{1}{3}$
$= 4\frac{3}{6} + 5\frac{1}{6} - 1\frac{2}{6} = 8\frac{2}{6} = \mathbf{8\frac{1}{3}}$

28. $(0.06 \div 5) \div 0.004$
$= (0.012) \div 0.004$
$= \mathbf{3}$

29. $9\frac{1}{2} = 9.5,\ 9.5 \times 9.2 = \mathbf{87.4}$

30. (a)
$$\begin{array}{r} \$15.00 \\ \times\ \ 0.06 \\ \hline \$0.9000 \end{array} = \$0.90 \qquad \begin{array}{r} \$15.00 \\ +\ \ \$0.90 \\ \hline \mathbf{\$15.90} \end{array}$$

(b)
$$\begin{array}{r} \$15.00 \\ \times\ \ 0.15 \\ \hline \mathbf{\$2.25} \end{array}$$

INVESTIGATION 7

1. **Subtract 18.**

2. **Subtract 18 from both sides of the equation.**

3. $45 - 18 = 27$
$x + 18 - 18 = x$
On the left side will be 27, and on the right side will be x.

4. **Yes, the equation would still be balanced.**
$\mathbf{x + 18 = 45}$

5. 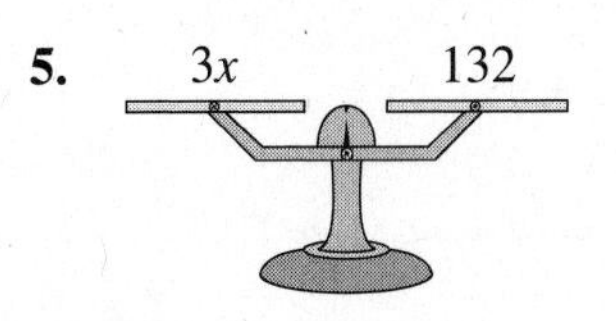

6. **Divide by 3.**

7. **Divide both sides of the equation by 3.**

8. $3x \div 3 = x$
$132 \div 3 = 44$

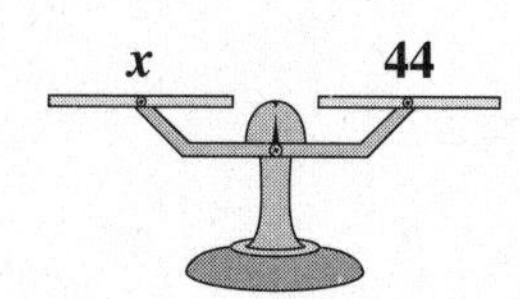

9. $$3x = 132$$
$$\frac{3x}{3} = \frac{132}{3}$$
$$1x = 44$$
$$\mathbf{x = 44}$$

10. $$3x = 132$$
$$3(44) = 132$$
$$132 = 132 \checkmark$$

11. **Multiply by $\frac{4}{3}$.**

12. **Multiply both sides of the equation by $\frac{4}{3}$.**

13. $$\frac{3}{4}x = \frac{9}{10}$$
$$\frac{4}{3} \cdot \frac{3}{4}x = \frac{4}{3} \cdot \frac{9}{10}$$
$$1x = \frac{36}{30}$$
$$\mathbf{x = \frac{6}{5}\left(\text{or } 1\frac{1}{5}\right)}$$

14. $$\frac{3}{4}x = \frac{9}{10}$$
$$\frac{3}{4} \cdot \frac{6}{5} = \frac{9}{10}$$
$$\frac{18}{20} = \frac{9}{10}$$
$$\frac{9}{10} = \frac{9}{10} \checkmark$$

15. (a) **Subtract 2.5.**

(b) **Subtract 2.5 from both sides of the equation.**

(c) $$x + 2.5 = 7$$
$$x + 2.5 - 2.5 = 7 - 2.5$$
$$x + 0 = 4.5$$
$$\mathbf{x = 4.5}$$

(d) $$x + 2.5 = 7$$
$$4.5 + 2.5 = 7$$
$$7 = 7 \checkmark$$

16. (a) **Subtract 2.**

(b) **Subtract 2 from both sides of the equation.**

(c) $$3.6 = y + 2$$
$$3.6 - 2 = y + 2 - 2$$
$$1.6 = y + 0$$
$$\mathbf{1.6 = y}$$

(d) $$3.6 = y + 2$$
$$3.6 = 1.6 + 2$$
$$3.6 = 3.6 \checkmark$$

17. (a) **Divide by 4.**

(b) **Divide both sides of the equation by 4.**

(c) $$4w = 132$$
$$\frac{4w}{4} = \frac{132}{4}$$
$$1w = 33$$
$$\mathbf{w = 33}$$

(d) $$4w = 132$$
$$4(33) = 132$$
$$132 = 132 \checkmark$$

18. (a) **Divide by 1.2.**

(b) **Divide both sides of the equation by 1.2.**

(c) $$1.2m = 1.32$$
$$\frac{1.2m}{1.2} = \frac{1.32}{1.2}$$
$$1m = 1.1$$
$$\mathbf{m = 1.1}$$

(d) $$1.2m = 1.32$$
$$1.2(1.1) = 1.32$$
$$1.32 = 1.32 \checkmark$$

19. (a) **Subtract $\frac{3}{4}$.**

(b) **Subtract $\frac{3}{4}$ from both sides of the equation.**

(c) $$x + \frac{3}{4} = \frac{5}{6}$$
$$x + \frac{3}{4} - \frac{3}{4} = \frac{5}{6} - \frac{3}{4}$$
$$x + 0 = \frac{10}{12} - \frac{9}{12}$$
$$\mathbf{x = \frac{1}{12}}$$

(d) $$x + \frac{3}{4} = \frac{5}{6}$$
$$\frac{1}{12} + \frac{3}{4} = \frac{5}{6}$$
$$\frac{1}{12} + \frac{9}{12} = \frac{5}{6}$$
$$\frac{10}{12} = \frac{5}{6}$$
$$\frac{5}{6} = \frac{5}{6} \checkmark$$

20. (a) **Multiply by $\frac{4}{3}$.**

(b) **Multiply both sides of the equation by $\frac{4}{3}$.**

(c)
$$\frac{3}{4}x = \frac{5}{6}$$
$$\frac{4}{3} \cdot \frac{3}{4}x = \frac{4}{3} \cdot \frac{5}{6}$$
$$1x = \frac{20}{18}$$
$$\mathbf{x = \frac{10}{9}}$$

(d)
$$\frac{3}{4}x = \frac{5}{6}$$
$$\frac{3}{4} \cdot \frac{10}{9} = \frac{5}{6}$$
$$\frac{30}{36} = \frac{5}{6}$$
$$\frac{5}{6} = \frac{5}{6} \checkmark$$

21. **See student work.**

22. **See student work.**

Lesson 71, Warm-Up

a. **−15**

b. **0.0045**

c. **80**

d. **30**

e. **0.5 m**

f. **$27**

g. **$32.40**

Problem Solving
9 combinations:
A1, A2, A3, B1, B2, B3, C1, C2, C3

Lesson 71, Lesson Practice

a.

25 babies

$\frac{3}{5}$ were boys (15).
- 5 babies
- 5 babies
- 5 babies

$\frac{2}{5}$ were girls.
- 5 babies
- 5 babies

$15 \div 3 = 5$

5×5 babies = **25 babies**

b.

40 clowns

$\frac{5}{8}$ had happy faces.
- 5 clowns
- 5 clowns
- 5 clowns
- 5 clowns
- 5 clowns

$\frac{3}{8}$ did not have happy faces (15).
- 5 clowns
- 5 clowns
- 5 clowns

$15 \div 3 = 5$

5×8 clowns = **40 clowns**

c.

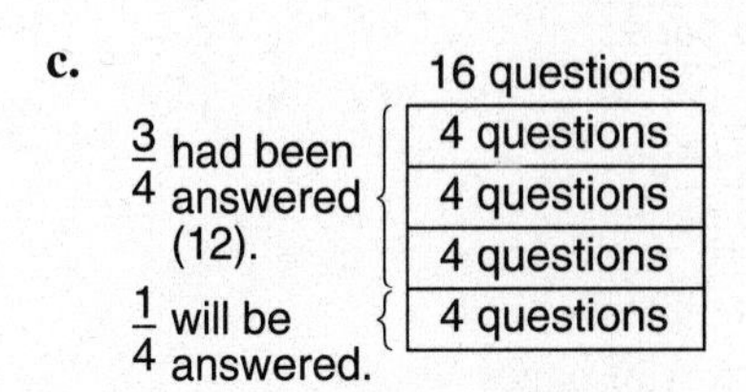

$12 \div 3 = 4$

4×4 questions = **16 questions**

Lesson 71, Mixed Practice

1. $9 \text{ seconds} \cdot \frac{331 \text{ meters}}{1 \text{ second}}$ (seconds cancelled)

= 2979 meters; **about 3 kilometers**

2.
$$\begin{array}{r} 3.33 \\ 3.45 \\ +\ 3.51 \\ \hline 10.29 \end{array} \qquad 3\overline{)10.29} = \mathbf{3.43}$$

3. 2(80 percent) + 3(90 percent)

= 430 percent

$\frac{430 \text{ percent}}{5}$ = **86 percent**

4.
$$\begin{array}{r} 20{,}000{,}000{,}000 \\ -\ 9{,}000{,}000{,}000 \\ \hline 11{,}000{,}000{,}000 \end{array}$$

$\mathbf{1.1 \times 10^{10}}$

5.
$$\begin{array}{r} 2 \\ 3 \\ 5 \\ 7 \\ +\ 11 \\ \hline \mathbf{28} \end{array}$$

6.

	Ratio	Actual Count
New ones	4	N
Used ones	7	U
Total	11	242

$$\frac{4}{11} = \frac{N}{242}$$
$11N = 968$
$N =$ **88 new ones**

7.

130 pages

$\frac{3}{5}$ read (78)	26 pages
	26 pages
	26 pages
$\frac{2}{5}$ not read	26 pages
	26 pages

(a) $78 \div 3 = 26$
$5 \times 26 =$ **130 pages**

(b) $2 \times 26 =$ **52 pages**

8. (4 1-inch cubes) (4 1-inch cubes)
= 16 1-inch cubes
$\frac{\text{16 1-inch cubes}}{\text{1 layer}} \cdot \text{4 layers}$
= **64 1-inch cubes**

9. $(4 \text{ in.})(4 \text{ in.}) = 16 \text{ in.}^2$
$6(16 \text{ in.}^2) =$ $\mathbf{96 \textbf{ in.}^2}$

10. (a) $C \approx 3.14(28 \text{ cm})$
$C \approx$ **87.92 cm**

(b) $C \approx \frac{22}{7}(28 \text{ cm})$
$C \approx$ **88 cm**

11. (a) $\mathbf{2.5 \times 10^7}$

(b) $\mathbf{2.5 \times 10^{-5}}$

12. (a) $0.1 = \mathbf{\frac{1}{10}}$

(b) $0.1 = \frac{1}{10} = \frac{10}{100} =$ **10%**

(c) $0.5\% = \frac{0.5}{100} = \mathbf{\frac{1}{200}}$

(d) $0.5\% = \frac{0.5}{100} =$ **0.005**

13. (a) $\mathbf{W_N = 0.35 \times 80}$
$W_N =$ **28**

(b) $\mathbf{\frac{3}{\cancel{4}_1} \times \overset{6}{\cancel{24}} = W_N}$

$3(6) = W_N$
$\mathbf{18} = W_N$

14. **Add 7 to the "in" number to find the "out" number.** $0 + 7 =$ **7**

15.

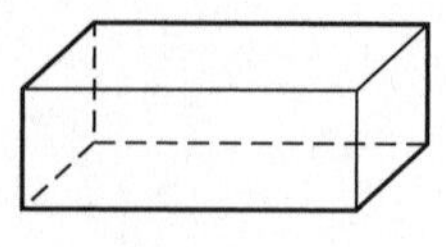

8 vertices

16. (a) Perimeter = 10 cm + 12 cm + 8 cm + 6 cm = **36 cm**

(b) $\text{Area} = \frac{(6 \text{ cm})(8 \text{ cm})}{2} =$ $\mathbf{24 \textbf{ cm}^2}$

(c) $\text{Area} = \frac{(12 \text{ cm})(8 \text{ cm})}{2} =$ $\mathbf{48 \textbf{ cm}^2}$

(d) $\text{Area} = 24 \text{ cm}^2 + 48 \text{ cm}^2 =$ $\mathbf{72 \textbf{ cm}^2}$

17.
$16.50
× 0.15
$2.475

About $2.50

18.

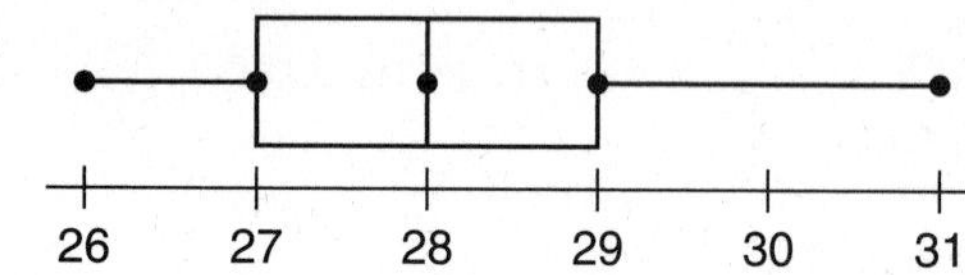

19.

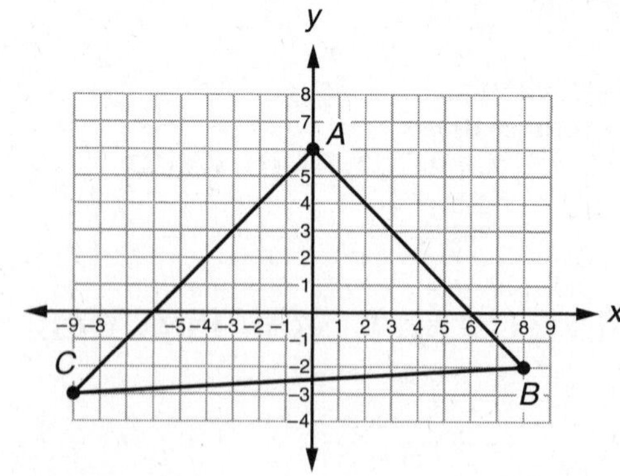

$\mathbf{m\angle A = 90°}$
$\mathbf{m\angle B = 48°}$
$\mathbf{m\angle C = 42°}$

20. $(0.01) - (0.1)(0.01)$
$= 0.01 - 0.001 =$ **0.009**

21.
$$m + 5.75 = 26.4$$
$$m + 5.75 - 5.75 = 26.4 - 5.75$$
$$m = \mathbf{20.65}$$
check: $20.65 + 5.75 = 26.4$
$$26.4 = 26.4$$

22.
$$\frac{3}{4}x = 48$$
$$\left(\frac{\cancel{4}^{1}}{\cancel{3}_{1}}\right)\frac{\cancel{3}^{1}}{\cancel{4}_{1}}x = \left(\frac{4}{\cancel{3}_{1}}\right)\cancel{48}^{16}$$
$$x = \mathbf{64}$$
check: $\frac{3}{\cancel{4}_{1}}(\cancel{64}^{16}) = 48$
$$3(16) = 48$$
$$48 = 48$$

23. **Rhombus**

24.
$$\frac{4^2 + \{20 - 2[6 - (5 - 2)]\}}{\sqrt{36}}$$
$$\frac{16 + \{20 - 2[6 - (3)]\}}{6}$$
$$\frac{16 + \{20 - 2[3]\}}{6}$$
$$\frac{16 + \{20 - 6\}}{6}$$
$$\frac{16 + \{14\}}{6}$$
$$\frac{30}{6}$$
$$\mathbf{5}$$

25.
$$\begin{array}{rrr} \cancel{1}^{0}\text{ yd} & (\cancel{3}^{2}\text{ ft}) & (12\text{ in.}) \\ - & 1\text{ ft} & 1\text{ in.} \\ \hline & \mathbf{1\text{ ft}} & \mathbf{11\text{ in.}} \end{array}$$

26. $3.5\ \cancel{\text{hr}} \cdot \frac{60\ \cancel{\text{min}}}{1\ \cancel{\text{hr}}} \cdot \frac{60\text{ s}}{1\ \cancel{\text{min}}} = \mathbf{12{,}600\text{ s}}$

27.
$$4\frac{1}{2} \cdot 2\frac{2}{3} = \frac{\cancel{9}^{3}}{\cancel{2}_{1}} \cdot \frac{\cancel{8}^{4}}{\cancel{3}_{1}} = 12$$
$$6\frac{2}{3} \div 12 = \frac{20}{3} \div \frac{12}{1}$$
$$= \frac{\cancel{20}^{5}}{3} \times \frac{1}{\cancel{12}_{3}} = \mathbf{\frac{5}{9}}$$

28.
$$\begin{array}{rcl} 7\frac{1}{2} & = & 7\frac{3}{6} \\ -\ 5\frac{1}{6} & = & 5\frac{1}{6} \\ \hline & & 2\frac{2}{6} = 2\frac{1}{3} \end{array}$$
$$\begin{array}{r} 2\frac{1}{3} \\ +\ 1\frac{1}{3} \\ \hline \mathbf{3\frac{2}{3}} \end{array}$$

29. (a)
$$(-5) + (-6) - |-7|$$
$$(-5) + (-6) + [-|-7|]$$
$$(-5) + (-6) + [-7]$$
$$\mathbf{-18}$$

(b)
$$(-15) - (-24) - (+8)$$
$$(-15) + [-(-24)] + [-(+8)]$$
$$(-15) + [24] + [-8]$$
$$\mathbf{1}$$

30.
$$1.5 = 1\frac{1}{2}$$
$$\begin{array}{rcl} 2\frac{2}{3} & = & 2\frac{4}{6} \\ -\ 1\frac{1}{2} & = & 1\frac{3}{6} \\ \hline & & \mathbf{1\frac{1}{6}} \end{array}$$

Lesson 72, Warm-Up

a. **7**

b. $\mathbf{\frac{4}{9}}$

c. **5**

d. **3.6**

e. **0.5 kg**

f. **$12**

g. **$4.50**

Problem Solving

$$\mathbf{3X + 6 = 18 + X}$$
$$3X + 6 - 6 = 18 - 6 + X$$
$$3X - X = 12 + X - X$$
$$2X \div 2 = 12 \div 2$$
$$X = \mathbf{6\text{ oz}}$$

Lesson 72, Lesson Practice

a. **Between 4 and 6 hours**

	Case 1	Case 2
km	30	75
Hours	2	h

$$\frac{30}{2} = \frac{75}{h}$$
$$30h = 2(75)$$
$$30h = 150$$
$$h = \frac{\overset{5}{\cancel{150}}}{\underset{1}{\cancel{30}}}$$
$$h = \textbf{5 hours}$$

b. $\overset{5}{\cancel{50}}\ \cancel{\text{head}} \cdot \dfrac{\overset{3}{\cancel{6}}\text{ bales}}{\underset{\underset{2}{\cancel{4}}}{\cancel{40}}\ \cancel{\text{head}}}$

$$= \frac{15\text{ bales}}{2} = \mathbf{7\frac{1}{2}}\textbf{ bales}$$

	Case 1	Case 2
Head of cattle	40	50
Bales	6	b

$$\frac{40}{6} = \frac{50}{b}$$
$$40b = 6(50)$$
$$40b = 300$$
$$b = \frac{\overset{\overset{15}{\cancel{30}}}{\cancel{300}}}{\underset{\underset{2}{\cancel{4}}}{\cancel{40}}}$$
$$b = \frac{15}{2}\text{ bales} = \mathbf{7\frac{1}{2}}\textbf{ bales}$$

c.

	Case 1	Case 2
First number	5	9
Second number	15	n

$$\frac{5}{15} = \frac{9}{n}$$
$$5n = 15(9)$$
$$5n = 135$$
$$n = \frac{\overset{27}{\cancel{135}}}{\underset{1}{\cancel{5}}}$$
$$n = \mathbf{27}$$

Lesson 72, Mixed Practice

1.
$$\begin{array}{r} 1821 \\ -\ 1769 \\ \hline \textbf{52 years} \end{array}$$

2.
$$\begin{aligned} 4(4\text{ points}) &+ 6(9\text{ points}) \\ &= 16\text{ points} + 54\text{ points} \\ &= 70\text{ points} \end{aligned}$$
$$\frac{70\text{ points}}{10} = \textbf{7 points}$$

3. $\mathbf{2.5}\ \cancel{\textbf{liters}} \cdot \dfrac{\textbf{1000 milliliters}}{\textbf{1}\ \cancel{\textbf{liter}}}$

$$= \textbf{2500 milliliters}$$

4.
$$\left(\frac{1}{2} + \frac{2}{5}\right) - \left(\frac{1}{2} \cdot \frac{2}{5}\right)$$
$$= \left(\frac{5}{10} + \frac{4}{10}\right) - \left(\frac{1}{5}\right)$$
$$= \frac{9}{10} - \frac{1}{5} = \frac{9}{10} - \frac{2}{10} = \mathbf{\frac{7}{10}}$$

5.

	Ratio	Actual Count
Carnivores	2	126
Herbivores	7	H

$$\frac{2}{7} = \frac{126}{H}$$
$$2H = 882$$
$$H = \textbf{441 herbivores}$$

6.

	Case 1	Case 2
Books	4	14
Pounds	9	P

$$\frac{4}{9} = \frac{14}{p}$$
$$4p = 9(14)$$
$$4p = 126$$
$$p = \frac{\overset{63}{\cancel{126}}}{\underset{2}{\cancel{4}}}$$
$$p = \mathbf{31\frac{1}{2}}\textbf{ pounds}$$

7. (a) $\dfrac{\mathbf{2}}{\underset{1}{\cancel{\mathbf{5}}}} \times \overset{12}{\cancel{\mathbf{60}}} = W_N$

$$\mathbf{24} = W_N$$

(b) $M = \mathbf{0.75} \times \mathbf{\$24}$
$M = \mathbf{\$18}$

8. $C \approx 3.14(20\text{ in.})$
$C \approx 62.8\text{ in.} \approx \textbf{63 in.}$

9.

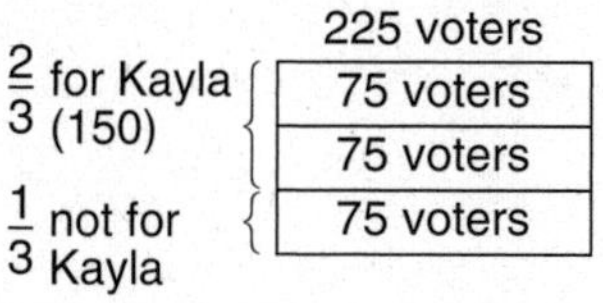

(a) $\dfrac{150 \text{ votes}}{2} = 75 \text{ votes}$

$3(75 \text{ votes}) = \mathbf{225 \text{ votes}}$

(b) $1(75 \text{ votes}) = \mathbf{75 \text{ votes}}$

10. $(10 \text{ ice cubes})(8 \text{ ice cubes})$

$= 80 \text{ ice cubes}$

$\dfrac{80 \text{ ice cubes}}{1 \cancel{\text{layer}}} \cdot 6 \cancel{\text{layers}}$

$= \mathbf{480 \text{ ice cubes}}$

11. $(10 \text{ in.})(8 \text{ in.}) = 80 \text{ in.}^2$

$(6 \text{ in.})(10 \text{ in.}) = 60 \text{ in.}^2$

$(6 \text{ in.})(8 \text{ in.}) = 48 \text{ in.}^2$

$2(80 \text{ in.}^2) + 2(60 \text{ in.}^2) + 2(48 \text{ in.}^2)$

$= 160 \text{ in.}^2 + 120 \text{ in.}^2 + 96 \text{ in.}^2$

$= \mathbf{376 \text{ in.}^2}$

12. (a) $\mathbf{6 \times 10^5}$

(b) $\mathbf{6 \times 10^{-7}}$

13.

$$\begin{array}{r} 1.35 \\ 1.44 \\ +\ 1.59 \\ \hline 4.38 \end{array} \qquad \begin{array}{r} \mathbf{1.46} \\ 3\overline{)4.38} \end{array}$$

14. (a) $\begin{array}{r} \mathbf{0.6} \\ 3\overline{)5.0} \end{array}$

(b) $\dfrac{3}{5} \times 100\% = \dfrac{300\%}{5} = \mathbf{60\%}$

(c) $2.5\% = \dfrac{2.5}{100} = \mathbf{\dfrac{1}{40}}$

(d) $2.5\% = \dfrac{2.5}{100} = \mathbf{0.025}$

15. (a) $\mathbf{2^2 \cdot 3^4 \cdot 5^2}$

(b) **90**

16. (a) $\text{Area} = (8 \text{ in.})(6 \text{ in.}) = \mathbf{48 \text{ in.}^2}$

(b) $\text{Area} = \dfrac{(8 \text{ in.})(6 \text{ in.})}{2} = \mathbf{24 \text{ in.}^2}$

(c) $180° - 72° = \mathbf{108°}$

17. (a) **Sphere**

(b) **Triangular prism**

(c) **Cylinder**

Only the triangular prism is a polyhedron, because it is the only figure whose faces are polygons.

18. (a) $C = \pi(60 \text{ mm})$

$C = \mathbf{60\pi \text{ mm}}$

(b) $C \approx 3.14(60 \text{ mm})$

$C \approx \mathbf{188.4 \text{ mm}}$

19. $\dfrac{2}{3}$ ⓧ 0.667 — $\dfrac{2}{3} \; \textcircled{<} \; 0.667$

20.

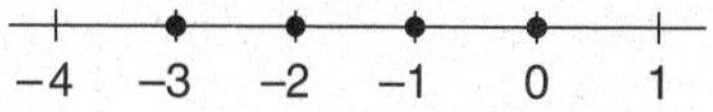

21. (a) $(5)^2 - (4)^2 = 25 - 16 = \mathbf{9}$

(b) $(5)^0 - (4)^{-1} = 1 - \dfrac{1}{4}$

$= \dfrac{4}{4} - \dfrac{1}{4} = \mathbf{\dfrac{3}{4}}$

22.

$$m - \frac{2}{3} = 1\frac{3}{4}$$

$$m - \frac{2}{3} + \frac{2}{3} = 1\frac{3}{4} + \frac{2}{3}$$

$$m = 1\frac{9}{12} + \frac{8}{12}$$

$$m = 1\frac{17}{12} = \mathbf{2\frac{5}{12}}$$

check:

$$2\frac{5}{12} - \frac{2}{3} = 1\frac{3}{4}$$

$$2\frac{5}{12} - \frac{8}{12} = 1\frac{3}{4}$$

$$1\frac{17}{12} - \frac{8}{12} = 1\frac{3}{4}$$

$$1\frac{9}{12} = 1\frac{3}{4}$$

$$1\frac{3}{4} = 1\frac{3}{4}$$

23.

$$\frac{2}{3}w = 24$$

$$\left(\frac{\cancel{3}^{1}}{\cancel{2}_{1}}\right)\frac{\cancel{2}^{1}}{\cancel{3}_{1}}w = \left(\frac{3}{\cancel{2}_{1}}\right)\cancel{24}^{12}$$

$$w = 3(12)$$

$$w = \mathbf{36}$$

check:

$$\left(\frac{2}{\cancel{3}_{1}}\right)\cancel{36}^{12} = 24$$

$$2(12) = 24$$

$$24 = 24$$

24. $\dfrac{[30 - 4(5 - 2)] + 5(3^3 - 5^2)}{\sqrt{9} + \sqrt{16}}$

$\dfrac{[30 - 12] + 5(27 - 25)}{3 + 4}$

$\dfrac{18 + 5(2)}{7}$

$\dfrac{18 + 10}{7}$

$\dfrac{28}{7}$

4

25.

$$\begin{array}{r} \overset{1}{\cancel{2}}\text{ gal} \quad 1\text{ qt} \;(4\text{ qt}) \\ -\ 1\text{ gal} \quad 1\text{ qt} \quad 1\text{ pt} \\ \hline \end{array} \longrightarrow$$

$$\begin{array}{r} \overset{1}{\cancel{2}}\text{ gal} \quad \overset{4}{\cancel{5}}\;\cancel{1}\text{ qt} \;(2\text{ pt}) \\ -\ 1\text{ gal} \quad 1\text{ qt} \quad 1\text{ pt} \\ \hline \end{array} \longrightarrow$$

$$\begin{array}{r} \overset{1}{\cancel{2}}\text{ gal} \quad \overset{4}{\cancel{5}}\;\cancel{1}\text{ qt} \quad 2\text{ pt} \\ -\ 1\text{ gal} \quad 1\text{ qt} \quad 1\text{ pt} \\ \hline \textbf{3 qt} \quad \textbf{1 pt} \end{array}$$

26. $\dfrac{1}{\underset{1}{\cancel{2}}}\cancel{\text{mi}} \cdot \dfrac{\overset{880}{\cancel{\overset{1760}{\cancel{5280}}}}\ \cancel{\text{ft}}}{1\ \cancel{\text{mi}}} \cdot \dfrac{1\text{ yd}}{\underset{1}{\cancel{3}}\ \cancel{\text{ft}}}$

$= \textbf{880 yd}$

27. $4\dfrac{1}{2} \cdot 6\dfrac{2}{3} = \dfrac{\overset{3}{\cancel{9}}}{\underset{1}{\cancel{2}}} \cdot \dfrac{\overset{10}{\cancel{20}}}{\underset{1}{\cancel{3}}} = 30$

$\left(2\dfrac{1}{2}\right)^2 \div 30 = \left(\dfrac{5}{2}\right)^2 \div 30$

$= \dfrac{25}{4} \div \dfrac{30}{1} = \dfrac{\overset{5}{\cancel{25}}}{4} \times \dfrac{1}{\underset{6}{\cancel{30}}}$

$= \mathbf{\dfrac{5}{24}}$

28.

$$\begin{array}{rcl} 5\frac{1}{6} & = & 5\frac{1}{6} \\ +\ 1\frac{1}{3} & = & 1\frac{2}{6} \\ \hline & & 6\frac{3}{6} = 6\frac{1}{2} \end{array}$$

$$\begin{array}{r} 7\frac{1}{2} \\ -\ 6\frac{1}{2} \\ \hline \mathbf{1} \end{array}$$

29. (a) $(-7) + |+5| + (-9)$
$= (-7) + 5 + (-9) = \mathbf{-11}$

(b) $(16) + (-24) - (-18)$
$(16) + (-24) + [-(-18)]$
$(16) + (-24) + [18]$
10

30. $5\dfrac{1}{4} = 5.25$

$$\begin{array}{r} 5.25 \\ +\ 1.9 \\ \hline \mathbf{7.15} \end{array}$$

Lesson 73, Warm-Up

a. **−10**

b. **8750**

c. **36**

d. **63**

e. **0.5 L**

f. **$24**

g. **6**

Problem Solving

We can use 20×14, or 280 tiles, to cover most of the floor, leaving a 6-in. strip along two walls. To cover the untiled strip, we can use a 6-in.-square piece in the corner and 6-by-12-in. pieces along the rest. Each 6-by-12-in. piece is half of a whole tile. We need $20 + 14$, or 34 such half pieces, in addition to the 6-in.-square piece. So if all cut-off portions of tiles may be used, we need $280 + (34 \div 2) + 1$, or 298 tiles. **Fifteen boxes** (with a total of 300 tiles) would be needed.

However, if we may use only one portion from each tile that is cut, then we need $280 + 34 + 1$, or 315 tiles. **Sixteen boxes** (with a total of 320 tiles) would be needed.

Lesson 73, Lesson Practice

a. $(-7)(3) = \mathbf{-21}$

b. $(+4)(-8) = \mathbf{-32}$

c. $(8)(+5) = \mathbf{40}$

d. $(-8)(-3) = \mathbf{24}$

e. $\dfrac{25}{-5} = \mathbf{-5}$

f. $\dfrac{-27}{-3} = \mathbf{9}$

g. $\dfrac{-28}{4} = \mathbf{-7}$

h. $\dfrac{+30}{6} = \mathbf{5}$

i. $\dfrac{+45}{-3} = \mathbf{-15}$

Lesson 73, Mixed Practice

1.

	Case 1	Case 2
Packages	12	p
Minutes	5	60

$$\frac{12}{5} = \frac{p}{60}$$
$$5p = 12(60)$$
$$p = \frac{720}{5}$$
$$p = \mathbf{144\ packages}$$

2. 5(30 minutes) + 3(46 minutes)
= 150 minutes + 138 minutes
= 288 minutes

$\dfrac{288 \text{ minutes}}{8} = \mathbf{36\ minutes}$

3. $\dfrac{(0.2 + 0.5)}{(0.2)(0.5)} = \dfrac{0.7}{0.1} = \mathbf{7}$

4. $23\ \cancel{\text{cm}} \cdot \dfrac{10 \text{ mm}}{1\ \cancel{\text{cm}}} = \mathbf{230\ mm}$

5.

	Ratio	Actual Count
Paperback books	3	P
Hardback books	11	9240
Total	14	T

$$\frac{11}{14} = \frac{9240}{T}$$
$$11T = 14(9240)$$
$$T = \frac{129{,}360}{11}$$
$$T = \mathbf{11{,}760\ books}$$

6. (a) $\mathbf{2.4 \times 10^{-4}}$

(b) $\mathbf{2.4 \times 10^{8}}$

7.

120 questions

$\frac{1}{4}$ were true-false. { 30 questions

$\frac{3}{4}$ were not true-false. { 30 questions / 30 questions / 30 questions

(a) 4(30 questions) = **120 questions**

(b) 3(30 questions) = **90 questions**

8. (a) $\dfrac{5}{\cancel{9}_{1}} \times \cancel{45}^{5} = W_N$

$25 = W_N$

(b) $W_N = 0.8 \times 760$

$W_N = \mathbf{608}$

9. (a) $\dfrac{-36}{9} = \mathbf{-4}$

(b) $\dfrac{-36}{-6} = \mathbf{6}$

(c) $9(-3) = \mathbf{-27}$

(d) $(+8)(+7) = \mathbf{56}$

10. $\dfrac{\text{composite numbers}}{\text{total numbers}} = \mathbf{\dfrac{3}{8}}$

11.

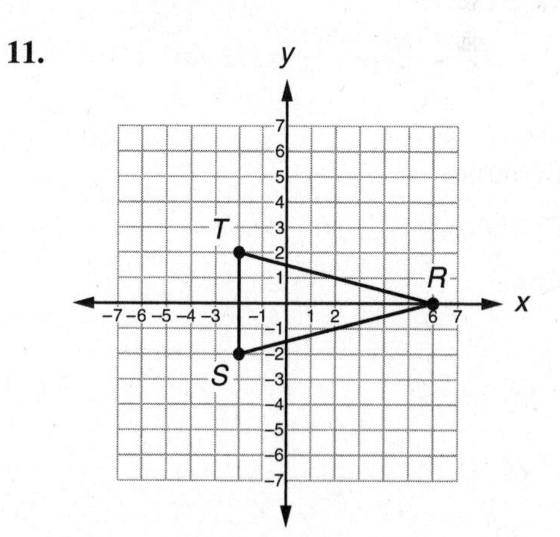

(a) **(−2, 2)**

(b) **Isosceles triangle**

(c) $m\angle S = \dfrac{180° - 28°}{2} = \mathbf{76°}$

12. If the signs of the two factors are the same—both positive or both negative—then the product is positive. If the signs of the two factors are different, the product is negative.

13. (4 1-ft cubes)(3 1-ft cubes)
$= 12$ 1-ft cubes
$\frac{12 \text{ 1-ft cubes}}{1 \text{ layer}} \cdot 8 \text{ layers}$
$=$ **96 1-ft cubes**

14. (a) $C \approx 3.14(42 \text{ m})$
$C \approx$ **131.88 m**

(b) $C \approx \frac{22}{7}(42 \text{ m})$
$C \approx$ **132 m**

15. (a) $2.5 = 2\frac{5}{10} = \mathbf{2\frac{1}{2}}$

(b) $2.5 \times 100\% =$ **250%**

(c) $0.2\% = \frac{0.2}{100} = \mathbf{\frac{1}{500}}$

(d) $0.2\% = \frac{0.2}{100} =$ **0.002**

16. **Right; scalene**
$\text{Area} = \frac{(6 \text{ cm})(8 \text{ cm})}{2} = \mathbf{24 \text{ cm}^2}$

17. **Obtuse; scalene**
$\text{Area} = \frac{(4 \text{ cm})(5 \text{ cm})}{2} = \mathbf{10 \text{ cm}^2}$

18. **Acute; isosceles**
$\text{Area} = \frac{(6 \text{ cm})(4 \text{ cm})}{2} = \mathbf{12 \text{ cm}^2}$

19. (a) **Pyramid**
(b) **Cylinder**
(c) **Cone**

20. $\frac{2}{3} \times 96 = 64$
$\frac{5}{6} \times 84 = 70$
$64 < 70$
$\frac{2}{3}$ of 96 $\textcircled{<}$ $\frac{5}{6}$ of 84

21. $\left(\frac{5}{6}\right)\left(\frac{3}{4}\right) - \left(\frac{5}{6} - \frac{3}{4}\right)$
$= \frac{5}{8} - \left(\frac{10}{12} - \frac{9}{12}\right) = \frac{5}{8} - \frac{1}{12}$
$= \frac{15}{24} - \frac{2}{24} = \mathbf{\frac{13}{24}}$

22. $\frac{3}{5}w = 15$
$\left(\frac{\cancel{5}^1}{\cancel{3}_1}\right)\left(\frac{\cancel{3}^1}{\cancel{5}_1}\right)w = \left(\frac{5}{\cancel{3}_1}\right)\cancel{15}^5$
$w = 5(5)$
$w = \mathbf{25}$
check: $\frac{3}{\cancel{5}_1}(\cancel{25}^5) = 15$
$3(5) = 15$
$15 = 15$

23. $b - 1.6 = (0.4)^2$
$b - 1.6 + 1.6 = (0.4)^2 + 1.6$
$b = 0.16 + 1.6$
$b = \mathbf{1.76}$
check: $1.76 - 1.6 = (0.4)^2$
$0.16 = 0.16$

24. $20w = 5.6$
$\frac{\cancel{20}^1 w}{\cancel{20}_1} = \frac{5.6}{20}$
$w =$ **0.28**
check: $20(0.28) = 5.6$
$5.6 = 5.6$

25.

	2 yd	1 ft	7 in.
+	1 yd	2 ft	8 in.
	3 yd	3 ft	15 in.

15 in. = 1 ft 3 in.

	1 ft	3 in.
+	3 ft	
	4 ft	3 in.

4 ft = 1 yd 1 ft

	1 yd	1 ft	3 in.
+	3 yd		
	4 yd	**1 ft**	**3 in.**

26. $0.5 \cancel{\text{m}} \cdot \frac{100 \cancel{\text{cm}}}{1 \cancel{\text{m}}} \cdot \frac{10 \text{ mm}}{1 \cancel{\text{cm}}} =$ **500 mm**

27. $12\frac{1}{2} \cdot 4\frac{1}{5} \cdot 2\frac{2}{3} = \frac{\cancel{25}^5}{\cancel{2}_1} \cdot \frac{\cancel{21}^7}{\cancel{5}_1} \cdot \frac{\cancel{8}^4}{\cancel{3}_1} =$ **140**

28. $6\frac{2}{3} \cdot 1\frac{1}{5} = \frac{\cancel{20}^4}{\cancel{3}_1} \cdot \frac{\cancel{6}^2}{\cancel{5}_1} = 8$
$7\frac{1}{2} \div 8 = \frac{15}{2} \div \frac{8}{1} = \frac{15}{2} \times \frac{1}{8}$
$= \mathbf{\frac{15}{16}}$

29. (a) $(-8) + (-7) - (-15)$
$(-8) + (-7) + [-(-15)]$
$(-8) + (-7) + [15]$
$\mathbf{0}$

(b) $(-15) + (+11) - |+24|$
$(-15) + (+11) + [-|+24|]$
$(-15) + (+11) + [-24]$
$\mathbf{-28}$

30. $2.25 = 2\frac{1}{4}$

$2\frac{1}{4} \times 1\frac{1}{3} = \frac{\overset{3}{\cancel{9}}}{\underset{1}{\cancel{4}}} \times \frac{\overset{1}{\cancel{4}}}{\underset{1}{\cancel{3}}} = \mathbf{3}$

Lesson 74, Warm-Up

a. **4**

b. **0.045**

c. **7**

d. **$3.00**

e. **0.4 km**

f. **$20**

g. **1**

Problem Solving

The riders travel about 30×3.14, or 94.2 feet, in one turn of the Gravitron. In 30 turns, the riders travel 94.2×30, or **2826 feet, which is more than $\frac{1}{2}$ mile.** (One mile is 5280 feet, so $\frac{1}{2}$ mile is 2640 feet.)

Lesson 74, Lesson Practice

a. $\mathbf{W_F \times 130 = 80}$

$\frac{W_F \times 130}{130} = \frac{80}{130}$

$W_F = \mathbf{\frac{8}{13}}$

b. $\mathbf{75 = W_D \times 300}$

$\frac{75}{300} = \frac{W_D \times 300}{300}$

$\mathbf{0.25} = W_D$

c. $\mathbf{80 = 0.4 \times W_N}$

$\frac{80}{0.4} = \frac{0.4 \times W_N}{0.4}$

$\mathbf{200} = W_N$

d. $\mathbf{60 = \frac{5}{6} \times W_N}$

$\frac{6}{5} \times 60 = \frac{6}{5} \times \frac{5}{6} \times W_N$

$6(12) = W_N$

$\mathbf{72} = W_N$

e. $\mathbf{60 = W_F \times 90}$

$\frac{60}{90} = \frac{W_F \times 90}{90}$

$\mathbf{\frac{2}{3}} = W_F$

f. $\mathbf{W_D \times 80 = 60}$

$\frac{W_D \times 80}{80} = \frac{60}{80}$

$W_D = \mathbf{0.75}$

g. $\mathbf{40 = 0.08 \times W_N}$

$\frac{40}{0.08} = \frac{0.08 \times W_N}{0.08}$

$\mathbf{500} = W_N$

h. $\mathbf{\frac{6}{5} \times W_N = 60}$

$\frac{5}{6} \times \frac{6}{5} \times W_N = \frac{5}{6} \times 60$

$W_N = 5(10)$

$W_N = \mathbf{50}$

Lesson 74, Mixed Practice

1. $3(28 \text{ pages}) + 4(42 \text{ pages})$
$= 84 \text{ pages} + 168 \text{ pages}$
$= 252 \text{ pages}$

$\frac{252 \text{ pages}}{7} = \mathbf{36 \text{ pages}}$

2. $\frac{\$1.14}{12 \text{ ounces}} = \frac{\$0.095}{1 \text{ ounce}}$

$\frac{\$1.28}{16 \text{ ounces}} = \frac{\$0.08}{1 \text{ ounce}}$

$\$0.095 - \$0.08 = \$0.015$

1.5¢ per ounce

3. $4\frac{1}{2}\ \cancel{\text{feet}} \cdot \frac{12\text{ inches}}{1\ \cancel{\text{foot}}} = \textbf{54 inches}$

4.

	Ratio	A. C.
Black-and-white photographs	2	B
Color photographs	3	18
Total	5	T

$\frac{3}{5} = \frac{18}{T}$

$3T = 90$

$T = \textbf{30 photographs}$

5.

	Case 1	Case 2
Pounds	5	8
Cost	$1.40	C

$\frac{5}{\$1.40} = \frac{8}{C}$

$5C = (\$1.40)(8)$

$C = \frac{\$11.20}{5}$

$C = \textbf{\$2.24}$

6.

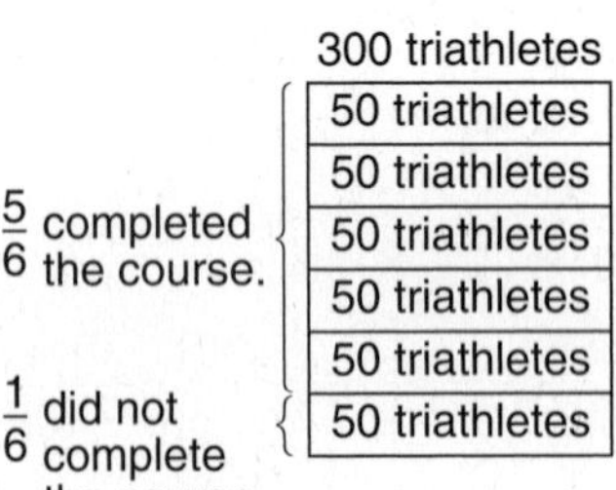

(a) **250 triathletes**

(b) $\frac{\text{completed the course}}{\text{did not complete the course}} = \mathbf{\frac{5}{1}}$

7. $\mathbf{15 = \frac{3}{8} \times W_N}$

$\frac{8}{3} \times 15 = \frac{8}{3} \times \frac{3}{8} \times W_N$

$8(5) = W_N$

$\mathbf{40} = W_N$

8. $\mathbf{70 = W_D \times 200}$

$\frac{70}{200} = \frac{W_D \times 200}{200}$

$\mathbf{0.35} = W_D$

9. $\mathbf{\frac{2}{5} \times W_N = 120}$

$\frac{5}{2} \times \frac{2}{5} \times W_N = \frac{5}{2} \times 120$

$W_N = 5(60)$

$W_N = \mathbf{300}$

10. $\mathbf{P = 0.6 \times \$180}$

$P = \mathbf{\$108}$

11. $\mathbf{W_N = 0.2 \times \$35}$

$W_N = \mathbf{\$7}$

12. (a) $(3\text{ in.})(3\text{ in.}) = 9\text{ in.}^2$

$\frac{9\text{ in.}^2}{1\ \cancel{\text{layer}}} \cdot (1\text{ in.})(3\ \cancel{\text{layers}})$

$= \mathbf{27\ in.^3}$

(b) $6(9\text{ in.}^2) = \mathbf{54\ in.^2}$

13. (a) $C \approx \frac{22}{7}(14\text{ m})$

$C \approx \mathbf{44\ m}$

(b) $C = \pi(14\text{ m})$

$C = \mathbf{14\pi\ m}$

14. (a) $3\frac{1}{2} = \mathbf{3.5}$

(b) $3\frac{1}{2} = \frac{7}{2} \times 100\% = \frac{700\%}{2} = \mathbf{350\%}$

(c) $35\% = \frac{35}{100} = \mathbf{\frac{7}{20}}$

(d) $35\% = \frac{35}{100} = \mathbf{0.35}$

15. **Multiply the "in" number by 3; then add 1 to find the "out" number.**

$5(3) + 1 = \mathbf{16}$

$25 - 1 = 24$

$24 \div 3 = \mathbf{8}$

16. $\mathbf{4.25 \times 10^8}$

17. (a) **Parallelogram**

(b) **Trapezoid**

(c) **Isosceles triangle**

18. (a) $m\angle ABC = 180° - 100° = \mathbf{80°}$

(b) $m\angle BCE = m\angle A = \mathbf{100°}$

(c) $m\angle ECD = 180° - 100° = \mathbf{80°}$

(d) $m\angle EDC = m\angle ECD = \mathbf{80°}$

(e) $m\angle DEC = 180° - (80° + 80°)$

$= 180° - 160° = \mathbf{20°}$

(f) $m\angle DEA = 20° + 80° = \mathbf{100°}$

19. **0.0103, 0.013, 0.021, 0.1023**

20. $\left(1\frac{1}{2} + 2\frac{2}{3}\right) - \left(1\frac{1}{2}\right)\left(2\frac{2}{3}\right)$

$= \left(\frac{3}{2} + \frac{8}{3}\right) - \left(\frac{3}{2}\right)\left(\frac{8}{3}\right)$

$= \left(\frac{3}{2} + \frac{8}{3}\right) - (4)$

$= \frac{9}{6} + \frac{16}{6} - \frac{24}{6}$

$= \frac{25}{6} - \frac{24}{6} = \mathbf{\frac{1}{6}}$

21. $p + 3\frac{1}{5} = 7\frac{1}{2}$

$p + 3\frac{1}{5} - 3\frac{1}{5} = 7\frac{1}{2} - 3\frac{1}{5}$

$p = \frac{15}{2} - \frac{16}{5}$

$p = \frac{75}{10} - \frac{32}{10}$

$p = \frac{43}{10} = \mathbf{4\frac{3}{10}}$

check: $4\frac{3}{10} + 3\frac{1}{5} = 7\frac{1}{2}$

$4\frac{3}{10} + 3\frac{2}{10} = 7\frac{1}{2}$

$7\frac{5}{10} = 7\frac{1}{2}$

$7\frac{1}{2} = 7\frac{1}{2}$

22. $3n = 0.138$

$\frac{3n}{3} = \frac{0.138}{3}$

$n = \mathbf{0.046}$

check: $3(0.046) = 0.138$

$0.138 = 0.138$

23. $n - 0.36 = 4.8$

$n - 0.36 + 0.36 = 4.8 + 0.36$

$n = \mathbf{5.16}$

check: $5.16 - 0.36 = 4.8$

$4.8 = 4.8$

24. $\frac{2}{3}x = \frac{8}{9}$

$\left(\frac{\cancel{3}^{1}}{\cancel{2}_{1}}\right)\left(\frac{\cancel{2}^{1}}{\cancel{3}_{1}}x\right) = \left(\frac{\cancel{3}^{1}}{\cancel{2}_{1}}\right)\left(\frac{\cancel{8}^{4}}{\cancel{9}_{3}}\right)$

$x = \mathbf{\frac{4}{3}}$

check: $\left(\frac{2}{3}\right)\left(\frac{4}{3}\right) = \frac{8}{9}$

$\frac{8}{9} = \frac{8}{9}$

25. $\sqrt{49} + \{5[3^2 - (2^3 - \sqrt{25})] - 5^2\}$

$7 + \{5[9 - (8 - 5)] - 25\}$

$7 + \{5[9 - (3)] - 25\}$

$7 + \{5[6] - 25\}$

$7 + \{30 - 25\}$

$7 + \{5\}$

12

26.

$\begin{array}{r} \overset{3}{\cancel{4}}\text{ hr } \overset{4}{\cancel{5}}\text{ min } 15\text{ s} \\ -\ 1\text{ hr } 15\text{ min } 30\text{ s} \\ \hline \end{array}$ (60 min) (60 s) $\longrightarrow$

$\begin{array}{r} \overset{3}{\cancel{4}}\text{ hr } \overset{64}{\cancel{5}}\text{ min } \overset{75}{\cancel{15}}\text{ s} \\ -\ 1\text{ hr } 15\text{ min } 30\text{ s} \\ \hline \mathbf{2\text{ hr } 49\text{ min } 45\text{ s}} \end{array}$

27. (a) $(-9) + (-11) - (+14)$

$(-9) + (-11) + [-(+14)]$

$(-9) + (-11) + [-14]$

−34

(b) $(26) + (-43) - |-36|$

$(26) + (-43) + [-|-36|]$

$(26) + (-43) + [-36]$

−53

28. (a) $(-3)(12) = \mathbf{-36}$

(b) $(-3)(-12) = \mathbf{36}$

(c) $\frac{-12}{3} = \mathbf{-4}$

(d) $\frac{-12}{-3} = \mathbf{4}$

29. $7.5 = 7\frac{1}{2}$

$8\frac{1}{3} + 7\frac{1}{2} = 8\frac{2}{6} + 7\frac{3}{6} = \mathbf{15\frac{5}{6}}$

30. **South**

LESSON 75, WARM-UP

a. **−35**

b. **225**

c. **9**

d. **$60.00**

e. **0.25 g**

f. **$3.50**

g. **1.8**

Problem Solving

$$\frac{\overset{1}{\cancel{5}}}{\underset{\underset{1}{\cancel{2}}}{\cancel{6}}} \times \frac{\overset{1}{\cancel{3}}}{\underset{\underset{1}{\cancel{2}}}{\cancel{10}}} \times \frac{\overset{\overset{4}{\cancel{8}}}{\cancel{16}}}{3} \times \frac{?}{?} = 1$$

After canceling, we have

$\frac{1}{1} \times \frac{1}{1} \times \frac{4}{3} \times \frac{?}{?} = 1.$

The unknown fraction must be the reciprocal of $\frac{4}{3}$, which is $\mathbf{\frac{3}{4}}$.

Lesson 75, Lesson Practice

a.

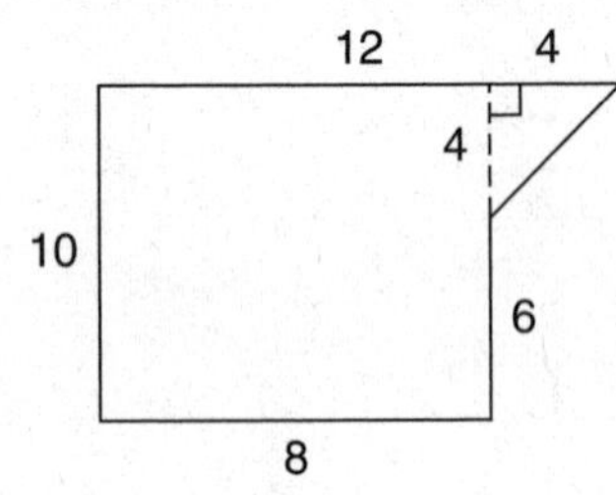

$$\text{Area of rectangle} = (8 \text{ cm})(10 \text{ cm})$$
$$= 80 \text{ cm}^2$$
$$\text{Area of triangle} = \frac{(4 \text{ cm})(4 \text{ cm})}{2}$$
$$= 8 \text{ cm}^2$$
$$\text{Area of figure} = 80 \text{ cm}^2 + 8 \text{ cm}^2$$
$$= \mathbf{88 \text{ cm}^2}$$

b.

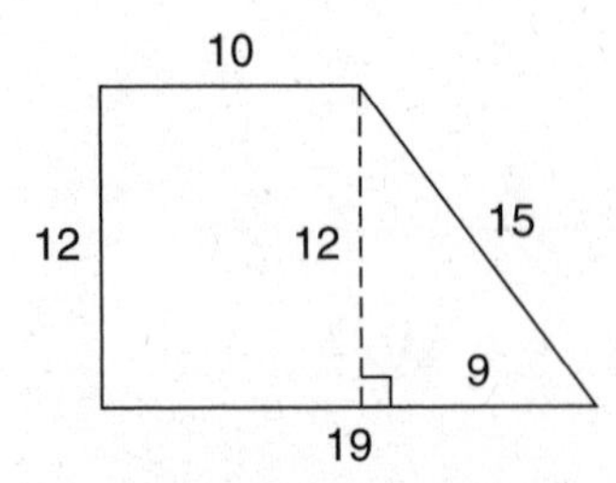

$$\text{Area of rectangle} = (10 \text{ cm})(12 \text{ cm})$$
$$= 120 \text{ cm}^2$$
$$\text{Area of triangle} = \frac{(9 \text{ cm})(12 \text{ cm})}{2}$$
$$= 54 \text{ cm}^2$$
$$\text{Area of figure} = 120 \text{ cm}^2 + 54 \text{ cm}^2$$
$$= \mathbf{174 \text{ cm}^2}$$

c.

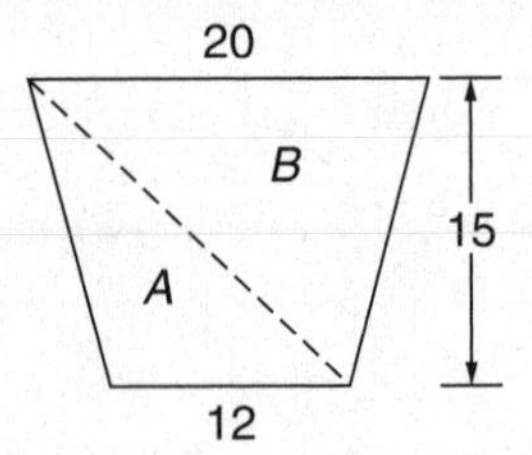

$$\text{Area of triangle } A = \frac{(12 \text{ cm})(15 \text{ cm})}{2}$$
$$= 90 \text{ cm}^2$$
$$\text{Area of triangle } B = \frac{(20 \text{ cm})(15 \text{ cm})}{2}$$
$$= 150 \text{ cm}^2$$
$$\text{Area of figure} = 90 \text{ cm}^2 + 150 \text{ cm}^2$$
$$= \mathbf{240 \text{ cm}^2}$$

Lesson 75, Mixed Practice

1. 5(72 seconds) + 3(80 seconds)
= 360 seconds + 240 seconds = 600 seconds

$$\frac{600 \text{ seconds}}{8} = \mathbf{75 \text{ seconds}}$$

2. $\frac{\$2.49}{30 \text{ ounces}} = \frac{\$0.083}{1 \text{ ounce}}$

8.3¢ per ounce

3. **1.5 kilometers**

4.
$$\left(\frac{1}{2} + \frac{3}{5}\right) - \left(\frac{1}{2}\right)\left(\frac{3}{5}\right)$$
$$= \left(\frac{5}{10} + \frac{6}{10}\right) - \left(\frac{3}{10}\right)$$
$$= \frac{11}{10} - \frac{3}{10} = \frac{8}{10}$$
$$= \mathbf{\frac{4}{5}}$$

5. $\frac{3}{2} = \frac{60 \text{ years}}{C}$

$3C = 2(60 \text{ years})$

$C = 40 \text{ years}$

60 years
− 40 years
20 years

6. 12.5×10^{-4} ⊜ 1.25×10^{-3}

7.

	Case 1	Case 2
Miles	40	100
Hours	3	h

$$\frac{40}{3} = \frac{100}{h}$$
$$40h = 3(100)$$
$$h = \frac{300}{40}$$
$$h = \frac{15}{2} = \mathbf{7\frac{1}{2}\ hours}$$

8.

	21,000 books
$\frac{2}{5}$ were checked out.	4200 books
	4200 books
$\frac{3}{5}$ were not checked out.	4200 books
	4200 books
	4200 books

(a) **8400 books**

(b) **12,600 books**

9.
$$\mathbf{60 = \frac{5}{12} \times W_N}$$
$$\frac{12}{\cancel{5}_1}(\cancel{60}^{12}) = \frac{\cancel{12}^1}{\cancel{5}_1} \times \frac{\cancel{5}^1}{\cancel{12}_1} \times W_N$$
$$\mathbf{144} = W_N$$

10. $\mathbf{0.7 \times \$35.00 = M}$
$$\mathbf{\$24.50} = M$$

11. $\mathbf{35 = W_F \times 80}$
$$\frac{35}{80} = \frac{W_F \times 80}{80}$$
$$\mathbf{\frac{7}{16}} = W_F$$

12. $\mathbf{56 = W_D \times 70}$
$$\frac{56}{70} = \frac{W_D \times 70}{70}$$
$$\mathbf{0.8} = W_D$$

13. (a) $\frac{-120}{4} = \mathbf{-30}$

(b) $(-12)(11) = \mathbf{-132}$

(c) $\frac{-120}{-5} = \mathbf{24}$

(d) $12(+20) = \mathbf{240}$

14. $(20\text{ cm})(15\text{ cm}) = 300\text{ cm}^2$
$$\frac{300\text{ cm}^2}{1\ \cancel{\text{layer}}} \cdot (1\text{ cm})(10\ \cancel{\text{layers}})$$
$$\mathbf{3000\ cm^3}$$

15.
$$C \approx \frac{22}{7}(11\text{ inches})$$
$$C \approx \frac{242}{7}\text{ inches}$$
$$C \approx 34\frac{4}{7}\text{ inches} \approx \mathbf{34\frac{1}{2}\ inches}$$

16.

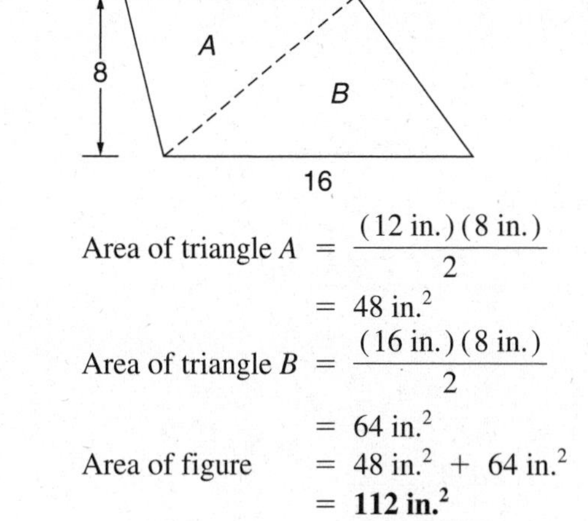

$$\text{Area of triangle } A = \frac{(12\text{ in.})(8\text{ in.})}{2}$$
$$= 48\text{ in.}^2$$
$$\text{Area of triangle } B = \frac{(16\text{ in.})(8\text{ in.})}{2}$$
$$= 64\text{ in.}^2$$
$$\text{Area of figure} = 48\text{ in.}^2 + 64\text{ in.}^2$$
$$= \mathbf{112\ in.^2}$$

17.

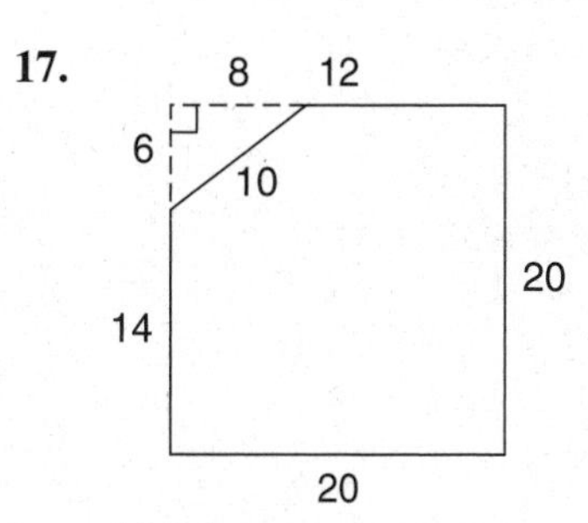

(a) **20 cm**

(b) Perimeter $= 12\text{ cm} + 20\text{ cm} + 20\text{ cm} + 14\text{ cm} + 10\text{ cm} = \mathbf{76\ cm}$

(c) Area of rectangle $= (20\text{ cm})(20\text{ cm})$
$$= 400\text{ cm}^2$$
Area of triangle $= \frac{(8\text{ cm})(6\text{ cm})}{2}$
$$= 24\text{ cm}^2$$
Area of figure $= 400\text{ cm}^2 - 24\text{ cm}^2$
$$= \mathbf{376\ cm^2}$$

18. (a) $125\% = \frac{125}{100} = \mathbf{1\frac{1}{4}}$

(b) $125\% = \frac{125}{100} = \mathbf{1.25}$

(c) $8\overline{)1.000}$ with quotient **0.125**

(d) $\frac{1}{8} \times 100\% = \frac{100\%}{8} = \mathbf{12\frac{1}{2}\%}$

or **12.5%**

19.

$12.50	$12.50
× 0.20	+ $ 2.50
$ 2.50	**$15.00**

20. $(2)^3 - (2)(0.5) - \frac{2}{0.5}$
$= 8 - 1 - 4 = \mathbf{3}$

21. $\frac{5}{8}x = 40$

$\left(\frac{8}{5}\right)\frac{5}{8}x = \left(\frac{8}{5}\right)40$

$x = \mathbf{64}$

check: $\frac{5}{8}(64) = 40$

$5(8) = 40$

$40 = 40$

22. $1.2w = 26.4$

$\frac{1.2w}{1.2} = \frac{26.4}{1.2}$

$w = \mathbf{22}$

check: $1.2(22) = 26.4$

$26.4 = 26.4$

23. $y + 3.6 = 8.47$

$y + 3.6 - 3.6 = 8.47 - 3.6$

$y = \mathbf{4.87}$

check: $4.87 + 3.6 = 8.47$

$8.47 = 8.47$

24. $9^2 - [3^3 - (9 \cdot 3 - \sqrt{9})]$

$81 - [27 - (27 - 3)]$

$81 - [27 - (24)]$

$81 - [3]$

78

25.

```
   47 ──→(60 s)
 2 hr 48 min 20 s
− 1 hr 23 min 48 s   ──→
────────────────────
      47     80
 2 hr 48 min 20 s
− 1 hr 23 min 48 s
────────────────────
 1 hr 24 min 32 s
```

26. $100 \text{ yd} \cdot \frac{3 \text{ ft}}{1 \text{ yd}} \cdot \frac{12 \text{ in.}}{1 \text{ ft}}$
$= \mathbf{3600 \text{ in.}}$

27. $3 \div 1\frac{1}{3} = \frac{3}{1} \div \frac{4}{3}$

$= \frac{3}{1} \times \frac{3}{4} = \frac{9}{4}$

$5\frac{1}{3} \cdot \frac{9}{4} = \frac{16}{3} \cdot \frac{9}{4} = \mathbf{12}$

28. $3\frac{1}{5} + 2\frac{1}{2} - 1\frac{1}{4}$

$= 3\frac{4}{20} + 2\frac{10}{20} - 1\frac{5}{20}$

$= 5\frac{14}{20} - 1\frac{5}{20} = \mathbf{4\frac{9}{20}}$

29. (a) $(-26) + (-15) - (-40)$

$(-26) + (-15) + [-(-40)]$

$(-26) + (-15) + [40]$

$\mathbf{-1}$

(b) $(-5) + (-4) - (-3) - (+2)$

$(-5) + (-4) + [-(-3)] + [-(+2)]$

$(-5) + (-4) + [3] + [-2]$

$\mathbf{-8}$

30. (a) 5^7

(b) 5^3

(c) 5^0

(d) 5^{-3}

Lesson 76, Warm-Up

a. **−12**

b. **0.0625**

c. **4**

d. **18**

e. **0.5 cm**

f. **8**

g. **$26.50**

Problem Solving

Kim: **7, 7, 3, 1, 1, 1**
7, 3, 3, 3, 3, 1
Shell: **7, 7, 3, 3; 4 attempts**

Lesson 76, Lesson Practice

a. $\frac{37\frac{1}{2}}{100} = \frac{\frac{75}{2}}{\frac{100}{1}}$

$\frac{\frac{75}{2}}{\frac{100}{1}} \cdot \frac{\frac{1}{100}}{\frac{1}{100}} = \frac{\frac{75}{200}}{1} = \frac{75}{200}$

$= \mathbf{\frac{3}{8}}$

b. $\dfrac{12}{\frac{5}{6}} = \dfrac{\frac{12}{1}}{\frac{5}{6}}$

$\dfrac{\frac{12}{1}}{\frac{5}{6}} \cdot \dfrac{\frac{6}{5}}{\frac{6}{5}} = \dfrac{\frac{72}{5}}{1} = \dfrac{72}{5} = \mathbf{14\frac{2}{5}}$

c. $\dfrac{\frac{2}{5}}{\frac{2}{3}} \cdot \dfrac{\frac{3}{2}}{\frac{3}{2}} = \dfrac{\frac{6}{10}}{1} = \dfrac{6}{10} = \mathbf{\frac{3}{5}}$

d. $66\frac{2}{3}\% = \dfrac{66\frac{2}{3}}{100} = \dfrac{\frac{200}{3}}{\frac{100}{1}}$

$\dfrac{\frac{200}{3}}{\frac{100}{1}} \cdot \dfrac{\frac{1}{100}}{\frac{1}{100}} = \dfrac{\frac{200}{300}}{1} = \dfrac{200}{300} = \mathbf{\frac{2}{3}}$

e. $8\frac{1}{3}\% = \dfrac{8\frac{1}{3}}{100} = \dfrac{\frac{25}{3}}{\frac{100}{1}}$

$\dfrac{\frac{25}{3}}{\frac{100}{1}} \cdot \dfrac{\frac{1}{100}}{\frac{1}{100}} = \dfrac{\frac{25}{300}}{1} = \dfrac{25}{300} = \mathbf{\frac{1}{12}}$

f. $4\frac{1}{6}\% = \dfrac{4\frac{1}{6}}{100} = \dfrac{\frac{25}{6}}{\frac{100}{1}}$

$\dfrac{\frac{25}{6}}{\frac{100}{1}} \cdot \dfrac{\frac{1}{100}}{\frac{1}{100}} = \dfrac{\frac{25}{600}}{1} = \dfrac{25}{600} = \mathbf{\frac{1}{24}}$

Lesson 76, Mixed Practice

1. $\dfrac{42 \text{ kilometers}}{1.75 \text{ hours}} = \mathbf{24 \text{ kilometers per hour}}$

2.
$$\begin{array}{r} 7.9 \\ 8.3 \\ 8.1 \\ 8.1 \\ +\ 8.2 \\ \hline 40.6 \end{array} \qquad 5\overline{)40.60} = \mathbf{8.12}$$

3.

	Ratio	Actual Count
Good guys	2	G
Bad guys	5	B
Total	7	35

$\dfrac{2}{7} = \dfrac{G}{35}$

$7G = 35(2)$

$G = \dfrac{70}{7}$

$G = \mathbf{10\ guys}$

4. $\mathbf{3.5\ \cancel{grams} \cdot \dfrac{1000\ milligrams}{1\ \cancel{gram}}}$

$= \mathbf{3500\ milligrams}$

5. $16\frac{2}{3}\% = \dfrac{16\frac{2}{3}}{100} = \dfrac{\frac{50}{3}}{\frac{100}{1}}$

$\dfrac{\frac{50}{3}}{\frac{100}{1}} \cdot \dfrac{\frac{1}{100}}{\frac{1}{100}} = \dfrac{\frac{50}{300}}{1} = \dfrac{50}{300} = \mathbf{\frac{1}{6}}$

6. **East**

7. $\dfrac{1}{\cancel{6}_1} \times \overset{24}{\cancel{144}}$ grams $=$ **24 grams**

8. (a) $\sqrt{2(2)^3} = \sqrt{2^4} = 2^2 = \mathbf{4}$

(b) $(2)^{-1} \cdot (2)^{-2} = \dfrac{1}{2} \cdot \dfrac{1}{4} = \mathbf{\frac{1}{8}}$

9. (a) $\dfrac{-60}{-12} = \mathbf{5}$

(b) $(-8)(6) = \mathbf{-48}$

(c) $\dfrac{40}{-8} = \mathbf{-5}$

(d) $(-5)(-15) = \mathbf{75}$

10. $C = \pi(30 \text{ cm})$
$C = \mathbf{30\pi\ cm}$

11.

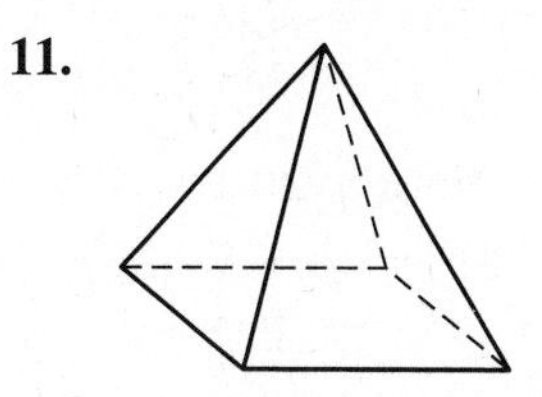

(a) **5 faces**

(b) **8 edges**

(c) **5 vertices**

12. $\mathbf{W_N = 0.1 \times \$37.50}$
$W_N = \mathbf{\$3.75}$

13. $\mathbf{W_N = \dfrac{5}{\cancel{8}_1} \times \overset{9}{\cancel{72}}}$

$W_N = \mathbf{45}$

14. $\mathbf{25 = W_F \times 60}$

$\dfrac{25}{60} = \dfrac{W_F \times 60}{60}$

$\mathbf{\dfrac{5}{12}} = W_F$

15.
$$60 = W_D \times 80$$
$$\frac{60}{80} = \frac{W_D \times 80}{80}$$
$$\mathbf{0.75} = W_D$$

16. (a) $m\angle ACB = 180° - 115° = \mathbf{65°}$

(b) $m\angle ABC = m\angle ACB = \mathbf{65°}$

(c) $m\angle CAB = 180° - (65° + 65°)$
$= 180° - 130° = \mathbf{50°}$

17. (a) $6\overline{)5.00\ldots}$ quotient $0.83\ldots$ $\quad \mathbf{0.8\overline{3}}$

(b) $\frac{5}{6} \times 100\% = \frac{500\%}{6} = \mathbf{83\frac{1}{3}\%}$

(c) $0.1\% = \frac{0.1}{100} = \mathbf{\frac{1}{1000}}$

(d) $0.1\% = \frac{0.1}{100} = \mathbf{0.001}$

18.

4 5
3 5
9
6
9

(a) Perimeter = 5 in. + 9 in. + 9 in. + 6 in. + 5 in. = **34 in.**

(b) Area of rectangle = (9 in.)(9 in.)
= 81 in.^{2}

Area of triangle = $\frac{(4 \text{ in.})(3 \text{ in.})}{2}$
= 6 in.^{2}

Area of figure = 81 in.^{2} − 6 in.^{2}
= **75 in.^{2}**

19. **Rectangle**

20.
$$\begin{array}{r} 212°\text{F} \\ +\ 32\ \text{F} \\ \hline 244°\text{F} \end{array} \qquad \frac{244°\text{F}}{2} = \mathbf{122°F}$$

21.
$$x - 25 = 96$$
$$x - 25 + 25 = 96 + 25$$
$$x = \mathbf{121}$$

check: $121 - 25 = 96$
$96 = 96$

22.
$$\frac{2}{3}m = 12$$
$$\left(\frac{\cancel{3}^1}{\cancel{2}_1}\right)\frac{\cancel{2}^1}{\cancel{3}_1}m = \left(\frac{3}{\cancel{2}_1}\right)\cancel{12}^6$$
$$m = \mathbf{18}$$

check: $\frac{2}{\cancel{3}_1}(\cancel{18}^6) = 12$
$2(6) = 12$
$12 = 12$

23.
$$2.5p = 6.25$$
$$\frac{2.5p}{2.5} = \frac{6.25}{2.5}$$
$$p = \mathbf{2.5}$$

check: $(2.5)(2.5) = 6.25$
$6.25 = 6.25$

24.
$$10 = f + 3\frac{1}{3}$$
$$10 - 3\frac{1}{3} = f + 3\frac{1}{3} - 3\frac{1}{3}$$
$$9\frac{3}{3} - 3\frac{1}{3} = f$$
$$\mathbf{6\frac{2}{3}} = f$$

check: $10 = 6\frac{2}{3} + 3\frac{1}{3}$
$10 = 9\frac{3}{3}$
$10 = 10$

25. $\sqrt{13^2 - 5^2} = \sqrt{169 - 25}$
$= \sqrt{144} = \mathbf{12}$

26. 1 ton = 2000 lb
2000 lb − 400 lb = **1600 lb**

27. $3\frac{3}{4} \times 4\frac{1}{6} \times (0.4)^2$
$$= \frac{15}{4} \times \frac{25}{6} \times \left(\frac{4}{10}\right)^2$$
$$= \frac{\cancel{15}^5}{\cancel{4}_1} \times \frac{\cancel{25}^1}{\cancel{6}_2} \times \frac{\cancel{16}^{\cancel{4}\,1}}{\cancel{100}_{\cancel{4}\,1}}$$
$$= \frac{5}{2} = \mathbf{2\frac{1}{2}}$$

28. $3\frac{1}{8} + 6.7 + 8\frac{1}{4}$
$= 3.125 + 6.7 + 8.25 = \mathbf{18.075}$

29. (a)
$$(-3) + (-5) - (-3) - |+5|$$
$$(-3) + (-5) + [-(-3)] + [-|+5|]$$
$$(-3) + (-5) + [3] + [-5]$$
$$\mathbf{-10}$$

(b)
$$(-73) + (-24) - (-50)$$
$$(-73) + (-24) + [-(-50)]$$
$$(-73) + (-24) + [50]$$
$$\mathbf{-47}$$

30. The quotient is a little more than 1 because the dividend is slightly greater than the divisor.

$$\frac{\frac{5}{6}}{\frac{2}{3}} \cdot \frac{\frac{3}{2}}{\frac{3}{2}} = \frac{\frac{15}{12}}{1} = \frac{15}{12} = \frac{5}{4} = \mathbf{1\frac{1}{4}}$$

Lesson 77, Warm-Up

a. **30**

b. **4,000,000**

c. **45**

d. **$9**

e. **0.25 m**

f. **16**

g. **21**

Problem Solving

$$\mathbf{4X + 25 = X + 100}$$
$$4X - X + 25 = X - X + 100$$
$$3X + 25 - 25 = 100 - 25$$
$$3X \div 3 = 75 \div 3$$
$$X = \mathbf{25\ g}$$

Lesson 77, Lesson Practice

a.
$$24 = W_P \times 40$$
$$\frac{\overset{3}{\cancel{24}}}{\underset{5}{\cancel{40}}} = \frac{W_P \times \overset{1}{\cancel{40}}}{\underset{1}{\cancel{40}}}$$
$$\frac{3}{5} = W_P$$
$$W_P = \frac{3}{5} \times 100\% = \mathbf{60\%}$$

b.
$$W_P \times 6 = 2$$
$$\frac{W_P \times \overset{1}{\cancel{6}}}{\underset{1}{\cancel{6}}} = \frac{\overset{1}{\cancel{2}}}{\underset{3}{\cancel{6}}}$$
$$W_P = \frac{1}{3}$$
$$W_P = \frac{1}{3} \times 100\% = \mathbf{33\frac{1}{3}\%}$$

c.
$$\frac{15}{100} \times W_N = 45$$
$$\frac{\overset{1}{\cancel{15}}}{\underset{1}{\cancel{100}}} \cdot \frac{\overset{1}{\cancel{100}}}{\underset{1}{\cancel{15}}} \times W_N = \overset{3}{\cancel{45}} \cdot \frac{100}{\underset{1}{\cancel{15}}}$$
$$W_N = \mathbf{300}$$

d.
$$W_P \times 4 = 6$$
$$\frac{W_P \times 4}{4} = \frac{\overset{3}{\cancel{6}}}{\underset{2}{\cancel{4}}}$$
$$W_P = \frac{3}{2} \times 100\% = \mathbf{150\%}$$

e.
$$24 = 120\% \times W_N$$
$$24 = \frac{120}{100} \times W_N$$
$$\frac{\overset{20}{\cancel{100}}}{\underset{1}{\cancel{120}}} \times \overset{1}{\cancel{24}} = \frac{\overset{1}{\cancel{100}}}{\underset{1}{\cancel{120}}} \times \frac{\overset{1}{\cancel{120}}}{\underset{1}{\cancel{100}}} \times W_N$$
$$\mathbf{20} = W_N$$

f.
$$60 = \frac{\overset{3}{\cancel{150}}}{\underset{2}{\cancel{100}}} \times W_N$$
$$60 = \mathbf{\frac{3}{2}} \times W_N$$
$$\frac{2}{\underset{1}{\cancel{3}}} \times \overset{20}{\cancel{60}} = \left(\frac{\overset{1}{\cancel{2}}}{\underset{1}{\cancel{3}}}\right)\left(\frac{\overset{1}{\cancel{3}}}{\underset{1}{\cancel{2}}}\right) \times W_N$$
$$\mathbf{40} = W_N$$

g.
$$W_P \times \$5.00 = \$0.35$$
$$\frac{W_P \times \overset{1}{\cancel{\$5.00}}}{\underset{1}{\cancel{\$5.00}}} = \frac{\$0.35}{\$5.00}$$
$$W_P = 0.07$$
$$W_P = 0.07 \times 100\% = \mathbf{7\%}$$

LESSON 77, MIXED PRACTICE

1.

	Ratio	Actual Count
Nickels	2	70
Pennies	5	P
Total	7	T

$$\frac{2}{7} = \frac{70}{T}$$
$$2T = 7(70)$$
$$T = \frac{490}{2}$$
$$T = \textbf{245 coins}$$

2. 0.8(50 questions) = **40 questions**

3.

$$\begin{array}{r} 80\% \\ 75\% \\ 80\% \\ 95\% \\ 80\% \\ +\ 100\% \\ \hline 510\% \end{array} \qquad 6\overline{)510\%} = \textbf{85\%}$$

4. (a) **80%**

(b) 100% − 75% = **25%**

5. (a) **Sphere**

(b) **Cylinder**

(c) **Rectangular prism**

6.

	Case 1	Case 2
Inches	100	250
Centimeters	254	C

$$\frac{100}{254} = \frac{250}{C}$$
$$100C = (254)(250)$$
$$C = \frac{63500}{100}$$
$$C = \textbf{635 centimeters}$$

7.

30 people

$\frac{3}{5}$ agreed.	6 people
	6 people
	6 people
$\frac{2}{5}$ disagreed.	6 people
	6 people

(a) $\mathbf{\frac{2}{5}}$

(b) 12 ÷ 2 = 6, 6(5 people) = **30 people**

(c) 3(6 people) = **18 people**

(d) $\frac{\text{agreed}}{\text{disagreed}} = \frac{18}{12} = \mathbf{\frac{3}{2}}$

8.

$$\mathbf{40 = \frac{4}{25} \times W_N}$$
$$\frac{25}{\cancel{4}_1} \times \cancel{40}^{10} = \frac{\cancel{25}^1}{\cancel{4}_1} \times \frac{\cancel{4}^1}{\cancel{25}_1} \times W_N$$
$$\mathbf{250} = W_N$$

9. $\mathbf{0.24 \times 10{,}000 = W_N}$

$$\mathbf{2400} = W_N$$

10. $\mathbf{0.12 \times W_N = 240}$

$$\frac{\cancel{0.12}^1 \times W_N}{\cancel{0.12}_1} = \frac{\cancel{240}^{2000}}{\cancel{0.12}_1}$$
$$W_N = \mathbf{2000}$$

11. $\mathbf{20 = W_P \times 25}$

$$\frac{\cancel{20}^4}{\cancel{25}_5} = \frac{W_P \times \cancel{25}^1}{\cancel{25}_1}$$
$$\frac{4}{5} = W_P$$
$$W_P = \frac{4}{5} \times 100\% = \mathbf{80\%}$$

12. (a) $(25)(-5) = \mathbf{-125}$

(b) $(-15)(-5) = \mathbf{75}$

(c) $\frac{-250}{-5} = \mathbf{50}$

(d) $\frac{-225}{15} = \mathbf{-15}$

13. (a) $0.2 = \frac{2}{10} = \mathbf{\frac{1}{5}}$

(b) $0.2 = \frac{2}{10} = \frac{20}{100} = \mathbf{20\%}$

(c) $2\% = \frac{2}{100} = \mathbf{\frac{1}{50}}$

(d) $2\% = \frac{2}{100} = \mathbf{0.02}$

14.

$$\begin{array}{r} \$21.00 \\ \times\ \ 0.075 \\ \hline \$1.575 \end{array} \longrightarrow \$1.58$$

$$\begin{array}{r} \$21.00 \\ +\ \$\ 1.58 \\ \hline \mathbf{\$22.58} \end{array}$$

15. (a) $$\frac{14\frac{2}{7}}{100} = \frac{\frac{100}{7}}{\frac{100}{1}}$$

$$\frac{\frac{100}{7}}{\frac{100}{1}} \cdot \frac{\frac{1}{100}}{\frac{1}{100}} = \frac{\frac{100}{700}}{1} = \frac{100}{700} = \mathbf{\frac{1}{7}}$$

(b) $$\frac{60}{\frac{2}{3}} = \frac{\frac{60}{1}}{\frac{2}{3}}$$

$$\frac{\frac{60}{1}}{\frac{2}{3}} \cdot \frac{\frac{3}{2}}{\frac{3}{2}} = \frac{\frac{180}{2}}{1} = \frac{180}{2} = \mathbf{90}$$

16.

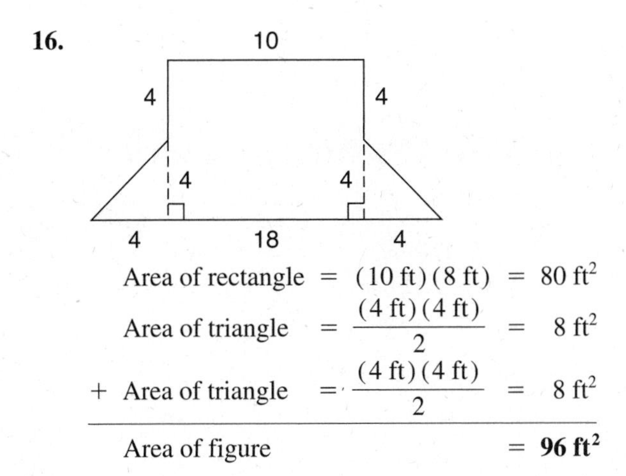

$$\text{Area of rectangle} = (10\text{ ft})(8\text{ ft}) = 80\text{ ft}^2$$
$$\text{Area of triangle} = \frac{(4\text{ ft})(4\text{ ft})}{2} = 8\text{ ft}^2$$
$$+\ \text{Area of triangle} = \frac{(4\text{ ft})(4\text{ ft})}{2} = 8\text{ ft}^2$$
$$\text{Area of figure} = \mathbf{96\text{ ft}^2}$$

17.

2 cm
2 cm
2 cm

(a) $(2\text{ cm})(2\text{ cm}) = 4\text{ cm}^2$

$$\frac{4\text{ cm}^2}{1\text{ layer}} \cdot 1\text{ cm}(2\text{ layers}) = \mathbf{8\text{ cm}^3}$$

(b) **One way to find the surface area of a cube is to find the area of one face of the cube and then multiply that area by 6.**

18. $\mathbf{1.2 \times 10^{10}}$

19. (a) $C = \pi(20\text{ mm})$
$C = \mathbf{20\pi\text{ mm}}$

(b) $C \approx 3.14(20\text{ mm})$
$C \approx \mathbf{62.8\text{ mm}}$

20. $3x = 26.7$

$$\frac{\overset{1}{\cancel{3}}x}{\underset{1}{\cancel{3}}} = \frac{\overset{8.9}{\cancel{26.7}}}{\underset{1}{\cancel{3}}}$$

$x = \mathbf{8.9}$

check: $3(8.9) = 26.7$
$26.7 = 26.7$

21. $$y - 3\frac{1}{3} = 7$$
$$y - 3\frac{1}{3} + 3\frac{1}{3} = 7 + 3\frac{1}{3}$$
$$y = \mathbf{10\frac{1}{3}}$$

check: $10\frac{1}{3} - 3\frac{1}{3} = 7$
$7 = 7$

22. $$\frac{2}{3}x = 48$$

$$\left(\frac{\overset{1}{\cancel{3}}}{\underset{1}{\cancel{2}}}\right)\left(\frac{\overset{1}{\cancel{2}}}{\underset{1}{\cancel{3}}}\right)x = \left(\frac{3}{\underset{1}{\cancel{2}}}\right)\overset{24}{\cancel{48}}$$

$x = \mathbf{72}$

check: $\frac{2}{\underset{1}{\cancel{3}}}(\overset{24}{\cancel{72}}) = 48$
$2(24) = 48$
$48 = 48$

23. **Multiply the "in" number by 4; then add 1 to find the "out" number.**

$3(4) + 1 = \mathbf{13}$

$1 - 1 = 0,\ \frac{0}{4} = \mathbf{0}$

24. $5^2 - \{2^3 + 3[4^2 - (4)(\sqrt{9})]\}$
$25 - \{8 + 3[16 - (4)(3)]\}$
$25 - \{8 + 3[16 - 12]\}$
$25 - \{8 + 3[4]\}$
$25 - \{8 + 12\}$
$25 - 20$
$\mathbf{5}$

25.

4 gal	3 qt	1 pt
+ 1 gal	2 qt	1 pt
5 gal	5 qt	2 pt

2 pt = 1 qt, 5qt + 1 qt = 6 qt
6 qt = 1 gal 2 qt

1 gal	2 qt
+ 5 gal	
6 gal	**2 qt**

26. $$1\ \cancel{\text{ft}^2} \cdot \frac{12\text{ in.}}{1\ \cancel{\text{ft}}} \cdot \frac{12\text{ in.}}{1\ \cancel{\text{ft}}}$$
$= \mathbf{144\text{ in.}^2}$

27. $$1\frac{1}{3} \div 3 = \frac{4}{3} \div \frac{3}{1} = \frac{4}{3} \times \frac{1}{3} = \frac{4}{9}$$

$$5\frac{1}{3} \div \frac{4}{9} = \frac{16}{3} \div \frac{4}{9} = \frac{\overset{4}{\cancel{16}}}{\underset{1}{\cancel{3}}} \times \frac{\overset{3}{\cancel{9}}}{\underset{1}{\cancel{4}}} = \mathbf{12}$$

28. $3\frac{1}{5} - 2\frac{1}{2} + 1\frac{1}{4} = \left(\frac{16}{5} - \frac{5}{2}\right) + \frac{5}{4}$

$= \left(\frac{64}{20} - \frac{50}{20}\right) + \frac{25}{20} = \frac{14}{20} + \frac{25}{20}$

$= \frac{39}{20} = \mathbf{1\frac{19}{20}}$

29. $2.5 = 2\frac{1}{2}$

$3\frac{1}{3} \div 2\frac{1}{2} = \frac{10}{3} \div \frac{5}{2}$

$= \frac{\overset{2}{\cancel{10}}}{3} \times \frac{2}{\underset{1}{\cancel{5}}} = \frac{4}{3} = \mathbf{1\frac{1}{3}}$

30. (a) $(-3) + (-4) - (+5)$

$(-3) + (-4) + [-(+5)]$

$(-3) + (-4) + [-5]$

$\mathbf{-12}$

(b) $(-6) - (-16) - (+30)$

$(-6) + [-(-16)] + [-(+30)]$

$(-6) + [16] + [-30]$

$\mathbf{-20}$

Lesson 78, Warm-Up

a. **8**

b. **2**

c. **6**

d. **$32.00**

e. **0.75 kg**

f. **72**

g. **$4.50 to $4.80**

Problem Solving

6 dots (1, 2, and 3)
7 dots (1, 2, and 4)
9 dots (1, 3, and 5)
10 dots (1, 4, and 5)
11 dots (2, 3, and 6)
12 dots (2, 4, and 6)
14 dots (3, 5, and 6)
15 dots (4, 5, and 6)

Lesson 78, Lesson Practice

a.

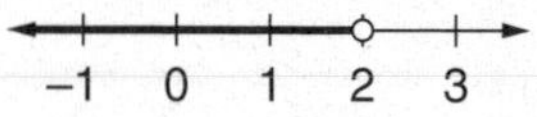

b.

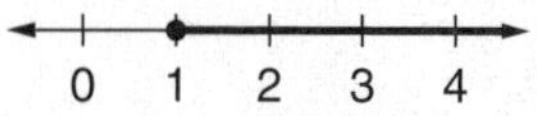

c.

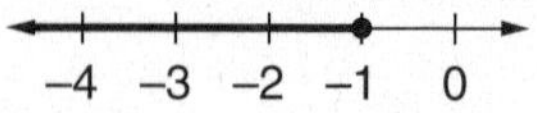

d.

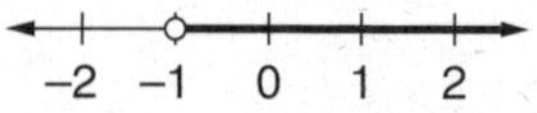

Lesson 78, Mixed Practice

1.

	Case 1	Case 2
Cartons	4	C
Hungry children	30	75

$\frac{4}{30} = \frac{C}{75}$

$30C = 4(75)$

$C = \frac{300}{30}$

$C = \mathbf{10\ cartons}$

2. $4(88) = 352$

$6(90) = 540$

$540 - 352 = 188$

$188 \div 2 = \mathbf{94}$

3. $\left(\frac{2}{3} + \frac{3}{4}\right) \div \left(\frac{2}{3} \times \frac{3}{4}\right)$

$= \left(\frac{8}{12} + \frac{9}{12}\right) \div \left(\frac{1}{2}\right)$

$= \frac{17}{12} \div \frac{1}{2}$

$= \frac{17}{\underset{6}{\cancel{12}}} \times \frac{\overset{1}{\cancel{2}}}{1} = \frac{17}{6} = \mathbf{2\frac{5}{6}}$

4.

	Ratio	Actual Count
Monocotyledons	3	M
Dicotyledons	4	84

$\frac{3}{4} = \frac{M}{84}$

$4M = 3(84)$

$M = \frac{252}{4}$

$M = \mathbf{63\ monocotyledons}$

5. $C \approx 3.14(21 \text{ millimeters})$

$C \approx 65.94 \text{ millimeters}$

$C \approx$ **66 millimeters**

6. (a) number line: 1 2 3 4 5 (open circle at 2, shaded to the right)

(b) number line: −2 −1 0 1 2 (closed circle at 1, shaded to the left)

7. $\mathbf{1.5\ \cancel{kg} \cdot \frac{1000\ g}{1\ \cancel{kg}} = 1500\ g}$

8. $\frac{5}{\cancel{6}_1} \times \cancel{30}^5 = 25$

(a) $25 - 5 = 20$

20 more people

(b) $\frac{\text{preferred Brand } Y}{\text{preferred Brand } X} = \frac{5}{25} = \mathbf{\frac{1}{5}}$

9. $\mathbf{42 = \frac{7}{10} \times W_N}$

$\left(\frac{10}{\cancel{7}_1}\right)\cancel{42}^6 = \left(\frac{\cancel{10}^1}{\cancel{7}_1}\right)\frac{\cancel{7}^1}{\cancel{10}_1} \times W_N$

$\mathbf{60} = W_N$

10. $\mathbf{1.5 \times W_N = 600}$

$\frac{\cancel{1.5}^1 \times W_N}{\cancel{1.5}_1} = \frac{600}{1.5}$

$W_N = \mathbf{400}$

11. $\mathbf{0.4 \times 50 = W_N}$

$\mathbf{20} = W_N$

12. $\mathbf{40 = W_P \times 50}$

$\frac{\cancel{40}^4}{\cancel{50}_5} = \frac{W_P \times \cancel{50}^1}{\cancel{50}_1}$

$\frac{4}{5} = W_P$

$W_P = \frac{4}{5} \times 100\% = \frac{400\%}{5} = \mathbf{80\%}$

13. (a) **0.0015**

(b) $\mathbf{2.5 \times 10^7}$

14. (a) $\frac{-45}{9} = \mathbf{-5}$

(b) $\frac{-450}{15} = \mathbf{-30}$

(c) $15(-20) = \mathbf{-300}$

(d) $(-15)(-12) = \mathbf{180}$

15. (a) $50\% = \frac{50}{100} = \mathbf{\frac{1}{2}}$

(b) $50\% = \frac{50}{100} = \mathbf{0.5}$

(c) $12\overline{)1.000\ldots}$ quotient $\mathbf{0.08\overline{3}}$

(d) $\frac{1}{12} \times 100\% = \frac{100\%}{12} = \mathbf{8\frac{1}{3}\%}$

16. $\frac{83\frac{1}{3}}{100} = \frac{\frac{250}{3}}{\frac{100}{1}}$

$\frac{\frac{250}{3}}{\frac{100}{1}} \cdot \frac{\frac{1}{100}}{\frac{1}{100}} = \frac{\frac{250}{300}}{1}$

$= \frac{250}{300} = \mathbf{\frac{5}{6}}$

17. Trapezoid figure: top 40, bottom 24, height 20, side 23, triangles A and B

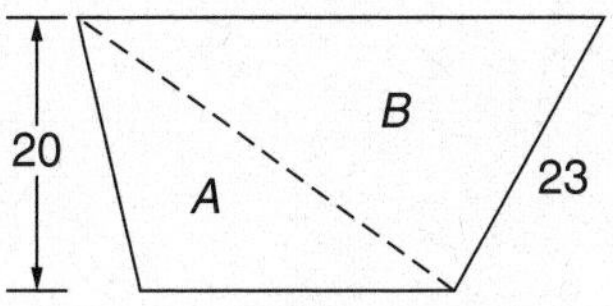

Area of triangle $A = \frac{(24 \text{ mm})(20 \text{ mm})}{2}$

$= 240 \text{ mm}^2$

Area of triangle $B = \frac{(40 \text{ mm})(20 \text{ mm})}{2}$

$= 400 \text{ mm}^2$

Area of figure $= 240 \text{ mm}^2 + 400 \text{ mm}^2$

$= \mathbf{640\ mm^2}$

18.

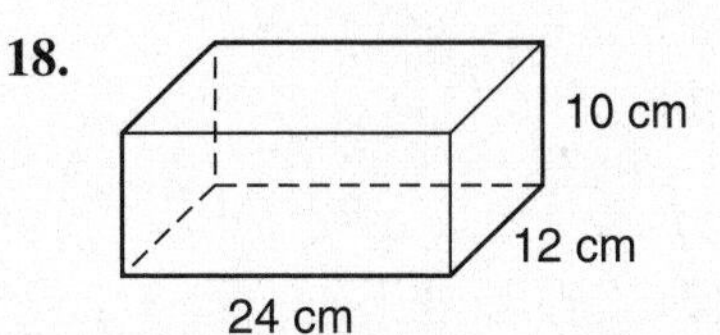

$(24 \text{ cm})(12 \text{ cm}) = 288 \text{ cm}^2,$

$\frac{288 \text{ cm}^2}{1 \cancel{\text{layer}}} \cdot (1 \text{ cm})(10 \cancel{\text{layers}})$

$\mathbf{2880\ cm^3}$

19. One possibility:

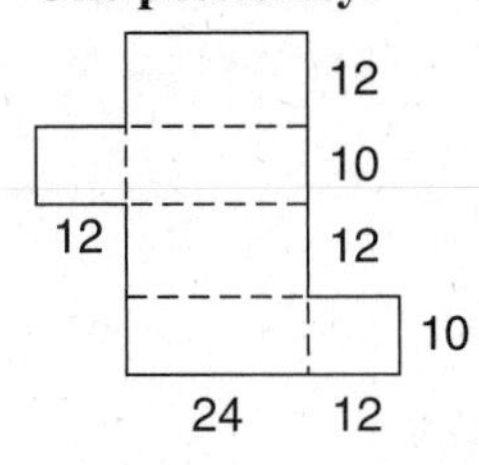

20. Multiply the "in" number by 5; then subtract 1 to find the "out" number.

$4(5) - 1 = \mathbf{19}$

$4 + 1 = 5,\ 5 \div 5 = \mathbf{1}$

21.

$$\begin{array}{r} \$18.50 \\ \times \quad 0.30 \\ \hline \$5.5500 \end{array} = \mathbf{\$5.55}$$

22.

$$\begin{aligned} m + 8.7 &= 10.25 \\ m + 8.7 - 8.7 &= 10.25 - 8.7 \\ m &= \mathbf{1.55} \end{aligned}$$

check:

$$\begin{aligned} 1.55 + 8.7 &= 10.25 \\ 10.25 &= 10.25 \end{aligned}$$

23.

$$\frac{4}{3}w = 36$$

$$\left(\frac{\cancel{3}^{1}}{\cancel{4}_{1}}\right)\frac{\cancel{4}^{1}}{\cancel{3}_{1}}w = \left(\frac{3}{\cancel{4}_{1}}\right)\cancel{36}^{9}$$

$$w = \mathbf{27}$$

check:

$$\left(\frac{4}{\cancel{3}_{1}}\right)\cancel{27}^{9} = 36$$

$$(4)9 = 36$$

$$36 = 36$$

24.

$$0.7y = 48.3$$

$$\frac{0.7y}{0.7} = \frac{48.3}{0.7}$$

$$y = \mathbf{69}$$

check:

$$\begin{aligned} 0.7(69) &= 48.3 \\ 48.3 &= 48.3 \end{aligned}$$

25.

$$\{4^2 + 10[2^3 - (3)(\sqrt{4})]\} - \sqrt{36}$$
$$\{16 + 10[8 - (3)(2)]\} - 6$$
$$\{16 + 10[8 - 6]\} - 6$$
$$\{16 + 10[2]\} - 6$$
$$\{16 + 20\} - 6$$
$$\{36\} - 6$$
$$\mathbf{30}$$

26.

$$|5 - 3| - |3 - 5|$$
$$= |2| - |-2| = 2 - 2 = \mathbf{0}$$

27.

$$1\,\cancel{m^2} \cdot \frac{100\text{ cm}}{1\,\cancel{m}} \cdot \frac{100\text{ cm}}{1\,\cancel{m}} = \mathbf{10{,}000\ cm^2}$$

28.

$$7\frac{1}{2} \cdot 3 \cdot \left(\frac{2}{3}\right)^2 =$$

$$\frac{\cancel{15}^{5}}{\cancel{2}_{1}} \cdot \frac{\cancel{3}^{1}}{1} \cdot \frac{\cancel{4}^{2}}{\cancel{9}_{\cancel{3}_{1}}} = \mathbf{10}$$

29.

$$3\frac{1}{5} - \left(2\frac{1}{2} - 1\frac{1}{4}\right)$$
$$= 3\frac{4}{20} - \left(2\frac{10}{20} - 1\frac{5}{20}\right)$$
$$= 3\frac{4}{20} - 1\frac{5}{20} = 2\frac{24}{20} - 1\frac{5}{20}$$
$$= \mathbf{1\frac{19}{20}}$$

30. (a)

$$(-10) - (-8) - (+6)$$
$$(-10) + [-(-8)] + [-(+6)]$$
$$(-10) + [8] + [-6]$$
$$\mathbf{-8}$$

(b)

$$(+10) + (-20) - (-30)$$
$$(+10) + (-20) + [-(-30)]$$
$$(+10) + (-20) + [30]$$
$$\mathbf{20}$$

Lesson 79, Warm-Up

a. **−40**

b. **3750**

c. **250**

d. **\$1.80**

e. **1.2 L**

f. **25**

g. **25**

Problem Solving

	A	B	C
M	Not		Not
S		Not	
H			Not

We see that the B must be in math and not in history. That means the A must be in history and the C must be in science.

Lesson 79, Lesson Practice

a. $x \; \textcircled{>} \; y$

b. $m \; \textcircled{=} \; n$

c. **Insufficient information**

d. **Insufficient information** (Both x and y could be zero, or y could be greater.)

Lesson 79, Mixed Practice

1. $4(33.5 \text{ cassettes}) = 134 \text{ cassettes}$

$$5\overline{)134.0} = \mathbf{26.8 \text{ cassettes}}$$

2. $\dfrac{315 \text{ kilometers}}{35 \text{ liters}} = \mathbf{9 \dfrac{kilometers}{liter}}$

3.

	Ratio	Actual Count
Winners	7	W
Losers	5	L
Total	12	1260

$$\frac{7}{12} = \frac{W}{1260}$$
$$12W = 7(1260)$$
$$W = \frac{8820}{12}$$
$$W = 735 \text{ winners}$$
$$\frac{5}{12} = \frac{L}{1260}$$
$$12L = 5(1260)$$
$$L = \frac{6300}{12}$$
$$L = 525 \text{ losers}$$

$$\begin{array}{r} 735 \\ -\ 525 \\ \hline \mathbf{210} \end{array}$$ **more winners**

4. (a) $\mathbf{3.75 \times 10^{-5}}$

(b) $\mathbf{3.75 \times 10^{7}}$

5. **Insufficient information** ($x < y$ if both are positive; $x > y$ if both are negative.)

6. (a)

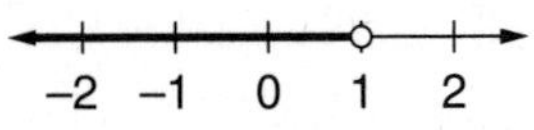

(b)

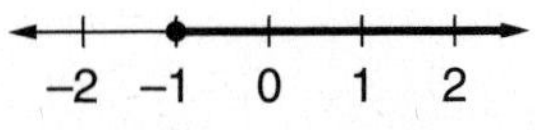

7.

	Case 1	Case 2
Inches	4	12
Hours	3	h

$$\frac{4}{3} = \frac{12}{h}$$
$$4h = 3(12)$$
$$h = \frac{36}{4}$$
$$h = \mathbf{9 \text{ hours}}$$

8.

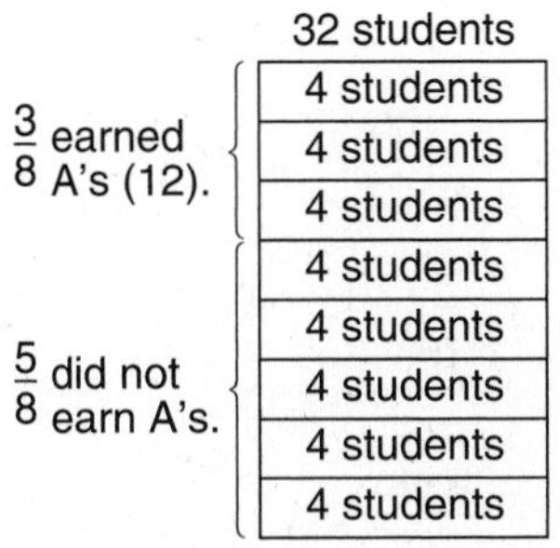

(a) $\dfrac{12 \text{ students}}{3} = 4 \text{ students}$

$5(4 \text{ students}) = \mathbf{20 \text{ students}}$

(b) $\dfrac{5}{8} \times 100\% = \dfrac{500\%}{8} = \mathbf{62\frac{1}{2}\%}$

9.
$$\mathbf{35 = 0.7 \times W_N}$$
$$\frac{35}{0.7} = \frac{0.7 \times W_N}{0.7}$$
$$\mathbf{50} = W_N$$

10.
$$\mathbf{W_P \times 20 = 17}$$
$$\frac{W_P \times 20}{20} = \frac{17}{20}$$
$$W_P = \frac{17}{20}$$
$$W_P = \frac{17}{20} \times 100\% = \frac{1700\%}{20} = \mathbf{85\%}$$

11.
$$\mathbf{W_P \times 20 = 25}$$
$$\frac{W_P \times 20}{20} = \frac{\overset{5}{\cancel{25}}}{\underset{4}{\cancel{20}}}$$
$$W_P = \frac{5}{4}$$
$$W_P = \frac{5}{4} \times 100\% = \frac{500\%}{4} = \mathbf{125\%}$$

12.
$$\mathbf{360 = \frac{3}{4} \times W_N}$$
$$\left(\frac{4}{\underset{1}{\cancel{3}}}\right)\overset{120}{\cancel{360}} = \left(\frac{\overset{1}{\cancel{4}}}{\underset{1}{\cancel{3}}}\right)\frac{\overset{1}{\cancel{3}}}{\underset{1}{\cancel{4}}} \times W_N$$
$$(4)120 = W_N$$
$$\mathbf{480} = W_N$$

13. (a) $\dfrac{144}{-8} = \mathbf{-18}$

(b) $\dfrac{-144}{+6} = \mathbf{-24}$

(c) $-12(12) = \mathbf{-144}$

(d) $-16(-9) = \mathbf{144}$

14. (a) $25\overline{)1.00}$ quotient **0.04**

(b) $\dfrac{1}{25} \times 100\% = \dfrac{100\%}{25} = \mathbf{4\%}$

(c) $8\% = \dfrac{8}{100} = \mathbf{\dfrac{2}{25}}$

(d) $8\% = \dfrac{8}{100} = \mathbf{0.08}$

15.
$$\begin{array}{r} \$4500 \\ \times \quad 0.05 \\ \hline \$225.00 \end{array} = \mathbf{\$225}$$

16. $\dfrac{62\frac{1}{2}}{100} = \dfrac{\frac{125}{2}}{\frac{100}{1}}$

$\dfrac{\frac{125}{2}}{\frac{100}{1}} \cdot \dfrac{\frac{1}{100}}{\frac{1}{100}} = \dfrac{\frac{125}{200}}{1}$

$= \dfrac{125}{200} = \mathbf{\dfrac{5}{8}}$

17.

7, 3, 4, 5, 10, 6, 10

(a) Perimeter $= 10 \text{ in.} + 7 \text{ in.} + 5 \text{ in.} + 6 \text{ in.} + 10 \text{ in.} = \mathbf{38 \text{ in.}}$

(b) Area of square $= (10 \text{ in.})(10 \text{ in.})$
$= 100 \text{ in.}^2$

Area of triangle $= \dfrac{(3 \text{ in.})(4 \text{ in.})}{2}$
$= 6 \text{ in.}^2$

Area of figure $= 100 \text{ in.}^2 - 6 \text{ in.}^2$
$= \mathbf{94 \text{ in.}^2}$

18. (a) $(6 \text{ cubes})(3 \text{ cubes}) = 18 \text{ cubes}$

$\dfrac{18 \text{ cubes}}{1 \text{ layer}} \cdot 4 \text{ layers} = 72 \text{ cubes}$

$\mathbf{72 \text{ cm}^3}$

(b) $(6 \text{ cm})(3 \text{ cm}) = 18 \text{ cm}^2$
$(4 \text{ cm})(3 \text{ cm}) = 12 \text{ cm}^2$
$(6 \text{ cm})(4 \text{ cm}) = 24 \text{ cm}^2$
$2(18 \text{ cm}^2) + 2(12 \text{ cm}^2) + 2(24 \text{ cm}^2)$
$= 36 \text{ cm}^2 + 24 \text{ cm}^2 + 48 \text{ cm}^2$
$= \mathbf{108 \text{ cm}^2}$

19. (a) $C \approx 3.14(1 \text{ m})$
$C \approx \mathbf{3.14 \text{ m}}$

(b) $C = \pi(1 \text{ m})$
$C = \boldsymbol{\pi} \textbf{ m}$

20. (a) **Right; scalene**

(b) **Obtuse; isosceles**

(c) **Acute; equilateral**

21. $1.2x = 2.88$

$\dfrac{1.2x}{1.2} = \dfrac{2.88}{1.2}$

$x = \mathbf{2.4}$

check: $1.2(2.4) = 2.88$
$2.88 = 2.88$

22. $3\frac{1}{3} = x + \frac{5}{6}$

$3\frac{1}{3} - \frac{5}{6} = x + \frac{5}{6} - \frac{5}{6}$

$3\frac{2}{6} - \frac{5}{6} = x$

$2\frac{8}{6} - \frac{5}{6} = x$

$2\frac{3}{6} = x$

$\mathbf{2\frac{1}{2}} = x$

check: $3\frac{1}{3} = 2\frac{1}{2} + \frac{5}{6}$

$3\frac{1}{3} = 2\frac{3}{6} + \frac{5}{6}$

$3\frac{1}{3} = 2\frac{8}{6}$

$3\frac{1}{3} = 3\frac{2}{6}$

$3\frac{1}{3} = 3\frac{1}{3}$

23. $\frac{3}{2}w = \frac{9}{10}$

$\left(\frac{\overset{1}{\cancel{2}}}{\underset{1}{\cancel{3}}}\right)\frac{\overset{1}{\cancel{3}}}{\underset{1}{\cancel{2}}}w = \left(\frac{\overset{1}{\cancel{2}}}{\underset{1}{\cancel{3}}}\right)\frac{\overset{3}{\cancel{9}}}{\underset{5}{\cancel{10}}}$

$w = \mathbf{\frac{3}{5}}$

check: $\left(\frac{3}{2}\right)\frac{3}{5} = \frac{9}{10}$

$\frac{9}{10} = \frac{9}{10}$

24. $\frac{\sqrt{100} + 5[3^3 - 2(3^2 + 3)]}{5}$

$\frac{10 + 5[27 - 2(9 + 3)]}{5}$

$\frac{10 + 5[27 - 2(12)]}{5}$

$\frac{10 + 5[27 - 24]}{5}$

$\frac{10 + 5[3]}{5}$

$\frac{10 + 15}{5}$

$\frac{25}{5}$

$\mathbf{5}$

25. $\overset{2}{\cancel{3}}$ hr $\overset{14}{\cancel{15}}$ min 24 s (60 min) (60 s)
$-$ 2 hr 45 min 30 s ⟶

$\overset{2}{\cancel{3}}$ hr $\overset{74}{\cancel{\overset{14}{\cancel{15}}}}$ min $\overset{84}{\cancel{24}}$ s
$-$ 2 hr 45 min 30 s
29 min 54 s

26. $1\ \cancel{yd^2} \cdot \frac{3\text{ ft}}{1\ \cancel{yd}} \cdot \frac{3\text{ ft}}{1\ \cancel{yd}} = \mathbf{9\ ft^2}$

27. $7\frac{1}{2} \cdot \left(3 \div \frac{5}{9}\right) = \frac{15}{2} \cdot \left(\frac{3}{1} \div \frac{5}{9}\right)$

$= \frac{15}{2} \cdot \left(\frac{3}{1} \times \frac{9}{5}\right) = \frac{\overset{3}{\cancel{15}}}{2} \cdot \frac{27}{\underset{1}{\cancel{5}}}$

$= \frac{81}{2} = \mathbf{40\frac{1}{2}}$

28. $4\frac{5}{6} + 3\frac{1}{3} + 7\frac{1}{4}$

$= 4\frac{10}{12} + 3\frac{4}{12} + 7\frac{3}{12}$

$= 14\frac{17}{12} = \mathbf{15\frac{5}{12}}$

29. $3\frac{3}{4} = 3.75,\ 3.75 \div 1.5 = \mathbf{2.5}$

30. (a) $(-10) - (+20) - (-30)$

$(-10) + [-(+20)] + [-(-30)]$

$(-10) + [-20] + [30]$

$\mathbf{0}$

(b) $(-10) - |(-20) - (+30)|$

$(-10) + [-|(-20) + [-(+30)]|]$

$(-10) + [-|(-20) + [-30]|]$

$(-10) + [-|-50|]$

$(-10) + [-(+50)]$

$(-10) + [-50]$

$\mathbf{-60}$

Lesson 80, Warm-Up

a. **−75**

b. **1600**

c. **8**

d. **25**

e. **1.5 km**

f. **9**

g. **6 m; 2.25 m²**

Problem Solving

$\sqrt{1^3} = \sqrt{1} = \mathbf{1}$

$\sqrt{1^3 + 2^3} = \sqrt{9} = \mathbf{3}$

$\sqrt{1^3 + 2^3 + 3^3} = \sqrt{36} = \mathbf{6}$

$\mathbf{\sqrt{1^3 + 2^3 + 3^3 + 4^3} = \sqrt{100} = 10}$

$\mathbf{\sqrt{1^3 + 2^3 + 3^3 + 4^3 + 5^3} = \sqrt{225} = 15}$

Lesson 80, Lesson Practice

a. **See lesson.**

b. **$W'(3, -4), X'(1, -4), Y'(1, -1), Z'(3, -1)$**

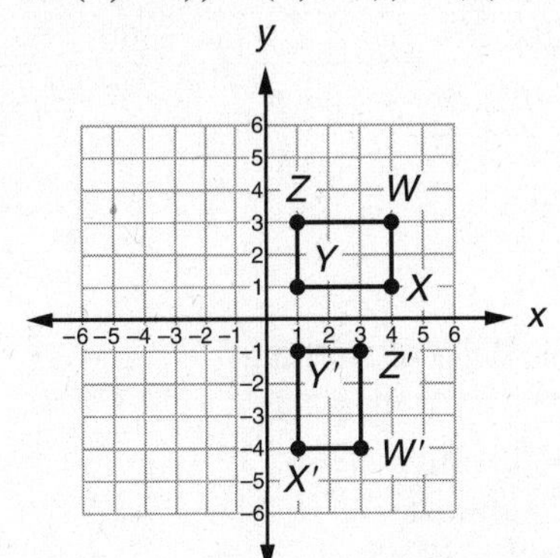

c. $J'(-1, -1), K'(-3, -2), L'(-1, -3)$

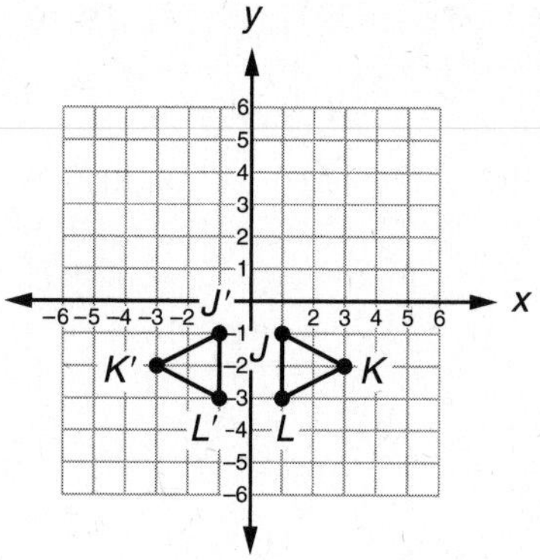

d. $P'(6, 0), Q'(5, -2), R'(2, -2), S'(3, 0)$

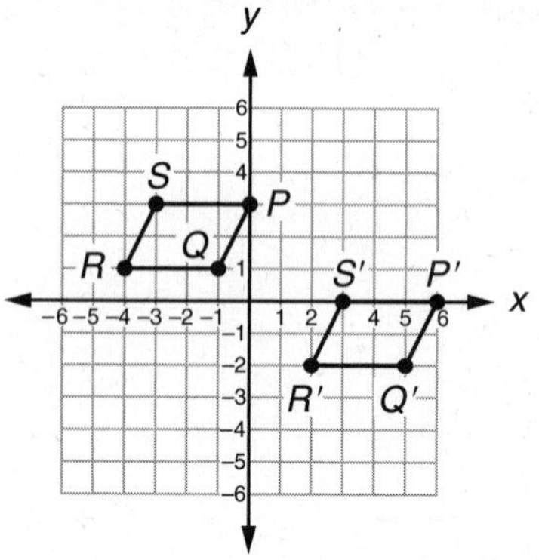

Lesson 80, Mixed Practice

1. $4(\$7.00) + 3(\$6.30)$
$= \$28.00 + \$18.90 = \$46.90$
\$6.70 per hour
$7\overline{)\$46.90}$

2. $(4) + [(4)^2 - (4)(3)] - 3$
$= 4 + [16 - 12] - 3$
$= 4 + [4] - 3 = 8 - 3 = \mathbf{5}$

3. **Insufficient information**

4.

	Ratio	Actual Count
Clean clothes	2	C
Dirty clothes	3	D
Total	5	30

$\frac{2}{5} = \frac{C}{30}$
$5C = 2(30)$
$C = \frac{60}{5}$
$C =$ **12 articles of clothing**

5. $C \approx 3.14(30 \text{ millimeters})$
$C \approx 94.2$ millimeters
$C \approx$ **94 millimeters**

6. $1\frac{1}{2}\ \cancel{\text{quarts}} \cdot \frac{2 \text{ pints}}{1\ \cancel{\text{quart}}} =$ **3 pints**

7. (a)

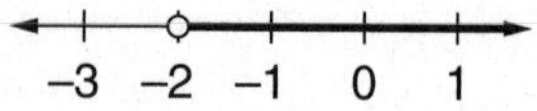

(b)

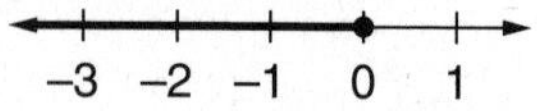

8.

	Case 1	Case 2
Minutes	25	60
Customers	400	c

$\frac{25}{400} = \frac{60}{c}$
$25c = 400(60)$
$c = \frac{24{,}000}{25}$
$c =$ **960 customers**

9.

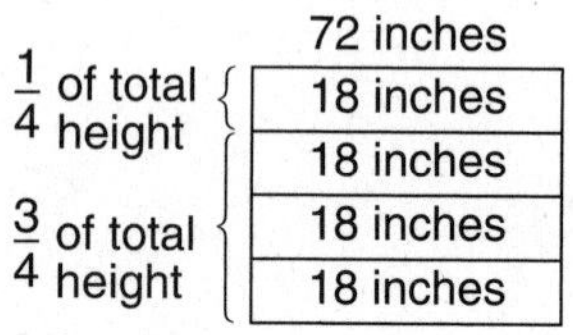

(a) $4(18 \text{ inches}) =$ **72 inches**

(b) $\overset{6}{\cancel{72}}\ \cancel{\text{inches}} \cdot \frac{1 \text{ foot}}{\underset{1}{\cancel{12}}\ \cancel{\text{inches}}}$
$=$ **6 feet**

10. $\mathbf{600 = \frac{5}{9} \times W_N}$
$\frac{9}{\cancel{5}}(\overset{120}{\cancel{600}}) = \left(\frac{\overset{1}{\cancel{9}}}{\underset{1}{\cancel{5}}}\right)\frac{\overset{1}{\cancel{5}}}{\underset{1}{\cancel{9}}} \times W_N$
$9(120) = W_N$
$\mathbf{1080} = W_N$

11. $\mathbf{280 = W_P \times 400}$
$\frac{\overset{7}{\cancel{280}}}{\underset{10}{\cancel{400}}} = \frac{W_P \times \overset{1}{\cancel{400}}}{\underset{1}{\cancel{400}}}$
$\frac{7}{10} = W_P$
$W_P = \frac{7}{10} \times 100\% = \frac{700\%}{10} = \mathbf{70\%}$

12. $\mathbf{W_N = 0.04 \times 400}$
$W_N = \mathbf{16}$

13. $\mathbf{60 = 0.6 \times W_N}$
$\frac{60}{0.6} = \frac{\overset{1}{\cancel{0.6}} \times W_N}{\underset{1}{\cancel{0.6}}}$
$\mathbf{100} = W_N$

14. (a) $\frac{600}{-15} = \mathbf{-40}$

(b) $\frac{-600}{-12} = \mathbf{50}$

(c) $20(-30) = \mathbf{-600}$

(d) $+15(40) = \mathbf{600}$

15.
$$\begin{array}{r} \$850 \\ \times\ 0.06 \\ \hline \$51.00 \end{array} = \mathbf{\$51}$$

16. (a) $0.3 = \mathbf{\frac{3}{10}}$

(b) $0.3 = \frac{3}{10} = \frac{30}{100} = \mathbf{30\%}$

(c) $12\overline{)5.000\ldots}$ $\mathbf{0.41\overline{6}}$

(d) $\frac{5}{12} \times 100\% = \frac{500\%}{12} = \mathbf{41\frac{2}{3}\%}$

17. (a) $\mathbf{3 \times 10^7}$

(b) $\mathbf{3 \times 10^{-5}}$

18.

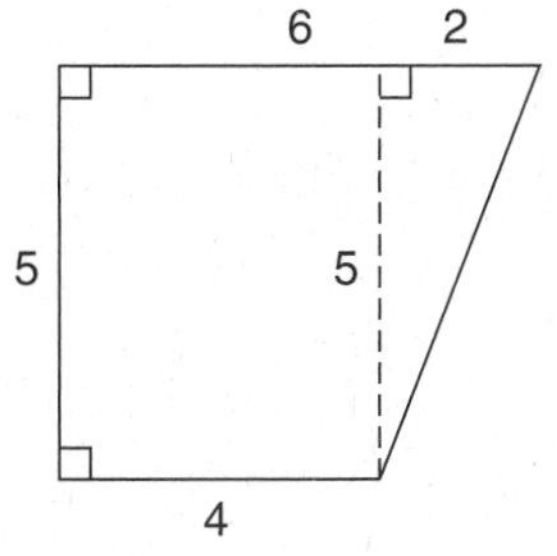

Figure: a rectangle with top side 6, left side 5, bottom side 4, plus a right triangle with top 2 and height 5 (dashed).

Area of rectangle	=	(4 m)(5 m)	=	20 m^2
+ Area of triangle	=	$\frac{(2\text{ m})(5\text{ m})}{2}$	=	5 m^2
Area of figure			=	**25 m^2**

19. (a) $(5\text{ in.})(5\text{ in.}) = 25\text{ in.}^2$
$(5\text{ in.})(25\text{ in.}^2) = \mathbf{125\text{ in.}^3}$

(b) $6(25\text{ in.}^2) = \mathbf{150\text{ in.}^2}$

20. $\frac{\text{green marbles}}{\text{total marbles}} = \frac{40}{100} = \mathbf{\frac{2}{5}}$

21.
$17a = 408$

$\frac{\overset{1}{\cancel{17}}a}{\underset{1}{\cancel{17}}} = \frac{\overset{24}{\cancel{408}}}{\underset{1}{\cancel{17}}}$

$a = \mathbf{24}$

check: $17(24) = 408$
$408 = 408$

22.
$\frac{3}{8}m = 48$

$\left(\frac{\overset{1}{\cancel{8}}}{\underset{1}{\cancel{3}}}\right)\frac{\overset{1}{\cancel{3}}}{\underset{1}{\cancel{8}}}m = \left(\frac{8}{\cancel{3}}\right)\overset{16}{\cancel{48}}$

$m = (8)16$

$m = \mathbf{128}$

check: $\frac{3}{\underset{1}{\cancel{8}}}(\overset{16}{\cancel{128}}) = 48$

$3(16) = 48$

$48 = 48$

23.
$1.4 = x - 0.41$

$1.4 + 0.41 = x - 0.41 + 0.41$

$\mathbf{1.81} = x$

check: $1.4 = 1.81 - 0.41$
$1.4 = 1.4$

24.
$$\frac{2^3 + 4 \cdot 5 - 2 \cdot 3^2}{\sqrt{25} \cdot \sqrt{4}}$$
$$\frac{8 + 4 \cdot 5 - 2 \cdot 9}{5 \cdot 2}$$
$$\frac{8 + 20 - 2 \cdot 9}{10}$$
$$\frac{8 + 20 - 18}{10}$$
$$\frac{28 - 18}{10}$$
$$\frac{10}{10}$$
$$\mathbf{1}$$

25. $7\frac{1}{7} \times 1.4 = 7\frac{1}{7} \times 1\frac{4}{10}$

$= \frac{\overset{5}{\cancel{50}}}{\underset{1}{\cancel{7}}} \times \frac{\overset{2}{\cancel{14}}}{\underset{1}{\cancel{10}}} = \mathbf{10}$

26.
$$\begin{array}{rr} \overset{9}{\cancel{10}}\text{ lb} \to (16\text{ oz}) & 6\text{ oz} \\ -\ 7\text{ lb} & 11\text{ oz} \\ \hline \end{array} \longrightarrow$$

$$\begin{array}{rr} \overset{9}{\cancel{10}}\text{ lb} & \overset{22}{\cancel{6}}\text{ oz} \\ -\ 7\text{ lb} & 11\text{ oz} \\ \hline \mathbf{2\text{ lb}} & \mathbf{11\text{ oz}} \end{array}$$

27. $1\ \cancel{\text{cm}^2} \cdot \frac{10\text{ mm}}{1\ \cancel{\text{cm}}} \cdot \frac{10\text{ mm}}{1\ \cancel{\text{cm}}} = \mathbf{100\text{ mm}^2}$

28.
$\overset{1}{\cancel{3}} \cdot \frac{5}{\underset{3}{\cancel{9}}} = \frac{5}{3}$

$7\frac{1}{2} \div \frac{5}{3} = \frac{15}{2} \div \frac{5}{3}$

$= \frac{\overset{3}{\cancel{15}}}{2} \times \frac{3}{\underset{1}{\cancel{5}}} = \frac{9}{2} = \mathbf{4\frac{1}{2}}$

29. $2^{-4} + 4^{-2} = \frac{1}{2^4} + \frac{1}{4^2}$
$= \frac{1}{16} + \frac{1}{16} = \frac{2}{16} = \mathbf{\frac{1}{8}}$

30.

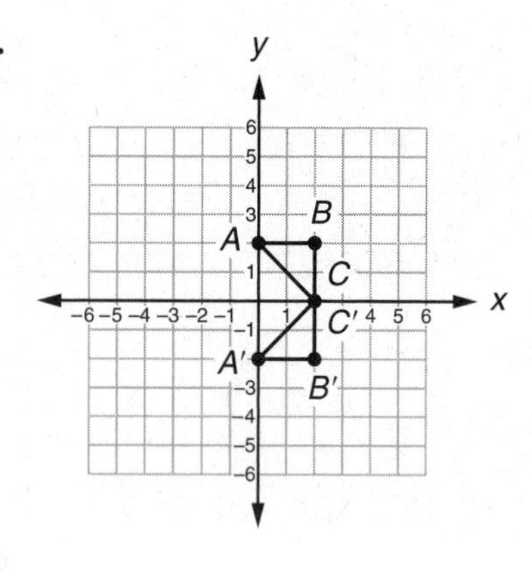

INVESTIGATION 8

1. **3 cm; 3 cm**

2. **90°**

3. **The line is perpendicular to the segment and divides the segment in half.**

4. **90°**

5. (a) $A = \frac{1}{2}bh$
$A = \frac{1}{2} \times 2 \times 2$
$A = \mathbf{2\ cm^2}$

(b) **Add the areas of the four small right triangles. The area of the square is 8 cm².**

6. **See student work. Each of the smaller angles should be half the measure of the larger angle.**

7. **The ray divides the original angle into two congruent angles. That is, the ray divides the angle in half.**

8. **45°**

9. **Both triangles are isosceles.**

10. (a) **45°**

(b) **45°**

11. (a) $\mathbf{67\frac{1}{2}°}$

(b) $\mathbf{67\frac{1}{2}°}$

(c) $\mathbf{67\frac{1}{2}°}$

(d) $\mathbf{67\frac{1}{2}°}$

12. (a) **135°**

(b) **135°**

LESSON 81, WARM-UP

a. **+100**

b. **0.0012**

c. **60**

d. **$3.60**

e. **2.5 cm**

f. **36**

g. **$27**

Problem Solving

Each term is found by adding the two previous terms.

1, 1, 2, 3, 5, 8, 13, **21, 34, 55**

LESSON 81, LESSON PRACTICE

a. Estimate: $\frac{21}{70} = \frac{3}{10} = 30\%$

$100\% - 30\% = \mathbf{70\%}$

	Percent	Actual Count
Planted with alfalfa	P_P	21
Not planted with alfalfa	P_N	49
Total	100	70

$\frac{P_N}{100} = \frac{49}{70}$

$70P_N = 4900$

$P_N = \mathbf{70\%}$

b. Estimate: $60 \times 3 = \mathbf{180}$ **pages**

	Percent	Actual Count
Pages read	40	120
Pages left to read	60	P
Total	100	T

$$\frac{40}{60} = \frac{120}{P}$$
$$40P = 7200$$
$$P = \mathbf{180\ pages}$$

c. Estimate: $\frac{26}{30} = \frac{13}{15} \approx \mathbf{87\%}$

	Percent	Actual Count
Missed	P_M	4
Correct	P_C	26
Total	100	30

$$\frac{P_C}{100} = \frac{26}{30}$$
$$30P_C = 2600$$
$$P_C = \mathbf{86\frac{2}{3}\%}$$

Lesson 81, Mixed Practice

1.

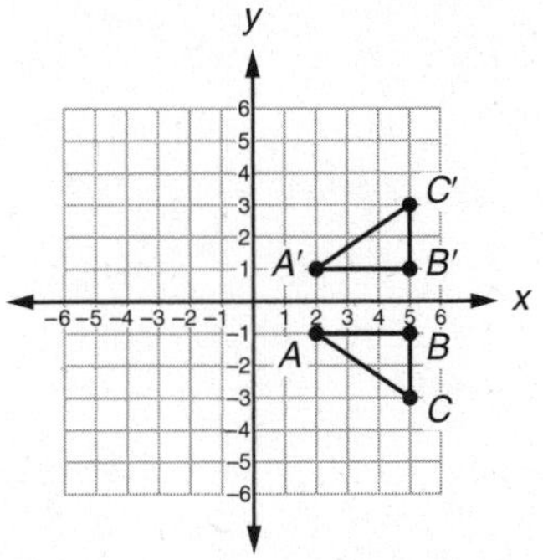

$\mathbf{A'(2, 1), B'(5, 1), C'(5, 3)}$

2. (a)

$$\begin{array}{r} 70 \\ 85 \\ 80 \\ 85 \\ 90 \\ 80 \\ 85 \\ 80 \\ 90 \\ 95 \\ 85 \\ 90 \\ 100 \\ 85 \\ +\ 90 \\ \hline 1290 \end{array} \qquad \begin{array}{r} \mathbf{86} \\ 15\overline{)1290} \end{array}$$

(b) **85**

3. (a) **85**

(b) $100 - 70 = \mathbf{30}$

4.

$$\begin{array}{rr} \overset{5}{\cancel{6}}\text{ ft} \longrightarrow & \overset{(12\text{ in.})}{1\text{ in.}} \longrightarrow \\ -\ 5\text{ ft} & 6\frac{1}{2}\text{ in.} \\ \hline \end{array}$$

$$\begin{array}{rr} \overset{5}{\cancel{6}}\text{ ft} & \overset{13}{\cancel{1}}\text{ in.} \\ -\ 5\text{ ft} & 6\frac{1}{2}\text{ in.} \\ \hline & \mathbf{6\frac{1}{2}\ inches} \end{array}$$

5.

	Case 1	Case 2
Pencils	5	12
Price	75¢	p

$$\frac{5}{\$0.75} = \frac{12}{p}$$
$$5p = \$9.00$$
$$p = \mathbf{\$1.80}$$

6. (a) $x < 4$

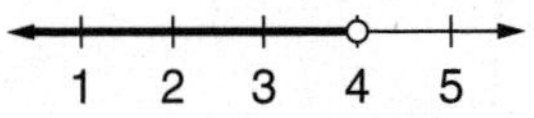

(b) $x \geq -2$

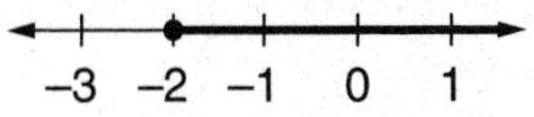

7.

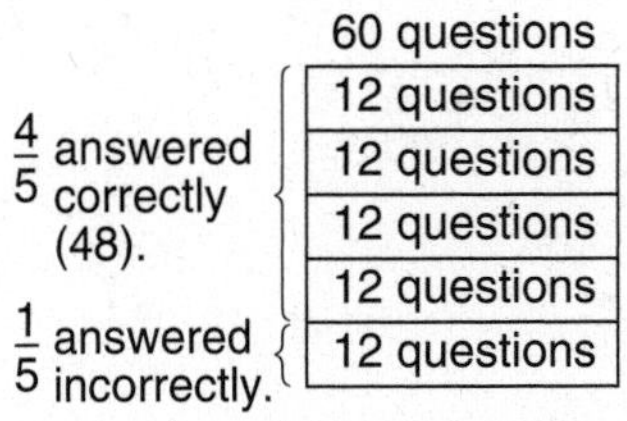

(a) $48 \div 4 = 12$

$5(12\text{ questions}) = \mathbf{60\ questions}$

(b) $\frac{\text{correct answers}}{\text{incorrect answers}} = \frac{48}{12} = \mathbf{\frac{4}{1}}$

8. $3\frac{3}{4} \div 2 = \frac{15}{4} \times \frac{1}{2}$

$= \frac{15}{8} = \mathbf{1\frac{7}{8}\ inches}$

9.

	Ratio	Actual Count
Gleeps	9	G
Bobbles	5	B
Total	14	2800

$$\frac{9}{14} = \frac{G}{2800}$$
$$14G = 25{,}200$$
$$G = \mathbf{1800\ gleeps}$$

10. $(9)^2 + \sqrt{9} = 81 + 3 = \mathbf{84}$

11. **Insufficient information** ($m < n$ if both are positive; $m > n$ if both are negative.)

12. (a) $2\frac{1}{4} = 2\frac{25}{100} = \mathbf{2.25}$

(b) $2.25 \times 100\% = \mathbf{225\%}$

(c) $2\frac{1}{4}\% = \frac{2\frac{1}{4}}{100} = \frac{\frac{9}{4}}{100} \cdot \frac{\frac{1}{100}}{\frac{1}{100}} = \mathbf{\frac{9}{400}}$

(d) $2\frac{1}{4}\% = 2.25\% = \mathbf{0.0225}$

13. $\mathbf{p = 0.4 \times \$12}$
$p = \mathbf{\$4.80}$

14. $\mathbf{0.5 \times W_N = 0.4}$

$\frac{0.5 \times W_N}{0.5} = \frac{0.4}{0.5}$

$W_N = \mathbf{0.8}$

15. $\frac{16\frac{2}{3}}{100} = \frac{\frac{50}{3}}{\frac{100}{1}}$

$\frac{\frac{50}{3}}{\frac{100}{1}} \cdot \frac{\frac{1}{100}}{\frac{1}{100}} = \frac{\frac{50}{300}}{1} = \frac{50}{300}$

$= \mathbf{\frac{1}{6}}$

16.

	Percent	Actual Count
Correct	P_C	21
Incorrect	P_I	4
Total	100	25

$\frac{P_C}{100} = \frac{21}{25}$

$25P_C = 2100$

$P_C = \mathbf{84\%}$

17.

	Percent	Actual Count
Left fallow	20	F
Not left fallow	80	N
Total	100	4000

$\frac{80}{100} = \frac{N}{4000}$

$100N = 320{,}000$

$N = \mathbf{3200\ acres}$

18. (a) **Angles ABC and CBD are supplementary (total 180°), so m∠CBD = 40°.**

(b) **Angles CBD and DBE are complementary (total 90°), so m∠DBE = 50°.**

(c) **Angles DBE and EBA are supplementary (total 180°), so m∠EBA = 130°.**

(d) **360°**

19. $\frac{3000}{6300} = \frac{\cancel{2}^1 \cdot \cancel{2}^1 \cdot 2 \cdot \cancel{3}^1 \cdot \cancel{5}^1 \cdot \cancel{5}^1 \cdot 5}{\cancel{2}_1 \cdot \cancel{2}_1 \cdot \cancel{3}_1 \cdot 3 \cdot \cancel{5}_1 \cdot \cancel{5}_1 \cdot 7}$

$= \mathbf{\frac{10}{21}}$

20. (a)

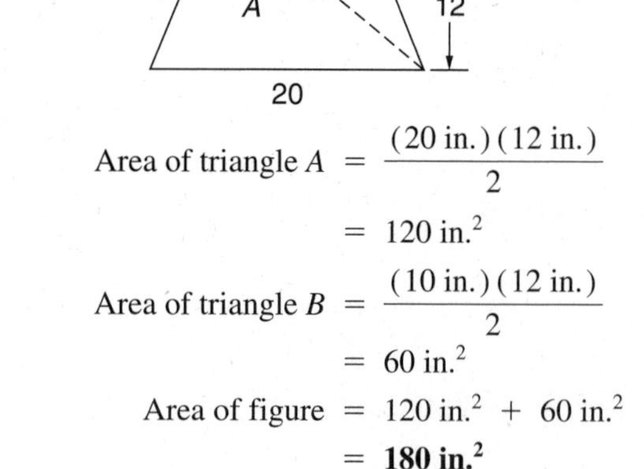

Area of triangle $A = \frac{(20\text{ in.})(12\text{ in.})}{2}$

$= 120\text{ in.}^2$

Area of triangle $B = \frac{(10\text{ in.})(12\text{ in.})}{2}$

$= 60\text{ in.}^2$

Area of figure $= 120\text{ in.}^2 + 60\text{ in.}^2$

$= \mathbf{180\ in.^2}$

(b)

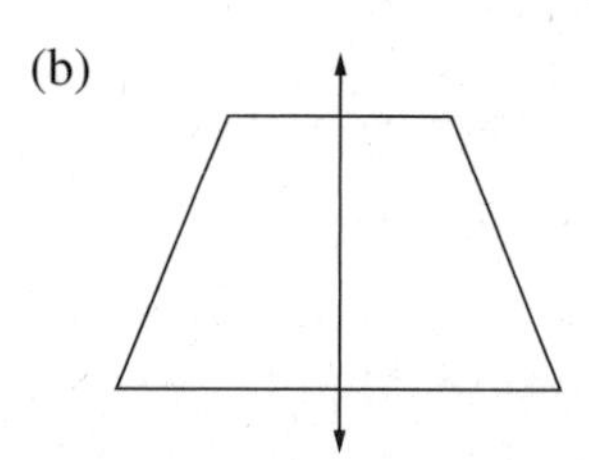

21. **Double the "in" number, then subtract 1 to find the "out" number.**
$4 \cdot 2 - 1 = \mathbf{7}$
$-1 + 1 = 0,\ 0/2 = \mathbf{0}$

22. (a) $\mathbf{5.6 \times 10^8}$

(b) $\mathbf{5.6 \times 10^{-6}}$

23. $5x = 16.5$

$\frac{5x}{5} = \frac{16.5}{5}$

$x = \mathbf{3.3}$

check: $5(3.3) = 16.5$
$16.5 = 16.5$

24.

$$3\frac{1}{2} + a = 5\frac{3}{8}$$
$$3\frac{1}{2} - 3\frac{1}{2} + a = 5\frac{3}{8} - 3\frac{1}{2}$$
$$a = 5\frac{3}{8} - 3\frac{4}{8}$$
$$a = 4\frac{11}{8} - 3\frac{4}{8}$$
$$a = \mathbf{1\frac{7}{8}}$$

check:
$$3\frac{1}{2} + 1\frac{7}{8} = 5\frac{3}{8}$$
$$3\frac{4}{8} + 1\frac{7}{8} = 5\frac{3}{8}$$
$$4\frac{11}{8} = 5\frac{3}{8}$$
$$5\frac{3}{8} = 5\frac{3}{8}$$

25.
$$3^2 + 5[6 - (10 - 2^3)]$$
$$9 + 5[6 - (10 - 8)]$$
$$9 + 5[6 - 2]$$
$$9 + 5[4]$$
$$9 + 20$$
$$\mathbf{29}$$

26.
$$\sqrt{2^2 \cdot 3^4 \cdot 5^2} = 2 \cdot 3^2 \cdot 5$$
$$= 2 \cdot 9 \cdot 5 = \mathbf{90}$$

27.
$$2\frac{2}{3} \times 4\frac{1}{2} \div 6$$
$$= \left(\frac{\overset{4}{\cancel{8}}}{\underset{1}{\cancel{3}}} \times \frac{\overset{3}{\cancel{9}}}{\underset{1}{\cancel{2}}}\right) \div 6 = 12 \div 6 = \mathbf{2} \text{ or } \mathbf{\frac{2}{1}}$$

28.
$$(3.5)^2 - (5 - 3.4)$$
$$= (12.25) - (1.6) = \mathbf{10.65}$$

29. (a) **108**

(b) **−75**

(c) **−20**

(d) **−5**

30. (a)
$$(-3) + |-4| - (-5)$$
$$(-3) + (4) + [-(-5)]$$
$$(-3) + (4) + [5]$$
$$\mathbf{6}$$

(b)
$$(-18) - (+20) + (-7)$$
$$(-18) + [-(+20)] + (-7)$$
$$(-18) + [-20] + [-7]$$
$$\mathbf{-45}$$

Lesson 82, Warm-Up

a. **18**

b. $\mathbf{\frac{1}{100}}$ **or 0.01**

c. **9**

d. **31.4 ft**

e. **1.5 m**

f. **32**

g. **60**

Problem Solving

$$\mathbf{24 + 4W = 60 + 2W}$$
$$24 + 4W - 2W = 60 + 2W - 2W$$
$$24 - 24 + 2W = 60 - 24$$
$$2W \div 2 = 36 \div 2$$
$$W = \mathbf{18 \text{ ounces}}$$

Lesson 82, Lesson Practice

a.
$$A = \pi r^2$$
$$A \approx 3.14(36 \text{ ft}^2)$$
$$A \approx \mathbf{113 \text{ ft}^2}$$

b.
$$A = \pi r^2$$
$$A \approx 3.14(16 \text{ cm}^2)$$
$$A \approx \mathbf{50.24 \text{ cm}^2}$$

c.
$$A = \pi r^2$$
$$A = \pi(16 \text{ cm}^2)$$
$$A = \mathbf{16\pi \text{ cm}^2}$$

d.
$$A = \pi r^2$$
$$A \approx \frac{22}{7}(16 \text{ cm})^2$$
$$A \approx \frac{352}{7} \text{ cm}^2$$
$$A \approx \mathbf{50\frac{2}{7} \text{ cm}^2}$$

Lesson 82, Mixed Practice

1.
$$(2 \text{ ft})(4 \text{ ft}) = 8 \text{ ft}^2$$
$$\frac{8 \text{ ft}^2}{1 \cancel{\text{layer}}}(1 \text{ ft})(2.5 \cancel{\text{layers}})$$
$$= \mathbf{20 \text{ ft}^3}$$

2.

$$\begin{aligned} 6'\,3'' &= 75'' \\ 6'\,5'' &= 77'' \\ 5'\,11'' &= 71'' \\ 6'\,2'' &= 74'' \\ +\ 6'\,1'' &= 73'' \\ \hline & \ 370'' \end{aligned}$$

$$5\overline{)370''} = 74''$$

$74'' =$ **6' 2"**

3.

	Ratio	Actual Count
Patients	20	P
Doctors	1	48

$$\frac{20}{1} = \frac{P}{48}$$

$P =$ **960 patients**

4. $2.54 \text{ ~~centimeters~~} \cdot \dfrac{1 \text{ meter}}{100 \text{ ~~centimeters~~}}$

$=$ **0.0254 meter**

5. (a)

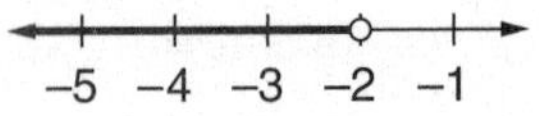

(b)

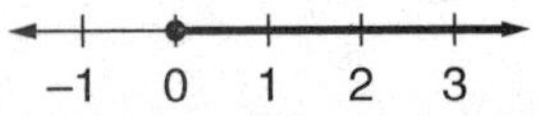

6.

	Case 1	Case 2
Beats	225	b
Minutes	3	5

$$\frac{225}{3} = \frac{b}{5}$$

$3b = 1125$

$b =$ **375 times**

7.

25 children

$\frac{2}{5}$ were boys. { 5 children | 5 children }

$\frac{3}{5}$ were girls (15). { 5 children | 5 children | 5 children }

(a) $15 \div 3 = 5$

5(5 children) = **25 children**

(b) $\dfrac{\text{girls}}{\text{boys}} = \dfrac{15}{10} = \mathbf{\dfrac{3}{2}}$

8. $x^2 - y^2 = 5^2 - 3^2 = 25 - 9 = 16$

$(x + y)(x - y) = (5 + 3)(5 - 3)$

$= (8)(2) = 16, 16 = 16$

$x^2 - y^2 \ \textcircled{=}\ (x + y)(x - y)$

9. Percent unshaded: 25% + 23% = 48%

Percent shaded: 100% − 48% = **52%**

10. $a \ \textcircled{<}\ b$

11. (a) $C \approx 3.14\,(14 \text{ cm})$

$C \approx$ **43.96 cm**

(b) $C \approx \dfrac{22}{\cancel{7}_1}(\cancel{14}^{\,2} \text{ cm})$

$C \approx$ **44 cm**

12. (a) $A = \pi r^2$

$A \approx 3.14\,(49 \text{ cm}^2)$

$A \approx$ **153.86 cm²**

(b) $A = \pi r^2$

$A \approx \dfrac{22}{\cancel{7}_1}(\cancel{49}^{\,7} \text{ cm}^2)$

$A \approx$ **154 cm²**

13. (a) $1.6 = 1\dfrac{6}{10} = \mathbf{1\dfrac{3}{5}}$

(b) 1.6 × 100% = **160%**

(c) $1.6\% = \dfrac{1.6}{100} \cdot \dfrac{10}{10} = \dfrac{16}{1000} = \mathbf{\dfrac{2}{125}}$

(d) 1.6% = **0.016**

14. M **= 0.064 × $25**

$M =$ **$1.60**

15. (a) $\mathbf{1.2 \times 10^6}$

(b) $\mathbf{1.2 \times 10^{-4}}$

16.

	Percent	Actual Count
Correctly described	64	C
Incorrectly described	36	63
Total	100	T

$$\frac{64}{36} = \frac{C}{63}$$

$36C = 4032$

$C =$ **112 students**

17.

	Percent	Actual Count
Pages read	60	180
Pages left to read	40	P
Total	100	T

$$\frac{60}{40} = \frac{180}{P}$$

$60P = 7200$

$P =$ **120 pages**

18.

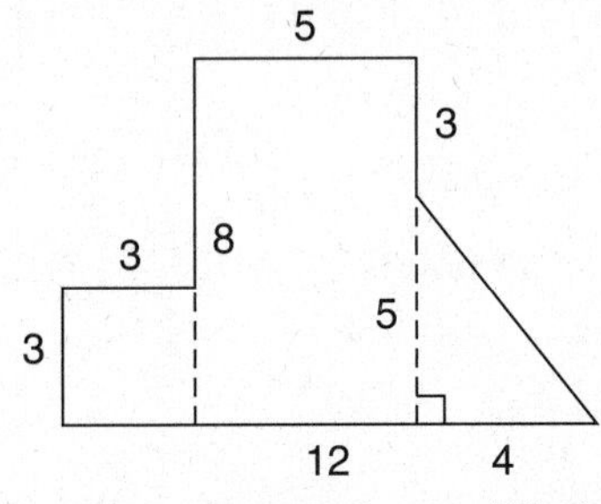

Area of rectangle	= (5 in.)(8 in.)	=	40 in.^{2}
Area of square	= (3 in.)(3 in.)	=	9 in.^{2}
+ Area of triangle	= $\frac{(5\text{ in.})(4\text{ in.})}{2}$	=	10 in.^{2}
Area of figure		=	**59 in.^{2}**

19.

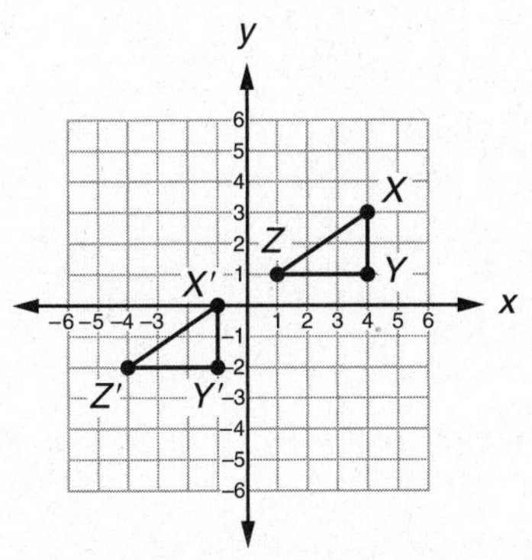

$X'(-1, 0), Y'(-1, -2), Z'(-4, -2)$

20. $\frac{240}{816} = \frac{\overset{1}{\cancel{2}} \cdot \overset{1}{\cancel{2}} \cdot \overset{1}{\cancel{2}} \cdot \overset{1}{\cancel{2}} \cdot \overset{1}{\cancel{3}} \cdot 5}{\underset{1}{\cancel{2}} \cdot \underset{1}{\cancel{2}} \cdot \underset{1}{\cancel{2}} \cdot \underset{1}{\cancel{2}} \cdot \underset{1}{\cancel{3}} \cdot 17}$

$= \mathbf{\frac{5}{17}}$

21. (a) **3 chords**

(b) **6 chords**

(c) **60°**

(d) **120°**

22. $\mathbf{1 \times 10^8}$

23. $\frac{3}{4}x = 36$

$\left(\frac{\overset{1}{\cancel{4}}}{\underset{1}{\cancel{3}}}\right)\frac{\overset{1}{\cancel{3}}}{\underset{1}{\cancel{4}}}x = \left(\frac{4}{\underset{1}{\cancel{3}}}\right)\overset{12}{\cancel{36}}$

$x = \mathbf{48}$

check: $\frac{3}{\underset{1}{\cancel{4}}}(\overset{12}{\cancel{48}}) = 36$

$3(12) = 36$

$36 = 36$

24. $3.2 + a = 3.46$

$3.2 - 3.2 + a = 3.46 - 3.2$

$a = \mathbf{0.26}$

check: $3.2 + 0.26 = 3.46$

$3.46 = 3.46$

25. $\frac{\sqrt{3^2 + 4^2}}{5} = \frac{\sqrt{9 + 16}}{5}$

$= \frac{\sqrt{25}}{5} = \frac{5}{5} = \mathbf{1}$

26. $(8 - 3)^2 - (3 - 8)^2$

$= (5)^2 - (-5)^2 = 25 - 25 = \mathbf{0}$

27. $3.5 \div (7 \div 0.2)$

$= 3.5 \div 35 = \mathbf{0.1}$

28. $4\frac{1}{2} + 2\frac{2}{3} - 3$

$= 4\frac{3}{6} + 2\frac{4}{6} - 2\frac{6}{6}$

$= 6\frac{7}{6} - 2\frac{6}{6} = \mathbf{4\frac{1}{6}}$

29. (a) **−6**

(b) **24**

30. (a) $(-3) + (-4) - (-2)$

$(-3) + (-4) + [-(-2)]$

$(-3) + (-4) + [2]$

−5

(b) $(-20) + (+30) - |-40|$

$(-20) + (+30) + [-|-40|]$

$(-20) + (+30) + [-40]$

−30

Lesson 83, Warm-Up

a. **−20**

b. **6,750,000**

c. **20**

d. **$18**

e. **0.5 g**

f. **64**

g. **150 mi**

Problem Solving

10 triangles

(one whole figure, four small triangles, three triangles composed of pairs of small triangles, and two triangles composed of triplets of small triangles)

Lesson 83, Lesson Practice

a. $(4.2 \times 1.4) \times (10^6 \times 10^3)$
$= \mathbf{5.88 \times 10^9}$

b. $(5 \times 3) \times (10^5 \times 10^7) = 15 \times 10^{12}$
$= \mathbf{1.5 \times 10^{13}}$

c. $(4 \times 2.1) \times (10^{-3} \times 10^{-7})$
$= \mathbf{8.4 \times 10^{-10}}$

d. $(6 \times 7) \times (10^{-2} \times 10^{-5}) = 42 \times 10^{-7}$
$= \mathbf{4.2 \times 10^{-6}}$

Lesson 83, Mixed Practice

1. $\dfrac{\$1.12}{16 \text{ ounces}} = \dfrac{\$0.07}{1 \text{ ounce}}$

$\dfrac{\$1.32}{24 \text{ ounces}} = \dfrac{\$0.055}{1 \text{ ounce}}$

$\$0.070$
$- \$0.055$
$0.015 more per ounce

2.

	Ratio	Actual Count
Good apples	5	G
Bad apples	2	B
Total	7	70

$\dfrac{5}{7} = \dfrac{G}{70}$

$7G = 350$

$G =$ **50 apples**

3. $15(82) + 5(90) = 1230 + 450 = 1680$

$20\overline{)1680}$ = **84**

4. $\$6(2.5) =$ **$15**

5. $24 \text{ ~~shillings~~} \cdot \dfrac{12 \text{ pence}}{1 \text{ ~~shilling~~}} =$ **288 pence**

6.
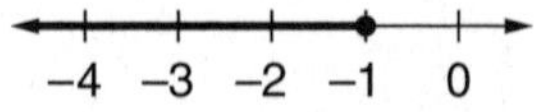

7.

	Case 1	Case 2
First number	5	20
Second number	12	n

$\dfrac{5}{12} = \dfrac{20}{n}$

$5n = 240$

$n =$ **48**

8. $4(1.5) + 5 = 6 + 5 = \mathbf{11}$

9. $\dfrac{30 \text{ points}}{5} = 6 \text{ points}$

$4(6 \text{ points}) =$ **24 points**

10. $x(x + y) \; \textcircled{=} \; x^2 + xy$

11. (a) $C = \pi(28 \text{ cm})$
$C = \mathbf{28\pi \text{ cm}}$

(b) $C \approx \dfrac{22}{\not7_{1}}(\not{28}^{4} \text{ cm})$

$C \approx$ **88 cm**

12. (a) $A = \pi r^2$
$A = \pi(196 \text{ cm}^2)$
$A = \mathbf{196\pi \text{ cm}^2}$

(b) $A = \pi r^2$
$A \approx \dfrac{22}{\not7_{1}}(\not{196}^{28} \text{ cm}^2)$

$A \approx \mathbf{616 \text{ cm}^2}$

13. (a) $(10 \text{ cm})(10 \text{ cm}) = 100 \text{ cm}^2$
$(100 \text{ cm}^2)(10 \text{ cm}) = \mathbf{1000 \text{ cm}^3}$

(b) $6(100 \text{ cm}^2) = \mathbf{600 \text{ cm}^2}$

14. (a) $250\% = \dfrac{250}{100} = 2\dfrac{50}{100} = \mathbf{2\dfrac{1}{2}}$

(b) $250\% = \dfrac{250}{100} = \mathbf{2.5}$

(c) $12\overline{)7.0000}$ = $0.5833\ldots = \mathbf{0.58\overline{3}}$

```
   0.5833...
12)7.0000
   6 0
   1 00
     96
      40
      36
       40
       36
        4
```

(d) $\dfrac{7}{12} \times 100\% = \dfrac{700\%}{12} = \mathbf{58\dfrac{1}{3}\%}$

$12\overline{)700}$ = $58\frac{4}{12} = 58\frac{1}{3}\%$

```
   58 4/12
12)700
   60
   100
    96
     4
```

15.
$\$8.50$
$\times\ 0.065$
$\$0.5525 \longrightarrow$ **$0.55**

16.

	Percent	Actual Count
Commercial time	P_C	12
Other	P_O	48
Total	100	60

$$\frac{P_C}{100} = \frac{12}{60}$$
$$60P_C = 1200$$
$$P_C = \mathbf{20\%}$$

17.

	Percent	Actual Count
Steam-powered	30	S
Not steam-powered	70	42
Total	100	T

$$\frac{70}{100} = \frac{42}{T}$$
$$70T = 4200$$
$$T = \textbf{60 boats}$$

18. $$\frac{420}{630} = \frac{2 \cdot \overset{1}{\cancel{2}} \cdot \overset{1}{\cancel{3}} \cdot \overset{1}{\cancel{5}} \cdot \overset{1}{\cancel{7}}}{\underset{1}{\cancel{2}} \cdot 3 \cdot \underset{1}{\cancel{3}} \cdot \underset{1}{\cancel{5}} \cdot \underset{1}{\cancel{7}}} = \mathbf{\frac{2}{3}}$$

19.

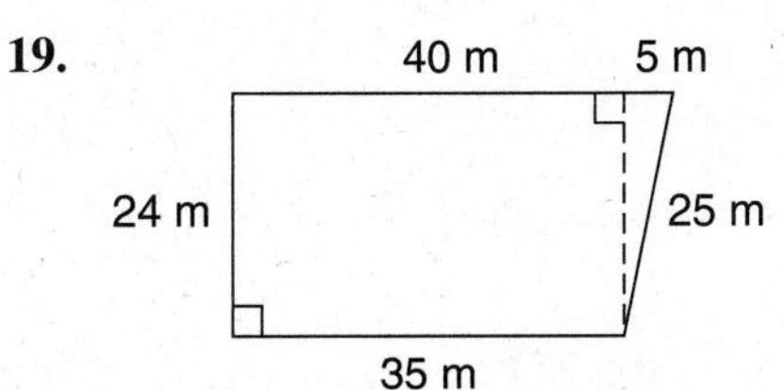

Area of rectangle $= (35 \text{ m})(24 \text{ m})$
$= 840 \text{ m}^2$

Area of triangle $= \frac{(24 \text{ m})(5 \text{ m})}{2}$
$= 60 \text{ m}^2$

Area of figure $= 840 \text{ m}^2 + 60 \text{ m}^2$
$= \mathbf{900 \text{ m}^2}$

20. (a) $m\angle ECD = 180° - (90° + 54°)$
$= 180° - 144° = \mathbf{36°}$

(b) $m\angle ECB = 180° - 36° = \mathbf{144°}$

(c) $m\angle ACB = m\angle ECD = \mathbf{36°}$

(d) $m\angle BAC = \frac{180° - 36°}{2} = \frac{144°}{2} = \mathbf{72°}$

21. **Multiply the "in" number by 2; then add 1 to find the "out" number.**

$21 - 1 = 20, \frac{20}{2} = \mathbf{10}$

$-5(2) + 1 = -10 + 1 = \mathbf{-9}$

22. (a) $(3 \times 6) \times (10^4 \times 10^5) = 18 \times 10^9$
$= \mathbf{1.8 \times 10^{10}}$

(b) $(1.2 \times 4) \times (10^{-3} \times 10^{-6})$
$= \mathbf{4.8 \times 10^{-9}}$

23.
$$b - 1\frac{2}{3} = 4\frac{1}{2}$$
$$b - 1\frac{2}{3} + 1\frac{2}{3} = 4\frac{1}{2} + 1\frac{2}{3}$$
$$b = 4\frac{3}{6} + 1\frac{4}{6}$$
$$b = 5\frac{7}{6}$$
$$b = \mathbf{6\frac{1}{6}}$$

check:
$$6\frac{1}{6} - 1\frac{2}{3} = 4\frac{1}{2}$$
$$6\frac{1}{6} - 1\frac{4}{6} = 4\frac{1}{2}$$
$$5\frac{7}{6} - 1\frac{4}{6} = 4\frac{1}{2}$$
$$4\frac{3}{6} = 4\frac{1}{2}$$
$$4\frac{1}{2} = 4\frac{1}{2}$$

24.
$$0.4y = 1.44$$
$$\frac{0.4y}{0.4} = \frac{1.44}{0.4}$$
$$y = \mathbf{3.6}$$

check:
$$0.4(3.6) = 1.44$$
$$1.44 = 1.44$$

25. $2^3 + 2^2 + 2^1 + 2^0 + 2^{-1}$
$= 8 + 4 + 2 + 1 + \frac{1}{2} = \mathbf{15\frac{1}{2}}$

26. $\frac{6}{10} \times 3\frac{1}{3} \div 2$
$= \left(\frac{6}{10} \times \frac{10}{3}\right) \div 2 = 2 \div 2 = \mathbf{1}$

27. (a) **4**

(b) **−60**

28.
$$\frac{5}{24} = \frac{25}{120}$$
$$-\frac{7}{60} = \frac{14}{120}$$
$$\mathbf{\frac{11}{120}}$$

29. (a) $(-3) + (-4) - (-5)$
$(-3) + (-4) + [-(-5)]$
$(-3) + (-4) + [5]$
$\mathbf{-2}$

(b) $(-15) - (+14) + (+10)$
$(-15) + [-(+14)] + (+10)$
$(-15) + [-14] + (+10)$
$\mathbf{-19}$

30.

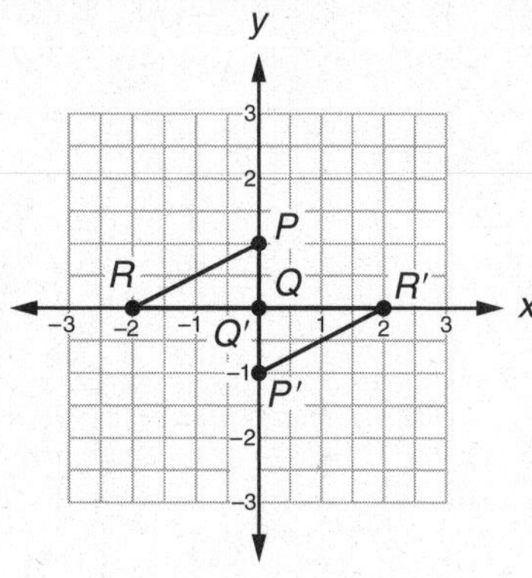

***P′*(0, −1), *Q′*(0, 0), *R′*(2, 0)**

Lesson 84, Warm-Up

a. **0**

b. **625**

c. **7**

d. **94.2 cm**

e. **15 mm**

f. **36**

g. **3**

Problem Solving

The left edge hangs over 4 in. more than the right edge, so the tablecloth should be shifted **2 in. to the right.**

The front edge hangs over 4 in. more than the back edge, so the tablecloth should be shifted **2 in. to the back.**

Lesson 84, Lesson Practice

a. **Binomial**

b. **Trinomial**

c. **Monomial**

d. **Binomial**

e. $3a + 2a^2 - a + a^2$
$2a^2 + a^2 + 3a - a$
$\mathbf{3a^2 + 2a}$

f. $5xy - x + xy - 2x$
$5xy + xy - x - 2x$
$\mathbf{6xy - 3x}$

g. $3 + x^2 + x - 5 + 2x^2$
$x^2 + 2x^2 + x + 3 - 5$
$\mathbf{3x^2 + x - 2}$

h. $3\pi + 1.4 - \pi + 2.8$
$3\pi - \pi + 1.4 + 2.8$
$\mathbf{2\pi + 4.2}$

Lesson 84, Mixed Practice

1. **18°F**

2. $2xy + xy - 3x + x$
$\mathbf{3xy - 2x}$

3. (a) 79°F − 68°F = **11°F**

(b) **Thursday**

(c)
68°F
72°F
70°F
76°F
+ 79°F
365°F

$5\overline{)365°F}$ = 73°F

73°F − 70°F = **3°F**

4. (a)
90
90
100
95
95
85
100
100
80
+ 100
935

$10\overline{)935.0}$ = **93.5**

(b) **95**

(c) **100**

(d) 100 − 80 = **20**

5.

	Ratio	Actual Count
Rowboats	3	*R*
Sailboats	7	*S*
Total	10	210

$\frac{7}{10} = \frac{S}{210}$

$10S = 1470$

$S =$ **147 sailboats**

6. **2nd**

7. $\frac{\$1.40}{4} = \frac{c}{10}$

$4c = \$14$

$c = \mathbf{\$3.50}$

8. $36 \div 3 = 12$

$5(12 \text{ members}) = \mathbf{60\ members}$

9. (a) $(5)^2 - 2(5) + 1 = 25 - 10 + 1$

$= \mathbf{16}$

(b) $(5 - 1)^2 = (4)^2 = \mathbf{16}$

10. $f \ \textcircled{=} \ g$

11. (a) $C \approx 3.14(6 \text{ in.})$

$C \approx \mathbf{18.84\ in.}$

(b) $A = \pi r^2$

$A \approx 3.14(9 \text{ in.}^2)$

$A \approx \mathbf{28.26\ in.^2}$

12. $\mathbf{4.8\ \cancel{meters}} \cdot \dfrac{\mathbf{100\ centimeters}}{\mathbf{1\ \cancel{meter}}}$

$= \mathbf{480\ centimeters}$

13.

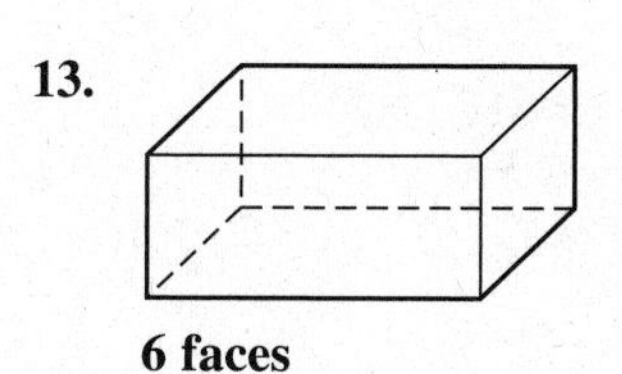

6 faces

14. (a) $1\frac{4}{5} = 1\frac{8}{10} = \mathbf{1.8}$

(b) $1.8 \times 100\% = \mathbf{180\%}$

(c) $1.8\% = \frac{1.8}{100} \cdot \frac{10}{10} = \frac{18}{1000} = \mathbf{\frac{9}{500}}$

(d) $1.8\% = \mathbf{0.018}$

15. $\mathbf{p = 0.3 \times \$18.00}$

$p = \mathbf{\$5.40}$

16. $\dfrac{12\frac{1}{2}}{100} = \dfrac{\frac{25}{2}}{\frac{100}{1}}$

$\dfrac{\frac{25}{2}}{\frac{100}{1}} \cdot \dfrac{\frac{1}{100}}{\frac{1}{100}} = \dfrac{\frac{25}{200}}{1} = \dfrac{25}{200}$

$= \mathbf{\frac{1}{8}}$

17.

	Percent	Actual Count
Flew the coop	40	36
Stayed	60	S
Total	100	T

$\frac{40}{100} = \frac{36}{T}$

$40T = 3600$

$T = \mathbf{90\ pigeons}$

18.

	Percent	Actual Count
3 feet tall or less	60	L
More than 3 feet tall	40	M
Total	100	300

$\frac{40}{100} = \frac{M}{300}$

$100M = 12{,}000$

$M = \mathbf{120\ saplings}$

19.

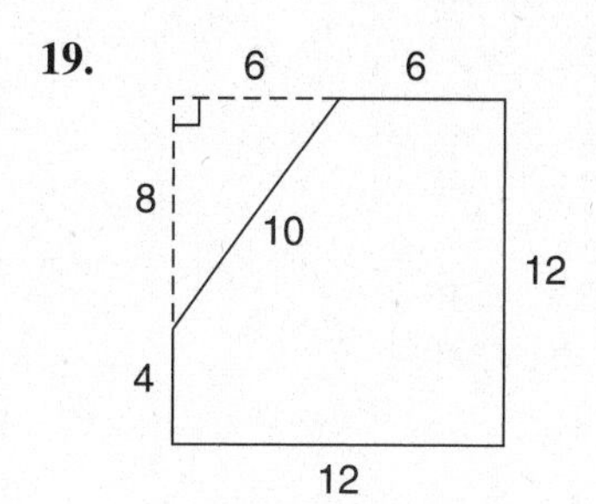

(a) Perimeter $= 12 \text{ in.} + 12 \text{ in.} + 4 \text{ in.}$

$+ 10 \text{ in.} + 6 \text{ in.}$

$= \mathbf{44\ in.}$

(b) Area of square $= (12 \text{ in.})(12 \text{ in.})$

$= 144 \text{ in.}^2$

Area of triangle $= \dfrac{(6 \text{ in.})(8 \text{ in.})}{2}$

$= 24 \text{ in.}^2$

Area of figure $= 144 \text{ in.}^2 - 24 \text{ in.}^2$

$= \mathbf{120\ in.^2}$

20. (a) $\frac{60°}{360°}$ $\mathbf{\frac{1}{6}}$

(b) $\frac{45°}{360°}$ $\mathbf{\frac{1}{8}}$

(c) $\frac{75°}{360°}$ $\mathbf{\frac{5}{24}}$

21. **To find a term in the sequence, double the preceding term and add 1.**

Note: **Other rule descriptions are possible, including "The value of the *n*th term is $2^n - 1$."**

$31 \times 2 + 1 = \mathbf{63}$

$63 \times 2 + 1 = \mathbf{127}$

$127 \times 2 + 1 = \mathbf{255}$

22. (a) $(1.5 \times 3) \times (10^{-3} \times 10^{6})$
$= \mathbf{4.5 \times 10^{3}}$

(b) $(3 \times 5) \times (10^{4} \times 10^{5}) = 15 \times 10^{9}$
$= \mathbf{1.5 \times 10^{10}}$

23. (a) 10^{6}

(b) 10^{-4}

24.
$$b - 4.75 = 5.2$$
$$b - 4.75 + 4.75 = 5.2 + 4.75$$
$$b = \mathbf{9.95}$$
check: $9.95 - 4.75 = 5.2$
$5.2 = 5.2$

25.
$$\frac{2}{3}y = 36$$
$$\left(\frac{\cancel{3}^{1}}{\cancel{2}_{1}}\right)\frac{\cancel{2}^{1}}{\cancel{3}_{1}}y = \left(\frac{3}{\cancel{2}_{1}}\right)\cancel{36}^{18}$$
$$y = \mathbf{54}$$
check: $\frac{2}{\cancel{3}_{1}}(\cancel{54}^{18}) = 36$
$2(18) = 36$
$36 = 36$

26. $\sqrt{5^{2} - 4^{2}} + 2^{3}$
$= \sqrt{25 - 16} + 8 = \sqrt{9} + 8$
$= 3 + 8 = \mathbf{11}$

27. 1 m = 1000 mm
1000 mm − 45 mm = **955 mm**

28. $0.9 \div 2.25 \times 24 = 0.4 \times 24 = \mathbf{9.6}$

29. (a) **4**

(b) **−30**

30. (a) $(+30) - (-50) - (+20)$
$(+30) + [-(-50)] + [-(+20)]$
$(+30) + [50] + [-20]$
60

(b) $(-3) - (-4) - (5)$
$(-3) + [-(-4)] + [-(5)]$
$(-3) + [4] + [-5]$
−4

LESSON 85, WARM-UP

a. **−72**

b. $\mathbf{8 \times 10^{9}}$

c. **120**

d. **$48**

e. **800 m**

f. **$3**

g. **10 m; 6.25 m²**

Problem Solving
$$n + 2n + 3n = 180$$
$$6n = 180$$
$$6n \div 6 = 180 \div 6$$
$$n = 30$$
30; 60; 90

LESSON 85, LESSON PRACTICE

a. $(-3) + (-3)(-3) - \frac{(-3)}{(+3)}$
$(-3) + 9 - (-1)$
$(-3) + 9 + [-(-1)]$
$(-3) + 9 + [1]$
7

b. $(-3) - [(-4) - (-5)(-6)]$
$(-3) - [(-4) - (+30)]$
$(-3) - [-34]$
$(-3) + 34$
31

c. $(-2)[(-3) - (-4)(-5)]$
$(-2)[(-3) - (+20)]$
$(-2)[-23]$
46

d. $(-5) - (-5)(-5) + |-5|$
$(-5) - (+25) + 5$
$(-5) + [-(+25)] + 5$
$(-5) + [-25] + 5$
−25

e. $y = 3(3) - 1$
$y = 9 - 1$
$y = \mathbf{8;}$
$y = 3(1) - 1$
$y = 3 - 1$
$y = \mathbf{2;}$
$y = 3(0) - 1$
$y = 0 - 1$
$y = \mathbf{-1}$

f. $y = \frac{1}{2}x;$

$y = \frac{1}{2}(6)$

$y = \mathbf{3;}$

$y = \frac{1}{2}(0)$

$y = \mathbf{0;}$

$4 = \frac{1}{2}x$

$\left(\frac{2}{1}\right)4 = \left(\frac{2}{1}\right)\left(\frac{1}{2}\right)x$

$\mathbf{8} = x$

g. $y = 8 - x$

$y = 8 - 7$

$y = \mathbf{1;}$

$y = 8 - 1$

$y = \mathbf{7;}$

$y = 8 - 4$

$y = \mathbf{4}$

h. $y = x^2$

x	*y*
1	**1**
2	**4**
3	**9**

Lesson 85, Mixed Practice

1. (a)

```
   70
   80
   90
   80
   70
   90
   75
   95
  100
+  90
-----
  840
```

$10\overline{)840}$ = **84**

(b) **85**

(c) **90**

(d) **30**

2.

	Ratio	Actual Count
Won	3	*W*
Lost	1	*L*
Total	4	24

$\frac{1}{4} = \frac{L}{24}$

$4L = 24$

$L = \mathbf{6\ games}$

3.

	Ratio	Actual Count
Dandelions	11	*D*
Marigolds	4	44

$\frac{11}{4} = \frac{D}{44}$

$4D = 484$

$D = \mathbf{121\ dandelions}$

4.

	Case 1	Case 2
Miles	2	*m*
Seconds	10	60

$\frac{2}{10} = \frac{m}{60}$

$10m = 120$

$m = \mathbf{12\ miles}$

5. **0.98 ~~liter~~ · $\frac{\text{1000 milliliters}}{\text{1 \sout{liter}}}$ = 980 milliliters**

6. 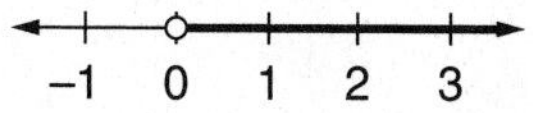

7. 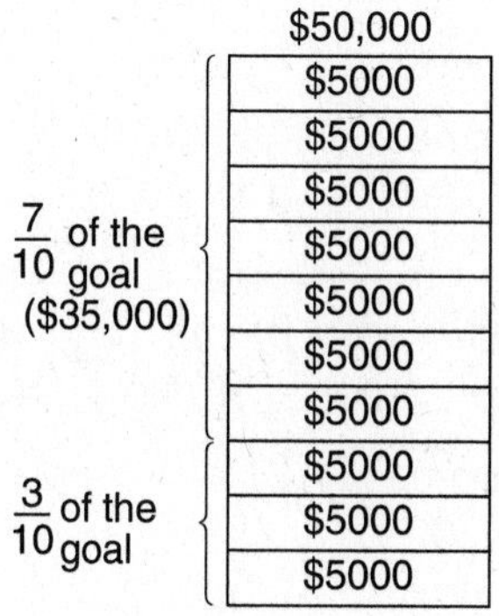

(a) $\frac{\$35{,}000}{7} = \$5000$

$10 \times \$5000 = \mathbf{\$50{,}000}$

(b) **30%**

8. **Insufficient information**

9. (a) $C \approx 3.14(8 \text{ m})$
$C \approx \mathbf{25.12 \text{ m}}$

(b) $A = \pi r^2$
$A \approx 3.14(16 \text{ m}^2)$
$A \approx \mathbf{50.24 \text{ m}^2}$

10. $1 - \frac{8}{20} - \frac{5}{20} = \mathbf{\frac{7}{20}}$

11.

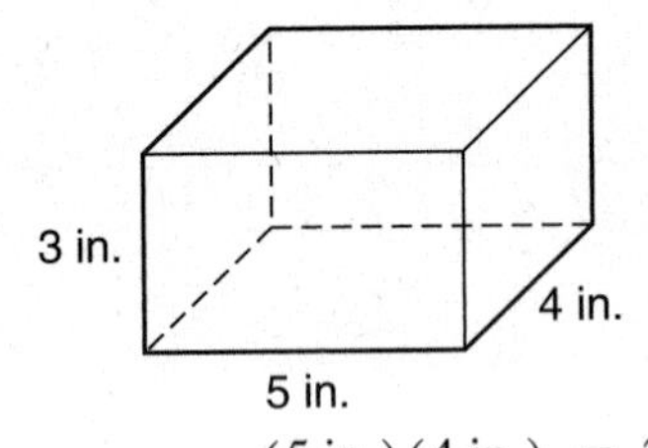

$(5 \text{ in.})(4 \text{ in.}) = 20 \text{ in.}^2$

$\frac{20 \text{ in.}^2}{1 \text{ layer}} \cdot 1 \text{ in.}(3 \text{ layers}) = \mathbf{60 \text{ in.}^3}$

12. One possibility:

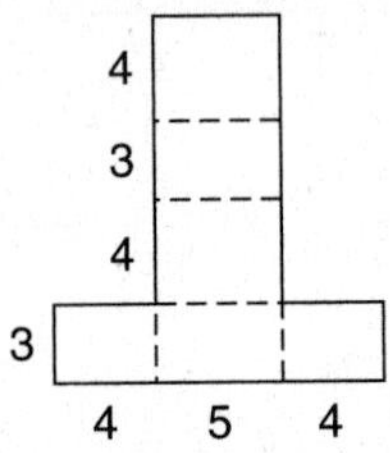

$2(15 \text{ in.}^2) + 2(20 \text{ in.}^2) + 2(12 \text{ in.}^2)$
$= 30 \text{ in.}^2 + 40 \text{ in.}^2 + 24 \text{ in.}^2$
$= \mathbf{94 \text{ in.}^2}$

13. (a) $\frac{1}{40} \cdot \frac{25}{25} = \frac{25}{1000} = \mathbf{0.025}$

(b) $\frac{1}{40} \times 100\% = \frac{100\%}{40} = \mathbf{2\frac{1}{2}\%}$

$$40\overline{)100} = 2\frac{20}{40} = 2\frac{1}{2}\%$$
$$\begin{array}{r} 80 \\ \hline 20 \end{array}$$

(c) $0.25\% = \frac{0.25}{100} \cdot \frac{100}{100} = \frac{25}{10{,}000} = \mathbf{\frac{1}{400}}$

(d) $0.25\% = \mathbf{0.0025}$

14.
$$\begin{array}{r} \$180{,}000 \\ \times \quad 0.06 \\ \hline \mathbf{\$10{,}800} \end{array}$$

15. (a) $\mathbf{2^3 \cdot 3^2 \cdot 5 \cdot 7^2}$

(b) **The exponents of the prime factors of 17,640 are not all even numbers.**

16. $\frac{8\frac{1}{3}}{100} = \frac{\frac{25}{3}}{\frac{100}{1}}$

$\frac{\frac{25}{3}}{\frac{100}{1}} \cdot \frac{\frac{1}{100}}{\frac{1}{100}} = \frac{\frac{25}{300}}{1} = \frac{25}{300}$

$= \mathbf{\frac{1}{12}}$

17.

	Percent	Actual Count
Correct	P_C	38
Incorrect	P_I	2
Total	100	40

$\frac{P_C}{100} = \frac{38}{40}$
$40P_C = 3800$
$P_C = \mathbf{95\%}$

18.

	Percent	Actual Count
Happy faces	35	H
Not happy faces	65	91
Total	100	T

$\frac{65}{100} = \frac{91}{T}$
$65T = 9100$
$T = \mathbf{140 \text{ children}}$

19.

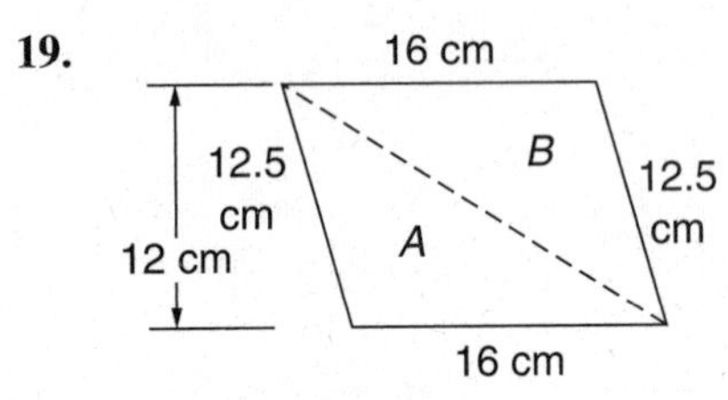

(a) **Parallelogram**

(b) Perimeter $= 12.5 \text{ cm} + 16 \text{ cm} + 12.5 \text{ cm} + 16 \text{ cm} = \mathbf{57 \text{ cm}}$

(c) Area $= (12 \text{ cm})(16 \text{ cm})$
$= \mathbf{192 \text{ cm}^2}$

(d)

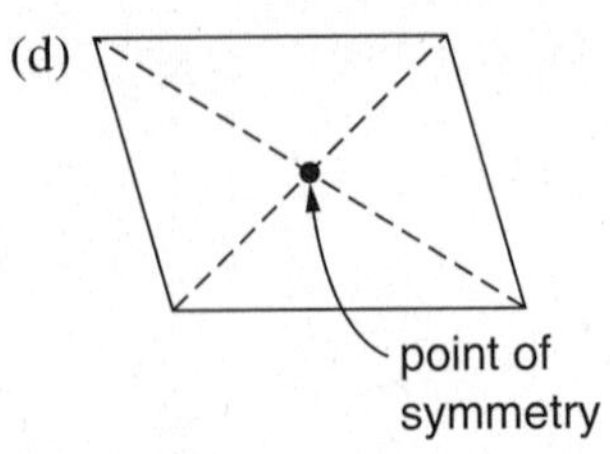

20. (a) $m\angle TOS = \mathbf{90°}$

(b) $m\angle QOT = \mathbf{180°}$

(c) $m\angle QOR = \frac{90°}{3} = \mathbf{30°}$

(d) $m\angle TOR = 90° + 60° = \mathbf{150°}$

21. $y = 2(5) - 1$
$y = 10 - 1$
$y = \mathbf{9};$
$y = 2(3) - 1$
$y = 6 - 1$
$y = \mathbf{5};$
$y = 2(1) - 1$
$y = 2 - 1$
$y = \mathbf{1}$

22. $30 \times 10^5 = (5 \times 10^{-3})(6 \times 10^8)$
$= (5 \times 10^8)(6 \times 10^{-3})$

23. $13.2 = 1.2w$
$\frac{13.2}{1.2} = \frac{1.2w}{1.2}$
$\mathbf{11} = w$
check: $13.2 = 1.2(11)$
$13.2 = 13.2$

24. $c + \frac{5}{6} = 1\frac{1}{4}$
$c + \frac{5}{6} - \frac{5}{6} = 1\frac{1}{4} - \frac{5}{6}$
$c = 1\frac{3}{12} - \frac{10}{12}$
$c = \frac{15}{12} - \frac{10}{12}$
$c = \mathbf{\frac{5}{12}}$
check: $\frac{5}{12} + \frac{5}{6} = 1\frac{1}{4}$
$\frac{5}{12} + \frac{10}{12} = 1\frac{1}{4}$
$\frac{15}{12} = 1\frac{1}{4}$
$1\frac{3}{12} = 1\frac{1}{4}$
$1\frac{1}{4} = 1\frac{1}{4}$

25. $3\{20 - [6^2 - 3(10 - 4)]\}$
$3\{20 - [36 - 3(6)]\}$
$3\{20 - [36 - 18]\}$
$3\{20 - [18]\}$
$3\{2\}$
6

26.
(60 min) (60 s)
~~3~~ 2 hr ~~15~~ 14 min 25 s
− 2 hr 45 min 30 s

~~3~~ 2 hr ~~15~~ ~~14~~ 74 min ~~25~~ 85 s
− 2 hr 45 min 30 s
29 min 55 s

27. $1 + 0.2 + 0.25 = \mathbf{1.45}$

28. (a) $(-2) + (-2)(+2) - \frac{(-2)}{(-2)}$
$(-2) + (-4) - 1$
$(-2) + (-4) + [-(1)]$
$(-2) - (-4) + (-1)$
$\mathbf{-7}$

(b) $(-3) - [(-2) - (+4)(-5)]$
$(-3) - [(-2) - (-20)]$
$(-3) - [(-2) + 20]$
$(-3) - [18]$
$(-3) + [-(18)]$
$(-3) + (-18)$
$\mathbf{-21}$

29. $x^2 + 6x - 2x - 12$
$\mathbf{x^2 + 4x - 12}$

30. (a) ***D*(1, −1)**

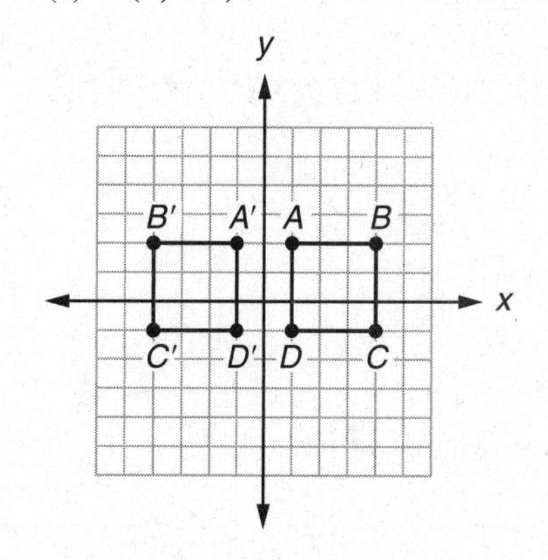

(b) ***A'*(−1, 2), *B'*(−4, 2), *C'*(−4, −1), *D'*(−1, −1)**

Lesson 86, Warm-Up

a. **−58**

b. $\mathbf{9 \times 10^{-6}}$

c. **8**

d. **$2.70**

e. **200 mL**

f. **$90**

g. **$214**

Problem Solving

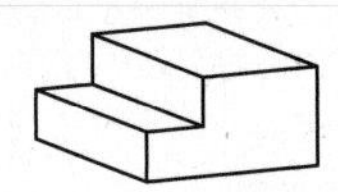

LESSON 86, LESSON PRACTICE

a. 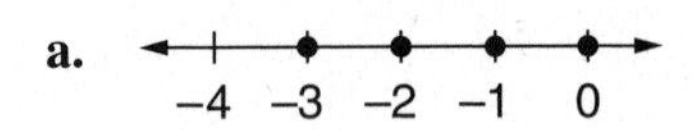

b. −1 0 1 2 3 4

c. **False**

d. **True**

LESSON 86, MIXED PRACTICE

1. $\frac{\$28.50}{3 \text{ ounces}} = \frac{\$9.50}{1 \text{ ounce}}$

$\frac{\$4.96}{8 \text{ ounces}} = \frac{\$0.62}{1 \text{ ounce}}$

$$\begin{array}{r} \$9.50 \\ -\ \$0.62 \\ \hline \$8.88 \end{array}$$

$8.88 more per ounce

2.

	Ratio	Actual Count
Rookies	2	R
Veterans	7	V
Total	9	252

$\frac{2}{9} = \frac{R}{252}$

$9R = 504$

$R = \mathbf{56 \text{ rookies}}$

3. (a) **213 lb**

(b) **213 lb**

(c)
$$\begin{array}{r} 197 \\ 213 \\ 246 \\ 205 \\ 238 \\ 213 \\ +\ 207 \\ \hline 1519 \end{array}$$

$1519 \div 7 = \mathbf{217 \text{ lb}}$

(d) **49 lb**

4. $12 \text{ bushels} \cdot \frac{4 \text{ pecks}}{1 \text{ bushel}} = \mathbf{48 \text{ pecks}}$

5. $\frac{468 \text{ miles}}{9 \text{ hours}} = \mathbf{52 \text{ miles per hour}}$

6. −1 0 1 2 3 4

7.

	Case 1	Case 2
First number	9	n
Second number	6	30

$\frac{9}{6} = \frac{n}{30}$

$6n = 270$

$n = \mathbf{45}$

8. (a) $1800 \div 10 = 180$

$9(180 \text{ employees}) = \mathbf{1620 \text{ employees}}$

(b) $\frac{1}{10} \times 100\% = \mathbf{10\%}$

9. $\sqrt{(5)^2 - 4(1)(4)}$

$= \sqrt{25 - 16} = \sqrt{9} = \mathbf{3}$

10. **Insufficient information**

11. (a) $C = \pi(24 \text{ in.})$

$C = \mathbf{24\pi \text{ in.}}$

(b) $A = \pi r^2$

$A = \pi(144 \text{ in.}^2)$

$A = \mathbf{144\pi \text{ in.}^2}$

12. (a) $\mathbf{10^5}$

(b) $\mathbf{10^{-3}}$

13.

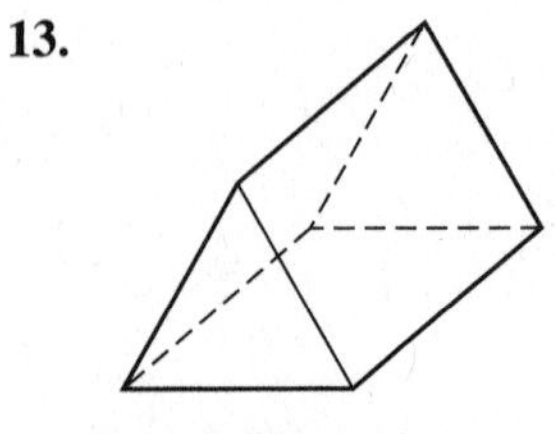

(a) **5 faces**

(b) **9 edges**

(c) **6 vertices**

14. (a) $0.9 = \mathbf{\frac{9}{10}}$

(b) $0.9 = \mathbf{90\%}$

(c) $\frac{11}{12} = \mathbf{0.91\overline{6}}$

$$12\overline{)11.0000\ldots} = 0.9166\ldots = 0.91\overline{6}$$

```
   0.9166
12)11.0000 . . .  = 0.916̄
   10 8
      20
      12
       80
       72
        80
        72
         8
```

(d) $\frac{11}{12} \times 100\% = \frac{1100\%}{12} = \mathbf{91\frac{2}{3}\%}$

```
   91 8/12 = 91 2/3%
12)1100
   108
     20
     12
      8
```

15. **North**

16.

	Percent	Actual Count
Sale price	60	\$24
Regular price	100	R

$\frac{60}{100} = \frac{\$24}{R}$

$60R = \$2400$

$R = \mathbf{\$40}$

17.

	Percent	Actual Count
Sprouted seeds	75	48
Unsprouted seeds	25	U
Total	100	T

$\frac{75}{25} = \frac{48}{U}$

$75U = 1200$

$U = \mathbf{16\ seeds}$

18. $\mathbf{30 = W_P \times 20}$

$\frac{30}{20} = \frac{W_P \times 20}{20}$

$\frac{3}{2} = W_P$

$W_P = \frac{3}{2} \times 100\% = \mathbf{150\%}$

19.

15 mm
20 mm
25 mm
15 mm
30 mm

(a) **Trapezoid**

(b) Perimeter = 15 mm + 25 mm + 30 mm + 20 mm = **90 mm**

(c) Area of rectangle = (15 mm)(20 mm)
= 300 mm^2

Area of triangle = $\frac{(15\text{ mm})(20\text{ mm})}{2}$
= 150 mm^2

Area of figure = 300 mm^2 + 150 mm^2
= **450 mm^2**

20. (a) **120°**

(b) **165°**

(c) 165° − 30° = **135°**

21.
$y = 3(4) + 1$
$y = 12 + 1$
$y = \mathbf{13;}$
$y = 3(7) + 1$
$y = 21 + 1$
$y = \mathbf{22;}$
$y = 3(0) + 1$
$y = 0 + 1$
$y = \mathbf{1}$

22. (a) $(1.2 \times 1.2) \times (10^5 \times 10^{-8})$
$= \mathbf{1.44 \times 10^{-3}}$

(b) $(6 \times 7) \times (10^{-3} \times 10^{-4})$
$= 42 \times 10^{-7}$
$= \mathbf{4.2 \times 10^{-6}}$

23.
$56 = \frac{7}{8}w$

$\left(\frac{8}{7}\right)56 = \left(\frac{8}{7}\right)\frac{7}{8}w$

$\mathbf{64} = w$

check: $56 = \frac{7}{8}(64)$
$56 = 7(8)$
$56 = 56$

24.
$4.8 + c = 7.34$
$4.8 - 4.8 + c = 7.34 - 4.8$
$c = \mathbf{2.54}$

check: $4.8 + 2.54 = 7.34$
$7.34 = 7.34$

25. $\sqrt{10^2 - 6^2} - \sqrt{10^2 - 8^2}$
$\sqrt{100 - 36} - \sqrt{100 - 64}$
$\sqrt{64} - \sqrt{36}$
$8 - 6$
$\mathbf{2}$

26.
```
   5 lb   9 oz
 + 4 lb   7 oz
   9 lb  16 oz
```
16 oz = 1 lb
1 lb + 9 lb = **10 lb**

27. $1.4 \div 3.5 \times 1000$
$= 0.4 \times 1000 = \mathbf{400}$

28. (a) $(-4)(-5) - (-4)(+3)$
$+20 - (-12)$
$20 + [-(-12)]$
$+20 + 12$
$\mathbf{32}$

(b) $(-2)[(-3) - (-4)(+5)]$
$(-2)[(-3) - (-20)]$
$(-2)[(-3) + (20)]$
$(-2)[+17]$
$\mathbf{-34}$

29. $x^2 + 3xy + 2x^2 - xy$
$x^2 + 2x^2 + 3xy - xy$
$\mathbf{3x^2 + 2xy}$

30. $\mathbf{3 \cdot 3 \cdot x \cdot y \cdot y}$

Lesson 87, Warm-Up

a. **−30**

b. $\mathbf{8 \times 10^2}$

c. **40**

d. **62.8 ft**

e. **0.75 kg**

f. **$125**

g. $\mathbf{2\frac{1}{2}}$

Problem Solving

$\mathbf{3M + 250 = M + 1000}$
$3M - M + 250 = M - M + 1000$
$2M + 250 - 250 = 1000 - 250$
$2M \div 2 = 750 \div 2$
$M = \mathbf{375\ g}$

Lesson 87, Lesson Practice

a. $(-3x)(-2xy)$
$= (-3) \cdot x \cdot (-2) \cdot x \cdot y$
$= (-3)(-2) \cdot x \cdot x \cdot y$
$= \mathbf{6x^2y}$

b. $3x^2(xy^3)$
$= (3) \cdot x \cdot x \cdot x \cdot y \cdot y \cdot y$
$= \mathbf{3x^3y^3}$

c. $(2a^2b)(-3ab^2)$
$= (2) \cdot a \cdot a \cdot b \cdot (-3) \cdot a \cdot b \cdot b$
$= (2)(-3) \cdot a \cdot a \cdot a \cdot b \cdot b \cdot b$
$= \mathbf{-6a^3b^3}$

d. $(-5x^2y)(-4x)$
$= (-5) \cdot x \cdot x \cdot y \cdot (-4) \cdot x$
$= (-5)(-4) \cdot x \cdot x \cdot x \cdot y$
$= \mathbf{20x^3y}$

e. $(-xy^2)(xy)(2y)$
$= (-1) \cdot x \cdot y \cdot y \cdot x \cdot y \cdot (2) \cdot y$
$= (-1)(2) \cdot x \cdot x \cdot y \cdot y \cdot y \cdot y$
$= \mathbf{-2x^2y^4}$

f. $(-3m)(-2mn)(m^2n)$
$= (-3) \cdot m \cdot (-2) \cdot m \cdot n \cdot m \cdot m \cdot n$
$= (-3)(-2) \cdot m \cdot m \cdot m \cdot m \cdot n \cdot n$
$= \mathbf{6m^4n^2}$

g. $(4wy)(3wx)(-w^2)(x^2y)$
$= (4) \cdot w \cdot y \cdot (3) \cdot w \cdot x \cdot (-1) \cdot w \cdot w \cdot x \cdot x \cdot y$
$= (4)(3)(-1) \cdot w \cdot w \cdot w \cdot w \cdot x \cdot x \cdot x \cdot y \cdot y$
$= \mathbf{-12w^4x^3y^2}$

h. $5d(-2df)(-3d^2fg)$
$= (5) \cdot d \cdot (-2) \cdot d \cdot f \cdot (-3) \cdot d \cdot d \cdot f \cdot g$
$= (5)(-2)(-3) \cdot d \cdot d \cdot d \cdot d \cdot f \cdot f \cdot g$
$= \mathbf{30d^4f^2g}$

Lesson 87, Mixed Practice

1. $2.5\ \text{hours} \cdot \dfrac{450\ \text{miles}}{1\ \text{hour}}$
$= \mathbf{1125\ miles}$

2. $\mathbf{12.5\ centimeters \cdot \dfrac{1\ meter}{100\ centimeters}}$
$= \mathbf{0.125\ meter}$

3.

	Ratio	Actual Count
Girls	4	240
Boys	3	180
Total	7	420

$\frac{\text{boys}}{\text{girls}} = \mathbf{\frac{3}{4}}$

4.

$$\begin{aligned} 18'\ 3'' &= 219'' \\ 17'\ 10'' &= 214'' \\ +\ 17'\ 11'' &= 215'' \\ & \overline{\ \ 648''} \end{aligned}$$

$$3\overline{)648''} = 216''$$

$216'' = \mathbf{18'}$

5. $\frac{468 \text{ miles}}{18 \text{ gallons}} = \mathbf{26 \text{ miles per gallon}}$

6.

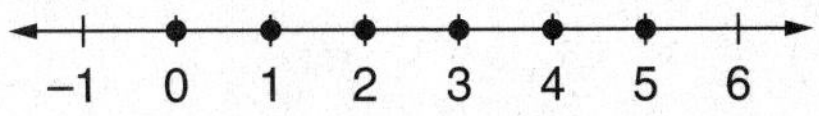

7.

	Case 1	Case 2
Yards	100	1500
Feet	36	*f*

$$\frac{100}{36} = \frac{1500}{f}$$
$$100f = 54{,}000$$
$$f = \mathbf{540 \text{ feet}}$$

8. (a) $m\angle a = 180° - 105° = \mathbf{75°}$

(b) $m\angle b = 180° - 75° = \mathbf{105°}$

(c) $m\angle c = 180° - 105° = \mathbf{75°}$

(d) $m\angle d = m\angle a = \mathbf{75°}$

(e) $m\angle e = m\angle b = \mathbf{105°}$

9.
$$y = 3(-4) - 1$$
$$y = -12 - 1$$
$$y = \mathbf{-13}$$

10. (a) $C \approx \frac{22}{\cancel{7}_1}(\cancel{140}^{20} \text{ mm})$

$C \approx \mathbf{440 \text{ mm}}$

(b) $A = \pi r^2$

$A \approx \frac{22}{\cancel{7}_1}(\cancel{4900}^{700} \text{ mm}^2)$

$A \approx \mathbf{15{,}400 \text{ mm}^2}$

11.

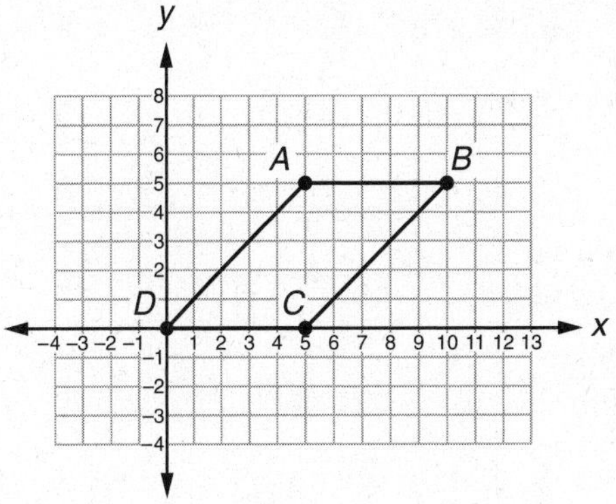

(a) Area $= (5 \text{ units})(5 \text{ units})$
$= \mathbf{25 \text{ units}^2}$

(b) $m\angle A = \mathbf{135°}$
$m\angle B = \mathbf{45°}$
$m\angle C = \mathbf{135°}$
$m\angle D = \mathbf{45°}$

12. Bottom rectangular prism: (6 cubes) (3 cubes)
= 18 cubes

$\frac{18 \text{ cubes}}{1 \cancel{\text{layer}}} \cdot 2 \cancel{\text{layers}} = 36 \text{ cubes}$

36 in.^{3}

Top rectangular prism: (2 cubes) (3 cubes)
= 6 cubes

$\frac{6 \text{ cubes}}{1 \cancel{\text{layer}}} \cdot 2 \cancel{\text{layers}} = 12 \text{ cubes, } 12 \text{ in.}^3$

$36 \text{ in.}^3 + 12 \text{ in.}^3 = \mathbf{48 \text{ in.}^3}$

13. (a) $12\frac{1}{2}\% = \frac{12\frac{1}{2}}{100} = \frac{\frac{25}{2}}{100}$

$= \frac{\frac{1}{100}}{\frac{1}{100}} = \frac{25}{200} = \mathbf{\frac{1}{8}}$

(b) $12\frac{1}{2}\% = 12.5\% = \mathbf{0.125}$

(c) $\frac{7}{8} = \mathbf{0.875}$

$$\begin{array}{r} 0.875 \\ 8\overline{)7.000} \\ \underline{6\,4} \\ 60 \\ \underline{56} \\ 40 \\ \underline{40} \\ 0 \end{array}$$

(d) $\frac{7}{8} \times 100\% = \frac{700\%}{8} = \mathbf{87\frac{1}{2}\%}$

$$\begin{array}{r} 87\,\frac{4}{8} = 87\tfrac{1}{2}\% \\ 8\overline{)700} \\ \underline{64} \\ 60 \\ \underline{56} \\ 4 \end{array}$$

14. $\mathbf{W_N = \frac{1}{4} \times 4}$

$W_N = \mathbf{1}$

15.

	Percent	Actual Count
Sale price	80	$24
Regular price	100	P

$$\frac{80}{100} = \frac{\$24}{P}$$
$$80P = \$2400$$
$$P = \mathbf{\$30}$$

16.

	Percent	Actual Count
Meters ran	60	M
Meters left	40	2000
Total	100	T

$$\frac{40}{100} = \frac{2000}{T}$$
$$40T = 200{,}000$$
$$T = \mathbf{5000\ meters}$$

17. $\mathbf{100 = W_P \times 80}$

$$\frac{100}{80} = \frac{W_P \times 80}{80}$$
$$\frac{5}{4} = W_P$$
$$W_P = \frac{5}{4} \times 100\% = \mathbf{125\%}$$

18.

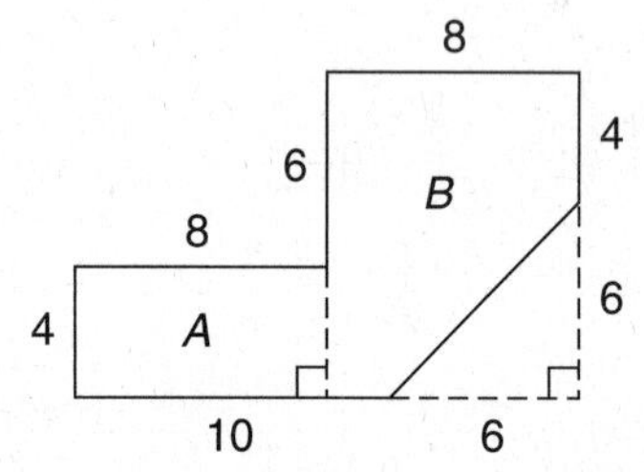

Area of rectangle A = (4 cm)(8 cm) = 32 cm^2

Area of rectangle B = (8 cm)(10 cm) = 80 cm^2

Area of $A + B$ = 32 cm^2 + 80 cm^2 = 112 cm^2

Area of triangle $= \frac{(6\text{ cm})(6\text{ cm})}{2}$ = 18 cm^2

112 cm^2 − 18 cm^2 = **94 cm^2**

19. (a) $m\angle AOB = \frac{90°}{3} = \mathbf{30°}$

(b) $m\angle AOC = 30° + 30° = \mathbf{60°}$

(c) $m\angle EOC = 90° + 30° = \mathbf{120°}$

(d) $\mathbf{\angle COA}$ or $\mathbf{\angle AOC}$

20.
$$\frac{66\frac{2}{3}}{100} = \frac{\frac{200}{3}}{\frac{100}{1}}$$
$$\frac{\frac{200}{3}}{\frac{100}{1}} \cdot \frac{\frac{1}{100}}{\frac{1}{100}} = \frac{\frac{200}{300}}{1} = \frac{200}{300}$$
$$= \mathbf{\frac{2}{3}}$$

21. $y = \frac{24}{3}$

$y = \mathbf{8;}$

$y = \frac{24}{4}$

$y = \mathbf{6;}$

$y = \frac{24}{12}$

$y = \mathbf{2}$

22. (a) $(4 \times 2.1) \times (10^{-5} \times 10^{-7}) = \mathbf{8.4 \times 10^{-12}}$

(b) $(4 \times 6) \times (10^5 \times 10^7) = 24 \times (10^{12}) = \mathbf{2.4 \times 10^{13}}$

23.
$$d - 8.47 = 9.1$$
$$d - 8.47 + 8.47 = 9.1 + 8.47$$
$$d = \mathbf{17.57}$$

check: $17.57 - 8.47 = 9.1$

$9.1 = 9.1$

24.
$$0.25m = 3.6$$
$$\frac{0.25m}{0.25} = \frac{3.6}{0.25}$$
$$m = \mathbf{14.4}$$

check: $0.25(14.4) = 3.6$

$3.6 = 3.6$

25.
$$\frac{3 + 5.2 - 1}{4 - 3 + 2}$$
$$\frac{8.2 - 1}{1 + 2}$$
$$\frac{7.2}{3}$$
$$\mathbf{2.4}$$

26. 1 kg = 1000 g, 1000 g − 75 g = **925 g**

27. 3.7 + 2.625 + 15 = **21.325**

28. (a) $(-5) - (-2)[(-3) - (+4)]$

$(-5) - (-2)[-7]$

$(-5) - (+14)$

$(-5) + [-(+14)]$

$(-5) + [-14]$

$\mathbf{-19}$

(b) $\dfrac{(-3) + (-3)(+4)}{(+3) + (-4)}$

$\dfrac{(-3) + (-12)}{-1}$

$\dfrac{-15}{-1}$

$\mathbf{15}$

29. (a) $(3x)(4y)$

$= (3) \cdot x \cdot (4) \cdot y$

$= (3)(4) \cdot x \cdot y = \mathbf{12xy}$

(b) $(6m)(-4m^2n)(-mnp)$

$= (6) \cdot m \cdot (-4) \cdot m \cdot m \cdot n \cdot (-1) \cdot m \cdot n \cdot p$

$= (6)(-4)(-1) \cdot m \cdot m \cdot m \cdot m \cdot n \cdot n \cdot p$

$= \mathbf{24m^4n^2p}$

30. $3ab + a - ab - 2ab + a$

$3ab - ab - 2ab + a + a$

$\mathbf{2a}$

Lesson 88, Warm-Up

a. **−75**

b. $\mathbf{3 \times 10^9}$

c. **12**

d. **$15**

e. **25.4 mm**

f. **$2.50**

g. **1250 mi**

Problem Solving

Vincent offered 10 skillings and 2 ore, which is 42 ore.

André wanted 2 gilders, which is 12 skillings, which is 48 ore.

André wanted 6 ore more than Vincent's offer. One skilling is 4 ore, so André wanted **1 skilling, 2 ore** more than Vincent's offer.

Lesson 88, Lesson Practice

a. $5\text{ yd} \cdot \dfrac{3\text{ ft}}{1\text{ yd}} \cdot \dfrac{12\text{ in.}}{1\text{ ft}} = \mathbf{180\text{ in.}}$

b. $1\frac{1}{2}\text{ hr} \cdot \dfrac{60\text{ min}}{1\text{ hr}} \cdot \dfrac{60\text{ s}}{1\text{ min}}$

$= \mathbf{5400\text{ s}}$

c. $15\text{ yd}^2 \cdot \dfrac{3\text{ ft}}{1\text{ yd}} \cdot \dfrac{3\text{ ft}}{1\text{ yd}}$

$= \mathbf{135\text{ ft}^2}$

d. $20\text{ cm}^2 \cdot \dfrac{10\text{ mm}}{1\text{ cm}} \cdot \dfrac{10\text{ mm}}{1\text{ cm}}$

$= \mathbf{2000\text{ mm}^2}$

Lesson 88, Mixed Practice

1. $\$6(3.25) = \mathbf{\$19.50}$

2. $4(93) = 372$

$10(84) = 840$

$840 - 372 = 468$

$468 \div 6 = \mathbf{78}$

3. $6\text{ ft}^2 \cdot \dfrac{12\text{ in.}}{1\text{ ft}} \cdot \dfrac{12\text{ in.}}{1\text{ ft}} = \mathbf{864\text{ in.}^2}$

4.

	Ratio	Actual Count
Woodwinds	3	15
Brass instruments	2	B

$\dfrac{3}{2} = \dfrac{15}{B}$

$3B = 30$

$B =$ **10 brass instruments**

5. 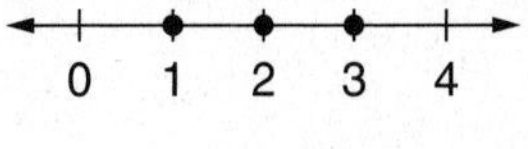

6.

	Case 1	Case 2
Artichokes	8	36
Price	$2	p

$\dfrac{8}{\$2} = \dfrac{36}{p}$

$8p = \$72$

$p = \mathbf{\$9}$

7.

	27 lights
$\frac{2}{3}$ were on (18).	9 lights
	9 lights
$\frac{1}{3}$ were off.	9 lights

(a) $18 \div 2 = 9,\ 1(9 \text{ lights}) = \mathbf{9\ lights}$

(b) $\frac{2}{3} \times 100\% = \mathbf{66\frac{2}{3}\%}$

8. $(5) - [(3) - (5 - 3)]$
$= 5 - [3 - 2] = 5 - [1] = \mathbf{4}$

9. $x \textcircled{<} y$

10. (a) $C = \pi(60 \text{ ft})$
$C \approx 3.14(60 \text{ ft})$
$C \approx \mathbf{188.4\ ft}$

(b) $A = \pi r^2$
$A \approx 3.14(900 \text{ ft}^2)$
$A \approx \mathbf{2826\ ft^2}$

11. $1 - \left(\frac{4}{12} + \frac{5}{12}\right) = \frac{3}{12} = \frac{1}{4}$

$\frac{1}{4} \times 100\% = \mathbf{25\%}$

12.

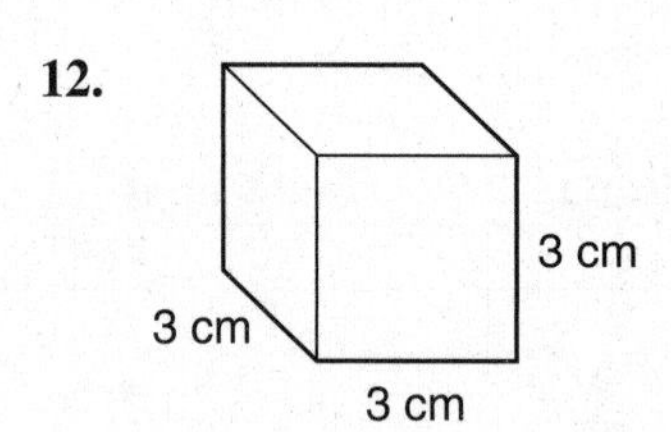

(a) $(3 \text{ cm})(3 \text{ cm}) = 9 \text{ cm}^2$

$\frac{9 \text{ cm}^2}{1 \cancel{\text{layer}}} \cdot (1 \text{ cm})(3 \cancel{\text{layers}}) = \mathbf{27\ cm^3}$

(b) $6(9 \text{ cm}^2) = \mathbf{54\ cm^2}$

13. $2x + 3y - 5 + x - y - 1$
$2x + x + 3y - y - 5 - 1$
$\mathbf{3x + 2y - 6}$

14. $x^2 + 2x - x - 2$
$\mathbf{x^2 + x - 2}$

15. (a) $0.125 = \frac{125}{100} = \mathbf{\frac{1}{8}}$

(b) $0.125 \times 100\% = 12.5\%$ or $\mathbf{12\frac{1}{2}\%}$

(c) $\frac{3}{8} = \mathbf{0.375}$

$$\begin{array}{r} 0.375 \\ 8\overline{)3.000} \\ \underline{2\,4} \\ 60 \\ \underline{56} \\ 40 \\ \underline{40} \\ 0 \end{array}$$

(d) $\frac{3}{8} \times 100\% = \frac{300\%}{8} = \mathbf{37\frac{1}{2}\%}$

$$\begin{array}{r} 37\frac{4}{8} = 37\frac{1}{2}\% \\ 8\overline{)300} \\ \underline{24} \\ 60 \\ \underline{56} \\ 4 \end{array}$$

16. $\frac{60}{1\frac{1}{4}} = \frac{\frac{60}{1}}{\frac{5}{4}}$

$\frac{\frac{60}{1}}{\frac{5}{4}} \cdot \frac{\frac{4}{5}}{\frac{4}{5}} = \frac{\frac{240}{5}}{1} = \frac{240}{5}$

$= \mathbf{48}$

17.

	Percent	Actual Count
Sale price	P_S	\$18
Regular price	100	\$24

$\frac{P_S}{100} = \frac{\$18}{\$24}$
$24P_S = 1800$
$P_S = \mathbf{75\%}$

18.

	Percent	Actual Count
With seats	30	375
Without seats	70	W
Total	100	T

$\frac{30}{70} = \frac{375}{W}$
$30W = 26{,}250$
$W = \mathbf{875}$

19. $\mathbf{24 = \frac{1}{4} \times W_N}$

$\left(\frac{4}{1}\right)24 = \left(\frac{\cancel{4}^1}{\cancel{1}_1}\right)\left(\frac{\cancel{1}^1}{\cancel{4}_1}\right) \times W_N$

$\mathbf{96} = W_N$

20.

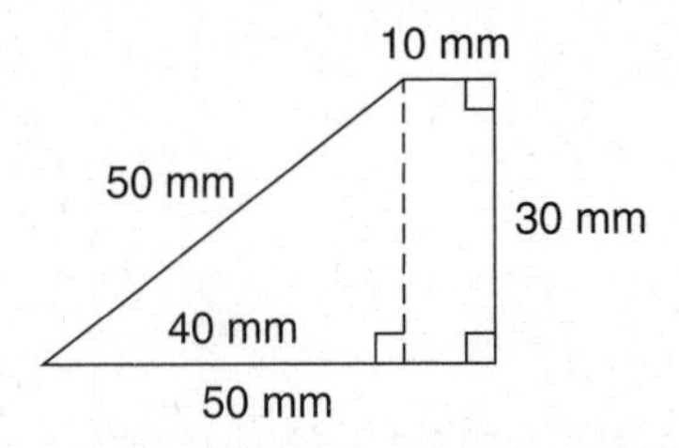

(a) **Trapezoid**

(b) Perimeter $= 10 \text{ mm} + 30 \text{ mm} + 50 \text{ mm} + 50 \text{ mm} = \mathbf{140 \text{ mm}}$

(c) Area of rectangle $= (10 \text{ mm})(30 \text{ mm}) = 300 \text{ mm}^2$

Area of triangle $= \dfrac{(40 \text{ mm})(30 \text{ mm})}{2} = 600 \text{ mm}^2$

Area of figure $= 300 \text{ mm}^2 + 600 \text{ mm}^2 = \mathbf{900 \text{ mm}^2}$

21. $y = (10) - 5$
$y = \mathbf{5;}$
$y = (7) - 5$
$y = \mathbf{2;}$
$y = (5) - 5$
$y = \mathbf{0}$

22. (a) $(9 \times 4) \times (10^{-6} \times 10^{-8})$
$= 36 \times (10^{-14})$
$= \mathbf{3.6 \times 10^{-13}}$

(b) $(9 \times 4) \times (10^{6} \times 10^{8})$
$= 36 \times (10^{14})$
$= \mathbf{3.6 \times 10^{15}}$

23.
$$8\frac{5}{6} = d - 5\frac{1}{2}$$
$$8\frac{5}{6} + 5\frac{1}{2} = d - 5\frac{1}{2} + 5\frac{1}{2}$$
$$8\frac{5}{6} + 5\frac{3}{6} = d$$
$$13\frac{8}{6} = d$$
$$14\frac{2}{6} = d$$
$$\mathbf{14\frac{1}{3}} = d$$

check:
$$8\frac{5}{6} = 14\frac{1}{3} - 5\frac{1}{2}$$
$$8\frac{5}{6} = 14\frac{2}{6} - 5\frac{3}{6}$$
$$8\frac{5}{6} = 13\frac{8}{6} - 5\frac{3}{6}$$
$$8\frac{5}{6} = 8\frac{5}{6}$$

24.
$$\frac{5}{6}m = 90$$
$$\left(\frac{\cancel{6}^1}{\cancel{5}_1}\right)\frac{\cancel{5}^1}{\cancel{6}_1}m = \left(\frac{6}{\cancel{5}_1}\right)\cancel{90}^{18}$$
$$m = \mathbf{108}$$

check:
$$\frac{5}{\cancel{6}_1}(\cancel{108}^{18}) = 90$$
$$90 = 90$$

25.

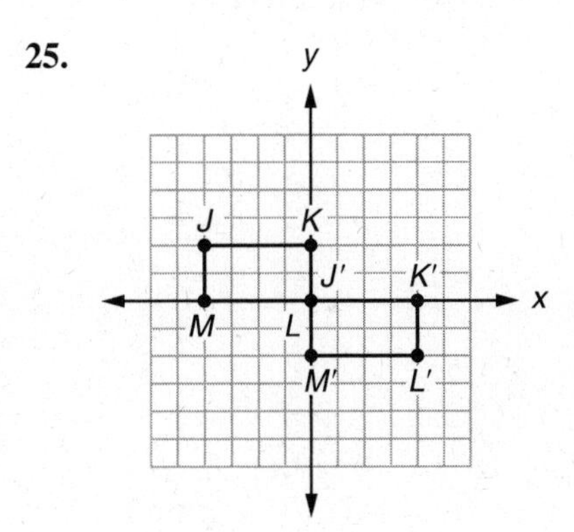

(a) $\mathbf{M(-4, 0)}$

(b) $\mathbf{J'(0, 0), K'(4, 0), L'(4, -2), M'(0, -2)}$

26. **D.** $\mathbf{4^2 + 4}$

27. $\frac{2}{3}(0.12) = 0.08$
$0.5(0.08) = \mathbf{0.04}$

28. $6\{5 \cdot 4 - 3[6 - (3 - 1)]\}$
$6\{20 - 3[6 - 2]\}$
$6\{20 - 3[4]\}$
$6\{20 - 12\}$
$6\{8\}$
$\mathbf{48}$

29. (a)
$$\frac{(-3)(-4) - (-3)}{(-3) - (+4)(+3)}$$
$$\frac{12 - (-3)}{-3 - (12)}$$
$$\frac{12 + 3}{-3 - 12}$$
$$\frac{15}{-15}$$
$$\mathbf{-1}$$

(b) $(+5) + (-2)[(+3) - (-4)]$
$(+5) + (-2)[+3 + (+4)]$
$(+5) + (-2)[7]$
$(+5) + (-14)$
$\mathbf{-9}$

30. (a) $(-2x)(-3x)$
$= (-2) \cdot x \cdot (-3) \cdot (x)$
$= (-2)(-3) \cdot x \cdot x = \mathbf{6x^2}$

(b) $(ab)(2a^2b)(-3a)$
$= a \cdot b \cdot (2) \cdot a \cdot a \cdot b \cdot (-3) \cdot a$
$= (2)(-3) \cdot a \cdot a \cdot a \cdot a \cdot b \cdot b$
$= \mathbf{-6a^4b^2}$

Lesson 89, Warm-Up

a. **+20**

b. **7.5×10^4**

c. **10**

d. **$0.64**

e. **187 cm**

f. **$2.50**

g. **8**

Problem Solving

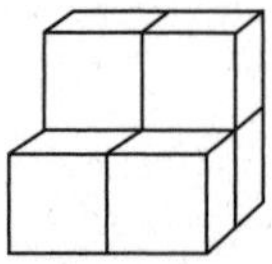

6 in.^{3}

The bottom "stair" is 2 in.-by-2 in.-by-1-in., which is 4 in.^{3}.
The top "stair" is 2 in.-by-1 in.-by-1 in., which is 2 in.^{3}.
4 in.^{3} + 2 in.^{3} = **6 in.^{3}**

Lesson 89, Lesson Practice

a. **For answers, see solutions to examples 1–4.**

b. **One possibility:**

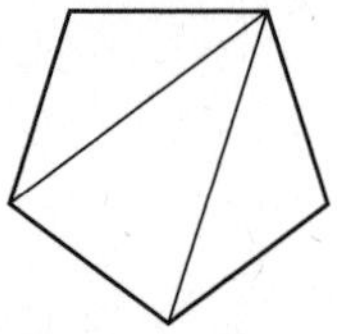

2 diagonals

c. **3 triangles**

d. $3 \times 180° = \mathbf{540°}$

e. $\dfrac{540°}{5} = \mathbf{108°}$

f. $\dfrac{360°}{5} = \mathbf{72°}$

g. $108° + 72° = \mathbf{180°}$

Lesson 89, Mixed Practice

1.

	Case 1	Case 2
Feet	440	5280
Seconds	10	s

$\dfrac{440}{10} = \dfrac{5280}{s}$
$440s = 52{,}800$
$s =$ **120 seconds** or **2 minutes**

2.

	Ratio	Actual Count
Lions	3	18
Tigers	2	T

$\dfrac{3}{2} = \dfrac{18}{T}$
$3T = 36$
$T = 12$ tigers

	Ratio	Actual Count
Tigers	3	12
Bears	4	B

$\dfrac{3}{4} = \dfrac{12}{B}$
$3B = 48$
$B =$ **16 bears**

3. $(30\text{ cm})(15\text{ cm}) = 450\text{ cm}^2$
$\dfrac{450\text{ cm}^2}{1\ \cancel{\text{layer}}} \cdot (1\text{ cm}) \cdot 12\ \cancel{\text{layers}} = \mathbf{5400\ cm^3}$

4. $\dfrac{24}{61} = 0.393442\ldots$ **0.393**

5. $\overset{2}{\cancel{\overset{6}{\cancel{18}}}}\ \cancel{\text{ft}^2} \cdot \dfrac{1\text{ yd}}{\underset{1}{\cancel{3}}\ \cancel{\text{ft}}} \cdot \dfrac{1\text{ yd}}{\underset{1}{\cancel{3}}\ \cancel{\text{ft}}} =$ **2 square yards**

6.

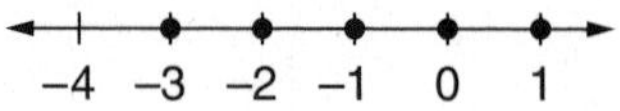

7.

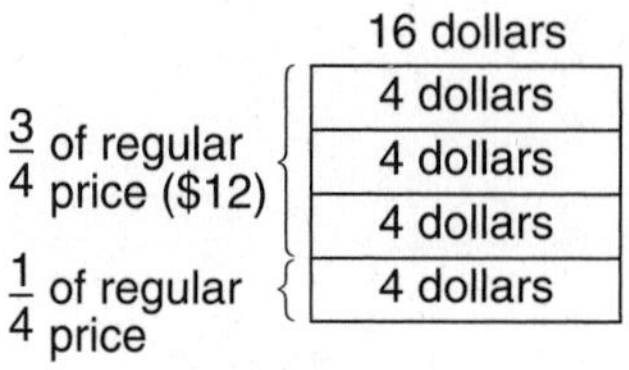

(a) $\$12 \div 3 = \$4$
$(\$4)(4) = \mathbf{\$16}$

(b) $\dfrac{3}{4} \times 100\% = \mathbf{75\%}$

8. (a) $m\angle a = 180° - (90° + 35°) = \mathbf{55°}$

(b) $m\angle b = 180° - 55° = \mathbf{125°}$

(c) $m\angle c = m\angle a = \mathbf{55°}$

(d) $m\angle d = 180° - (55° + 70°) = \mathbf{55°}$

9. (a) $C \approx \frac{22}{\cancel{7}_1}(\cancel{42}^{6}\text{in.})$

$C \approx \mathbf{132\ in.}$

(b) $A = \pi r^2$

$A \approx \frac{22}{\cancel{7}_1}(\cancel{441}^{63}\text{ in.}^2)$

$A \approx \mathbf{1386\ in.^2}$

10. $\frac{91\frac{2}{3}}{100} = \frac{\frac{275}{3}}{\frac{100}{1}}$

$\frac{\frac{275}{3}}{\frac{100}{1}} \cdot \frac{\frac{1}{100}}{\frac{1}{100}} = \frac{\frac{275}{300}}{1} = \frac{275}{300}$

$= \mathbf{\frac{11}{12}}$

11. $\frac{(10)(5) + (10)}{(10) + (5)} = \frac{50 + 10}{15}$

$= \frac{60}{15} = \mathbf{4}$

12. 0.25 ◯ 0.5

a^2 ⓒ a — $a^2 < a$ (circled)

13. (a) $\frac{7}{8} = \mathbf{0.875}$

```
   0.875
8)7.000
  6 4
    60
    56
     40
     40
      0
```

(b) $\frac{7}{8} \times 100\% = \frac{700\%}{8} = \mathbf{87\frac{1}{2}\%}$

```
  87 4/8 = 87 1/2%
8)700
  64
   60
   56
    4
```

(c) $875\% = \frac{875}{100} = 8\frac{75}{100} = \mathbf{8\frac{3}{4}}$

(d) $875\% = \mathbf{8.75}$

14. (a) **4:00, 8:00**

(b) **120°**

15.

	Percent	Actual Count
Ordered a hamburger	45	H
Other customers	55	C
Total	100	3000

$\frac{45}{100} = \frac{H}{3000}$

$100H = 135{,}000$

$H = \mathbf{1350\ customers}$

16.

	Percent	Actual Count
Sale price	75	$24
Regular price	100	R

$\frac{75}{100} = \frac{\$24}{R}$

$75R = \$2400$

$R = \$32$

$\$32 - \$24 = \mathbf{\$8}$

17. $\mathbf{20 = W_P \times 200}$

$\frac{20}{200} = \frac{W_P \times 200}{200}$

$\frac{1}{10} = W_P$

$W_P = \frac{1}{10} \times 100\% = \mathbf{10\%}$

18. (a)

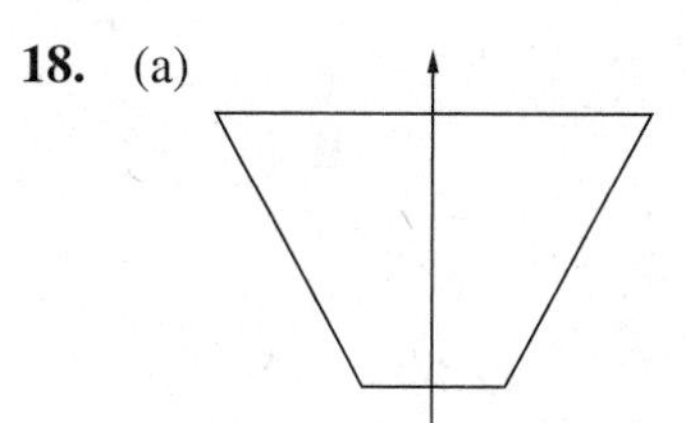

(b)

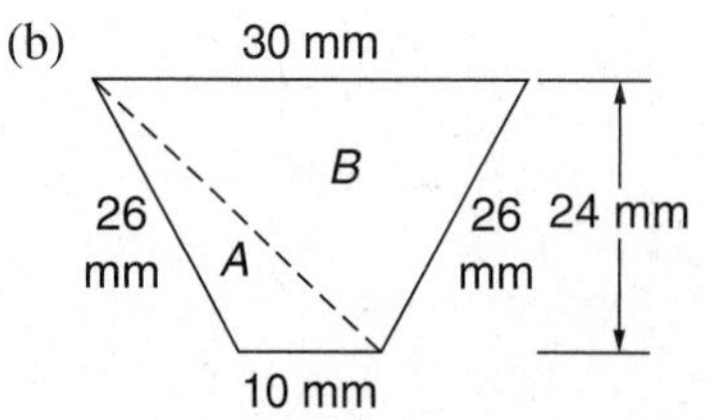

Area of triangle $A = \frac{(10\text{ mm})(24\text{ mm})}{2}$

$= 120\text{ mm}^2$

Area of triangle $B = \frac{(30\text{ mm})(24\text{ mm})}{2}$

$= 360\text{ mm}^2$

Area of figure $= 120\text{ mm}^2 + 360\text{ mm}^2$

$= \mathbf{480\ mm^2}$

19. $\frac{360°}{3} = \mathbf{120°}$

20.
$$y = \frac{1}{3}(12)$$
$$y = \mathbf{4;}$$
$$y = \frac{1}{3}(9)$$
$$y = \mathbf{3;}$$
$$6 = \frac{1}{3}x$$
$$\left(\frac{3}{1}\right)6 = \left(\frac{3}{1}\right)\frac{1}{3}x$$
$$\mathbf{18} = x$$

21.
$$(1.25 \times 8) \times (10^{-3} \times 10^{-5})$$
$$= 10 \times 10^{-8}$$
$$= \mathbf{1 \times 10^{-7}}$$

22. (a)

10 cm, 10 cm, 4 cm

Perimeter = 10 cm + 10 cm + 4 cm
= **24 cm**

(b) **There can only be one answer. A triangle with side lengths of 4 cm, 4 cm, and 10 cm cannot exist.**

23.
$$\frac{4}{9}p = 72$$
$$\left(\frac{\cancel{9}^{1}}{\cancel{4}_{1}}\right)\frac{\cancel{4}^{1}}{\cancel{9}_{1}}p = \left(\frac{9}{\cancel{4}_{1}}\right)\cancel{72}^{18}$$
$$p = \mathbf{162}$$
check:
$$\frac{4}{\cancel{9}_{1}}(\cancel{162}^{18}) = 72$$
$$4(18) = 72$$
$$72 = 72$$

24.
$$12.3 = 4.56 + f$$
$$12.3 - 4.56 = 4.56 - 4.56 + f$$
$$\mathbf{7.74} = f$$
check:
$$12.3 = 4.56 + 7.74$$
$$12.3 = 12.3$$

25.
$$2x + 3y - 4 + x - 3y - 1$$
$$2x + x + 3y - 3y - 4 - 1$$
$$\mathbf{3x - 5}$$

26.
$$\frac{9 \cdot 8 - 7 \cdot 6}{6 \cdot 5}$$
$$\frac{72 - 42}{30}$$
$$\frac{30}{30}$$
$$\mathbf{1}$$

27.
$$3\frac{2}{10} \times \frac{1}{4^2} \times 10^2$$
$$3\frac{2}{10} \times \frac{1}{16} \times 100$$
$$= \frac{\cancel{32}^{2}}{\cancel{10}_{1}} \times \frac{1}{\cancel{16}_{1}} \times \frac{\cancel{100}^{10}}{1} = \mathbf{20}$$

28.
$$4.75 + \frac{3}{4} = 4\frac{3}{4} + \frac{3}{4}$$
$$= 4\frac{6}{4} = 5\frac{2}{4} = 5\frac{1}{2}$$
$$13\frac{1}{3} - 5\frac{1}{2} = 13\frac{2}{6} - 5\frac{3}{6}$$
$$= 12\frac{8}{6} - 5\frac{3}{6} = \mathbf{7\frac{5}{6}}$$

29. (a)
$$\frac{(+3) + (-4)(-6)}{(-3) + (-4) - (-6)}$$
$$\frac{(+3) + (24)}{-7 - (-6)}$$
$$\frac{27}{-1}$$
$$\mathbf{-27}$$

(b)
$$(-5) - (+6)(-2) + (-2)(-3)(-1)$$
$$(-5) - (-12) + (-6)$$
$$(-5) + (+12) + (-6)$$
$$\mathbf{1}$$

30. (a)
$$(3x^2)(2x)$$
$$= (3) \cdot x \cdot x \cdot (2) \cdot x$$
$$= (3)(2) \cdot x \cdot x \cdot x$$
$$= \mathbf{6x^3}$$

(b)
$$(-2ab)(-3b^2)(-a)$$
$$= (-2) \cdot a \cdot b \cdot (-3) \cdot b \cdot b \cdot (-1) \cdot a$$
$$= (-2)(-3)(-1) \cdot a \cdot a \cdot b \cdot b \cdot b$$
$$= \mathbf{-6a^2b^3}$$

Lesson 90, Warm-Up

a. **−20**

b. **8.4×10^{-10}**

c. **11**

d. **$3.60**

e. **0.8 kg**

f. **$1.25**

g. **1000 in.³**

Problem Solving

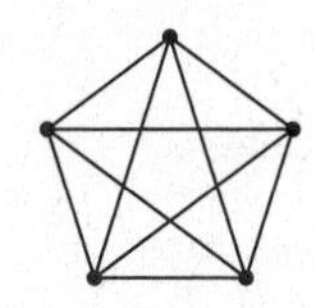

10 handshakes

LESSON 90, LESSON PRACTICE

a.
$$1\frac{1}{8}x = 36$$
$$\frac{9}{8}x = 36$$
$$\left(\frac{\cancel{8}^{1}}{\cancel{9}_{1}}\right)\left(\frac{\cancel{9}^{1}}{\cancel{8}_{1}}x\right) = \left(\frac{8}{\cancel{9}_{1}}\right) \cdot \cancel{36}^{4}$$
$$x = \mathbf{32}$$

b.
$$3\frac{1}{2}a = 490$$
$$\frac{7}{2}a = 490$$
$$\left(\frac{\cancel{2}^{1}}{\cancel{7}_{1}}\right)\left(\frac{\cancel{7}^{1}}{\cancel{2}_{1}}a\right) = \left(\frac{2}{\cancel{7}_{1}}\right)\cancel{490}^{70}$$
$$a = \mathbf{140}$$

c.
$$2\frac{3}{4}w = 6\frac{3}{5}$$
$$\frac{11}{4}w = \frac{33}{5}$$
$$\left(\frac{\cancel{4}^{1}}{\cancel{11}_{1}}\right)\left(\frac{\cancel{11}^{1}}{\cancel{4}_{1}}w\right) = \left(\frac{4}{\cancel{11}_{1}}\right)\left(\frac{\cancel{33}^{3}}{5}\right)$$
$$w = \mathbf{\frac{12}{5}}$$

d.
$$2\frac{2}{3}y = 1\frac{4}{5}$$
$$\frac{8}{3}y = \frac{9}{5}$$
$$\left(\frac{\cancel{3}^{1}}{\cancel{8}_{1}}\right)\left(\frac{\cancel{8}^{1}}{\cancel{3}_{1}}y\right) = \left(\frac{3}{8}\right)\left(\frac{9}{5}\right)$$
$$y = \mathbf{\frac{27}{40}}$$

e.
$$-3x = 0.45$$
$$\frac{-3x}{-3} = \frac{0.45}{-3}$$
$$x = \mathbf{-0.15}$$

f.
$$-\frac{3}{4}m = \frac{2}{3}$$
$$\left(-\frac{\cancel{4}^{1}}{\cancel{3}_{1}}\right)\left(-\frac{\cancel{3}^{1}}{\cancel{4}_{1}}m\right) = \left(-\frac{4}{3}\right)\left(\frac{2}{3}\right)$$
$$m = \mathbf{-\frac{8}{9}}$$

g.
$$-10y = -1.6$$
$$\frac{-10y}{-10} = \frac{-1.6}{-10}$$
$$y = \mathbf{0.16}$$

h.
$$-2\frac{1}{2}w = 3\frac{1}{3}$$
$$-\frac{5}{2}w = \frac{10}{3}$$
$$\left(-\frac{\cancel{2}^{1}}{\cancel{5}_{1}}\right)\left(-\frac{\cancel{5}^{1}}{\cancel{2}_{1}}w\right) = \left(-\frac{2}{\cancel{5}_{1}}\right)\left(\frac{\cancel{10}^{2}}{3}\right)$$
$$w = \mathbf{-\frac{4}{3}}$$

LESSON 90, MIXED PRACTICE

1. $(0.8 + 0.9) - (0.8)(0.9)$
$= 1.7 - 0.72 = 0.98$
Ninety-eight hundredths

2. (a)

```
      8
      6
      9
     10
      8
      7
      9
     10
      8
     10
      9
  +   8
  -----
    102
```

$$12\overline{)102.0} = \mathbf{8.5}$$

(b) **8.5**

(c) **8**

(d) $10 - 6 = \mathbf{4}$

3. $\frac{\$1.20}{24 \text{ ounces}} = \frac{\$0.05}{1 \text{ ounce}}$

$\frac{\$1.44}{32 \text{ ounces}} = \frac{\$0.045}{1 \text{ ounce}}$

$\$0.050 - \$0.045 = \$0.005$

0.5¢ more per ounce

4. (a) $\frac{360°}{10} = \mathbf{36°}$

(b) $180° - 36° = \mathbf{144°}$

5. $x^2 + 2xy + y^2 + x^2 - y^2$

$x^2 + x^2 + 2xy + y^2 - y^2$

$\mathbf{2x^2 + 2xy}$

6.

	Percent	Actual Count
Sale price	90	\$36
Regular price	100	R

$\frac{90}{100} = \frac{\$36}{R}$

$90R = \$3600$

$R = \mathbf{\$40}$

7.

	Percent	Actual Count
Voted for Graham	75	V
Did not vote for Graham	25	D
Total	100	800

$\frac{25}{100} = \frac{D}{800}$

$100D = 20{,}000$

$D = \mathbf{200 \text{ citizens}}$

8. (a) $\mathbf{24 = W_P \times 30}$

$\frac{24}{30} = \frac{W_P \times 30}{30}$

$\frac{4}{5} = W_P$

$W_P = \frac{4}{5} \times 100\% = \mathbf{80\%}$

(b) $\mathbf{30 = W_P \times 24}$

$\frac{30}{24} = \frac{W_P \times 24}{24}$

$\frac{5}{4} = W_P$

$W_P = \frac{5}{4} \times 100\% = \mathbf{125\%}$

9. $2 \cancel{\text{ft}^2} \cdot \frac{12 \text{ in.}}{1 \cancel{\text{ft}}} \cdot \frac{12 \text{ in.}}{1 \cancel{\text{ft}}} = \mathbf{288 \text{ square inches}}$

10.

750 doctors

$\frac{2}{5}$ of doctors (300) did.
- 150 doctors
- 150 doctors

$\frac{3}{5}$ of doctors did not.
- 150 doctors
- 150 doctors
- 150 doctors

(a) $300 \div 2 = 150$

$5(150 \text{ doctors}) = \mathbf{750 \text{ doctors}}$

(b) $3(150 \text{ doctors}) = \mathbf{450 \text{ doctors}}$

11. $y = 2(4.5) + 1$

$y = 9 + 1$

$y = \mathbf{10}$

12. a ⓪ ab — $a \; \text{(>)} \; ab$

13. $\frac{12 \text{ inches}}{4} = 3 \text{ inches}$

$(3 \text{ inches})(3 \text{ inches}) = \mathbf{9 \text{ square inches}}$

14. (a) $1.75 = 1\frac{75}{100} = \mathbf{1\frac{3}{4}}$

(b) $1.75 \times 100\% = \mathbf{175\%}$

15. $\$325 \times 0.06 = \$19.50$

$\$325 + \$19.50 = \mathbf{\$344.50}$

16. $(6 \times 8) \times (10^4 \times 10^{-7})$

$= (48 \times (10^{-3})$

$= \mathbf{4.8 \times 10^{-2}}$

17. (a) $(8 \text{ in.})(3 \text{ in.}) = 24 \text{ in.}^2$

$\frac{24 \text{ in.}^2}{1 \cancel{\text{layer}}} \cdot (1 \text{ in.}) \cdot (12 \cancel{\text{layers}}) = \mathbf{288 \text{ in.}^3}$

(b) $2(8 \text{ in.} \times 12 \text{ in.}) + 2(8 \text{ in.} \times 3 \text{ in.})$

$+ 2(12 \text{ in.} \times 3 \text{ in.})$

$= 192 \text{ in}^2 + 48 \text{ in.}^2 + 72 \text{ in.}^2$

$= \mathbf{312 \text{ in.}^2}$

18. (a) $C \approx 3.14(100 \text{ mm})$

$C \approx \mathbf{314 \text{ mm}}$

(b) $A = \pi r^2$

$A \approx 3.14(2500 \text{ mm}^2)$

$A \approx \mathbf{7850 \text{ mm}^2}$

19. **0**

20.

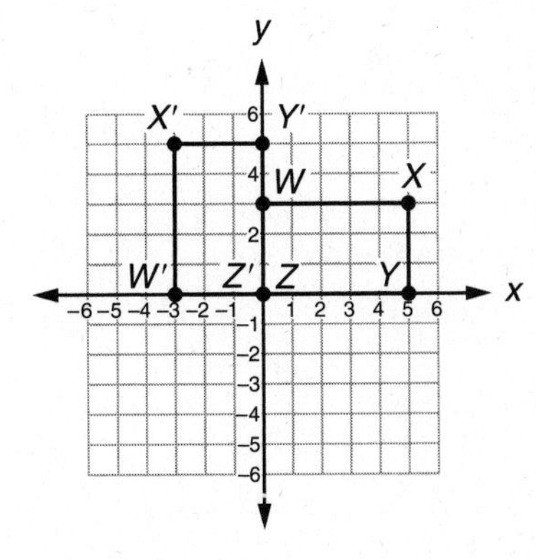

(a) **Z (0, 0)**

(b) **W′(−3, 0), X′(−3, 5), Y′(0, 5), Z′(0, 0)**

21. $\frac{2}{3} \times 20 = \frac{40}{3} = \mathbf{13\frac{1}{3}}$

22. Number line from −1 to 5 with a ray starting at 4 pointing left.

23.
$$x + 3.5 = 4.28$$
$$x + 3.5 - 3.5 = 4.28 - 3.5$$
$$x = \mathbf{0.78}$$

24.
$$2\frac{2}{3}w = 24$$
$$\frac{8}{3}w = 24$$
$$\left(\frac{\cancel{3}^{1}}{\cancel{8}_{1}}\right)\frac{\cancel{8}^{1}}{\cancel{3}_{1}}w = \left(\frac{3}{\cancel{8}_{1}}\right)\cancel{24}^{3}$$
$$w = \mathbf{9}$$

25.
$$-4y = 1.4$$
$$\frac{-4y}{-4} = \frac{1.4}{-4}$$
$$y = \mathbf{-0.35}$$

26. $10^1 + 10^0 + 10^{-1}$
$$= 10 + 1 + \frac{1}{10} = 11 + 0.1$$
$$= \mathbf{11.1}$$

27. $(-2x^2)(-3xy)(-y)$
$$= (-2) \cdot x \cdot x \cdot (-3) \cdot x \cdot y \cdot (-1) \cdot y$$
$$= (-2)(-3)(-1) \cdot x \cdot x \cdot x \cdot y \cdot y$$
$$= \mathbf{-6x^3y^2}$$

28.
$$\frac{8}{75} = \frac{32}{300}$$
$$-\ \frac{9}{100} = \frac{27}{300}$$
$$\frac{5}{300} = \mathbf{\frac{1}{60}}$$

29. (a) $(-3) + (-4)(-5) - (-6)$
$$(-3) + (20) + (+6)$$
$$(-3) + (26)$$
$$\mathbf{23}$$

(b) $\frac{(-2)(-4)}{(-4) - (-2)}$
$$\frac{8}{-2}$$
$$\mathbf{-4}$$

30.
$$10^2 - 5^2 = 100 - 25 = 75$$
$$(10 + 5)(10 - 5) = (15)(5) = 75$$
$$75 = 75$$
$$x^2 - y^2 \;\textcircled{=}\; (x + y)(x - y)$$

INVESTIGATION 9

1. **The student should select two or three points such as (−1, 0), (5, 6), and ($1\frac{1}{2}, 2\frac{1}{2}$). Substitute the (*x*, *y*) numbers in *y* = *x* + 1 and simplify.**

2.

$y = \frac{1}{2}x - 1$

(a) **The student should select a point on the line and determine its coordinates.**

(b) **No. For this equation, if *x* is negative, then *y* must also be negative. Thus the graph does not enter the quadrant where *x* is negative and *y* is positive.**

3. **The endpoint of the ray is (0, 0). If the graph were a line, there would be negative numbers for the length of the side and for the perimeter. Lengths cannot be negative, so the graph cannot be a line.**

4. **Doubling the side lengths doubles the perimeter.** ***Note:*** **This is true of any polygon.**

5. $p = 3s$

s	p
1	3
2	6
3	9

p = perimeter of triangle
s = length of side

Perimeter of an Equilateral Triangle

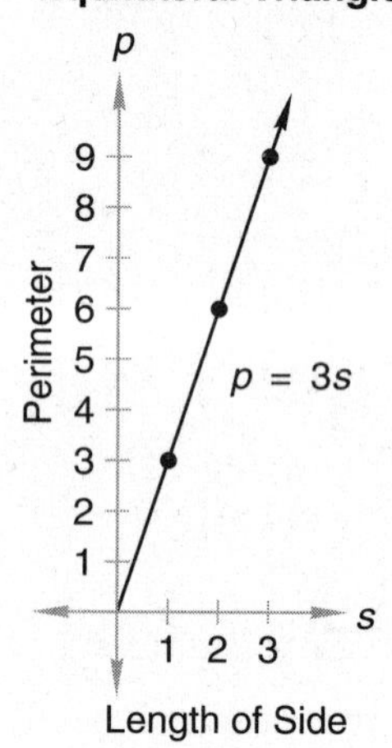

6. $d = 6h$

h	d
1	6
2	12

d = distance in miles
h = time in hours

Distance Sam Jogged at 6 Miles per Hour

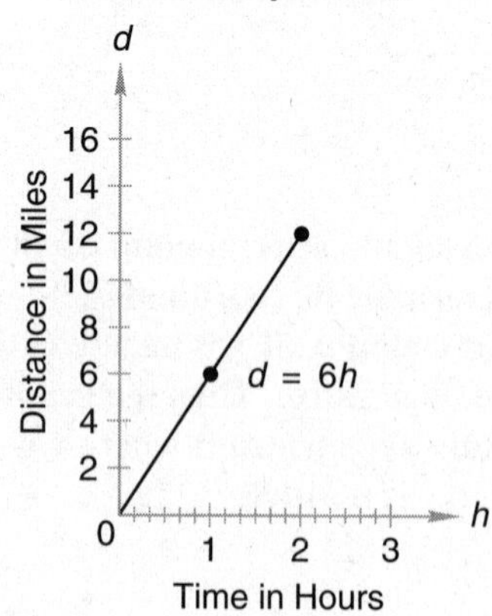

7. **4 miles**

8. $\mathbf{1\frac{1}{2}}$ **hours**

9. **Sam did not continue jogging forever. He stopped after 2 hours.**

10. **Doubling the side lengths of a square quadruples the area.** ***Note:*** **This is true of any polygon.**

11. $\frac{1}{2}(3)^2 = \mathbf{4\frac{1}{2}}$
$\frac{1}{2}(4)^2 = \mathbf{8}$

12. **Area of Half of a Square**

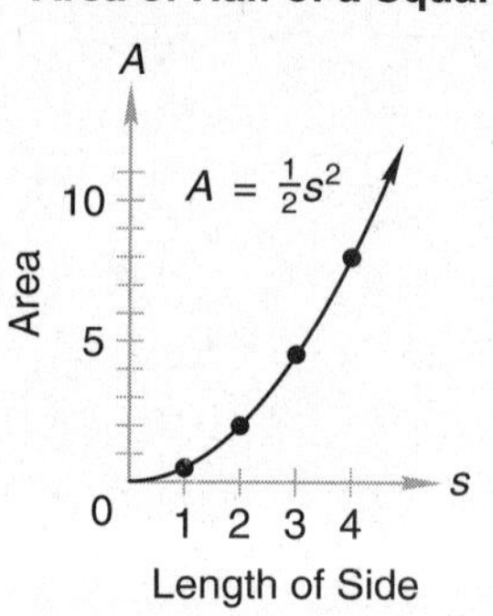

Extensions

a. $V = e^3$

e	V
0	0
1	1
2	8
3	27
4	64

V = volume of cube
e = length of edge

b. **Volume (V) of a Cube with Edge Length e**

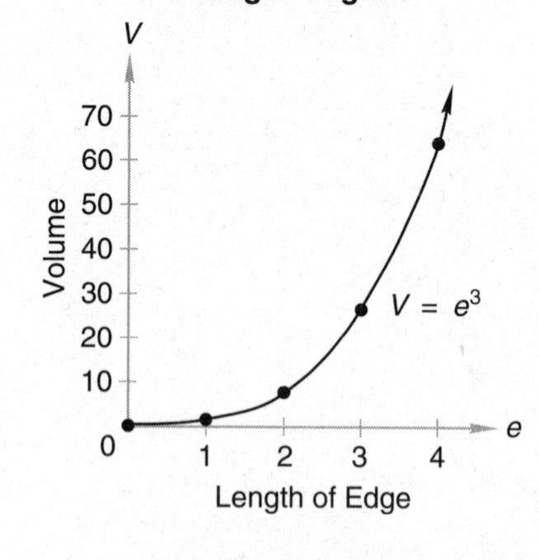

LESSON 91, WARM-UP

a. **−134**

b. $\mathbf{1.44 \times 10^6}$

c. **60**

d. **15**

e. **1500 mL**

f. **$200**

g. **15**

Problem Solving

15 ft = 5 yd
12 ft = 4 yd
5 yd × 4 yd = **20 sq. yd**

Lesson 91, Lesson Practice

a. $(3) + (3)(-2) - (-2)$
$(3) + (-6) - (-2)$
$(-3) - (-2)$
$\mathbf{-1}$

b. $-(-2) + (-5) - (-2)(-5)$
$-(-2) + (-5) - (+10)$
$(-3) - (+10)$
$\mathbf{-13}$

c. $-3 + 4 - 5 - 2$
$+4 \underbrace{- 3 - 5 - 2}$
$+4 \quad -10$
$\mathbf{-6}$

d. $-2 + 3(-4) - 5(-2)$
$-2 + (-12) - (-10)$
$-14 - (-10)$
$\mathbf{-4}$

e. $-3(-2) - 5(2) + 3(-4)$
$(+6) - (+10) + (-12)$
$\mathbf{-16}$

f. $-4(-3)(-2) - 6(-4)$
$-4(+6) - (-24)$
$(-24) - (-24)$
$\mathbf{0}$

Lesson 91, Mixed Practice

1. $6(86) + 4(94) = 516 + 376$
$= 892$

$892.0 \div 10 = \mathbf{89.2}$

2. Median $= \dfrac{7 + 9}{2} = 8$ $\qquad 11 - 8 = \mathbf{3}$

Mean $= 88 \div 8 = 11$

3. $\dfrac{130 \text{ miles}}{2.5 \text{ hours}} = \mathbf{52 \text{ miles per hour}}$

4.

	Ratio	Actual Count
Laborers	3	L
Supervisors	5	S
Total	8	120

$\dfrac{3}{8} = \dfrac{L}{120}$
$8L = 360$
$L = \mathbf{45 \text{ laborers}}$

5.

	Case 1	Case 2
Notebooks	3	5
Price	$8.55	p

$\dfrac{3}{\$8.55} = \dfrac{5}{p}$
$3p = \$42.75$
$p = \mathbf{\$14.25}$

6.

	Percent	Actual Count
Sale price	90	S
Regular price	100	$36

$\dfrac{90}{100} = \dfrac{S}{\$36}$
$100S = \$3240$
$S = \mathbf{\$32.40}$

7.

	Percent	Actual Count
People who came	80	40
Invited people	100	I

$\dfrac{80}{100} = \dfrac{40}{I}$
$80I = 4000$
$I = \mathbf{50 \text{ people}}$

8. (a) $\mathbf{20 = 0.4 \times W_N}$
$\dfrac{20}{0.4} = \dfrac{0.4 \times W_N}{0.4}$
$\mathbf{50} = W_N$

(b) $\mathbf{20 = W_P \times 40}$
$\dfrac{20}{40} = \dfrac{W_P \times 40}{40}$
$\dfrac{1}{2} = W_P$
$W_P = \dfrac{1}{2} \times 100\% = \mathbf{50\%}$

9. $\overset{25}{\cancel{3600}} \cancel{\text{in.}^2} \cdot \dfrac{1 \text{ foot}}{\underset{1}{\cancel{12}} \cancel{\text{in.}}} \cdot \dfrac{1 \text{ foot}}{\underset{1}{\cancel{12}} \cancel{\text{in.}}}$
$= \mathbf{25 \text{ square feet}}$

10. 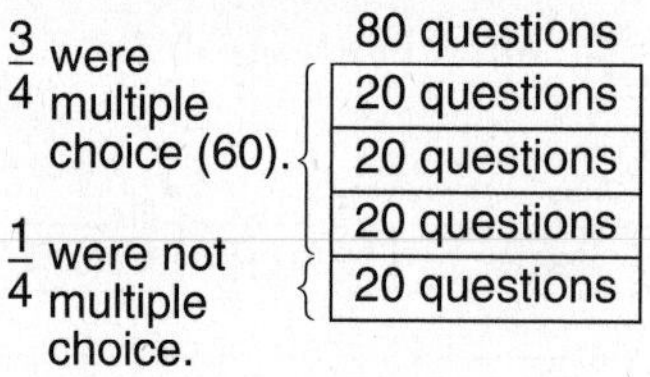

(a) $60 \div 3 = 20$

$4(20 \text{ questions}) = \mathbf{80 \text{ questions}}$

(b) **25%**

11. $(-3) - (-2) - (-3)(-2)$

$(-3) - (-2) - (+6)$

$\mathbf{-7}$

12. **Insufficient information**

13. 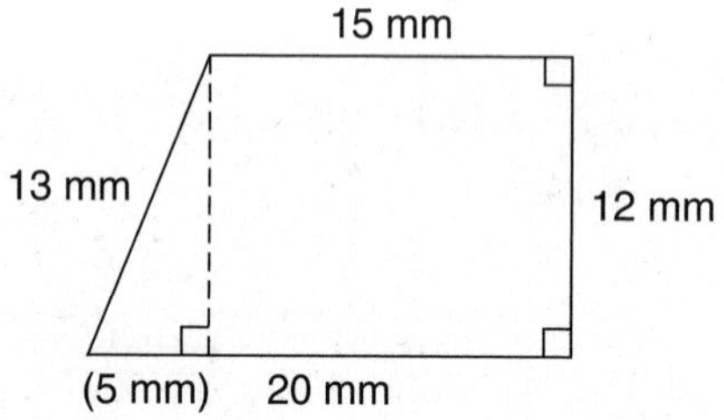

(a) **Trapezoid**

(b) Perimeter $= 15 \text{ mm} + 12 \text{ mm} + 20 \text{ mm} + 13 \text{ mm} = \mathbf{60 \text{ mm}}$

(c) Area of rectangle $= (15 \text{ mm})(12 \text{ mm}) = 180 \text{ mm}^2$

Area of triangle $= \dfrac{(5 \text{ mm})(12 \text{ mm})}{2} = 30 \text{ mm}^2$

Area of figure $= 180 \text{ mm}^2 + 30 \text{ mm}^2 = \mathbf{210 \text{ mm}^2}$

(d) $180° - 75° = \mathbf{105°}$

14. (a) **Associative property of addition**

(b) **Commutative property of multiplication**

(c) **Distributive property**

15. 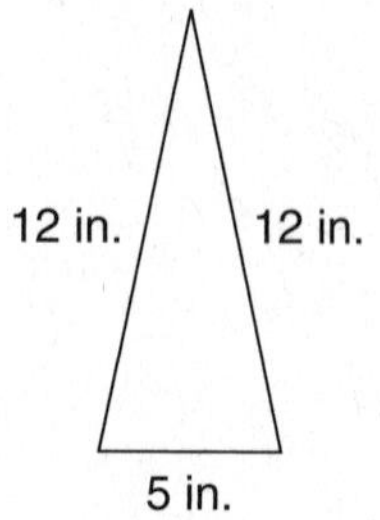

Perimeter $= 12 \text{ in.} + 12 \text{ in.} + 5 \text{ in.} = \mathbf{29 \text{ in.}}$

16. $(2.4 \times 10^{-4})(5 \times 10^{-7})$

12×10^{-11}

$(1.2 \times 10^{1}) \times 10^{-11}$

$\mathbf{1.2 \times 10^{-10}}$

17. (a) **5 faces**

(b) **8 edges**

(c) **5 vertices**

18. (a) $C \approx 3.14(8 \text{ cm})$

$C \approx \mathbf{25.12 \text{ cm}}$

(b) $A \approx 3.14(16 \text{ cm}^2)$

$A \approx \mathbf{50.24 \text{ cm}^2}$

19. 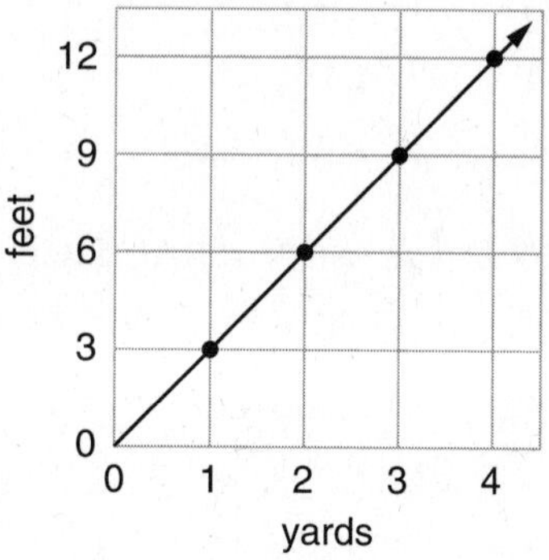

20. (a) $m\angle x = 180° - (90° + 30°) = \mathbf{60°}$

(b) $m\angle y = m\angle x = \mathbf{60°}$

(c) $m\angle A = 180° - (60° + 65°) = \mathbf{55°}$

(d) **No. The triangles do not have the same shape, nor do they have matching angles.**

21. (a) $-3x - 3 - x - 1$

$-3x - x - 3 - 1$

$\mathbf{-4x - 4}$

(b) $(-3x)(-3)(-x)(-1)$

$(-3) \cdot x \cdot (-3) \cdot (-1) \cdot x \cdot (-1)$

$(-3)(-3)(-1)(-1) \cdot x \cdot x$

$\mathbf{9x^2}$

22. 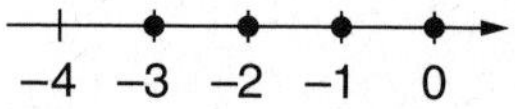

23. AB is 60 mm

BC is 40 mm

$60 \text{ mm} - 40 \text{ mm} = \mathbf{20 \text{ mm}}$

24. $5 = y - 4.75$

$5 + 4.75 = y - 4.75 + 4.75$

$\mathbf{9.75} = y$

25. $3\frac{1}{3}y = 7\frac{1}{2}$

$\frac{10}{3}y = \frac{15}{2}$

$\left(\frac{\cancel{3}^{1}}{\cancel{10}_{1}}\right)\frac{\cancel{10}^{1}}{\cancel{3}_{1}}y = \left(\frac{3}{\cancel{10}_{2}}\right)\frac{\cancel{15}^{3}}{2}$

$y = \mathbf{\frac{9}{4}}$

26. $-9x = 414$

$$\frac{-\cancel{9}x}{-\cancel{9}} = \frac{\overset{-46}{\cancel{414}}}{\underset{1}{-\cancel{9}}}$$

$x = \mathbf{-46}$

27. $\frac{32\text{ ft}}{1\text{ s}} \cdot \frac{60\text{ s}}{1\text{ min}} = \mathbf{1920\frac{ft}{min}}$

28. $5\frac{1}{3} + 2\frac{1}{2} + \frac{1}{6} = 5\frac{2}{6} + 2\frac{3}{6} + \frac{1}{6}$

$= 7\frac{6}{6} = \mathbf{8}$

29. $\frac{2.75 + 3.5}{2.5} = \frac{6.25}{2.5} = \mathbf{2.5}$

30. (a) $\frac{(-3) - (-4)(+5)}{-2}$

$\frac{(-3) - (-20)}{-2}$

$\frac{17}{-2}$

$\mathbf{-8\frac{1}{2}}$

(b) $-3(+4) - 5(+6) - 7$

$(-12) - (+30) - 7$

$\mathbf{-49}$

Lesson 92, Warm-Up

a. **−25**

b. $\mathbf{2.25 \times 10^{-10}}$

c. **16**

d. **20**

e. **30 cm**

f. **$60**

g. **$36.00**

Problem Solving

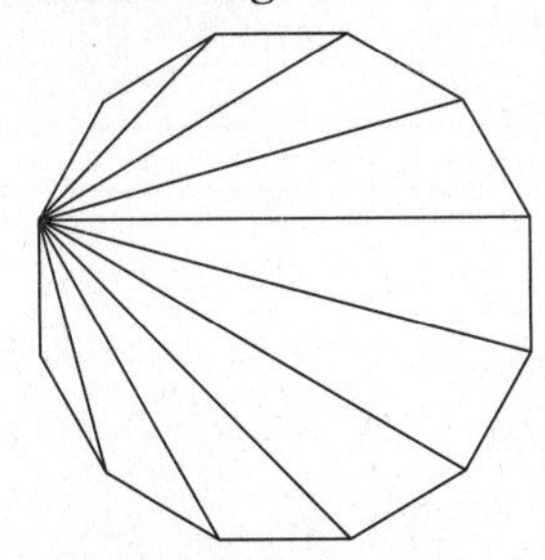

10 triangles

$10 \times 180° = \mathbf{1800°}$

Lesson 92, Lesson Practice

a.

	Percent	Actual Count
Original	100	$24.50
− Change	30	C
New	70	N

$\frac{100}{70} = \frac{\$24.50}{N}$

$100N = \$1715$

$N = \mathbf{\$17.15}$

b.

	Percent	Actual Count
Original	100	O
+ Change	20	C
New	120	60

$\frac{100}{120} = \frac{O}{60}$

$120O = 6000$

$O = \mathbf{50\ books}$

c.

	Percent	Actual Count
Original	100	O
− Change	20	C
New	80	$120

$\frac{20}{80} = \frac{C}{\$120}$

$80C = \$2400$

$C = \mathbf{\$30}$

d.

	Percent	Actual Count
Original	100	$15
+ Change	80	C
New	180	N

$\frac{100}{180} = \frac{\$15}{N}$

$100N = 2700$

$N = \mathbf{\$27}$

Lesson 92, Mixed Practice

1. $(7 + 11 + 13) - (2 \times 3 \times 5)$

$= 31 - 30$

$= \mathbf{1}$

2. $5(88) = 440$

$7(90) = 630$

$630 - 440 = 190$

$190 \div 2 = \mathbf{95}$

3. $\frac{2\text{ miles}}{0.25\text{ hour}} = \mathbf{8\ miles\ per\ hour}$

4.

	Ratio	Actual Count
Girls	9	45
Boys	7	35
Total	16	80

$$\frac{35}{45} = \mathbf{\frac{7}{9}}$$

5.

	Case 1	Case 2
Sparklers	24	60
Price	$3.60	p

$$\frac{24}{\$3.60} = \frac{60}{p}$$
$$24p = \$216$$
$$p = \mathbf{\$9.00}$$

6.

	Percent	Actual Count
Original	100	340,000
+ Change	20	C
New	120	N

$$\frac{100}{120} = \frac{340{,}000}{N}$$
$$100N = 40{,}800{,}000$$
$$N = \mathbf{408{,}000}$$

7.

	Percent	Actual Count
Original	100	O
+ Change	50	C
New	150	96

$$\frac{100}{150} = \frac{O}{96}$$
$$150O = 9600$$
$$O = \mathbf{64¢ \text{ per pound}}$$

8. (a) $\mathbf{60 = W_P \times 75}$

$$\frac{60}{75} = \frac{W_P \times 75}{75}$$
$$\frac{12}{15} = W_P$$
$$W_P = \frac{12}{15} \times 100\% = \mathbf{80\%}$$

(b) $\mathbf{75 = W_P \times 60}$

$$\frac{75}{60} = \frac{W_P \times 60}{60}$$
$$\frac{15}{12} = W_P$$
$$W_P = \frac{15}{12} \times 100\% = \mathbf{125\%}$$

9. $100\ \cancel{\text{cm}^2} \cdot \frac{10 \text{ mm}}{1\ \cancel{\text{cm}}} \cdot \frac{10 \text{ mm}}{1\ \cancel{\text{cm}}}$

$= \mathbf{10{,}000 \text{ square millimeters}}$

10.

256 trees

32 trees
32 trees
32 trees
32 trees
32 trees
32 trees
32 trees
32 trees

$\frac{5}{8}$ were deciduous (160).

$\frac{3}{8}$ were not deciduous.

(a) $160 \div 5 = 32$

$8(32 \text{ trees}) = \mathbf{256 \text{ trees}}$

(b) $3(32 \text{ trees}) = \mathbf{96 \text{ trees}}$

11.
$$y = 3(-5) - 1$$
$$y = -15 - 1$$
$$y = \mathbf{-16}$$

12.
$$30\% \times 20 = \frac{3}{10} \times 20 = 6$$
$$20\% \times 30 = \frac{2}{10} \times 30 = 6$$
$$6 = 6$$

30% of 20 (=) 20% of 30

13.

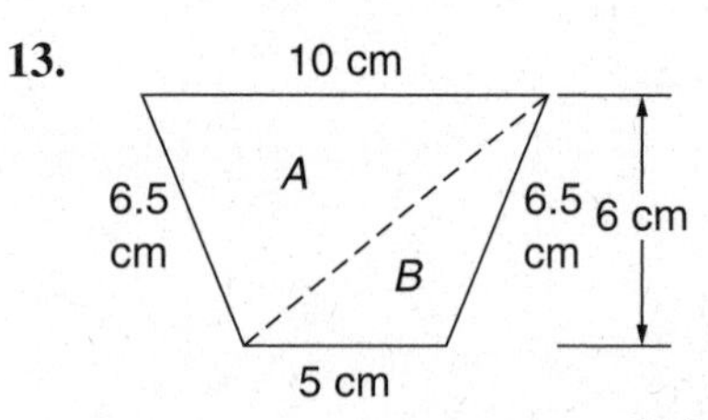

(a) Area of triangle $A = \frac{(10 \text{ cm})(6 \text{ cm})}{2}$

$= 30 \text{ cm}^2$

Area of triangle $B = \frac{(5 \text{ cm})(6 \text{ cm})}{2}$

$= 15 \text{ cm}^2$

Area of figure $= 30 \text{ cm}^2 + 15 \text{ cm}^2$

$= \mathbf{45 \text{ cm}^2}$

(b)

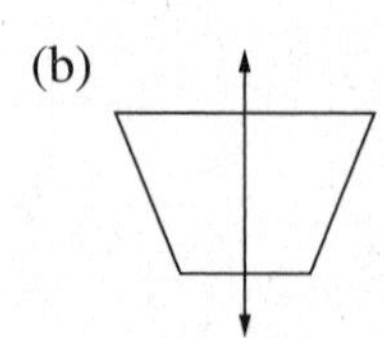

14.

	Percent	Actual Count
Original	100	$90.00
+ Change	75	C
New	175	N

(a) $\frac{100}{175} = \frac{\$90}{N}$

$100N = \$15{,}750$

$N = \mathbf{\$157.50}$

(b)
$$\begin{array}{r} \$157.50 \\ \times\ \ 0.06 \\ \hline \$9.45 \end{array} \qquad \begin{array}{r} \$157.50 \\ +\ \ \$9.45 \\ \hline \mathbf{\$166.95} \end{array}$$

15. $(8 \times 10^{-5})(3 \times 10^{12})$
24×10^7
$(2.4 \times 10^1) \times 10^7$
$\mathbf{2.4 \times 10^8}$

16. (a) $\mathbf{2.\overline{3}}$

(b) $\mathbf{233\frac{1}{3}\%}$

(c) $\mathbf{\frac{1}{30}}$

(d) $\mathbf{0.0\overline{3}}$

17. $\mathbf{\frac{3}{51} = \frac{1}{17}}$

18. $W_N = 2.5 \times 60$
$W_N = \mathbf{150}$

19. (a) **2**

(b) **3**

20. $C \approx 3.14(24 \text{ inches})$
$C \approx 75.36 \text{ inches}$
$C \approx$ **75 inches**

21. $y = 2(0) + 1$
$y = 0 + 1$
$y =$ **1;**
$y = 2(3) + 1$
$y = 6 + 1$
$y =$ **7;**
$y = 2(-2) + 1$
$y = -4 + 1$
$y =$ **−3**

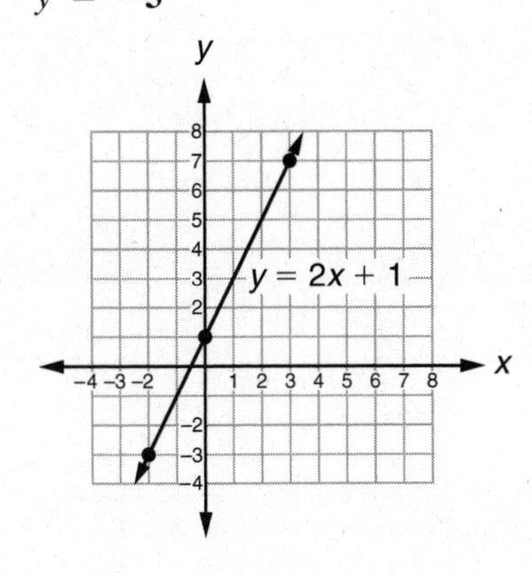

22. $180° \div 9 = 20°$
$4(20°) = \mathbf{80°}$

23. (a) $x + y + 3 + x - y - 1$
$x + x + y - y + 3 - 1$
$\mathbf{2x + 2}$

(b) $(3x)(2x) + (3x)(2)$
$[(3) \cdot x \cdot (2) \cdot x] + [(3) \cdot x \cdot (2)]$
$[(3)(2) \cdot x \cdot x] + [(3)(2) \cdot x]$
$\mathbf{6x^2 + 6x}$

24. **One possibility:**

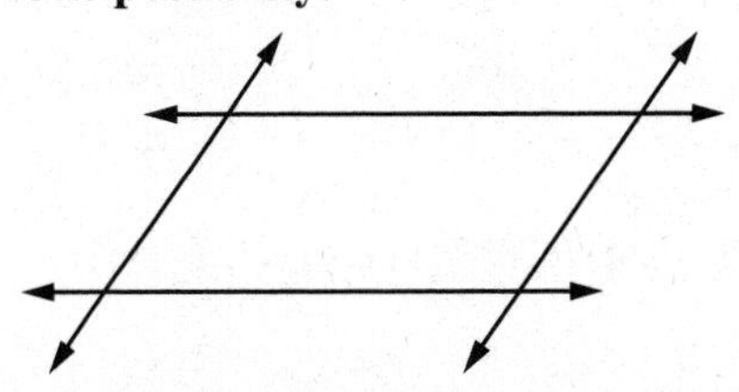

Parallelogram

25. $3\frac{1}{7}x = 66$

$\frac{22}{7}x = 66$

$\left(\frac{\overset{1}{\cancel{7}}}{\underset{1}{\cancel{22}}}\right)\frac{\overset{1}{\cancel{22}}}{\underset{1}{\cancel{7}}}x = \left(\frac{7}{\cancel{22}}\right)\overset{3}{\cancel{66}}$

$x = \mathbf{21}$

26. $w - 0.15 = 4.9$
$w - 0.15 + 0.15 = 4.9 + 0.15$
$w = \mathbf{5.05}$

27. $-8y = 600$

$\frac{-8y}{-8} = \frac{\overset{-75}{\cancel{600}}}{\underset{1}{\cancel{-8}}}$

$y = \mathbf{-75}$

28. $(2 \cdot 3)^2 - 2(3^2)$
$= (6)^2 - 2(9) = 36 - 18 = \mathbf{18}$

29. $5 - \left(3\frac{1}{3} - 1\frac{1}{2}\right)$

$= 4\frac{6}{6} - \left(3\frac{2}{6} - 1\frac{3}{6}\right)$

$= 4\frac{6}{6} - \left(2\frac{8}{6} - 1\frac{3}{6}\right)$

$= 4\frac{6}{6} - 1\frac{5}{6} = \mathbf{3\frac{1}{6}}$

30. (a) $\dfrac{(-8)(-6)(-5)}{(-4)(-3)(-2)}$

$$\frac{(48)(-5)}{(12)(-2)}$$

$$\frac{-240}{-24}$$

$$\mathbf{10}$$

(b) $-6 - 5(-4) - 3(-2)(-1)$

$$-6 - (-20) - 3(2)$$

$$-6 - (-20) - (6)$$

$$\mathbf{8}$$

Lesson 93, Warm-Up

a. **200**

b. **7.5×10^4**

c. **250**

d. **288 in.^{2}**

e. **18**

f. **54**

g. **15 cm**

Problem Solving

"Double the Celsius temperature":
$20° \times 2 = 40°$
"Subtract 10%": $40° - 4° = 36°$
"Add 32°": $36° + 32° =$ **68°F**

Lesson 93, Lesson Practice

a.
$$\begin{aligned} 8x - 15 &= 185 \\ 8x - 15 + 15 &= 185 + 15 \\ 8x &= 200 \\ \frac{8x}{8} &= \frac{200}{8} \\ x &= \mathbf{25} \end{aligned}$$

b.
$$\begin{aligned} 0.2y + 1.5 &= 3.7 \\ 0.2y + 1.5 - 1.5 &= 3.7 - 1.5 \\ 0.2y &= 2.2 \\ \frac{0.2y}{0.2} &= \frac{2.2}{0.2} \\ y &= \mathbf{11} \end{aligned}$$

c.
$$\begin{aligned} \frac{3}{4}m - \frac{1}{3} &= \frac{1}{2} \\ \frac{3}{4}m - \frac{1}{3} + \frac{1}{3} &= \frac{1}{2} + \frac{1}{3} \\ \frac{3}{4}m &= \frac{5}{6} \\ \left(\frac{4}{3}\right)\frac{3}{4}m &= \left(\frac{4}{3}\right)\frac{5}{6} \\ m &= \frac{20}{18} \\ m &= \mathbf{\frac{10}{9}} \end{aligned}$$

d.
$$\begin{aligned} 1\frac{1}{2}n + 3\frac{1}{2} &= 14 \\ 1\frac{1}{2}n + 3\frac{1}{2} - 3\frac{1}{2} &= 14 - 3\frac{1}{2} \\ 1\frac{1}{2}n &= 14 - 3\frac{1}{2} \\ 1\frac{1}{2}n &= 10\frac{1}{2} \\ \left(\frac{2}{3}\right)\frac{3}{2}n &= \left(\frac{2}{3}\right)\frac{21}{2} \\ n &= \frac{42}{6} \\ n &= \mathbf{7} \end{aligned}$$

e.
$$\begin{aligned} -6p + 36 &= 12 \\ -6p + 36 - 36 &= 12 - 36 \\ -6p &= -24 \\ \frac{-6p}{-6} &= \frac{-24}{-6} \\ p &= \mathbf{4} \end{aligned}$$

f.
$$\begin{aligned} 38 &= 4w - 26 \\ 38 + 26 &= 4w - 26 + 26 \\ 64 &= 4w \\ \mathbf{16} &= w \end{aligned}$$

g.
$$\begin{aligned} -\frac{5}{3}m + 15 &= 60 \\ -\frac{5}{3}m + 15 - 15 &= 60 - 15 \\ -\frac{5}{3}m &= 45 \\ \left(-\frac{3}{5}\right)\left(-\frac{5}{3}m\right) &= \left(-\frac{3}{5}\right)45 \\ m &= \mathbf{-27} \end{aligned}$$

h.
$$\begin{aligned} 4.5 &= 0.6d - 6.3 \\ 4.5 + 6.3 &= 0.6d - 6.3 + 6.3 \\ 10.8 &= 0.6d \\ \frac{10.8}{0.6} &= \frac{0.6d}{0.6} \\ \mathbf{18} &= d \end{aligned}$$

i.
$$\begin{aligned} 2x + 5 &\ge 1 \\ 2x + 5 - 5 &\ge 1 - 5 \\ 2x &\ge -4 \\ \frac{2x}{2} &\ge \frac{-4}{2} \\ \mathbf{x} &\ \mathbf{\ge -2} \end{aligned}$$

$x \ge -2$

−3 −2 −1 0 1

j.
$$\begin{aligned} 2x - 5 &< 1 \\ 2x - 5 + 5 &< 1 + 5 \\ 2x &< 6 \\ \frac{2x}{2} &< \frac{6}{2} \\ \mathbf{x} &\ \mathbf{< 3} \end{aligned}$$

$x < 3$

0 1 2 3 4

Lesson 93, Mixed Practice

1. $\frac{60 \text{ kilometers}}{2.5 \text{ hours}} = $ **24 kilometers per hour**

2. (a)

$$\begin{array}{r} 3 \\ 9 \\ 7 \\ 5 \\ 10 \\ 4 \\ 5 \\ 8 \\ 5 \\ 4 \\ 8 \\ +\ 40 \\ \hline 108 \end{array} \qquad 12\overline{)108} = 9$$

(b) $40 - 3 = $ **37**

3.

	Ratio	Actual Count
Red marbles	7	R
Blue marbles	5	B
Total	12	600

(a)
$$\begin{aligned} \frac{5}{12} &= \frac{B}{600} \\ 12B &= 3000 \\ B &= \mathbf{250\ marbles} \end{aligned}$$

(b) $\mathbf{\frac{5}{12}}$

4.

	Case 1	Case 2
Pterodactyls	500	p
Minutes	20	90

$$\begin{aligned} \frac{500}{20} &= \frac{p}{90} \\ 20p &= 45{,}000 \\ p &= \mathbf{2250\ plastic\ pterodactyls} \end{aligned}$$

5. (a)

	Percent	Actual Count
Original	100	\$24
− Change	25	C
New	75	N

$$\begin{aligned} \frac{100}{75} &= \frac{\$24}{N} \\ 100N &= \$1800 \\ N &= \mathbf{\$18} \end{aligned}$$

(b)

	Percent	Actual Count
Original	100	O
− Change	25	C
New	75	\$24

$$\begin{aligned} \frac{100}{75} &= \frac{O}{\$24} \\ 75O &= \$2400 \\ O &= \mathbf{\$32} \end{aligned}$$

6.
$$(-3x^2)(2xy)(-x)(3y^2)$$
$$(-3) \cdot x \cdot x \cdot (2) \cdot x \cdot y \cdot (-1) \cdot x \cdot (3) \cdot y \cdot y$$
$$(-3)(2)(-1)(3) \cdot x \cdot x \cdot x \cdot x \cdot y \cdot y \cdot y$$
$$\mathbf{18x^4y^3}$$

7. $\mathbf{\frac{2}{50} = \frac{1}{25}}$

8. $7 \text{ days} \cdot \frac{24 \text{ hours}}{1 \text{ day}} \cdot \frac{60 \text{ minutes}}{1 \text{ hour}}$ (days, day, hours, hour crossed out)
= **10,080 minutes**

9.

45 cars

5 cars
5 cars
5 cars
5 cars
5 cars
5 cars
5 cars
5 cars
5 cars

$\frac{5}{9}$ were not cattle cars. (top five rows)
$\frac{4}{9}$ were cattle cars. (bottom four rows)

(a) **20 cattle cars**

(b) $\mathbf{55\frac{5}{9}\%}$

10. $\frac{1}{3}$ (>) 33%

11. $(-3)(-1) - (-3) - (-1)$
$(3) - (-3) - (-1)$
$\mathbf{7}$

12.
$$\begin{array}{r} \$7.95 \\ \$0.90 \\ \underline{\$2.35} \\ \$11.20 \end{array} \qquad \begin{array}{r} \$11.20 \\ \underline{\times \quad 0.05} \\ \$0.56 \end{array} \qquad \begin{array}{r} \$11.20 \\ \underline{+ \quad \$0.56} \\ \mathbf{\$11.76} \end{array}$$

13.

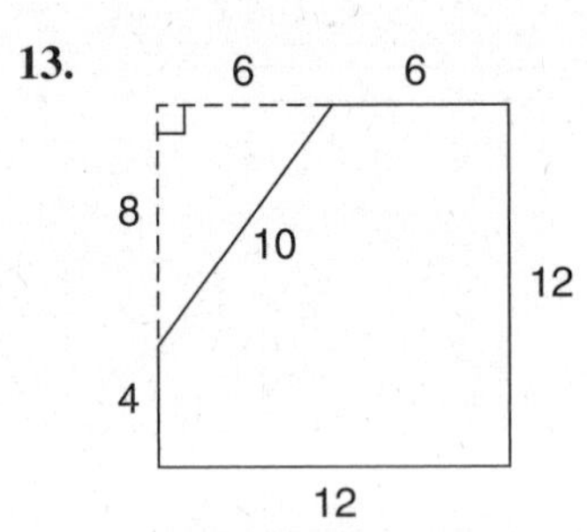

(a) Perimeter $= 6 \text{ in.} + 12 \text{ in.} + 12 \text{ in.} + 4 \text{ in.} + 10 \text{ in.} = \mathbf{44\ in.}$

(b) Area of square $= (12 \text{ in.})(12 \text{ in.}) = 144 \text{ in.}^2$

Area of triangle $= \dfrac{(6 \text{ in.})(8 \text{ in.})}{2} = 24 \text{ in.}^2$

Area of figure $= 144 \text{ in.}^2 - 24 \text{ in.}^2 = \mathbf{120\ in.^2}$

14. (a) $\mathbf{\dfrac{2}{25}}$

(b) **8%**

(c) $\mathbf{\dfrac{1}{12}}$

(d) $\mathbf{0.08\overline{3}}$

15.

	Percent	Actual Count
Original	100	\$3.60
+ Change	120	C
New	220	N

$\dfrac{100}{220} = \dfrac{\$3.60}{N}$
$100N = \$792$
$N = \mathbf{\$7.92}$

16. $(8 \times 10^{-3})(6 \times 10^{7})$
48×10^{4}
$(4.8 \times 10^{1})10^{4}$
$\mathbf{4.8 \times 10^{5}}$

17. (a) $(10 \text{ cm})(10 \text{ cm})(10 \text{ cm}) = \mathbf{1000\ cm^3}$

(b) $6(100 \text{ cm}^2) = \mathbf{600\ cm^2}$

18. (a) $A \approx 3.14(100 \text{ cm}^2)$
$A \approx \mathbf{314\ cm^2}$

(b) $C \approx 3.14(20 \text{ cm})$
$C \approx \mathbf{62.8\ cm}$

19. $-x + 2x^2 - 1 + x - x^2$
$2x^2 - x^2 - x + x - 1$
$\mathbf{x^2 - 1}$

20. $y = 2(1) + 3$
$y = 2 + 3$
$y = \mathbf{5;}$
$y = 2(0) + 3$
$y = 0 + 3$
$y = \mathbf{3;}$
$y = 2(-2) + 3$
$y = -4 + 3$
$y = \mathbf{-1}$

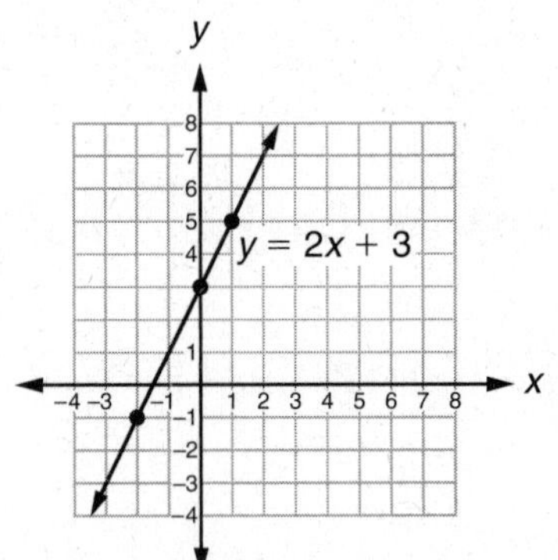

21. $\mathbf{60 = \dfrac{3}{8} \times W_N}$

$\dfrac{8}{3} \cdot 60 = \left(\dfrac{8}{3}\right)\dfrac{3}{8} \times W_N$

$\mathbf{160 = W_N}$

22. $2x - 5 > -1$
$2x - 5 + 5 > -1 + 5$
$2x > 4$
$\dfrac{2x}{2} > \dfrac{4}{2}$
$\mathbf{x > 2}$

$x > 2$

0 1 2 3 4

23. (a) $m\angle x = 180° - (90° + 50°) = \mathbf{40°}$
$m\angle y = m\angle x = \mathbf{40°}$
$m\angle z = 180° - (90° + 40°) = \mathbf{50°}$

(b) **Yes. The triangles have the same shape. Their corresponding angles are congruent.**

24.

$$\begin{array}{r} 0.42 \\ +\ 0.45 \\ \hline \mathbf{0.87} \end{array}$$

25.

$$\begin{aligned} 3x + 2 &= 9 \\ 3x + 2 - 2 &= 9 - 2 \\ 3x &= 7 \\ \frac{3x}{3} &= \frac{7}{3} \\ x &= \mathbf{\frac{7}{3}} \end{aligned}$$

26.

$$\begin{aligned} \frac{2}{3}w + 4 &= 14 \\ \frac{2}{3}w + 4 - 4 &= 14 - 4 \\ \frac{2}{3}w &= 10 \\ \left(\frac{3}{2}\right)\frac{2}{3}w &= \left(\frac{3}{2}\right)10 \\ w &= \mathbf{15} \end{aligned}$$

27.

$$\begin{aligned} 0.2y - 1 &= 7 \\ 0.2y - 1 + 1 &= 7 + 1 \\ 0.2y &= 8 \\ \frac{0.2y}{0.2} &= \frac{8}{0.2} \\ y &= \mathbf{40} \end{aligned}$$

28.

$$\begin{aligned} -\frac{2}{3}m &= -6 \\ \left(-\frac{3}{2}\right)\left(-\frac{2}{3}m\right) &= \left(-\frac{3}{2}\right)(-6) \\ m &= \mathbf{9} \end{aligned}$$

29. $3(2^3 + \sqrt{16}) - 4^0 - 8 \cdot 2^{-3}$

$$\begin{gathered} 3(8 + 4) - 1 - 8 \cdot \frac{1}{8} \\ 3(12) - 1 - 1 \\ 36 - 2 \\ \mathbf{34} \end{gathered}$$

30. (a) $\dfrac{(-9)(+6)(-5)}{(-4) - (-1)}$

$$\begin{gathered} \frac{(-54)(-5)}{-3} \\ \frac{270}{-3} \\ \mathbf{-90} \end{gathered}$$

(b) $-3(4) + 2(3) - 1$

$$\begin{gathered} (-12) + (6) - 1 \\ \mathbf{-7} \end{gathered}$$

Lesson 94, Warm-Up

a. **24**

b. $\mathbf{6 \times 10^{-5}}$

c. **0.6**

d. **86°F**

e. **16**

f. **55**

g. **5**

Problem Solving

$$\begin{aligned} \mathbf{4X + 1.7} &= \mathbf{2X + 4.3} \\ 4X - 2X + 1.7 &= 2X - 2X + 4.3 \\ 2X + 1.7 - 1.7 &= 4.3 - 1.7 \\ 2X \div 2 &= 2.6 \div 2 \\ X &= \mathbf{1.3\ lb} \end{aligned}$$

Lesson 94, Lesson Practice

a.

$$\begin{aligned} p(4, 5) &= \frac{1}{6} \cdot \frac{1}{6} = \frac{1}{36} \\ p(3, 6) &= \frac{1}{6} \cdot \frac{1}{6} = \frac{1}{36} \\ p(5, 4) &= \frac{1}{6} \cdot \frac{1}{6} = \frac{1}{36} \\ p(6, 3) &= \frac{1}{6} \cdot \frac{1}{6} = \frac{1}{36} \end{aligned}$$

$$\frac{1}{36} + \frac{1}{36} + \frac{1}{36} + \frac{1}{36} = \mathbf{\frac{4}{36}} = \mathbf{\frac{1}{9}}$$

b.

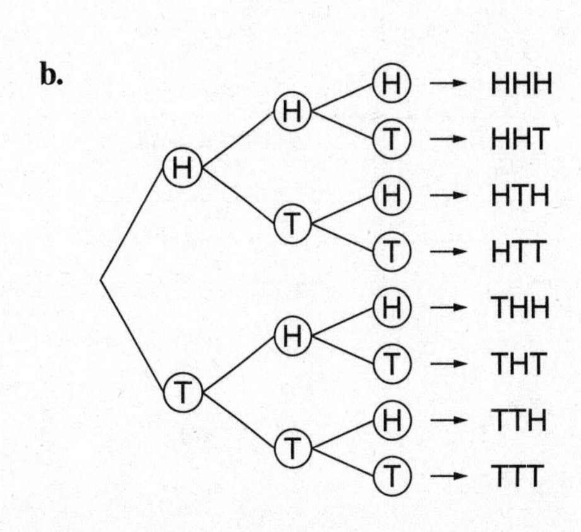

c. $\frac{1}{3} \cdot \frac{1}{4} = \frac{1}{12}$

d. Second Draw (columns); First Draw (rows)

	Red	White	Blue
Red	R, R	R, W	R, B
White	W, R	W, W	W, B
Blue	B, R	B, W	B, B

e. Second Draw (columns); First Draw (rows)

	Red	White	Blue
Red		R, W	R, B
White	W, R		W, B
Blue	B, R	B, W	

f. **Independent; dependent**

LESSON 94, MIXED PRACTICE

1. $$\begin{array}{r} 21{,}000{,}000{,}000 \\ -\ 9{,}800{,}000{,}000 \\ \hline 11{,}200{,}000{,}000 \end{array}$$ $\mathbf{1.12 \times 10^{10}}$

2. $\frac{96 \text{ miles} + 240 \text{ miles}}{2 \text{ hours} + 4 \text{ hours}}$

$= \frac{336 \text{ miles}}{6 \text{ hours}} =$ **56 miles per hour**

3. $\frac{\$8.40}{10 \text{ pounds}} = \frac{\$0.84}{1 \text{ pound}}$

$\frac{\$10.50}{15 \text{ pounds}} = \frac{\$0.70}{1 \text{ pound}}$

$$\begin{array}{r} \$0.84 \\ -\ \$0.70 \\ \hline \end{array}$$

10-pound box; $0.14 per pound more

4. $\frac{6}{12} = \mathbf{\frac{1}{2}}$

5.

	Ratio	Actual Count
Won	3	12
Lost	2	L
Total	5	T

$\frac{3}{5} = \frac{12}{T}$

$3T = 5(12)$

$3T = 60$

$T =$ **20 games**

6.

	Case 1	Case 2
First number	24	42
Second number	36	n

$\frac{24}{36} = \frac{42}{n}$

$24n = 1512$

$n = \mathbf{63}$

7. $100\% - 20\% = \mathbf{80\%}$

8.

	Percent	Actual Count
Original	100	O
− Change	20	C
New	80	$20

$\frac{100}{80} = \frac{O}{\$20}$

$80O = \$2000$

$O =$ **$25**

9. (a) $12 \cancel{\text{ft}} \cdot \frac{12 \text{ inches}}{1 \cancel{\text{ft}}} \cdot \frac{12 \text{ inches}}{1 \cancel{\text{ft}}}$

= **1728 square inches**

(b) $1 \cancel{\text{km}} \cdot \frac{1000 \cancel{\text{m}}}{1 \cancel{\text{km}}} \cdot \frac{1000 \text{ mm}}{1 \cancel{\text{m}}}$

= **1,000,000 millimeters**

10. 300 male serfs

$\frac{2}{5}$ were conscripted (120).
- 60 male serfs
- 60 male serfs

$\frac{3}{5}$ were not conscripted.
- 60 male serfs
- 60 male serfs
- 60 male serfs

(a) $120 \div 2 = 60$

$5(60 \text{ male serfs}) =$ **300 male serfs**

(b) $3(60 \text{ male serfs}) =$ **180 male serfs**

11. (a) **0**

(b) $p(1, 1) = \frac{1}{6} \cdot \frac{1}{6} = \mathbf{\frac{1}{36}}$

(c) $p(1, 2) = \frac{1}{6} \cdot \frac{1}{6} = \frac{1}{36}$

$p(2, 1) = \frac{1}{6} \cdot \frac{1}{6} = \frac{1}{36}$

$\frac{1}{36} + \frac{1}{36} = \mathbf{\frac{2}{36}} = \mathbf{\frac{1}{18}}$

12. $y = 4(-2) - 3$

$y = -8 - 3$

$y = \mathbf{-11}$

13. $\frac{4 \text{ yards}}{4} = 1 \text{ yard} = 3 \text{ feet}$

$(3 \text{ feet})(3 \text{ feet})$

= **9 square feet**

14. (a)
$$\begin{array}{r} \$14{,}500 \\ \times \quad 0.065 \\ \hline \mathbf{\$942.50} \end{array}$$

(b)
$$\begin{array}{r} \$14{,}500 \\ + \quad 942.50 \\ \hline \mathbf{\$15{,}442.50} \end{array}$$

(c)
$$\begin{array}{r} \$14{,}500 \\ \times \quad 0.02 \\ \hline \mathbf{\$290} \end{array}$$

15. (a) $\mathbf{\frac{2}{3}}$

(b) $\mathbf{0.\overline{6}}$

(c) **1.75**

(d) **175%**

16. (a) $2 \times \$7.50 = \mathbf{\$15.00}$

(b) $\frac{100}{300} = \frac{\$7.50}{P}$

$P = \mathbf{\$22.50}$

17. $(2 \times 10^8)(8 \times 10^2)$
16×10^{10}
$(1.6 \times 10^1) \times 10^{10}$
$\mathbf{1.6 \times 10^{11}}$

18. (8 cubes)(6 cubes)(2 cubes)
= **96 cubes**

19. Area of square = (14 in.)(14 in.)
= 196 in.^{2}

Area of circle $\approx \frac{22}{7}$ (49 in.^{2})
$\approx$ 154 in.^{2}

$$\begin{array}{r} 196 \text{ in.}^2 \\ -\ 154 \text{ in.}^2 \\ \hline \mathbf{42 \text{ in.}^2} \end{array}$$

20.
$$\begin{array}{r} 65.4545\ldots \\ 0.11\overline{)7.200000} \\ \underline{6\,6} \\ 60 \\ \underline{55} \\ 50 \\ \underline{44} \\ 60 \\ \underline{55} \\ 50 \\ \underline{44} \\ 60 \\ \underline{55} \\ 5 \end{array}$$

$65.4545\ldots = \mathbf{65.\overline{45}}$

21. $y = 3(3) - 2$
$y = 9 - 2$
$y = \mathbf{7}$;
$y = 3(0) - 2$
$y = 0 - 2$
$y = \mathbf{-2}$;
$y = 3(-1) - 2$
$y = -3 - 2$
$y = \mathbf{-5}$

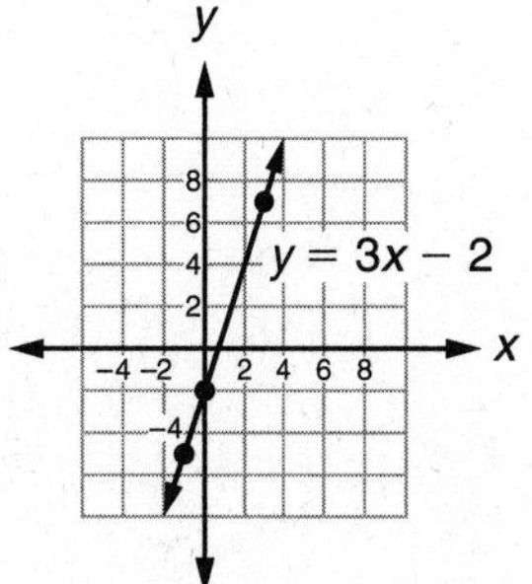

22.
$2x - 5 < -1$
$2x - 5 + 5 < -1 + 5$
$2x < 4$
$\frac{2x}{2} < \frac{4}{2}$
$\mathbf{x < 2}$

x < 2

−1 0 1 2 3

23. (a) $m\angle AOB = \frac{90°}{3} = \mathbf{30°}$

(b) $m\angle EOC = 90° + 45° = \mathbf{135°}$

24. AB is $1\frac{3}{4}$ in.

BC is $1\frac{1}{2}$ in.

$$\begin{array}{rcr} 1\frac{3}{4} \text{ in.} & \longrightarrow & 1\frac{3}{4} \text{ in.} \\ -\ 1\frac{1}{2} \text{ in.} & \longrightarrow & -\ 1\frac{2}{4} \text{ in.} \\ \hline & & \mathbf{\frac{1}{4} \text{ in.}} \end{array}$$

25.
$1.2p + 4 = 28$
$1.2p + 4 - 4 = 28 - 4$
$1.2p = 24$
$\frac{1.2p}{1.2} = \frac{24}{1.2}$
$p = \mathbf{20}$

26.
$$-6\frac{2}{3}m = 1\frac{1}{9}$$
$$-\frac{20}{3}m = \frac{10}{9}$$
$$\left(-\frac{3}{20}\right)\left(-\frac{20}{3}m\right) = \left(-\frac{3}{20}\right)\left(\frac{10}{9}\right)$$
$$m = \mathbf{-\frac{1}{6}}$$

27. (a) $6x^2 + 3x - 2x - 1$
$\mathbf{6x^2 + x - 1}$

(b)
$$(5x)(3x) - (5x)(-4)$$
$$[(5) \cdot x \cdot (3) \cdot x] - [(5) \cdot x \cdot (-4)]$$
$$[(5)(3) \cdot x \cdot x] - [(5)(-4) \cdot x]$$
$$15x^2 - (-20x)$$
$$\mathbf{15x^2 + 20x}$$

28. (a) $\dfrac{-8 - (-6) - (4)}{-3}$

$\dfrac{-8 + 6 - 4}{-3}$

$\dfrac{-6}{-3}$

2

(b) $-5(-4) - 3(-2) - 1$
$20 - (-6) - 1$
25

29. $(-2)^2 - 4(-1)(3)$
$4 - 4(-3)$
$4 - (-12)$
16

30. $\mathbf{Q_H D_H}$
$\mathbf{Q_H D_T}$
$\mathbf{Q_T D_H}$
$\mathbf{Q_T D_T}$

Lesson 95, Warm-Up

a. **−28**

b. $\mathbf{6.25 \times 10^{12}}$

c. **0.9**

d. **77°F**

e. **22**

f. **90**

g. **10**

Problem Solving

Because both factors have only two digits and the product has a 9 in the thousands place, both factors must be close to 100. We write a 9 in the tens place of both factors. We write a 7 in the ones place of the bottom factor because $3 \times 7 = 21$. To find the remaining missing digits, we multiply 93 by 97.

```
   93
 × 97
  651
 837
 9021
```

Lesson 95, Lesson Practice

a. Area of base $= \dfrac{(8\text{ cm})(6\text{ cm})}{2} = 24\text{ cm}^2$
Volume $= (24\text{ cm}^2)(12\text{ cm}) = \mathbf{288\text{ cm}^3}$

b. Area of base $= \dfrac{(10\text{ cm})(6\text{ cm})}{2}$
$= 30\text{ cm}^2$
Volume $= (30\text{ cm}^2)(12\text{ cm}) = \mathbf{360\text{ cm}^3}$

c. Area of base $= \pi(3\text{ cm})^2 = 9\pi\text{ cm}^2$
Volume $= (9\pi\text{ cm}^2)(10\text{ cm}) = \mathbf{90\pi\text{ cm}^3}$

d. Area of base $= (7\text{ cm})(2\text{ cm})$
$+ (3\text{ cm})(3\text{ cm})$
$= 14\text{ cm}^2 + 9\text{ cm}^2 = 23\text{ cm}^2$
Volume $= (23\text{ cm}^2)(10\text{ cm}) = \mathbf{230\text{ cm}^3}$

e. Area of base $= \pi(1\text{ cm})^2 = \pi\text{ cm}^2$
Volume $= (\pi\text{ cm}^2)(10\text{ cm}) = \mathbf{10\pi\text{ cm}^3}$

Lesson 95, Mixed Practice

1. $\$1.40 + 40(\$0.35)$
$= \$1.40 + \$14.00 = \$15.40$
$\dfrac{\$15.40}{4\text{ miles}} = \mathbf{\$3.85\text{ per mile}}$

2.

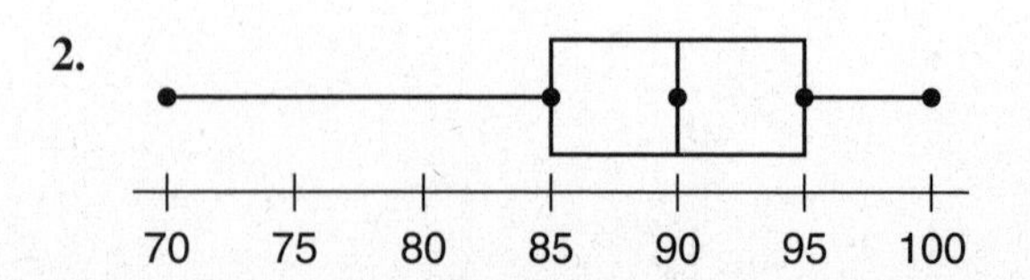

3.

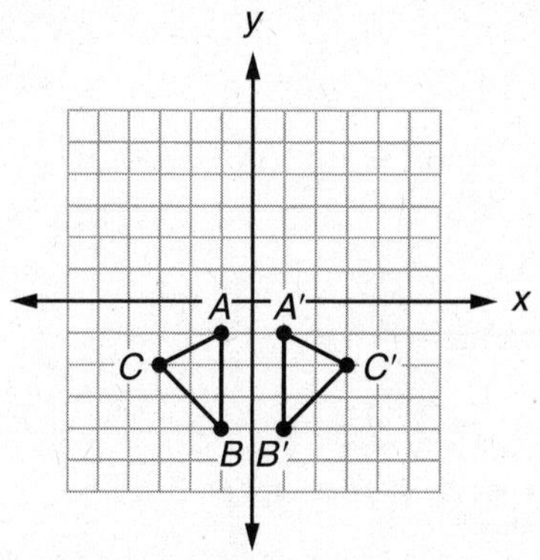

$\mathbf{A'(1, -1), B'(1, -4), C'(3, -2)}$

4. $\$6\left(4\frac{1}{3}\right) = \$6\left(\frac{13}{3}\right) = \mathbf{\$26}$

5.
$$\text{Area of rectangle} = (12)(8) = 96$$
$$\text{Area of triangle} = \frac{(6)(8)}{2} = 24$$
$$\frac{24}{96} = \frac{1}{4} = \frac{\text{shaded area}}{\text{total area}}$$
$$\frac{\text{unshaded area}}{\text{total area}} = \frac{3}{4}$$
$$\frac{\text{shaded area}}{\text{unshaded area}} = \mathbf{\frac{1}{3}}$$

6.
$$1 \text{ ton} = 2000 \text{ pounds}$$
$$\frac{600}{\$7.20} = \frac{2000}{p}$$
$$600p = \$14{,}400$$
$$p = \mathbf{\$24.00}$$

7.

	Percent	Actual Count
Original	100	*O*
+ Change	30	*C*
New	130	$3.90

$$\frac{100}{130} = \frac{O}{\$3.90}$$
$$130O = \$390$$
$$O = \mathbf{\$3 \text{ per unit}}$$

8.

	Percent	Actual Count
Original	100	$3.90
+ Change	30	*C*
New	130	*N*

$$\frac{100}{130} = \frac{\$3.90}{N}$$
$$100N = \$507$$
$$N = \mathbf{\$5.07}$$

9. $\overset{10}{\cancel{1000}}\ \cancel{\text{mm}^2} \cdot \frac{1 \text{ cm}}{\cancel{10}\ \cancel{\text{mm}}} \cdot \frac{1 \text{ cm}}{\cancel{10}\ \cancel{\text{mm}}}$
$= \mathbf{10 \text{ cm}^2}$

10.

150 Lilliputians

$\frac{3}{5}$ believed.
- 30 Lilliputians
- 30 Lilliputians
- 30 Lilliputians

$\frac{2}{5}$ did not believe (60).
- 30 Lilliputians
- 30 Lilliputians

(a) $60 \div 2 = 30$
$5(30 \text{ Lilliputians}) = \mathbf{150 \text{ Lilliputians}}$

(b) $3(30 \text{ Lilliputians}) = \mathbf{90 \text{ Lilliputians}}$

11. **Insufficient information**

12.
$$(-2)[(-2) + (-3)]$$
$$-2[-5]$$
$$\mathbf{10}$$

13. (a)
$$p(1, 6) = \frac{1}{6} \cdot \frac{1}{6} = \frac{1}{36}$$
$$p(3, 4) = \frac{1}{6} \cdot \frac{1}{6} = \frac{1}{36}$$
$$p(6, 1) = \frac{1}{6} \cdot \frac{1}{6} = \frac{1}{36}$$
$$p(4, 3) = \frac{1}{6} \cdot \frac{1}{6} = \frac{1}{36}$$
$$p(2, 5) = \frac{1}{6} \cdot \frac{1}{6} = \frac{1}{36}$$
$$p(5, 2) = \frac{1}{6} \cdot \frac{1}{6} = \frac{1}{36}$$
$$\frac{1}{36} + \frac{1}{36} + \frac{1}{36} + \frac{1}{36} + \frac{1}{36} + \frac{1}{36}$$
$$= \mathbf{\frac{6}{36} = \frac{1}{6}}$$

(b)
$$p(1, 1 \text{ or } 2 \text{ or } 3 \text{ or } 4 \text{ or } 5) = \frac{1}{6} \cdot \frac{5}{6} = \frac{5}{36}$$
$$p(2, 1 \text{ or } 2 \text{ or } 3 \text{ or } 4) = \frac{1}{6} \cdot \frac{4}{6} = \frac{4}{36}$$
$$p(3, 1 \text{ or } 2 \text{ or } 3) = \frac{1}{6} \cdot \frac{3}{6} = \frac{3}{36}$$
$$p(4, 1 \text{ or } 2) = \frac{1}{6} \cdot \frac{2}{6} = \frac{2}{36}$$
$$p(5, 1) = \frac{1}{6} \cdot \frac{1}{6} = \frac{1}{36}$$
$$\frac{5}{36} + \frac{4}{36} + \frac{3}{36} + \frac{2}{36} + \frac{1}{36}$$
$$= \mathbf{\frac{15}{36} = \frac{5}{12}}$$

14.
$$\text{Area of base} = \frac{(30 \text{ mm})(40 \text{ mm})}{2}$$
$$= 600 \text{ mm}^2$$
$$\text{Volume} = (600 \text{ mm}^2)(50 \text{ mm})$$
$$= \mathbf{30{,}000 \text{ mm}^3}$$

15.
$$\text{Area of base} \approx 3.14(9 \text{ cm}^2)$$
$$\approx 28.26 \text{ cm}^2$$
$$\text{Volume} \approx (28.26 \text{ cm}^2)(10 \text{ cm})$$
$$\approx \mathbf{282.6 \text{ cm}^3}$$

16. $3(\$1.25) + 2(\$0.95) + \$1.30$
$= \$3.75 + \$1.90 + \$1.30 = \$6.95$

$$\begin{array}{r} \$6.95 \\ \times \ \ 0.06 \\ \hline \$0.417 \end{array} \longrightarrow \$0.42 \qquad \begin{array}{r} \$6.95 \\ + \ \$0.42 \\ \hline \mathbf{\$7.37} \end{array}$$

17.

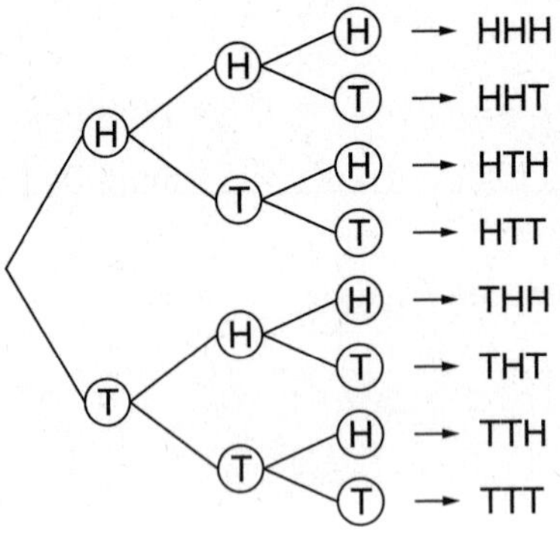

18. (a) $(-2xy)(-2x)(x^2y)$
$(-2) \cdot x \cdot y \cdot (-2) \cdot x \cdot x \cdot x \cdot y$
$(-2)(-2) \cdot x \cdot x \cdot x \cdot x \cdot y \cdot y$
$\mathbf{4x^4y^2}$

(b) $6x - 4y + 3 - 6x - 5y - 8$
$6x - 6x - 4y - 5y + 3 - 8$
$\mathbf{-9y - 5}$

19. 6
32×10^{-2}
$(3.2 \times 10^1) \times 10^{-2}$
$\mathbf{3.2 \times 10^{-1}}$

20. (a) $y = \frac{1}{2}(6) + 1$
$y = 3 + 1$
$y = \mathbf{4;}$
$y = \frac{1}{2}(4) + 1$
$y = 2 + 1$
$y = \mathbf{3;}$
$y = \frac{1}{2}(-2) + 1$
$y = -1 + 1$
$y = \mathbf{0}$

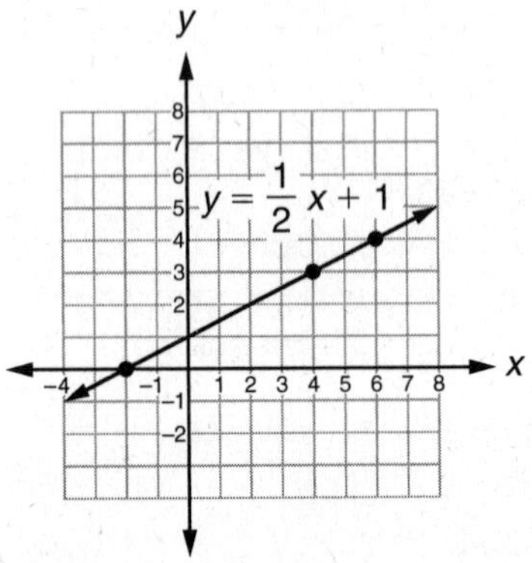

(b) **(0, 1)**

21. (a) $m\angle x = 180° - (90° + 55°) = \mathbf{35°}$

(b) $m\angle y = 180° - (90° + 35°) = \mathbf{55°}$

(c) $m\angle A = 180° - (110° + 55°)$
$= \mathbf{15°}$

22.

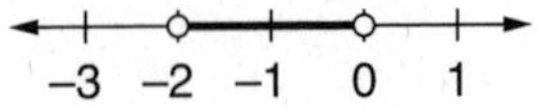

23. $(1.52 + 1.56) \div 2 = \mathbf{1.54}$

24.
$$-5w + 11 = 51$$
$$-5w + 11 - 11 = 51 - 11$$
$$-5w = 40$$
$$\frac{-5w}{-5} = \frac{40}{-5}$$
$$w = \mathbf{-8}$$

25.
$$\frac{4}{3}x - 2 = 14$$
$$\frac{4}{3}x - 2 + 2 = 14 + 2$$
$$\frac{4}{3}x = 16$$
$$\left(\frac{3}{4}\right)\frac{4}{3}x = \left(\frac{3}{4}\right)16$$
$$x = \mathbf{12}$$

26.
$$0.9x + 1.2 \le 3$$
$$0.9x + 1.2 - 1.2 \le 3 - 1.2$$
$$0.9x \le 1.8$$
$$\frac{0.9x}{0.9} \le \frac{1.8}{0.9}$$
$$x \le \mathbf{2}$$

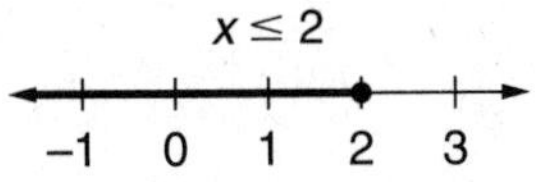

27.
$$\frac{10^3 \cdot 10^2}{10^5} - 10^{-1}$$
$$= \frac{10^5}{10^5} - \frac{1}{10}$$
$$= 1 - \frac{1}{10} = \frac{10}{10} - \frac{1}{10} = \mathbf{\frac{9}{10}} \text{ or } \mathbf{0.9}$$

28. $\sqrt{1^3 + 2^3} + (1 + 2)^3$
$= \sqrt{1 + 8} + (3)^3 = \sqrt{9} + 27$
$= 3 + 27 = \mathbf{30}$

29. $5 - 2\frac{2}{3}\left(1\frac{3}{4}\right) = 5 - \frac{8}{3}\left(\frac{7}{4}\right)$
$= 5 - \frac{14}{3} = 4\frac{3}{3} - 4\frac{2}{3} = \mathbf{\frac{1}{3}}$

30. (a) $\frac{(-10) + (-8) - (-6)}{(-2)(+3)}$

$\frac{-18 - (-6)}{-6}$

$\frac{-12}{-6}$

2

(b) $-8 + 3(-2) - 6$

$-8 + (-6) - 6$

$\mathbf{-20}$

Lesson 96, Warm-Up

a. **23**

b. **1×10^{13}**

c. **20**

d. **59°F**

e. **15**

f. **$100**

g. **$7.50**

Problem Solving

$\frac{1}{4} = \frac{3}{12}; \frac{1}{6} = \frac{2}{12}$

$\frac{3}{12} + \frac{2}{12} + \frac{1}{12} = \frac{6}{12} = \frac{1}{2}$

$\frac{1}{2} \div 3 = \mathbf{\frac{1}{6}}$

Lesson 96, Lesson Practice

a. $4 \times 6° = \mathbf{24°}$

b. $20 \times 6° = \mathbf{120°}$

c. $7 \times 6° = \mathbf{42°}$

d. **See student work; 55°**

e. **See student work; 15°**

f. **See student work; 45°**

g. **See student work; 145°**

h. $x(x - y)$

$\mathbf{x^2 - xy}$

i. $-3(2x - 1)$

$\mathbf{-6x + 3}$

j. $-x(x - 2)$

$\mathbf{-x^2 + 2x}$

k. $-2(4 - 3x)$

$\mathbf{-8 + 6x}$

l. $x^2 + 2x - 3(x + 2)$

$x^2 + 2x - 3x - 6$

$\mathbf{x^2 - x - 6}$

m. $x^2 - 2x - 3(x - 2)$

$x^2 - 2x - 3x + 6$

$\mathbf{x^2 - 5x + 6}$

Lesson 96, Mixed Practice

1. $3(\$280) + 5(\$240)$

$= \$840 + \$1200 = \$2040$

$8\overline{)\$2040}$ = **$255 per ton**

2. $\frac{9^2}{\sqrt{9}} = \frac{81}{3} = \mathbf{27}$

3. $\frac{2000 \text{ miles}}{25 \text{ miles per gallon}} = 80 \text{ gallons}$

$\frac{80 \text{ gallons}}{16 \text{ gallons per tank}} = \mathbf{5\ tanks}$

4. $\frac{\text{vertices}}{\text{edges}} = \mathbf{\frac{2}{3}}$

5.

	Case 1	Case 2
Dollars	12	d
Yuan	100	475

$\frac{12}{100} = \frac{d}{475}$

$100d = 5700$

$d = \mathbf{57\ dollars}$

6.

	Percent	Actual Count
Original	100	O
− Change	20	C
New	80	60

$\frac{100}{80} = \frac{O}{60}$

$80O = 6000$

$O = \mathbf{75}$

7.

	Percent	Actual Count
Original	100	120
+ Change	25	C
New	125	N

$$\frac{100}{125} = \frac{120}{N}$$
$$100N = 15{,}000$$
$$N = \textbf{150 customers per day}$$

8. (a) $\mathbf{60 = W_P \times 50}$
$$\frac{60}{50} = \frac{W_P \times 50}{50}$$
$$\frac{6}{5} = W_P$$
$$W_P = \frac{6}{5} \times 100\% = \mathbf{120\%}$$

(b) $\mathbf{50 = W_P \times 60}$
$$\frac{50}{60} = \frac{W_P \times 60}{60}$$
$$\frac{5}{6} = W_P$$
$$W_P = \frac{5}{6} \times 100\% = \mathbf{83\frac{1}{3}\%}$$

9. $1.2\,\cancel{m^2} \cdot \frac{100\text{ cm}}{1\,\cancel{m}} \cdot \frac{100\text{ cm}}{1\,\cancel{m}}$
$$= \mathbf{12{,}000\text{ cm}^2}$$

10. (a) **Angles: $\angle A$ and $\angle E$, $\angle B$ and $\angle D$, $\angle ACB$ (or $\angle BCA$) and $\angle ECD$ (or $\angle DCE$)**
Sides: $\overline{AB}$ (or $\overline{BA}$) and $\overline{ED}$ (or $\overline{DE}$), $\overline{BC}$ (or $\overline{CB}$) and $\overline{DC}$ (or $\overline{CD}$), $\overline{AC}$ (or $\overline{CA}$) and $\overline{EC}$ (or $\overline{CE}$)

(b) $m\angle ECD = 90° - 53° = \mathbf{37°}$

11. $x + y \; \textcircled{>} \; x - y$

12. $(-2)[(-4) + (-3)]$
$$-2[-7]$$
$$\mathbf{14}$$

13. 1 yard = 36 inches
Area = (9 inches)(9 inches)
= **81 square inches**

14. (a) $p(3, 3) = \frac{1}{4} \cdot \frac{1}{4} = \mathbf{\frac{1}{16}}$

(b) $p(1, 1, 1, 1) = \frac{2}{4} \cdot \frac{2}{4} \cdot \frac{2}{4} \cdot \frac{2}{4}$
$$= \frac{16}{256} = \mathbf{\frac{1}{16}}$$

15. (a) Area of base = (3 cm)(3 cm)
= 9 cm^2
Volume = (9 cm^2)(3 cm) = **27 cm^3**

(b) Area of base $= \frac{(4\text{ cm})(6\text{ cm})}{2}$
= 12 cm^2
Volume = (12 cm^2)(5 cm) = **60 cm^3**

16.
$$\begin{array}{r} \$14.50 \\ \times \quad 20 \\ \hline \$290 \end{array} \qquad \begin{array}{r} \$290 \\ \times \; 0.07 \\ \hline \$20.30 \end{array} \qquad \begin{array}{r} \$290 \\ + \; \$20.30 \\ \hline \mathbf{\$310.30} \end{array}$$

17. (a) $\mathbf{\frac{3}{80}}$

(b) **0.0375**

18.

	Percent	Actual Count
Original	100	\$24
– Change	$33\frac{1}{3}$	C
New	$66\frac{2}{3}$	N

(a) $\frac{100}{33\frac{1}{3}} = \frac{\$24}{C}$
$$100C = \$24\left(33\frac{1}{3}\right)$$
$$100C = \frac{\$2400}{3}$$
$$C = \mathbf{\$8}$$

(b)
$$\begin{array}{r} \$24 \\ - \;\; \$8 \\ \hline \mathbf{\$16} \end{array}$$

19. $24 \times 10^{-5} = \mathbf{2.4 \times 10^{-4}}$

20. (a) $C = \pi(12\text{ m})$
$C = \mathbf{12\pi\text{ m}}$

(b) $A = \pi(36\text{ m}^2)$
$A = \mathbf{36\pi\text{ m}^2}$

21. (a) $15 \times 6° = \mathbf{90°}$

(b) $25 \times 6° = \mathbf{150°}$

(c) $8 \times 6° = \mathbf{48°}$

22. (a) $\frac{360°}{8} = \mathbf{45°}$

(b) $180° - 45° = \mathbf{135°}$

23. **$Q'(8, -4)$, $R'(4, 0)$, $S'(0, -4)$, $T'(4, -8)$**

24.
$$0.8x + 1.5 < 4.7$$
$$0.8x + 1.5 - 1.5 < 4.7 - 1.5$$
$$0.8x < 3.2$$
$$\frac{0.8x}{0.8} < \frac{3.2}{0.8}$$
$$\mathbf{x < 4}$$

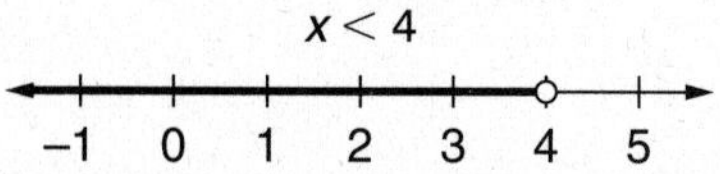

25.
$$2\frac{1}{2}x - 7 = 13$$
$$2\frac{1}{2}x - 7 + 7 = 13 + 7$$
$$2\frac{1}{2}x = 20$$
$$\left(\frac{2}{5}\right)\frac{5}{2}x = \left(\frac{2}{5}\right)20$$
$$x = \mathbf{8}$$

26.
$$-3x + 8 = -10$$
$$-3x + 8 - 8 = -10 - 8$$
$$-3x = -18$$
$$\frac{-3x}{-3} = \frac{-18}{-3}$$
$$x = \mathbf{6}$$

27. (a) $-3(x - 4)$
$\mathbf{-3x + 12}$

(b) $x(x + y)$
$\mathbf{x^2 + xy}$

28. (a)
$$\frac{(-4) - (-8)(-3)(-2)}{-2}$$
$$\frac{(-4) - (24)(-2)}{-2}$$
$$\frac{(-4) - (-48)}{-2}$$
$$\frac{44}{-2}$$
$$\mathbf{-22}$$

(b) $(-3)^2 + 3^2 = 9 + 9 = \mathbf{18}$

29. (a) $(-4ab^2)(-3b^2c)(5a)$
$(-4) \cdot a \cdot b \cdot b \cdot (-3) \cdot b \cdot b \cdot c \cdot (5) \cdot a$
$(-4)(-3)(5) \cdot a \cdot a \cdot b \cdot b \cdot b \cdot b \cdot c$
$\mathbf{60\,a^2b^4c}$

(b) $a^2 + ab - ab - b^2$
$\mathbf{a^2 - b^2}$

30. **A, A; A, B; A, C; B, A; B, B; B, C; C, A; C, B; C, C**

Lesson 97, Warm-Up

a. **−125**

b. **3×10^{12}**

c. **0.9**

d. **41°F**

e. **65**

f. **$60**

g. **$3\frac{1}{2}$**

Problem Solving

hexagon: **3; 4 × 180°**

n-gon: ***n* − 3; (*n* − 2) × 180°**

Lesson 97, Lesson Practice

a. **Corresponding angles: $\angle W$ and $\angle R$; $\angle Y$ and $\angle Q$; $\angle X$ and $\angle P$**
Corresponding sides: $\overline{YW}$ and $\overline{QR}$; $\overline{WX}$ and $\overline{RP}$; $\overline{XY}$ and $\overline{PQ}$

b. **See student work.**

c.
$$\frac{6}{9} = \frac{x}{12}$$
$$9x = 72$$
$$x = \mathbf{8}$$

d. **See student work.**

e.
$$\frac{6}{9} = \frac{12}{y}$$
$$6y = 108$$
$$y = \mathbf{18}$$

f.
$$\frac{H_T}{6\text{ ft}} = \frac{18\text{ ft}}{9\text{ ft}}$$
$$9\text{ ft} \cdot H_T = 18\text{ ft} \cdot 6\text{ ft}$$
$$H_T = \frac{\overset{6}{\cancel{18}}\text{ ft} \cdot \overset{2}{\cancel{6}}\,\cancel{\text{ft}}}{\underset{\underset{1}{\cancel{3}}}{\cancel{9}}\,\cancel{\text{ft}}}$$
$$H_T = \mathbf{12\text{ ft}}$$

g. **About 11 ft**

LESSON 97, MIXED PRACTICE

1. $\$8.95 \times 0.06 = \$0.537 \longrightarrow \$0.54$

$\$8.95 + \$0.54 = \$9.49$

$\$10.00 - \$9.49 = \mathbf{\$0.51}$

2. $2,000,000,000,000 - 300,000,000,000 = 1,700,000,000,000$ $\mathbf{1.7 \times 10^{12}}$

3. (a) **90**

(b) **90**

(c) **20**

4.

	Case 1	Case 2
Miles	24	m
Minutes	60	5

$$\frac{24}{60} = \frac{m}{5}$$
$$60m = 120$$
$$m = \mathbf{2\ miles}$$

5.

	Case 1	Case 2
Yards	3520	y
Minutes	5	8

$$\frac{3520}{5} = \frac{y}{8}$$
$$5y = 28{,}160$$
$$y = \mathbf{5632\ yards}$$

6. **Translation of 5 units to the right**

7. 1 yard = 36 inches

$$\frac{3}{\cancel{4}_1} \times \overset{9}{\cancel{36}} \text{ inches} = \mathbf{27\ inches}$$

8.

	Ratio	Actual Count
Leeks	5	L
Radishes	7	420

$$\frac{5}{7} = \frac{L}{420}$$
$$7L = 2100$$
$$L = \mathbf{300\ leeks}$$

9. $\mathbf{40 = 2.5 \times W_N}$

$$\frac{40}{2.5} = \frac{2.5 \times W_N}{2.5}$$
$$\mathbf{16} = W_N$$

10. $\mathbf{40 = W_P \times 60}$

$$\frac{40}{60} = \frac{W_P \times 60}{60}$$
$$\frac{2}{3} = W_P$$
$$W_P = \frac{2}{3} \times 100\% = \mathbf{66\frac{2}{3}\%}$$

11. $W_D = 0.4 \times 6$

$W_D = \mathbf{2.4}$

12.

	Percent	Actual Count
Original	100	O
+ Change	10	C
New	110	\$17,600

$$\frac{100}{110} = \frac{O}{\$17{,}600}$$
$$110O = \$1{,}760{,}000$$
$$O = \mathbf{\$16{,}000}$$

13. $$\frac{1.78 + 2.04}{2} = \mathbf{1.91}$$

14. (a) $\mathbf{3\frac{1}{4}}$

(b) **325%**

(c) $\mathbf{0.1\overline{6}}$

(d) $\mathbf{16\frac{2}{3}\%}$

15. $x + y$ ⓧ$<$ $x - y$ (circled: $<$)

16. $(5.4 \times 6) \times (10^8 \times 10^{-4})$
$= 32.4 \times 10^4 = \mathbf{3.24 \times 10^5}$

17. (a) $C \approx 3.14(20\text{ mm})$
$C \approx \mathbf{62.8\ mm}$

(b) $A \approx 3.14(100\text{ mm}^2)$
$A \approx \mathbf{314\ mm^2}$

18.

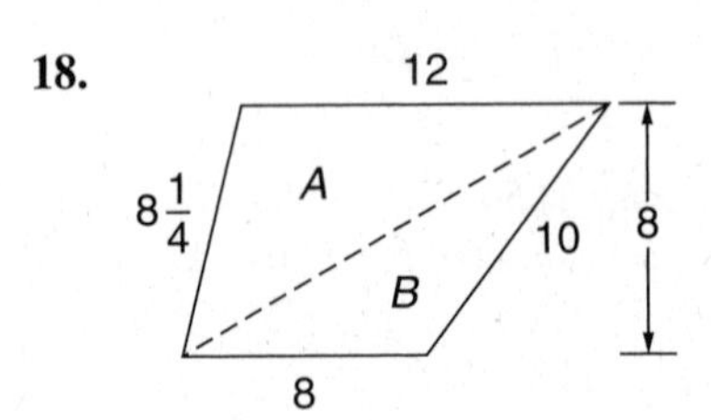

$$\text{Area of triangle } A = \frac{(12\text{ ft})(8\text{ ft})}{2} = 48\text{ ft}^2$$
$$+\ \text{Area of triangle } B = \frac{(8\text{ ft})(8\text{ ft})}{2} = 32\text{ ft}^2$$
$$\text{Area of figure} = \mathbf{80\ ft^2}$$

19. (a) Area of base $= \dfrac{(1 \text{ m})(2 \text{ m})}{2} = 1 \text{ m}^2$

Volume $= (1 \text{ m}^2)(2 \text{ m}) = \mathbf{2 \text{ m}^3}$

(b) Area of base $= \pi\,(1 \text{ m}^2) = \pi \text{ m}^2$

Volume $= (\pi \text{ m}^2)(1 \text{ m}) = \boldsymbol{\pi} \mathbf{\text{ m}^3}$

20. (a) $m\angle X = 180° - (120° + 25°)$
$= \mathbf{35°}$

(b) $m\angle Y = 180° - (90° + 35°) = \mathbf{55°}$

(c) $m\angle A = 180° - (60° + 55°)$
$= \mathbf{65°}$

21. $\dfrac{x}{12 \text{ cm}} = \dfrac{10 \cancel{\text{cm}}}{15 \cancel{\text{cm}}}$

$15 \cdot x = 120 \text{ cm}$

$x = \mathbf{8 \text{ cm}}$

22. See student diagram.

$\dfrac{H_P}{3 \text{ ft}} = \dfrac{72 \text{ ft}}{4 \text{ ft}}$

$4 \text{ ft} \times H_P = 3 \text{ ft} \times 72 \text{ ft}$

$H_P = \dfrac{3 \text{ ft} \times \overset{18}{\cancel{72}}\, \cancel{\text{ft}}}{\underset{1}{\cancel{4}}\, \cancel{\text{ft}}}$

$H_P = \mathbf{54 \text{ ft}}$

23. $\dfrac{(40{,}000)(600)}{80} = \mathbf{300{,}000}$

24.

$1.2m + 0.12 = 12$

$1.2m + 0.12 - 0.12 = 12 - 0.12$

$1.2m = 11.88$

$\dfrac{1.2m}{1.2} = \dfrac{11.88}{1.2}$

$m = \mathbf{9.9}$

25.

$1\frac{3}{4}y - 2 = 12$

$1\frac{3}{4}y - 2 + 2 = 12 + 2$

$1\frac{3}{4}y = 14$

$\left(\dfrac{\overset{1}{\cancel{4}}}{\underset{1}{\cancel{7}}}\right)\dfrac{\overset{1}{\cancel{7}}}{\underset{1}{\cancel{4}}}y = \left(\dfrac{4}{\underset{1}{\cancel{7}}}\right)\overset{2}{\cancel{14}}$

$y = \mathbf{8}$

26. $3x - y + 8 + x + y - 2$

$3x + x - y + y + 8 - 2$

$\mathbf{4x + 6}$

27. (a) $3(x - y)$
$\mathbf{3x - 3y}$

(b) $x(y - 3)$
$\mathbf{xy - 3x}$

28. $4\frac{1}{2} \div 1\frac{1}{8} = \dfrac{\overset{1}{\cancel{9}}}{\underset{1}{\cancel{2}}} \times \dfrac{\overset{4}{\cancel{8}}}{\underset{1}{\cancel{9}}} = 4$

$3\frac{1}{3} \div 4 = \dfrac{10}{3} \times \dfrac{1}{4} = \dfrac{10}{12} = \mathbf{\dfrac{5}{6}}$

29. $\dfrac{(-2) - (+3) + (-4)(-3)}{(-2) + (+3) - (+4)}$

$\dfrac{(-2) - (+3) + (+12)}{(-2) + (+3) - (+4)}$

$\dfrac{(-5) + (12)}{(1) - (+4)}$

$\dfrac{7}{-3}$

$\mathbf{-2\frac{1}{3}}$

30. A, B
A, C
B, A
B, C
C, A
C, B

Lesson 98, Warm-Up

a. **−45**

b. **1 × 10⁶**

c. $\mathbf{2\frac{1}{2}}$

d. **20 g**

e. **15**

f. **$80**

g. **21 mi**

Problem Solving

"Double the Celsius number":
$-10° \times 2 = -20°$
"Subtract 10%": $-20° - (-2°) = -18°$
"Add 32°": $-18° + 32° =$ **14°F**

Lesson 98, Lesson Practice

a. $\frac{6 \cdot \overset{2}{\cancel{24}}}{\underset{1}{\cancel{12}}} = \textbf{12 feet}$

b. $\frac{54}{\underset{3}{\cancel{36}}} \cdot \overset{1}{\cancel{12}} = \textbf{18 inches}$

c. $5 \times 3 = 15$
$7 \times 3 = \textbf{21}$

d. $3 \times 7 = 21$
$42 \div 7 = \textbf{6}$

e. $10f = 25$
$f = \frac{25}{10}$
$f = \textbf{2.5}$

f. $(2.5)^2 = \textbf{6.25}$

g.

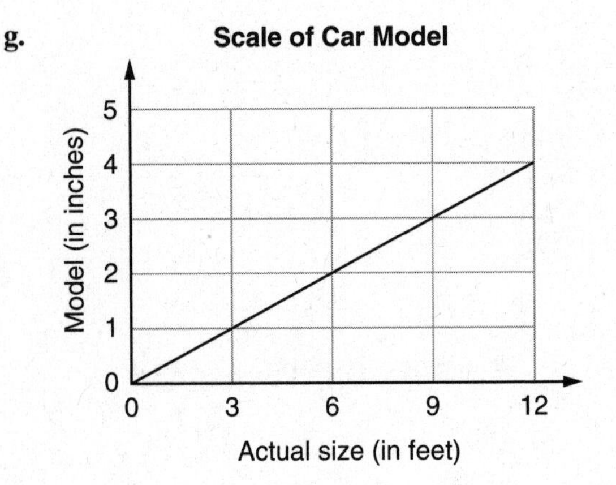

h. **C. 9 ft**

Lesson 98, Mixed Practice

1. (a) $p(2, 6) = \frac{1}{6} \cdot \frac{1}{6} = \frac{1}{36}$

$p(3, 5 \text{ or } 6) = \frac{1}{6} \cdot \frac{2}{6} = \frac{2}{36}$

$p(4, 4 \text{ or } 5 \text{ or } 6) = \frac{1}{6} \cdot \frac{3}{6} = \frac{3}{36}$

$p(5, 3 \text{ or } 4 \text{ or } 5 \text{ or } 6) = \frac{1}{6} \cdot \frac{4}{6} = \frac{4}{36}$

$p(6, 2 \text{ or } 3 \text{ or } 4 \text{ or } 5 \text{ or } 6) = \frac{1}{6} \cdot \frac{5}{6} = \frac{5}{36}$

$\mathbf{\frac{15}{36} = \frac{5}{12}}$

(b) **0**

2.

	Percent	Actual Count
Original	100	$45
− Change	20	C
New	80	N

$\frac{100}{80} = \frac{\$45}{N}$
$100N = \$3600$
$N = \mathbf{\$36}$

3.

	Case 1	Case 2
Dollars	5	d
Kroner	40	100

$\frac{5}{40} = \frac{d}{100}$
$40d = 500$
$d = \mathbf{\$12.50}$

4.

	Percent	Actual Count
Original	100	O
+ Change	25	20
New	125	N

$\frac{25}{125} = \frac{20}{N}$
$25N = 2500$
$N = \textbf{100 club members}$

5. $(3x)(x) - (x)(2x)$
$[(3) \cdot x \cdot x] - [x \cdot (2) \cdot x]$
$[(3) \cdot x \cdot x] - [(2) \cdot x \cdot x]$
$3x^2 - 2x^2$
$\mathbf{x^2}$

6. $\frac{6(10) + 9(15)}{15}$
$= \frac{60 + 135}{15} = \textbf{13 points per game}$

7. (a) 44,010 miles
− 43,764 miles
246 miles

$\frac{246 \text{ miles}}{12 \text{ gallons}} = \textbf{20.5 miles per gallon}$

(b) $\frac{246 \text{ miles}}{5 \text{ hours}} = \textbf{49.2 miles per hour}$

8. $\mathbf{\frac{3}{5} \times W_N = 60}$

$\left(\frac{\overset{1}{\cancel{5}}}{\underset{1}{\cancel{3}}}\right)\left(\frac{\overset{1}{\cancel{3}}}{\underset{1}{\cancel{5}}} \times W_N\right) = \left(\frac{5}{\underset{1}{\cancel{3}}}\right)\overset{20}{\cancel{60}}$

$W_N = \textbf{100}$

9.

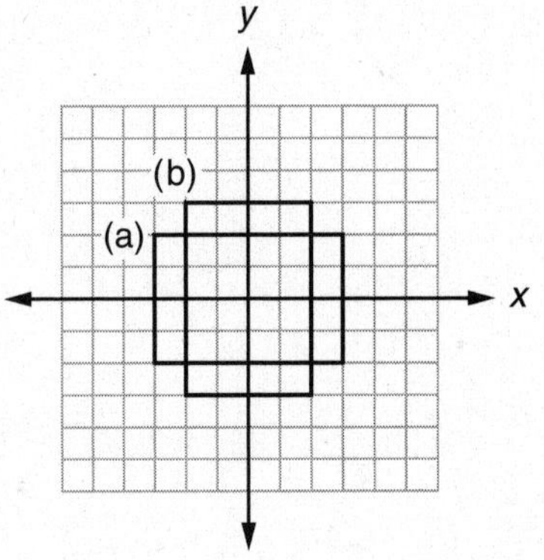

(a) **(3, −2)**

(b) **(2, 3), (2, −3), (−2, −3), (−2, 3)**

10. $\frac{2}{6} = \mathbf{\frac{1}{3}}$

11. $a = 9$, $9^2 = \mathbf{81}$

12.
$$\mathbf{40 = W_P \times 250}$$
$$\frac{40}{250} = \frac{W_P \times 250}{250}$$
$$\frac{4}{25} = W_P$$
$$W_P = \frac{4}{25} \times 100\% = \mathbf{16\%}$$

13.
$$\mathbf{0.4 \times W_N = 60}$$
$$\frac{0.4 \times W_N}{0.4} = \frac{60}{0.4}$$
$$W_N = \mathbf{150}$$

14.

	Percent	Actual Count
Original	100	$40
+ Change	60	*C*
New	160	*N*

$$\frac{100}{160} = \frac{\$40}{N}$$
$$100N = \$6400$$
$$N = \mathbf{\$64}$$

15. $2\frac{1}{8} - 1\frac{3}{8} = \frac{17}{8} - \frac{11}{8} = \frac{6}{8} = \mathbf{\frac{3}{4}}$ **inch**

16.

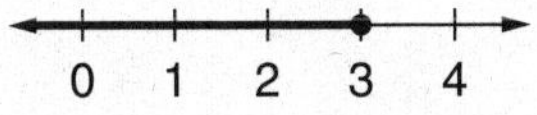

17. (a) $\mathbf{\frac{7}{500}}$

(b) **0.014**

18. $(1.4 \times 10^{-6})(5 \times 10^{4})$
$\mathbf{7.0 \times 10^{-2}}$

19.
$$y = -2(3)$$
$$y = \mathbf{-6;}$$
$$y = -2(0)$$
$$y = \mathbf{0;}$$
$$y = -2(-2)$$
$$y = \mathbf{4}$$

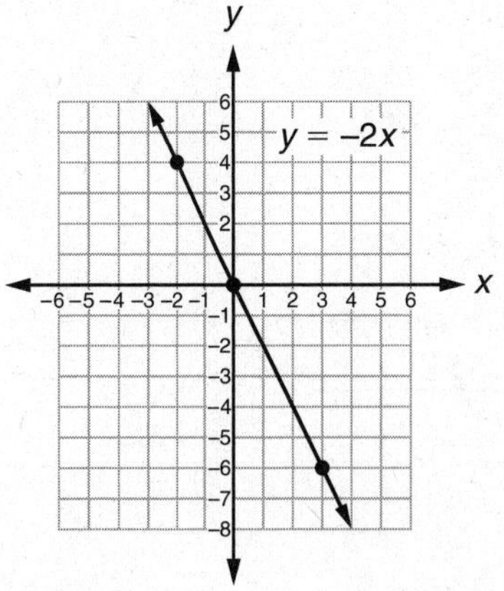

(0,0)

20. (a) $C \approx 3.14(2\text{ ft})$
$C \approx \mathbf{6.28\text{ ft}}$

(b) $A \approx 3.14(1\text{ ft}^2)$
$A \approx \mathbf{3.14\text{ ft}^2}$

21. **See student work; 40°.**

22.
$$\frac{x}{30} = \frac{20}{50}$$
$$50x = 600$$
$$\mathbf{x = 12\text{ in.}}$$
$$\frac{y}{16} = \frac{50}{20}$$
$$20y = 800$$
$$\mathbf{y = 40\text{ in.}}$$
$$\textbf{Area} = \frac{(16\text{ in.})(12\text{ in.})}{2} = \mathbf{96\text{ in.}^2}$$

23.
$$20f = 50$$
$$f = \frac{50}{20}$$
$$f = \mathbf{2.5}$$

24.
$$-\frac{3}{5}m + 8 = 20$$
$$-\frac{3}{5}m + 8 - 8 = 20 - 8$$
$$-\frac{3}{5}m = 12$$
$$\left(-\frac{\overset{1}{\cancel{5}}}{\underset{1}{\cancel{3}}}\right)\left(-\frac{\overset{1}{\cancel{3}}}{\underset{1}{\cancel{5}}}m\right) = \left(-\frac{5}{\underset{1}{\cancel{3}}}\right)\cancel{12}^{4}$$
$$m = \mathbf{-20}$$

25.
$$\begin{aligned} 0.3x - 2.7 &= 9 \\ 0.3x - 2.7 + 2.7 &= 9 + 2.7 \\ 0.3x &= 11.7 \\ \frac{0.3x}{0.3} &= \frac{11.7}{0.3} \\ x &= \mathbf{39} \end{aligned}$$

26. $\sqrt{5^3 - 5^2} = \sqrt{125 - 25} = \sqrt{100}$
$= \mathbf{10}$

27.
$$\begin{array}{rrr} \overset{0}{\cancel{1}}\text{ gal} & \overset{0}{\cancel{1}}\text{ qt} & (2\text{ pt}) \\ - & 1\text{ qt} & 1\text{ pt} \\ \hline \end{array} \longrightarrow$$
(1 gal → (4 qt); 1 qt → (2 pt))

$$\begin{array}{rrr} \overset{0}{\cancel{1}}\text{ gal} & \overset{4}{\overset{0}{\cancel{1}}}\text{ qt} & 2\text{ pt} \\ - & 1\text{ qt} & 1\text{ pt} \\ \hline & \mathbf{3\ qt} & \mathbf{1\ pt} \end{array}$$

28. $(0.25)(1.25 - 1.2)$
$= (0.25)(0.05) = \mathbf{0.0125}$

29. $7\frac{1}{3} - \left(1\frac{3}{4} \div 3\frac{1}{2}\right)$
$= 7\frac{1}{3} - \left(\frac{7}{4} \div \frac{7}{2}\right) = 7\frac{1}{3} - \left(\frac{\overset{1}{\cancel{7}}}{\underset{2}{\cancel{4}}} \times \frac{\overset{1}{\cancel{2}}}{\underset{1}{\cancel{7}}}\right)$
$= 7\frac{1}{3} - \frac{1}{2} = 7\frac{2}{6} - \frac{3}{6} = 6\frac{8}{6} - \frac{3}{6}$
$= \mathbf{6\frac{5}{6}}$

30. $\dfrac{(-2)(3) - (3)(-4)}{(-2)(-3) - (4)}$
$\dfrac{(-6) - (-12)}{(6) - (4)}$
$\dfrac{6}{2}$
$\mathbf{3}$

Lesson 99, Warm-Up

a. **3**

b. $\mathbf{3.2 \times 10^{11}}$

c. **1**

d. **5°F**

e. **28**

f. **$40**

g. **10**

Problem Solving

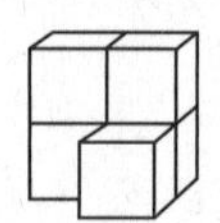

The back part of the object is 2 in.-by-2 in.-by-1 in., which is 4 in.^{3}.
The front part of the object is 1 in.-by-1 in.-by-1 in., which is 1 in.^{3}.
4 in.^{3} + 1 in.^{3} = **5 in.^{3}**

Lesson 99, Lesson Practice

a.

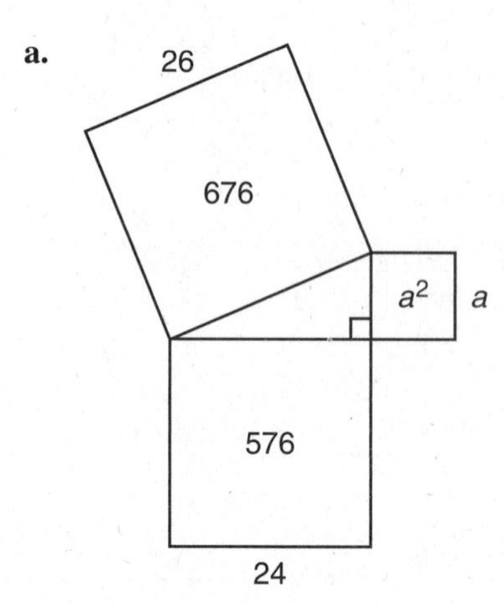

$$\begin{aligned} a^2 + 576 &= 676 \\ a^2 &= 100 \\ a &= \mathbf{10} \end{aligned}$$

b.

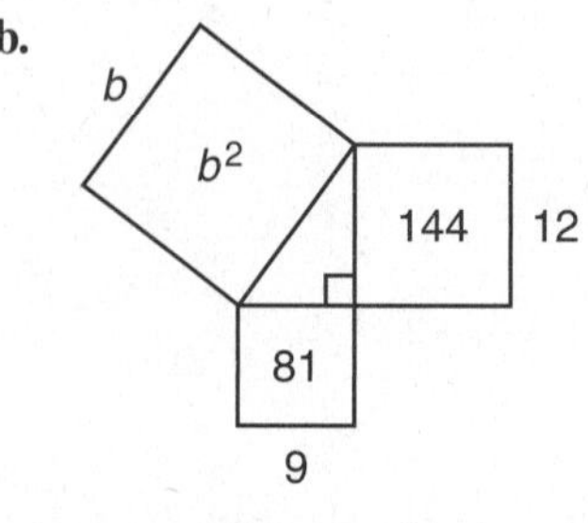

$$\begin{aligned} 81 + 144 &= b^2 \\ 225 &= b^2 \\ \mathbf{15} &= b \end{aligned}$$

c. 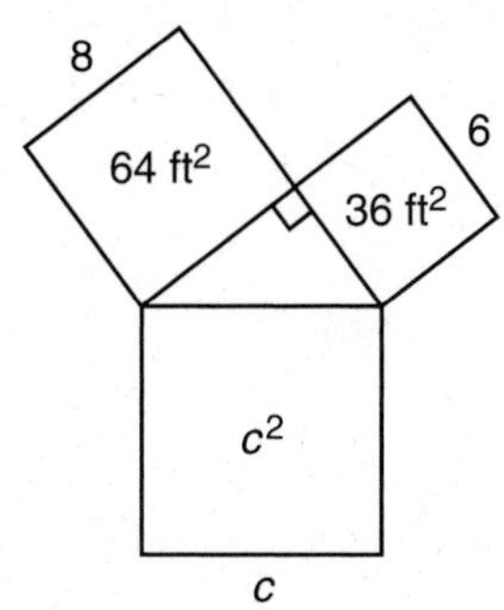

$$64 \text{ ft}^2 + 36 \text{ ft}^2 = c^2$$
$$100 = c^2$$
$$10 = c$$
Perimeter $= 8 \text{ ft} + 6 \text{ ft} + 10 \text{ ft}$
$= \mathbf{24 \text{ ft}}$

LESSON 99, MIXED PRACTICE

1. $\$15.00 \times 0.15 = \mathbf{\$2.25}$

2. $0.0024\overset{4}{\cancel{5}}{}^{1}00 - 0.000020 = 0.002480$ **2.48×10^{-3}**

3. (a) $\dfrac{4(30) + 7(31) + 29}{12} = \mathbf{30.5 \text{ days}}$

(b) **31 days**

(c) **31 days**

(d) **2 days**

4. 2 pounds = 32 ounces

$$\frac{\$2.72}{32 \text{ ounces}} = \frac{\$0.085}{1 \text{ ounce}}$$

$$\frac{\$3.60}{48 \text{ ounces}} = \frac{\$0.075}{1 \text{ ounce}}$$

$\$0.085 - \$0.075 = \$0.01$

1¢ more per ounce

5.

	Case 1	Case 2
Pounds	80	300
Price	$96	p

$$\frac{80}{\$96} = \frac{300}{p}$$
$$80p = \$28{,}800$$
$$p = \mathbf{\$360}$$

6.

	Ratio	Actual Count
Stalactites	9	C
Stalagmites	5	G
Total	14	1260

$$\frac{5}{14} = \frac{G}{1260}$$
$$14G = 6300$$
$$G = \mathbf{450 \text{ stalagmites}}$$

7. $\dfrac{5}{\cancel{8}_1}(\overset{2}{\cancel{16}} \text{ ounces}) = \mathbf{10 \text{ ounces}}$

8. (a) $\mathbf{0.1 \times W_N = 20}$

$$\frac{0.1 \times W_N}{0.1} = \frac{20}{0.1}$$
$$W_N = \mathbf{200}$$

(b) $\mathbf{20 = W_P \times 60}$

$$\frac{20}{60} = \frac{W_P \times 60}{60}$$
$$\frac{1}{3} = W_P$$
$$W_P = \frac{1}{3} \times 100\% = \mathbf{33\frac{1}{3}\%}$$

9. (a) **Right triangle**

(b) **Equilateral triangle**

(c) **Isosceles triangle**

10.

	Percent	Actual Count
Original	100	$3.40
− Change	20	C
New	80	N

$$\frac{100}{80} = \frac{\$3.40}{N}$$
$$100N = \$272$$
$$N = \mathbf{\$2.72}$$

11. $100\% - 20\% = \mathbf{80\%}$

12. Area $= \dfrac{(6 \text{ units})(3 \text{ units})}{2} = \mathbf{9 \text{ units}^2}$

13.

	Scale	Measure
Model	1	8
Object	60	m

$$\frac{1}{60} = \frac{8}{m}$$
$$m = \mathbf{480 \text{ inches}}$$
$$\overset{40}{\cancel{480}} \cancel{\text{inches}} \cdot \frac{1 \text{ foot}}{\cancel{12} \cancel{\text{inches}}} = \mathbf{40 \text{ feet}}$$

14. (a) $\mathbf{1.\overline{3}}$

(b) $\mathbf{133\frac{1}{3}\%}$

(c) $\mathbf{\frac{1}{75}}$

(d) $\mathbf{0.01\overline{3}}$

15. (a) $(ax^2)(-2ax)(-a^2)$
$a \cdot x \cdot x \cdot (-2) \cdot a \cdot x \cdot (-1) \cdot a \cdot a$
$(-2)(-1) \cdot a \cdot a \cdot a \cdot a \cdot x \cdot x \cdot x$
$\mathbf{2a^4x^3}$

(b) $\frac{1}{2}\pi + \frac{2}{3}\pi - \pi = \frac{3}{6}\pi + \frac{4}{6}\pi - \frac{6}{6}\pi$
$= \frac{7}{6}\pi - \frac{6}{6}\pi = \mathbf{\frac{1}{6}\pi}$

16. $(8.1 \times 9) \times (10^{-6} \times 10^{10}) = 72.9 \times 10^4$
$= \mathbf{7.29 \times 10^5}$

17. $\sqrt{(15)^2 - (12)^2} = \sqrt{225 - 144}$
$= \sqrt{81} = \mathbf{9}$

18.
$$\begin{aligned} 5^2 + c^2 &= 13^2 \\ 25 + c^2 &= 169 \\ c^2 &= 169 - 25 \\ c^2 &= 144 \\ c &= \mathbf{12} \end{aligned}$$

19. Area of base $\approx 3.14(100\text{ cm}^2)$
$\approx 314\text{ cm}^2$
Volume $\approx (314\text{ cm}^2)(10\text{ cm}) = \mathbf{3140\text{ cm}^3}$

20. (a) $m\angle X = 180° - 138° = \mathbf{42°}$

(b) $m\angle Y = 180° - (100° + 42°) = \mathbf{38°}$

(c) $m\angle Z = 180° - (90° + 38°) = \mathbf{52°}$

21. (a)
$$\begin{aligned} \frac{x}{6} &= \frac{12}{8} \\ 8x &= 72 \\ x &= \mathbf{9\text{ inches}} \end{aligned}$$

(b)
$$\begin{aligned} 6f &= 9 \\ f &= \frac{9}{6} \\ f &= \mathbf{1.5} \end{aligned}$$

(c) $(1.5)(1.5) = $ **2.25 times**

22. $\frac{(40{,}000)(400)}{80} = \mathbf{200{,}000}$

23.
$$\begin{aligned} 4n + 1.64 &= 2 \\ 4n + 1.64 - 1.64 &= 2 - 1.64 \\ 4n &= 0.36 \\ \frac{4n}{4} &= \frac{0.36}{4} \\ n &= \mathbf{0.09} \end{aligned}$$

24.
$$\begin{aligned} 3\frac{1}{3}x - 1 &= 49 \\ 3\frac{1}{3}x - 1 + 1 &= 49 + 1 \\ 3\frac{1}{3}x &= 50 \\ \left(\frac{3}{10}\right)\left(\frac{10}{3}x\right) &= \left(\frac{3}{10}\right)50 \\ x &= \mathbf{15} \end{aligned}$$

25.
$$\begin{aligned} \frac{17}{25} &= \frac{m}{75} \\ 25m &= 1275 \\ m &= \mathbf{51} \end{aligned}$$

26. $3^3 + 4^2 - \sqrt{225}$
$= 27 + 16 - 15 = \mathbf{28}$

27. $\sqrt{225} - 15^0 + 10^{-1} = 15 - 1 + \frac{1}{10}$
$= 14 + \frac{1}{10} = \mathbf{14\frac{1}{10}}$ or **14.1**

28. $\left(3\frac{1}{3}\right)\left(\frac{3}{4}\right)\left(\frac{40}{1}\right)$
$= \left(\frac{10}{\cancel{3}_1}\right)\left(\frac{\cancel{3}^1}{\cancel{4}_1}\right)\left(\frac{\cancel{40}^{10}}{1}\right) = \mathbf{100}$

29. $\frac{-12 - (6)(-3)}{(-12) - (-6) + (3)}$

$\frac{-12 - (-18)}{(-6) + 3}$

$\frac{6}{-3}$

$\mathbf{-2}$

30. $3(x - 2) = 3(x) - 3(2) = \mathbf{3x - 6}$

Lesson 100, Warm-Up

a. **6**

b. $\mathbf{2.5 \times 10^{-9}}$

c. **24**

d. **−4°F**

e. **75**

f. **$16**

g. **3**

Problem Solving

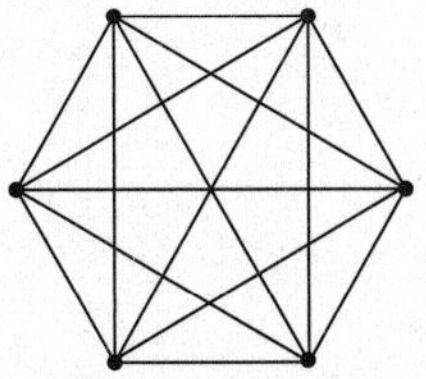

15 handshakes

Lesson 100, Lesson Practice

a. **2 and 3**

b. **8 and 9**

c. **26 and 27**

d.
$$x^2 = 1^2 + 1^2$$
$$x^2 = 1 + 1$$
$$x^2 = 2$$
$$x = \mathbf{\sqrt{2}}$$

e.
$$2^2 = 1^2 + y^2$$
$$4 = 1 + y^2$$
$$3 = y^2$$
$$\mathbf{\sqrt{3}} = y$$

f.

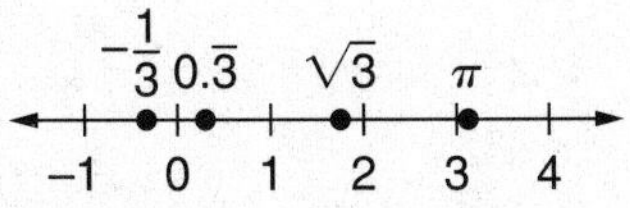

$\mathbf{\sqrt{3}, \pi}$

Lesson 100, Mixed Practice

1. $2.5(\$2.60) + 2(\$1.49)$
$= \$6.50 + \$2.98 = \$9.48$

$$\begin{array}{r} \$20.00 \\ -\ \ \$9.48 \\ \hline \mathbf{\$10.52} \end{array}$$

2. (a)
$$p(6) = \frac{1}{5}$$
$$p(4) = \frac{1}{5}$$
$$\frac{1}{5} + \frac{1}{5} = \mathbf{\frac{2}{5}}$$

(b)
$$p(3, 3 \text{ or } 5 \text{ or } 7) = \frac{1}{5} \cdot \frac{3}{5} = \frac{3}{25}$$
$$p(5, 3 \text{ or } 5 \text{ or } 7) = \frac{1}{5} \cdot \frac{3}{5} = \frac{3}{25}$$
$$p(7, 3 \text{ or } 5 \text{ or } 7) = \frac{1}{5} \cdot \frac{3}{5} = \frac{3}{25}$$
$$\frac{3}{25} + \frac{3}{25} + \frac{3}{25} = \mathbf{\frac{9}{25}}$$

3. Average $= (1 + 2 + 3 + 4 + 5 + 6 + 7 + 8 + 9 + 10) \div 10$
$$= \frac{55}{10} = \mathbf{5.5}$$

4. $375 \text{ ~~miles~~} \cdot \dfrac{1 \text{ hour}}{50 \text{ ~~miles~~}} = \mathbf{7\frac{1}{2} \text{ hours}}$

5.

	Case 1	Case 2
Kilometers	300	500
Hours	4	h

$$\frac{300}{4} = \frac{500}{h}$$
$$300h = 2000$$
$$h = 6\frac{2}{3} \text{ hours}$$
$$6\frac{2}{3} \text{ hours} = \textbf{6 hours 40 minutes}$$

6.

	Ratio	Actual Count
Winners	1	W
Losers	15	L
Total	16	800

$$\frac{1}{16} = \frac{W}{800}$$
$$16W = 800$$
$$W = \textbf{50 winners}$$

7.

	Percent	Actual Count
Original	100	O
− Change	30	C
New	70	350

$$\frac{100}{70} = \frac{O}{350}$$
$$70O = 35{,}000$$
$$O = \mathbf{500}$$

8. $\frac{3}{4} \cdot n = 36$

$\left(\frac{\overset{1}{\cancel{4}}}{\underset{1}{\cancel{3}}}\right)\left(\frac{\overset{1}{\cancel{3}}}{\underset{1}{\cancel{4}}}n\right) = \left(\frac{4}{\underset{1}{\cancel{3}}}\right)\overset{12}{\cancel{36}}$

$n = 48$

$\frac{1}{\underset{1}{\cancel{2}}} \cdot \overset{24}{\cancel{48}} = \mathbf{24}$

9. (a) $\mathbf{300 = 0.06 \times W_N}$

$\frac{300}{0.06} = \frac{0.06 \times W_N}{0.06}$

$\mathbf{5000} = W_N$

(b) $\mathbf{20 = W_P \times 10}$

$\frac{20}{10} = \frac{W_P \times 10}{10}$

$2 = W_P$

$W_P = 2 \times 100\% = \mathbf{200\%}$

10. $\$40.00 \times 0.065 = \$2.60$ $\$40.00 + \$2.60 = \mathbf{\$42.60}$

11. $x(x + 3) = x(x) + x(3)$

$= \mathbf{x^2 + 3x}$

12.

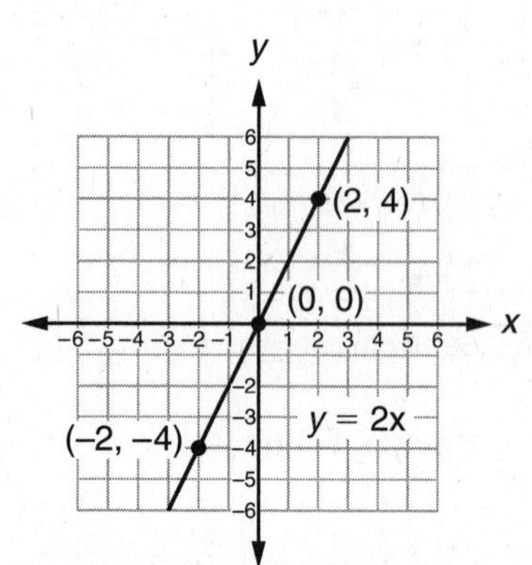

Another point from the 3rd quadrant could be (−1, −2) or (−3, −6).

13. (a) $\mathbf{2\frac{1}{2}}$ **inches**

(b) $120f = 5$

$f = \frac{5}{120}$

$f = \mathbf{\frac{1}{24}}$

(c) **14 feet**

14. (a) $2f = 6$

$f = \frac{6}{2}$

$f = \mathbf{3}$

(b) $(3)(3) = \mathbf{9}$ **times**

(c) $3^3 = \mathbf{27}$ **times**

15. **Insufficient information**

16. (a) $\mathbf{\frac{18}{25}}$

(b) **0.72**

17. $(4.5 \times 6) \times (10^6 \times 10^3) = 27 \times 10^9$

$= \mathbf{2.7 \times 10^{10}}$

18. (a) $6^2 = 36, 7^2 = 49$

6 and 7

(b) $4^2 = 16, 5^2 = 25$

4 and 5

19. (a) $C \approx \frac{22}{\underset{1}{\cancel{7}}}(\overset{2}{\cancel{14}} \text{ in.})$

$C \approx \mathbf{44 \text{ in.}}$

(b) $A \approx \frac{22}{\underset{1}{\cancel{7}}}(\overset{7}{\cancel{49}} \text{ in.}^2)$

$A \approx \mathbf{154 \text{ in.}^2}$

20. $15^2 + a^2 = 17^2$

$225 + a^2 = 289$

$a^2 = 289 - 225$

$a^2 = 64$

$a = \mathbf{8 \text{ cm}}$

21. Area of base $= \frac{(3 \text{ cm})(4 \text{ cm})}{2}$

$= 6 \text{ cm}^2$

Volume $= (6 \text{ cm}^2)(6 \text{ cm}) = \mathbf{36 \text{ cm}^3}$

22. Area of base $\approx 3.14(4 \text{ cm}^2)$

$\approx 12.56 \text{ cm}^2$

Volume $\approx (12.56 \text{ cm}^2)(10 \text{ cm})$

$\approx \mathbf{125.6 \text{ cm}^3}$

23. $m\angle a = 180° - 48° = \mathbf{132°}$

$m\angle b = 180° - 132° = \mathbf{48°}$

$m\angle c = 180° - (90° + 48°) = \mathbf{42°}$

24. $-4\frac{1}{2}x + 8^0 = 4^3$

$-4\frac{1}{2}x + 1 - 1 = 64 - 1$

$-4\frac{1}{2}x = 63$

$\left(-\frac{2}{9}\right)\left(-\frac{9}{2}x\right) = \left(-\frac{2}{9}\right)63$

$x = \mathbf{-14}$

25.
$$\frac{15}{w} = \frac{45}{3.3}$$
$$(45)w = 15(3.3)$$
$$\frac{45w}{45} = \frac{49.5}{45}$$
$$w = \mathbf{1.1}$$

26. $\sqrt{6^2 + 8^2} = \sqrt{36 + 64} = \sqrt{100} = \mathbf{10}$

27.
$$3\frac{1}{3}\left(7\frac{2}{10} \div \frac{3}{5}\right) = 3\frac{1}{3}\left(\frac{\overset{12}{\cancel{\overset{24}{\cancel{72}}}}}{\underset{1}{\cancel{\underset{2}{\cancel{10}}}}} \times \frac{\overset{1}{\cancel{5}}}{\underset{1}{\cancel{3}}}\right)$$
$$= 3\frac{1}{3}(12) = \frac{10}{\underset{1}{\cancel{3}}}(\overset{4}{\cancel{12}}) = \mathbf{40}$$

28.
$$8\frac{5}{6} - 2\frac{1}{2} - 1\frac{1}{3}$$
$$= 8\frac{5}{6} - 2\frac{3}{6} - 1\frac{2}{6} = 6\frac{2}{6} - 1\frac{2}{6}$$
$$= \mathbf{5}$$

29.
$$\frac{|-18| - (2)(-3)}{(-3) + (-2) - (-4)}$$
$$\frac{18 - (-6)}{(-5) - (-4)}$$
$$\frac{24}{-1}$$
$$\mathbf{-24}$$

30.

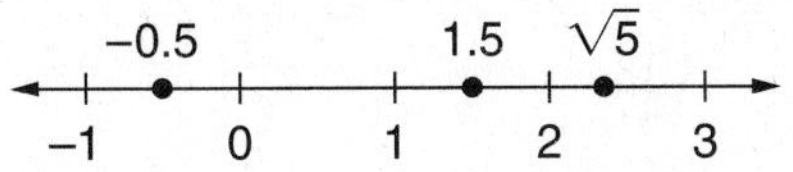

Investigation 10

1. There are two possible outcomes for a flipped coin: heads or tails. The decimal probability that a flipped coin will land heads up is **0.5.**

2.
```
     0.027
36)1.000
    72
    280
    252
     28
```
0.027 rounds to **0.03**

3. **The ratio $\frac{1}{13}$ is more precise than 0.08 because the decimal form is rounded.**

4. $100\% - 40\% = \mathbf{60\%}$

5. There are two possible outcomes for a flipped coin: heads or tails. The chance that a flipped coin will land heads up is **50%.**

6. There are four suits in a normal deck of cards: heart, spade, club, and diamond. The chance of not selecting a heart by drawing one card is **75%.**

7. There are 4 white marbles. There are 12 marbles altogether. Therefore, the probability of drawing a white marble is $\mathbf{\frac{1}{3}}$.

8. There are 3 red marbles. There are 12 marbles altogether. Therefore, the probability of drawing a red marble is $\frac{1}{4}$. The probability of not drawing a red marble is equal to $1 - \frac{1}{4}$, or $\mathbf{\frac{3}{4}}$.

9. There are 5 blue marbles. There are 7 marbles that are not blue. Therefore, the odds of drawing a blue marble are **5 to 7.**

10. Darcy made 64 free throws. She shot $64 + 32$, or 96 free throws. Therefore, the statistical probability that Darcy will make her next free throw is $\mathbf{\frac{2}{3}}$.

11. $\mathbf{\frac{3}{200}}$ **or 0.015**

12. **The insurance rates (premiums) for younger drivers are much higher than the rates for older drivers.**

Extension

See student work.

Lesson 101, Warm-Up

a. **6**

b. **34**

c. **16**

d. $\mathbf{3 \times 10^{-10}}$

e. **0.06**

f. **−13°F**

g. **\$60**

h. **\$60**

i. **\$4.50**

Problem Solving

There are $360° \div 12$, or $30°$, between each hour mark. So, between the 12 and the 5, there are $5 \times 30°$, or **150°**.

Lesson 101, Lesson Practice

a.
$$\mathbf{3x + 6 = 30}$$
$$3x = 24$$
$$x = \frac{24}{3}$$
$$x = \mathbf{8}$$

b.
$$\mathbf{\frac{1}{2}x - 10 = 30}$$
$$\frac{1}{2}x = 40$$
$$\left(\frac{2}{1}\right)\left(\frac{1}{2}x\right) = \left(\frac{2}{1}\right)40$$
$$x = \mathbf{80}$$

c.
$$\mathbf{3x + 2x = 90°}$$
$$5x = 90°$$
$$x = \frac{90°}{5}$$
$$x = 18°$$
$$2(18°) = \mathbf{36°}$$

d.
$$\mathbf{x + 2x + 3x = 180°}$$
$$6x = 180°$$
$$x = \frac{180°}{6}$$
$$x = \mathbf{30°}$$
$$2x = 2(30°) = \mathbf{60°}$$
$$3x = 3(30°) = \mathbf{90°}$$

Lesson 101, Mixed Practice

1. (a) **11.9**

(b) **11.8 and 11.9**

(c) **0.7**

2. $\$6(3.75) = \mathbf{\$22.50}$

3.

	Case 1	Case 2
Flour	3	15
Eggs	2	*e*

$$\frac{3}{2} = \frac{15}{e}$$
$$3e = 30$$
$$e = \mathbf{10\ eggs}$$

4.

	Case 1	Case 2
Words	48	*w*
Seconds	60	90

$$\frac{48}{60} = \frac{w}{90}$$
$$60w = 4320$$
$$w = \mathbf{72\ words}$$

5.

	Percent	Actual Count
Surpassed the quota	40	10
Did not surpass the quota	60	*N*
Total	100	*T*

$$\frac{40}{100} = \frac{10}{T}$$
$$40T = 1000$$
$$T = \mathbf{25\ salespeople}$$

6.

	Percent	Actual Count
Original	100	\$24
− Change	40	*C*
New	60	*N*

$$\frac{100}{60} = \frac{\$24}{N}$$
$$100N = \$1440$$
$$N = \mathbf{\$14.40}$$

7.
$$3(x - 4) - x$$
$$3x - 12 - x$$
$$3x - x - 12$$
$$\mathbf{2x - 12}$$

8. $3\ \cancel{\text{gallons}} \cdot \frac{4\ \cancel{\text{quarts}}}{1\ \cancel{\text{gallon}}} \cdot \frac{2\ \text{pints}}{1\ \cancel{\text{quart}}} = \mathbf{24\ pints}$

9.

24 games

$\frac{5}{6}$ won (20).	4 games
	4 games
	4 games
	4 games
	4 games
$\frac{1}{6}$ lost.	4 games

(a) $20 \div 5 = 4$, $6(4\text{ games}) = \mathbf{24\ games}$

(b) $\frac{\text{won}}{\text{lost}} = \frac{20}{4} = \mathbf{\frac{5}{1}}$

10. **14 and 15**

11. $w \; \textcircled{<} \; m$

12.

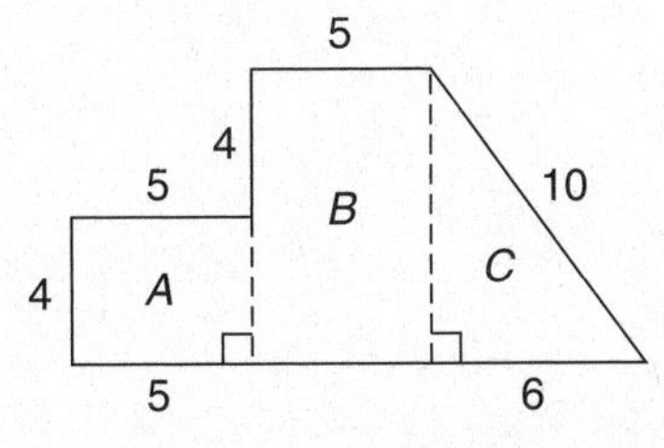

$$\begin{aligned} \text{Area } A &= (4\text{ cm})(5\text{ cm}) = 20\text{ cm}^2 \\ \text{Area } B &= (5\text{ cm})(8\text{ cm}) = 40\text{ cm}^2 \\ + \text{ Area } C &= \frac{(6\text{ cm})(8\text{ cm})}{2} = 24\text{ cm}^2 \\ \hline \text{Area of figure} &= \mathbf{84\text{ cm}^2} \end{aligned}$$

13. $\mathbf{6n - 3 = 45}$

$6n = 48$

$n = \mathbf{8}$

14. $(8 \times 4) \times (10^8 \times 10^{-2}) = 32 \times 10^6$

$= \mathbf{3.2 \times 10^7}$

15. (a) $\mathbf{\frac{1}{50}}$

(b) **2%**

(c) $\mathbf{\frac{1}{500}}$

(d) **0.002**

16. $y = 2(-1) + 1$
$y = -2 + 1$
$y = \mathbf{-1;}$
$y = 2(0) + 1$
$y = 0 + 1$
$y = \mathbf{1;}$
$y = 2(1) + 1$
$y = 2 + 1$
$y = \mathbf{3;}$
$y = 2(2) + 1$
$y = 4 + 1$
$y = \mathbf{5}$

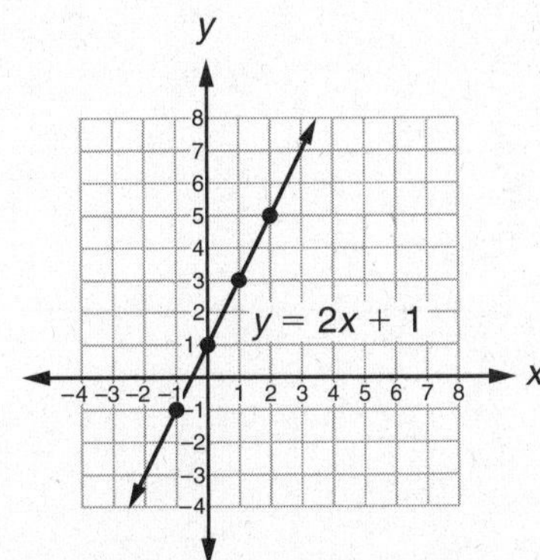

17. (a) Area of base $= (4\text{ in.})(4\text{ in.}) = 16\text{ in.}^2$

Volume $= (16\text{ in.}^2)(4\text{ in.}) = \mathbf{64\text{ in.}^3}$

(b) Surface area $= 6(16\text{ in.}^2) = \mathbf{96\text{ in.}^2}$

18. (a) $C = \pi(18\text{ cm})$

$C = \mathbf{18\pi\text{ cm}}$

(b) $A = \pi(81\text{ cm}^2)$

$A = \mathbf{81\pi\text{ cm}^2}$

19. **1 to 1**

20.
$$\begin{aligned} 2x + x &= 90° \\ 3x &= 90° \\ x &= \frac{90°}{3} \\ x &= \mathbf{30°} \\ 2x = 2(30°) &= \mathbf{60°} \end{aligned}$$

21. (a)
$$\begin{array}{r} 0.1366\ldots \\ 9\overline{)1.2300} \\ \underline{9} \\ 33 \\ \underline{27} \\ 60 \\ \underline{54} \\ 60 \\ \underline{54} \\ 6 \end{array}$$

$\mathbf{0.1366\ldots = 0.13\overline{6}}$

(b) 0.1366 … rounded to three decimal places is **0.137**

22.
$$\begin{aligned} (AB)^2 &= (9\text{ cm})^2 + (12\text{ cm})^2 \\ (AB)^2 &= 81\text{ cm}^2 + 144\text{ cm}^2 \\ (AB)^2 &= 225\text{ cm}^2 \\ AB &= \mathbf{15\text{ cm}} \end{aligned}$$

23. (a) Perimeter $= 2(9\text{ cm}) + 2(12\text{ cm}) + 2(15\text{ cm})$

$= 18\text{ cm} + 24\text{ cm} + 30\text{ cm}$

$= \mathbf{72\text{ cm}}$

(b) Area $= \frac{(18\text{ cm})(24\text{ cm})}{2} = \mathbf{216\text{ cm}^2}$

24. $p(B, B) = \frac{4}{10} \cdot \frac{3}{9} = \frac{12}{90} = \mathbf{\frac{2}{15}}$

25.
$$3\frac{1}{7}d = 88$$
$$\left(\frac{7}{22}\right)\frac{22}{7}d = \left(\frac{7}{22}\right)88$$
$$d = \mathbf{28}$$

26.
$$\begin{aligned} 3x + 20 &\geq 14 \\ 3x + 20 - 20 &\geq 14 - 20 \\ 3x &\geq -6 \\ \frac{3x}{3} &\geq \frac{-6}{3} \\ \mathbf{x} &\geq \mathbf{-2} \end{aligned}$$

$x \geq -2$

−3 −2 −1 0 1 2 3

27. $5^2 + (3^3 - \sqrt{81})$
$= 25 + (27 - 9) = 25 + 18 = \mathbf{43}$

28. $3x + 2(x - 1)$
$3x + 2x - 2$
$\mathbf{5x - 2}$

29. $\left(4\frac{4}{9}\right)\left(2\frac{7}{10}\right)\left(1\frac{1}{3}\right) = \frac{\overset{4}{\cancel{40}}}{\underset{1}{\cancel{9}}} \cdot \frac{\overset{\overset{1}{\cancel{3}}}{\cancel{27}}}{\underset{1}{\cancel{10}}} \cdot \frac{4}{\underset{1}{\cancel{3}}} = \mathbf{16}$

30. $(-2)(-3) - (-4)(-5)$
$(6) - (20)$
$\mathbf{-14}$

Lesson 102, Warm-Up

a. **11**

b. **94**

c. **−1**

d. $\mathbf{6 \times 10^8}$

e. $\mathbf{2\frac{1}{2}}$

f. **0.5 L**

g. **$60**

h. **$100**

i. **2**

Problem Solving

The number of triangles a polygon can be divided into equals the number of sides minus 2. So a 22-gon can be divided into 22 − 2, or **20 triangles.**

$20 \times 180° = \mathbf{3600°}$

Lesson 102, Lesson Practice

a. $\boldsymbol{\angle s}$ **and** $\boldsymbol{\angle u}$**,** $\boldsymbol{\angle t}$ **and** $\boldsymbol{\angle v}$**,** $\boldsymbol{\angle w}$ **and** $\boldsymbol{\angle y}$**,** $\boldsymbol{\angle x}$ **and** $\boldsymbol{\angle z}$

b. $\boldsymbol{\angle t}$ **and** $\boldsymbol{\angle y}$**,** $\boldsymbol{\angle x}$ **and** $\boldsymbol{\angle u}$

c. $\boldsymbol{\angle s}$ **and** $\boldsymbol{\angle z}$**,** $\boldsymbol{\angle w}$ **and** $\boldsymbol{\angle v}$

d. $\mathbf{m\angle t = m\angle v = m\angle y = 80°}$
$\mathbf{m\angle s = m\angle u = m\angle x = m\angle z = 100°}$

e.
$$\begin{aligned} 3w - 10 + w &= 90 \\ 4w - 10 &= 90 \\ 4w - 10 + 10 &= 90 + 10 \\ 4w &= 100 \\ w &= \frac{100}{4} \\ w &= \mathbf{25} \end{aligned}$$

f.
$$\begin{aligned} x + x + 10 + 2x - 10 &= 180 \\ 4x + 10 - 10 &= 180 \\ 4x &= 180 \\ x &= \frac{180}{4} \\ x &= \mathbf{45} \end{aligned}$$

g.
$$\begin{aligned} 3y + 5 &= y - 25 \\ 3y + 5 - y &= y - 25 - y \\ 2y + 5 &= -25 \\ 2y + 5 - 5 &= -25 - 5 \\ 2y &= -30 \\ y &= \frac{-30}{2} \\ y &= \mathbf{-15} \end{aligned}$$

h.
$$\begin{aligned} 4n - 5 &= 2n + 3 \\ 4n - 5 - 2n &= 2n + 3 - 2n \\ 2n - 5 &= 3 \\ 2n - 5 + 5 &= 3 + 5 \\ 2n &= 8 \\ n &= \frac{8}{2} \\ n &= \mathbf{4} \end{aligned}$$

i.
$$\begin{aligned} 3x - 2(x - 4) &= 32 \\ 3x - 2x + 8 &= 32 \\ x + 8 &= 32 \\ x + 8 - 8 &= 32 - 8 \\ x &= \mathbf{24} \end{aligned}$$

j.
$$\begin{aligned} 3x &= 2(x - 4) \\ 3x &= 2x - 8 \\ 3x - 2x &= 2x - 8 - 2x \\ x &= \mathbf{-8} \end{aligned}$$

Lesson 102, Mixed Practice

1. (a)

$$p(1,4) = \frac{1}{6} \cdot \frac{1}{6} = \frac{1}{36}$$
$$p(4,1) = \frac{1}{6} \cdot \frac{1}{6} = \frac{1}{36}$$
$$p(2,3) = \frac{1}{6} \cdot \frac{1}{6} = \frac{1}{36}$$
$$p(3,2) = \frac{1}{6} \cdot \frac{1}{6} = \frac{1}{36}$$
$$\frac{1}{36} + \frac{1}{36} + \frac{1}{36} + \frac{1}{36} = \mathbf{\frac{4}{36}} = \mathbf{\frac{1}{9}}$$

(b) **8 to 1**

2.

$$\mathbf{3x - 12 = 36}$$
$$3x - 12 + 12 = 36 + 12$$
$$3x = 48$$
$$x = \frac{48}{3}$$
$$x = \mathbf{16}$$

3. (a) $\frac{360°}{10} = \mathbf{36°}$

(b) $180° - 36° = \mathbf{144°}$

4.

	Ratio	Actual Count
Youths	3	Y
Adults	7	A
Total	10	4500

$$\frac{7}{10} = \frac{A}{4500}$$
$$10A = 31{,}500$$
$$A = \textbf{3150 adults}$$

5.

	Case 1	Case 2
Over	2	8
Up	1	u

$$\frac{2}{1} = \frac{8}{u}$$
$$2u = 8$$
$$u = \mathbf{4}$$

6.

	Percent	Actual Count
People invited	100	40
People who came	80	P

$$\frac{100}{80} = \frac{40}{P}$$
$$100P = 3200$$
$$P = 32 \text{ people came}$$
$$40 - 32 = \textbf{8 people}$$

7.

	Percent	Actual Count
Sale price	60	\$24
Regular price	100	R

$$\frac{60}{100} = \frac{\$24}{R}$$
$$60R = \$2400$$
$$R = \mathbf{\$40}$$

8.

$$\mathbf{2n + 3 = -13}$$
$$2n + 3 - 3 = -13 - 3$$
$$2n = -16$$
$$n = \frac{-16}{2}$$
$$n = \mathbf{-8}$$

9.

$$(3x - 25) + (x + 5) = 180°$$
$$4x - 20 = 180°$$
$$4x - 20 + 20 = 180° + 20$$
$$4x = 200°$$
$$x = \frac{200°}{4}$$
$$x = 50°$$
$$\mathbf{x + 5} = 50° + 5 = \mathbf{55°}$$
$$\mathbf{3x - 25} = 3(50°) - 25 = \mathbf{125°}$$

10.

2000 voters

$\frac{7}{10}$ voted for incumbent (1400).
200 voters
200 voters
200 voters
200 voters
200 voters
200 voters
200 voters

$\frac{3}{10}$ did not vote for incumbent.
200 voters
200 voters
200 voters

(a) $1400 \div 7 = 200$
$10(200 \text{ voters}) = \textbf{2000 voters}$

(b) **30%**

11. $(3) + (3)(-2) - (3)(-2)$
$3 + (-6) - (-6)$
3

12. a ⓛ $a - a$ (circled: $<$)

13. $\frac{1 \text{ meter}}{4} = \frac{100 \text{ cm}}{4} = 25 \text{ cm}$

$(25 \text{ cm})(25 \text{ cm}) = \mathbf{625\ cm^2}$

14.
$$\begin{array}{r} \$12.95 \\ \$7.85 \\ +\ \$49.50 \\ \hline \$70.30 \end{array} \qquad \begin{array}{r} \$70.30 \\ \times\ \ 0.07 \\ \hline \$4.921 \end{array} \longrightarrow \$4.92$$

$$\begin{array}{r} \$70.30 \\ +\ \ \$4.92 \\ \hline \mathbf{\$75.22} \end{array}$$

15. $(3.5 \times 3) \times (10^5 \times 10^6) = 10.5 \times 10^{11}$
$= \mathbf{1.05 \times 10^{12}}$

16. (a) $\angle \boldsymbol{g}$

(b) $\angle \boldsymbol{d}$

(c) $\angle \boldsymbol{a}$

(d) **70°**

17. (a) $1.25 \times 84 = W_N$
$\mathbf{105} = W_N$

(b) $1.25 \times 84 = \mathbf{105}$

18. **3 to 2**

19. Area of base $= (6 \text{ ft})(4 \text{ ft}) = 24 \text{ ft}^2$
Volume $= (24 \text{ ft}^2)(3 \text{ ft}) = \mathbf{72 \text{ ft}^3}$

20. (a) $C \approx \frac{22}{7}(14 \text{ m})$
$C \approx \mathbf{44 \text{ m}}$

(b) $A \approx \frac{22}{7}(49 \text{ m}^2)$
$A \approx \mathbf{154 \text{ m}^2}$

21. $y = 3(2)$
$y = \mathbf{6;}$
$y = 3(-1)$
$y = \mathbf{-3;}$
$y = 3(0)$
$y = \mathbf{0}$

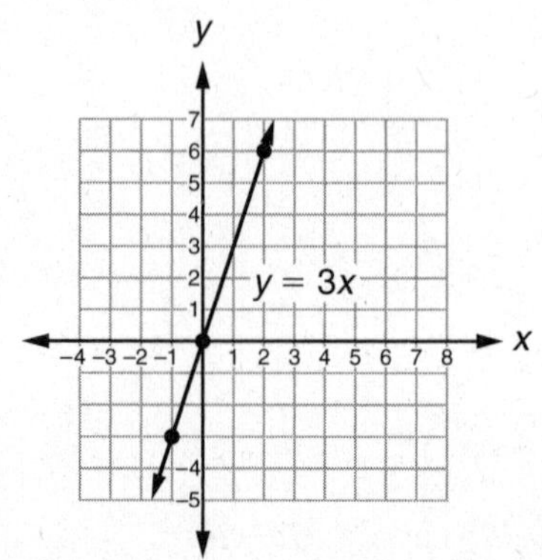

22. (a) $m\angle a = m\angle c = 180° - (90° + 34°)$
$= \mathbf{56°}$

(b) $m\angle b = 90° - 56° = \mathbf{34°}$

(c) $m\angle c = 90° - 34° = \mathbf{56°}$

23. −3 −2 −1 0 1

24. 1/4 0.4 √4
0 1 2

All three numbers are rational.

25.
$$\begin{aligned} 3x + x + 3^0 &= 49 \\ 4x + 1 &= 49 \\ 4x + 1 - 1 &= 49 - 1 \\ 4x &= 48 \\ x &= \frac{48}{4} \\ x &= \mathbf{12} \end{aligned}$$

26.
$$\begin{aligned} 3y + 2 &= y + 32 \\ 3y + 2 - y &= y + 32 - y \\ 2y + 2 &= 32 \\ 2y + 2 - 2 &= 32 - 2 \\ 2y &= 30 \\ \frac{2y}{2} &= \frac{30}{2} \\ y &= \mathbf{15} \end{aligned}$$

27.
$$\begin{aligned} x + 2(x + 3) &= 36 \\ x + 2x + 6 &= 36 \\ 3x + 6 - 6 &= 36 - 6 \\ 3x &= 30 \\ \frac{3x}{3} &= \frac{30}{3} \\ x &= \mathbf{10} \end{aligned}$$

28. (a) $(3x^2y)(-2x)(xy^2)$
$(3) \cdot x \cdot x \cdot y \cdot (-2) \cdot x \cdot x \cdot y \cdot y$
$\mathbf{-6x^4y^3}$

(b) $-3x + 2y - x - y$
$-3x - x + 2y - y$
$\mathbf{-4x + y}$

29. $\left(4\frac{1}{2}\right)\left(\frac{2}{10}\right)\left(\frac{100}{1}\right) = \frac{9}{\cancel{2}_1} \cdot \frac{\cancel{2}^1}{\cancel{10}_1} \cdot \frac{\cancel{100}^{10}}{1} = \mathbf{90}$

30.
$$\begin{gathered} \frac{(-4)(+3)}{(-2)} - (-1) \\ \frac{(-12)}{(-2)} + 1 \\ 6 + 1 \\ \mathbf{7} \end{gathered}$$

Lesson 103, Warm-Up

a. **3**

b. **81**

c. **0.5**

d. **1.2×10^{-6}**

e. **15**

f. **−22°F**

g. **$200**

h. **$400**

i. **9×10^{6}**

Problem Solving

$\sqrt[3]{64} = \mathbf{4}$

$\sqrt[3]{125} = \mathbf{5}$

Lesson 103, Lesson Practice

a. $(-5)(-4)(-3)(-2)(-1) = \mathbf{-120}$

b. $(+5)(-4)(+3)(-2)(+1) = \mathbf{120}$

c. $(-2)^3 = (-2)(-2)(-2) = \mathbf{-8}$

d. $(-3)^4 = (-3)(-3)(-3)(-3) = \mathbf{81}$

e. $(-9)^2 = (-9)(-9) = \mathbf{81}$

f. $(-1)^5 = (-1)(-1)(-1)(-1)(-1) = \mathbf{-1}$

g. $$\frac{6a^2b^3c}{3ab} = \frac{2 \cdot \overset{1}{\cancel{3}} \cdot \overset{1}{\cancel{a}} \cdot a \cdot \overset{1}{\cancel{b}} \cdot b \cdot b \cdot c}{\underset{1}{\cancel{3}} \cdot \underset{1}{\cancel{a}} \cdot \underset{1}{\cancel{b}}}$$
$$= \mathbf{2ab^2c}$$

h. $$\frac{8xy^3z^2}{6x^2y} = \frac{\overset{1}{\cancel{2}} \cdot 2 \cdot 2 \cdot \overset{1}{\cancel{x}} \cdot \overset{1}{\cancel{y}} \cdot y \cdot y \cdot z \cdot z}{\underset{1}{\cancel{2}} \cdot 3 \cdot \underset{1}{\cancel{x}} \cdot x \cdot \underset{1}{\cancel{y}}}$$
$$= \mathbf{\frac{4y^2z^2}{3x}}$$

i. $$\frac{15mn^2p}{25m^2n^2} = \frac{3 \cdot \overset{1}{\cancel{5}} \cdot \overset{1}{\cancel{m}} \cdot \overset{1}{\cancel{n}} \cdot \overset{1}{\cancel{n}} \cdot p}{\underset{1}{\cancel{5}} \cdot 5 \cdot \underset{1}{\cancel{m}} \cdot m \cdot \underset{1}{\cancel{n}} \cdot \underset{1}{\cancel{n}}} = \mathbf{\frac{3p}{5m}}$$

Lesson 103, Mixed Practice

1.
$$\begin{array}{r} \$25.00 \\ \times \quad 0.15 \\ \hline \mathbf{\$3.75} \end{array}$$

2. (a) **85**

(b) **90**

3.

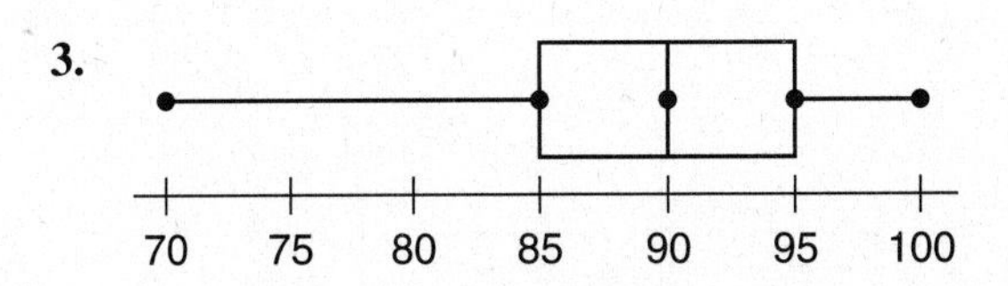

4. $\dfrac{1280 \text{ kilometers}}{2.5 \text{ hours}} = $ **512 kilometers per hour**

5.

	Case 1	Case 2
Dollars	\$25	*d*
Hours	4	7

$$\frac{\$25}{4} = \frac{d}{7}$$
$$4d = \$175$$
$$d = \mathbf{\$43.75}$$

6.

	Percent	Ratio
Lights on	40	2
Lights off	60	3

$$\frac{40}{60} = \mathbf{\frac{2}{3}}$$

7.

	Percent	Actual Count
Original	100	*O*
− Change	20	\$25
New	80	*N*

$$\frac{100}{20} = \frac{O}{\$25}$$
$$20O = \$2500$$
$$O = \mathbf{\$125}$$

8.

	Percent	Actual Count
Original	100	\$30
+ Change	60	*C*
New	160	*N*

$$\frac{100}{60} = \frac{\$30}{C}$$
$$100C = \$1800$$
$$C = \mathbf{\$18}$$

9. (a) $20\,(54 \text{ inches}) = \mathbf{1080\ inches}$

(b) $\overset{90}{\cancel{1080}}\ \cancel{\text{inches}} \cdot \dfrac{1 \text{ foot}}{\underset{1}{\cancel{12}}\ \cancel{\text{inches}}} = \mathbf{90\ feet}$

10. $20^3 = \mathbf{8000\ times}$

11. $\dfrac{8}{5}\,(1000 \text{ meters}) = \mathbf{1600\ meters}$

12.

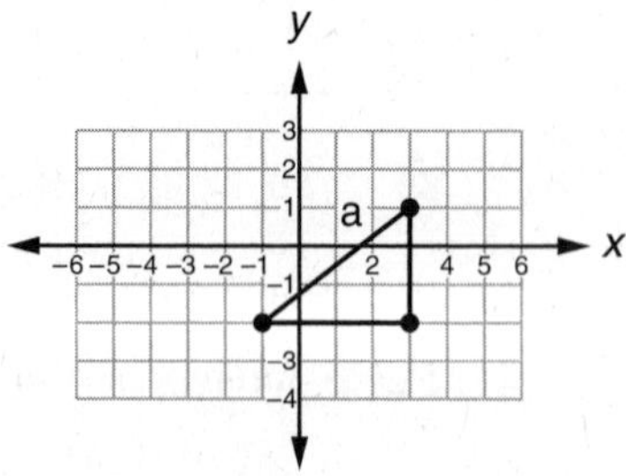

$$a^2 = (4 \text{ units})^2 + (3 \text{ units})^2$$
$$a^2 = 16 \text{ units}^2 + 9 \text{ units}^2$$
$$a^2 = 25 \text{ units}^2$$
$$a = \mathbf{5\ units}$$

13. $\dfrac{35}{50} = \mathbf{70\%}$

14.

$$W_P \times 25 = 20$$
$$\frac{W_P \times 25}{25} = \frac{20}{25}$$
$$W_P = \frac{4}{5}$$
$$W_P = \frac{4}{5} \times 100\% = \mathbf{80\%}$$

15. (a) $\dfrac{120°}{360°} = \mathbf{\dfrac{1}{3}}$

(b) $\dfrac{2}{3} \times 100\% = \mathbf{66\dfrac{2}{3}\%}$

16. $y = \mathbf{-2;}$
$y = -0$
$y = \mathbf{0;}$
$y = -(-1)$
$y = \mathbf{1}$

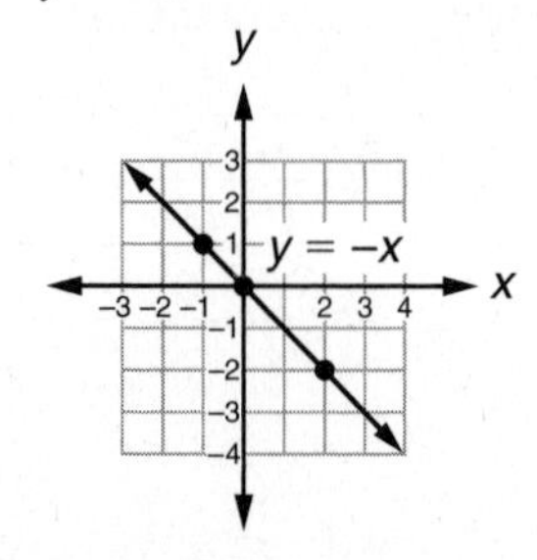

17.

$$\mathbf{2x - 3 = -7}$$
$$2x - 3 + 3 = -7 + 3$$
$$2x = -4$$
$$\frac{2x}{2} = \frac{-4}{2}$$
$$x = \mathbf{-2}$$

18. (a) $m\angle CAB = 90° - 36° = \mathbf{54°}$

(b) $m\angle CAD = m\angle ACB = \mathbf{36°}$

(c) $m\angle ACD = m\angle CAB = \mathbf{54°}$

19. (a) **See student work;**

$$\frac{x}{12} = \frac{12}{16}$$
$$16x = 144$$
$$x = \mathbf{9}$$

(b)
$$12f = 9$$
$$\frac{12f}{12} = \frac{9}{12}$$
$$f = \frac{3}{4} = \mathbf{0.75}$$

20. See student work; 50°

21. (a) $C \approx 3.14\,(2 \text{ ft})$
$C \approx \mathbf{6.28\ ft}$

(b) $A \approx 3.14(1 \text{ ft}^2)$
$A \approx \mathbf{3.14\ ft^2}$

22. $\sqrt{144} = 12$ and $\sqrt{169} = 13$

C. $\sqrt{150}$ is between $\sqrt{144}$ and $\sqrt{169}$.

23. Area of base $= \dfrac{(6 \text{ units})(3 \text{ units})}{2} = 9 \text{ units}^2$

Volume $= (9 \text{ units}^2)(6 \text{ units}) = \mathbf{54\ units^3}$

24. Area of base $\approx 3.14(9 \text{ units}^2) \approx 28.26 \text{ units}^2$

Volume $\approx (28.26 \text{ units}^2)(3 \text{ units})$
$\approx \mathbf{84.78\ units^3}$

25.

$$3x + x - 5 = 2(x - 2)$$
$$4x - 5 = 2x - 4$$
$$4x - 5 - 2x = 2x - 4 - 2x$$
$$2x - 5 = -4$$
$$2x - 5 + 5 = -4 + 5$$
$$2x = 1$$
$$\frac{2x}{2} = \frac{1}{2}$$
$$x = \mathbf{\frac{1}{2}}$$

26. $$6\frac{2}{3}f - 5 = 5$$
$$6\frac{2}{3}f - 5 + 5 = 5 + 5$$
$$6\frac{2}{3}f = 10$$
$$\left(\frac{3}{20}\right)\left(\frac{20}{3}f\right) = \left(\frac{3}{20}\right)10$$
$$f = \mathbf{\frac{3}{2}}$$

27. $$10\frac{1}{2} \cdot 1\frac{3}{7} \cdot 5^{-2} = \frac{\overset{3}{\cancel{21}}}{\underset{1}{\cancel{2}}} \cdot \frac{\overset{\overset{1}{\cancel{5}}}{\cancel{10}}}{\underset{1}{\cancel{7}}} \cdot \frac{1}{\underset{5}{\cancel{25}}} = \mathbf{\frac{3}{5}}$$

28. $$12\frac{1}{2} - 8\frac{1}{3} + 1\frac{1}{6}$$
$$= 12\frac{3}{6} - 8\frac{2}{6} + 1\frac{1}{6} = 5\frac{2}{6} = \mathbf{5\frac{1}{3}}$$

29. (a) $$\frac{(-3)(-2)(-1)}{-|(-3)(+2)|}$$
$$\frac{(6)(-1)}{-|(-6)|}$$
$$\frac{-6}{-(+6)}$$
$$\mathbf{1}$$

(b) $3^2 - (-3)^2 = 9 - (9) = \mathbf{0}$

30. (a) $$\frac{6a^3b^2c}{2abc} = \frac{\overset{1}{\cancel{2}} \cdot 3 \cdot \overset{1}{\cancel{a}} \cdot a \cdot a \cdot \overset{1}{\cancel{b}} \cdot b \cdot \overset{1}{\cancel{c}}}{\underset{1}{\cancel{2}} \cdot \underset{1}{\cancel{a}} \cdot \underset{1}{\cancel{b}} \cdot \underset{1}{\cancel{c}}}$$
$$= \mathbf{3a^2b}$$

(b) $$\frac{8x^2yz^3}{12xy^2z} = \frac{\overset{1}{\cancel{2}} \cdot \overset{1}{\cancel{2}} \cdot 2 \cdot \overset{1}{\cancel{x}} \cdot x \cdot \overset{1}{\cancel{y}} \cdot \overset{1}{\cancel{z}} \cdot z \cdot z}{\underset{1}{\cancel{2}} \cdot \underset{1}{\cancel{2}} \cdot 3 \cdot \underset{1}{\cancel{x}} \cdot \underset{1}{\cancel{y}} \cdot y \cdot \underset{1}{\cancel{z}}}$$
$$= \mathbf{\frac{2xz^2}{3y}}$$

Lesson 104, Warm-Up

a. **15**

b. **190**

c. **0.5**

d. **2.5×10^{11}**

e. **25**

f. **9 sq. ft**

g. **$50**

h. **$75**

i. **9 miles**

Problem Solving

$$\mathbf{4X = X + 60}$$
$$4X - X = X - X + 60$$
$$3X \div 3 = 60 \div 3$$
$$X = \textbf{20 ounces}$$

Lesson 104, Lesson Practice

a. $$C \approx \frac{3.14(6 \text{ cm})}{2}$$
$$C \approx 9.42 \text{ cm}$$
Perimeter $\approx$ 3 cm + 4 cm + 9.42 cm + 3 cm + 10 cm $\approx$ **29.42 cm**

b.

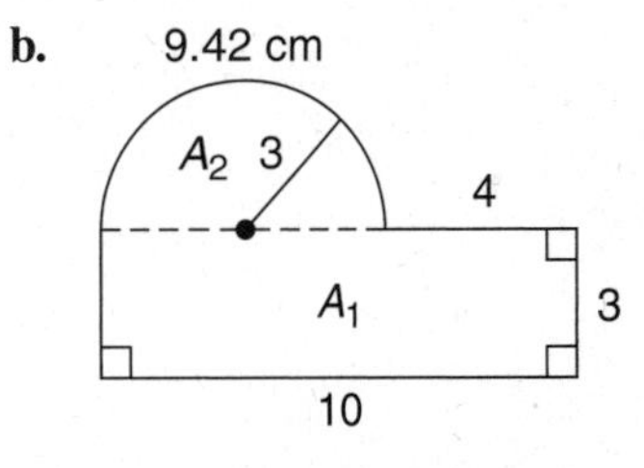

$$A_1 = (3 \text{ cm})(10 \text{ cm}) = 30 \text{ cm}^2$$
$$A_2 \approx \frac{3.14(9 \text{ cm}^2)}{2} \approx 14.13 \text{ cm}^2$$
$$A_1 + A_2 \approx \mathbf{44.13 \text{ cm}^2}$$

c. $$\frac{45°}{360°} = \frac{1}{8}$$
$$A = \frac{\pi(16 \text{ cm}^2)}{8}$$
$$A = \frac{16\pi \text{ cm}^2}{8}$$
$$A = \mathbf{2\pi \text{ cm}^2}$$

d. $$C \approx \frac{3.14(8 \text{ cm})}{8}$$
$$C \approx 3.14 \text{ cm}$$
Perimeter $\approx$ 3.14 cm + 4 cm + 4 cm $\approx$ **11.14 cm**

Lesson 104, Mixed Practice

1.
$$\begin{array}{r} \$12.50 \\ \times \quad 0.4 \\ \hline \mathbf{\$5.00} \end{array}$$

2.
$$p(1, 1 \text{ or } 2 \text{ or } 4 \text{ or } 6) = \frac{1}{6} \cdot \frac{4}{6} = \frac{4}{36}$$
$$p(2, 1 \text{ or } 3 \text{ or } 5) = \frac{1}{6} \cdot \frac{3}{6} = \frac{3}{36}$$
$$p(3, 2 \text{ or } 4) = \frac{1}{6} \cdot \frac{2}{6} = \frac{2}{36}$$
$$p(4, 1 \text{ or } 3) = \frac{1}{6} \cdot \frac{2}{6} = \frac{2}{36}$$
$$p(5, 2 \text{ or } 6) = \frac{1}{6} \cdot \frac{2}{6} = \frac{2}{36}$$
$$p(6, 1 \text{ or } 5) = \frac{1}{6} \cdot \frac{2}{6} = \frac{2}{36}$$
$$\frac{4}{36} + \frac{3}{36} + \frac{2}{36} + \frac{2}{36} + \frac{2}{36} + \frac{2}{36}$$
$$= \mathbf{\frac{15}{36}} = \mathbf{\frac{5}{12}}$$

3.
$$10(88) = 880$$
$$880 - 70 = 810$$
$$\frac{810}{9} = \mathbf{90}$$

4.
$$\frac{\$3.42}{36 \text{ ounces}} = \frac{\$0.095}{1 \text{ ounce}}$$
$$\frac{\$3.84}{48 \text{ ounces}} = \frac{\$0.08}{1 \text{ ounce}}$$
$$\begin{array}{r} \$0.095 \\ - \ \$0.08 \\ \hline \$0.015 \end{array}$$

1.5¢ more per ounce

5.

	Case 1	Case 2
Pages	18	180
Minutes	30	m

$$\frac{18}{30} = \frac{180}{m}$$
$$18m = 5400$$
$$m = 300 \text{ minutes}$$
$$\overset{5}{\cancel{300}} \ \cancel{\text{minutes}} \cdot \frac{1 \text{ hour}}{\underset{1}{\cancel{60}} \ \cancel{\text{minutes}}} = \mathbf{5 \text{ hours}}$$

6.
$$\frac{5}{6} \times W_N = 75$$
$$\frac{6}{5}\left(\frac{5}{6} \times W_N\right) = \left(\frac{6}{5}\right)75$$
$$W_N = 90$$
$$\frac{3}{5}(90) = \mathbf{54}$$

7.

	Ratio	Actual Count
Crawfish	2	C
Tadpoles	21	1932

$$\frac{2}{21} = \frac{C}{1932}$$
$$21C = 3864$$
$$C = \mathbf{184 \text{ crawfish}}$$

8. (a)
$$\mathbf{W_P \times \$60 = \$45}$$
$$\frac{W_P \times \$60}{\$60} = \frac{\$45}{\$60}$$
$$W_P = \frac{3}{4}$$
$$W_P = \frac{3}{4} \times 100\% = \mathbf{75\%}$$

(b)
$$\mathbf{M = 0.45 \times \$60}$$
$$M = \mathbf{\$27}$$

9. (a)
$$A = \pi(144 \text{ units}^2)$$
$$A = \mathbf{144\pi \text{ units}^2}$$

(b)
$$C = \pi(24 \text{ units})$$
$$\mathbf{C = 24\pi \text{ units}}$$

10.
$$360° - (60° + 180°) = 120°$$
$$\frac{120°}{360°} = \frac{1}{3}$$
$$A = \frac{144\pi \text{ units}^2}{3}$$
$$A = \mathbf{48\pi \text{ units}^2}$$

11. (a)
$$360° - 120° = \mathbf{240°}$$

(b)
$$\frac{240°}{360°} = \frac{2}{3}$$
$$C = \frac{2(24\pi \text{ units})}{3}$$
$$C = \mathbf{16\pi \text{ units}}$$

12. $y = 2x - 1$

x	y
−1	−3
0	−1
1	1

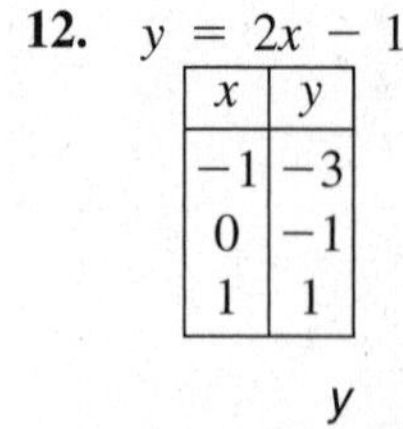

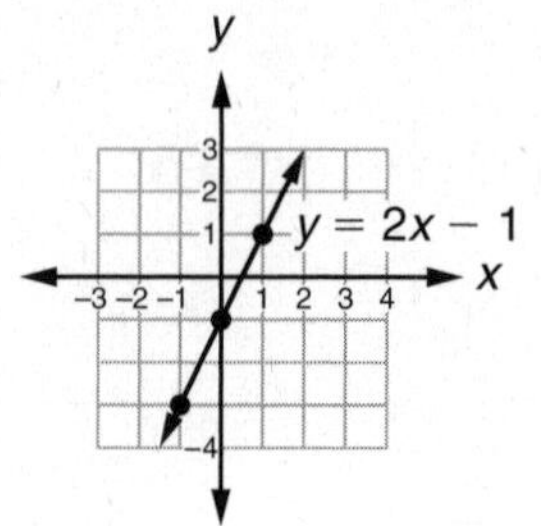

13. (a) $\mathbf{\frac{11}{500}}$

(b) **0.022**

14. (a) **75 miles**

(b) **The car traveled about 60 miles in 1 hour, so its speed was about 60 miles per hour.**

15. $ab \; (<) \; a - b$

16. $(3.6 \times 10^{-4})(9 \times 10^{8})$

32.4×10^{4}

$(3.24 \times 10^{1}) \times 10^{4}$

$\mathbf{3.24 \times 10^{5}}$

17.

3

A_1 3 A_2 6

$$A_1 \approx \frac{3.14(9 \text{ cm}^2)}{2}$$

$$A_1 \approx 14.13 \text{ cm}^2$$

$$A_2 = (3 \text{ cm})(6 \text{ cm}) = 18 \text{ cm}^2$$

$$A_1 + A_2 \approx \mathbf{32.13 \text{ cm}^2}$$

18. $C \approx \dfrac{3.14(6 \text{ cm})}{2}$

$C \approx 9.42$ cm

Perimeter $\approx$ 3 cm + 9.42 cm + 3 cm + 6 cm $\approx$ **21.42 cm**

19. (a) Area of base = (1 ft)(1 ft) = 1 ft^2

Volume = (1 ft^2)(1 ft) = 1 ft^3

$$1 \cancel{\text{ft}^3} \cdot \frac{12 \text{ in.}}{1 \cancel{\text{ft}}} \cdot \frac{12 \text{ in.}}{1 \cancel{\text{ft}}} \cdot \frac{12 \text{ in.}}{1 \cancel{\text{ft}}} = \mathbf{1728 \text{ in.}^3}$$

(b) $6(1 \text{ ft}^2) = \mathbf{6 \text{ ft}^2}$

20. $5(30°) = \mathbf{150°}$

21. $180° - 139° = 41°$

$m\angle x = 180° - (90° + 41°) = \mathbf{49°}$

22. (a) $\dfrac{x}{13} = \dfrac{6}{12}$

$12x = 78$

$x = \dfrac{78}{12}$

$x = \mathbf{6\frac{1}{2}}$

(b) $6f = 12$

$\dfrac{6f}{6} = \dfrac{12}{6}$

$f = \mathbf{2}$

(c) $(2)(2) = $ **4 times**

23.

$$12^2 + y^2 = 13^2$$
$$144 + y^2 = 169$$
$$144 + y^2 - 144 = 169 - 144$$
$$y^2 = 25$$
$$y = \mathbf{5}$$

24.

$$2\frac{3}{4}w + 4 = 48$$
$$2\frac{3}{4}w + 4 - 4 = 48 - 4$$
$$2\frac{3}{4}w = 44$$
$$\left(\frac{4}{11}\right)\left(\frac{11}{4}w\right) = \left(\frac{4}{11}\right)44$$
$$w = \mathbf{16}$$

25.

$$2.4n + 1.2n - 0.12 = 7.08$$
$$3.6n - 0.12 + 0.12 = 7.08 + 0.12$$
$$3.6n = 7.2$$
$$\frac{3.6n}{3.6} = \frac{7.2}{3.6}$$
$$n = \mathbf{2}$$

26. $\sqrt{(3^2)(10^2)} = \sqrt{(9)(100)} = \sqrt{900} = \mathbf{30}$

27. (a) $$\frac{24x^2y}{8x^3y^2} = \frac{\overset{1}{\cancel{2}} \cdot \overset{1}{\cancel{2}} \cdot \overset{1}{\cancel{2}} \cdot 3 \cdot \overset{1}{\cancel{x}} \cdot \overset{1}{\cancel{x}} \cdot \overset{1}{\cancel{y}}}{\underset{1}{\cancel{2}} \cdot \underset{1}{\cancel{2}} \cdot \underset{1}{\cancel{2}} \cdot \underset{1}{\cancel{x}} \cdot \underset{1}{\cancel{x}} \cdot x \cdot \underset{1}{\cancel{y}} \cdot y} = \mathbf{\frac{3}{xy}}$$

(b) $3x^2 + 2x(x - 1)$

$3x^2 + 2x^2 - 2x$

$\mathbf{5x^2 - 2x}$

28.

$$12\frac{1}{2} - \left(8\frac{1}{3} + 1\frac{1}{6}\right)$$
$$= 12\frac{3}{6} - \left(8\frac{2}{6} + 1\frac{1}{6}\right)$$
$$= 12\frac{3}{6} - 9\frac{3}{6} = \mathbf{3}$$

29.

$$\left(4\frac{1}{6} \div 3\frac{3}{4}\right) \div 2.5$$
$$= \left(\frac{\overset{5}{\cancel{25}}}{\underset{3}{\cancel{6}}} \times \frac{\overset{2}{\cancel{4}}}{\underset{3}{\cancel{15}}}\right) \div 2\frac{1}{2} = \frac{10}{9} \div \frac{5}{2}$$
$$= \frac{\overset{2}{\cancel{10}}}{9} \times \frac{2}{\underset{1}{\cancel{5}}} = \mathbf{\frac{4}{9}}$$

30. (a) $\frac{(-3)(4)}{-2} - \frac{(-3)(-4)}{-2}$

$\frac{(-12)}{-2} - \frac{(12)}{-2}$

$6 - (-6)$

12

(b) $\frac{(-2)^3}{(-2)^2} = \frac{\cancel{(-2)}^{1}\cancel{(-2)}^{1}(-2)}{\cancel{(-2)}_{1}\cancel{(-2)}_{1}} = \mathbf{-2}$

Lesson 105, Warm-Up

a. **16**

b. **1209**

c. **16**

d. $\mathbf{1.6 \times 10^{-7}}$

e. **12**

f. **−31°F**

g. **$100**

h. **$100**

i. **2**

Problem Solving

The bottom "stair" is 2 in.-by-2 in.-by-1 in., which is 4 in.^{3}.

The top "stair" is 2 in.-by-1 in.-by-1 in., which is 2 in.^{3}.

4 in.^{3} + 2 in.^{3} = **6 in.^{3}**

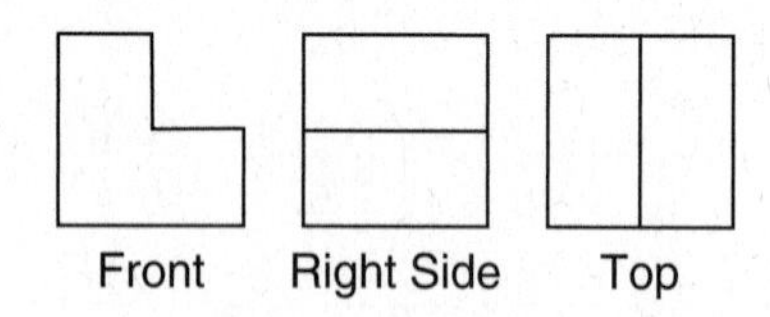

Lesson 105, Lesson Practice

a.
Area of rectangle	= (10 m)(4 m)	=	40 m^2
Area of rectangle	= (10 m)(4 m)	=	40 m^2
Area of rectangle	= (10 m)(4 m)	=	40 m^2
Area of rectangle	= (10 m)(4 m)	=	40 m^2
Area of square	= (4 m)(4 m)	=	16 m^2
+ Area of square	= (4 m)(4 m)	=	16 m^2
Total surface area		=	**192 m^2**

b.
Area of triangle $= \frac{(6 \text{ in.})(8 \text{ in.})}{2}$ = 24 in.^{2}

Area of triangle $= \frac{(6 \text{ in.})(8 \text{ in.})}{2}$ = 24 in.^{2}

Area of rectangle = (10 in.)(10 in.) = 100 in.^{2}

Area of rectangle = (10 in.)(8 in.) = 80 in.^{2}

+ Area of rectangle = (10 in.)(6 in.) = 60 in.^{2}

Total surface area = 24 in.^{2} + 24 in.^{2} + 100 in.^{2} + 80 in.^{2} + 60 in.^{2} = **288 in.^{2}**

c. Area = $\pi d \cdot$ height
Area ≈ (3.14)(10 cm)(4 cm)
Area ≈ **125.6 cm^2**

d. $A \approx 3.14(25 \text{ cm}^2) \approx 78.5 \text{ cm}^2$

Area of top	=	78.5 cm^2
Area of bottom	=	78.5 cm^2
+ Area of lateral surface	=	125.6 cm^2
Total surface area	=	**282.6 cm^2**

e. $A = 4\pi r^2$
$A \approx 4(3.14)(4 \text{ cm}^2)$
$A \approx 50.24 \text{ cm}^2$
$A \approx$ **50 cm^2**

f. **4, −4**

g. $\sqrt[3]{125} = \sqrt[3]{5 \cdot 5 \cdot 5} = \mathbf{5}$

h. $\sqrt[3]{-8} = \sqrt[3]{(-2) \cdot (-2) \cdot (-2)} = \mathbf{-2}$

Lesson 105, Mixed Practice

1.
$$\begin{array}{r} \overset{1}{\cancel{2}}\,\overset{9}{\cancel{0}},{}^{1}000{,}000{,}000 \\ -\quad 900{,}000{,}000 \\ \hline 19{,}100{,}000{,}000 \end{array}$$
$\mathbf{1.91 \times 10^{10}}$

2. Mean = 12.95 ÷ 5 = 2.59
Median − mean = 3.1 − 2.59 = **0.51**

3. $\sqrt{(10)^2 - (8)^2} = \sqrt{100 - 64} = \sqrt{36} = \mathbf{6}$

4. $6.5(\$8.50) = \mathbf{\$55.25}$

5.

	Case 1	Case 2
Cost	\$2.48	c
Kilograms	6	45

$$\frac{\$2.48}{6} = \frac{c}{45}$$
$$6c = \$111.60$$
$$c = \mathbf{\$18.60}$$

6.

	Percent	Actual Count
Regular price	100	\$30
Sale price	75	S

(a) $\frac{100}{75} = \frac{\$30}{S}$
$$100S = \$2250$$
$$S = \mathbf{\$22.50}$$

(b) **75%**

7.

	Ratio	Actual Count
Whigs	7	W
Tories	3	T
Total	10	210

$$\frac{3}{10} = \frac{T}{210}$$
$$10T = 630$$
$$T = \mathbf{63\ Tories}$$

8.

	Percent	Actual Count
Original	100	\$60
− Change	30	C
New	70	N

$$\frac{100}{70} = \frac{\$60}{N}$$
$$100N = \$4200$$
$$N = \mathbf{\$42}$$

9. $60\ \text{feet} \cdot \frac{12\ \text{inches}}{1\ \text{foot}} = 720\ \text{inches}$

	Scale	Actual Count
Model	1	m
Airplane	36	720

$$\frac{1}{36} = \frac{m}{720}$$
$$36m = 720$$
$$m = \mathbf{20\ inches}$$

10. $4x + 5x = 180°$
$$9x = 180°$$
$$\frac{9x}{9} = \frac{180°}{9}$$
$$x = \mathbf{20°}$$

11.
$$\mathbf{W_P \times \$60 = \$3}$$
$$\frac{W_P \times \$60}{\$60} = \frac{\$3}{\$60}$$
$$W_P = \frac{1}{20}$$
$$W_P = \frac{1}{20} \times 100\% = \mathbf{5\%}$$

12. $\mathbf{W_F = \frac{1}{10} \times 4}$
$$W_F = \mathbf{\frac{2}{5}}$$

13.
$$\mathbf{2x - 12 = 86}$$
$$2x - 12 + 12 = 86 + 12$$
$$2x = 98$$
$$\frac{2x}{2} = \frac{98}{2}$$
$$x = \mathbf{49}$$

14.

$$a^2 = (4\ \text{units})^2 + (3\ \text{units})^2$$
$$a^2 = 16\ \text{units}^2 + 9\ \text{units}^2$$
$$a^2 = 25\ \text{units}^2$$
$$a = \mathbf{5\ units}$$

15. a^3 (<) a^2

16. $\frac{13}{52} \cdot \frac{12}{51} = \frac{156}{2652} = \mathbf{\frac{1}{17}}$

17. $A = 4\pi r^2$
$$A \approx 4(3.14)(100\ \text{in.}^2)$$
$$A \approx \mathbf{1256\ in.^2}$$

18. $(8 \times 3.2) \times (10^{-4} \times 10^{-10})$
$$= 25.6 \times 10^{-14} = \mathbf{2.56 \times 10^{-13}}$$

19.
$$C \approx \frac{3.14(20\ \text{m})}{2}$$
$$C \approx 31.4\ \text{m}$$
$$\text{Perimeter} \approx 31.4\ \text{m} + 5\ \text{m} + 30\ \text{m} + 5\ \text{m} + 10\ \text{m} \approx \mathbf{81.4\ m}$$

20. $y = -2(3) - 1$
$y = \mathbf{-7;}$
$y = -2(-2) - 1$
$y = 4 - 1$
$y = \mathbf{3;}$
$y = -2(0) - 1$
$y = \mathbf{-1}$

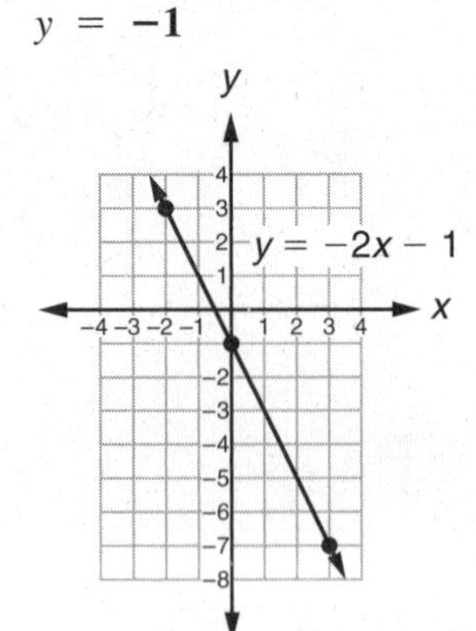

21. (a) Volume $= (5\text{ mm})(5\text{ mm})(5\text{ mm})$
$= \mathbf{125\ mm^3}$

(b) Surface area $= 6(25\text{ mm}^2) = \mathbf{150\ mm^2}$

22. Area of base $\approx 3.14(100\text{ cm}^2) \approx 314\text{ cm}^2$
Volume $\approx (314\text{ cm}^2)(10\text{ cm}) \approx \mathbf{3140\ cm^3}$

23. Lateral surface area $\approx$
$(3.14)(20\text{ cm})(10\text{ cm}) \approx 628\text{ cm}^2$

Area of top	=	314 cm²
Area of bottom	=	314 cm²
+ Lateral surface area	=	628 cm²
Total surface area	=	**1256 cm²**

24. $m\angle a = 180° - (90° + 30°) = 60°$
$m\angle y = m\angle a = 60°$
$m\angle b = 180° - (90° + 60°) = \mathbf{30°}$

25. (a) **See student work;**
$\frac{x}{6} = \frac{8}{12}$
$12x = 48$
$x = \mathbf{4}$

(b) $8f = 12$
$f = \frac{12}{8}$
$f = \frac{3}{2}$
$f = 1\frac{1}{2} = \mathbf{1.5}$

(c) $(1.5)^2 = $ **2.25 times**

26. $4\frac{1}{2}x + 4 = 48 - x$
$4\frac{1}{2}x + 4 + x = 48 - x + x$
$5\frac{1}{2}x + 4 = 48$
$5\frac{1}{2}x + 4 - 4 = 48 - 4$
$5\frac{1}{2}x = 44$
$\left(\frac{2}{11}\right)\left(\frac{11}{2}x\right) = \left(\frac{2}{11}\right)44$
$x = \mathbf{8}$

27. $\frac{3.9}{75} = \frac{c}{25}$
$75c = 97.5$
$\frac{75c}{75} = \frac{97.5}{75}$
$c = \mathbf{1.3}$

28. $3.2 \div \left(2\frac{1}{2} \div \frac{5}{8}\right)$
$= 3\frac{2}{10} \div \left(\frac{5}{2} \times \frac{8}{5}\right) = \frac{32}{10} \div \frac{4}{1}$
$= \frac{32}{10} \times \frac{1}{4} = \mathbf{\frac{4}{5}}$ or **0.8**

29. (a) $\frac{(2xy)(4x^2y)}{8x^2y}$
$= \frac{\overset{1}{\cancel{2}} \cdot \overset{1}{\cancel{x}} \cdot \overset{1}{\cancel{y}} \cdot \overset{1}{\cancel{2}} \cdot \overset{1}{\cancel{2}} \cdot \overset{1}{\cancel{x}} \cdot x \cdot y}{\underset{1}{\cancel{2}} \cdot \underset{1}{\cancel{2}} \cdot \underset{1}{\cancel{2}} \cdot \underset{1}{\cancel{x}} \cdot \underset{1}{\cancel{x}} \cdot \underset{1}{\cancel{y}}} = \mathbf{xy}$

(b) $3(x - 3) - 3 = 3x - 9 - 3$
$= \mathbf{3x - 12}$

30. (a) $\frac{(-10)(-4) - (3)(-2)(-1)}{(-4) - (-2)}$
$\frac{(40) - (6)}{-2}$
$\frac{34}{-2}$
$\mathbf{-17}$

(b) $(-2)^4 - (-2)^2 + \sqrt[3]{-1} + 2^0$
$= (-2)(-2)(-2)(-2) - (-2)(-2)$
$+ \sqrt[3]{(-1)(-1)(-1)} + 1$
$= (16) - (4) + (-1) + 1 = \mathbf{12}$

Lesson 106, Warm-Up

a. **17**

b. **2001**

c. **–125**

d. **4×10^{-1}**

e. **1.8**

f. **18 sq. ft**

g. **$30**

h. **$30**

i. **$6.00**

Problem Solving

$\frac{1}{6} = \frac{2}{12}; \frac{1}{4} = \frac{3}{12}; \frac{1}{3} = \frac{4}{12}$

$\frac{1}{12} + \frac{2}{12} + \frac{3}{12} + \frac{4}{12} + \frac{5}{12} = \frac{15}{12}$

$\frac{15}{12} \div 5 = \frac{15}{60} = \mathbf{\frac{1}{4}}$

Lesson 106, Lesson Practice

a.
$$a + b = c$$
$$a + b - b = c - b$$
$$\mathbf{a = c - b}$$

b.
$$wx = y$$
$$\frac{wx}{x} = \frac{y}{x}$$
$$\mathbf{w = \frac{y}{x}}$$

c.
$$y - b = mx$$
$$y - b + b = mx + b$$
$$\mathbf{y = mx + b}$$

d.
$$A = bh$$
$$\frac{A}{h} = \frac{bh}{h}$$
$$\mathbf{\frac{A}{h} = b}$$

Lesson 106, Mixed Practice

1.
$$\begin{array}{r} \$1.85 \\ \$1.85 \\ \$1.85 \\ +\ \$8.95 \\ \hline \$14.50 \end{array} \qquad \begin{array}{r} \$14.50 \\ \times\ 0.06 \\ \hline \$0.87 \end{array} \qquad \begin{array}{r} \$14.50 \\ +\ \$0.87 \\ \hline \$15.37 \end{array}$$

$$\begin{array}{r} \$20.00 \\ -\ \$15.37 \\ \hline \mathbf{\$4.63} \end{array}$$

2. (a) **1 to 2**

(b) **25%**

3. $\frac{\$2.80}{16 \text{ ounces}} = \mathbf{17.5¢ \text{ per ounce}}$

4. $6(90) = 540,\ 540 - 75 = 465$

$\frac{465}{5} = \mathbf{93}$

5.

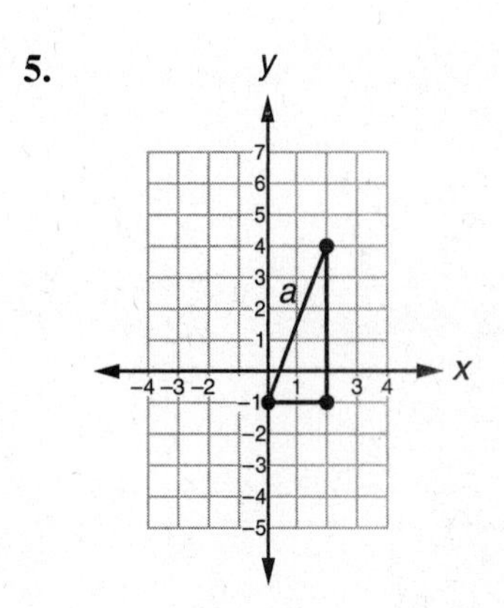

$$a^2 = (2 \text{ units})^2 + (5 \text{ units})^2$$
$$a^2 = 4 \text{ units}^2 + 25 \text{ units}^2$$
$$a^2 = 29 \text{ units}^2$$
$$a = \mathbf{\sqrt{29} \text{ units}}$$

6.

	Case 1	Case 2
Problems	3	27
Minutes	4	m

$$\frac{3}{4} = \frac{27}{m}$$
$$3m = 108$$
$$m = \mathbf{36 \text{ minutes}}$$

7.

	Ratio	Actual Count
Residents	2	R
Visitors	3	V
Total	5	60

$$\frac{3}{5} = \frac{V}{60}$$
$$5V = 180$$
$$V = \mathbf{36 \text{ visitors}}$$

8.

	Percent	Actual Count
Original	100	O
+ Change	25	C
New	125	80

$$\frac{100}{125} = \frac{O}{80}$$
$$125O = 8000$$
$$O = \mathbf{64 \text{ clients}}$$

9. (a) **8 and −8**

(b) $3\sqrt{-64} = \sqrt[3]{(-4)(-4)(-4)} = \mathbf{-4}$

10. $\mathbf{W_N = 2.25 \times 40}$
$\mathbf{W_N = 90}$

11. (a)

$\frac{2}{2}$ $\sqrt{2}$ $|-2|$ 2^2

0 1 2 3 4

(b) $\mathbf{|-2|, \frac{2}{2}, 2^2}$

12.

$$\mathbf{66 = \frac{2}{3} \times W_N}$$
$$\left(\frac{3}{2}\right)66 = \left(\frac{3}{2}\right)\frac{2}{3} \times W_N$$
$$\mathbf{99} = W_N$$

13.

$$\mathbf{0.75 \times W_N = 2.4}$$
$$\frac{0.75 \times W_N}{0.75} = \frac{2.4}{0.75}$$
$$W_N = \mathbf{3.2}$$

14. (a) $\mathbf{1\frac{1}{20}}$

(b) **1.05**

15. See student work;

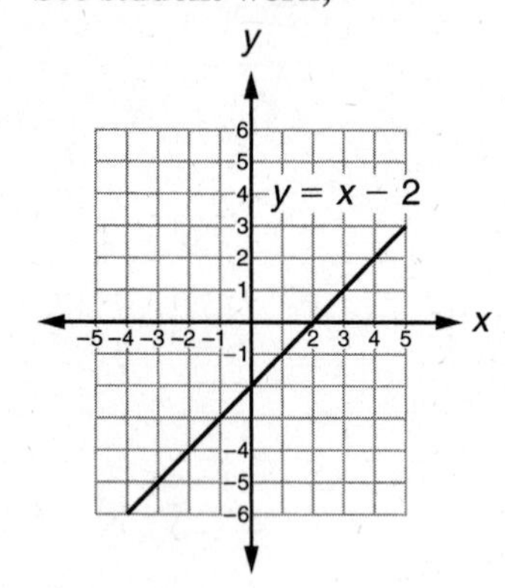

16.

$$81\overline{)6.7500} = 0.0833\ldots \rightarrow \mathbf{0.083}$$

6 48
270
243
270
243
27

17.

$$(4.8 \times 6) \times (10^{-10} \times 10^{-6})$$
$$= 28.8 \times 10^{-16}$$
$$= \mathbf{2.88 \times 10^{-15}}$$

18.

$$(-3)^2 + (-5)(-3) + (6)$$
$$9 + (15) + 6$$
$$\mathbf{30}$$

19.

A_1 10
8
4
A_2
28

$$A_1 \approx \frac{(3.14)(100 \text{ mm}^2)}{2}$$
$$A_1 \approx 157 \text{ mm}^2$$
$$A_2 = (4 \text{ mm})(28 \text{ mm}) = 112 \text{ mm}^2$$
$$A_1 + A_2 \approx 157 \text{ mm}^2 + 112 \text{ mm}^2 \approx \mathbf{269 \text{ mm}^2}$$

20.

Area of triangle $= \frac{(4 \text{ cm})(3 \text{ cm})}{2} = 6 \text{ cm}^2$

Area of triangle $= \frac{(4 \text{ cm})(3 \text{ cm})}{2} = 6 \text{ cm}^2$

Area of rectangle $= (10 \text{ cm})(5 \text{ cm}) = 50 \text{ cm}^2$

Area of rectangle $= (10 \text{ cm})(4 \text{ cm}) = 40 \text{ cm}^2$

+ Area of rectangle $= (10 \text{ cm})(3 \text{ cm}) = 30 \text{ cm}^2$

Total surface area $= \mathbf{132 \text{ cm}^2}$

21. Area of base $\approx 3.14(1 \text{ in.})^2 \approx 3.14 \text{ in.}^2$

Volume $\approx (3.14 \text{ in.}^2)(10 \text{ in.})$
$\approx \mathbf{31.4 \text{ in.}^3}$

22.

$$180° - 140° = 40°$$
$$180° - (90° + 40°) = 50°$$
$$m\angle b = 180° - (90° + 50°) = \mathbf{40°}$$

23. (a)

$$x + c = d$$
$$x + c - c = d - c$$
$$\mathbf{x = d - c}$$

(b)

$$an = b$$
$$\frac{an}{a} = \frac{b}{a}$$
$$\mathbf{n = \frac{b}{a}}$$

24.

$$6w - 2(4 + w) = w + 7$$
$$6w - 8 - 2w = w + 7$$
$$4w - 8 = w + 7$$
$$4w - 8 - w = w + 7 - w$$
$$3w - 8 = 7$$
$$3w - 8 + 8 = 7 + 8$$
$$3w = 15$$
$$\frac{3w}{3} = \frac{15}{3}$$
$$w = \mathbf{5}$$

25.
$$\begin{aligned} 6x + 8 &< 14 \\ 6x + 8 - 8 &< 14 - 8 \\ 6x &< 6 \\ \frac{6x}{6} &< \frac{6}{6} \\ \mathbf{x} &< \mathbf{1} \end{aligned}$$

$x < 1$

−2 −1 0 1 2

26. $37 = 3x - 5,\ 42 = 3x,\ x = \mathbf{14}$

27. $25 - [3^2 + 2(5 - 3)] = 25 - [9 + 4]$
$= 25 - 13 = \mathbf{12}$

28.
$$\frac{6x^2 + (5x)(2x)}{4x}$$
$$\frac{6x^2 + 10x^2}{4x}$$
$$\frac{16x^2}{4x}$$
$$\frac{\overset{1}{\cancel{4}} \cdot 4 \cdot \overset{1}{\cancel{x}} \cdot x}{\underset{1}{\cancel{4}} \cdot \underset{1}{\cancel{x}}} = \mathbf{4x}$$

29. $4^0 + 3^{-1} + 2^{-2}$
$$1 + \frac{1}{3} + \frac{1}{4} = \frac{12}{12} + \frac{4}{12} + \frac{3}{12} = \frac{19}{12}$$
$$= \mathbf{1\frac{7}{12}}$$

30.
$$(-3)(-2)(+4)(-1) + (-3)^2 + \sqrt[3]{-64} - (-2)^3$$
$$(6)(+4)(-1) + (-3)(-3) + \sqrt[3]{(-4)(-4)(-4)} - (-2)(-2)(-2)$$
$$-24 + (9) + (-4) - (-8)$$
$$\mathbf{-11}$$

Lesson 107, Warm-Up

a. **24**

b. **800**

c. **10**

d. $\mathbf{6.25 \times 10^{12}}$

e. **3**

f. **−58°F**

g. **$45**

h. **$105**

i. **6**

Problem Solving

There are sixteen 1-by-1 squares, nine 2-by-2 squares, four 3-by-3 squares, and one 4-by-4 square. The total number of squares is $16 + 9 + 4 + 1$, or **30.**

Lesson 107, Lesson Practice

a. **"Yards to feet":** $\frac{\text{rise}}{\text{run}} = \frac{3}{1} = \mathbf{3}$

"Feet to yards": $\frac{\text{rise}}{\text{run}} = \mathbf{\frac{1}{3}}$

b. Graph (a): $\frac{\text{rise}}{\text{run}} = \frac{1}{1} = \mathbf{1}$

Graph (c): $\frac{\text{rise}}{\text{run}} = \frac{1}{-2} = \mathbf{-\frac{1}{2}}$

c. $\frac{\text{rise}}{\text{run}} = \mathbf{\frac{1}{3}}$

$\frac{\text{rise}}{\text{run}} = \frac{2}{-3} = \mathbf{-\frac{2}{3}}$

$\frac{\text{rise}}{\text{run}} = \mathbf{0}$

$\frac{\text{rise}}{\text{run}} = \frac{2}{-1} = \mathbf{-2}$

d. $\mathbf{\frac{1}{3}; -\frac{2}{3}; 0; -2}$

Lesson 107, Mixed Practice

1. $\frac{2}{3} \times \$21 = \mathbf{\$14}$

2.
```
   0 9 9
   1̸, 0̸ 0̸¹0,000,000,000
 −    9 7 5,000,000,000
        2 5,000,000,000
```
$\mathbf{2.5 \times 10^{10}}$

3. (a) **11**

(b) **16**

4.

	Case 1	Case 2
Miles	18	m
Minutes	60	40

$$\frac{18}{60} = \frac{m}{40}$$
$$60m = 720$$
$$m = \mathbf{12\ miles}$$

5.

	Ratio	Actual Count
Earthworms	5	E
Cutworms	2	C
Total	7	140

$$\frac{5}{7} = \frac{E}{140}$$
$$7E = 700$$
$$E = \mathbf{100\ earthworms}$$

6.

	Percent	Actual Count
Original	100	\$16,550
+ Change	8	C
New	108	N

$$\frac{100}{108} = \frac{\$16{,}550}{N}$$
$$100N = \$1{,}787{,}400$$
$$N = \mathbf{\$17{,}874}$$

7.

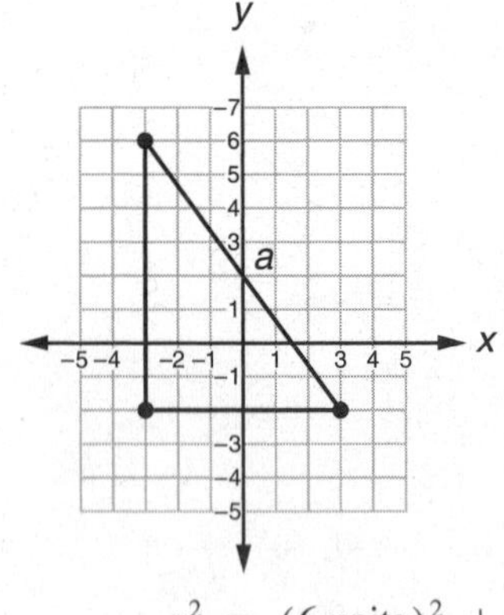

$$a^2 = (6\text{ units})^2 + (8\text{ units}^2)$$
$$a^2 = 36\text{ units}^2 + 64\text{ units}^2$$
$$a^2 = 100\text{ units}^2$$
$$a = 10\text{ units}$$
$$\text{Perimeter} = 6\text{ units} + 8\text{ units} + 10\text{ units} = \mathbf{24\ units}$$

8. (a) $m\angle ABD = 180° - (90° + 50°) = \mathbf{40°}$

(b) $m\angle DBC = 90° - 40° = \mathbf{50°}$

(c) $m\angle BCD = 180° - (90° + 50°) = \mathbf{40°}$

(d) **All three triangles are similar.**

9.
$$\mathbf{60 = 1.25 \times W_N}$$
$$\frac{60}{1.25} = \frac{1.25 \times W_N}{1.25}$$
$$\mathbf{48} = W_N$$

10.
$$\mathbf{60 = W_P \times 25}$$
$$\frac{60}{25} = \frac{W_P \times 25}{25}$$
$$\frac{12}{5} = W_P$$
$$W_P = \frac{12}{5} \times 100\% = \mathbf{240\%}$$

11.
$$\mathbf{60 = 2n + 4}$$
$$60 - 4 = 2n + 4 - 4$$
$$56 = 2n$$
$$\frac{56}{2} = \frac{2n}{2}$$
$$\mathbf{28} = n$$

12. (a) $\dfrac{\text{not red marbles}}{\text{total marbles}} = \dfrac{\mathbf{80}}{\mathbf{100}} = \mathbf{80\%}$

(b) $p(y, y) = \dfrac{\mathbf{10}}{\mathbf{100}} \cdot \dfrac{\mathbf{9}}{\mathbf{99}} = \dfrac{90}{9900} = \dfrac{\mathbf{1}}{\mathbf{110}}$

13. (a) $\mathbf{0.8\overline{3}}$

(b) $\mathbf{83\frac{1}{3}\%}$

14. $(x - y)^2 \ (=)\ (y - x)^2$

15. $(1.8 \times 9) \times (10^{10} \times 10^{-6})$
$$= 16.2 \times 10^4$$
$$= \mathbf{1.62 \times 10^5}$$

16. (a) **24 and 25**

(b) $\mathbf{\sqrt{10}}$ **and** $\mathbf{-\sqrt{10}}$

17. **See student work;**

(a)

y

y = x + 1

x

(b) $\dfrac{\text{rise}}{\text{run}} = \dfrac{1}{1} = \mathbf{1}$

18. $360° - 120° = 240°, \quad \frac{240°}{360°} = \frac{2}{3}$

$$A = \frac{2\pi(36\text{ cm}^2)}{3}$$

$$A = \mathbf{24\pi\text{ cm}^2}$$

19.

Area of square	=	(4 in.)(4 in.)	=	16 in.^{2}
Area of square	=	(4 in.)(4 in.)	=	16 in.^{2}
Area of rectangle	=	(4 in.)(8 in.)	=	32 in.^{2}
Area of rectangle	=	(4 in.)(8 in.)	=	32 in.^{2}
Area of rectangle	=	(4 in.)(8 in.)	=	32 in.^{2}
+ Area of rectangle	=	(4 in.)(8 in.)	=	32 in.^{2}
Total surface area			=	**160 in.^{2}**

20. Area of base $\approx 3.14(16\text{ cm}^2)$

$\approx 50.24\text{ cm}^2$

Volume $\approx (50.24\text{ cm}^2)(10\text{ cm})$

$\approx \mathbf{502.4\text{ cm}^3}$

21. $C \approx 3.14(8\text{ cm})$

$C \approx 25.12\text{ cm}$

Area of lateral surface	=	(25.12 cm)(10 cm)
	=	251.2 cm^2
Area of base	=	50.24 cm^2
+ Area of top	=	50.24 cm^2
Total surface area	=	**351.68 cm^2**

22. $180° - (90° + 60°) = 30°$

$90° - 30° = 60°$

$180° - (90° + 50°) = \mathbf{30°} = m\angle X$

23. $\frac{\text{rise}}{\text{run}} = \frac{2}{1} = \mathbf{2}$

24. (a)

$$x - y = z$$
$$x - y + y = z + y$$
$$\boldsymbol{x = z + y}$$

(b)

$$w = xy$$
$$\frac{w}{y} = \frac{xy}{y}$$
$$\boldsymbol{\frac{w}{y} = x}$$

25.

$$\frac{a}{21} = \frac{1.5}{7}$$
$$7a = 31.5$$
$$\frac{7a}{7} = \frac{31.5}{7}$$
$$a = \mathbf{4.5}$$

26.

$$6x + 5 = 7 + 2x$$
$$6x + 5 - 2x = 7 + 2x - 2x$$
$$4x + 5 = 7$$
$$4x + 5 - 5 = 7 - 5$$
$$4x = 2$$
$$\frac{4x}{4} = \frac{2}{4}$$
$$x = \mathbf{\frac{1}{2}}$$

27.

$$62 + 5\{20 - [4^2 + 3(2 - 1)]\}$$
$$62 + 5\{20 - [16 + 3(1)]\}$$
$$62 + 5\{20 - [19]\}$$
$$62 + 5\{1\}$$
$$62 + (5)$$
$$\mathbf{67}$$

28.

$$\frac{(6x^2y)(2xy)}{4xy^2} = \frac{\overset{1}{\cancel{2}} \cdot 3 \cdot \overset{1}{\cancel{x}} \cdot x \cdot \overset{1}{\cancel{y}} \cdot \overset{1}{\cancel{2}} \cdot x \cdot \overset{1}{\cancel{y}}}{\underset{1}{\cancel{2}} \cdot \underset{1}{\cancel{2}} \cdot \underset{1}{\cancel{x}} \cdot \underset{1}{\cancel{y}} \cdot \underset{1}{\cancel{y}}}$$

$$= \mathbf{3x^2}$$

29.

$$5\frac{1}{6} + 3\frac{1}{2} - \frac{1}{3} = 5\frac{1}{6} + 3\frac{3}{6} - \frac{2}{6}$$
$$= 8\frac{4}{6} - \frac{2}{6} = 8\frac{2}{6} = \mathbf{8\frac{1}{3}}$$

30.

$$\frac{(5)(-3)(2)(-4) + (-2)(-3)}{|-6|}$$
$$\frac{(-30)(-4) + (6)}{6}$$
$$\frac{120 + 6}{6}$$
$$\mathbf{21}$$

Lesson 108, Warm-Up

a. **14**

b. **45**

c. **0**

d. $\mathbf{1 \times 10^{-7}}$

e. **8**

f. **27 sq. ft**

g. **$180**

h. **$180**

i. **210 mi**

Problem Solving

$\sqrt[3]{1{,}000{,}000} = \mathbf{100}$

LESSON 108, LESSON PRACTICE

a.
$$20 = b(4)$$
$$\frac{20}{4} = b$$
$$\mathbf{5} = b$$

b.
$$20 = \frac{1}{2}(b)(4)$$
$$20 = 2b$$
$$\frac{20}{2} = b$$
$$\mathbf{10} = b$$

c.
$$F = 1.8(-40) + 32$$
$$F = \mathbf{-40}$$

LESSON 108, MIXED PRACTICE

1.
$$\begin{array}{r} \$8.35 \\ \$1.25 \\ +\ \$2.40 \\ \hline \$12.00 \end{array} \qquad \begin{array}{r} \$12.00 \\ \times\ \ 0.15 \\ \hline \mathbf{\$1.80} \end{array}$$

2.
$$\begin{array}{r} 0.000120 \\ -\ 0.000020 \\ \hline 0.000100 \end{array} \qquad \mathbf{1 \times 10^{-4}}$$

3. **4, 7, 8, 8, 8, 9, 9, 10, 12, 15**
Median = 8.5
Mode = 8

4. $p(5, 5) = \mathbf{\frac{4}{52} \cdot \frac{3}{51} = \frac{1}{221}}$

5.

	Case 1	Case 2
Dollars	$200	d
Francs	300	240

$$\frac{\$200}{300} = \frac{d}{240}$$
$$300d = \$48{,}000$$
$$d = \mathbf{\$160}$$

6.

	Ratio	Actual Count
Red beans	5	175
Brown beans	7	B
Total	12	T

$$\frac{5}{12} = \frac{175}{T}$$
$$5T = 2100$$
$$T = \mathbf{420\ beans}$$

7.

	Percent	Actual Count
Original	100	$90
− Change	35	C
New	65	N

$$\frac{100}{65} = \frac{\$90}{N}$$
$$100N = \$5850$$
$$N = \$58.50$$
$$2(\$58.50) = \mathbf{\$117}$$

8.
$$1 \text{ ton} = 2000 \text{ pounds}$$
$$\frac{3}{8}(2000 \text{ pounds}) = \mathbf{750\ pounds}$$

9.
$$\mathbf{W_N = 0.025 \times 800}$$
$$W_N = \mathbf{20}$$

10.
$$\mathbf{0.1 \times W_N = \$2500}$$
$$\frac{0.1 \times W_N}{0.1} = \frac{\$2500}{0.1}$$
$$W_N = \mathbf{\$25{,}000}$$

11.
$$\mathbf{56 = 2x - 8}$$
$$56 + 8 = 2x - 8 + 8$$
$$64 = 2x$$
$$\frac{64}{2} = \frac{2x}{2}$$
$$\mathbf{32} = x$$

12. $\frac{\text{rise}}{\text{run}} = \frac{2}{-3} = \mathbf{-\frac{2}{3}}$

13. (a)

	Scale	Measure
Model	1	$6/7\frac{1}{2}$
Object	24	O_1/O_2

$$\frac{1}{24} = \frac{6}{O_1} \qquad \frac{1}{24} = \frac{7.5}{O_2}$$
$$O_1 = 144 \qquad O_2 = 180$$
$$O_1 = 12 \text{ ft} \qquad O_2 = 15 \text{ ft}$$
$$12 \text{ ft} \times 15 \text{ ft} = \mathbf{180\ ft^2}$$

(b) **9 in.; We round 17 ft $9\frac{1}{2}$ in. to 18 ft. Since every 2 ft is 1 in. in the floor plan, we estimate by dividing 18 by 2.**

14. $4x = \mathbf{180°}$

$\frac{4x}{4} = \frac{180°}{4}$

$x = \mathbf{45°}$

$2x = 2(45°) = \mathbf{90°}$

15. $(2.8 \times 8) \times (10^5 \times 10^{-8})$

$= 22.4 \times 10^{-3}$

$= \mathbf{2.24 \times 10^{-2}}$

16. $c = 2.54(12)$

$c = \mathbf{30.48\ cm}$

17. **See student work;**

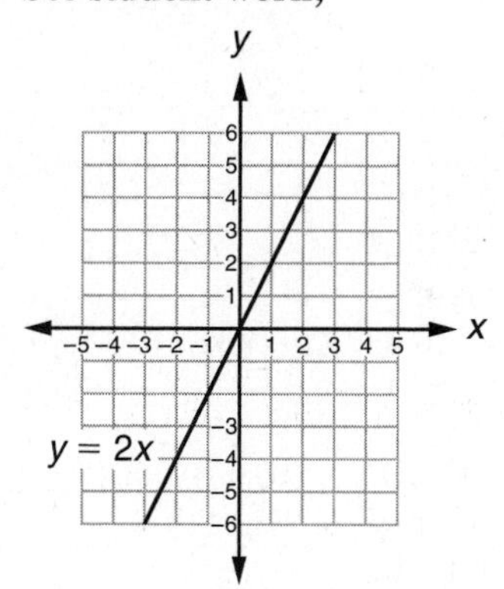

18.

5

4

2

$C \approx \frac{3.14(4\text{ in.})}{2}$

$C \approx 6.28\text{ in.}$

Perimeter $\approx$ 4 in. + 5 in. + 6.28 in. + 5 in. $\approx$ **20.28 in.**

19. $6(100\text{ in.}^2) = \mathbf{600\ in.^2}$

20. Area of base $\approx 3.14(25\text{ cm}^2)$

$\approx 78.5\text{ cm}^2$

Volume $\approx (78.5\text{ cm}^2)(5\text{ cm})$

$\approx \mathbf{392.5\ cm^3}$

21. $m\angle x = 180° - 150° = \mathbf{30°}$

22. (a) $\frac{y}{10} = \frac{9}{6}$

$6y = 90$

$y = \mathbf{15\ cm}$

(b) $6f = 9$

$f = \frac{9}{6}$

$f = 1\frac{1}{2} = \mathbf{1.5}$

(c) $(1.5)(1.5) =$ **2.25 times**

23. $x^2 + (6\text{ cm})^2 = (10\text{ cm})^2$

$x^2 + 36\text{ cm}^2 = 100\text{ cm}^2$

$x^2 + 36\text{ cm}^2 - 36\text{ cm}^2 = 100\text{ cm}^2 - 36\text{ cm}^2$

$x^2 = 64\text{ cm}^2$

$x = \mathbf{8\ cm}$

24. Surface area $= 4\pi r^2$

$\approx 4(3.14)(25\text{ in.}^2)$

$\approx \mathbf{314\ in.^2}$

25. $1\frac{2}{3}x = 32 - x$

$1\frac{2}{3}x + x = 32 - x + x$

$2\frac{2}{3}x = 32$

$\left(\frac{3}{8}\right)\left(\frac{8}{3}x\right) = \left(\frac{3}{8}\right)32$

$x = \mathbf{12}$

26. $x^2 + x(x + 2)$

$x^2 + x^2 + 2x$

$\mathbf{2x^2 + 2x}$

27. $\frac{\cancel{(-2)}^1 \cdot 2 \cdot a \cdot \cancel{x}^1 \cdot \cancel{3}^1 \cdot \cancel{x}^1 \cdot y}{\cancel{(-2)}_1 \cdot \cancel{3}_1 \cdot \cancel{x}_1 \cdot \cancel{x}_1}$

$= \mathbf{2ay}$

28. $1.1\{1.1[1.1(1000)]\}$

$1.1\{1.1[1100]\}$

$1.1\{1210\}$

1331

29. $3\frac{3}{4} \cdot 2\frac{2}{3} \div 10$

$= \frac{\cancel{15}^5}{\cancel{4}_1} \cdot \frac{\cancel{8}^2}{\cancel{3}_1} \div \frac{10}{1} = \cancel{10}^1 \times \frac{1}{\cancel{10}_1} = \mathbf{1}$

30. (a) $(-6) - (7)(-4) + \sqrt[3]{125} + \frac{(-8)(-9)}{(-3)(-2)}$

$(-6) - (-28) + (5) + \frac{72}{6}$

$22 + (5) + 12$

39

(b) $(-1) + (-1)^2 + (-1)^3 + (-1)^4$

$(-1) + (1) + (-1) + (1)$

0

Lesson 109, Warm-Up

a. **18**

b. **1960**

c. $\frac{1}{4}$

d. **1.44 × 10^24**

e. **1**

f. **1.5 m**

g. **$10**

h. **$70**

i. **40**

Problem Solving

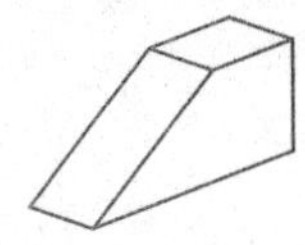

Lesson 109, Lesson Practice

a.
$$\begin{aligned} 3x^2 - 8 &= 100 \\ 3x^2 - 8 + 8 &= 100 + 8 \\ 3x^2 &= 108 \\ x^2 &= 36 \\ x &= \mathbf{6, -6} \end{aligned}$$

b.
$$\begin{aligned} x^2 + x^2 &= 12 \\ 2x^2 &= 12 \\ x^2 &= 6 \\ x &= \mathbf{\sqrt{6}, -\sqrt{6}} \end{aligned}$$

c.
$$\begin{aligned} 157 &= 2(-x)^2 - 5 \\ 157 &= 2x^2 - 5 \\ 157 + 5 &= 2x^2 - 5 + 5 \\ 162 &= 2x^2 \\ 81 &= x^2 \\ 9, -9 &= x \\ &\mathbf{-9} \end{aligned}$$

d.
$$\begin{aligned} 7x^2 &= 21 \\ x^2 &= 3 \\ x &= \sqrt{3}, -\sqrt{3} \\ &\mathbf{\sqrt{3}} \end{aligned}$$

e.
$$\begin{aligned} \frac{w}{4} &= \frac{9}{w} \\ w^2 &= 36 \\ w &= \mathbf{6, -6} \end{aligned}$$

Lesson 109, Mixed Practice

1. $\frac{(0.2)(0.05)}{0.2 + 0.05} = \frac{0.01}{0.25} = \mathbf{0.04}$

2. (a) $\angle z$

(b) $\angle w$

(c) $\angle y$

(d)
$$\begin{aligned} 3m + m &= 180^\circ \\ 4m &= 180^\circ \\ m &= 45^\circ \\ 3m &= 3(45^\circ) = \mathbf{135^\circ} \end{aligned}$$

3.
$$\begin{aligned} 20 &= 5 + (10 \times W_D) \\ 20 - 5 &= 5 + (10 \times W_D) - 5 \\ 15 &= 10 \times W_D \\ \frac{15}{10} &= \frac{10 \times W_D}{10} \\ \mathbf{1.5} &= W_D \end{aligned}$$

4.
$$\begin{aligned} &1\ \cancel{\text{km}^2} \cdot \frac{1000\text{ m}}{1\ \cancel{\text{km}}} \cdot \frac{1000\text{ m}}{1\ \cancel{\text{km}}} \\ &= \mathbf{1{,}000{,}000\ m^2} \end{aligned}$$

5.
$$\begin{aligned} 4(5) &= 20 \\ 10(5) &= 50 \\ \frac{\text{quarters}}{\text{dimes}} &= \frac{20}{50} = \mathbf{\frac{2}{5}} \end{aligned}$$

6.

	Case 1	Case 2
Meters	3000	5000
Minutes	9	m

$$\begin{aligned} \frac{3000}{9} &= \frac{5000}{m} \\ 3000m &= 45{,}000 \\ m &= \mathbf{15\ minutes} \end{aligned}$$

7.

	Percent	Actual Count
Original	100	O
+ Change	20	C
New	120	60

$$\begin{aligned} \frac{100}{120} &= \frac{O}{60} \\ 120O &= 6000 \\ O &= \mathbf{50} \end{aligned}$$

8.

	Percent	Actual Count
Original	100	\$36
− Change	25	C
New	75	N

$$\frac{100}{75} = \frac{\$36}{N}$$
$$100N = \$2700$$
$$N = \mathbf{\$27}$$

9. $\mathbf{60 = 1.5 \times W_N}$

$$\frac{60}{1.5} = \frac{1.5 \times W_N}{1.5}$$
$$\mathbf{40} = W_N$$

10.

702 cards	
$\frac{2}{3}$ kept	234 cards
	234 cards
$\frac{1}{3}$ given away (234)	234 cards

(a) 3(234 cards) = **702 cards**

(b) 2(234 cards) = **468 cards**

11. $a - b$ ⓻ $b - a$ (circled: $>$)

12. $p(c, c, c, c, c) = \frac{1}{2} \cdot \frac{1}{2} \cdot \frac{1}{2} \cdot \frac{1}{2} \cdot \frac{1}{2}$

$$= \left(\frac{1}{2}\right)^5 = \mathbf{\frac{1}{32}}$$

13.

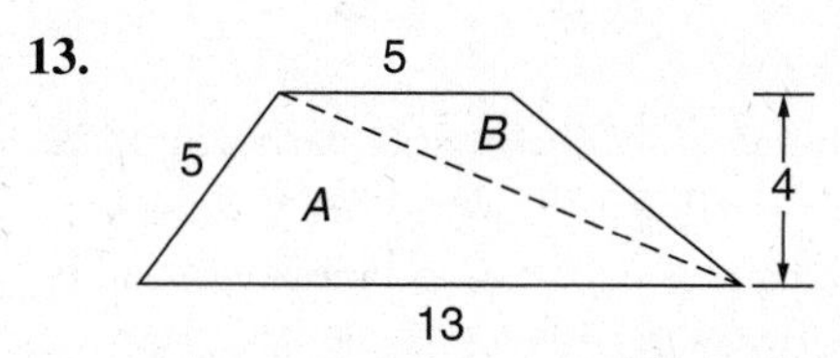

$$\text{Area of triangle } A = \frac{(13 \text{ cm})(4 \text{ cm})}{2} = 26 \text{ cm}^2$$
$$+ \text{ Area of triangle } B = \frac{(5 \text{ cm})(4 \text{ cm})}{2} = 10 \text{ cm}^2$$
$$\text{Area of figure} = \mathbf{36 \text{ cm}^2}$$

14. $\text{Area of base} = \frac{(6 \text{ in.})(3 \text{ in.})}{2} = 9 \text{ in.}^2$

$$\text{Volume} = (9 \text{ in.}^2)(6 \text{ in.}) = \mathbf{54 \text{ in.}^3}$$

15. $C \approx 3.14(6 \text{ in.})$

$$C \approx 18.84 \text{ in.}$$
$$\text{Area} \approx (18.84 \text{ in.})(3 \text{ in.}) \approx \mathbf{56.52 \text{ in.}^2}$$

16. (a)

$$\begin{array}{r} \$36.00 \\ \times\ 0.065 \\ \hline \mathbf{\$2.34} \end{array}$$

(b)

$$\begin{array}{r} \$36.00 \\ +\ \$2.34 \\ \hline \mathbf{\$38.34} \end{array}$$

17. (a) $\mathbf{\frac{1}{200}}$

(b) **0.005**

18. $\frac{100}{166\frac{2}{3}} = \frac{\frac{100}{1}}{\frac{500}{3}} \cdot \frac{\frac{3}{500}}{\frac{3}{500}} = \frac{\frac{300}{500}}{1} = \frac{3}{5}$

$$\frac{3}{5} = \frac{48}{n}$$
$$3n = 240$$
$$n = \mathbf{80}$$

19. $(6 \times 10^{-8})(8 \times 10^4)$

$$48 \times 10^{-4}$$
$$(4.8 \times 10^1) \times 10^{-4}$$
$$\mathbf{4.8 \times 10^{-3}}$$

20. $y = \frac{2}{3}(6) - 1$

$y = 2(2) - 1$
$y = 4 - 1$
$y = \mathbf{3;}$
$y = \frac{2}{3}(0) - 1$
$y = 0 - 1$
$y = \mathbf{-1;}$
$y = \frac{2}{3}(-3) - 1$
$y = 2(-1) - 1$
$y = -2 - 1$
$y = \mathbf{-3}$

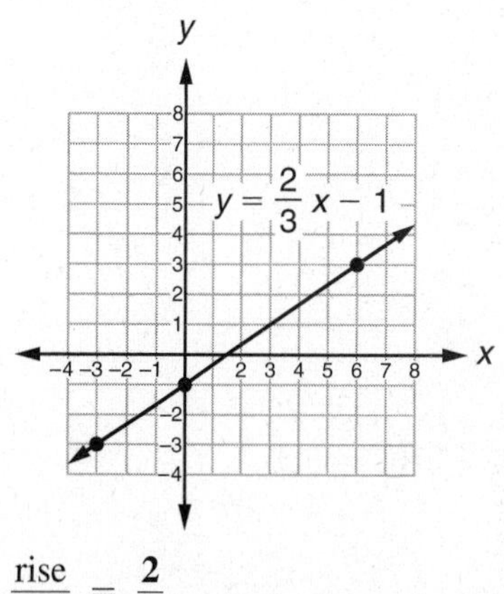

$$\frac{\text{rise}}{\text{run}} = \mathbf{\frac{2}{3}}$$

21.

	Ratio	Actual Count
Angle 1	7	A_1
Angle 2	8	A_2
Total	15	90°

$$\frac{7}{15} = \frac{A_1}{90°}$$
$$15A_1 = 630°$$
$$A_1 = \mathbf{42°}$$

22. $4x + 3x + 2x + x = 360°$
$10x = 360°$
$x = \mathbf{36°}$

23. $d^2 = (2 \text{ units})^2 + (5 \text{ units})^2$
$d^2 = 4 \text{ units}^2 + 25 \text{ units}^2$
$d^2 = 29 \text{ units}^2$
$d = \mathbf{\sqrt{29} \text{ units}}$

24. $3m^2 + 2 = 50$
$3m^2 + 2 - 2 = 50 - 2$
$3m^2 = 48$
$m^2 = 16$
$m = \mathbf{4, -4}$

25. $7(y - 2) = 4 - 2y$
$7y - 14 = 4 - 2y$
$7y - 14 + 2y = 4 - 2y + 2y$
$9y - 14 = 4$
$9y - 14 + 14 = 4 + 14$
$9y = 18$
$\frac{9y}{9} = \frac{18}{9}$
$y = \mathbf{2}$

26. $\sqrt{144} - (\sqrt{36})(\sqrt{4})$
$= 12 - (6)(2) = 12 - 12 = \mathbf{0}$

27. $x^2y + xy^2 + x(xy - y^2)$
$= x^2y + xy^2 + x^2y - xy^2$
$= \mathbf{2x^2y}$

28. $\left(1\frac{5}{9}\right)\left(1\frac{1}{2}\right) \div 2\frac{2}{3}$
$= \left(\frac{14}{9}\right)\left(\frac{3}{2}\right) \div \frac{8}{3} = \frac{7}{3} \div \frac{8}{3}$
$= \frac{7}{3} \times \frac{3}{8} = \mathbf{\frac{7}{8}}$

29. $9.5 - (4.2 - 3.4)$
$= 9.5 - 0.8 = \mathbf{8.7}$

30. (a) $\dfrac{(-18) + (-12) - (-6)(3)}{-3}$
$\dfrac{(-30) - (-18)}{-3}$
$\dfrac{-12}{-3}$
$\mathbf{4}$

(b) $\sqrt[3]{1000} - \sqrt[3]{125} = 10 - 5 = \mathbf{5}$

(c) $2^2 + 2^1 + 2^0 + 2^{-1}$
$= 4 + 2 + 1 + \frac{1}{2} = \mathbf{7\frac{1}{2}}$ or **7.5**

Lesson 110, Warm-Up

a. **30**

b. **678**

c. $\mathbf{\frac{1}{4}}$

d. $\mathbf{5.4 \times 10^{16}}$

e. $\mathbf{\frac{5}{2}}$

f. **1500 mL**

g. **$90**

h. **$90**

i. **10**

Problem Solving

Recall that there are 30° between each hour mark on a clock. At 10:30 ($1\frac{1}{2}$ hours after 9 o'clock) the minute hand is on the 6, and the hour hand is halfway between the 10 and the 11. The total angle measure is $4 \times 30° + 15°$, or **135°**.

Lesson 110, Lesson Practice

a. **$132,528.15**

b. $\$6000 \times 0.08 = \$480$
$\frac{8}{12} \times \$480 = \$320$
$\$6000 + \$320 = \mathbf{\$6320}$
$\mathbf{\frac{8}{12}}$ or $\mathbf{\frac{2}{3}}$

c. 80% of 80% of $300
$= 0.8 \times 0.8 \times \$300$
$= 0.64 \times \$300 = \mathbf{\$192}$

LESSON 110, MIXED PRACTICE

1. $3(\$5.95) = \$17.85$

$$\begin{array}{r} \$17.85 \\ \times \ \ 0.06 \\ \hline \$1.071 \end{array} \longrightarrow \$1.07 \qquad \begin{array}{r} \$17.85 \\ + \ \$1.07 \\ \hline \$18.92 \end{array}$$

$$\begin{array}{r} \$20.00 \\ - \ \$18.92 \\ \hline \mathbf{\$1.08} \end{array}$$

2. 90% of 75% of \$24
$= 0.9 \times 0.75 \times \$24 = 0.675 \times \$24$
$= \mathbf{\$16.20}$

3.

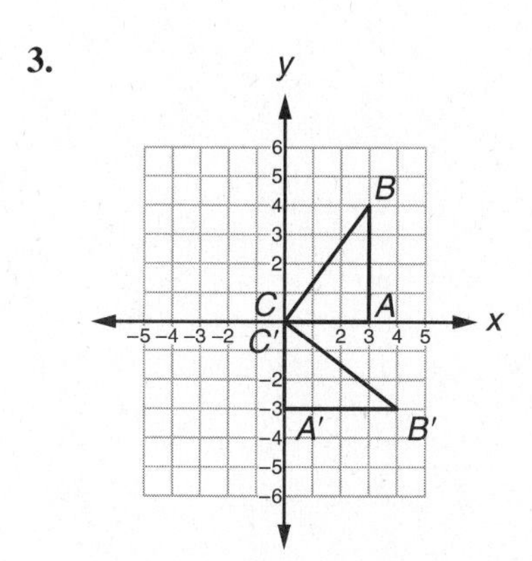

$A'(0, -3), B'(4, -3), C'(0, 0)$

4.

	Case 1	Case 2
Miles	1	m
Calories	100	350

$$\frac{1}{100} = \frac{m}{350}$$
$100m = 350$
$m = \mathbf{3.5\ miles}$

5.

	Case 1	Case 2
Roses	12	30
Dollars	\$4.90	d

$$\frac{12}{\$4.90} = \frac{30}{d}$$
$12d = \$147$
$d = \mathbf{\$12.25}$

6.
$$C \approx \frac{\frac{22}{7}(7\text{ in.})}{2}$$
$$C \approx \frac{22\text{ in.}}{2}$$
$C \approx 11$ in.
Perimeter ≈ 11 in. $+ 7$ in. $+ 7$ in.
$+ 7$ in. $\approx$ **32 in.**

7.
$8(4) = 32$
$32 - (2 + 4 + 6) = \mathbf{20}$

8. $\mathbf{150 = W_P \times 60}$
$$\frac{150}{60} = \frac{W_P \times 60}{60}$$
$$\frac{5}{2} = W_P$$
$$W_P = \frac{5}{2} \times 100\% = \mathbf{250\%}$$

9. $\mathbf{0.6 \times W_N = 150}$
$$\frac{0.6 \times W_N}{0.6} = \frac{150}{0.6}$$
$W_N = \mathbf{250}$

10. $\mathbf{(-x)^2 + 6 = 150}$
$x^2 + 6 = 150$
$x^2 + 6 - 6 = 150 - 6$
$x^2 = 144$
$x = 12, -12$
$\mathbf{-12}$

11.

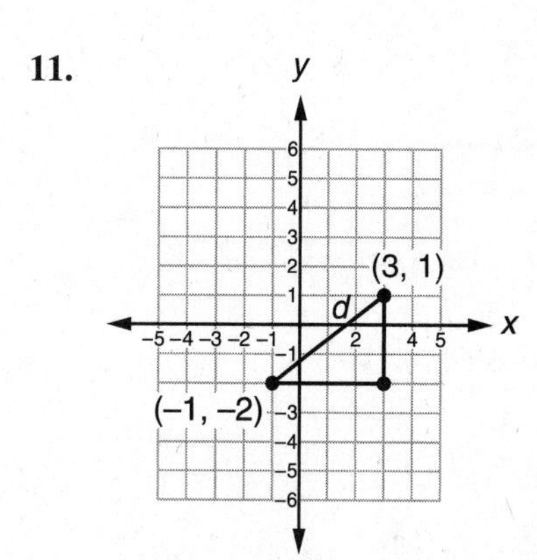

$d^2 = (4\text{ units})^2 + (3\text{ units})^2$
$d^2 = 16\text{ units}^2 + 9\text{ units}^2$
$d^2 = 25\text{ units}^2$
$d = \mathbf{5\ units}$

12.

	Percent	Actual Count
Original	100	O
− Change	40	C
New	60	\$48

$$\frac{100}{60} = \frac{O}{\$48}$$
$60O = \$4800$
$O = \mathbf{\$80}$

13.

	Scale	Actual Size
Model	1	m
Car	36	180

$$\frac{1}{36} = \frac{m}{180}$$
$36m = 180$
$m = \mathbf{5\ inches}$

14. **8 and 9**

15. **See student work;**

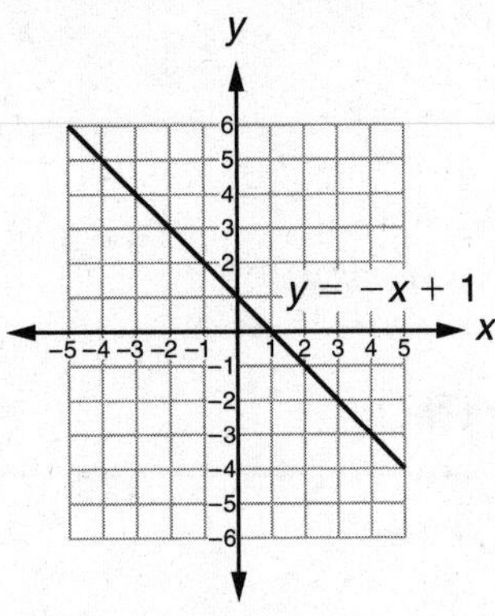

$\frac{\text{rise}}{\text{run}} = \frac{1}{-1} = \mathbf{-1}$

16.
$$5x + 12 \geq 2$$
$$5x + 12 - 12 \geq 2 - 12$$
$$5x \geq -10$$
$$\frac{5x}{5} \geq \frac{-10}{5}$$
$$\mathbf{x \geq -2}$$

$x \geq -2$

−3 −2 −1 0 1

17. $(6.3 \times 9) \times (10^7 \times 10^{-3})$
$$= 56.7 \times 10^4$$
$$= \mathbf{5.67 \times 10^5}$$

18.
$$\frac{1}{2}y = x + 2$$
$$\left(\frac{2}{1}\right)\frac{1}{2}y = \frac{2}{1}(x + 2)$$
$$\mathbf{y = 2x + 4}$$

19.

$$\begin{array}{r} \$4000 \\ \times\ 1.09 \\ \hline \$4360 \\ \times\ 1.09 \\ \hline \$4752.40 \\ \times\ 1.09 \\ \hline \$5180.116 \end{array} \longrightarrow \mathbf{\$5180.12}$$

20. (a) **See student work;**
$$\frac{x}{8} = \frac{3}{4}$$
$$4x = 24$$
$$x = \mathbf{6\ in.}$$

(b) $4f = 3$
$$f = \frac{3}{4} = \mathbf{0.75}$$

21. Area of base $= \frac{(16\text{ in.})(12\text{ in.})}{2}$
$$= 96\text{ in.}^2$$
Volume $= (96\text{ in.}^2)(10\text{ in.}) = \mathbf{960\ in.^3}$

22.

Area of triangle		=	96 in.^{2}
Area of triangle		=	96 in.^{2}
Area of rectangle	= (10 in.)(20 in.)	=	200 in.^{2}
Area of rectangle	= (10 in.)(16 in.)	=	160 in.^{2}
+ Area of rectangle	= (10 in.)(12 in.)	=	120 in.^{2}
Total surface area		=	**672 in.^{2}**

23. $m\angle x = 180° - (80° + 30°) = 70°$
$$180° - 70° = \mathbf{110°}$$

24.
$$\frac{w}{2} = \frac{18}{w}$$
$$w^2 = 36$$
$$w = \mathbf{6, -6}$$

25.
$$3\frac{1}{3}w^2 - 4 = 26$$
$$3\frac{1}{3}w^2 - 4 + 4 = 26 + 4$$
$$\frac{10}{3}w^2 = 30$$
$$\left(\frac{3}{10}\right)\frac{10}{3}w^2 = \left(\frac{3}{10}\right)30$$
$$w^2 = 9$$
$$w = \mathbf{3, -3}$$

26. $16 - \{27 - 3[8 - (3^2 - 2^3)]\}$
$$16 - \{27 - 3[8 - (9 - 8)]\}$$
$$16 - \{27 - 3[8 - (1)]\}$$
$$16 - \{27 - 3[7]\}$$
$$16 - \{27 - 21\}$$
$$16 - \{6\}$$
$$\mathbf{10}$$

27.
$$\frac{(6ab^2)(8ab)}{12a^2b^2}$$
$$= \frac{\overset{1}{\cancel{2}} \cdot \overset{1}{\cancel{3}} \cdot \overset{1}{\cancel{a}} \cdot \overset{1}{\cancel{b}} \cdot \overset{1}{\cancel{b}} \cdot \overset{1}{\cancel{2}} \cdot 2 \cdot 2 \cdot \overset{1}{\cancel{a}} \cdot b}{\underset{1}{\cancel{2}} \cdot \underset{1}{\cancel{2}} \cdot \underset{1}{\cancel{3}} \cdot \underset{1}{\cancel{a}} \cdot \underset{1}{\cancel{a}} \cdot \underset{1}{\cancel{b}} \cdot \underset{1}{\cancel{b}}}$$
$$= \mathbf{4b}$$

28. $3\frac{1}{3} + 1\frac{1}{2} + 4\frac{5}{6}$
$$= 3\frac{2}{6} + 1\frac{3}{6} + 4\frac{5}{6} = 8\frac{10}{6} = 9\frac{4}{6}$$
$$= \mathbf{9\frac{2}{3}}$$

29. $20 \div \left(3\frac{1}{3} \div 1\frac{1}{5}\right) = 20 \div \left(\frac{\overset{5}{\cancel{10}}}{3} \times \frac{5}{\underset{3}{\cancel{6}}}\right)$
$$= 20 \div \frac{25}{9} = \frac{\overset{4}{\cancel{20}}}{1} \times \frac{9}{\underset{5}{\cancel{25}}} = \frac{36}{5} = \mathbf{7\frac{1}{5}}$$

30. $(-3)^2 + (-2)^3 = 9 + (-8) = \mathbf{1}$

INVESTIGATION 11

Activity

1-cm cube

Edge length: **1 cm**

Surface area: $1 \times 1 \times 6 =$ **6 cm^2**

Volume: $1 \times 1 \times 1 =$ **1 cm^3**

2-cm cube

Edge length: **2 cm**

Surface area: $2 \times 2 \times 6 =$ **24 cm^2**

Volume: $2 \times 2 \times 2 =$ **8 cm^3**

3-cm cube

Edge length: **3 cm**

Surface area: $3 \times 3 \times 6 =$ **54 cm^2**

Volume: $3 \times 3 \times 3 =$ **27 cm^3**

4-cm cube

Edge length: **4 cm**

Surface area: $4 \times 4 \times 6 =$ **96 cm^2**

Volume: $4 \times 4 \times 4 =$ **64 cm^3**

1. $2 \div 1 =$ **2 times**

2. $24 \div 6 =$ **4 times**

3. $8 \div 1 =$ **8 times**

4. $4 \div 2 =$ **2 times**

5. $96 \div 24 =$ **4 times**

6. $64 \div 8 =$ **8 times**

7. $3 \div 1 =$ **3 times**

8. $54 \div 6 =$ **9 times**

9. $27 \div 1 =$ **27 times**

10. **3 times**

11. **9 times**

12. **27 times**

13. (a) $6 \times 6 \times 6 =$ **216 cm^2**
    (b) $6 \times 6 \times 6 =$ **216 cm^3**

14. $216 \div 24 =$ **9 times**

15. $216 \div 8 =$ **27 times**

16. (a) $24 \div 6 =$ **4**
    (b) $4^2 =$ **16**
    (c) $4^3 =$ **64**

17. (a) $3 \div 1\frac{1}{2} = 2$
    $2^3 =$ **8**
    (b) $9 \div 3 = 3$
    $3^2 =$ **9**

18. (a) $5 \div 2\frac{1}{2} =$ **2**
    (b) $2^2 =$ **4**

19. $15 \div 10 = 1.5$
    $1.5^2 = 2.25$
    $\$10.00 \times 2.25 =$ **$22.50**

20. (a) $230 \div 2.3 =$ **100**
    (b) $100^2 =$ **10,000**
    (c) $100^3 =$ **1,000,000**

21. The edge length of the large cube is 4 cm.
    $4 \text{ cm} \times 4 \text{ cm} \times 6 =$ **96 cm^2**

22. The edge length of each cube is 2 cm.
    $2 \text{ cm} \times 2 \text{ cm} \times 6 = 24 \text{ cm}^2$
    There are 8 smaller cubes in all.
    $8 \times 24 \text{ cm}^2 =$ **192 cm^2**

23. The edge length of each cube is 1 cm.
    $1 \text{ cm} \times 1 \text{ cm} \times 6 = 6 \text{ cm}^2$
    There are 64 small cubes in all.
    $64 \times 6 \text{ cm}^2 =$ **384 cm^2**

24. **Although the volumes are the same, the small cubes will melt sooner than the large block because a much greater surface area is exposed to the warmer surroundings.**

25. **Because the surface area of a smaller animal is greater relative to its volume than the surface area for a larger animal, smaller animals work harder to regulate body temperature than larger animals in the same environment. So smaller animals, like sparrows, need to eat a greater percentage of their weight in food than do larger animals, like hawks.**

Extensions

a. **See student work.**

b. **See student work.**

c. **See student work.**

LESSON 111, WARM-UP

a. **23**

b. **321**

c. **0.05**

d. **6.4×10^{-7}**

e. **60**

f. **144 in.^{2}**

g. **\$50**

h. **\$100**

i. **\$16.00**

Problem Solving

36 ft = 12 yd
12 yd × 12 yd = 144 sq. yd
144 × \$25 = **\$3600**

LESSON 111, LESSON PRACTICE

a. $2\overline{)3.6}$ = 1.8 **1.8×10^6**

b. $25\overline{)75}$ = 3 **3×10^{-6}**

c. $3\overline{)4.5}$ = 1.5 **1.5×10^{-4}**

d. $15\overline{)60}$ = 4 **4×10^4**

e. $8\overline{)4.0}$ = 0.5
$0.5 \times 10^8 \longrightarrow$ **5×10^7**

f. $3\overline{)1.5}$ = 0.5
$0.5 \times 10^{-8} \longrightarrow$ **5×10^{-9}**

g. $6\overline{)3.6}$ = 0.6
$0.6 \times 10^{-6} \longrightarrow$ **6×10^{-7}**

h. $9\overline{)1.8}$ = 0.2
$0.2 \times 10^6 \longrightarrow$ **2×10^5**

LESSON 111, MIXED PRACTICE

1.
$$\begin{array}{r} 1909 \\ -\ 1859 \\ \hline 50 \end{array}$$
50 years + 1 year = **51 years (The year 1859 should be counted.)**

2.
$$15y = 600$$
$$\frac{15y}{15} = \frac{600}{15}$$
$$y = 40$$
$$40 + 15 = \mathbf{55}$$

3. (a) $70\% = \frac{70}{100} = \mathbf{\frac{7}{10}}$

(b) $\frac{\text{agreed}}{\text{disagreed}} = \mathbf{\frac{3}{7}}$

4.

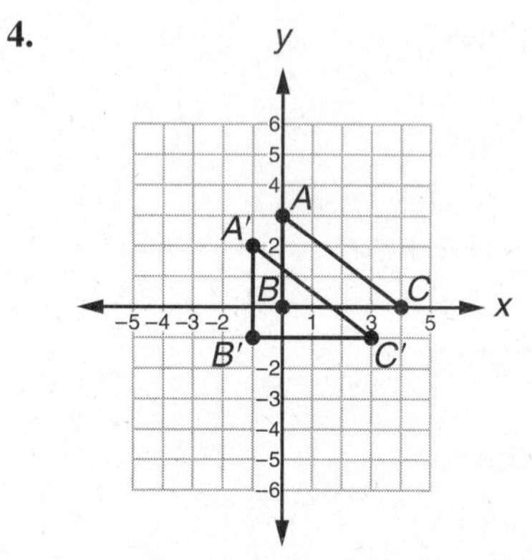

$A'(-1, 2)$, $B'(-1, -1)$, $C'(3, -1)$

5. (a) **2^{10}**

(b) **32**

6. (a) $180° - 150° = \mathbf{30°}$

(b) $\frac{360°}{30°} = 12$, **12 sides**

(c) **Dodecagon**

7.
$$\frac{100}{60} = \frac{p}{\$48}$$
$$60p = \$4800$$
$$p = \mathbf{\$80}$$

8. (a) $P(R, W, B) = \frac{\mathbf{3}}{\mathbf{12}} \cdot \frac{\mathbf{4}}{\mathbf{11}} \cdot \frac{\mathbf{5}}{\mathbf{10}} = \frac{\mathbf{1}}{\mathbf{22}}$

(b) $P(B, W, R) = \frac{\mathbf{5}}{\mathbf{12}} \cdot \frac{\mathbf{4}}{\mathbf{11}} \cdot \frac{\mathbf{3}}{\mathbf{10}} = \frac{\mathbf{1}}{\mathbf{22}}$

9.
$$\begin{aligned} \mathbf{2x + 6} &= \mathbf{36} \\ 2x + 6 - 6 &= 36 - 6 \\ 2x &= 30 \\ \frac{2x}{2} &= \frac{30}{2} \\ x &= \mathbf{15} \end{aligned}$$

10.
$$\begin{aligned} 2x + x + x &= 180° \\ 4x &= 180° \\ \frac{4x}{4} &= \frac{180°}{4} \\ x &= \mathbf{45°} \end{aligned}$$

11.
$$\begin{aligned} c^2 - b^2 &= a^2 \\ c^2 - b^2 + b^2 &= a^2 + b^2 \\ \mathbf{c^2} &= \mathbf{a^2 + b^2} \end{aligned}$$

12. (a) $m\angle a = m\angle h = \mathbf{105°}$

(b) $m\angle b = 180° - 105° = \mathbf{75°}$

(c) $m\angle c = m\angle b = \mathbf{75°}$

(d) $m\angle d = m\angle a = \mathbf{105°}$

13.
$$\begin{aligned} F &= 1.8C + 32 \\ F &= 1.8(17) + 32 \\ F &= 62.6° \end{aligned}$$
63°F

14. $\frac{45°}{360°} = \frac{1}{8}$

$$A \approx \frac{3.14(144 \text{ in.}^2)}{8}$$

$A \approx 56.52 \text{ in.}^2 \approx \mathbf{57\ in.^2}$

15. **See student work;**

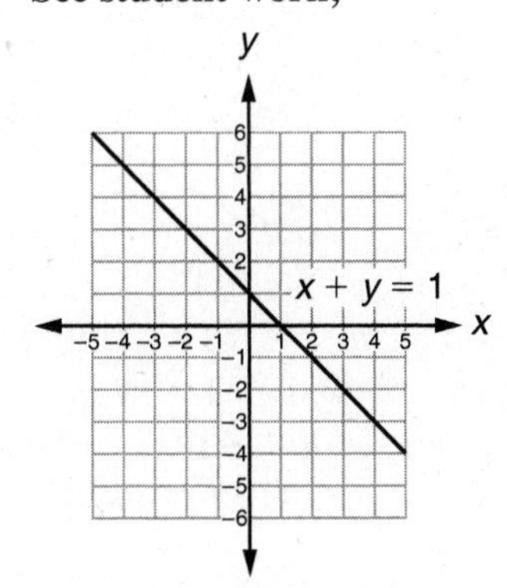

16. (a) $\frac{\text{rise}}{\text{run}} = \frac{1}{-1} = \mathbf{-1}$

(b) **(0, 1)**

17. 24 in. = 2 ft; 36 in. = 3 ft
$$\begin{aligned} C &\approx 3.14(2 \text{ ft}) \\ C &\approx 6.28 \text{ ft} \\ A &\approx (6.28 \text{ ft})(3 \text{ ft}) \\ A &\approx \mathbf{18.84\ ft^2} \end{aligned}$$

18.
$$\begin{aligned} \text{Area of base} &\approx 3.14(1 \text{ ft}^2) \\ &\approx 3.14 \text{ ft}^2 \\ \text{Volume} &\approx (3.14 \text{ ft}^2)(3 \text{ ft}) \\ &\approx \mathbf{9.42\ ft^3} \end{aligned}$$

19. (a)
$$\begin{aligned} 2x^2 + 1 &= 19 \\ 2x^2 + 1 - 1 &= 19 - 1 \\ 2x^2 &= 18 \\ \frac{2x^2}{2} &= \frac{18}{2} \\ x^2 &= 9 \\ x &= \mathbf{3, -3} \end{aligned}$$

(b)
$$\begin{aligned} 2x^2 - 1 &= 19 \\ 2x^2 - 1 + 1 &= 19 + 1 \\ 2x^2 &= 20 \\ \frac{2x^2}{2} &= \frac{20}{2} \\ x^2 &= 10 \\ x &= \mathbf{\sqrt{10}, -\sqrt{10}} \end{aligned}$$

20.

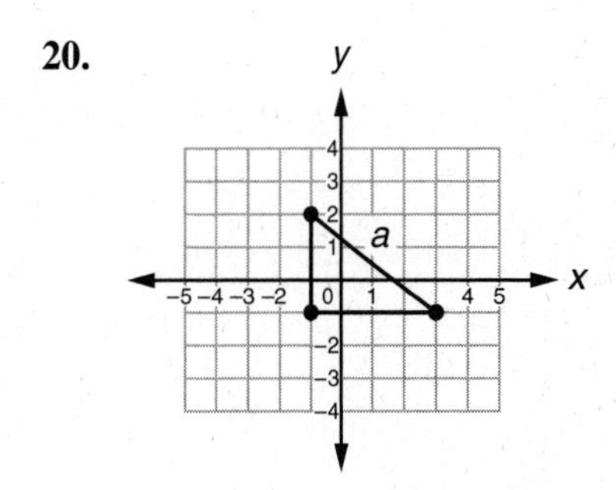

$$\begin{aligned} a^2 &= (3 \text{ units})^2 + (4 \text{ units})^2 \\ a^2 &= 9 \text{ units}^2 + 16 \text{ units}^2 \\ a^2 &= 25 \text{ units}^2 \\ a &= 5 \text{ units} \\ \text{Perimeter} &= 5 \text{ units} + 3 \text{ units} \\ &\quad + 4 \text{ units} = \mathbf{12\ units} \end{aligned}$$

21.
$$\begin{array}{r} \$5000 \\ \times \quad 1.05 \\ \hline \$5250 \\ \times \quad 1.05 \\ \hline \$5512.50 \\ \times \quad 1.05 \\ \hline \$5788.125 \\ \times \quad 1.05 \\ \hline \$6077.53125 \\ \times \quad 1.05 \\ \hline \$6381.407813 \end{array} \longrightarrow \mathbf{\$6381.41}$$

22. (a)
$$AB^2 = (15\text{ cm})^2 + (20\text{ cm})^2$$
$$AB^2 = 225\text{ cm}^2 + 400\text{ cm}^2$$
$$AB^2 = 625\text{ cm}^2$$
$$AB = \mathbf{25\text{ cm}}$$

(b)
$$\frac{20\text{ cm}}{25\text{ cm}} = \frac{CD}{15\text{ cm}}$$
$$(25\text{ cm})CD = 300\text{ cm}^2$$
$$CD = \mathbf{12\text{ cm}}$$

23. (a) $6\overline{)3.6}$ = 0.6 $\quad 0.6 \times 10^2 \longrightarrow \mathbf{6 \times 10^1}$

(b) $12\overline{)36}$ = 3 $\quad \mathbf{3 \times 10^{-2}}$

24.
$$180° - 140° = 40°$$
$$180° - (90° + 40°) = 50°$$
$$m\angle y = 180° - 50° = \mathbf{130°}$$

25.
$$5x + 3x = 18 + 2x$$
$$8x - 2x = 18 + 2x - 2x$$
$$6x = 18$$
$$\frac{6x}{6} = \frac{18}{6}$$
$$x = \mathbf{3}$$

26.
$$\frac{3.6}{x} = \frac{4.5}{0.06}$$
$$4.5x = 0.216$$
$$\frac{4.5x}{4.5} = \frac{0.216}{4.5}$$
$$x = \mathbf{0.048}$$

27. (a) $(-1)^6 + (-1)^5 = 1 + (-1) = \mathbf{0}$

(b) $(-10)^6 \div (-10)^5 = (-10)^1 = \mathbf{-10}$

28. (a)
$$\frac{(4a^2b)(9ab^2c)}{6abc}$$
$$= \frac{\overset{1}{\cancel{2}} \cdot 2 \cdot \overset{1}{\cancel{3}} \cdot 3 \cdot \overset{1}{\cancel{a}} \cdot a \cdot a \cdot \overset{1}{\cancel{b}} \cdot b \cdot b \cdot \overset{1}{\cancel{c}}}{\underset{1}{\cancel{2}} \cdot \underset{1}{\cancel{3}} \cdot \underset{1}{\cancel{a}} \cdot \underset{1}{\cancel{b}} \cdot \underset{1}{\cancel{c}}}$$
$$= \mathbf{6a^2b^2}$$

(b)
$$x(x - c) + cx$$
$$x^2 - cx + cx$$
$$\mathbf{x^2}$$

29.
$$(-3) + (+2)(-4) - (-6)(-2) - (-8)$$
$$(-3) + (-8) - (12) - (-8)$$
$$\mathbf{-15}$$

30.
$$3\frac{1}{3} \cdot 1\frac{4}{5} = \frac{\overset{2}{\cancel{10}}}{\underset{1}{\cancel{3}}} \cdot \frac{\overset{3}{\cancel{9}}}{\underset{1}{\cancel{5}}} = 6$$
$$6 + 1.5 = 7.5$$
$$\frac{7.5}{0.03} = \mathbf{250}$$

Lesson 112, Warm-Up

a. **32**

b. **99**

c. **–900**

d. $\mathbf{2 \times 10^3}$

e. **10, –10**

f. **122°F**

g. **\$500**

h. **\$2500**

i. $\mathbf{2\frac{1}{2}}$

Problem Solving

One third of the book is 20 pages more than one fourth of the book, so one twelfth of the book $(\frac{1}{3} - \frac{1}{4})$ is 20 pages. So the whole book is 20×12, or 240 pages long. If Mariabella has read three fourths of the book, she has one fourth of the book left to read. One fourth of the book is $240 \times \frac{1}{4}$, or **60 pages.**

Lesson 112, Lesson Practice

a.

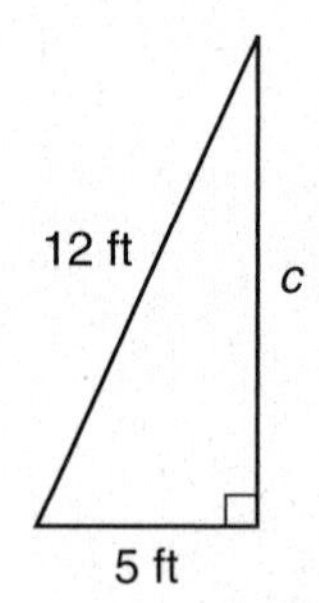

$$(5\text{ ft})^2 + c^2 = (12\text{ ft})^2$$
$$25\text{ ft}^2 + c^2 = 144\text{ ft}^2$$
$$25\text{ ft}^2 + c^2 - 25\text{ ft}^2 = 144\text{ ft}^2 - 25\text{ ft}^2$$
$$c^2 = 119\text{ ft}^2$$
$$c = \sqrt{119\text{ ft}^2} \approx 10.9\text{ ft}$$
$$c \approx \mathbf{10\text{ feet }11\text{ inches}}$$

b.
$$(AC)^2 = (400\text{ ft})^2 + (300\text{ ft})^2$$
$$(AC)^2 = 160{,}000\text{ ft}^2 + 90{,}000\text{ ft}^2$$
$$(AC)^2 = 250{,}000\text{ ft}^2$$
$$AC = 500\text{ feet}$$
$$(300\text{ feet} + 400\text{ feet}) - 500\text{ feet} = \mathbf{200\text{ feet}}$$

Lesson 112, Mixed Practice

1.

$$\begin{array}{r} \$3000 \\ \times \quad 1.08 \\ \hline \$3240 \\ \times \quad 1.08 \\ \hline \$3499.20 \\ \times \quad\quad 1.08 \\ \hline \$3799.1360 \end{array} \longrightarrow \mathbf{\$3799.14}$$

2. $\sqrt{(3)^2 + (4)^2} = \sqrt{9 + 16} = \sqrt{25} = \mathbf{5}$

3. (a) **90**

(b) **95**

4. $\dfrac{840 \text{ kilometers}}{10.5 \text{ hours}} = $ **80 kilometers per hour**

5.

	Case 1	Case 2
Hours	6	9
Earnings	\$28	e

$$\frac{6}{\$28} = \frac{9}{e}$$
$$6e = \$252$$
$$e = \mathbf{\$42}$$

6.

	Percent	Actual Count
Original	100	O
− Change	25	C
New	75	\$48

$$\frac{100}{75} = \frac{O}{\$48}$$
$$75O = \$4800$$
$$O = \mathbf{\$64}$$

7.

	Percent	Actual Count
Original	100	\$6
+ Change	25	C
New	125	N

$$\frac{100}{25} = \frac{\$6}{C}$$
$$100C = \$150$$
$$C = \mathbf{\$1.50}$$

8. 50% of 50% of \$1.00
$= (0.5)(0.5) \times \$1.00 = 0.25 \times \$1.00$
$=$ **\$0.25**

9. $60\% = \dfrac{60}{100} = \dfrac{6}{10} = \dfrac{3}{5} = \dfrac{\text{boys}}{\text{total}}$

$\dfrac{\text{boys}}{\text{girls}} = \mathbf{\dfrac{3}{2}}$

10.

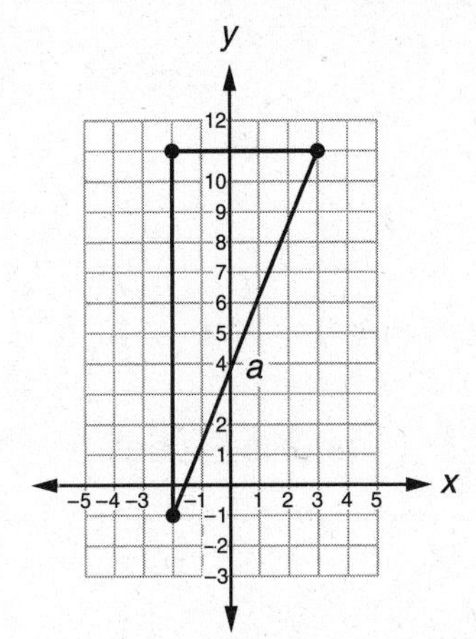

$$a^2 = (12 \text{ units})^2 + (5 \text{ units})^2$$
$$a^2 = 144 \text{ units}^2 + 25 \text{ units}^2$$
$$a^2 = 169 \text{ units}^2$$
$$a = \mathbf{13 \text{ units}}$$

11.

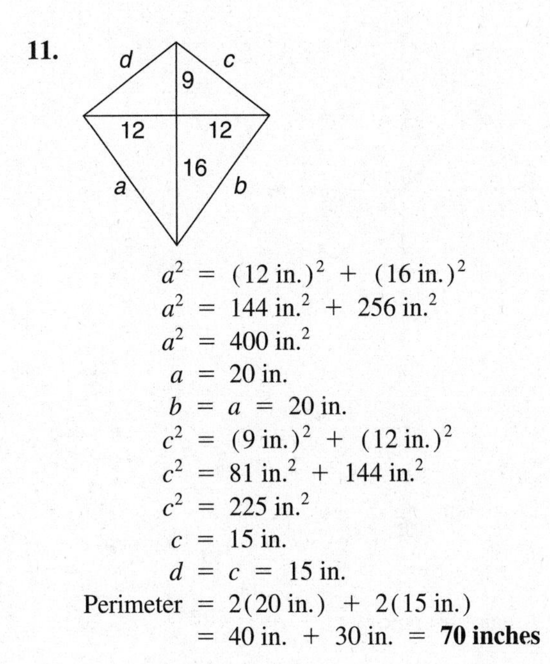

$$a^2 = (12 \text{ in.})^2 + (16 \text{ in.})^2$$
$$a^2 = 144 \text{ in.}^2 + 256 \text{ in.}^2$$
$$a^2 = 400 \text{ in.}^2$$
$$a = 20 \text{ in.}$$
$$b = a = 20 \text{ in.}$$
$$c^2 = (9 \text{ in.})^2 + (12 \text{ in.})^2$$
$$c^2 = 81 \text{ in.}^2 + 144 \text{ in.}^2$$
$$c^2 = 225 \text{ in.}^2$$
$$c = 15 \text{ in.}$$
$$d = c = 15 \text{ in.}$$
$$\text{Perimeter} = 2(20 \text{ in.}) + 2(15 \text{ in.})$$
$$= 40 \text{ in.} + 30 \text{ in.} = \mathbf{70 \text{ inches}}$$

12. $\mathbf{W_P \times 2.5 = 2}$

$$\frac{W_P \times 2.5}{2.5} = \frac{2}{2.5}$$
$$W_P = 0.8$$
$$W_P = 0.8 \times 100\% = \mathbf{80\%}$$

13. $2 \times 2 \times 2 \times 2 = 16$ possible outcomes
1 favorable outcome
15 unfavorable outcomes
Odds = favorable to unfavorable
= **1 to 15**

14.

$$\begin{array}{r} \$4000 \\ \times \quad 0.09 \\ \hline \$360 \end{array} \qquad \frac{6}{12} = \frac{1}{2}$$

$$\left(\frac{1}{2}\right)\$360 = \mathbf{\$180}$$

15. (a) **0.625**

(b) **62.5%**

16. (a) $2\overline{)5.0}$ with quotient 2.5 $\quad$ $\mathbf{2.5 \times 10^4}$

(b) $4\overline{)1.2}$ with quotient 0.3 $\quad$ $0.3 \times 10^{-4} \longrightarrow \mathbf{3 \times 10^{-5}}$

17. $300 \text{ kg} \cdot \dfrac{1000 \text{ g}}{1 \text{ kg}} = \mathbf{300{,}000 \text{ g}}$

18.
$$d = rt$$
$$\frac{d}{r} = \frac{rt}{r}$$
$$\boldsymbol{\frac{d}{r} = t}$$

19. **See student work;**

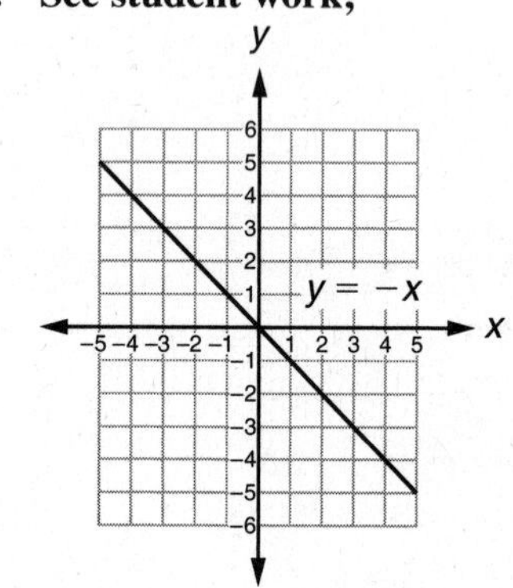

20.
$$C \approx \frac{3.14(20 \text{ cm})}{2}$$
$$C \approx 31.4 \text{ cm}$$
$$\text{Perimeter} \approx 31.4 \text{ cm} + 20 \text{ cm} + 20 \text{ cm} + 20 \text{ cm} \approx \mathbf{91.4 \text{ cm}}$$

21.

Area of triangle	=	$\dfrac{(9 \text{ ft})(12 \text{ ft})}{2}$	=	54 ft^2
Area of triangle	=	$\dfrac{(9 \text{ ft})(12 \text{ ft})}{2}$	=	54 ft^2
Area of rectangle	=	$(10 \text{ ft})(12 \text{ ft})$	=	120 ft^2
Area of rectangle	=	$(10 \text{ ft})(15 \text{ ft})$	=	150 ft^2
+ Area of rectangle	=	$(10 \text{ ft})(9 \text{ ft})$	=	90 ft^2
Total surface area			=	$\mathbf{468 \text{ ft}^2}$

22. (a) $\mathbf{2^{12} \cdot 5^{12}}$

(b) **1,000,000**

23. (a)
$$\frac{x}{9} = \frac{8}{12}$$
$$12x = 72$$
$$x = \mathbf{6 \text{ cm}}$$

(b)
$$8f = 12$$
$$f = \frac{12}{8}$$
$$f = \frac{3}{2} = \mathbf{1.5}$$

24.
$$\frac{16}{2.5} = \frac{48}{f}$$
$$16f = 120$$
$$f = \mathbf{7.5}$$

25.
$$2\frac{2}{3}x - 3 = 21$$
$$2\frac{2}{3}x - 3 + 3 = 21 + 3$$
$$2\frac{2}{3}x = 24$$
$$\left(\frac{3}{8}\right)\frac{8}{3}x = \left(\frac{3}{8}\right)24$$
$$x = \mathbf{9}$$

26.
$$10^2 - [10 - 10(10^0 - 10^{-1})]$$
$$100 - \left[10 - 10\left(1 - \frac{1}{10}\right)\right]$$
$$100 - \left[10 - 10\left(\frac{9}{10}\right)\right]$$
$$100 - [10 - 9]$$
$$100 - [1]$$
$$\mathbf{99}$$

27.
$$2\frac{3}{4} - \left(1\frac{1}{2} - \frac{1}{6}\right)$$
$$= \frac{11}{4} - \left(\frac{9}{6} - \frac{1}{6}\right) = \frac{11}{4} - \frac{8}{6}$$
$$= \frac{33}{12} - \frac{16}{12} = \frac{17}{12} = \mathbf{1\frac{5}{12}}$$

28.
$$3\frac{1}{2} \div 1\frac{2}{5} \div 3$$
$$= \left(\frac{7}{2} \times \frac{5}{7}\right) \div 3 = \frac{5}{2} \times \frac{1}{3} = \mathbf{\frac{5}{6}}$$

29.
$$|-4| - (-3)(-2)(-1) + \frac{(-5)(4)(-3)(2)}{-1}$$
$$4 - (-6) + \frac{(-20)(-6)}{-1}$$
$$4 - (-6) + \frac{120}{-1}$$
$$4 - (-6) - 120$$
$$10 + (-120)$$
$$\mathbf{-110}$$

30.
$$\text{Surface area} = 4\pi r^2$$
$$\approx 4\left(\frac{22}{7}\right)(49 \text{ cm}^2)$$
$$\approx 4(22)(7 \text{ cm}^2)$$
$$\approx 616 \text{ cm}^2$$
$$\approx \mathbf{600 \text{ cm}^2}$$

Lesson 113, Warm-Up

a. **31**

b. **1941**

c. **−0.25**

d. $\mathbf{2 \times 10^{-4}}$

e. **0.6**

f. **288 in.^{2}**

g. **$1600**

h. **$800**

i. $\mathbf{2\frac{1}{2}}$ **hr**

Problem Solving

5 and 12

Lesson 113, Lesson Practice

a. $\mathbf{\frac{1}{3}}$

b. $\mathbf{\frac{2}{3}}$

c. $\mathbf{\frac{2}{3}}$

d. $\mathbf{\frac{1}{3}}$

e. **All of the box would be filled.** $\mathbf{\left(\frac{1}{3} + \frac{2}{3} = 1\right)}$

f. Volume of box $= (12\text{ in.})(12\text{ in.})(12\text{ in.})$
$= \mathbf{1728\text{ in.}^3}$
Volume of pyramid $= \frac{1}{3}(1728\text{ in.}^3)$
$= \mathbf{576\text{ in.}^3}$

g. Area of base $= \pi(9\text{ in.}^2)$
$= 9\pi\text{ in.}^2$
Volume $= \frac{1}{3}(9\pi\text{ in.}^2)(6\text{ in.})$
$= \mathbf{18\pi\text{ in.}^3}$

h. Volume $= \frac{4}{3}\pi(3\text{ in.})^3$
$= \frac{4}{3}\pi(27\text{ in.}^3) = \mathbf{36\pi\text{ in.}^3}$

Lesson 113, Mixed Practice

1. 75% of 75% of $24
$= (0.75)(0.75) \times \$24 = \mathbf{\$13.50}$

2.
```
 0 9 9 1
 1̸0̸,0̸0 0,0 0 0,0 0 0
−    9 8 0,0 0 0,0 0 0
 ─────────────────────
   9,0 2 0,0 0 0,0 0 0
```
$\mathbf{9.02 \times 10^9}$

3. Mean $= 8.45 \div 5 = 1.69$
Mean − median $= 1.69 - 0.75$
$= \mathbf{0.94}$

4. (a) $\frac{\$24}{5\text{ hours}} = \mathbf{\$4.80\text{ per hour}}$

(b) $\frac{\$33}{6\text{ hours}} = \mathbf{\$5.50\text{ per hour}}$

(c) $5.50 − $4.80 = **$0.70 more per hour**

5.

	Case 1	Case 2
Kilograms	24	42
Cost	$31	c

$\frac{24}{\$31} = \frac{42}{c}$
$24c = \$1302$
$c = \mathbf{\$54.25}$

6. $\frac{5}{8}(1760\text{ yards}) =$ **about 1100 yards**

7. $P(H, H) = \frac{13}{52} \cdot \frac{13}{52} = \frac{1}{4} \cdot \frac{1}{4} = \mathbf{\frac{1}{16}}$

8. $\mathbf{W_P \times \$30 = \$1.50}$
$\frac{W_P \times \$30}{\$30} = \frac{\$1.50}{\$30}$
$W_P = 0.05$
$W_P = 0.05 \times 100\% = \mathbf{5\%}$

9. $\mathbf{\frac{1}{2} \times W_N = 2\frac{1}{2}}$
$\frac{2}{1}\left(\frac{1}{2} \times W_N\right) = \left(\frac{2}{1}\right)\frac{5}{2}$
$\mathbf{W_N = 5}$

10.
$$\begin{array}{r} \$5000 \\ \times\ \ 1.08 \\ \hline \$5400 \\ \times\ \ 1.08 \\ \hline \$5832 \\ \times\ \ 1.08 \\ \hline \$6298.56 \\ -\ \$5000.00 \\ \hline \mathbf{\$1298.56} \end{array}$$

11.

	Percent	Actual Count
Original	100	\$12
− Change	20	C
New	80	N

$$\frac{100}{80} = \frac{\$12}{N}$$
$$100N = \$960$$
$$N = \mathbf{\$9.60}$$

12.

	Scale	Actual Size
Model	24	6 ft = 72 in.
Figurine	1	F

$$\frac{24}{1} = \frac{72 \text{ in.}}{F}$$
$$24F = 72 \text{ in.}$$
$$F = \textbf{3 inches}$$

13.

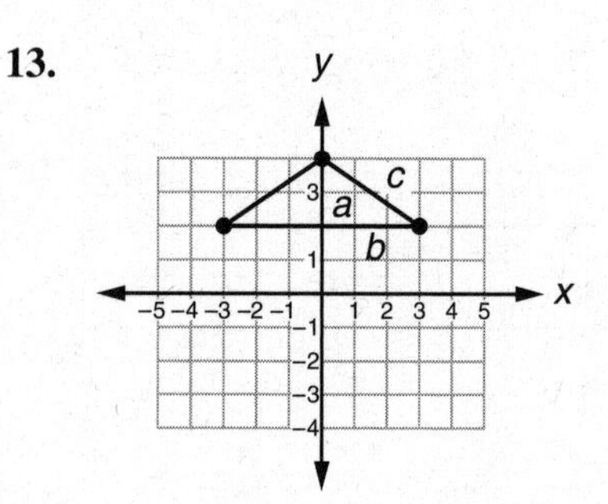

$$A = \frac{(6 \text{ units})(2 \text{ units})}{2} = \textbf{6 units}^{\mathbf{2}}$$

14.
$$(2 \text{ units})^2 + (3 \text{ units})^2 = c^2$$
$$4 \text{ units}^2 + 9 \text{ units}^2 = c^2$$
$$13 \text{ units}^2 = c^2$$
$$\mathbf{\sqrt{13}\ \textbf{units}} = c$$

15.
$$\text{Volume} \approx \frac{4}{3}(3)(3 \text{ cm})^3$$
$$\approx \frac{4}{3}(3)(27 \text{ cm}^3)$$
$$\approx 4(27 \text{ cm}^3)$$
$$\approx 108 \text{ cm}^3$$
$$\approx \textbf{110 cm}^{\mathbf{3}}$$

16.
$$(6.3 \times 7) \times (10^6 \times 10^{-3})$$
$$= 44.1 \times 10^3$$
$$= \mathbf{4.41 \times 10^4}$$

17.
$$s^2 = (40 \text{ yd})^2 + (30 \text{ yd})^2$$
$$s^2 = 1600 \text{ yd}^2 + 900 \text{ yd}^2$$
$$s^2 = 2500 \text{ yd}^2$$
$$s = 50 \text{ yd}$$
$$(40 \text{ yd} + 30 \text{ yd}) - 50 \text{ yd} = \textbf{20 yards}$$

18. (a)
$$A = \frac{1}{2}bh$$
$$\left(\frac{2}{1}\right)A = \frac{2}{1}\left(\frac{1}{2}bh\right)$$
$$2A = bh$$
$$\frac{2A}{b} = \frac{bh}{b}$$
$$\mathbf{\frac{2A}{b} = h}$$

(b)
$$h = \frac{2(16)}{8}$$
$$h = 2(2)$$
$$h = \mathbf{4}$$

19. See student work;

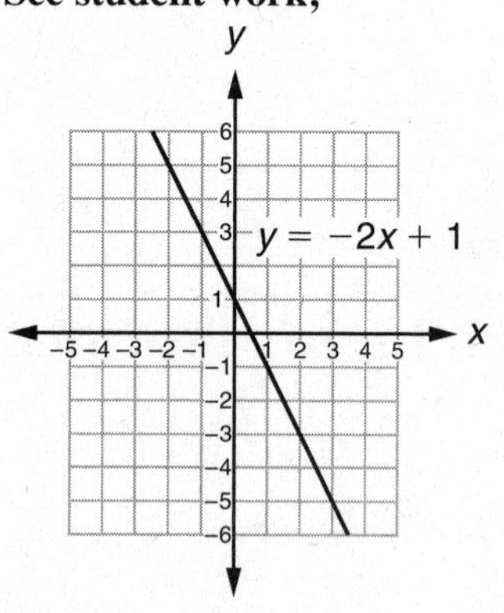

$$\frac{\text{rise}}{\text{run}} = \frac{2}{-1} = \mathbf{-2}$$

20.
$$\text{Area of base} = (40 \text{ m})(40 \text{ m}) = 1600 \text{ m}^2$$
$$\text{Volume} = \frac{1}{3}(1600 \text{ m}^2)(30 \text{ m})$$
$$= \textbf{16,000 m}^{\mathbf{3}}$$

21.
$$\text{Area of base} \approx 3.14(100 \text{ cm}^2) \approx 314 \text{ cm}^2$$
$$\text{Volume} \approx \frac{1}{3}(314 \text{ cm}^2)(60 \text{ cm})$$
$$\approx \textbf{6280 cm}^{\mathbf{3}}$$

22. (a) $m\angle D = 180° - (90° + 30°) = \mathbf{60°}$

(b) $m\angle E = m\angle D = \mathbf{60°}$

(c) $m\angle A = m\angle C = \mathbf{30°}$

23.
$$\frac{4 \text{ cm}}{6 \text{ cm}} = \frac{8 \text{ cm}}{CD}$$
$$(4 \text{ cm})CD = 48 \text{ cm}^2$$
$$CD = \textbf{12 cm}$$

24. $\dfrac{7.5}{d} = \dfrac{25}{16}$

$25d = 120$

$d = \mathbf{4.8}$

25. $1\frac{3}{5}w + 17 = 49$

$1\frac{3}{5}w + 17 - 17 = 49 - 17$

$1\frac{3}{5}w = 32$

$\left(\frac{5}{8}\right)\frac{8}{5}w = \left(\frac{5}{8}\right)32$

$w = \mathbf{20}$

26. $5^2 - \{4^2 - [3^2 - (2^2 - 1^2)]\}$

$25 - \{16 - [9 - (4 - 1)]\}$

$25 - \{16 - [9 - (3)]\}$

$25 - \{16 - [6]\}$

$25 - \{10\}$

15

27. $\dfrac{\overset{22}{\cancel{440}}\ \text{yd}}{1\ \cancel{\text{min}}} \cdot \dfrac{1\ \cancel{\text{min}}}{\underset{\underset{1}{\cancel{20}}}{\cancel{60}}\ \text{s}} \cdot \dfrac{\overset{1}{\cancel{3}}\ \text{ft}}{1\ \cancel{\text{yd}}} = \mathbf{22\ \dfrac{ft}{s}}$

28. $1\frac{3}{4} + 2\frac{2}{3} - 3\frac{5}{6}$

$= 1\frac{9}{12} + 2\frac{8}{12} - 3\frac{10}{12} = \mathbf{\frac{7}{12}}$

29. $\left(1\frac{3}{4}\right)\left(2\frac{2}{3}\right) \div 3\frac{5}{6}$

$= \left(\frac{7}{4} \cdot \frac{8}{3}\right) \div \frac{23}{6} = \frac{14}{3} \times \frac{6}{23}$

$= \frac{28}{23} = \mathbf{1\frac{5}{23}}$

30. $(-7) + |-3| - (2)(-3) + (-4) - (-3)(-2)(-1)$

$(-7) + 3 - (-6) + (-4) - (-6)$

$-4 - (-6) + (-4) - (-6)$

$2 + (-4) - (-6)$

$-2 - (-6)$

4

Lesson 114, Warm-Up

a. 9

b. 29

c. –18

d. 1.6×10^{17}

e. 9

f. 140°F

g. $4500

h. $7500

i. 2

Problem Solving

$\mathbf{3X + 250 = X + 500}$

$3X - X + 250 = X - X + 500$

$2X + 250 - 250 = 500 - 250$

$2X \div 2 = 250 \div 2$

$X = \mathbf{125\ g}$

Lesson 114, Lesson Practice

a.

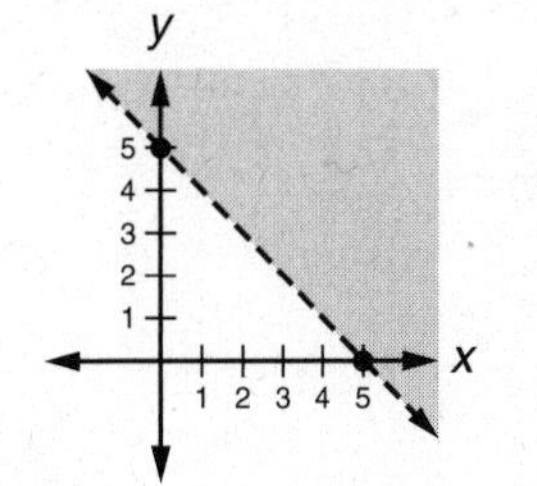

$x + y < 5$

b.

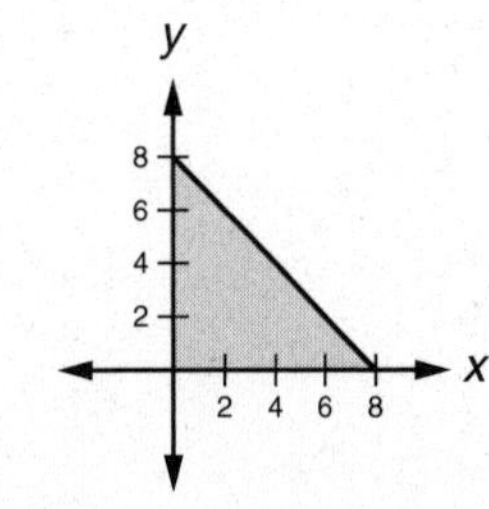

$x + y \geq 0$
$x + y \leq 8$

c.

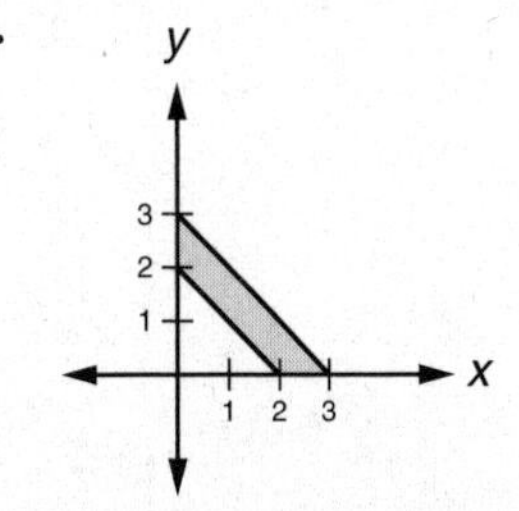

$x + y \geq 2$
$x + y \leq 3$

LESSON 114, MIXED PRACTICE

1.

	Percent	Actual Count
Original	100	$72.50
− Change	20	C
New	80	N

$$\frac{100}{80} = \frac{\$72.50}{N}$$
$$100N = \$5800$$
$$N = \$58$$

$$\$58.00 \times 0.07 = \$4.06 \qquad \$58.00 + \$4.06 = \mathbf{\$62.06}$$

2. $4(87) = 348,\ 6(90) = 540$

$540 - 348 = 192,\ \frac{192}{2} = \mathbf{96}$

3. (a) $P(B) = \frac{12}{27} = \mathbf{\frac{4}{9}}$

(b) $\mathbf{33\frac{1}{3}\%}$

(c) **7 to 2**

4. $\frac{\$10.80}{144 \text{ pencils}} = \frac{\$0.075}{1 \text{ pencil}}$

$7\frac{1}{2}$¢ per pencil

5. $\$5000 \times 0.08 = \$400$

$\frac{6}{12} = \frac{1}{2},\ \frac{\$400}{2} = \mathbf{\$200}$

6. (a) **Trees in the Park**

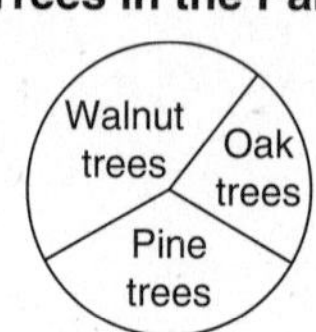

(b) $6(4 \text{ trees}) = 24 \text{ trees}$

$\frac{24 \text{ trees}}{3} = 8 \text{ trees}$

$24 \text{ trees} - (8 \text{ trees} + 6 \text{ trees}) = \mathbf{10 \text{ trees}}$

7.

	Ratio	Actual Count
Cars	5	C
Trucks	2	T
Total	7	3500

$$\frac{5}{7} = \frac{C}{3500}$$
$$7C = 17{,}500$$
$$C = \mathbf{2500 \text{ cars}}$$

8. Volume $\approx \frac{4}{3}(3)(2 \text{ ft})^3$

$\approx 4(8 \text{ ft}^3)$

$\approx \mathbf{32 \text{ ft}^3}$

9. $\mathbf{W_N = 1.2 \times \$240}$

$\mathbf{W_N = \$288}$

10. $\mathbf{60 = W_P \times 150}$

$$\frac{60}{150} = \frac{W_P \times 150}{150}$$
$$\frac{2}{5} = W_P$$
$$W_P = \frac{2}{5} \times 100\% = \mathbf{40\%}$$

11.

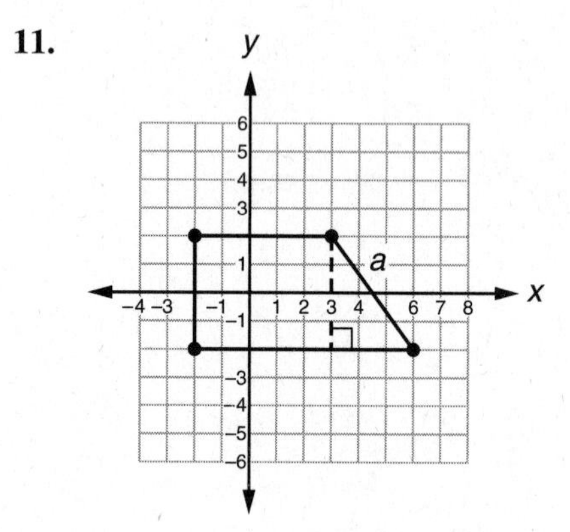

(a) Area of triangle $= \frac{(3 \text{ units})(4 \text{ units})}{2}$

$= 6 \text{ units}^2$

Area of rectangle $= (5 \text{ units})(4 \text{ units})$

$= 20 \text{ units}^2$

Area of figure $= 6 \text{ units}^2 + 20 \text{ units}^2$

$= \mathbf{26 \text{ units}^2}$

(b) $a^2 = (4 \text{ units})^2 + (3 \text{ units})^2$

$a^2 = 16 \text{ units}^2 + 9 \text{ units}^2$

$a^2 = 25 \text{ units}^2$

$a = 5 \text{ units}$

Perimeter $= 5 \text{ units} + 8 \text{ units} + 4 \text{ units} + 5 \text{ units}$

$= \mathbf{22 \text{ units}}$

12. (a) $\mathbf{-6, 0.6, \sqrt{6}, 6^2}$

(b) $\mathbf{6^2, -6, 0.6}$

13. (a) **1.8**

(b) **180%**

14. (a) $2\overset{\displaystyle 2.5}{\overline{)5.0}}$ $\quad \mathbf{2.5 \times 10^{-3}}$

(b) $5\overset{\displaystyle 0.4}{\overline{)2.0}}$ $\quad 0.4 \times 10^3 \longrightarrow \mathbf{4 \times 10^2}$

15. $10 \times 10^{-1} = \mathbf{1}$

16. $$12\ \cancel{\text{inches}} \cdot \frac{2.54 \text{ centimeters}}{1\ \cancel{\text{inch}}} = \mathbf{30.48 \text{ centimeters}}$$

17. (a)
$$\begin{aligned} C &= \pi d \\ \frac{C}{\pi} &= \frac{\pi d}{\pi} \\ \frac{C}{\pi} &= d \end{aligned}$$

(b)
$$\begin{aligned} \frac{62.8}{3.14} &\approx d \\ \mathbf{20} &\approx d \end{aligned}$$

18.

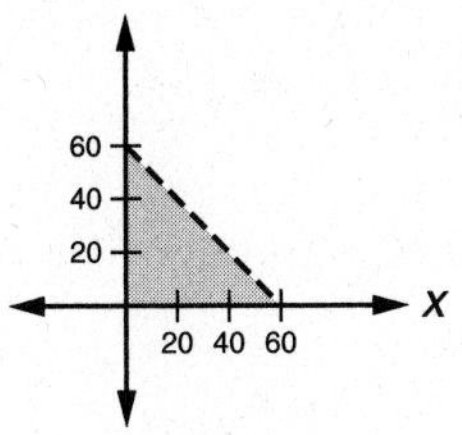

$\mathbf{x + y < 60}$

19.
$$\begin{aligned} C &\approx \frac{3.14(6 \text{ cm})}{2} \\ C &\approx 9.42 \text{ cm} \end{aligned}$$

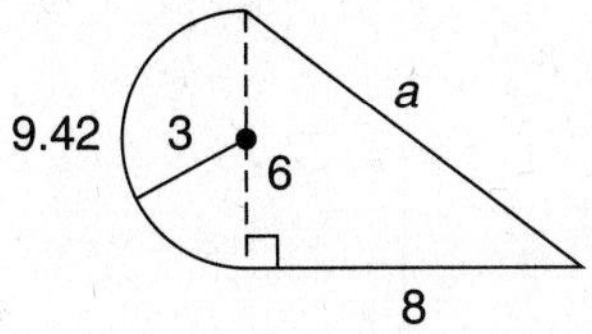

$$\begin{aligned} a^2 &= (6 \text{ cm})^2 + (8 \text{ cm})^2 \\ a^2 &= 36 \text{ cm}^2 + 64 \text{ cm}^2 \\ a^2 &= 100 \text{ cm}^2 \\ a &= 10 \text{ cm} \\ \text{Perimeter} &\approx 9.42 \text{ cm} + 10 \text{ cm} + 8 \text{ cm} \\ &\approx \mathbf{27.42 \text{ cm}} \end{aligned}$$

20. (a)
$$\begin{aligned} \text{Surface area} &= 6(3 \text{ ft} \times 3 \text{ ft}) \\ &= 6(9 \text{ ft}^2) = \mathbf{54 \text{ ft}^2} \end{aligned}$$

(b) $$\text{Volume} = \frac{1}{3}(9 \text{ ft}^2)(3 \text{ ft}) = \mathbf{9 \text{ ft}^3}$$

21.
$$\begin{aligned} \text{Area of base} &\approx 3.14(25 \text{ m}^2) \\ &\approx 78.5 \text{ m}^2 \\ \text{Volume} &\approx (78.5 \text{ m}^2)(3 \text{ m}) \approx \mathbf{235.5 \text{ m}^3} \end{aligned}$$

22. (a) $m\angle ACB = 180° - 130° = \mathbf{50°}$

(b) $m\angle CAB = 180° - (90° + 30°) = \mathbf{40°}$

(c)
$$\begin{aligned} m\angle CED &= m\angle ECD \\ &= 180° - 130° = 50° \\ m\angle CDE &= 180° - (50° + 50°) = \mathbf{80°} \end{aligned}$$

23.
$$\begin{aligned} \text{Volume} &= (40 \text{ cm})(10 \text{ cm})(20 \text{ cm}) \\ &= \mathbf{8000 \text{ cm}^3} \end{aligned}$$

24.
$$\begin{aligned} 0.8m - 1.2 &= 6 \\ 0.8m - 1.2 + 1.2 &= 6 + 1.2 \\ 0.8m &= 7.2 \\ \frac{0.8m}{0.8} &= \frac{7.2}{0.8} \\ m &= \mathbf{9} \end{aligned}$$

25.
$$\begin{aligned} 3(x - 4) &< x - 8 \\ 3x - 12 &< x - 8 \\ 3x - 12 + 12 &< x - 8 + 12 \\ 3x &< x + 4 \\ 3x - x &< x + 4 - x \\ 2x &< 4 \\ \frac{2x}{2} &< \frac{4}{2} \\ \mathbf{x} &\mathbf{< 2} \end{aligned}$$

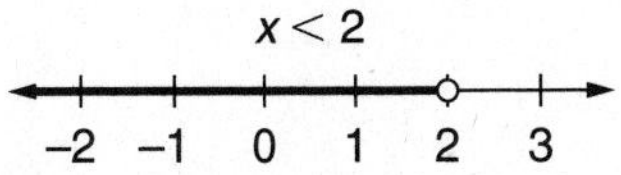

26. $$4^2 \cdot 2^{-3} \cdot 2^{-1} = \overset{\overset{1}{\cancel{2}}}{\cancel{16}} \cdot \frac{1}{\underset{1}{\cancel{8}}} \cdot \frac{1}{\underset{1}{\cancel{2}}} = \mathbf{1}$$

27.
$$\begin{aligned} 1 \text{ kilogram} &= 1000 \text{ grams} \\ 1000 \text{ grams} - 50 \text{ grams} &= \mathbf{950 \text{ grams}} \end{aligned}$$

28.
$$\begin{aligned} 1\frac{2}{10}\left(3\frac{3}{4}\right) \div 4\frac{1}{2} &= \left(\frac{\overset{3}{\cancel{12}}}{\underset{2}{\cancel{10}}} \cdot \frac{\overset{3}{\cancel{15}}}{\underset{1}{\cancel{4}}}\right) \div \frac{9}{2} \\ &= \frac{\overset{1}{\cancel{9}}}{\underset{1}{\cancel{2}}} \cdot \frac{\overset{1}{\cancel{2}}}{\underset{1}{\cancel{9}}} = \mathbf{1} \end{aligned}$$

29. $2\frac{3}{4} - 1\frac{1}{2} - \frac{1}{6} = 2\frac{9}{12} - 1\frac{6}{12} - \frac{2}{12}$

$= \mathbf{1\frac{1}{12}}$

30.

$(-3)(-2) - (2)(-3) - (-8) + (-2)(-3) + |-5|$

$(6) - (-6) - (-8) + (6) + (5)$

$12 - (-8) + 6 + (5)$

$20 + 6 + 5$

31

Lesson 115, Warm-Up

a. **22**

b. **154**

c. $\mathbf{\frac{1}{100}}$

d. **1**

e. **1**

f. **2.5 m**

g. **$800**

h. **$800**

i. **100¢**

Problem Solving

In Lesson 113 we learned that the volume of a sphere is $\frac{2}{3}$ the volume of a cylinder with the same radius and height. This cylinder has the same radius as one tennis ball and the same height as three tennis balls. Therefore, all three tennis balls occupy $\mathbf{\frac{2}{3}}$ **of the container.**

Lesson 115, Lesson Practice

a. **2 kg or 2000 g**

b. **3000 cm³**

c. **1000 milliliters**

d. Volume $= (25 \text{ cm})(10 \text{ cm})(8 \text{ cm})$

$= 2000 \text{ cm}^3 = 2000 \text{ mL}$

$=$ **2 liters**

Lesson 115, Mixed Practice

1.

$$\begin{array}{r} \$7000 \\ \times \quad 0.08 \\ \hline \$560 \end{array}$$

$\frac{9}{12} = \frac{3}{4}, \frac{3}{4}(\$560) = \mathbf{\$420}$

2. (a) $P(H, H) = \frac{1}{2} \cdot \frac{1}{2} = \mathbf{\frac{1}{4}}$

(b) **25%**

(c) **1 to 3**

3. $4(410) = 1640$

1640 miles + 600 miles = 2240 miles

$\frac{2240 \text{ miles}}{5 \text{ days}} = \mathbf{448 \frac{mi}{day}}$

4. $\frac{\$2.16}{18 \text{ ounces}} = \frac{\$0.12}{1 \text{ ounce}}$

$\frac{\$3.36}{32 \text{ ounces}} = \frac{\$0.105}{1 \text{ ounce}}$

$$\begin{array}{r} \$0.120 \\ - \$0.105 \\ \hline \$0.015 \end{array}$$

$\mathbf{1\frac{1}{2}}$**¢ more per ounce**

5.

	Case 1	Case 2
Words	160	800
Minutes	5	m

$\frac{160}{5} = \frac{800}{m}$

$160m = 4000$

$m =$ **25 minutes**

6.

	Ratio	Actual Count
Guinea pigs	7	G
Rats	5	R
Total	12	120

$\frac{7}{12} = \frac{G}{120}$

$12G = 840$

$G =$ **70 guinea pigs**

7.
$$\frac{3}{4}x = 48$$
$$\left(\frac{4}{3}\right)\frac{3}{4}x = \left(\frac{4}{3}\right)48$$
$$x = 64$$
$$\frac{5}{8}(64) = \mathbf{40}$$

8.

	Percent	Actual Count
Original	100	\$1500
+ Change	40	C
New	140	N

$$\frac{100}{140} = \frac{\$1500}{N}$$
$$100N = \$210{,}000$$
$$N = \$2100$$
$$\begin{array}{r} \$2100 \\ \times\ \ 0.08 \\ \hline \$168 \end{array} \qquad \begin{array}{r} \$2100 \\ +\ \ \$168 \\ \hline \mathbf{\$2268} \end{array}$$

9. 80% of 75% of \$80 $= (0.8)(0.75) \times \$80$
$= \mathbf{\$48}$

10.

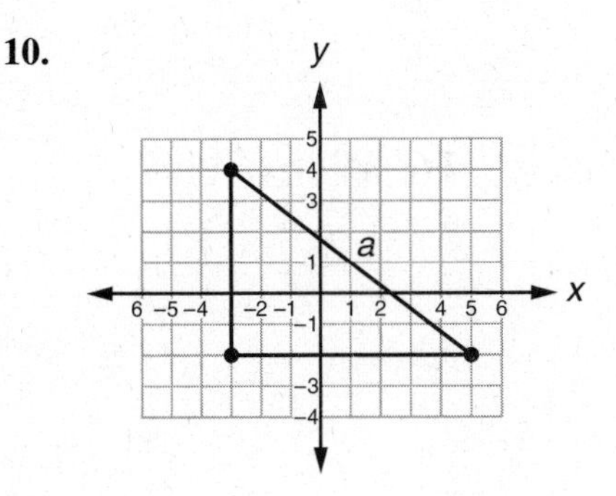

(a) Area $= \dfrac{(8 \text{ units})(6 \text{ units})}{2} = \mathbf{24\ units^2}$

(b)
$$a^2 = 64 \text{ units}^2 + 36 \text{ units}^2$$
$$a^2 = 100 \text{ units}^2$$
$$a = 10 \text{ units}$$
Perimeter $= 10$ units $+ 8$ units $+ 6$ units $=$ **24 units**

11. Volume $= (25 \text{ cm})(20 \text{ cm})(10 \text{ cm})$
$= 5000 \text{ cm}^3 = 5000 \text{ g} = 5 \text{ kg}$
$5 \text{ kg} + 5 \text{ kg} =$ **10 kg**

12. (a) $\mathbf{\dfrac{7}{8}}$

(b) $\mathbf{87\frac{1}{2}\%}$

13. $a \div b$ (<) $a - b$

14. (a) $(6.4 \times 10^6)(8 \times 10^{-8})$
51.2×10^{-2}
$(5.12 \times 10^1) \times 10^{-2}$
$\mathbf{5.12 \times 10^{-1}}$

(b) $\dfrac{6.4 \times 10^6}{8 \times 10^{-8}}$

$$8\overline{)6.4} = 0.8$$

$10^6 \div 10^{-8} = 10^{14} \longleftarrow [6 - (-8) = 14]$
0.8×10^{14}
$(8 \times 10^{-1}) \times 10^{14}$
$\mathbf{8 \times 10^{13}}$

15. 36 ~~inches~~ $\cdot \dfrac{\textbf{2.54 centimeters}}{\textbf{1 ~~inch~~}}$
= 91.44 centimeters

16. (a)
$$A = \frac{1}{2}bh$$
$$\left(\frac{2}{1}\right)A = \left(\frac{2}{1}\right)\frac{1}{2}bh$$
$$2A = bh$$
$$\frac{2A}{h} = \frac{bh}{h}$$
$$\frac{2A}{h} = b$$

(b)
$$\frac{2(24)}{6} = b$$
$$\mathbf{8 = b}$$

17. See student work;

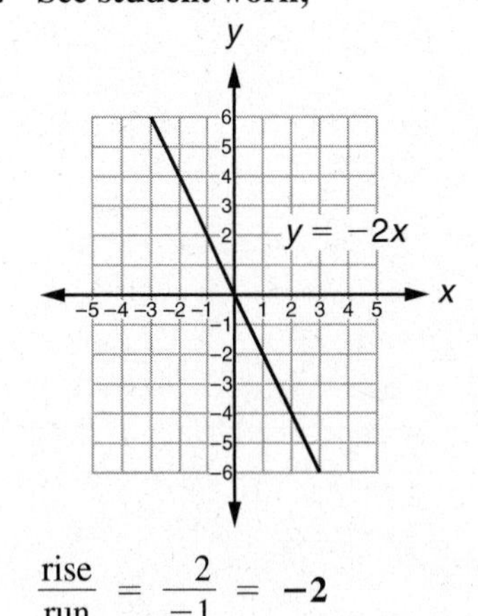

$$\frac{\text{rise}}{\text{run}} = \frac{2}{-1} = \mathbf{-2}$$

18. $A \approx \dfrac{3.14(4 \text{ mm}^2)}{2}$
$A \approx 6.28 \text{ mm}^2$

Area of square	$=$	$(6 \text{ mm})(6 \text{ mm})$
	$=$	36 mm^2
− Area of semicircle	$\approx$	6.28 mm^2
Area of figure	$\approx$	$\mathbf{29.72\ mm^2}$

19. (a) $6(100 \text{ cm} \times 100 \text{ cm}) = 6(10{,}000 \text{ cm}^2) = \mathbf{60{,}000 \text{ cm}^2}$

(b) $\text{Volume} = (100 \text{ cm})(100 \text{ cm})(100 \text{ cm}) = \mathbf{1{,}000{,}000 \text{ cm}^3}$

(c) **1 m**

20. (a) $\text{Area of base} = \pi(15 \text{ in.})^2 = 225\pi \text{ in.}^2$

$\text{Volume} = (225\pi \text{ in.}^2)(30 \text{ in.}) = \mathbf{6750\pi \text{ in.}^3}$

(b) $\text{Volume} = \frac{4}{3}\pi(15 \text{ in.})^3 = \mathbf{4500\pi \text{ in.}^3}$

21. (a) $m\angle YXZ = 180° - (90° + 35°) = \mathbf{55°}$

(b) $m\angle WXV = 180° - (90° + 55°) = \mathbf{35°}$

(c) $m\angle WVX = 180° - (90° + 35°) = \mathbf{55°}$

22.
$$\frac{21}{14} = \frac{12}{WV}$$
$$21(WV) = 168$$
$$WV = \mathbf{8 \text{ cm}}$$

23. $\text{Volume} = \frac{1}{3}(6 \text{ in.} \times 6 \text{ in.} \times 6 \text{ in.}) = \mathbf{72 \text{ in.}^3}$

24.
$$0.4n + 5.2 = 12$$
$$0.4n + 5.2 - 5.2 = 12 - 5.2$$
$$0.4n = 6.8$$
$$\frac{0.4n}{0.4} = \frac{6.8}{0.4}$$
$$n = \mathbf{17}$$

25.
$$\frac{18}{y} = \frac{36}{28}$$
$$36y = 504$$
$$\frac{36y}{36} = \frac{504}{36}$$
$$y = \mathbf{14}$$

26. $\sqrt{5^2 - 3^2} + \sqrt{5^2 - 4^2}$
$= \sqrt{25 - 9} + \sqrt{25 - 16} = \sqrt{16} + \sqrt{9}$
$= 4 + 3 = \mathbf{7}$

27.
$$\begin{array}{rrr} \overset{2}{\cancel{3}} \text{ yd} & (\overset{2}{\cancel{3}} \text{ ft}) & (12 \text{ in.}) \\ - & 2 \text{ ft} & 1 \text{ in.} \\ \hline \end{array} \longrightarrow \begin{array}{rrr} \overset{2}{\cancel{3}} \text{ yd} & \overset{2}{\cancel{3}} \text{ ft} & 12 \text{ in.} \\ - & 2 \text{ ft} & 1 \text{ in.} \\ \hline \mathbf{2 \text{ yd}} & & \mathbf{11 \text{ in.}} \end{array}$$

28. $3\frac{1}{2} \div \left(1\frac{2}{5} \div 3\right)$

$= \frac{7}{2} \div \left(\frac{7}{5} \times \frac{1}{3}\right) = \frac{7}{2} \div \frac{7}{15}$

$= \frac{\overset{1}{\cancel{7}}}{2} \times \frac{15}{\underset{1}{\cancel{7}}} = \frac{15}{2} = \mathbf{7\frac{1}{2}}$ or **7.5**

29. $3.5 + 2^{-2} - 2^{-3} = 3\frac{1}{2} + \frac{1}{4} - \frac{1}{8}$

$= 3\frac{4}{8} + \frac{2}{8} - \frac{1}{8} = \mathbf{3\frac{5}{8}}$ or **3.625**

30.
$$\frac{(3)(-2)(4)}{(-6)(2)} + (-8) + (-4)(+5) - (2)(-3)$$
$$\frac{(-6)(4)}{-12} + (-8) + (-20) - (-6)$$
$$\frac{-24}{-12} + (-8) + (-20) - (-6)$$
$$2 + (-8) + (-20) - (-6)$$
$$-6 + (-20) - (-6)$$
$$\mathbf{-20}$$

LESSON 116, WARM-UP

a. **42**

b. **812**

c. $\mathbf{4\frac{1}{4}}$

d. $\mathbf{2.5 \times 10^3}$

e. **0.4**

f. $\mathbf{10{,}000 \text{ cm}^2}$

g. **$500**

h. **$500**

i. **$3.60**

Problem Solving

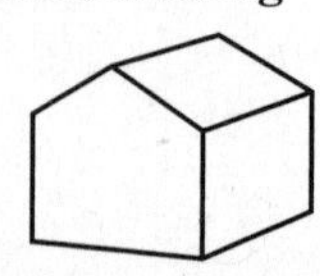

Lesson 116, Lesson Practice

a. $8m^2n = \mathbf{(2)(2)(2)mmn}$

b. $12mn^2 = \mathbf{(2)(2)(3)mnn}$

c. $18x^3y^2 = \mathbf{(2)(3)(3)xxxyy}$

d. $\frac{8m^2n}{4mn} + \frac{12mn^2}{4mn}$

$2m + 3n$

$\mathbf{4mn(2m + 3n)}$

e. $\frac{8xy^2}{4xy} - \frac{4xy}{4xy}$

$2y - 1$

$\mathbf{4xy(2y - 1)}$

f. $\frac{6a^2b^3}{3a^2b^2} + \frac{9a^3b^2}{3a^2b^2} + \frac{3a^2b^2}{3a^2b^2}$

$2b + 3a + 1$

$\mathbf{3a^2b^2(2b + 3a + 1)}$

Lesson 116, Mixed Practice

1. (a) $p(1, 4) = \frac{1}{6} \cdot \frac{1}{6} = \frac{1}{36}$

$p(4, 1) = \frac{1}{6} \cdot \frac{1}{6} = \frac{1}{36}$

$p(2, 3) = \frac{1}{6} \cdot \frac{1}{6} = \frac{1}{36}$

$p(3, 2) = \frac{1}{6} \cdot \frac{1}{6} = \frac{1}{36}$

$4\left(\frac{1}{36}\right) = \frac{1}{9} = 0.\overline{11}$, **0.11**

(b) **25%**

(c) **1 to 35**

2. 2^{10} bytes = **1024 bytes**

3. **The better sale seems to be the "40% of" sale, which is 60% off the regular price. Sixty percent off is better than forty percent off.**

4. (a) $\mathbf{1\frac{3}{4}}$

(b) **1.75**

(c) $\mathbf{0.08\overline{3}}$

(d) $\mathbf{8\frac{1}{3}\%}$

5.

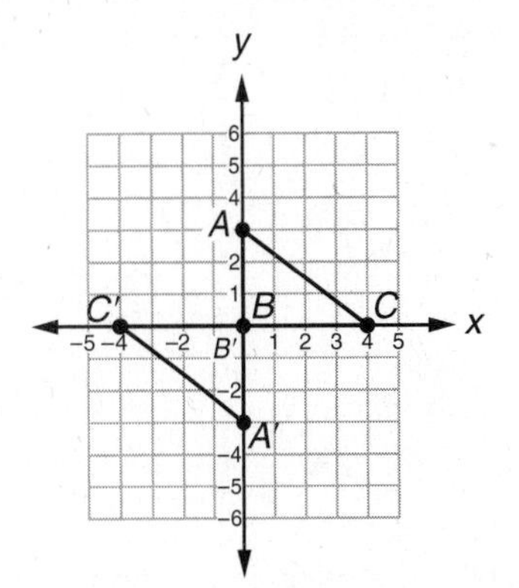

$\mathbf{A'(0, -3), B'(0, 0), C'(-4, 0)}$

6. Exterior angle: $\frac{360°}{20} = \mathbf{18°}$

Interior angle: $180° - 18° = \mathbf{162°}$

7.

	Percent	Actual Count
Original	100	O
− Change	30	C
New	70	\$42

$\frac{30}{70} = \frac{C}{\$42}$

$70C = \$1260$

$C = \mathbf{\$18}$

8. (a) Volume = (40 cm)(20 cm)(30 cm)

= 24,000 cm³ = 24,000 mL

24,000 mL = **24 liters**

(b) **24 kg**

9. **24 kg · $\frac{2.2 \text{ lb}}{1 \text{ kg}}$ = 52.8 lb**

10. $\mathbf{2x - 6 = 48}$

$2x - 6 + 6 = 48 + 6$

$2x = 54$

$\frac{2x}{2} = \frac{54}{2}$

$x = \mathbf{27}$

11. $(8x - 8) + (7x + 8) + (6x + 12) = 180$

$21x + 12 = 180$

$21x + 12 - 12 = 180 - 12$

$21x = 168$

$\frac{21x}{21} = \frac{168}{21}$

$x = 8°$

$7(8°) + 8 = \mathbf{64°}$

12.
$$\begin{aligned} F &= 1.8C + 32 \\ F - 32 &= 1.8C + 32 - 32 \\ F - 32 &= 1.8C \\ \frac{F - 32}{1.8} &= \frac{1.8C}{1.8} \\ \mathbf{\frac{F - 32}{1.8}} &\mathbf{= C} \end{aligned}$$

13.
$$\begin{aligned} C &\approx \frac{3.14(40 \text{ in.})}{2} \\ C &\approx 62.8 \text{ in.} \\ \text{Perimeter} &\approx 62.8 \text{ in.} + 66 \text{ in.} + 66 \text{ in.} \\ &\approx 194.8 \text{ in.} \approx \mathbf{195 \text{ in.}} \end{aligned}$$

14.
$$\begin{aligned} a^2 &= (15 \text{ cm})^2 + (20 \text{ cm})^2 \\ a^2 &= 225 \text{ cm}^2 + 400 \text{ cm}^2 \\ a^2 &= 625 \text{ cm}^2 \\ a &= 25 \text{ cm} \end{aligned}$$

$$\begin{aligned} \text{Area of triangle} &= \frac{(20 \text{ cm})(15 \text{ cm})}{2} \\ &= 150 \text{ cm}^2 \\ \text{Area of triangle} &= \frac{(20 \text{ cm})(15 \text{ cm})}{2} \\ &= 150 \text{ cm}^2 \\ \text{Area of rectangle} &= (20 \text{ cm})(20 \text{ cm}) \\ &= 400 \text{ cm}^2 \\ \text{Area of rectangle} &= (15 \text{ cm})(20 \text{ cm}) \\ &= 300 \text{ cm}^2 \\ + \text{ Area of rectangle} &= (20 \text{ cm})(25 \text{ cm}) \\ &= 500 \text{ cm}^2 \\ \hline \text{Total surface area} &= \mathbf{1500 \text{ cm}^2} \end{aligned}$$

15.
$$\begin{aligned} \text{Volume} &= (150 \text{ cm}^2)(20 \text{ cm}) \\ &= \mathbf{3000 \text{ cm}^3} \end{aligned}$$

16. (a) $\frac{\text{rise}}{\text{run}} = \frac{1}{1} = \mathbf{1;}$

(0, −2)

(b) $\frac{\text{rise}}{\text{run}} = \frac{2}{-1} = \mathbf{-2;}$

(0, 4)

17.
$$\begin{aligned} A &= \frac{1}{2}(12 \text{ cm} + 18 \text{ cm})8 \text{ cm} \\ &= \frac{1}{2}(30 \text{ cm})8 \text{ cm} \\ &= (15 \text{ cm})\ 8 \text{ cm} \\ &= \mathbf{120 \text{ cm}^2} \end{aligned}$$

18.
$$\begin{aligned} 3x^2 - 5 &= 40 \\ 3x^2 - 5 + 5 &= 40 + 5 \\ 3x^2 &= 45 \\ \frac{3x^2}{3} &= \frac{45}{3} \\ x^2 &= 15 \\ \mathbf{x} &\mathbf{= \sqrt{15}, -\sqrt{15}} \end{aligned}$$

19. (a) $4\overline{)8}$ with quotient 2 $\mathbf{2 \times 10^{-12}}$

(b) $8\overline{)4.0}$ with quotient 0.5 $0.5 \times 10^{12} \longrightarrow \mathbf{5 \times 10^{11}}$

20. The product is 1 because the numbers are reciprocals.

21. (a) $9x^2y = \mathbf{(3)(3)xxy}$

(b) $\frac{10a^2b}{5ab} + \frac{15a^2b^2}{5ab} + \frac{20abc}{5ab}$

$2a + 3ab + 4c$

$\mathbf{5ab(2a + 3ab + 4c)}$

22.
$$\begin{aligned} \text{Volume} &\approx \frac{4}{3}(3.14)(6 \text{ in.})^3 \\ &\approx \frac{4}{3}(3.14)(216 \text{ in.}^3) \\ &\approx 904.32 \text{ in.}^3 \\ &\approx \mathbf{904 \text{ in.}^3} \end{aligned}$$

23. (a) $m\angle BCD = 180° - (90° + 25°)$

$= 180° - 115° = \mathbf{65°}$

(b) $m\angle BAC = 180° - (25° + 90°) = \mathbf{65°}$

(c) $m\angle ACD = 180° - (90° + 65°) = \mathbf{25°}$

(d) **The three triangles are similar.**

24. $\mathbf{\frac{BD}{BC} = \frac{CD}{CA}}$

25.
$$\begin{aligned} x - 15 &= x + 2x + 1 \\ x - 15 &= 3x + 1 \\ x - 15 - x &= 3x + 1 - x \\ -15 &= 2x + 1 \\ -15 - 1 &= 2x + 1 - 1 \\ -16 &= 2x \\ \frac{-16}{2} &= \frac{2x}{2} \\ \mathbf{-8} &= x \end{aligned}$$

26.
$$\begin{aligned} 0.12(m - 5) &= 0.96 \\ 0.12m - 0.6 &= 0.96 \\ 0.12m - 0.6 + 0.6 &= 0.96 + 0.6 \\ 0.12m &= 1.56 \\ \frac{0.12m}{0.12} &= \frac{1.56}{0.12} \\ m &= \mathbf{13} \end{aligned}$$

27. $a(b - c) + b(c - a)$

$ab - ac + bc - ba$

$\mathbf{-ac + bc}$ or $\mathbf{bc - ac}$

28. $\dfrac{(8x^2y)(12x^3y^2)}{(4xy)(6y^2)}$

$$\frac{\cancel{2}\cdot\cancel{2}\cdot\cancel{2}\cdot 2\cdot 2\cdot\cancel{3}\cdot\cancel{x}\cdot x\cdot x\cdot x\cdot x\cdot\cancel{y}\cdot\cancel{y}\cdot\cancel{y}}{\cancel{2}\cdot\cancel{2}\cdot\cancel{2}\cdot\cancel{3}\cdot\cancel{x}\cdot\cancel{y}\cdot\cancel{y}\cdot\cancel{y}}$$

$\mathbf{4x^4}$

29. (a) $(-3)^2 + (-2)(-3) - (-2)^3$
$9 + (6) - (-8)$
$15 - (-8)$
23

(b) $\sqrt[3]{-8} + \sqrt[3]{8} = -2 + 2 = \mathbf{0}$

30. AD = 1.2 units − 0.75 unit
= **0.45 unit**

Lesson 117, Warm-Up

a. **64**

b. **1492**

c. **–6**

d. **3.5×10^2**

e. **5, –5**

f. **212°F**

g. **\$500**

h. **\$3500**

i. **–1**

Problem Solving

There are thirty-six 1-by-1 squares, twenty-five 2-by-2 squares, sixteen 3-by-3 squares, nine 4-by-4 squares, four 5-by-5 squares, and one 6-by-6 square. The total number of squares is 36 + 25 + 16 + 9 + 4 + 1, or **91.**

Lesson 117, Lesson Practice

a.
$2x + y = 3$
$2x + y - 2x = 3 - 2x$
$y = 3 - 2x$
$\mathbf{y = -2x + 3}$

b.
$y - 3 = x$
$y - 3 + 3 = x + 3$
$\mathbf{y = x + 3}$

c.
$2x + y - 3 = 0$
$2x + y - 3 + 3 = 0 + 3$
$2x + y = 3$
$2x + y - 2x = 3 - 2x$
$y = 3 - 2x$
$\mathbf{y = -2x + 3}$

d.
$x + y = 4 - x$
$x + y - x = 4 - x - x$
$y = 4 - 2x$
$\mathbf{y = -2x + 4}$

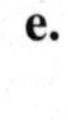

e.

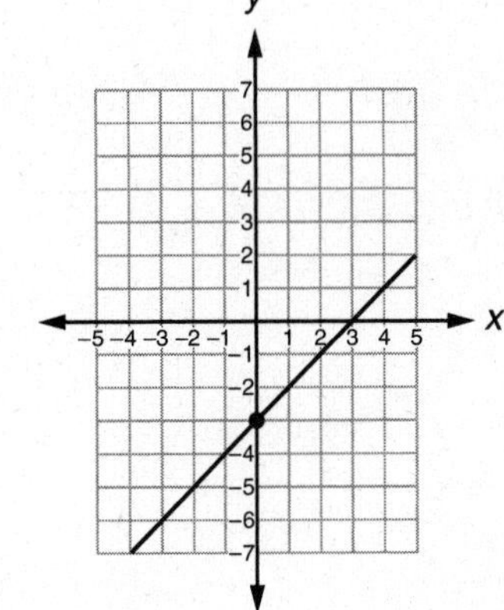

f.

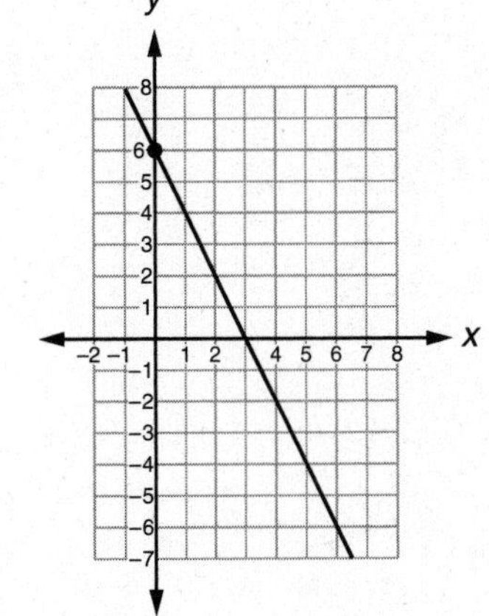

g.

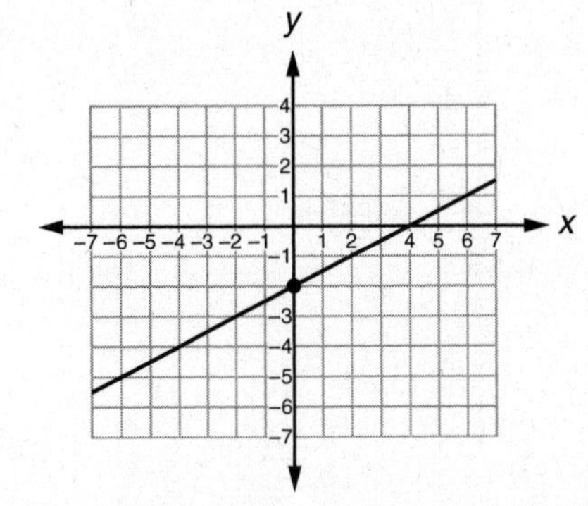

h.

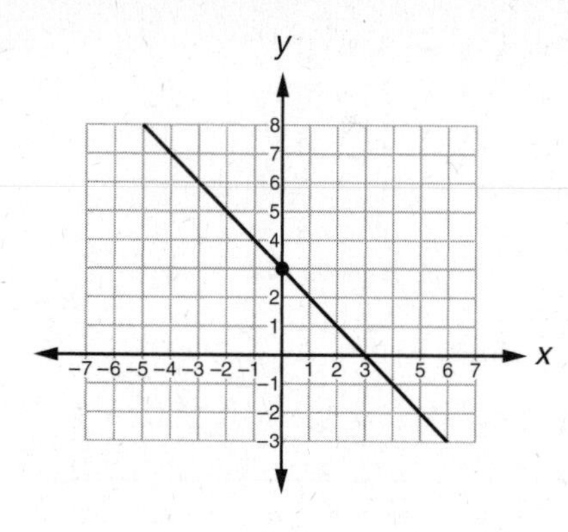

LESSON 117, MIXED PRACTICE

1.
$$\begin{array}{r} \$10{,}000 \\ \times \quad 1.07 \\ \hline \$10{,}700 \\ \times \quad 1.07 \\ \hline \$11{,}449 \\ \times \quad 1.07 \\ \hline \$12{,}250.43 \\ \times \quad 1.07 \\ \hline \$13{,}107.9601 \end{array}$$

$$\begin{array}{r} \$13{,}107.96 \\ -\ \$10{,}000.00 \\ \hline \mathbf{\$3107.96} \end{array}$$

2. (a) $\mathbf{\frac{1}{4}}$

(b) **1 to 3**

3. $4(75\%) + 6(85\%) = 810\%$

$$\frac{810\%}{10} = \mathbf{81\%}$$

4. (a) $\mathbf{1\frac{2}{5}}$

(b) **140%**

(c) $\mathbf{0.91\overline{6}}$

(d) $\mathbf{91\frac{2}{3}\%}$

5.

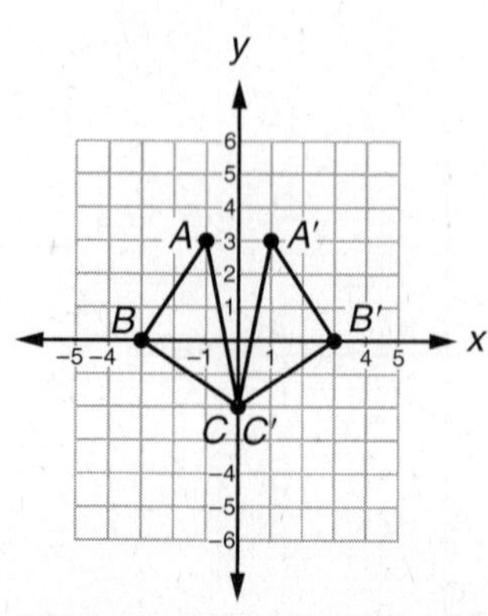

$\mathbf{A'(1, 3), B'(3, 0), C'(0, -2)}$

6. (a) $\frac{360°}{8} = \mathbf{45°}$

(b) $180° - 45° = \mathbf{135°}$

(c) **5 diagonals**

7.
$$\frac{100}{N} = \frac{1.2}{0.3}$$
$$1.2N = 30$$
$$N = \mathbf{25\%}$$

8. (a) **500 cubic centimeters**

(b) 500 mL = 500 g = **0.5 kilogram**

9. $\overset{\overset{60}{\cancel{180}}}{\cancel{540}}\ \text{ft}^2 \cdot \frac{1\ \text{yd}}{\underset{1}{\cancel{3}}\ \cancel{\text{ft}}} \cdot \frac{1\ \text{yd}}{\underset{1}{\cancel{3}}\ \cancel{\text{ft}}} = \mathbf{60\ yd^2}$

10.
$$3x^2 + 6 = 81$$
$$3x^2 + 6 - 6 = 81 - 6$$
$$3x^2 = 75$$
$$\frac{3x^2}{3} = \frac{75}{3}$$
$$x^2 = 25$$
$$x = \mathbf{5, -5}$$

11.
$$(3x + 5) + (x - 5) = 180°$$
$$4x = 180°$$
$$\frac{4x}{4} = \frac{180°}{4}$$
$$x = 45°$$

$$2x + (x - 5) = 2(45°) + (45° - 5)$$
$$= 90° + 40° = 130°$$
$$m\angle y = 180 - 130° = \mathbf{50°}$$

12.
$$c^2 - a^2 = b^2$$
$$c^2 - a^2 + a^2 = b^2 + a^2$$
$$\mathbf{c^2 = b^2 + a^2} \text{ or } \mathbf{c^2 = a^2 + b^2}$$

13. (a) **0.25**

(b) $\mathbf{33\frac{1}{3}\%}$

(c) **1 to 5**

14.

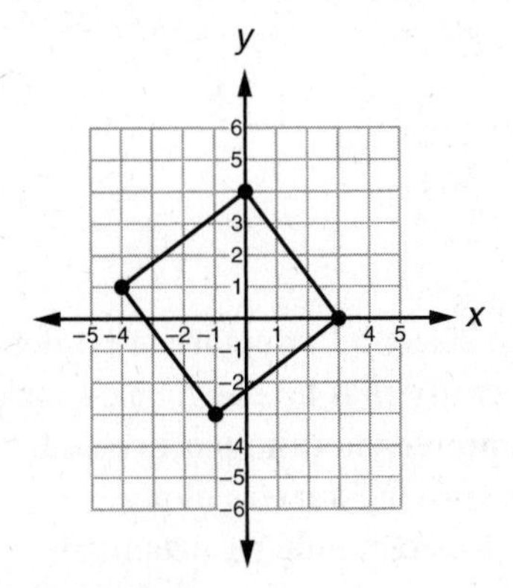

(a) **5 units**

(b) Perimeter $= 4(5 \text{ units}) = \mathbf{20 \text{ units}}$

(c) Area $= (5 \text{ units})(5 \text{ units}) = \mathbf{25 \text{ units}^2}$

15. (a)
$$\begin{aligned} \text{Area of base} &= \pi(3 \text{ in.})^2 \\ &= 9\pi \text{ in.}^2 \\ \text{Volume} &= (9\pi \text{ in.}^2)(8 \text{ in.}) \\ &= \mathbf{72\pi \text{ in.}^3} \end{aligned}$$

(b) Volume $= \frac{1}{3}(72\pi \text{ in.}^3) = \mathbf{24\pi \text{ in.}^3}$

16. (a)
$$\begin{aligned} V &= lwh \\ \frac{V}{lw} &= \frac{lwh}{lw} \\ \frac{V}{lw} &= h \end{aligned}$$

(b) $h = \dfrac{\overset{10}{\cancel{6000}} \text{ cm}^{\cancel{3}}}{(\underset{1}{\cancel{20}} \cancel{\text{cm}})(\underset{1}{\cancel{30}} \cancel{\text{cm}})} = \mathbf{10 \text{ cm}}$

17. (a) $\dfrac{\text{rise}}{\text{run}} = \dfrac{2}{1} = \mathbf{2}$

(b) **(0, 4)**

(c) $\mathbf{y = 2x + 4}$

18. (a)
$$\begin{aligned} y + 5 &= x \\ y + 5 - 5 &= x - 5 \\ \mathbf{y} &\mathbf{= x - 5} \end{aligned}$$

(b)
$$\begin{aligned} 2x + y &= 4 \\ 2x + y - 2x &= 4 - 2x \\ \mathbf{y} &\mathbf{= -2x + 4} \end{aligned}$$

19. (a) $24xy^2 = \mathbf{(2)(2)(2)(3)}xyy$

(b)
$$\frac{3x^2}{3x} + \frac{6xy}{3x} - \frac{9x}{3x}$$
$$x + 2y - 3$$
$$\mathbf{3x(x + 2y - 3)}$$

20. (a)
$$\begin{aligned} &(5 \times 10^3 \text{ mm})(5 \times 10^3 \text{ mm}) \\ &= 25 \times 10^6 \text{ mm}^2 \\ &= \mathbf{2.5 \times 10^7 \text{ mm}^2} \end{aligned}$$

(b) **25,000,000 mm²**

21. $\overset{25}{\cancel{25{,}000{,}000}} \cancel{\text{mm}^2} \cdot \dfrac{1 \text{ m}}{\underset{1}{\cancel{1000}} \cancel{\text{mm}}} \cdot \dfrac{1 \text{ m}}{\underset{1}{\cancel{1000}} \cancel{\text{mm}}}$

$= \mathbf{25 \text{ m}^2}$

22. (a) **Side *BD***

(b) **Side *AD***

23. $\dfrac{5}{3} \text{ in.} - \dfrac{5}{4} \text{ in.} = \dfrac{20}{12} \text{ in.} - \dfrac{15}{12} \text{ in.} = \mathbf{\dfrac{5}{12} \text{ in.}}$

24.
$$\begin{aligned} \frac{3}{4}x + 12 &< 15 \\ \frac{3}{4}x + 12 - 12 &< 15 - 12 \\ \frac{3}{4}x &< 3 \\ \left(\frac{4}{3}\right)\frac{3}{4}x &< \left(\frac{4}{3}\right)3 \\ \mathbf{x} &\mathbf{< 4} \end{aligned}$$

x < 4

0 1 2 3 4 5

25.
$$\begin{aligned} 6w - 3w + 18 &= 9(w - 4) \\ 3w + 18 &= 9w - 36 \\ 3w + 18 - 18 &= 9w - 36 - 18 \\ 3w &= 9w - 54 \\ 3w - 9w &= 9w - 54 - 9w \\ -6w &= -54 \\ \frac{-6w}{-6} &= \frac{-54}{-6} \\ w &= \mathbf{9} \end{aligned}$$

26.
$$3x(x - 2y) + 2xy(x + 3)$$
$$3x^2 - 6xy + 2x^2y + 6xy$$
$$\mathbf{3x^2 + 2x^2y}$$

27.
$$\begin{aligned} &2^{-2} + 4^{-1} + \sqrt[3]{27} + (-1)^3 \\ &= \frac{1}{4} + \frac{1}{4} + 3 + (-1) = \frac{2}{4} + 2 \\ &= \mathbf{2\frac{1}{2}} \end{aligned}$$

28.
$$(-3) + (-2)[(-3)(-2) - (+4)] - (-3)(-4)$$
$$(-3) + (-2)[(6) - (+4)] - (12)$$
$$(-3) + (-2)[2] - (12)$$
$$(-3) + (-4) - (12)$$
$$\mathbf{-19}$$

29. $4\overline{)1.2}$ with quotient 0.3; $\quad 0.3 \times 10^{-9} \longrightarrow \mathbf{3 \times 10^{-10}}$

30.
$$\frac{36a^2b^3c}{12ab^2c}$$
$$\frac{\overset{1}{\cancel{2}} \cdot \overset{1}{\cancel{2}} \cdot \overset{1}{\cancel{3}} \cdot 3 \cdot \overset{1}{\cancel{a}} \cdot a \cdot \overset{1}{\cancel{b}} \cdot \overset{1}{\cancel{b}} \cdot b \cdot \overset{1}{\cancel{c}}}{\underset{1}{\cancel{2}} \cdot \underset{1}{\cancel{2}} \cdot \underset{1}{\cancel{3}} \cdot \underset{1}{\cancel{a}} \cdot \underset{1}{\cancel{b}} \cdot \underset{1}{\cancel{b}} \cdot \underset{1}{\cancel{c}}}$$
$$\mathbf{3ab}$$

Lesson 118, Warm-Up

a. **43**

b. **1776**

c. $9\frac{1}{9}$

d. 1.5×10^{-3}

e. **22**

f. **7.5 kg**

g. **$6000**

h. **$10,000**

i. **1 hr 20 min**

Problem Solving

The frame will not lie flat in the box, but it will fit slanted (see picture below). By the Pythagorean theorem, we find that the length of the diagonal across the box is 15 in., which is long enough to hold the frame despite its thickness.

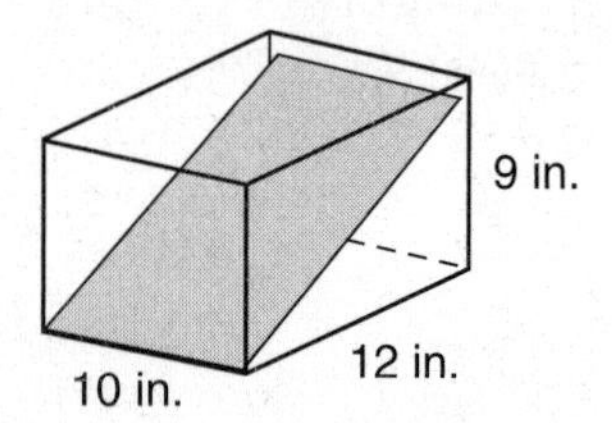

Lesson 118, Lesson Practice

a. **See student work.**

b. **See student work.**

Lesson 118, Mixed Practice

1. $$\frac{100}{C} = \frac{\$180{,}000}{\$9000}$$
$$\$180{,}000C = \$900{,}000$$
$$C = \mathbf{5\%}$$

2. $$\frac{40 \text{ cm}}{100 \text{ cm}} = \frac{600 \text{ cm}}{l}$$
$$(40 \text{ cm})l = 60{,}000 \text{ cm}^2$$
$$l = 1500 \text{ cm}$$
About 15 meters

3. **Armando can select a Pythagorean triplet like 3-4-5 to verify that he has formed a right angle. For example, he can measure and mark from a corner 3 meters along one line and 4 meters along a proposed perpendicular line. Then he can check to see whether it is 5 meters between the marks.**

4. $$15 \text{ meters} \cdot \frac{3.28 \text{ feet}}{1 \text{ meter}} \approx 49.2 \text{ feet}$$
$$\approx \mathbf{49 \text{ feet}}$$

5. $$\frac{1}{120} = \frac{1.5}{x}$$
$$(1)x = (1.5)(120)$$
$$x = (1.5)(120)$$
$$x = 180 \text{ in.} = 15 \text{ ft}$$
$$A = (15 \text{ ft})(15 \text{ ft}) = \mathbf{225 \text{ ft}^2}$$

6. (a) $\frac{1}{9}$

(b) $8\frac{1}{3}\%$

(c) **1 to 17**

7.

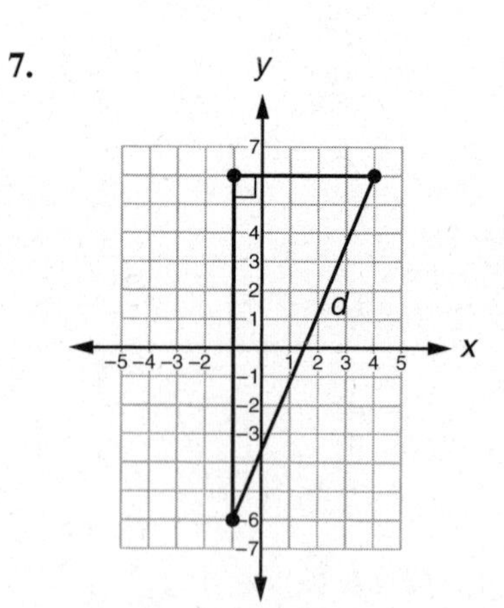

$$d^2 = (12 \text{ units})^2 + (5 \text{ units})^2$$
$$d^2 = 144 \text{ units}^2 + 25 \text{ units}^2$$
$$d^2 = 169 \text{ units}^2$$
$$d = \mathbf{13 \text{ units}}$$

8. (a) 2 L = 2000 mL
= **2000 cubic centimeters**

(b) **2 kilograms**

9. $\mathbf{\frac{1}{2}x - \frac{2}{3} = \frac{5}{6}}$

$$\frac{1}{2}x - \frac{2}{3} + \frac{2}{3} = \frac{5}{6} + \frac{2}{3}$$
$$\frac{1}{2}x = \frac{5}{6} + \frac{4}{6}$$
$$\frac{1}{2}x = \frac{9}{6}$$
$$\left(\frac{2}{1}\right)\frac{1}{2}x = \left(\frac{2}{1}\right)\left(\frac{9}{6}\right)$$
$$x = \frac{9}{3} = \mathbf{3}$$

10. $\frac{200°}{2} = 100°$

$m\angle g = 180° - 100° = \mathbf{80°}$

11. $3x + y = 6$

$$3x + y - 3x = 6 - 3x$$
$$y = 6 - 3x$$
$$\mathbf{y = -3x + 6}$$

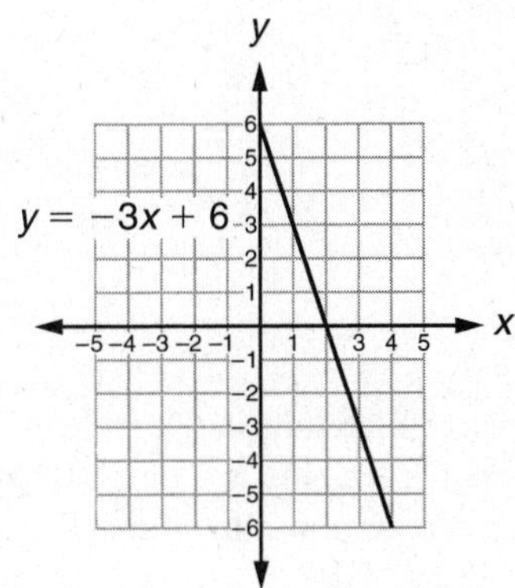

12. $(x + 30) + 3x + 2x = 180°$

$$6x + 30 = 180°$$
$$6x + 30 - 30 = 180° - 30$$
$$6x = 150°$$
$$\frac{6x}{6} = \frac{150°}{6}$$
$$x = 25°$$
$$2x = 2(25°) = \mathbf{50°}$$

13. Area of base $= (12 \text{ in.})(12 \text{ in.})$
$= 144 \text{ in.}^2$

Volume of pyramid $= \frac{1}{3}(144 \text{ in.}^2)(8 \text{ in.})$
$= 384 \text{ in.}^3$

Volume of cube $= (12 \text{ in.})(12 \text{ in.})(12 \text{ in.})$
$= 1728 \text{ in.}^3$

$384 \text{ in.}^3 + 1728 \text{ in.}^3 =$ **2112 cubic inches**

14. $AD = c - \mathbf{12}$

15. $\frac{x}{20} = \frac{20}{25}$

$$25x = 400$$
$$x = \mathbf{16}$$
$$y = 25 - 16$$
$$y = \mathbf{9}$$

16. $A \approx 4(3)(4 \text{ cm})^2$

$$\approx (12)(16 \text{ cm}^2)$$
$$\approx \mathbf{192 \text{ cm}^2}$$

17. $y - 2x + 5 = 1$

$$y - 2x + 2x + 5 = 1 + 2x$$
$$y + 5 = 1 + 2x$$
$$y + 5 - 5 = 1 + 2x - 5$$
$$\mathbf{y = 2x - 4}$$

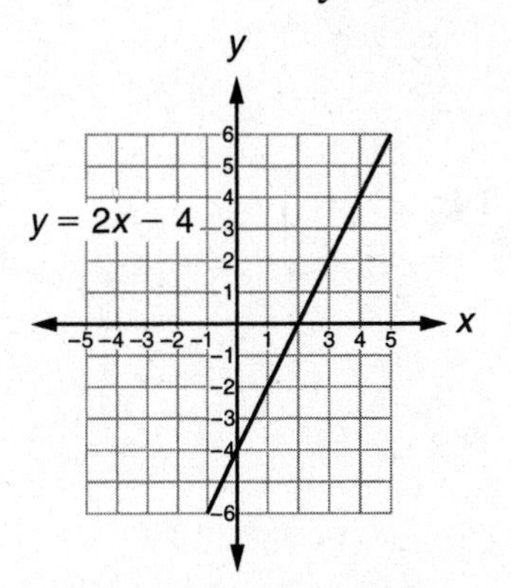

18. (a) $\frac{\text{rise}}{\text{run}} = \mathbf{\frac{1}{2}}$

(b) $\mathbf{-1}$

(c) $\mathbf{y = \frac{1}{2}x - 1}$

19. **See student work.**

20. $C \approx \frac{\frac{22}{7}(7 \text{ in.})}{2}$

$$C \approx \frac{22 \text{ in.}}{2}$$
$$C \approx 11 \text{ in.}$$

Perimeter $\approx 11 \text{ in.} + 4\frac{1}{4} \text{ in.} + 4\frac{1}{4} \text{ in.}$
$+ 11 \text{ in.} + 2 \text{ in.} + 2 \text{ in.}$
$\approx 26 \text{ in.} + 8\frac{2}{4} \text{ in.} \approx \mathbf{34\frac{1}{2} \text{ in.}}$

21. $\frac{1 \times 10^3}{1 \times 10^{-3}} = \mathbf{1 \times 10^6}$ **dimes**

22. (a) $\frac{x^2}{x} + \frac{x}{x}$

$x + 1$

$\mathbf{x(x + 1)}$

(b) $\frac{12m^2n^3}{6mn^2} + \frac{18mn^2}{6mn^2} - \frac{24m^2n^2}{6mn^2}$

$2mn + 3 - 4m$

$\mathbf{6mn^2(2mn + 3 - 4m)}$

23.

$$-2\frac{2}{3}w - 1\frac{1}{3} = 4$$
$$-2\frac{2}{3}w - 1\frac{1}{3} + 1\frac{1}{3} = 4 + 1\frac{1}{3}$$
$$-2\frac{2}{3}w = 5\frac{1}{3}$$
$$\left(-\frac{3}{8}\right)\left(-\frac{8}{3}w\right) = \left(-\frac{3}{8}\right)\left(\frac{16}{3}\right)$$
$$w = \mathbf{-2}$$

24.

$$5x^2 + 1 = 81$$
$$5x^2 + 1 - 1 = 81 - 1$$
$$5x^2 = 80$$
$$\frac{5x^2}{5} = \frac{80}{5}$$
$$x^2 = 16$$
$$x = \mathbf{4, -4}$$

25. $\left(\frac{1}{2}\right)^2 - 2^{-2} = \frac{1}{4} - \frac{1}{4} = \mathbf{0}$

26. $66\frac{2}{3}\%$ of $\frac{5}{6}$ of 0.144

$= 66\frac{2}{3}\%$ of 0.12

$= (0.66\bar{6})(0.12) = \mathbf{0.08}$ or $\mathbf{\frac{2}{25}}$

27. $[-3 + (-4)(-5)] - [-4 - (-5)(-2)]$

$[-3 + (20)] - [-4 - (10)]$

$[17] - [-14]$

31

28. $\frac{(5x^2yz)(6xy^2z)}{10\,xyz}$

$$\frac{\overset{1}{\cancel{2}} \cdot 3 \cdot \overset{1}{\cancel{5}} \cdot \overset{1}{\cancel{x}} \cdot x \cdot x \cdot \overset{1}{\cancel{y}} \cdot y \cdot y \cdot \overset{1}{\cancel{z}} \cdot z}{\underset{1}{\cancel{2}} \cdot \underset{1}{\cancel{5}} \cdot \underset{1}{\cancel{x}} \cdot \underset{1}{\cancel{y}} \cdot \underset{1}{\cancel{z}}}$$

$\mathbf{3x^2y^2z}$

29. $x(x + 2) + 2(x + 2)$

$x^2 + 2x + 2x + 4$

$\mathbf{x^2 + 4x + 4}$

30.

$$a^2 = (20\text{ mm})^2 + (10\text{ mm})^2$$
$$a^2 = 500\text{ mm}^2$$
$$a = \sqrt{500}\text{ mm}$$

22 mm and 23 mm

Lesson 119, Warm-Up

a. **27**

b. **1066**

c. **1**

d. $\mathbf{1 \times 10^{-4}}$

e. **0**

f. $\mathbf{500\text{ mm}^2}$

g. **$400**

h. **$400**

i. $\mathbf{\frac{1}{4}}$

Problem Solving

The perimeter of equilateral triangle ABC is 6 cm, which means each side is 2 cm. Using the Pythagorean theorem, we know that $(AC)^2 = (AD)^2 + (CD)^2$.

Substituting the information we know:

$$2^2 = (AD)^2 + 1^2$$
$$4 = (AD)^2 + 1$$
$$3 = (AD)^2$$
$$\mathbf{AD = \sqrt{3}\text{ cm}}$$

Lesson 119, Lesson Practice

a. **A typical error message display is [E 0]. Error messages vary.**

b. **0 ÷ 0 = 7 is not a fact, because division by zero is not possible.**

c. $\mathbf{w \neq 0}$

d. $\mathbf{x \neq 1}$

e. $\mathbf{w \neq 0}$

f. $y \neq 3$

g. $x \neq 2, -2$

h. $c \neq 0$

Lesson 119, Mixed Practice

1. $\frac{2}{50} = \frac{1}{25} = \mathbf{4\%}$

2. $\frac{100}{C} = \frac{\$20}{\$5}$
$\$20C = \$500$
$C = \mathbf{25\%}$

3. Number line from −4 to 3: closed dot at −3, open dot at 2, segment between them shaded.

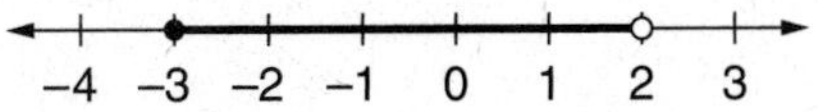

4. **360°**

5. (a) $\frac{1}{200}$
(b) **0.005**
(c) $0.\overline{8}$
(d) $88\frac{8}{9}\%$

6. (a) Right triangle with legs 10 cm and 10 cm
(b) **45°**
(c) **14 cm**

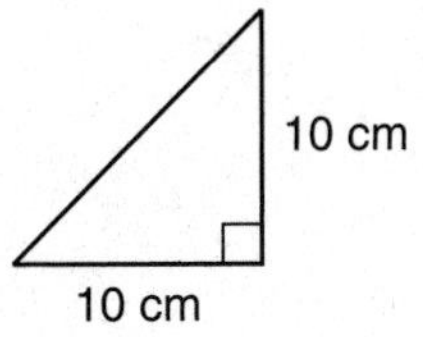

7. $\frac{(6 \times 10^5)(2 \times 10^6)}{(3 \times 10^4)} = \frac{12 \times 10^{11}}{3 \times 10^4}$
$3\overline{)12}$ = 4 $\mathbf{4 \times 10^7}$

8. (a) $\frac{2x^2}{x} + \frac{x}{x}$
$2x + 1$
$\mathbf{x(2x + 1)}$

(b) $\frac{3a^2b}{3a} - \frac{12a^2}{3a} + \frac{9ab^2}{3a}$
$ab - 4a + 3b^2$
$\mathbf{3a(ab - 4a + 3b^2)}$

9. Volume $= (3\text{ cm}^3) + 6(1\text{ cm}^3) + 9(1\text{ cm}^3)$
$= \mathbf{18\text{ cm}^3}$

10. Surface area $= 2(6\text{ cm}^2) + 6(3\text{ cm}^2) + 2(9\text{ cm}^2) = \mathbf{48\text{ cm}^2}$

11. $A = \frac{1}{2}bh$
$\left(\frac{2}{1}\right)A = \left(\frac{2}{1}\right)\frac{1}{2}bh$
$2A = bh$
$\frac{2A}{b} = \frac{bh}{b}$
$\mathbf{\frac{2A}{b} = h}$
$\frac{2(1.44\text{ m}^2)}{1.6\text{ m}} = h$
$\frac{2.88\text{ m}^2}{1.6\text{ m}} = h$
$\mathbf{1.8\text{ m}} = h$

12. $\frac{\text{boys}}{\text{total}} = \frac{3}{8} \times 100\% = \mathbf{37\frac{1}{2}\%}$

13. $(6\text{ ft})^2 + h^2 = (10\text{ ft})^2$
$36\text{ ft}^2 + h^2 = 100\text{ ft}^2$
$36\text{ ft}^2 + h^2 - 36\text{ ft}^2 = 100\text{ ft}^2 - 36\text{ ft}^2$
$h^2 = 64\text{ ft}^2$
$h = \mathbf{8\text{ ft}}$

14. (a) $\mathbf{y = 2x - 4}$
(b) $\mathbf{y = -\frac{1}{2}x + 1}$

15. **The product of the slopes is −1. The slopes are negative reciprocals.**

16.
$8000
× 1.06
$8480
× 1.06
$8988.80
× 1.06
$9528.1280
× 1.06
$10,099.81568 ⟶ **$10,099.82**

17. $1250\text{ sq. ft} \cdot \frac{1\text{ yd}}{3\text{ ft}} \cdot \frac{1\text{ yd}}{3\text{ ft}}$
≈ **139 square yards**

18. (a) $\mathbf{w \neq 0}$
(b) $\mathbf{m \neq -3}$

19. $DA = c - x$

20.
$$\begin{aligned}
\frac{y}{20} &= \frac{15}{25}\\
25y &= 300\\
y &= 12\text{ in.}\\
(12\text{ in.})^2 + z^2 &= (15\text{ in.})^2\\
144\text{ in.}^2 + z^2 &= 225\text{ in.}^2\\
144\text{ in.}^2 + z^2 - 144\text{ in.}^2 &= 225\text{ in.}^2 - 144\text{ in.}^2\\
z^2 &= 81\text{ in.}^2\\
z &= 9\text{ in.}\\
\text{Area} &= \frac{(12\text{ in.})(9\text{ in.})}{2} = \mathbf{54\text{ in.}^2}
\end{aligned}$$

21.
$$\begin{aligned}
\text{Volume} &\approx \frac{4}{3}(3.14)(15\text{ cm})^3\\
&\approx \frac{4}{3}(3.14)(3375\text{ cm}^3)\\
&\approx \mathbf{14{,}130\text{ cm}^3}
\end{aligned}$$

22. See student work.

23.
$$\begin{aligned}
\frac{2}{3}m + \frac{1}{4} &= \frac{7}{12}\\
\frac{2}{3}m + \frac{1}{4} - \frac{1}{4} &= \frac{7}{12} - \frac{1}{4}\\
\frac{2}{3}m &= \frac{7}{12} - \frac{3}{12}\\
\frac{2}{3}m &= \frac{4}{12}\\
\left(\frac{\cancel{3}^{1}}{\cancel{2}_{1}}\right)\left(\frac{\cancel{2}^{1}}{\cancel{3}_{1}}m\right) &= \left(\frac{\cancel{3}^{1}}{\cancel{2}_{1}}\right)\frac{\cancel{4}^{1}}{\cancel{12}_{\cancel{4}\,2}}\\
m &= \mathbf{\frac{1}{2}}
\end{aligned}$$

24.
$$\begin{aligned}
5(3 - x) &= 55\\
15 - 5x &= 55\\
15 - 5x - 15 &= 55 - 15\\
-5x &= 40\\
\frac{-5x}{-5} &= \frac{40}{-5}\\
x &= \mathbf{-8}
\end{aligned}$$

25.
$$\begin{aligned}
x + x + 12 &= 5x\\
2x + 12 &= 5x\\
2x + 12 - 12 &= 5x - 12\\
2x &= 5x - 12\\
2x - 5x &= 5x - 12 - 5x\\
-3x &= -12\\
\frac{-3x}{-3} &= \frac{-12}{-3}\\
x &= \mathbf{4}
\end{aligned}$$

26.
$$\begin{aligned}
10x^2 &= 100\\
\frac{10x^2}{10} &= \frac{100}{10}\\
x^2 &= 10\\
x &= \mathbf{\sqrt{10}, -\sqrt{10}}
\end{aligned}$$

27. 300

28.
$$\begin{aligned}
&x(x + 5) - 2(x + 5)\\
&x^2 + 5x - 2x - 10\\
&\mathbf{x^2 + 3x - 10}
\end{aligned}$$

29.
$$\frac{(12xy^2z)(9x^2y^2z)}{36xyz^2}$$

$$\frac{\cancel{2}\cdot\cancel{2}\cdot\cancel{3}\cdot\cancel{3}\cdot 3\cdot\cancel{x}\cdot x\cdot x\cdot\cancel{y}\cdot y\cdot y\cdot y\cdot\cancel{z}\cdot\cancel{z}}{\cancel{2}\cdot\cancel{2}\cdot\cancel{3}\cdot\cancel{3}\cdot\cancel{x}\cdot\cancel{y}\cdot\cancel{z}\cdot\cancel{z}}$$

$\mathbf{3x^2y^3}$

30.
$$\begin{aligned}
&33\tfrac{1}{3}\%\text{ of }0.12\text{ of }3\tfrac{1}{3}\\
&= \left(\frac{1}{3}\right) \times \left(\frac{\cancel{12}^{4}}{\cancel{100}_{10}} \cdot \frac{\cancel{10}^{1}}{\cancel{3}_{1}}\right)\\
&= \frac{1}{3} \times \frac{4}{10} = \mathbf{\frac{2}{15}}\text{ or }\mathbf{0.1\overline{3}}
\end{aligned}$$

LESSON 120, WARM-UP

a. **65**

b. **1969**

c. **1**

d. **2.5×10^{-9}**

e. **4, –4**

f. **32°F**

g. **$25**

h. **$275**

i. **1**

Problem Solving

In Lesson 113 we learned that the volume of a cone is $\frac{1}{3}$ the volume of a cylinder with the same height and diameter. Therefore, it will take **3 cones** of water to fill the beaker.

Lesson 120, Lesson Practice

a. **See student work.**

b. **See student work.**

c. **See student work.**

d. **See student work.**

Lesson 120, Mixed Practice

1. $P(\text{sum of } 7) = \frac{1}{6}$; **1 to 5**

2. $\frac{\$2.70}{1.08} = \mathbf{\$2.50}$

3. **Insufficient information**

4. **100°; 80°; 80°**

5. (a) $\mathbf{\frac{1}{1000}}$
   (b) **0.001**
   (c) **1.6**
   (d) **160%**

6. (a)
$$\begin{aligned} a^2 + (1 \text{ cm})^2 &= (2 \text{ cm})^2 \\ a^2 + 1 \text{ cm}^2 &= 4 \text{ cm}^2 \\ a^2 + 1 \text{ cm}^2 - 1 \text{ cm}^2 &= 4 \text{ cm}^2 - 1 \text{ cm}^2 \\ a^2 &= 3 \text{ cm}^2 \\ a &= \mathbf{\sqrt{3} \text{ cm}} \end{aligned}$$
   (b) **1.7 cm**

7. $\frac{(4 \times 10^{-5})(6 \times 10^{-4})}{8 \times 10^3} = \frac{24 \times 10^{-9}}{8 \times 10^3}$

   $8\overline{)24}$ = 3 $\quad \mathbf{3 \times 10^{-12}}$

8. (a) $\frac{3y^2}{y} - \frac{y}{y}$
   $3y - 1$
   $\mathbf{y(3y - 1)}$

   (b) $\frac{6w^2}{3w} + \frac{9wx}{3w} - \frac{12w}{3w}$
   $2w + 3x - 4$
   $\mathbf{3w(2w + 3x - 4)}$

9. $\mathbf{\frac{1}{3}}$

10.
$$\begin{aligned} C &\approx (3.14)(6 \text{ cm}) \\ C &\approx 18.84 \text{ cm} \\ A &\approx (18.84 \text{ cm})(6 \text{ cm}) \\ &\approx 113.04 \text{ cm}^2 \approx \mathbf{113 \text{ cm}^2} \end{aligned}$$

11.
$$\begin{aligned} E &= mc^2 \\ \frac{E}{c^2} &= \frac{mc^2}{c^2} \\ \mathbf{\frac{E}{c^2}} &\mathbf{= m} \end{aligned}$$

12. $\frac{60}{100} = \frac{3}{5} = \frac{\text{girls}}{\text{total}}$
   $\frac{\text{boys}}{\text{girls}} = \mathbf{\frac{2}{3}}$

13. line m: $\mathbf{y = -\frac{2}{3}x + 2}$

   line n: $\mathbf{y = \frac{3}{2}x - 2}$

14. **The product of the slopes is −1. The slopes are negative reciprocals.**

15.
$$\begin{array}{r} \$1000 \\ \times \quad 1.2 \\ \hline \$1200 \end{array} \text{ Year 1}$$
$$\begin{array}{r} \times \quad 1.2 \\ \hline \$1440 \end{array} \text{ Year 2}$$
$$\begin{array}{r} \times \quad 1.2 \\ \hline \$1728 \end{array} \text{ Year 3}$$
$$\begin{array}{r} \times \quad 1.2 \\ \hline \$2073.60 \end{array} \text{ Year 4}$$
   About 4 years

16.
$$\begin{aligned} d^2 &= (17 \text{ in.})^2 + (12 \text{ in.})^2 \\ d^2 &= 289 \text{ in.}^2 + 144 \text{ in.}^2 \\ d^2 &= 433 \text{ in.}^2 \\ d &= \sqrt{433 \text{ in.}^2} \\ d &\approx \mathbf{21 \text{ in.}} \end{aligned}$$

17. (a) Volume $= (36 \text{ feet})(21 \text{ feet})\left(\frac{1}{2} \text{ feet}\right)$
   $= \mathbf{378 \text{ cubic feet}}$

   (b) $\overset{14}{\cancel{378}} \text{ ft}^3 \cdot \frac{1 \text{ yd}}{\underset{1}{\cancel{3}} \cancel{\text{ft}}} \cdot \frac{1 \text{ yd}}{\underset{1}{\cancel{3}} \cancel{\text{ft}}} \cdot \frac{1 \text{ yd}}{\underset{1}{\cancel{3}} \cancel{\text{ft}}}$
   $= \mathbf{14 \text{ cubic yards}}$

18. (a) $\boldsymbol{m \neq 2}$

(b) $\boldsymbol{y \neq -5}$

19.

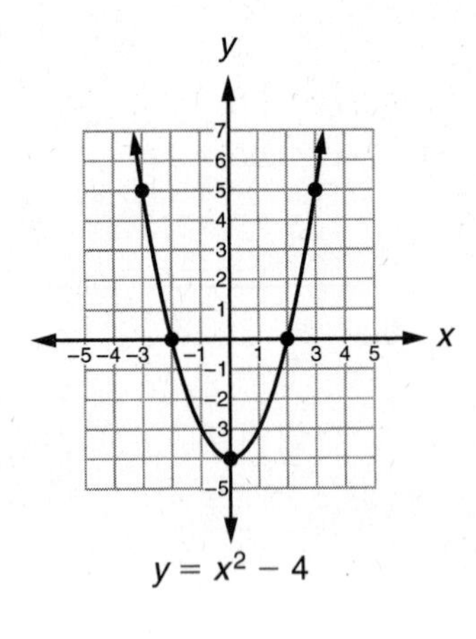

20. $\frac{c}{a} = \frac{a}{\boldsymbol{y}}$

21. Surface area $\approx 4(3.14)(3 \text{ in.})^2$
$\approx 4(3.14)(9 \text{ in.}^2)$
$\approx 113.04 \text{ in.}^2$
$\approx \mathbf{113 \text{ in.}^2}$

22. 250 cubic centimeters = 250 mL = **0.25 liter**

23.
$$\begin{aligned} 15 + x &= 3x - 17 \\ 15 + x - 3x &= 3x - 17 - 3x \\ 15 - 2x &= -17 \\ 15 - 2x - 15 &= -17 - 15 \\ -2x &= -32 \\ \frac{-2x}{-2} &= \frac{-32}{-2} \\ x &= \mathbf{16} \end{aligned}$$

24.
$$\begin{aligned} 3\tfrac{1}{3}x - 16 &= 74 \\ 3\tfrac{1}{3}x - 16 + 16 &= 74 + 16 \\ 3\tfrac{1}{3}x &= 90 \\ \left(\frac{\overset{1}{\cancel{3}}}{\underset{1}{\cancel{10}}}\right)\left(\frac{\overset{1}{\cancel{10}}}{\underset{1}{\cancel{3}}}x\right) &= \left(\frac{3}{\underset{1}{\cancel{10}}}\right)\overset{9}{\cancel{90}} \\ x &= \mathbf{27} \end{aligned}$$

25.
$$\begin{aligned} \frac{m^2}{4} &= 9 \\ m^2 &= 36 \\ m &= \mathbf{6, -6} \end{aligned}$$

26.
$$\begin{aligned} \frac{1.2}{m} &= \frac{0.04}{8} \\ 0.04m &= 9.6 \\ \frac{0.04m}{0.04} &= \frac{9.6}{0.04} \\ m &= \mathbf{240} \end{aligned}$$

27.
$$\begin{aligned} &x(x - 5) - 2(x - 5) \\ &x^2 - 5x - 2x + 10 \\ &\mathbf{x^2 - 7x + 10} \end{aligned}$$

28.
$$\frac{(3xy)(4x^2y)(5x^2y^2)}{10x^3y^3}$$

$$\frac{\overset{1}{\cancel{2}} \cdot 2 \cdot 3 \cdot \overset{1}{\cancel{5}} \cdot \overset{1}{\cancel{x}} \cdot \overset{1}{\cancel{x}} \cdot \overset{1}{\cancel{x}} \cdot x \cdot x \cdot \overset{1}{\cancel{y}} \cdot \overset{1}{\cancel{y}} \cdot \overset{1}{\cancel{y}} \cdot y}{\underset{1}{\cancel{2}} \cdot \underset{1}{\cancel{5}} \cdot \underset{1}{\cancel{x}} \cdot \underset{1}{\cancel{x}} \cdot \underset{1}{\cancel{x}} \cdot \underset{1}{\cancel{y}} \cdot \underset{1}{\cancel{y}} \cdot \underset{1}{\cancel{y}}}$$

$\mathbf{6x^2y}$

29.
$$\begin{aligned} &|-8| + 3(-7) - [(-4)(-5) - 3(-2)] \\ &8 + (-21) - [(20) - (-6)] \\ &8 + (-21) - [26] \\ &\mathbf{-39} \end{aligned}$$

30.
$$\begin{aligned} \frac{7\frac{1}{2} - \frac{2}{3}(0.9)}{0.03} &= \frac{7\frac{1}{2} - \frac{\overset{1}{\cancel{2}}}{\underset{1}{\cancel{3}}}\left(\frac{\overset{3}{\cancel{9}}}{\underset{5}{\cancel{10}}}\right)}{0.03} \\ &= \frac{7\frac{1}{2} - \frac{3}{5}}{0.03} = \frac{7\frac{5}{10} - \frac{6}{10}}{0.03} \\ &= \frac{6\frac{15}{10} - \frac{6}{10}}{0.03} = \frac{6\frac{9}{10}}{0.03} \\ &= \frac{6.9}{0.03} = \mathbf{230} \end{aligned}$$

INVESTIGATION 12

1.

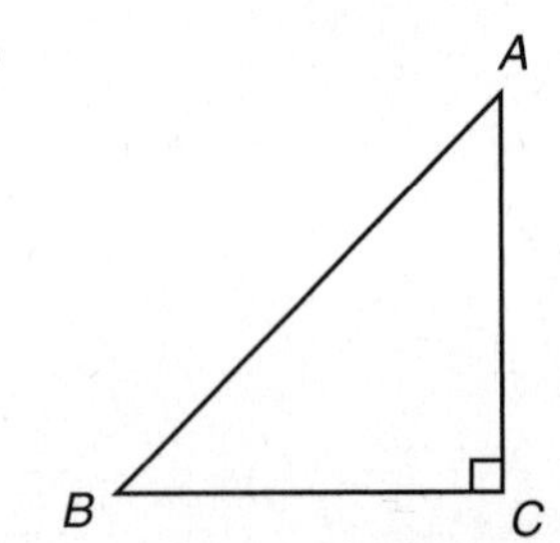

2.

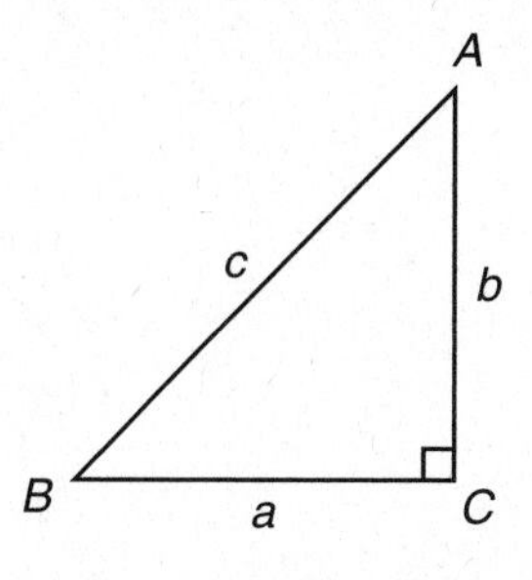

3.

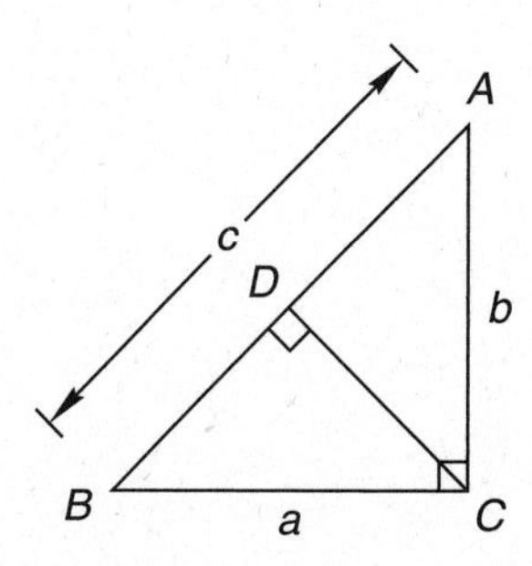

4.

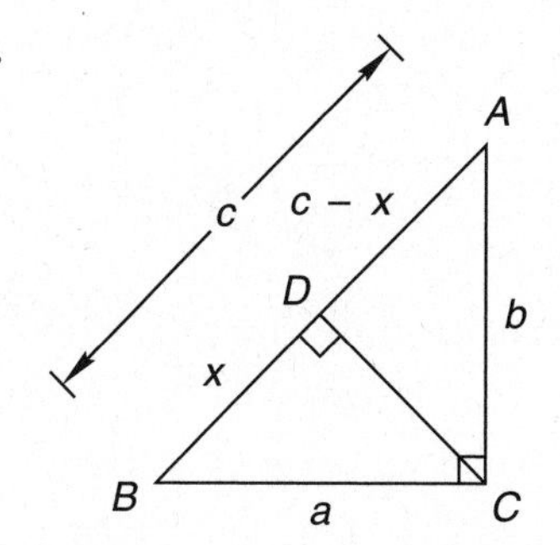

5. **90°; The sum of the measures of all three angles is 180°. The right angle of the triangle removes 90° from this total, leaving 90° to be shared by the remaining two acute angles.**

6. The measures of angle A and angle B should add to 90°. If angle B measures m degrees, then angle A measures **$90° - m$.**

7. (a) **$90° - m$**

   (b) **m**

8. **All three triangles are similar because each triangle has degree angle measures of 90, m, and $90 - m$. Since their corresponding angle measures are equal, their angles are congruent and the triangles are similar.**

9. $\frac{c}{a} = \frac{a}{x}$

10. $\frac{c}{b} = \frac{b}{c - x}$

11. From 9:

$$\frac{c}{a} = \frac{a}{x}$$
$$cx = a^2$$
or
$$a^2 = cx$$

From 10:

$$\frac{c}{b} = \frac{b}{c - x}$$
$$c^2 - cx = b^2$$
or
$$b^2 = c^2 - cx$$

12. **$c^2 - a^2 = b^2$ or $b^2 = c^2 - a^2$**

13. **$c^2 = a^2 + b^2$ or $a^2 + b^2 = c^2$; Pythagorean theorem**

Activity

One possibility:

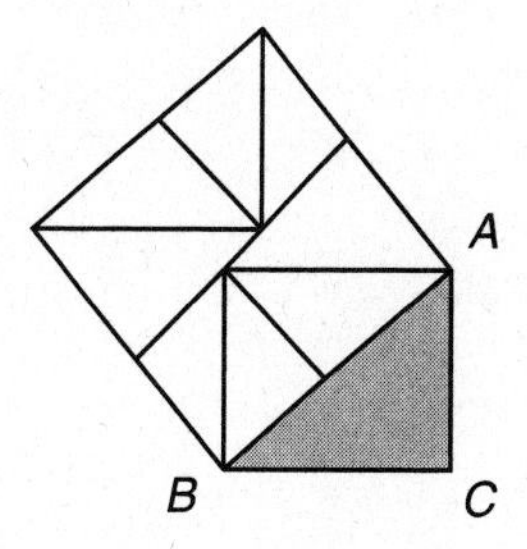

Solutions for

Appendix Topic

Topic A

a. 7

b. 8

c. 12

d. 17

e. 39

f. 64

g. 1919

h. 2002

Solutions for
Supplemental Practice

Supplemental Practice, Lesson 3

1. 26
2. 66
3. 12
4. 128
5. 54
6. 18
7. 8
8. 18
9. 9
10. 54
11. 6
12. 48
13. 7
14. 9

Supplemental Practice, Lesson 6

1. 1, 2, 3, 4, 6, 9
2. 1, 2, 3, 4, 5, 6, 8, 9, 10
3. 1, 2, 5, 7, 10
4. 1, 2, 3, 6
5. 1, 2, 3, 4, 5, 6, 8, 9, 10
6. 1, 2, 5, 10
7. 1, 2, 3, 4, 5, 6, 8, 10
8. 1, 2, 4, 5, 7, 10
9. 1, 2, 3, 5, 6, 9, 10
10. 1, 2, 3, 5, 6, 10
11. 1, 2, 4, 5, 8, 10
12. 1, 2, 3, 4, 5, 6, 7, 8, 9, 10

Supplemental Practice, Lesson 15

1. $\frac{3}{4}$
2. $\frac{1}{3}$
3. $\frac{3}{8}$
4. $\frac{2}{3}$
5. $\frac{4}{5}$
6. $\frac{1}{2}$
7. $\frac{2}{3}$
8. $\frac{4}{5}$
9. $3\frac{5}{6}$
10. $6\frac{3}{4}$
11. $8\frac{3}{5}$
12. $4\frac{9}{16}$

Supplemental Practice, Lesson 19

1. 64 in.
2. 58 in.
3. 82 in.
4. 86 in.

Supplemental Practice, Lesson 20

1. 64
2. 64

3. 27

4. 100,000

5. 17

6. 9

7. 64

8. 225

9. 10

10. 8

11. 500

12. 625

13. 9

14. 11

15. 7

16. 12

17. 30

18. 25

19. 14

20. 21

Supplemental Practice, Lesson 21

1. $3 \times 3 \times 3 \times 3$

2. $2 \times 2 \times 3 \times 5 \times 5$

3. $2 \times 2 \times 2 \times 2 \times 5 \times 5 \times 5$

4. $5 \times 5 \times 5 \times 5$

5. $2 \times 3 \times 3 \times 5 \times 5$

6. $2 \times 2 \times 2 \times 2 \times 3 \times 5 \times 5$

7. $2 \times 2 \times 2 \times 5 \times 11$

8. $2 \times 3 \times 5 \times 5 \times 5$

9. $2 \times 2 \times 2 \times 2 \times 5 \times 5 \times 5 \times 5$

10. $2 \times 2 \times 2 \times 2 \times 2 \times 2 \times 2$

11. $2 \times 2 \times 3 \times 5 \times 13$

12. $2 \times 2 \times 5 \times 7 \times 11$

Supplemental Practice, Lesson 23

1. $8\frac{2}{5}$

2. $8\frac{3}{4}$

3. 6

4. $9\frac{1}{2}$

5. $12\frac{1}{2}$

6. $11\frac{1}{3}$

7. $4\frac{3}{4}$

8. $1\frac{3}{4}$

9. $3\frac{2}{5}$

10. $3\frac{2}{3}$

11. $2\frac{3}{5}$

12. $3\frac{1}{3}$

Supplemental Practice, Lesson 26

1. $1\frac{1}{2}$

2. 7

3. 6

Supplemental Practice, Lesson 45

1. 0.3
2. 240
3. 45
4. 6.25
5. 312.5
6. 240
7. 430
8. 12.5
9. 5
10. 1.8
11. 200
12. 25

Supplemental Practice, Lesson 48

1. $0.8\overline{3}$
2. $83\frac{1}{3}\%$
3. $1\frac{1}{5}$
4. 120%
5. $\frac{2}{25}$
6. 0.08
7. 1.6
8. 160%
9. $\frac{3}{40}$
10. $7\frac{1}{2}\%$
11. $1\frac{1}{4}$
12. 1.25

Supplemental Practice, Lesson 49

1. 3 feet 4 inches
2. 3 minutes 20 seconds
3. 4 ft 9 in.
4. 3 hr 30 min
5. 5 yd 1 ft 3 in.
6. 8 hr 4 min 11 s
7. 12 lb 2 oz
8. 6 gal 2 qt

Supplemental Practice, Lesson 50

1. 288 in.
2. 8 yd
3. 5 hr
4. 18,000 s
5. 5 m
6. 5000 mm
7. 1600 oz
8. $\frac{1}{20}$ ton

Supplemental Practice, Lesson 52

1. 19
2. 33
3. 55
4. 15
5. 11
6. 1
7. 52

8. 29

9. 46

10. $\frac{3}{8}$

11. 0.55

12. -2

Supplemental Practice, Lesson 56

1. 1 ft 9 in.

2. 6 min 45 s

3. 3 yd 10 in.

4. 45 min 20 s

5. 6 yd 2 ft 7 in.

6. 28 min 47 s

Supplemental Practice, Lesson 57

1. $\frac{1}{16}$

2. $\frac{1}{8}$

3. 1

4. 3

5. $\frac{1}{32}$

6. $\frac{1}{2}$

7. $\frac{1}{10}$

8. $\frac{1}{9}$

9. 0.001

10. 2^2 or 4

Supplemental Practice, Lesson 60

1. $W_N = \frac{3}{4} \times 24$; 18

2. $\frac{3}{5} \times 60 = W_N$; 36

3. $W_N = 0.4 \times 80$; 32

4. $0.6 \times 60 = W_N$; 36

5. $W_N = 0.3 \times 120$; 36

6. $0.06 \times 250 = W_N$; 15

7. $W_N = \frac{5}{6} \times 300$; 250

8. $\frac{2}{3} \times 90 = W_N$; 60

9. $W_N = 0.5 \times 50$; 25

10. $0.7 \times 140 = W_N$; 98

11. $W_N = 0.75 \times 400$; 300

12. $0.8 \times 400 = W_N$; 320

Supplemental Practice, Lesson 64

1. 18

2. -41

3. -3

4. 13

5. -23

6. 26

7. 0

8. 5

9. $-5\frac{3}{4}$

10. $1\frac{1}{2}$

11. -6.7

12. -1.67

Supplemental Practice, Lesson 66

1. 125.6 cm
2. 44 cm
3. 12π cm
4. 31.4 cm
5. 88 cm
6. 15π cm

Supplemental Practice, Lesson 68

1. 5
2. 8
3. 23
4. −12
5. −2
6. 2
7. 5
8. 3
9. 9
10. 11
11. 4
12. −8

Supplemental Practice, Lesson 69

1. 1.5×10^{6}
2. 4.8×10^{-7}
3. 2×10^{6}
4. 7.2×10^{-5}
5. 1.25×10^{11}
6. 2.25×10^{-5}
7. 1.75×10^{11}
8. 3.75×10^{-9}

Supplemental Practice, Lesson 75

1. 94 cm^2
2. 108 cm^2
3. 160 cm^2
4. 168 cm^2

Supplemental Practice, Lesson 77

1. $W_P \times 75 = 60$; 80%
2. $60 = 0.75 \times W_N$; 80
3. $30 = W_P \times 90$; $33\frac{1}{3}\%$
4. $30 = 1.5 \times W_N$; 20
5. $W_P \times 40 = 50$; 125%
6. $0.2 \times W_N = 50$; 250
7. $W_P \times \$5.00 = \$3.50$; 70%
8. $12 = \frac{2}{3} \times W_N$; 18

Supplemental Practice, Lesson 82

1. 1256 cm^2
2. 616 cm^2
3. 16π cm^2
4. 314 cm^2
5. 154 cm^2
6. 64π cm^2

Supplemental Practice, Lesson 83

1. 3.6×10^{11}
2. 1.8×10^{10}
3. 1.05×10^{21}
4. 1.0×10^{13}
5. 8×10^{-11}
6. 2.4×10^{-11}
7. 1.3×10^{-11}
8. 2.4×10^{5}
9. 1.12×10^{-11}
10. 2.1×10^{-3}
11. 1.12×10^{3}
12. 3×10^{-1}

Supplemental Practice, Lesson 93

1. 15
2. 17
3. 24
4. 24
5. 9
6. 14
7. 3
8. 6
9. 96
10. 27
11. 3
12. −9

Supplemental Practice, Lesson 101

1. $2n + 6 = 72$; 33
2. $8n - 5 = 27$; 4
3. $n - 10 = 50$; 120
4. $n = (6 \times 4) + 12$; 36
5. $n + 6 = 12 - 5$; 1
6. $\frac{3}{4}n = 60 - 12$; 64

Supplemental Practice, Lesson 102

1. 12
2. 30
3. 16
4. 8
5. 40°
6. 90°

Supplemental Practice, Lesson 111

1. 2×10^{4}
2. 2×10^{-3}
3. 1.8×10^{-6}
4. 4×10^{3}
5. 3×10^{4}
6. 7.5×10^{1}
7. 6×10^{-4}
8. 5×10^{-5}
9. 3×10^{-5}
10. 1.5×10^{5}
11. 7×10^{3}
12. 8×10^{-5}

Solutions for
Facts Practice Tests

64 Multiplication Facts

Multiply.

<table>
<tr><td>6
× 8
48</td><td>5
× 7
35</td><td>3
× 3
9</td><td>6
× 2
12</td><td>4
× 7
28</td><td>9
× 3
27</td><td>8
× 5
40</td><td>2
× 4
8</td></tr>
<tr><td>7
× 2
14</td><td>4
× 5
20</td><td>8
× 2
16</td><td>8
× 6
48</td><td>2
× 9
18</td><td>5
× 6
30</td><td>9
× 7
63</td><td>4
× 9
36</td></tr>
<tr><td>8
× 9
72</td><td>7
× 9
63</td><td>2
× 6
12</td><td>3
× 8
24</td><td>7
× 8
56</td><td>9
× 6
54</td><td>3
× 2
6</td><td>6
× 7
42</td></tr>
<tr><td>5
× 2
10</td><td>3
× 7
21</td><td>8
× 7
56</td><td>6
× 3
18</td><td>2
× 2
4</td><td>7
× 7
49</td><td>9
× 8
72</td><td>4
× 3
12</td></tr>
<tr><td>7
× 6
42</td><td>8
× 8
64</td><td>4
× 8
32</td><td>3
× 5
15</td><td>8
× 3
24</td><td>9
× 5
45</td><td>2
× 7
14</td><td>5
× 8
40</td></tr>
<tr><td>6
× 6
36</td><td>2
× 3
6</td><td>4
× 4
16</td><td>5
× 3
15</td><td>9
× 9
81</td><td>3
× 9
27</td><td>8
× 4
32</td><td>7
× 3
21</td></tr>
<tr><td>4
× 6
24</td><td>7
× 5
35</td><td>3
× 6
18</td><td>6
× 9
54</td><td>5
× 4
20</td><td>9
× 4
36</td><td>2
× 5
10</td><td>6
× 4
24</td></tr>
<tr><td>5
× 9
45</td><td>3
× 4
12</td><td>6
× 5
30</td><td>2
× 8
16</td><td>7
× 4
28</td><td>4
× 2
8</td><td>5
× 5
25</td><td>9
× 2
18</td></tr>
</table>

B 30 Equations

Find the value of each variable.

$a + 12 = 20$ $a = \mathbf{8}$	$b - 8 = 10$ $b = \mathbf{18}$	$5c = 40$ $c = \mathbf{8}$
$\frac{d}{4} = 12$ $d = \mathbf{48}$	$11 + e = 24$ $e = \mathbf{13}$	$25 - f = 10$ $f = \mathbf{15}$
$10g = 60$ $g = \mathbf{6}$	$\frac{24}{h} = 6$ $h = \mathbf{4}$	$17 = j + 8$ $j = \mathbf{9}$
$20 = k - 5$ $k = \mathbf{25}$	$30 = 6m$ $m = \mathbf{5}$	$9 = \frac{n}{3}$ $n = \mathbf{27}$
$18 = 6 + p$ $p = \mathbf{12}$	$5 = 15 - q$ $q = \mathbf{10}$	$36 = 4r$ $r = \mathbf{9}$
$2 = \frac{16}{s}$ $s = \mathbf{8}$	$5 + 7 + t = 20$ $t = \mathbf{8}$	$u - 15 = 30$ $u = \mathbf{45}$
$8v = 48$ $v = \mathbf{6}$	$\frac{w}{3} = 6$ $w = \mathbf{18}$	$21 - x = 12$ $x = \mathbf{9}$
$y + 8 = 12$ $y = \mathbf{4}$	$36 = 3z$ $z = \mathbf{12}$	$\frac{48}{a} = 4$ $a = \mathbf{12}$
$b - 12 = 15$ $b = \mathbf{27}$	$75 = 3c$ $c = \mathbf{25}$	$\frac{d}{12} = 6$ $d = \mathbf{72}$
$36 = f + 24$ $f = \mathbf{12}$	$g - 24 = 24$ $g = \mathbf{48}$	$12h = 12$ $h = \mathbf{1}$

C

30 Improper Fractions and Mixed Numbers

Write each improper fraction as a mixed number or a whole number.

$\frac{5}{2} = \mathbf{2\frac{1}{2}}$	$\frac{6}{3} = \mathbf{2}$	$\frac{7}{4} = \mathbf{1\frac{3}{4}}$	$\frac{12}{5} = \mathbf{2\frac{2}{5}}$	$\frac{8}{2} = \mathbf{4}$
$\frac{10}{3} = \mathbf{3\frac{1}{3}}$	$\frac{15}{2} = \mathbf{7\frac{1}{2}}$	$\frac{21}{4} = \mathbf{5\frac{1}{4}}$	$\frac{15}{5} = \mathbf{3}$	$\frac{11}{8} = \mathbf{1\frac{3}{8}}$
$2\frac{3}{2} = \mathbf{3\frac{1}{2}}$	$4\frac{5}{4} = \mathbf{5\frac{1}{4}}$	$3\frac{6}{2} = \mathbf{6}$	$3\frac{7}{4} = \mathbf{4\frac{3}{4}}$	$6\frac{5}{2} = \mathbf{8\frac{1}{2}}$

Write each mixed number as an improper fraction.

$1\frac{1}{2} = \mathbf{\frac{3}{2}}$	$2\frac{2}{3} = \mathbf{\frac{8}{3}}$	$3\frac{3}{4} = \mathbf{\frac{15}{4}}$	$2\frac{1}{2} = \mathbf{\frac{5}{2}}$	$4\frac{1}{5} = \mathbf{\frac{21}{5}}$
$6\frac{2}{3} = \mathbf{\frac{20}{3}}$	$2\frac{3}{4} = \mathbf{\frac{11}{4}}$	$3\frac{1}{3} = \mathbf{\frac{10}{3}}$	$4\frac{1}{2} = \mathbf{\frac{9}{2}}$	$2\frac{4}{5} = \mathbf{\frac{14}{5}}$
$1\frac{5}{6} = \mathbf{\frac{11}{6}}$	$5\frac{3}{4} = \mathbf{\frac{23}{4}}$	$1\frac{7}{8} = \mathbf{\frac{15}{8}}$	$3\frac{1}{6} = \mathbf{\frac{19}{6}}$	$2\frac{3}{10} = \mathbf{\frac{23}{10}}$

40 Fractions to Reduce

Reduce each fraction to lowest terms.

$\frac{60}{100} = \mathbf{\frac{3}{5}}$	$\frac{2}{12} = \mathbf{\frac{1}{6}}$	$\frac{4}{16} = \mathbf{\frac{1}{4}}$	$\frac{2}{6} = \mathbf{\frac{1}{3}}$	$\frac{5}{10} = \mathbf{\frac{1}{2}}$
$\frac{50}{100} = \mathbf{\frac{1}{2}}$	$\frac{2}{16} = \mathbf{\frac{1}{8}}$	$\frac{8}{12} = \mathbf{\frac{2}{3}}$	$\frac{5}{100} = \mathbf{\frac{1}{20}}$	$\frac{3}{9} = \mathbf{\frac{1}{3}}$
$\frac{8}{16} = \mathbf{\frac{1}{2}}$	$\frac{2}{100} = \mathbf{\frac{1}{50}}$	$\frac{20}{100} = \mathbf{\frac{1}{5}}$	$\frac{6}{8} = \mathbf{\frac{3}{4}}$	$\frac{10}{100} = \mathbf{\frac{1}{10}}$
$\frac{2}{4} = \mathbf{\frac{1}{2}}$	$\frac{4}{10} = \mathbf{\frac{2}{5}}$	$\frac{90}{100} = \mathbf{\frac{9}{10}}$	$\frac{3}{12} = \mathbf{\frac{1}{4}}$	$\frac{6}{16} = \mathbf{\frac{3}{8}}$
$\frac{80}{100} = \mathbf{\frac{4}{5}}$	$\frac{9}{12} = \mathbf{\frac{3}{4}}$	$\frac{3}{6} = \mathbf{\frac{1}{2}}$	$\frac{12}{16} = \mathbf{\frac{3}{4}}$	$\frac{4}{8} = \mathbf{\frac{1}{2}}$
$\frac{6}{9} = \mathbf{\frac{2}{3}}$	$\frac{25}{100} = \mathbf{\frac{1}{4}}$	$\frac{4}{12} = \mathbf{\frac{1}{3}}$	$\frac{6}{10} = \mathbf{\frac{3}{5}}$	$\frac{40}{100} = \mathbf{\frac{2}{5}}$
$\frac{4}{100} = \mathbf{\frac{1}{25}}$	$\frac{2}{10} = \mathbf{\frac{1}{5}}$	$\frac{10}{16} = \mathbf{\frac{5}{8}}$	$\frac{10}{12} = \mathbf{\frac{5}{6}}$	$\frac{4}{6} = \mathbf{\frac{2}{3}}$
$\frac{14}{16} = \mathbf{\frac{7}{8}}$	$\frac{2}{8} = \mathbf{\frac{1}{4}}$	$\frac{6}{12} = \mathbf{\frac{1}{2}}$	$\frac{8}{10} = \mathbf{\frac{4}{5}}$	$\frac{75}{100} = \mathbf{\frac{3}{4}}$

E

Circles

Write the word that completes each sentence.

1. The distance around a circle is its **circumference**.
2. Every point on a circle is the same distance from the **center**.
3. The distance across a circle through its center is its **diameter**.
4. The distance from a circle to its center is its **radius**.
5. Two or more circles with the same center are **concentric** circles.
6. A segment between two points on a circle is a **chord**.
7. Part of a circumference is an **arc**.
8. A portion of a circle and its interior, bound by an arc and two radii, is a **sector**.
9. Half of a circle is a **semicircle**.
10. An angle whose vertex is the center of a circle is a **central** angle.
11. An angle whose vertex is on the circumference of a circle and whose sides include chords of the circle is an **inscribed** angle.
12. A polygon whose vertices are on a circle and whose edges are within the circle is an **inscribed** polygon.

Illustrate answers 1–12 below.

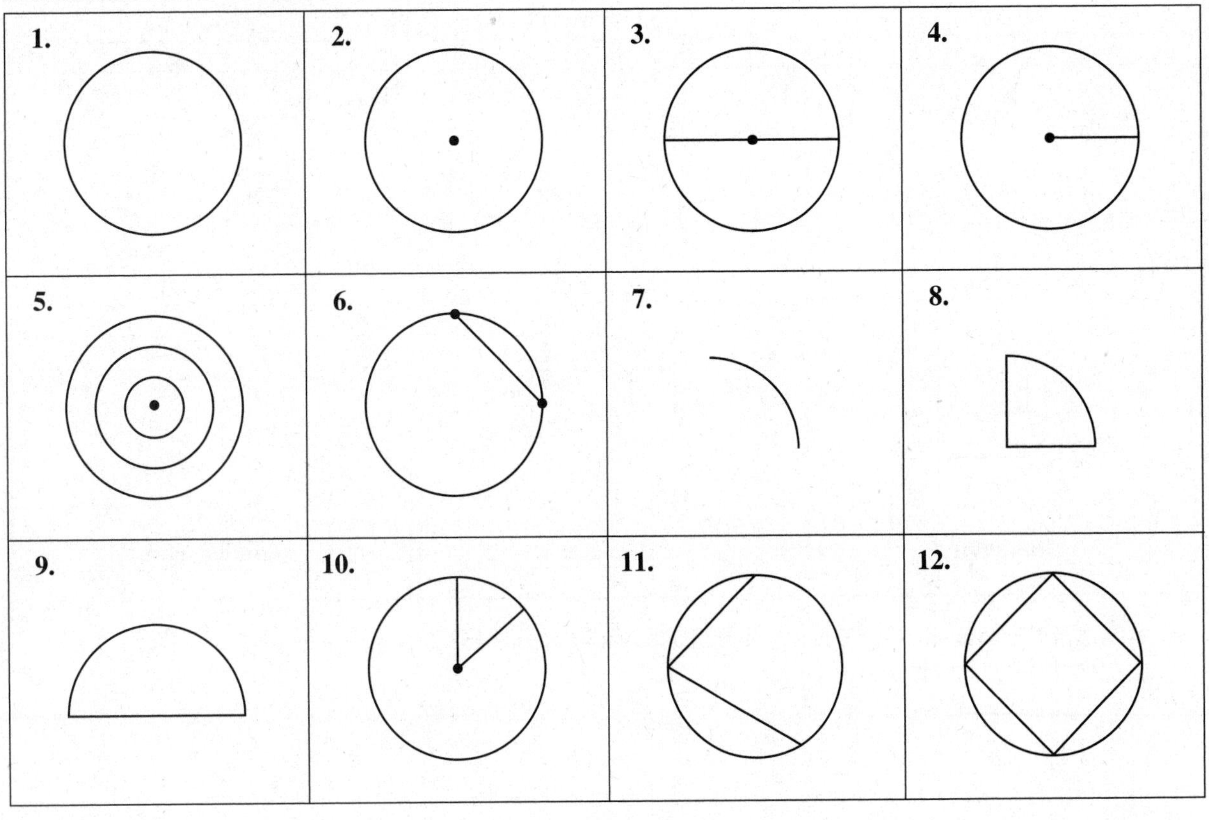

1.	2.	3.	4.
5.	6.	7.	8.
9.	10.	11.	12.

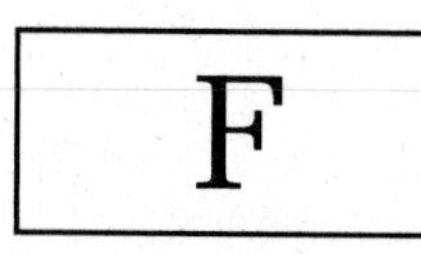

Lines, Angles, Polygons

Name each figure illustrated.

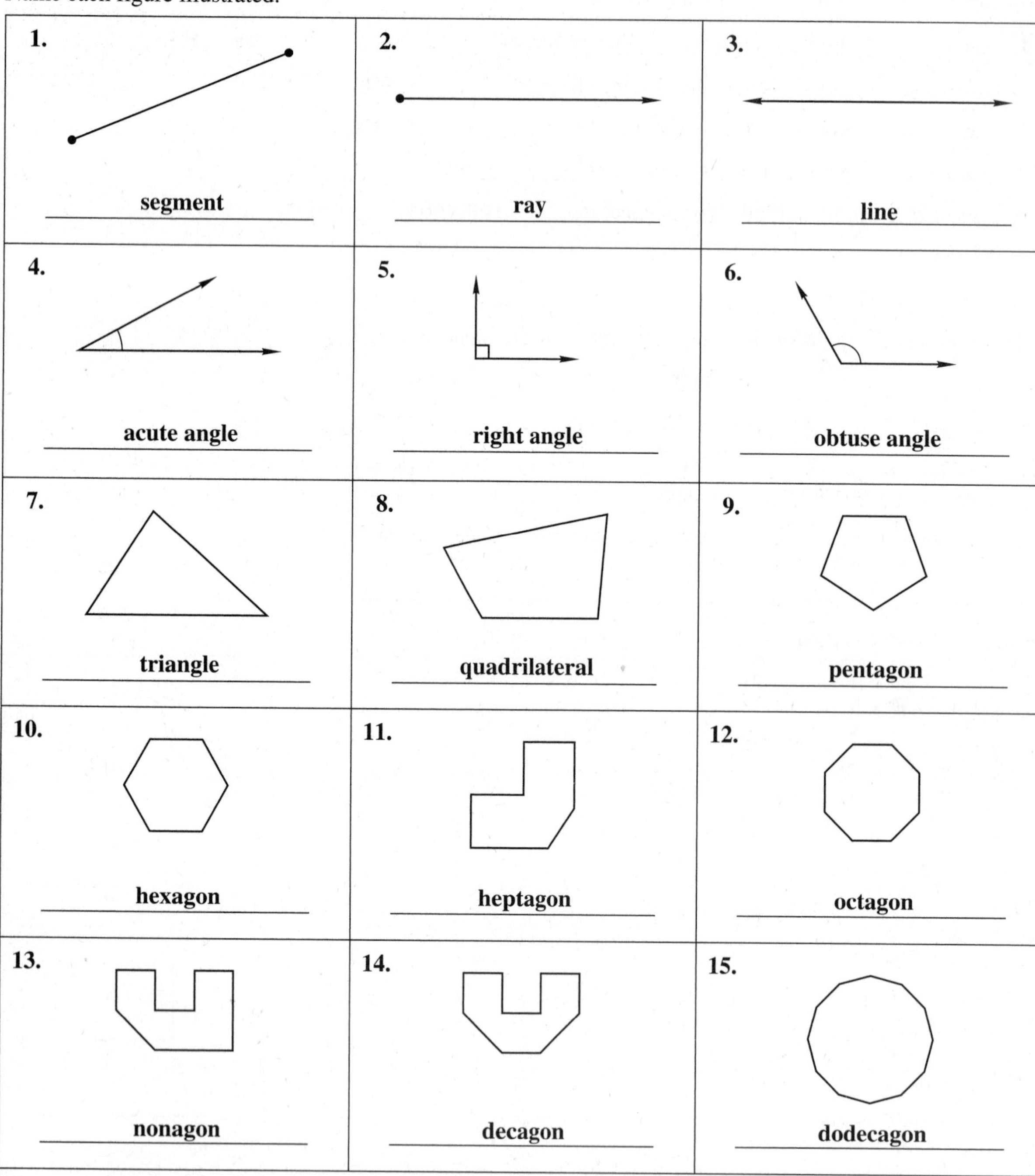

16. A polygon whose sides are equal in length and whose angles are equal in measure is a **regular** polygon.

G

+ − × ÷ Fractions

Simplify these expressions. Reduce the answers.

$\frac{2}{3} + \frac{2}{3} = \mathbf{1\frac{1}{3}}$	$\frac{2}{3} - \frac{2}{3} = \mathbf{0}$	$\frac{2}{3} \times \frac{2}{3} = \mathbf{\frac{4}{9}}$	$\frac{2}{3} \div \frac{2}{3} = \mathbf{1}$
$\frac{3}{4} + \frac{1}{4} = \mathbf{1}$	$\frac{3}{4} - \frac{1}{4} = \mathbf{\frac{1}{2}}$	$\frac{3}{4} \times \frac{1}{4} = \mathbf{\frac{3}{16}}$	$\frac{3}{4} \div \frac{1}{4} = \mathbf{3}$
$\frac{2}{3} + \frac{1}{2} = \mathbf{1\frac{1}{6}}$	$\frac{2}{3} - \frac{1}{2} = \mathbf{\frac{1}{6}}$	$\frac{2}{3} \times \frac{1}{2} = \mathbf{\frac{1}{3}}$	$\frac{2}{3} \div \frac{1}{2} = \mathbf{1\frac{1}{3}}$
$\frac{3}{4} + \frac{2}{3} = \mathbf{1\frac{5}{12}}$	$\frac{3}{4} - \frac{2}{3} = \mathbf{\frac{1}{12}}$	$\frac{3}{4} \times \frac{2}{3} = \mathbf{\frac{1}{2}}$	$\frac{3}{4} \div \frac{2}{3} = \mathbf{1\frac{1}{8}}$
$\frac{2}{5} + \frac{1}{4} = \mathbf{\frac{13}{20}}$	$\frac{2}{5} - \frac{1}{4} = \mathbf{\frac{3}{20}}$	$\frac{2}{5} \times \frac{1}{4} = \mathbf{\frac{1}{10}}$	$\frac{2}{5} \div \frac{1}{4} = \mathbf{1\frac{3}{5}}$
$\frac{1}{2} + \frac{5}{8} = \mathbf{1\frac{1}{8}}$	$\frac{5}{8} - \frac{1}{2} = \mathbf{\frac{1}{8}}$	$\frac{1}{2} \times \frac{5}{8} = \mathbf{\frac{5}{16}}$	$\frac{1}{2} \div \frac{5}{8} = \mathbf{\frac{4}{5}}$

H Measurement Facts

Write the number that makes each statement true.

Customary Units

Linear Measure

1. 1 foot = **12** inches

2. 1 yard = **36** inches

3. 1 yard = **3** feet

4. 1 mile = **5280** feet

5. 1 mile = **1760** yards

Area

6. 1 $foot^2$ = **144** $inches^2$

7. 1 $yard^2$ = **9** $feet^2$

Volume

8. 1 $yard^3$ = **27** $feet^3$

Weight

9. 1 pound = **16** ounces

10. 1 ton = **2000** pounds

Liquid Measure

11. 1 pint = **16** ounces

12. 1 pint = **2** cups

13. 1 quart = **2** pints

14. 1 gallon = **4** quarts

Temperature

15. Water freezes at **32** °F.

16. Water boils at **212** °F.

17. Normal body temperature is **98.6** °F.

Customary to Metric

18. 1 inch = **2.54** centimeters

Metric Units

Linear Measure

19. 1 centimeter = **10** millimeters

20. 1 meter = **100** centimeters

21. 1 meter = **1000** millimeters

22. 1 kilometer = **1000** meters

Area

23. 1 $meter^2$ = **10,000** $centimeters^2$

24. 1 $kilometer^2$ = **1,000,000** $meters^2$

Volume

25. 1 $meter^3$ = **1,000,000** $centimeters^3$

Mass

26. 1 gram = **1000** milligrams

27. 1 kilogram = **1000** grams

28. 1 metric ton = **1000** kilograms

Capacity

29. 1 liter = **1000** milliliters

30. One milliliter of water has a volume of **1 cm^3** and a mass of **1 gram**. One thousand cm^3 of water fills a **1**-liter container and has a mass of **1** kilogram.

Temperature

31. Water freezes at **0** °C.

32. Water boils at **100** °C.

33. Normal body temperature is **37** °C.

I

Proportions

Find the number that completes each proportion.

$\frac{3}{4} = \frac{a}{12}$ $a =$ **9**	$\frac{3}{4} = \frac{12}{b}$ $b =$ **16**	$\frac{c}{5} = \frac{12}{20}$ $c =$ **3**	$\frac{2}{d} = \frac{12}{24}$ $d =$ **4**
$\frac{4}{10} = \frac{e}{30}$ $e =$ **12**	$\frac{8}{12} = \frac{4}{f}$ $f =$ **6**	$\frac{g}{10} = \frac{10}{5}$ $g =$ **20**	$\frac{5}{h} = \frac{6}{18}$ $h =$ **15**
$\frac{15}{20} = \frac{i}{40}$ $i =$ **30**	$\frac{25}{100} = \frac{5}{j}$ $j =$ **20**	$\frac{k}{30} = \frac{3}{9}$ $k =$ **10**	$\frac{5}{m} = \frac{10}{100}$ $m =$ **50**
$\frac{50}{100} = \frac{n}{30}$ $n =$ **15**	$\frac{20}{15} = \frac{60}{p}$ $p =$ **45**	$\frac{q}{40} = \frac{75}{100}$ $q =$ **30**	$\frac{5}{r} = \frac{4}{16}$ $r =$ **20**
$\frac{2}{5} = \frac{s}{100}$ $s =$ **40**	$\frac{6}{8} = \frac{9}{t}$ $t =$ **12**	$\frac{u}{16} = \frac{8}{4}$ $u =$ **32**	$\frac{60}{v} = \frac{3}{2}$ $v =$ **40**
$\frac{8}{10} = \frac{w}{100}$ $w =$ **80**	$\frac{9}{12} = \frac{36}{x}$ $x =$ **48**	$\frac{y}{30} = \frac{6}{20}$ $y =$ **9**	$\frac{24}{z} = \frac{8}{6}$ $z =$ **18**

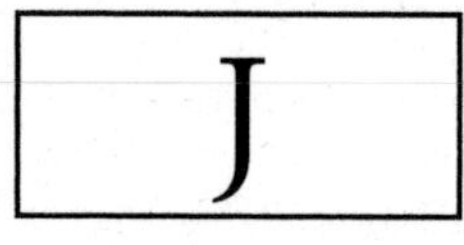

+ − × ÷ Decimals

Simplify these expressions.

0.8 + 0.4 = **1.2**	0.8 × 0.4 = **0.32**	0.8 ÷ 0.4 = **2**
1.2 − 0.4 = **0.8**	1.2 × 0.4 = **0.48**	1.2 ÷ 0.4 = **3**
1.2 + 0.04 = **1.24**	1.2 × 0.04 = **0.048**	1.2 ÷ 0.04 = **30**
1.2 + 4 = **5.2**	1.2 × 4 = **4.8**	1.2 ÷ 4 = **0.3**
6 − 0.3 = **5.7**	6 × 0.3 = **1.8**	6 ÷ 0.3 = **20**
0.3 + 6 = **6.3**	0.3 × 6 = **1.8**	0.3 ÷ 6 = **0.05**
0.01 − 0.01 = **0**	0.01 × 0.01 = **0.0001**	0.01 ÷ 0.01 = **1**

Powers and Roots

Simplify each power or root.

$\sqrt{100}$ = **10**	$\sqrt{16}$ = **4**	$\sqrt{81}$ = **9**	$\sqrt{4}$ = **2**
$\sqrt{144}$ = **12**	$\sqrt{1}$ = **1**	$\sqrt{64}$ = **8**	$\sqrt{49}$ = **7**
$\sqrt{25}$ = **5**	$\sqrt{121}$ = **11**	$\sqrt{9}$ = **3**	$\sqrt{36}$ = **6**
$\sqrt{169}$ = **13**	$\sqrt{225}$ = **15**	$\sqrt{196}$ = **14**	$\sqrt{625}$ = **25**
8^2 = **64**	5^2 = **25**	3^2 = **9**	12^2 = **144**
10^2 = **100**	2^3 = **8**	6^2 = **36**	3^3 = **27**
4^2 = **16**	10^3 = **1000**	7^2 = **49**	15^2 = **225**
5^3 = **125**	25^2 = **625**	4^3 = **64**	9^2 = **81**

L Fraction-Decimal-Percent Equivalents

Write each fraction as a decimal and as a percent. Write repeating decimals with a bar over the repetend.

Fraction	Decimal	Percent
$\frac{1}{2}$	**0.5**	**50%**
$\frac{1}{3}$	**$0.\overline{3}$**	**$33\frac{1}{3}\%$**
$\frac{2}{3}$	**$0.\overline{6}$**	**$66\frac{2}{3}\%$**
$\frac{1}{4}$	**0.25**	**25%**
$\frac{3}{4}$	**0.75**	**75%**
$\frac{1}{5}$	**0.2**	**20%**
$\frac{2}{5}$	**0.4**	**40%**
$\frac{3}{5}$	**0.6**	**60%**
$\frac{4}{5}$	**0.8**	**80%**
$\frac{1}{6}$	**$0.1\overline{6}$**	**$16\frac{2}{3}\%$**
$\frac{5}{6}$	**$0.8\overline{3}$**	**$83\frac{1}{3}\%$**
$\frac{1}{8}$	**0.125**	**$12\frac{1}{2}\%$**
$\frac{3}{8}$	**0.375**	**$37\frac{1}{2}\%$**

Fraction	Decimal	Percent
$\frac{5}{8}$	**0.625**	**$62\frac{1}{2}\%$**
$\frac{7}{8}$	**0.875**	**$87\frac{1}{2}\%$**
$\frac{1}{9}$	**$0.\overline{1}$**	**$11\frac{1}{9}\%$**
$\frac{1}{10}$	**0.1**	**10%**
$\frac{3}{10}$	**0.3**	**30%**
$\frac{7}{10}$	**0.7**	**70%**
$\frac{9}{10}$	**0.9**	**90%**
$\frac{1}{20}$	**0.05**	**5%**
$\frac{1}{25}$	**0.04**	**4%**
$\frac{1}{50}$	**0.02**	**2%**
$\frac{1}{100}$	**0.01**	**1%**
$1\frac{1}{2}$	**1.5**	**150%**

M

Metric Conversions

Complete each equivalence.

1. 2 meters = **200** centimeters

2. 1.5 kilometers = **1500** meters

3. 2.54 centimeters = **25.4** millimeters

4. 125 centimeters = **1.25** meters

5. 75 millimeters = **7.5** centimeters

6. 0.8 meter = **800** millimeters

7. 10 kilometers = **10,000** meters

8. 0.1 kilometer = **100** meters

9. 5000 meters = **5** kilometers

10. 50 centimeters = **0.5** meter

11. 50 centimeters = **500** millimeters

12. 2 liters = **2000** milliliters

13. 250 milliliters = **0.25** liter

14. 4 kilograms = **4000** grams

15. 2.5 grams = **2500** milligrams

16. 500 milligrams = **0.5** gram

17. 0.5 kilogram = **500** grams

18. Two liters of water has a volume of **2000** cubic centimeters and a mass of **2** kilograms.

Record the factor indicated by each prefix.

	Prefix	Factor
19.	kilo-	**1000**
20.	hecto-	**100**
21.	deka-	**10**
	(unit)	1
22.	deci-	$\frac{1}{10}$
23.	centi-	$\frac{1}{100}$
24.	milli-	$\frac{1}{1000}$

N + − × ÷ Mixed Numbers

Simplify these expressions. Reduce the answers.

$3 + 1\frac{2}{3} = \mathbf{4\frac{2}{3}}$	$3 - 1\frac{2}{3} = \mathbf{1\frac{1}{3}}$	$3 \times 1\frac{2}{3} = \mathbf{5}$	$3 \div 1\frac{2}{3} = \mathbf{1\frac{4}{5}}$
$1\frac{2}{3} + 1\frac{1}{2} = \mathbf{3\frac{1}{6}}$	$1\frac{2}{3} - 1\frac{1}{2} = \mathbf{\frac{1}{6}}$	$1\frac{2}{3} \times 1\frac{1}{2} = \mathbf{2\frac{1}{2}}$	$1\frac{2}{3} \div 1\frac{1}{2} = \mathbf{1\frac{1}{9}}$
$2\frac{1}{2} + 1\frac{2}{3} = \mathbf{4\frac{1}{6}}$	$2\frac{1}{2} - 1\frac{2}{3} = \mathbf{\frac{5}{6}}$	$2\frac{1}{2} \times 1\frac{2}{3} = \mathbf{4\frac{1}{6}}$	$2\frac{1}{2} \div 1\frac{2}{3} = \mathbf{1\frac{1}{2}}$
$4\frac{1}{2} + 2\frac{1}{4} = \mathbf{6\frac{3}{4}}$	$4\frac{1}{2} - 2\frac{1}{4} = \mathbf{2\frac{1}{4}}$	$4\frac{1}{2} \times 2\frac{1}{4} = \mathbf{10\frac{1}{8}}$	$4\frac{1}{2} \div 2\frac{1}{4} = \mathbf{2}$
$6\frac{2}{3} + 3\frac{3}{4} = \mathbf{10\frac{5}{12}}$	$6\frac{2}{3} - 3\frac{3}{4} = \mathbf{2\frac{11}{12}}$	$6\frac{2}{3} \times 3\frac{3}{4} = \mathbf{25}$	$3\frac{3}{4} \div 6\frac{2}{3} = \mathbf{\frac{9}{16}}$

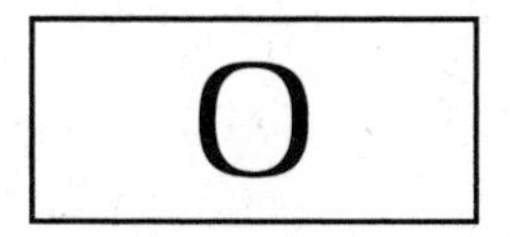

O Classifying Quadrilaterals and Triangles

Select from the words at the bottom of the page to describe each figure.

1.

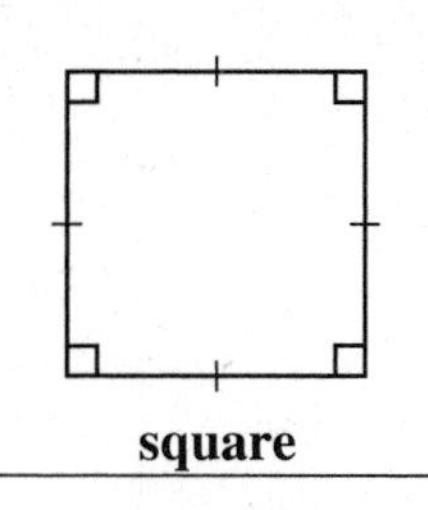

square

rectangle

rhombus

parallelogram

2.

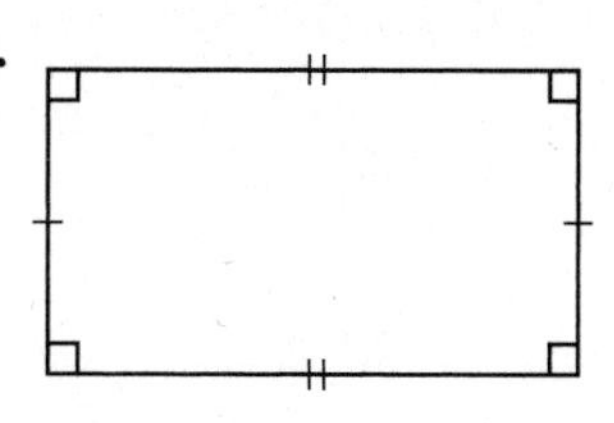

rectangle

parallelogram

3.

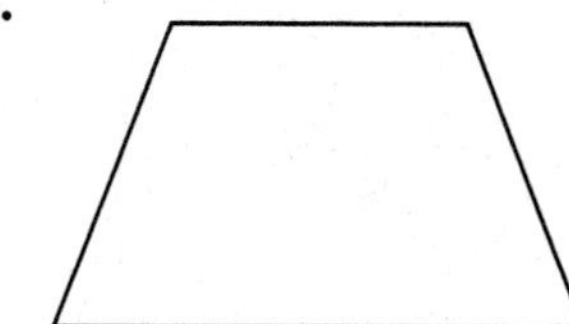

trapezoid

4.

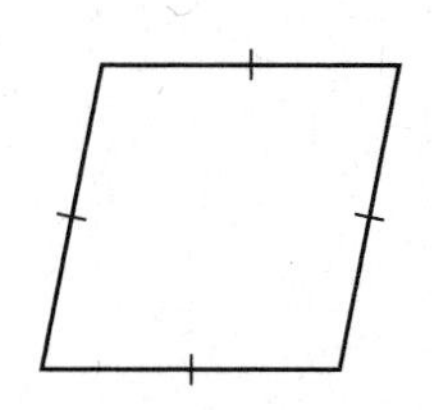

rhombus

parallelogram

5.

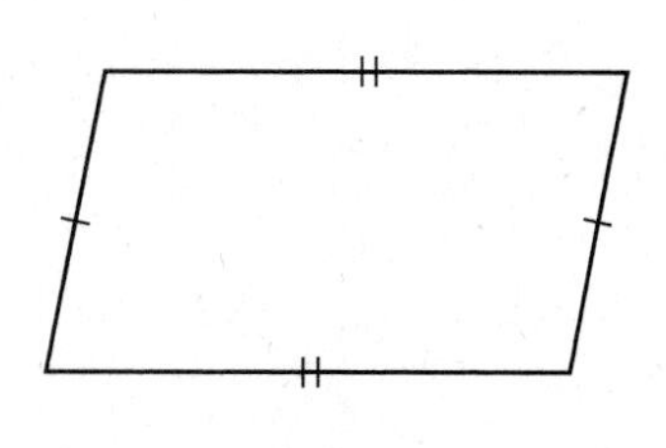

parallelogram

6.

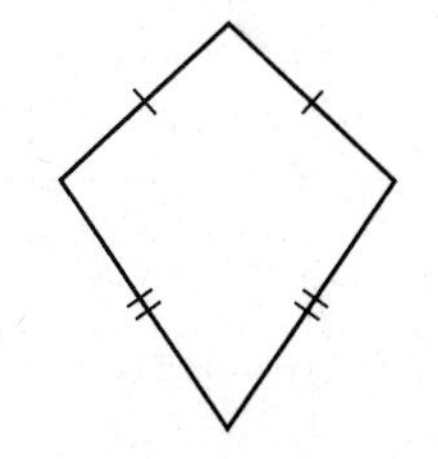

kite

7.

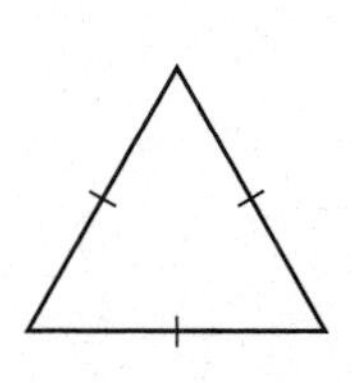

equilateral triangle

acute triangle

isosceles triangle

8.

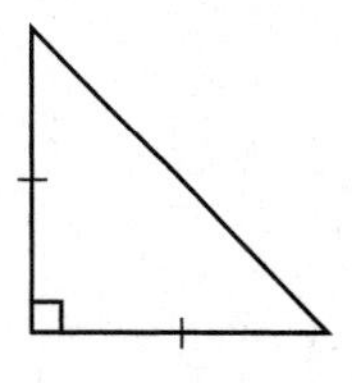

isosceles triangle

right triangle

9.

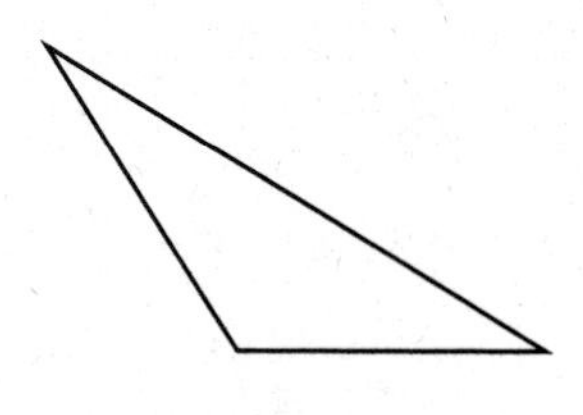

scalene triangle

obtuse triangle

kite	rectangle	isosceles triangle	right triangle
trapezoid	rhombus	scalene triangle	acute triangle
parallelogram	square	equilateral triangle	obtuse triangle

P

+ – × ÷ Integers

Simplify.

(–8) + (–2) = **–10**	(–8) – (–2) = **–6**	(–8)(–2) = **+16**	$\frac{-8}{-2}$ = **+4**
(–9) + (+3) = **–6**	(–9) – (+3) = **–12**	(–9)(+3) = **–27**	$\frac{-9}{+3}$ = **–3**
12 + (–2) = **+10**	12 – (–2) = **+14**	(12)(–2) = **–24**	$\frac{12}{-2}$ = **–6**
(+12) + (+6) = **+18**	(+12) – (+6) = **+6**	(+12)(+6) = **+72**	$\frac{+12}{+6}$ = **+2**
–20 + (+5) = **–15**	–20 – (+5) = **–25**	(–20)(+5) = **–100**	$\frac{-20}{+5}$ = **–4**
(–15) + (–3) = **–18**	(–15) – (–3) = **–12**	(–15)(–3) = **+45**	$\frac{-15}{-3}$ = **+5**
(+30) + (–6) = **+24**	(+30) – (–6) = **+36**	(+30)(–6) = **–180**	$\frac{+30}{-6}$ = **–5**
(–5) + (–6) + (–2) = **–13**	(–5) – (–6) – (–2) = **+3**	(–5)(–6)(–2) = **–60**	$\frac{(-5)(-6)}{(-2)}$ = **–15**

Percent-Decimal-Fraction Equivalents

Write each percent as a decimal and as a reduced fraction. Write repeating decimals with a bar over the repetend.

Percent	Decimal	Fraction
10%	**0.1**	$\frac{1}{10}$
90%	**0.9**	$\frac{9}{10}$
5%	**0.05**	$\frac{1}{20}$
40%	**0.4**	$\frac{2}{5}$
$12\frac{1}{2}$%	**0.125**	$\frac{1}{8}$
50%	**0.5**	$\frac{1}{2}$
2%	**0.02**	$\frac{1}{50}$
30%	**0.3**	$\frac{3}{10}$
$87\frac{1}{2}$%	**0.875**	$\frac{7}{8}$
25%	**0.25**	$\frac{1}{4}$
80%	**0.8**	$\frac{4}{5}$
$33\frac{1}{3}$%	$0.\overline{3}$	$\frac{1}{3}$
60%	**0.6**	$\frac{3}{5}$

Percent	Decimal	Fraction
$62\frac{1}{2}$%	**0.625**	$\frac{5}{8}$
20%	**0.2**	$\frac{1}{5}$
4%	**0.04**	$\frac{1}{25}$
75%	**0.75**	$\frac{3}{4}$
$66\frac{2}{3}$%	$0.\overline{6}$	$\frac{2}{3}$
$37\frac{1}{2}$%	**0.375**	$\frac{3}{8}$
70%	**0.7**	$\frac{7}{10}$
1%	**0.01**	$\frac{1}{100}$
$16\frac{2}{3}$%	$0.1\overline{6}$	$\frac{1}{6}$
$83\frac{1}{3}$%	$0.8\overline{3}$	$\frac{5}{6}$
$8\frac{1}{3}$%	$0.08\overline{3}$	$\frac{1}{12}$
$11\frac{1}{9}$%	$0.\overline{1}$	$\frac{1}{9}$

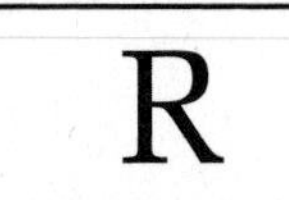

Area

Find the area of each figure. Angles that look like right angles are right angles.

1. 10 cm; 10 cm **100 cm²**	**2.** 8 in.; 4 in. **32 in.²**	**3.** 6 cm; 4 cm; 5 cm **24 cm²**
4. 7 in. Use $\frac{22}{7}$ for π. **154 in.²**	**5.** 20 cm Use 3.14 for π. **314 cm²**	**6.** 10 in. Leave π as π. **25π in.²**
7. 6 cm; 10 cm; 8 cm **24 cm²**	**8.** 10 in.; 6 in.; 6 in. **18 in.²**	**9.** 10 cm; 8 cm; 10 cm; 12 cm **48 cm²**
10. 7 cm; 5 cm; 4 cm; 10 cm **34 cm²**	**11.** 12 in.; 5 in.; 10 in.; 6 in. **90 in.²**	**12.** 4 cm; 12 cm; 6 cm; 10 cm **78 cm²**

S

Scientific Notation

Write each number in scientific notation.

186,000 = $\mathbf{1.86 \times 10^5}$	0.0005 = $\mathbf{5 \times 10^{-4}}$
30,500,000 = $\mathbf{3.05 \times 10^7}$	36×10^4 = $\mathbf{3.6 \times 10^5}$
0.35×10^5 = $\mathbf{3.5 \times 10^4}$	48×10^{-3} = $\mathbf{4.8 \times 10^{-2}}$
2.5 billion = $\mathbf{2.5 \times 10^9}$	15 thousandths = $\mathbf{1.5 \times 10^{-2}}$
12 million = $\mathbf{1.2 \times 10^7}$	$\frac{1}{1,000,000}$ = $\mathbf{1 \times 10^{-6}}$

Write each number in standard form.

1×10^6 = **1,000,000**	1×10^{-6} = **0.000001**
2.4×10^4 = **24,000**	5×10^{-4} = **0.0005**
4.75×10^5 = **475,000**	2.5×10^{-3} = **0.0025**
3.125×10^3 = **3125**	1.25×10^{-2} = **0.0125**
3.025×10^2 = **302.5**	1.05×10^{-1} = **0.105**

T Order of Operations

Simplify.

$6 + 6 \times 6 - 6 \div 6 = \mathbf{41}$	$5 + 5^2 + 5 \div 5 - 5 \times 5 = \mathbf{6}$
$3^2 + \sqrt{4} + 5(6) - 7 + 8 = \mathbf{42}$	$6 \times 4 \div 2 - 6 \div 2 \times 4 = \mathbf{0}$
$4 + 2(3 + 5) - 6 \div 2 = \mathbf{17}$	$8 + 7 \times 6 - (5 + 4) \div 3 + 2 = \mathbf{49}$
$2 + 2[3 + 4(7 - 5)] = \mathbf{24}$	$3[10 + (6 - 4) - 3(2 + 1)] = \mathbf{9}$
$\frac{(4)(3)(2)}{4 - 3 + 2} = \mathbf{8}$	$\sqrt{1^3 + 2^3 + 3^3} = \mathbf{6}$
$\frac{6 + 8(7 - 5) - 2}{4(3) - (4 + 3)} = \mathbf{4}$	$(2 + 3)^2 + 5[4^2 - 2(3)] = \mathbf{75}$
$(-3) + (-3)(-3) - (-3) = \mathbf{9}$	$\sqrt{-3 - (3)(-3) - (-3)} = \mathbf{3}$
$\frac{3(-3) - (-3)(-3)}{(-3) - 3(-3)} = \mathbf{-3}$	$\frac{(-3) - (-3) - \sqrt{3(3)}}{3^2 - 3(3) - 3} = \mathbf{1}$

Two-Step Equations

Complete each step to solve these equations.

$2x + 5 = 45$ $2x = \mathbf{40}$ $x = \mathbf{20}$	$3y + 4 = 22$ $3y = \mathbf{18}$ $y = \mathbf{6}$	$6w + 8 = 50$ $6w = \mathbf{42}$ $w = \mathbf{7}$
$5n - 3 = 32$ $5n = \mathbf{35}$ $n = \mathbf{7}$	$3m - 7 = 26$ $3m = \mathbf{33}$ $m = \mathbf{11}$	$8p - 9 = 47$ $8p = \mathbf{56}$ $p = \mathbf{7}$
$15 = 3a - 6$ $\mathbf{21} = 3a$ $\mathbf{7} = a$	$24 = 3b + 6$ $\mathbf{18} = 3b$ $\mathbf{6} = b$	$45 = 5c - 10$ $\mathbf{55} = 5c$ $\mathbf{11} = c$
$-2x + 9 = 25$ $-2x = \mathbf{16}$ $x = \mathbf{-8}$	$\frac{3}{4}m + 12 = 36$ $\frac{3}{4}m = \mathbf{24}$ $m = \mathbf{32}$	$0.5w - 1.5 = 4.5$ $0.5w = \mathbf{6}$ $w = \mathbf{12}$
$-\frac{2}{3}n - 6 = 18$ $-\frac{2}{3}n = \mathbf{24}$ $n = \mathbf{-36}$	$25 = 10 - 5y$ $\mathbf{15} = -5y$ $\mathbf{-3} = y$	$-0.3f + 1.2 = 4.8$ $-0.3f = \mathbf{3.6}$ $f = \mathbf{-12}$

V + − × ÷ Algebraic Terms

Simplify.

$6x + 2x = \mathbf{8x}$	$6x - 2x = \mathbf{4x}$	$(6x)(2x) = \mathbf{12x^2}$	$\frac{6x}{2x} = \mathbf{3}$
$6xy + 2xy = \mathbf{8xy}$	$6xy - 2xy = \mathbf{4xy}$	$6xy(2xy) = \mathbf{12x^2y^2}$	$\frac{6xy}{2xy} = \mathbf{3}$
$x + y + x = \mathbf{2x + y}$	$x + y - x = \mathbf{y}$	$(x)(y)(-x) = \mathbf{-x^2y}$	$\frac{xy}{x} = \mathbf{y}$
$3x + x + 3 = \mathbf{4x + 3}$	$3x - x - 3 = \mathbf{2x - 3}$	$(3x)(-x)(-3) = \mathbf{9x^2}$	$\frac{(2x)(8xy)}{4y} = \mathbf{4x^2}$
$3x + 2y + x - y = \mathbf{4x + y}$		$5xy - 2x + xy - x = \mathbf{6xy - 3x}$	

W Multiplying and Dividing in Scientific Notation

Simplify each expression. Write each answer in scientific notation.

$(1 \times 10^6)(1 \times 10^6) =$ **1×10^{12}**	$(3 \times 10^3)(3 \times 10^3) =$ **9×10^6**	$(4 \times 10^{-5})(2 \times 10^{-6}) =$ **8×10^{-11}**
$(5 \times 10^5)(5 \times 10^5) =$ **2.5×10^{11}**	$(6 \times 10^{-3})(7 \times 10^{-4}) =$ **4.2×10^{-6}**	$(3 \times 10^6)(2 \times 10^{-4}) =$ **6×10^2**
$(9 \times 10^{-6})(2 \times 10^2) =$ **1.8×10^{-3}**	$(5 \times 10^8)(4 \times 10^{-2}) =$ **2×10^7**	$(2.5 \times 10^{-6})(4 \times 10^{-4}) =$ **1×10^{-9}**
$\frac{8 \times 10^8}{2 \times 10^2} =$ **4×10^6**	$\frac{5 \times 10^6}{2 \times 10^3} =$ **2.5×10^3**	$\frac{9 \times 10^3}{3 \times 10^8} =$ **3×10^{-5}**
$\frac{7.5 \times 10^3}{2.5 \times 10^6} =$ **3×10^{-3}**	$\frac{2 \times 10^6}{4 \times 10^2} =$ **5×10^3**	$\frac{1 \times 10^3}{4 \times 10^8} =$ **2.5×10^{-6}**
$\frac{6 \times 10^4}{2 \times 10^{-4}} =$ **3×10^8**	$\frac{8 \times 10^{-8}}{2 \times 10^{-2}} =$ **4×10^{-6}**	$\frac{2.5 \times 10^{-4}}{5 \times 10^{-8}} =$ **5×10^3**

$8 \overline{)3000}$

Solutions for

Tests

Test 1

1. $\frac{12 \cdot 60}{12 + 36} = \frac{720}{48} = \mathbf{15}$

2. $\mathbf{4 \times 5 = 5 \times 4}$

3. $\mathbf{-5 < 3}$

4. **forty-three million, eighty thousand, seventy**

5. **25, 36, 49**

6. $\mathbf{(1 \times 100{,}000) + (3 \times 10{,}000) + (5 \times 100)}$

7. $-6 \mathbf{>} -8$

8. $ab = 14 \cdot 4 = \mathbf{56}$

9. **3,040,700**

10. $$\begin{array}{r} \$12.00 \\ -\ \$5.50 \\ \hline \mathbf{\$6.50} \end{array}$$

11. $$\begin{array}{r} 2{,}480 \\ +\ 9{,}630 \\ \hline \mathbf{12{,}110} \end{array}$$

12. $$\begin{array}{r} \mathbf{\$7.40} \\ 7\overline{)\$51.80} \\ 49 \\ \hline 2\,8 \\ 2\,8 \\ \hline 00 \\ 0 \\ \hline 0 \end{array}$$

13. $$\begin{array}{r} 6048 \\ -\ 2532 \\ \hline \mathbf{3516} \end{array}$$

14. $$\begin{array}{r} \mathbf{85} \\ 15\overline{)1275} \\ 120 \\ \hline 75 \\ 75 \\ \hline 0 \end{array}$$

15. $$\begin{array}{r} 18{,}400 \\ -\ 7{,}520 \\ \hline \mathbf{10{,}880} \end{array}$$

16. $$\begin{array}{r} 22 \\ \times\ 9 \\ \hline 198 \end{array} \qquad \begin{array}{r} 198 \\ \times\ 25 \\ \hline 990 \\ 396 \\ \hline \mathbf{4950} \end{array}$$

17. $$\begin{array}{r} 720 \\ -\ 38 \\ \hline 682 \end{array} \qquad \begin{array}{r} 1000 \\ -\ 682 \\ \hline \mathbf{318} \end{array}$$

18. $$\begin{array}{r} \mathbf{6{,}359} \\ 6\overline{)38{,}154} \\ 36\phantom{{,}154} \\ \hline 2\,1 \\ 1\,8 \\ \hline 35 \\ 30 \\ \hline 54 \\ 54 \\ \hline 0 \end{array}$$

19. $$\begin{array}{r} 170 \\ \times\ 18 \\ \hline 1360 \\ 170 \\ \hline \mathbf{3060} \end{array}$$

20. $$\begin{array}{r} \mathbf{\$4.13} \\ 10\overline{)\$41.30} \\ 40 \\ \hline 1\,3 \\ 1\,0 \\ \hline 30 \\ 30 \\ \hline 0 \end{array}$$

Test 2

1. (a) $\frac{3 \text{ dimes}}{10 \text{ dimes}} = \mathbf{\frac{3}{10}}$ (b) $\frac{30¢}{100¢} = \mathbf{30\%}$

2. $\overline{AC}$ (or $\overline{CA}$)

3. $\mathbf{\frac{8}{5}}$ or $\mathbf{1\frac{3}{5}}$

4. $9\frac{5}{8} = \frac{(8 \times 9) + 5}{8} = \mathbf{\frac{77}{8}}$

5. (a) $\mathbf{-8, 0, \frac{1}{8}, 8}$ (b) $\mathbf{\frac{1}{8}}$

6. $\mathbf{(2 + 6) + 8 = 2 + (6 + 8)}$

7. (a) $\mathbf{\frac{11}{15}}$ (b) $\mathbf{\frac{4}{15}}$

8.
$$\begin{array}{r} 300{,}000{,}000 \\ -\ \ 56{,}000{,}000 \\ \hline 244{,}000{,}000 \end{array}$$
two hundred forty-four million

9. (a) **1, 2, 4, 7, 14, 28**

(b) **1, 2, 3, 6, 7, 14, 21, 42**

(c) **1, 2, 7, 14**

(d) **14**

10. $\mathbf{4 \cdot 1 < 4 + 1}$

11.
$$\begin{array}{r} 7000 \\ -\ 3955 \\ \hline \mathbf{3045} \end{array}$$

12.
$$\begin{array}{r} \$6.45 \\ +\ \$4.85 \\ \hline \mathbf{\$11.30} \end{array}$$

13.
$$\begin{array}{r} \mathbf{23} \\ 34\overline{)782} \\ \underline{68} \\ 102 \\ \underline{102} \\ 0 \end{array}$$

14. $\frac{1}{11} + \frac{2}{11} = \mathbf{\frac{3}{11}}$

15. $\frac{7}{13} - \frac{6}{13} = \mathbf{\frac{1}{13}}$

16. $\frac{4}{7} \times \frac{4}{9} = \mathbf{\frac{16}{63}}$

17.
$$\begin{array}{r} \mathbf{4{,}324\ R\ 2} \\ 8\overline{)34{,}594} \\ \underline{32} \\ 2\,5 \\ \underline{2\,4} \\ 19 \\ \underline{16} \\ 34 \\ \underline{32} \\ 2 \end{array}$$

18.
$$\begin{array}{r} \$7.44 \\ \times\ \ \ \ 90 \\ \hline \mathbf{\$669.60} \end{array}$$

19. $\frac{3}{4} \cdot \frac{1}{8} \cdot \frac{5}{7} = \mathbf{\frac{15}{224}}$

20. (a) **line;** $\overleftrightarrow{XY}$ (or $\overleftrightarrow{YX}$)

(b) **segment;** $\overline{LM}$ (or $\overline{ML}$)

(c) **ray;** $\overrightarrow{BC}$

TEST 3

1.
$$\begin{array}{r} 22{,}374 \\ -\ 14{,}998 \\ \hline \mathbf{7{,}376} \end{array}$$

2.
$$\begin{array}{r} 13 \\ \times\ 18 \\ \hline 104 \\ \underline{13\ \ } \\ \mathbf{234} \end{array}$$
T-shirts

3. $(8 \cdot 4) - (8 + 4) = 32 - 12 = \mathbf{20}$

4.
$$\begin{array}{r} \$6.75 \\ \$3.95 \\ +\ \$0.95 \\ \hline \mathbf{\$11.65} \end{array}$$

5.
$$\begin{array}{r} 1728 \\ -\ 1556 \\ \hline \mathbf{172} \end{array}$$
years

6.
$$\begin{array}{r} 100\% \\ -\ \ 29\% \\ \hline \mathbf{71\%} \end{array}$$

7.

$1\frac{3}{8} = \frac{11}{8}$

8. (a) $\frac{2}{3} \cdot \frac{16}{16} = \mathbf{\frac{32}{48}}$ (b) $\frac{5}{8} \cdot \frac{6}{6} = \mathbf{\frac{30}{48}}$

9. $\frac{1}{5} \cdot \frac{2}{2} = \frac{2}{10};\ \frac{2}{10} - \frac{1}{10} = \mathbf{\frac{1}{10}}$

10. (a) **1, 3, 9, 27**

(b) **1, 3, 7, 9, 21, 63**

(c) **9**

11. $\overline{ST}$ (or $\overline{TS}$), $\overline{RS}$ (or $\overline{SR}$), $\overline{RT}$ (or $\overline{TR}$)

12. $\mathbf{2\frac{4}{7}}$

13. $\frac{11}{22} + \frac{13}{22} = \frac{24}{22} = \frac{12}{11} = \mathbf{1\frac{1}{11}}$

14. $\frac{3}{4} \cdot \frac{8}{5} = \frac{24}{20} = 1\frac{4}{20} = \mathbf{1\frac{1}{5}}$

15. $5\overline{)38{,}519}$ = **7,703 R 4**

$$\begin{array}{r} 35 \\ \hline 3\,5 \\ 3\,5 \\ \hline 01 \\ 0 \\ \hline 19 \\ 15 \\ \hline 4 \end{array}$$

16. $20\overline{)1060}$ = **53**

$$\begin{array}{r} 100 \\ \hline 60 \\ 60 \\ \hline 0 \end{array}$$

17.
$$\begin{array}{r} 122 \\ \times\ 84 \\ \hline 488 \\ 976 \\ \hline \mathbf{10{,}248} \end{array}$$

18. $(4 + 3)(3) = (7)(3) = \mathbf{21}$

19. $14\overline{)1820}$ = **130**

$$\begin{array}{r} 14 \\ \hline 42 \\ 42 \\ \hline 00 \\ 0 \\ \hline 0 \end{array}$$

20.
$$\begin{array}{r} \$20.00 \\ -\ \$4.52 \\ \hline \mathbf{\$15.48} \end{array}$$

Test 4

1.
$$\begin{array}{r} 2002 \\ -\ 84 \\ \hline \mathbf{1918} \end{array}$$

2. $40\overline{)5000}$ = **125 bushels**

3. $\frac{2 \text{ feet}}{3 \text{ feet}} = \frac{2}{3} = \mathbf{66\frac{2}{3}\%}$

4.
$$\begin{array}{r} 689 \text{ tickets} \\ -\ 39 \text{ tickets} \\ \hline \mathbf{650 \text{ tickets}} \end{array}$$

5.
$$\begin{array}{r} 1{,}000{,}000{,}000 \\ -\ 71{,}000{,}000 \\ \hline 929{,}000{,}000 \end{array}$$

nine hundred twenty-nine million

6. (a) $\frac{5}{16} + \left(\frac{3}{16} + \frac{7}{16}\right)$

$\mathbf{=} \left(\frac{5}{16} + \frac{3}{16}\right) + \frac{7}{16}$

(b) **associative property of addition**

7. $\mathbf{6 - 9 = -3}$

8. (a) Perimeter = 12 cm + 5 cm + 12 cm + 5 cm = **34 cm**

(b) Area = (12 cm)(5 cm) = $\mathbf{60\ cm^2}$

9. (a) $8\frac{10}{16} = \mathbf{8\frac{5}{8}}$ (b) $\frac{2}{16} = \mathbf{\frac{1}{8}}$

10. $2\frac{1}{4} = \frac{9}{4}; \frac{9}{4} \times \frac{1}{9} = \mathbf{\frac{1}{4}}$

11. (a) $\frac{3}{5} \cdot \frac{8}{8} = \mathbf{\frac{24}{40}}$ (b) $\frac{7}{10} \cdot \frac{4}{4} = \mathbf{\frac{28}{36}}$

12. 43°

13.
$$\begin{array}{r} 3446 \\ -\ 1428 \\ \hline \mathbf{2018} \end{array}$$

14.
$$\begin{array}{r} \mathbf{\$1.85} \\ 30\overline{)\$55.50} \\ \underline{30} \\ 25\ 5 \\ \underline{24\ 0} \\ 1\ 50 \\ \underline{1\ 50} \\ 0 \end{array}$$

15. $\frac{7}{3} = 2\frac{1}{3}$

D. $\mathbf{\frac{7}{3}}$

16. $\frac{3}{8} + \frac{3}{8} + \frac{3}{8} = \frac{9}{8} = \mathbf{1\frac{1}{8}}$

17. $\frac{11}{12} - \frac{5}{12} = \frac{6}{12} = \mathbf{\frac{1}{2}}$

18. $\left(\frac{1}{4}\right)^2 = \frac{1}{4} \cdot \frac{1}{4} = \mathbf{\frac{1}{16}}$

19. $\sqrt{256} = \mathbf{16}$

20. $14(10 + 11) = 14(21) = \mathbf{294}$

Test 5

1.
$$\begin{array}{r} \mathbf{23\ books} \\ 27\overline{)621} \\ \underline{54} \\ 81 \\ \underline{81} \\ 0 \end{array}$$

2.
$$\begin{array}{r} 1806 \\ -\ \ 800 \\ \hline \mathbf{1006\ years} \end{array}$$

3.
$$\begin{array}{r} \$20.00 \\ -\ \ \$8.24 \\ \hline \mathbf{\$11.76} \end{array}$$

4.
$$\begin{array}{r} 283 \\ -\ 197 \\ \hline \mathbf{86\ pages} \end{array}$$

5.

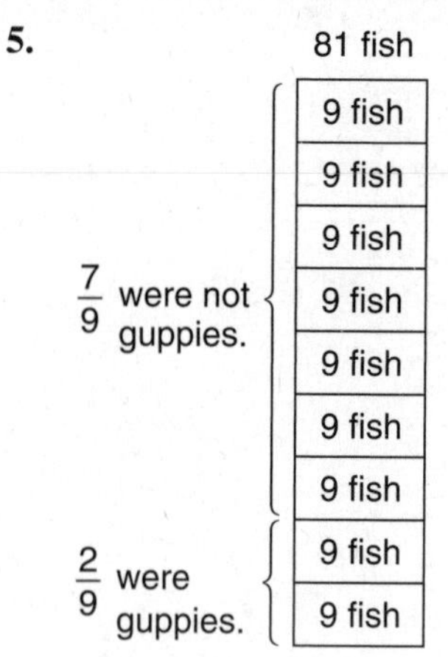

(a) 2×9 fish $=$ **18 fish**

(b) 7×9 fish $=$ **63 fish**

6.

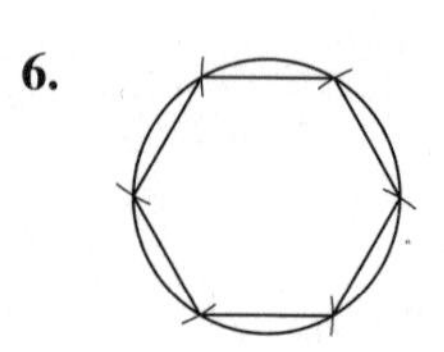

7.

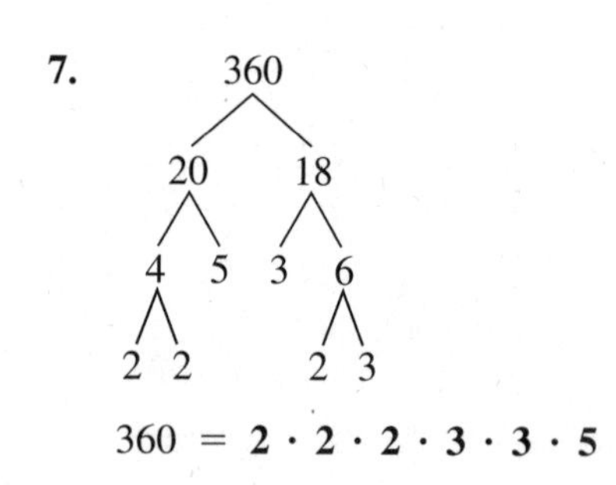

$360 = \mathbf{2 \cdot 2 \cdot 2 \cdot 3 \cdot 3 \cdot 5}$

8. (a) $\frac{105}{9} = 11\frac{6}{9} = \mathbf{11\frac{2}{3}}$

(b) $3\frac{8}{5} = \mathbf{4\frac{3}{5}}$

(c) $\frac{640}{780} \div \frac{20}{20} = \mathbf{\frac{32}{39}}$

9. (a) $\mathbf{\frac{7}{3}}$ (b) $\mathbf{\frac{3}{29}}$ (c) $\mathbf{\frac{1}{6}}$

10. (a) $\frac{3}{4} \cdot \frac{7}{7} = \mathbf{\frac{21}{28}}$ (b) $\frac{5}{7} \cdot \frac{4}{4} = \mathbf{\frac{20}{28}}$

11.
$$\begin{array}{r} 900 \\ -\ 610 \\ \hline \mathbf{290} \end{array}$$

12.
$$\begin{array}{r} 56 \\ +\ 57 \\ \hline \mathbf{113} \end{array}$$

13. $15\overline{)465}$ = **31**

14. $8 - 1\frac{2}{3} = 7\frac{3}{3} - 1\frac{2}{3} = \mathbf{6\frac{1}{3}}$

15. $6\frac{4}{7} + 5\frac{6}{7} = 11\frac{10}{7} = \mathbf{12\frac{3}{7}}$

16. $4\frac{1}{6} - 2\frac{5}{6} = 3\frac{7}{6} - 2\frac{5}{6} = 1\frac{2}{6} = \mathbf{1\frac{1}{3}}$

17. $\frac{2}{\cancel{5}_1} \cdot \frac{\cancel{5}^1}{\cancel{8}_1} \cdot \frac{\cancel{8}^1}{9} = \mathbf{\frac{2}{9}}$

18. $\frac{3}{4} \div \frac{2}{3} = \frac{3}{4} \cdot \frac{3}{2} = \frac{9}{8} = \mathbf{1\frac{1}{8}}$

19. $10^2 - \sqrt{36} = 100 - 6 = \mathbf{94}$

20. (a) **side *AD* (or *DA*)**

(b) Area = (28 mm)(14 mm) = **392 mm²**

TEST 6

1. $\frac{82 \text{ in.} + 74 \text{ in.} + 78 \text{ in.} + 80 \text{ in.} + 76 \text{ in.}}{5}$

$= \frac{390 \text{ in.}}{5} = \mathbf{78 \text{ in.}}$

2.
```
  $0.87        $10.00
×     9      −  $7.83
  $7.83         $2.17
```
(**$2.17**)

3.
```
  1634 miles
−  276 miles
  1358 miles
```
(**1358 miles**)

4. (a)

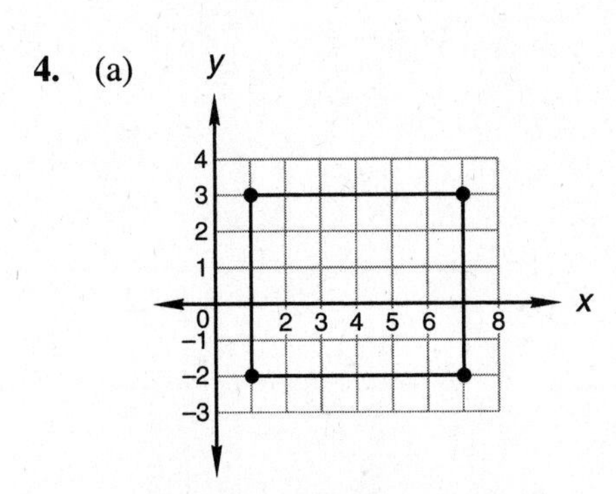

The fourth vertex is at **(1, 3)**.

(b) Area = (6 units)(5 units) = **30 units²**

5.

2540 miles

$\frac{1}{4}$ (or 25%) completed: 635 miles

$\frac{3}{4}$ (or 75%) left to drive: 635 miles, 635 miles, 635 miles

(a) **635 mi**

(b) 3 × 635 mi = **1905 mi**

6. 7 ft = 7 × 12 in. = 84 in.
84 in. ÷ 4 = **21 in.**

7.
$\frac{3}{5} \cdot \frac{3}{3} = \frac{9}{15}$

$+ \frac{2}{3} \cdot \frac{5}{5} = \frac{10}{15}$

$\frac{19}{15} = \mathbf{1\frac{4}{15}}$

8. (a) 68,②61 ⟶ **68,000**

(b) 68,2⑥1 ⟶ **68,300**

9. 40,000 ÷ 40 = **1000**

10. $\frac{160}{240} = \frac{\cancel{2} \cdot \cancel{2} \cdot \cancel{2} \cdot \cancel{2} \cdot 2 \cdot \cancel{5}}{\cancel{2} \cdot \cancel{2} \cdot \cancel{2} \cdot \cancel{2} \cdot 3 \cdot \cancel{5}} = \mathbf{\frac{2}{3}}$

11.
$\frac{7}{8} \cdot \frac{7}{7} = \frac{49}{56}$

$\frac{8}{7} \cdot \frac{8}{8} = \frac{64}{56}$

$\frac{49}{56} < \frac{64}{56}$

$\frac{7}{8} \mathbf{<} \frac{8}{7}$

12. Multiples of 8: 8, 16, 24, 32, 40, 48, …
Multiples of 10: 10, 20, 30, 40, …
The LCM of 8 and 10 is **40.**

13. (a) $\overline{ST}$ (or $\overline{TS}$)

(b) ∠***SRT*** (or ∠***TRS***)

14. (a) 324 = **2 · 2 · 3 · 3 · 3 · 3**

(b) $\sqrt{324} = 2 \cdot 3 \cdot 3 = \mathbf{18}$

15. $w = \frac{4 \cdot 36}{9} = \mathbf{16}$

16.
```
  971
− 287
  684
```
(**684**)

17. $\begin{array}{r} 94 \\ -\ 49 \\ \hline \mathbf{45} \end{array}$

18. $\begin{array}{r} \frac{3}{5} \cdot \frac{2}{2} = \frac{6}{10} \\ +\ \frac{1}{2} \cdot \frac{5}{5} = \frac{5}{10} \\ \hline \frac{11}{10} = \mathbf{1\frac{1}{10}} \end{array}$

19. $\left(\frac{2}{3} \cdot \frac{5}{6}\right) - \frac{2}{5} = \frac{5}{9} - \frac{2}{5} = \frac{25}{45} - \frac{18}{45}$
$= \mathbf{\frac{7}{45}}$

20. $5\frac{1}{3} \div 1\frac{7}{9} = \frac{16}{3} \div \frac{16}{9} = \frac{16}{3} \cdot \frac{9}{16} = \mathbf{3}$

TEST 7

1. $\begin{array}{r} \$113.96 \\ \$99.21 \\ \$93.20 \\ +\ \$128.95 \\ \hline \$435.32 \end{array}$ $\quad 4\overline{)\$435.32} = \mathbf{\$108.83}$

2. $\begin{array}{r} \$9521 \\ -\ \$3267 \\ \hline \mathbf{\$6254} \end{array}$

3. $\frac{\$41.40}{12} = \$3.45$
$\$4.25 - \$3.45 = \mathbf{\$0.80}$

4. 1 min 2 s = 62 s
62 s − 7 s = **55 s**

5. Perimeter of pentagon: 5 × 24 cm = 120 cm
Side of square: 120 cm ÷ 4 = **30 cm**

6. 45 fish

$\frac{2}{5}$ were guppies.	9 fish
	9 fish
$\frac{3}{5}$ were not guppies.	9 fish
	9 fish
	9 fish

(a) 2 × 9 fish = **18 fish**

(b) 3 × 9 fish = **27 fish**

7. Multiples of 5: 5, 10, 15, 20, 25, 30, 35, 40, ...
Multiples of 8: 8, 16, 24, 32, 40, 48, ...
Multiples of 10: 10, 20, 30, 40, ...
The LCM of 5, 8, and 10 is **40.**

8. (a) 1832.22④3 ⟶ **1832.22**

(b) 18③2.2243 ⟶ **1800**

9. (a) $\frac{30}{100} = \mathbf{\frac{3}{10}}$

(b) $\frac{30}{100} = \mathbf{30\%}$

10. **59.5**

11. (a)

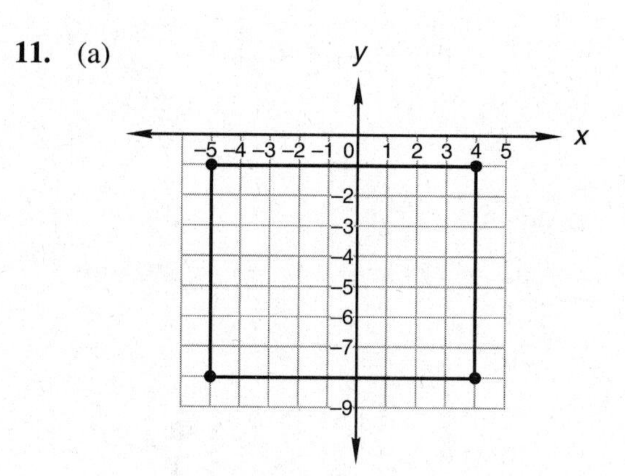

The fourth vertex is at **(−5, −8).**

(b) Area = (9 units)(7 units) = **63 units²**

12. **0.47**

13. $x = \frac{7 \cdot 10}{14} = \mathbf{5}$

14. **two hundred thirty-seven and two hundred eight thousandths**

15. $\begin{array}{r} 2.5 \\ \times\ 2.5 \\ \hline 125 \\ 50 \\ \hline \mathbf{6.25} \end{array}$

16. $\begin{array}{r} \mathbf{0.91} \\ 3\overline{)2.73} \\ 2\,7 \\ \hline 03 \\ 3 \\ \hline 0 \end{array}$

17.
$$\begin{array}{r} 8.5 \\ 1.83 \\ +\ 15. \\ \hline \mathbf{25.33} \end{array}$$

18.
$$\begin{array}{r} 31.740 \\ -\ 2.146 \\ \hline \mathbf{29.594} \end{array}$$

19. $4\frac{1}{3} - \left(\frac{1}{9} \cdot \frac{3}{4}\right) = 4\frac{1}{3} - \frac{1}{12}$
$= 4\frac{4}{12} - \frac{1}{12} = 4\frac{3}{12} = \mathbf{4\frac{1}{4}}$

20. $\left(3\frac{1}{3} + 1\frac{1}{4}\right) \div \left(6 - 4\frac{1}{6}\right)$
$= \left(3\frac{4}{12} + 1\frac{3}{12}\right) \div \left(5\frac{6}{6} - 4\frac{1}{6}\right)$
$= 4\frac{7}{12} \div 1\frac{5}{6} = \frac{55}{12} \div \frac{11}{6}$
$= \frac{55}{12} \cdot \frac{6}{11} = \frac{5}{2} = \mathbf{2\frac{1}{2}}$

Test 8

1.
$$\begin{array}{r} 7 \text{ red} \\ +\ 5 \text{ white} \\ \hline 12 \text{ total} \end{array} \qquad \frac{\text{white}}{\text{total}} = \mathbf{\frac{5}{12}}$$

2. (a) 6 minutes 50 seconds = **410 seconds**

(b) 410 seconds ÷ 5 = **82 seconds**

3. $27 \frac{\text{miles}}{\cancel{\text{gallon}}} \times 27 \cancel{\text{gallons}} = \mathbf{729\ miles}$

4.

116 adults

29 adults
29 adults
29 adults
29 adults

$\frac{3}{4}$ (or 75%) were 5 ft tall or taller. (first three rows)

$\frac{1}{4}$ (or 25%) were less than 5 ft tall. (last row)

(a) **29 adults**

(b) 3 × 29 adults = **87 adults**

5. Perimeter = 2 cm + 7 cm + 3 cm + 5 cm
+ 5 cm + 12 cm
= **34 cm**

6.

12 cm
2 cm | A
3 cm | B
(5 cm)

Area = area A + area B
= (12 cm × 2 cm) + (5 cm × 3 cm)
= 24 cm^2 + 15 cm^2 = **39 cm^2**

7. $AB + BC + CD = AD$
56 mm + BC + 14 mm = 98 mm
BC = **28 mm**

8. 56 mm = **5.6 cm**

9. 0.870⑤47 ⟶ **0.871**

10. 40° + 40° + a = 180°
The third angle measures **100°.**

11. **0.74**

12. (a) **48.09** (b) $\mathbf{48\frac{9}{100}}$

13. Area $= \frac{(24\text{ cm})(10\text{ cm})}{2} = \mathbf{120\ cm^2}$

14.
$$\begin{array}{r} 0.37 \\ \times\ 0.04 \\ \hline \mathbf{0.0148} \end{array}$$

15.
$$\begin{array}{r} \mathbf{0.026} \\ 6\overline{)0.156} \\ 12 \\ \hline 36 \\ 36 \\ \hline 0 \end{array}$$

16. $8\frac{1}{6} - 4\frac{2}{3} = 8\frac{1}{6} - 4\frac{4}{6} = 7\frac{7}{6} - 4\frac{4}{6}$
$= 3\frac{3}{6} = \mathbf{3\frac{1}{2}}$

17. $2\frac{1}{4} \cdot 2\frac{2}{3} = \frac{\overset{3}{\cancel{9}}}{\underset{1}{\cancel{4}}} \cdot \frac{\overset{2}{\cancel{8}}}{\underset{1}{\cancel{3}}} = \frac{6}{1} = \mathbf{6}$

18. $8 \div 2\frac{2}{3} = \frac{8}{1} \div \frac{8}{3} = \frac{8}{1} \cdot \frac{3}{8} = \mathbf{3}$

19.
$$\frac{8}{12} = \frac{w}{15}$$
$$12w = 8 \cdot 15$$
$$w = \frac{120}{12}$$
$$w = \mathbf{10}$$

20.
$$\begin{array}{r} 2.04 \\ -\ 0.34 \\ \hline \mathbf{1.70} \end{array}$$

Test 9

1. Primes: 2, 3, 5 Probability: $\frac{3}{6} = \mathbf{\frac{1}{2}}$

2. (a) $(100 + 98 + 91 + 84 + 93 + 88 + 97 + 91 + 87 + 91) \div 10 = \mathbf{92}$

(b) $\frac{91 + 91}{2} = \mathbf{91}$

3. $a(b + c) = 0.6(4.1 + 0.7) = 0.6(4.8) = \mathbf{2.88}$

4. (a) $22 - 14 = $ **8 votes**

(b) $\frac{16}{70} = \mathbf{\frac{8}{35}}$

5.

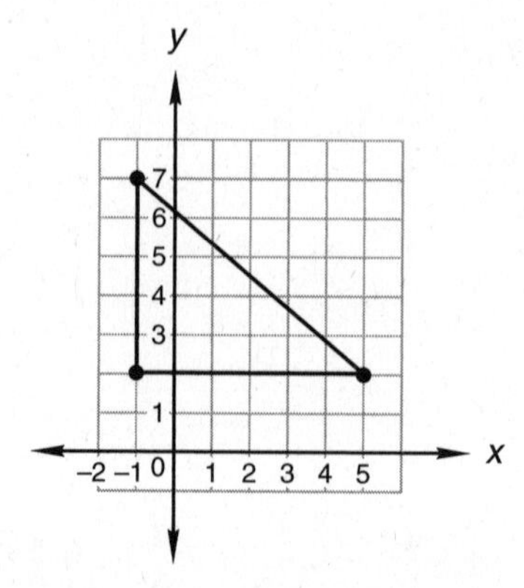

Area $= \frac{6 \cdot 5}{2} = $ **15 units²**

6. (Diagram not required.)

Those who rode the Giant Gyro

$\frac{5}{12}$ were euphoric.
$\frac{1}{12}$ of those who rode
$\frac{1}{12}$ of those who rode
$\frac{1}{12}$ of those who rode
$\frac{1}{12}$ of those who rode
$\frac{1}{12}$ of those who rode

$\frac{7}{12}$ were vertiginous.
$\frac{1}{12}$ of those who rode
$\frac{1}{12}$ of those who rode
$\frac{1}{12}$ of those who rode
$\frac{1}{12}$ of those who rode
$\frac{1}{12}$ of those who rode
$\frac{1}{12}$ of those who rode
$\frac{1}{12}$ of those who rode

(a) $\mathbf{\frac{7}{12}}$ **were vertiginous.**

(b) $\frac{\text{euphoric}}{\text{vertiginous}} = \mathbf{\frac{5}{7}}$

7. $l + 14 \text{ cm} = \text{half of the perimeter} = 32 \text{ cm}$
$l = 32 \text{ cm} - 14 \text{ cm} = $ **18 cm**

8. $m\angle a = 180° - (90° + 38°) = \mathbf{52°}$
$m\angle b = 180° - m\angle a$
$= 180° - 52° = \mathbf{128°}$
$m\angle c = m\angle a = \mathbf{52°}$

9. **3.75**

10. $5\overline{)7.0}$ = **1.4**

11. $67.\overline{24} = 67.242424\ldots$
$67.2424②4\ldots \longrightarrow$ **67.2424**

12.
$$\begin{array}{r} 0.355\ldots \\ 9\overline{)3.200\ldots} \\ \underline{2\ 7} \\ 50 \\ \underline{45} \\ 50 \\ \underline{45} \\ 5 \end{array}$$
$0.355\ldots = \mathbf{0.3\overline{5}}$

13. $\frac{14}{6} = \frac{7}{m}$

$14m = 6 \cdot 7$

$m = \frac{42}{14} = \mathbf{3}$

14. $\begin{array}{r} 8.00 \\ -\ 3.14 \\ \hline \mathbf{4.86} \end{array}$

15. $\begin{array}{r} 1.000 \\ -\ 0.072 \\ \hline \mathbf{0.928} \end{array}$

16. $5\frac{3}{4} + \frac{3}{5} + 3\frac{1}{2} = 5\frac{15}{20} + \frac{12}{20} + 3\frac{10}{20}$

$= 8\frac{37}{20} = \mathbf{9\frac{17}{20}}$

17. $4\frac{1}{6} - \left(3 - 1\frac{1}{4}\right) = 4\frac{1}{6} - \left(2\frac{4}{4} - 1\frac{1}{4}\right)$

$= 4\frac{1}{6} - 1\frac{3}{4} = 4\frac{2}{12} - 1\frac{9}{12}$

$= 3\frac{14}{12} - 1\frac{9}{12} = \mathbf{2\frac{5}{12}}$

18. $3\frac{1}{5} \cdot 4\frac{3}{8} \cdot 2 = \frac{\overset{2}{\cancel{16}}}{\underset{1}{\cancel{5}}} \cdot \frac{\overset{7}{\cancel{35}}}{\underset{1}{\cancel{8}}} \cdot \frac{2}{1} = \mathbf{28}$

19. $8 \div 10\frac{2}{3} = \frac{8}{1} \div \frac{32}{3} = \frac{\overset{1}{\cancel{8}}}{1} \cdot \frac{3}{\underset{4}{\cancel{32}}} = \mathbf{\frac{3}{4}}$

20. $\begin{array}{r} \mathbf{2.8} \\ 06.\overline{)16.8} \\ \underline{12} \\ 4\,8 \\ \underline{4\,8} \\ 0 \end{array}$

Test 10

1. Tax: $\begin{array}{r} \$19.98 \\ \times\ \ 0.07 \\ \hline \$1.3986 \end{array} \longrightarrow \$1.40$ Total: $\begin{array}{r} \$19.98 \\ +\ \ \$1.40 \\ \hline \mathbf{\$21.38} \end{array}$

2. $\begin{array}{r} \$0.21 \\ 13\overline{)\$2.73} \end{array}$ $\begin{array}{r} \$0.21 \\ +\ \$0.06 \\ \hline \$0.27 \end{array}$ $\begin{array}{r} \$0.27 \\ \times\ \ \ 16 \\ \hline \mathbf{\$4.32} \end{array}$

3. **8 to 17** or $\mathbf{\frac{8}{17}}$

4. $\begin{array}{r} \$119.97 \\ \$98.58 \\ \$99.18 \\ +\ \$105.15 \\ \hline \$422.88 \end{array}$ $\frac{\$422.88}{4} = \$105.72 \approx \mathbf{\$106.00}$

5. $\begin{array}{r} 3.50 \\ -\ 3.01 \\ \hline 0.49 \end{array}$ **forty-nine hundredths**

6.

80 boats

$\frac{3}{5}$ (60%) were for sale.

16 boats
16 boats
16 boats
16 boats
16 boats

$\frac{2}{5}$ (40%) were not for sale.

(a) $\mathbf{\frac{2}{5}}$

(b) 2×16 boats $=$ **32 boats**

7. Area $= (9\text{ cm} \cdot 9\text{ cm}) + \left(\frac{12\text{ cm} \cdot 9\text{ cm}}{2}\right)$

$= 81\text{ cm}^2 + 54\text{ cm}^2 = \mathbf{135\ cm^2}$

8. $16\% = \frac{16}{100} = \mathbf{\frac{4}{25}}$

9. $\begin{array}{r} 0.34545\ldots \\ 11\overline{)3.80000\ldots} \\ \underline{3\,3} \\ 50 \\ \underline{44} \\ 60 \\ \underline{55} \\ 50 \\ \underline{44} \\ 60 \\ \underline{55} \\ 5 \end{array}$ $= \mathbf{0.3\overline{45}}$

10. $\frac{630}{810} = \frac{\cancel{2} \cdot \cancel{3} \cdot \cancel{3} \cdot \cancel{5} \cdot 7}{\cancel{2} \cdot \cancel{3} \cdot \cancel{3} \cdot 3 \cdot 3 \cdot \cancel{5}} = \mathbf{\frac{7}{9}}$

11. Length of side $= 36$ in. $\div 4 = 9$ in.

Area $= 9$ in. $\times$ 9 in. $=$ **81 in.²**

12. $\frac{45}{72} = \frac{25}{f}$

$45f = 25 \cdot 72$

$f = \frac{\overset{5}{\cancel{25}} \cdot \overset{8}{\cancel{72}}}{\underset{1}{\cancel{\overset{\cancel{9}}{45}}}} = \mathbf{40}$

13. $4\overline{)9.6}$ = **2.4**

14. $9.00 - 1.83 = \mathbf{7.17}$

15. $7^2 - 3^3 = 49 - 27 = \mathbf{22}$

16.
```
      1 hr    1 min
   2 hr  47 min  50 s
 + 3 hr  34 min  45 s
   6 hr  22 min  35 s
```
6 hr 22 min 35 s

17. $15\frac{3}{12} - 7\frac{1}{8} = 15\frac{6}{24} - 7\frac{3}{24}$

$= 8\frac{3}{24} = \mathbf{8\frac{1}{8}}$

18. $6\frac{1}{4} \div 3\frac{1}{8} = \frac{25}{4} \div \frac{25}{8} = \frac{\overset{1}{\cancel{25}}}{\underset{1}{\cancel{4}}} \cdot \frac{\overset{2}{\cancel{8}}}{\underset{1}{\cancel{25}}} = \mathbf{2}$

19. $0.185 \times 10^4 = \mathbf{1850}$

20. $004.\overline{)021.060}$ = **5.265**
```
     5.265
004.)021.060
     20
      10
       8
       26
       24
        20
        20
         0
```

TEST 11

1. $\frac{3}{7} = \frac{54}{s}$

$3s = 7 \cdot 54$

$s = \frac{7 \cdot \overset{18}{\cancel{54}}}{\underset{1}{\cancel{3}}} = $ **126 sloops**

2. $91 \times 4 = 364$

$89 + 84 + 92 + f = 364$

The fourth number is **99.**

3. $12\overline{)\$9.36}$ = $0.78

$89¢ per quart − 78¢ per quart = **21¢ per quart**

4. $2\frac{1}{2}$ in. $- 1\frac{1}{4}$ in. $=$ **$1\frac{1}{4}$ in.**

5. 75 classic cars

$\frac{1}{5}$ were not hot rods.	15 classic cars
$\frac{4}{5}$ were hot rods.	15 classic cars
	15 classic cars
	15 classic cars
	15 classic cars

(a) 4×15 classic cars $=$ **60 classic cars**

(b) $\frac{4}{5} \times 100\% = \mathbf{80\%}$

6. (a) $34{,}000{,}000{,}000 = \mathbf{3.4 \times 10^{10}}$

(b) $6.51 \times 10^8 = \mathbf{651{,}000{,}000}$

7. $1.92 + 0.3 = 2.22$

$7 - 5.08 = 1.92$

$2.22 > 1.92$

$1.92 + 0.3 \mathbf{>} 7 - 5.08$

8. $710 \cancel{\text{mm}} \cdot \frac{1 \text{ cm}}{10 \cancel{\text{mm}}} = $ **71 cm**

9. (a) $450\% = \frac{450}{100} = \mathbf{4\frac{1}{2}}$

(b) $450\% = \frac{450}{100} = \mathbf{4.5}$

(c) $10\overline{)7.0}$ = **0.7**

(d) $\frac{7}{10} = \frac{70}{100} = \mathbf{70\%}$

10. $ab - bc = (5)(4) - (4)(3)$

$= 20 - 12 = \mathbf{8}$

11.

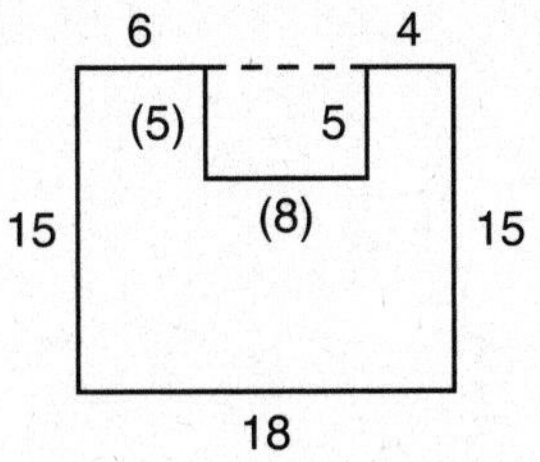

Diagram: top 6, 4; inner (5), 5; (8); sides 15, 15; bottom 18

Area = (18 in. · 15 in.) − (8 in. · 5 in.)
= 270 in.^{2} − 40 in.^{2}
= **230 in.^{2}**

12. Perimeter = 18 in. + 15 in. + 6 in. + 5 in. + 8 in. + 5 in. + 4 in. + 15 in. = **76 in.**

13.

$$\begin{array}{r} 10.00 \\ -\ 4.36 \\ \hline \mathbf{5.64} \end{array}$$

14. $\frac{a}{6} = \frac{35}{10}$

$10a = 6 \cdot 35$

$a = \frac{\overset{3}{\cancel{6}} \cdot \overset{7}{\cancel{35}}}{\underset{\underset{1}{\cancel{5}}}{\cancel{10}}} = \mathbf{21}$

15. $15^2 - 4^3 - 2^4 - \sqrt{225}$
$= 225 - 64 - 16 - 15 = \mathbf{130}$

16. $5 + 5 \cdot 5 - 5 \div 5 = 5 + 25 - 1 = \mathbf{29}$

17. $4\frac{3}{4} + 2\frac{1}{12} + 1\frac{1}{8}$
$= 4\frac{18}{24} + 2\frac{2}{24} + 1\frac{3}{24} = \mathbf{7\frac{23}{24}}$

18. $4\frac{4}{5} \cdot 3\frac{1}{8} \cdot 1\frac{9}{20}$

$= \frac{\overset{3}{\cancel{24}}}{\underset{1}{\cancel{5}}} \cdot \frac{\overset{\overset{1}{\cancel{5}}}{\cancel{25}}}{\underset{1}{\cancel{8}}} \cdot \frac{29}{\underset{4}{\cancel{20}}}$

$= \frac{87}{4} = \mathbf{21\frac{3}{4}}$

19. $0.8(0.25)(0.04) = (0.2)(0.04) = \mathbf{0.008}$

20.

$$\begin{array}{r} \mathbf{400} \\ 0018.\overline{)7200.} \\ \underline{72} \\ 00 \\ \underline{0} \\ 00 \\ \underline{0} \\ 0 \end{array}$$

Test 12

1. A half gallon is 4 pints.
$1.48 ÷ 4 = **$0.37 per pint**

2.

	Ratio	Actual Count
Oatmeal	4	3
Raisins	1	c

$\frac{4}{1} = \frac{3}{c}$

$4c = 3$

$c = \mathbf{\frac{3}{4}}$ **cups**

3.

$$\begin{array}{r} 53.4\text{ s} \\ \times\ 3 \\ \hline 160.2\text{ s} \end{array}$$

51.3 s + 56.4 s + t = 160.2 s
t = **52.5 s**

4. $4\frac{1}{2}$ mi + $4\frac{1}{2}$ mi = 9 mi

$\frac{9\text{ mi}}{60\text{ min}} = \frac{9\text{ mi}}{1\text{ hr}} =$ **9 mph**

5. $W_N = 30\% \cdot 50$
$W_N = 0.3 \cdot 50$
$W_N = \mathbf{15}$

6. (Diagram not required.)

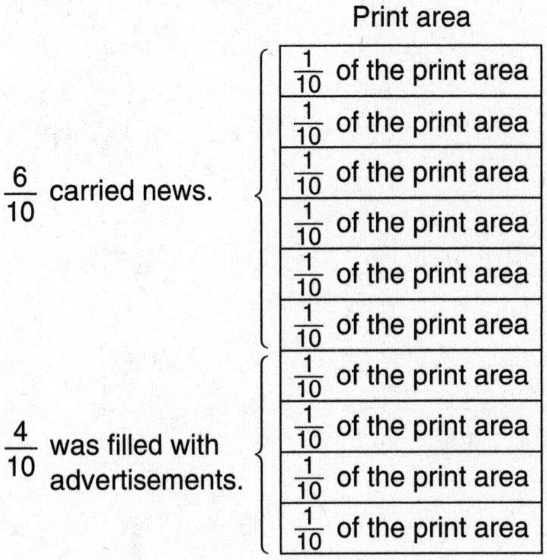

Print area

$\frac{6}{10}$ carried news. (six boxes: $\frac{1}{10}$ of the print area)

$\frac{4}{10}$ was filled with advertisements. (four boxes: $\frac{1}{10}$ of the print area)

(a) $\frac{4}{10} \times 100\% = \mathbf{40\%}$

(b) $\frac{\text{news area}}{\text{advertisement area}} = \frac{6}{4} = \mathbf{\frac{3}{2}}$

7. (a) $0.000309 = \mathbf{3.09 \times 10^{-4}}$

(b) $4.42 \times 10^{-6} = \mathbf{0.00000442}$

8. 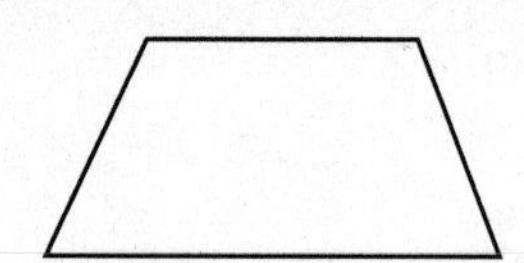 (sketches will vary)

trapezoid

9. $1050 \cancel{\text{yd}} \cdot \dfrac{3 \text{ ft}}{1 \cancel{\text{yd}}} = \mathbf{3150 \text{ ft}}$

10. (a) $18\% = \dfrac{18}{100} = \mathbf{\dfrac{9}{50}}$

(b) $18\% = \dfrac{18}{100} = \mathbf{0.18}$

(c) $50\overline{)1.00}$ with quotient $\mathbf{0.02}$

(d) $\dfrac{1}{50} = \dfrac{2}{100} = \mathbf{2\%}$

11. $\text{Perimeter} = 5 \text{ cm} + 3.5 \text{ cm} + 7 \text{ cm} + 1 \text{ cm} + 2 \text{ cm} + 2.5 \text{ cm}$
$= \mathbf{21 \text{ cm}}$

12.

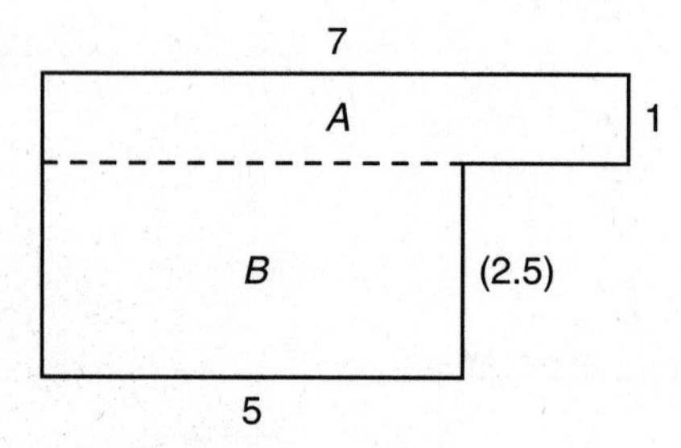

Area = area A + area B
$= (7 \text{ cm} \cdot 1 \text{ cm}) + (5 \text{ cm} \cdot 2.5 \text{ cm})$
$= 7 \text{ cm}^2 + 12.5 \text{ cm}^2 = \mathbf{19.5 \text{ cm}^2}$

13. $\dfrac{8}{18} = \dfrac{n}{54}$

$18n = 8 \cdot 54$

$n = \dfrac{\overset{4}{\cancel{8}} \cdot \overset{6}{\cancel{54}}}{\underset{1}{\cancel{\underset{\cancel{9}}{\cancel{18}}}}}$

$n = \mathbf{24}$

14.
$$\begin{array}{r} 8.0 \\ -\ 4.8 \\ \hline \mathbf{3.2} \end{array}$$

15. $3 + 3 \times 3 - 3 \div 3 = 3 + 9 - 1 = \mathbf{11}$

16. $10^3 - \sqrt{144} + 3^3 + 5^0$
$= 1000 - 12 + 27 + 1 = \mathbf{1016}$

17.
$$\begin{array}{rrr} \overset{6}{\cancel{7}} \text{ yd} & \overset{2}{\cancel{3}} \text{ ft} & 12 \text{ in.} \\ -\ 4 \text{ yd} & 2 \text{ ft} & 9 \text{ in.} \\ \hline \mathbf{2 \text{ yd}} & & \mathbf{3 \text{ in.}} \end{array}$$

18. $5\dfrac{2}{5} + \left(4\dfrac{1}{2} - 2\dfrac{5}{6}\right)$
$= 5\dfrac{12}{30} + \left(4\dfrac{15}{30} - 2\dfrac{25}{30}\right)$
$= 5\dfrac{12}{30} + \left(3\dfrac{45}{30} - 2\dfrac{25}{30}\right)$
$= 5\dfrac{12}{30} + 1\dfrac{20}{30} = 6\dfrac{32}{30}$
$= 7\dfrac{2}{30} = \mathbf{7\dfrac{1}{15}}$

19. $5\dfrac{5}{6} \div \left(2\dfrac{6}{7} \div 4\right)$
$= 5\dfrac{5}{6} \div \left(\dfrac{20}{7} \cdot \dfrac{1}{4}\right)$
$= 5\dfrac{5}{6} \div \dfrac{5}{7}$
$= \dfrac{\overset{7}{\cancel{35}}}{6} \cdot \dfrac{7}{\underset{1}{\cancel{5}}}$
$= \dfrac{49}{6} = \mathbf{8\dfrac{1}{6}}$

20. $5.3(0.03)(0.009) = (0.159)(0.009)$
$= \mathbf{0.001431}$

Test 13

1. Tax:
$$\begin{array}{r} \$20{,}000 \\ \times\ \ 0.085 \\ \hline \$1700.000 \end{array} = \$1700$$

Total:
$$\begin{array}{r} \$20{,}000 \\ +\ \ \$1{,}700 \\ \hline \mathbf{\$21{,}700} \end{array}$$

2. $\dfrac{\$60.75}{9 \text{ hr}} = \mathbf{\$6.75 \text{ per hour}}$

3.

	Ratio	Actual Count
Boys	4	b
Girls	5	g
Children	9	270

$\dfrac{5}{9} = \dfrac{g}{270}$

$9g = 1350$

$g = \mathbf{150 \text{ girls}}$

4. $\dfrac{3\frac{1}{2} + 4\frac{1}{3} + 3 + 5\frac{1}{6}}{4} = \dfrac{16}{4} = \mathbf{4}$

5. $W_N = 18\% \cdot 350$

$W_N = \frac{9}{50} \cdot 350$

$W_N = \mathbf{63}$

6.

228 sports cards

$\frac{3}{4}$ given to his brother	57 sports cards
	57 sports cards
	57 sports cards
$\frac{1}{4}$ left	57 sports cards

(a) $\frac{3}{4} = \mathbf{75\%}$

(b) $1 \cdot 57$ sports cards = **57 sports cards**

7. (a) $0.0002 = \mathbf{2 \times 10^{-4}}$

(b) $7.1 \times 10^{-2} = \mathbf{0.071}$

8. $m\angle D = 180° - m\angle C = 180° - 100°$
$= \mathbf{80°}$

9. $7.5 \text{ kg} \cdot \frac{1000 \text{ g}}{1 \text{ kg}} = 7500 \text{ g}$

$7.5 \text{ kg} = 7500 \text{ g}$

10. $033.\overline{)600.0}$ quotient **18.1**   18.1 rounds to **18**

11. $(-2) + (+7) + (-9) + (+3)$
$= (+5) + (-9) + (+3)$
$= (-4) + (+3) = \mathbf{-1}$

12. (a) $5\overline{)3.0}$ quotient **0.6**

(b) $\frac{3}{5} = \frac{60}{100} = \mathbf{60\%}$

(c) $0.16 = \frac{16}{100} = \mathbf{\frac{4}{25}}$

(d) $0.16 = \mathbf{16\%}$

13. Area = $20 \text{ cm} \times 9 \text{ cm} = \mathbf{180 \text{ cm}^2}$

14. $ab + a + b = \frac{3}{4} \cdot \frac{1}{2} + \frac{3}{4} + \frac{1}{2}$
$= \frac{3}{8} + \frac{6}{8} + \frac{4}{8} = \mathbf{\frac{13}{8}}$ or $\mathbf{1\frac{5}{8}}$

15. $\frac{w}{35} = \frac{16}{20}$

$20w = 16 \cdot 35$

$w = \frac{560}{20}$

$w = \mathbf{28}$

16. $16.\overline{)02.56}$ quotient **0.16**

17. $100 - 2[3(6 - 2)] = 100 - 2[3(4)]$
$= 100 - 2(12) = 100 - 24 = \mathbf{76}$

18. $2\frac{3}{4} + \left(4\frac{1}{6} - 3\frac{2}{3}\right)$

$= 2\frac{9}{12} + \left(4\frac{2}{12} - 3\frac{8}{12}\right)$

$= 2\frac{9}{12} + \left(3\frac{14}{12} - 3\frac{8}{12}\right)$

$= 2\frac{9}{12} + \frac{6}{12} = 2\frac{15}{12} = 3\frac{3}{12}$

$= \mathbf{3\frac{1}{4}}$

19. $6\frac{7}{8}\left(8 \div 2\frac{3}{4}\right) = 6\frac{7}{8}\left(\frac{8}{1} \cdot \frac{4}{11}\right)$

$= \frac{\overset{5}{\cancel{55}}}{\underset{1}{\cancel{8}}} \cdot \frac{\overset{4}{\cancel{32}}}{\underset{1}{\cancel{11}}} = \mathbf{20}$

20. $0.06(0.2)(2.4) = (0.012)(2.4) = \mathbf{0.0288}$

TEST 14

1. $280 \text{ km} + 280 \text{ km} = 560 \text{ km}$

$\frac{560 \text{ km}}{7 \text{ hr}} = \mathbf{80 \frac{km}{hr}}$

2.

	Ratio	Actual Count
Bunnies	3	b
Squirrels	5	s
Total	8	168

$\frac{5}{8} = \frac{s}{168}$

$8s = 840$

$s = $ **105 squirrels**

3. Susan's estimate was **a little too large, because $\pi > 3$.**

4. (a) $\frac{\$11.52}{4} = \mathbf{\$2.88}$

(b) $\$2.88 \times 10 = \mathbf{\$28.80}$

5. $(0.2 + 0.6) - (0.5 \times 0.3)$
$= 0.8 - 0.15$
$= \mathbf{0.65}$

6.

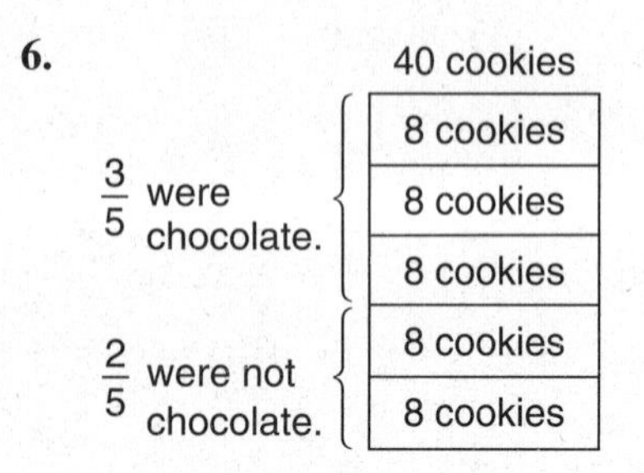

(a) 3(8 cookies) = **24 cookies**

(b) $\frac{2}{5} \times 100\% = \mathbf{40\%}$

7. (a) **6 faces**

(b) Volume = 4 in. · 4 in. · 4 in. = **64 in.^{3}**

8. (a) Circumference $= \pi d$
$= \pi \cdot 2(32 \text{ cm})$
$= \mathbf{64\pi \text{ cm}}$

(b) Circumference $= \pi d$
$\approx \frac{22}{\cancel{7}_1} \cdot \overset{7}{\cancel{49}} \text{ mm}$
$\approx$ **154 mm**

9. (a) $14 \times 10^{-7} = \mathbf{1.4 \times 10^{-6}}$

(b) $14 \times 10^{7} = \mathbf{1.4 \times 10^{8}}$

10. Area $= \frac{20 \text{ mm} \cdot 15 \text{ mm}}{2} = \mathbf{150 \text{ mm}^2}$

11. Area $= \frac{25 \text{ mm} \cdot 15 \text{ mm}}{2} = \mathbf{187.5 \text{ mm}^2}$

12. $ab - (a - b)$
$= (0.7)(0.6) - (0.7 - 0.6)$
$= 0.42 - 0.1 = \mathbf{0.32}$

13. $W_N = 15\% \cdot 1400$
$W_N = \frac{3}{20} \cdot 1400$
$W_N = \mathbf{210}$

14. (a) $8\overline{)7.000}$ = **0.875**

(b) $\frac{7}{8} = 0.875 = \mathbf{87.5\%}$

(c) $22\% = \frac{22}{100} = \mathbf{\frac{11}{50}}$

(d) $22\% = \mathbf{0.22}$

15. $6000 \cancel{g} \cdot \frac{1 \text{ kg}}{1000 \cancel{g}} = \mathbf{6 \text{ kg}}$

16.
$$\begin{array}{r} 50.5 \\ -\ 45. \\ \hline \mathbf{5.5} \end{array}$$

17. $\frac{50}{6} = \mathbf{8\frac{1}{3}}$

18. $3.2 \times 4\frac{1}{2} = 3.2 \times 4.5 = \mathbf{14.4}$

19. $(-4) + (-1) - (-7) - (+8)$
$= (-4) + (-1) + (+7) + (-8)$
$= \mathbf{-6}$

20. $2\frac{1}{4} \div \left(1\frac{1}{2} \cdot 3\right) = 2\frac{1}{4} \div \left(\frac{3}{2} \cdot \frac{3}{1}\right)$
$= 2\frac{1}{4} \div \frac{9}{2} = \frac{9}{4} \cdot \frac{2}{9}$
$= \mathbf{\frac{1}{2}}$

TEST 15

1.

	Case 1	Case 2
Earth	300	180
Moon	50	*m*

$\frac{300}{50} = \frac{180}{m}$
$300m = 9000$
$m =$ **30 pounds**

2. $\frac{3.2 + 5.4}{2} = \mathbf{4.3}$

3. Perimeter $= 9 \text{ in.} + 10 \text{ in.} + 13 \text{ in.} + 8 \text{ in.}$
$= \mathbf{40 \text{ in.}}$

4.

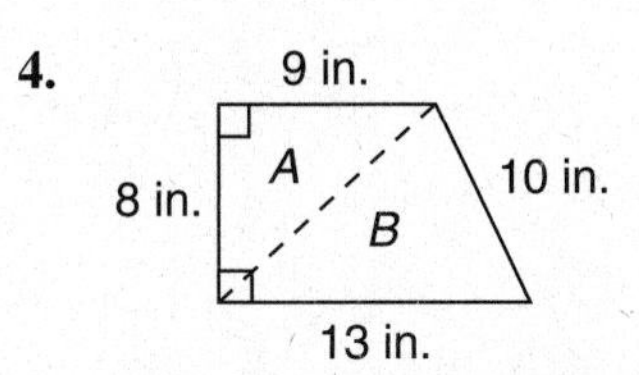

Area = area A + area B

$$= \frac{9 \text{ in.} \cdot 8 \text{ in.}}{2} + \frac{13 \text{ in.} \cdot 8 \text{ in.}}{2}$$

$$= 36 \text{ in.}^2 + 52 \text{ in.}^2 = \mathbf{88 \text{ in.}^2}$$

5. $4^2 - \sqrt{4} = 16 - 2 = \mathbf{14}$

6.

	Ratio	Actual Count
Boys	4	b
Girls	7	g
Children	11	550

$$\frac{7}{11} = \frac{g}{550}$$
$$11g = 3850$$
$$g = \mathbf{350 \text{ girls}}$$

7. $8.3 \text{ g} \cdot \dfrac{1000 \text{ mg}}{1 \text{ g}} = \mathbf{8300 \text{ mg}}$

8.

36 games

$\frac{1}{3}$ { 12 games

12 games

12 games

(a) total = 3(12 games) = **36 games**

(b) $w = 75\% \cdot 36$
$$w = \frac{3}{4} \cdot 36$$
$$w = \mathbf{27 \text{ games}}$$

9.
$$36 = \frac{4}{5} \times W_N$$
$$\frac{5}{4} \times 36 = \frac{5}{4} \times \frac{4}{5} \times W_N$$
$$\mathbf{45} = W_N$$

10.
$$\frac{1}{10} \times W_N = 291$$
$$10 \times \frac{1}{10} \times W_N = 10 \times 291$$
$$W_N = \mathbf{2910}$$

11. (a) $-8(-5) = \mathbf{40}$

(b) $-7(+2) = \mathbf{-14}$

(c) $\dfrac{-15}{-5} = \mathbf{3}$

(d) $\dfrac{18}{-3} = \mathbf{-6}$

12. Volume $= (9 \text{ cm})^3 = \mathbf{729 \text{ cm}^3}$

13. (a) Circumference $= \pi d$
$\approx 3.14 (52 \text{ in.})$
$\approx \mathbf{163.28 \text{ in.}}$

(b) Circumference $= \pi d$
$= \pi \cdot 2(26 \text{ m})$
$= \mathbf{52\pi \text{ m}}$

14. (a) $6\overline{)5.0000\ldots}$ with quotient $0.8333\ldots$; $0.8333\ldots = \mathbf{0.8\overline{3}}$

(b) $\dfrac{5}{6} \times 100\% = \mathbf{83\frac{1}{3}\%}$

(c) $0.65 = \dfrac{65}{100} = \mathbf{\dfrac{13}{20}}$

(d) $0.65 \times 100\% = \mathbf{65\%}$

15. $10m - (my - y^2)$
$= 10 \cdot 11 - (11 \cdot 8 - 8^2)$
$= 110 - (88 - 64) = \mathbf{86}$

16.
$$\frac{3}{4}y = 24$$
$$\frac{4}{3} \cdot \frac{3}{4}y = \overset{8}{\cancel{24}} \cdot \frac{4}{\underset{1}{\cancel{3}}}$$
$$y = \mathbf{32}$$

17.
$$s + 1.6 = 5$$
$$s + 1.6 - 1.6 = 5 - 1.6$$
$$s = \mathbf{3.4}$$

18. $5\frac{1}{9} \div \left(3\frac{1}{2} + 4\frac{1}{6}\right)$
$$= 5\frac{1}{9} \div \left(3\frac{3}{6} + 4\frac{1}{6}\right)$$
$$= 5\frac{1}{9} \div 7\frac{2}{3} = \frac{46}{9} \cdot \frac{3}{23} = \mathbf{\frac{2}{3}}$$

19. $(-6) - (-8) + (-7)$
$= (-6) + (+8) + (-7) = \mathbf{-5}$

20. $\dfrac{\$524}{1 \cancel{\text{wk}}} \cdot \dfrac{1 \cancel{\text{wk}}}{5 \cancel{\text{days}}} \cdot \dfrac{1 \cancel{\text{day}}}{8 \text{ hr}} = \dfrac{\$524}{40 \text{ hr}} = \mathbf{\dfrac{\$13.10}{\text{hr}}}$

Test 16

1. $\dfrac{(3 \times \$6.80) + (2 \times \$6.20)}{5 \text{ hr}}$

$= \dfrac{\$20.40 + \$12.40}{5 \text{ hr}} = \dfrac{\$32.80}{5 \text{ hr}} = \mathbf{\dfrac{\$6.56}{hr}}$

2. $x + (x^2 - xy) - y$

$= 8 + (8^2 - 8 \cdot 5) - 5$

$= 8 + 24 - 5 = \mathbf{27}$

3. If $a - b = 0$, then $a = b$.

$\mathbf{a = b}$

4.

	Ratio	Actual Count
Clean	3	c
Dirty	5	d
Total	8	48

$\dfrac{3}{8} = \dfrac{c}{48}$

$8c = 144$

$c =$ **18 articles of clothing**

5.

	Case 1	Case 2
Customers	600	c
Minutes	25	60

$\dfrac{600}{25} = \dfrac{c}{60}$

$25c = 36{,}000$

$c =$ **1440 customers**

6. Circumference $= \pi d$

$\approx 3.14 \cdot 18$ m

≈ 56.52 m

$\approx$ **57 m**

7.

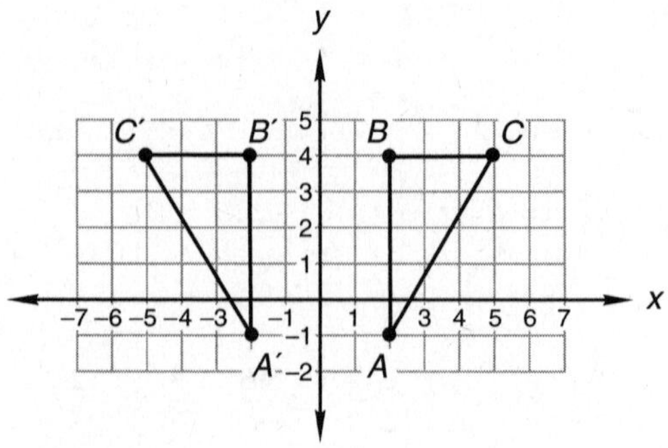

A′ (−2, −1)
B′ (−2, 4)
C′ (−5, 4)

8.

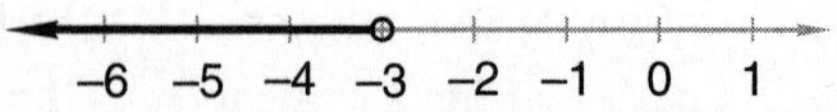

9. 28 in. $\times$ 4 = **112 in.**

10. (a) $\dfrac{350}{-7} = \mathbf{-50}$

(b) $\dfrac{-880}{-11} = \mathbf{80}$

(c) $14(-40) = \mathbf{-560}$

(d) $19(+60) = \mathbf{1140}$

11. (a) $9\overline{)4.000\ldots}$ = $0.444\ldots$ $0.444\ldots = \mathbf{0.\overline{4}}$

(b) $\dfrac{4}{9} \times 100\% = \mathbf{44\frac{4}{9}\%}$

(c) $0.85 = \dfrac{85}{100} = \mathbf{\dfrac{17}{20}}$

(d) $0.85 = \dfrac{85}{100} = \mathbf{85\%}$

12.

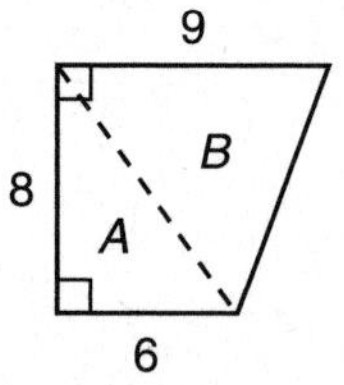

Area = area A + area B

$= \dfrac{6 \text{ m} \cdot 8 \text{ m}}{2} + \dfrac{9 \text{ m} \cdot 8 \text{ m}}{2}$

$= 24 \text{ m}^2 + 36 \text{ m}^2 = \mathbf{60 \text{ m}^2}$

13. $420 = \dfrac{3}{5} \times W_N$

$\dfrac{5}{\not{3}_1} \times \overset{140}{\not{420}} = \dfrac{5}{3} \times \dfrac{3}{5} \times W_N$

$\mathbf{700} = W_N$

14. $W_P \times 60 = 45$

$\dfrac{W_P \times 60}{60} = \dfrac{45}{60}$

$W_P = \dfrac{3}{4}$

$W_P = \mathbf{75\%}$

15. $\dfrac{4}{5}m = 52$

$\dfrac{5}{4} \cdot \dfrac{4}{5}m = \dfrac{5}{\not{4}_1} \times \overset{13}{\not{52}}$

$m = \mathbf{65}$

16. $3.5 = x - 0.09$

$3.5 + 0.09 = x - 0.09 + 0.09$

$\mathbf{3.59} = x$

17. $\dfrac{6\frac{1}{4}}{100} = \dfrac{25}{4} \cdot \dfrac{1}{100} = \dfrac{25}{400} = \mathbf{\dfrac{1}{16}}$

18. $\dfrac{3^3 + 4 \cdot 2 - 5 \cdot 2^2}{\sqrt{3^2 + 4^2}}$

$= \dfrac{27 + 8 - 20}{\sqrt{25}}$

$= \dfrac{15}{5} = \mathbf{3}$

19. $6\frac{1}{3} \div 1.9 = 6\frac{1}{3} \div 1\frac{9}{10}$

$= \dfrac{\overset{1}{\cancel{19}}}{3} \cdot \dfrac{10}{\underset{1}{\cancel{19}}} = \dfrac{10}{3} = \mathbf{3\frac{1}{3}}$

20. $-33 - (-23) + (+32)$

$= (-33) + (+23) + (+32)$

$= (-10) + (+32) = \mathbf{22}$

Test 17

1.

	Ratio	Actual Count
Won	2	w
Failed to win	5	f
Played	7	49

$\dfrac{5}{7} = \dfrac{f}{49}$

$7f = 245$

$f = $ **35 games**

2. (a) $(80 + 90 + 80 + 65 + 95 + 90 + 75 + 100 + 65 + 70) \div 10 = \mathbf{81}$

(b) $\dfrac{80 + 80}{2} = \mathbf{80}$

(c) **65, 80, 90**

(d) $100 - 65 = \mathbf{35}$

3.

	Ratio	Actual Count
Dandelions	12	d
Peonies	5	45

$\dfrac{12}{5} = \dfrac{d}{45}$

$5d = 12 \cdot 45$

$d = \dfrac{12 \cdot \overset{9}{\cancel{45}}}{\underset{1}{\cancel{5}}}$

$d = $ **108 dandelions**

4. $0.37\,\cancel{L} \cdot \dfrac{1000 \text{ mL}}{1\,\cancel{L}} = $ **370 mL**

5.

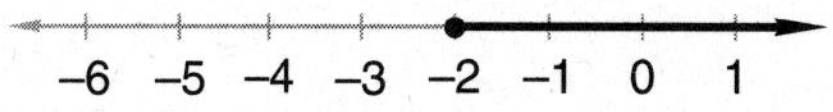

6. $6xy - xy - 5x + x = \mathbf{5xy - 4x}$

7.

	Case 1	Case 2
Distance	2 mi	d
Time	10 s	240 s

$\dfrac{2}{10} = \dfrac{d}{240}$

$10d = 480$

$d = $ **48 miles**

8.

	Ratio	Actual Count
Waved	65	w
Did not wave	35	133
Total	100	t

$\dfrac{35}{100} = \dfrac{133}{t}$

$35t = 13{,}300$

$t = $ **380 fans**

9.

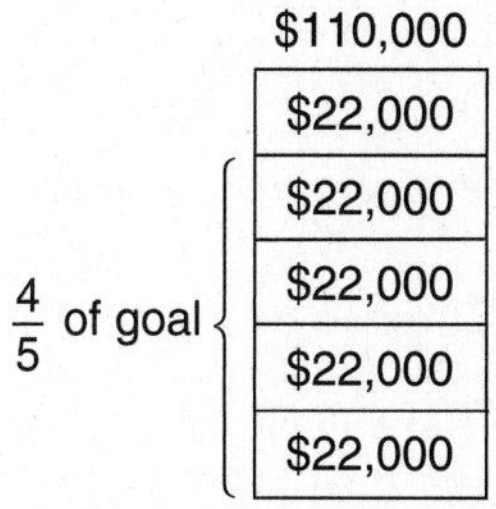

(a) goal $= 5 \times \$22{,}000 = $ **$110,000**

(b) The drive fell short by $\frac{1}{5}$, which is **20%.**

10.

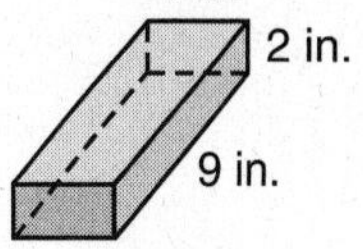

Volume $= 9 \text{ in.} \times 4 \text{ in.} \times 2 \text{ in.} = \mathbf{72 \text{ in.}^3}$

11. Area $= \pi r^2$

$\approx \dfrac{22}{7}(7 \text{ mm})^2$

$\approx \mathbf{154 \text{ mm}^2}$

12. (a) $20\overline{)11.00}$ → **0.55**

(b) $\frac{11}{20} = \frac{55}{100} = \mathbf{55\%}$

(c) $7\% = \mathbf{\frac{7}{100}}$

(d) $7\% = \mathbf{0.07}$

13. $(2.2 \times 10^5)(2.1 \times 10^6) = \mathbf{4.62 \times 10^{11}}$

14. **parallelogram**
Perimeter
$= 18\text{ cm} + 16\text{ cm} + 18\text{ cm} + 16\text{ cm}$
$= \mathbf{68\text{ cm}}$

15. $\text{Area} = 18\text{ cm} \cdot 15\text{ cm} = \mathbf{270\text{ cm}^2}$

16. $18.2 = 1.4p$
$\frac{18.2}{1.4} = \frac{1.4p}{1.4}$
$\mathbf{13} = p$

17. $z + \frac{5}{8} = 1\frac{1}{4}$
$z + \frac{5}{8} - \frac{5}{8} = 1\frac{1}{4} - \frac{5}{8}$
$z = \mathbf{\frac{5}{8}}$

18. $2\{45 - [6^2 - 3(11 - 7)]\}$
$= 2\{45 - [36 - 3(4)]\}$
$= 2[45 - (36 - 12)]$
$= 2(45 - 24) = 2(21) = \mathbf{42}$

19. $2.3 \div \left(6\frac{1}{2} - 3\frac{5}{8}\right)$
$= 2.3 \div \left(5\frac{12}{8} - 3\frac{5}{8}\right)$
$= 2.3 \div 2\frac{7}{8} = \frac{23}{10} \cdot \frac{8}{23} = \mathbf{\frac{4}{5}}$

20. $(-5) + (-9) - (-6) + (-1)$
$= (-5) + (-9) + (+6) + (-1)$
$= (-14) + (+6) + (-1)$
$= (-8) + (-1) = \mathbf{-9}$

TEST 18

1.

	Ratio	Actual Count
Lions	4	12
Tigers	5	t

$\frac{4}{5} = \frac{12}{t}$
$4t = 60$
$t = 15$ tigers

	Ratio	Actual Count
Tigers	3	15
Bears	5	b

$\frac{3}{5} = \frac{15}{b}$
$3b = 75$
$b = \mathbf{25\text{ bears}}$

2. Volume
$= 38\text{ cm} \cdot 20\text{ cm} \cdot 14\text{ cm} = \mathbf{10{,}640\text{ cm}^3}$

3. $\frac{28\text{ hits}}{76\text{ at-bats}} \approx \mathbf{0.368}$

4. $27\text{ ft}^2 \cdot \frac{1\text{ yd}}{3\text{ ft}} \cdot \frac{1\text{ yd}}{3\text{ ft}} = \mathbf{3\text{ yd}^2}$

5.

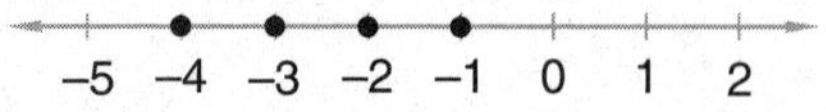

6. $\frac{1}{5} \cdot 360° = \mathbf{72°}$

7.

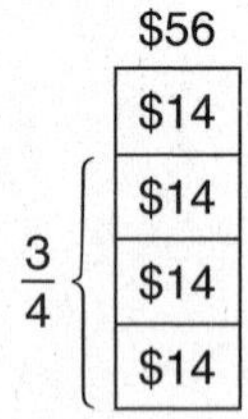

(a) regular price $= 4 \times \$14 = \mathbf{\$56}$

(b) $\frac{3}{4} \cdot 100\% = \mathbf{75\%}$

8. (a) $90° + 30° + m\angle w = 180°$
$m\angle w = \mathbf{60°}$

(b) $m\angle y = m\angle w$
$55° + m\angle y + m\angle z = 180°$
$55° + 60° + m\angle z = 180°$
$m\angle z = \mathbf{65°}$

9.

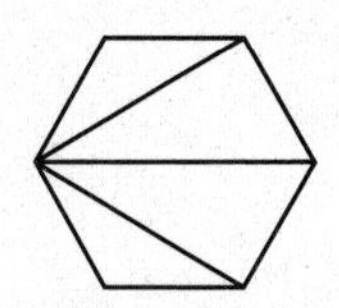

3 diagonals

10. Circumference $= \pi d$

$$\approx \frac{22}{7} \cdot 98 \text{ in.}$$

$$\approx \mathbf{308 \text{ in.}}$$

11.

18 m
13 m
10 m
B
A
32 m

Area = area A + area B

$$= \frac{32 \text{ m} \cdot 10 \text{ m}}{2} + \frac{18 \text{ m} \cdot 10 \text{ m}}{2}$$

$$= 160 \text{ m}^2 + 90 \text{ m}^2 = \mathbf{250 \text{ m}^2}$$

12. $a^2 = 0.7^2 = 0.49$

$a = 0.7$

$0.49 < 0.7$

$a^2 \mathbf{<} a$

13. (a) $0.05 = \frac{5}{100} = \mathbf{\frac{1}{20}}$

(b) $0.05 = 0.05 \cdot 100\% = \mathbf{5\%}$

14. $W_P \times 200 = 50$

$$\frac{W_P \times 200}{200} = \frac{50}{200}$$

$W_P = \mathbf{25\%}$

15.

	Ratio	Actual Count
Ordered hamburgers	25	h
Did not order hamburgers	75	n
Total	100	5000

$$\frac{75}{100} = \frac{n}{5000}$$

$100n = 375{,}000$

$n =$ **3750 customers**

16. $(2.75 \times 10^{-2})(6 \times 10^{-8})$

$= 16.50 \times 10^{-10}$

$= \mathbf{1.65 \times 10^{-9}}$

17. $7\frac{1}{2}y = 75$

$$\frac{2}{15} \cdot \frac{15}{2}y = \frac{2}{\cancel{15}_{3}} \cdot \cancel{75}^{15}$$

$y = \mathbf{10}$

18. $14.2 = 6.84 + f$

$14.2 - 6.84 = 6.84 + f - 6.84$

$\mathbf{7.36} = f$

19. $(-8x^2)(-2xy^2) = \mathbf{16x^3y^2}$

20. $(-6) - (-8)(+4) - (-2)(-5)$

$= -6 + 32 - 10 = \mathbf{16}$

Test 19

1.

$$\begin{array}{r} 83 \\ \times\ 4 \\ \hline 332 \end{array} \quad \begin{array}{r} 85 \\ \times\ 5 \\ \hline 425 \end{array} \quad \begin{array}{r} 425 \\ -\ 332 \\ \hline \mathbf{93} \end{array}$$

2.

$$\begin{array}{r} 75 \text{ girls} \\ +\ ?\text{ boys} \\ \hline 90 \text{ total} \end{array} \longrightarrow \begin{array}{r} 75 \text{ girls} \\ +\ 15 \text{ boys} \\ \hline 90 \text{ total} \end{array}$$

The ratio of boys to girls is $\frac{15}{75}$, which reduces to $\mathbf{\frac{1}{5}}$.

3. $72 \text{ \sout{juice bars}} \times \frac{\$5.60}{48 \text{ \sout{juice bars}}} = \frac{\$403.20}{48}$

$= \mathbf{\$8.40}$

4. $P(H, H, H) = \frac{1}{2} \cdot \frac{1}{2} \cdot \frac{1}{2} = \mathbf{\frac{1}{8}}$

5.

	%	Cost
Original	100	b
Increase	25	c
New	125	80¢

$$\frac{100}{125} = \frac{b}{80}$$

$125b = 8000$

$b =$ **64¢ per pound**

6. $30 = W_P \times 40$

$$\frac{30}{40} = \frac{W_P \times 40}{40}$$

$$\frac{3}{4} = W_P$$

$\mathbf{75\%} = W_P$

7. $2000\ \cancel{\text{cm}^2} \cdot \dfrac{10\text{ mm}}{1\ \cancel{\text{cm}}} \cdot \dfrac{10\text{ mm}}{1\ \cancel{\text{cm}}} = \mathbf{200{,}000\ mm^2}$

8. $y = 3x - 2$

$y = 3(-7) - 2$

$y = -21 - 2$

$y = \mathbf{-23}$

9. Volume = area of base × height

$= \left(\dfrac{6\text{ cm} \cdot 10\text{ cm}}{2}\right)(6\text{ cm}) = \mathbf{180\ cm^3}$

10. (a) $3\dfrac{3}{4} = 3\dfrac{75}{100} = \mathbf{3.75}$

(b) $3\dfrac{3}{4} \times 100\% = 3.75 \times 100\% = \mathbf{375\%}$

(c) $5\dfrac{1}{2}\% = \dfrac{5\frac{1}{2}}{100} = \dfrac{11}{2} \cdot \dfrac{1}{100} = \mathbf{\dfrac{11}{200}}$

(d) $5\dfrac{1}{2}\% = \dfrac{5.5}{100} = \mathbf{0.055}$

11. (a) Tax:
$$\begin{array}{r} \$85.00 \\ \times\quad 0.06 \\ \hline \mathbf{\$5.10} \end{array}$$

(b) Total:
$$\begin{array}{r} \$85.00 \\ +\quad \$5.10 \\ \hline \mathbf{\$90.10} \end{array}$$

12. $(2 \times 10^{-4})(9 \times 10^{7}) = 18 \times 10^{3}$

$= \mathbf{1.8 \times 10^4}$

13. 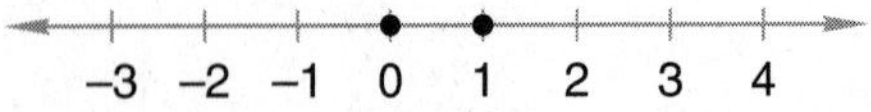

14. $4\dfrac{2}{3}x = 56$

$\dfrac{3}{14} \cdot \dfrac{14}{3}x = \dfrac{3}{\cancel{14}_{1}} \cdot \cancel{56}^{4}$

$x = \mathbf{12}$

15. $3m - 37 = 47$

$3m - 37 + 37 = 47 + 37$

$3m = 84$

$\dfrac{3m}{3} = \dfrac{84}{3}$

$m = \mathbf{28}$

16. $(4 \cdot 3)^2 - 4(3)^2 = 12^2 - 4(9)$

$= 144 - 36 = \mathbf{108}$

17. $(-3x^2)(3x^2y)(-2xy) = \mathbf{18x^5y^2}$

18. $4 - \left(7\dfrac{1}{2} - 5.6\right) = 4 - \left(7\dfrac{1}{2} - 5\dfrac{6}{10}\right)$

$= 4 - \left(6\dfrac{15}{10} - 5\dfrac{6}{10}\right) = 4 - 1\dfrac{9}{10}$

$= 3\dfrac{10}{10} - 1\dfrac{9}{10} = \mathbf{2\dfrac{1}{10}}$

19. $2x - y + x - y = 2x + x - y - y$

$= \mathbf{3x - 2y}$

20. $\dfrac{2 - 4 + 1 - 14 + 7(-6)}{3} = \dfrac{-57}{3} = \mathbf{-19}$

Test 20

1. $\dfrac{(4 \cdot 86) + (2 \cdot 92)}{6} = \mathbf{88}$

2.

	%	Pay
Original	100	\$6.90
Increase	20	*c*
New	120	*n*

$\dfrac{100}{120} = \dfrac{6.90}{n}$

$100n = 828$

$n = \mathbf{\$8.28\ per\ hour}$

3. $21 = W_P \times 14$

$\dfrac{21}{14} = \dfrac{W_P \times 14}{14}$

$\dfrac{3}{2} = W_P$

$\dfrac{150}{100} = W_P$

$\mathbf{150\%} = W_P$

4. $2.6\ \cancel{\text{m}^2} \cdot \dfrac{100\text{ cm}}{1\ \cancel{\text{m}}} \cdot \dfrac{100\text{ cm}}{1\ \cancel{\text{m}}} = \mathbf{26{,}000\ cm^2}$

5.

42 eggs

	7 eggs
$\frac{5}{6}$ were not cracked.	7 eggs
	7 eggs
	7 eggs
	7 eggs
$\frac{1}{6}$ were cracked.	7 eggs

(a) total = 6(7 eggs) = **42 eggs**

(b) If $\frac{1}{6}$ of the eggs were cracked, then $\frac{5}{6}$ of the eggs were not cracked.

$\frac{5}{6} \times 100\% = \frac{500\%}{6} = \mathbf{83\frac{1}{3}\%}$

6. $\frac{a + b}{c} = \frac{(-5) + (-4)}{(-6)} = \frac{-9}{-6} = \mathbf{\frac{3}{2}}$

7. Side length: 36 in. ÷ 4 = 9 in.

Area = $(9 \text{ in.})^2$ = **81 in.^{2}**

8. $P(C, C) = \frac{4}{6} \cdot \frac{4}{6} = \frac{16}{36} = \mathbf{\frac{4}{9}}$

9. Volume = area of base × height

$= \left(\frac{8 \text{ cm} \cdot 3 \text{ cm}}{2}\right)(9 \text{ cm}) = \mathbf{108 \text{ cm}^3}$

10. Area $= \pi r^2$

$\approx (3.14)(8 \text{ cm})^2$

$\approx \mathbf{200.96 \text{ cm}^2}$

11. Price:

$$\begin{array}{r} \$17.00 \\ \times \quad 20 \\ \hline \$340.00 \end{array}$$

Tax:

$$\begin{array}{r} \$340 \\ \times \ 0.07 \\ \hline \$23.80 \end{array}$$

Total:

$$\begin{array}{r} \$340.00 \\ + \quad 23.80 \\ \hline \mathbf{\$363.80} \end{array}$$

12. $W_N = 33\frac{1}{3}\% \times \$33.00$

$W_N = \frac{1}{3} \times \$33.00$

$W_N = \mathbf{\$11.00}$

13. $\frac{5}{12} \cdot 360° = \mathbf{150°}$

14. $(3 \times 10^4)(6 \times 10^{-7}) = 18 \times 10^{-3}$

$= \mathbf{1.8 \times 10^{-2}}$

15.

$$\begin{aligned} 0.5j - 1.2 &= 1.2 \\ 0.5j - 1.2 + 1.2 &= 1.2 + 1.2 \\ 0.5j &= 2.4 \\ \frac{0.5j}{0.5} &= \frac{2.4}{0.5} \\ j &= \mathbf{4.8} \end{aligned}$$

16.

$$\begin{aligned} \frac{2}{3}x - 3 &= 15 \\ \frac{2}{3}x - 3 + 3 &= 15 + 3 \\ \frac{2}{3}x &= 18 \\ \frac{3}{2} \cdot \frac{2}{3}x &= \frac{3}{2} \cdot 18 \\ x &= \mathbf{27} \end{aligned}$$

17. $4^3 - \sqrt{25} + 9 \cdot 2^4$

$= 64 - 5 + 144 = \mathbf{203}$

18.

$$\begin{array}{r} \scriptstyle 1 \text{ ft} \\ 5 \text{ yd } 1 \text{ ft } 11 \text{ in.} \\ + \qquad\qquad 6 \text{ in.} \\ \hline \mathbf{5 \text{ yd } 2 \text{ ft } 5 \text{ in.}} \end{array}$$

19. $3x + 8(x + 2) = 3x + 8x + 16$

$= \mathbf{11x + 16}$

20. $\frac{-4(-6) + 5(-1)(-3)}{(-3)} = \frac{24 + 15}{-3} = \mathbf{-13}$

TEST 21

1. 15% of \$21.00 = 0.15 × \$21.00 = **\$3.15**

2. $\frac{245 \text{ km}}{3.5 \text{ hr}} = \mathbf{\frac{70 \text{ km}}{\text{hr}}}$

3.

	Scale	Measure
Model	1	11
Object	48	h

$\frac{1}{48} = \frac{11}{h}$

h = **528 in.** or **44 ft**

4.

	%	Cost
Original	100	r
Discount	20	45
New	80	s

$$\frac{100}{20} = \frac{r}{45}$$
$$20r = 4500$$
$$r = \mathbf{225}$$

5.

	%	Cost
Original	100	30
Increase	60	c
New	160	p

$$\frac{100}{160} = \frac{30}{p}$$
$$100p = 4800$$
$$p = \mathbf{\$48}$$

6. $W_N = 6.5\% \times \$74.00$
$W_N = 0.065 \times \$74.00$
$W_N = \mathbf{\$4.81}$

7. Perimeter of semicircle $= \frac{\pi d}{2}$
$\approx \frac{(3.14)(12 \text{ cm})}{2}$
$\approx 18.84 \text{ cm}$

Perimeter $\approx 12 \text{ cm} + 2 \text{ cm} + 18.84 \text{ cm} + 2 \text{ cm} \approx \mathbf{34.84 \text{ cm}}$

8.
$$a^2 + b^2 = c^2$$
$$a^2 + (30 \text{ in.})^2 = (34 \text{ in.})^2$$
$$a^2 + 900 \text{ in.}^2 = 1156 \text{ in.}^2$$
$$a^2 = 256 \text{ in.}^2$$
$$a = \mathbf{16 \text{ in.}}$$

9.
Area of triangle $= \frac{6 \text{ in.} \cdot 8 \text{ in.}}{2} = 24 \text{ in.}^2$
Area of triangle $= \frac{6 \text{ in.} \cdot 8 \text{ in.}}{2} = 24 \text{ in.}^2$
Area of rectangle $= 6 \text{ in.} \cdot 6 \text{ in.} = 36 \text{ in.}^2$
Area of rectangle $= 8 \text{ in.} \cdot 6 \text{ in.} = 48 \text{ in.}^2$
+ Area of rectangle $= 10 \text{ in.} \cdot 6 \text{ in.} = 60 \text{ in.}^2$
Total surface area $= \mathbf{192 \text{ in.}^2}$

10. Volume = area of base × height
Area of base $= \pi r^2$
$\approx (3.14)(3 \text{ cm})^2$
$\approx 28.26 \text{ cm}^2$
Volume $\approx (28.26 \text{ cm}^2)(6 \text{ cm})$
$\approx \mathbf{169.56 \text{ cm}^3}$

11.
$$\frac{24}{8} = \frac{x}{6}$$
$$8x = 144$$
$$x = \mathbf{18}$$

12.
$$2x + 3x + 90° = 180°$$
$$5x + 90° = 180°$$
$$5x + 90° - 90° = 180° - 90°$$
$$5x = 90°$$
$$\frac{5x}{5} = \frac{90°}{5}$$
$$x = 18°$$
$$m\angle AOB = 3x = 3(18°) = \mathbf{54°}$$

13. (a) $\mathbf{-5, \sqrt{5}, 5, 5^2}$
(b) $\sqrt{5}$

14. $P(T, T) = \frac{1}{2} \cdot \frac{1}{2} = \frac{1}{4}$
The probability that the coin will land tails up twice is $\frac{1}{4}$, which is a **25%** chance.

15. Since $\sqrt{7} < 3$, the answer is not A.
Since $\sqrt[3]{125} = 5$, the answer is not B.
Since $\sqrt{36} = 6$ and $\sqrt{64} = 8$, the number between 6 and 8 is **C.** $\mathbf{\sqrt{57}}$**.**

16.
$$4x - 16 + x = 24$$
$$5x - 16 = 24$$
$$5x - 16 + 16 = 24 + 16$$
$$5x = 40$$
$$\frac{5x}{5} = \frac{40}{5}$$
$$x = \mathbf{8}$$

17.
$$\frac{16}{w} = \frac{94}{4.7}$$
$$94w = 16 \cdot 4.7$$
$$w = \frac{75.2}{94}$$
$$w = \mathbf{0.8}$$

18. $\frac{(4x^2y)(3x^2)}{6x^2} = \frac{12x^4y}{6x^2} = \mathbf{2x^2y}$

19. $(-3)^2 - 2^3 = 9 - 8 = \mathbf{1}$

20.
$$\frac{(-18) - (-4)(+5)}{(-4) - (+5) - (+5)}$$
$$= \frac{(-18) + (+20)}{(-4) + (-5) + (-5)}$$
$$= \frac{2}{-14} = \mathbf{-\frac{1}{7}}$$

Test 22

1. (a) $(60 + 46 + 64 + 69 + 72 + 66 + 72 + 59 + 68) \div 9 = \mathbf{64}$

(b) **66**

(c) **72**

(d) $72 - 46 = \mathbf{26}$

2. $P(C, C) = \frac{\overset{1}{\cancel{13}}}{\underset{\underset{1}{\cancel{4}}}{\cancel{52}}} \cdot \frac{\overset{\overset{1}{\cancel{4}}}{\cancel{12}}}{\underset{17}{\cancel{51}}} = \mathbf{\frac{1}{17}}$

3.

	Case 1	Case 2
Dollars	160	*d*
Francs	300	330

$$\frac{160}{300} = \frac{d}{330}$$
$$300d = 52{,}800$$
$$d = \mathbf{\$176}$$

4.

	Ratio	Actual Count
Ducks	4	*d*
Geese	9	*g*
Total	13	351

$$\frac{4}{13} = \frac{d}{351}$$
$$13d = 1404$$
$$d = \mathbf{108\ ducks}$$

5.

	%	$
Original	100	130
Decrease	30	*d*
New	70	*n*

$$\frac{100}{70} = \frac{130}{n}$$
$$100n = 9100$$
$$n = \mathbf{\$91\ per\ day}$$

6. Volume = area of base × height

Area of base $= \pi r^2$
$\approx (3.14)(5\text{ cm})^2$
$\approx 78.5\text{ cm}^2$

Volume $\approx (78.5\text{ cm}^2)(25\text{ cm})$
$\approx \mathbf{1962.5\ cm^3}$

7. $t = 1.04p$
$t = 1.04(7.5)$
$t = \mathbf{7.8}$

8. $y = 2x - 3$

x	*y*
0	–3
1	–1
2	1

Check student work. Number pairs will vary.

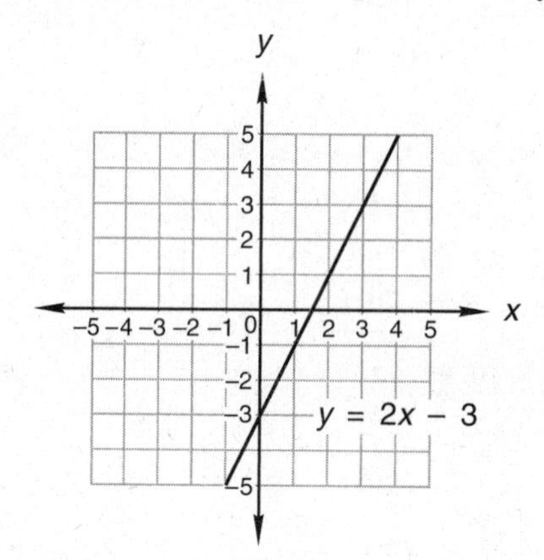

$$\text{Slope} = \frac{\text{rise}}{\text{run}} = \frac{+2}{1} = \mathbf{2}$$

9. $W_N = 6.5\% \times \$70.00$
$W_N = 0.065 \times \$70.00$
$W_N = \mathbf{\$4.55}$

10. $0.1 \times W_N = 220$
$W_N = \frac{220}{0.1}$
$W_N = \mathbf{2200}$

11. Perimeter of semicircle $= \frac{\pi d}{2}$
$\approx \frac{(3.14)(8\text{ cm})}{2}$
$\approx 12.56\text{ cm}$

Perimeter $\approx 8\text{ cm} + 3\text{ cm} + 12.56\text{ cm} + 3\text{ cm} \approx \mathbf{26.56\ cm}$

12. In the figure the acute angles are supplementary to the obtuse angles.

$m\angle h = 180° - m\angle a$
$m\angle h = 180° - 117°$
$m\angle h = \mathbf{63°}$

13. $d = rt$
$\frac{d}{r} = \frac{rt}{r}$
$\mathbf{\frac{d}{r} = t}$

14.

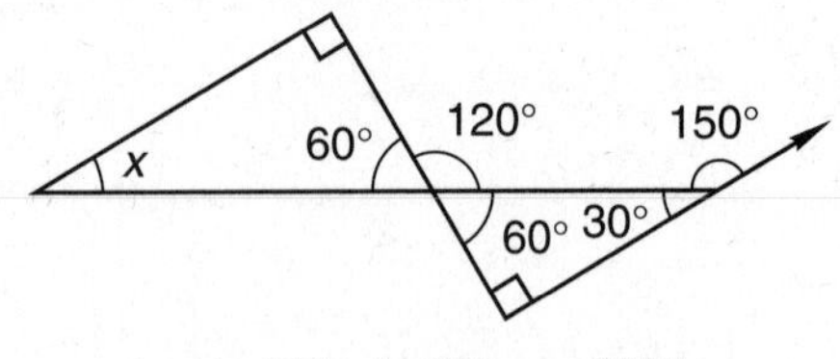

$$\begin{aligned} m\angle x + 90° + 60° &= 180° \\ m\angle x + 150° &= 180° \\ m\angle x + 150° - 150° &= 180° - 150° \\ m\angle x &= \mathbf{30°} \end{aligned}$$

15. (a)
$$\begin{aligned} \frac{8}{y} &= \frac{6}{15} \\ 6y &= 120 \\ y &= \mathbf{20} \end{aligned}$$

(b)
$$\begin{aligned} 6f &= 15 \\ f &= \frac{15}{6} \\ f &= \mathbf{\frac{5}{2}} \text{ or } \mathbf{2.5} \end{aligned}$$

16.
$$\begin{aligned} 1\frac{3}{7}x - 23 &= 27 \\ 1\frac{3}{7}x - 23 + 23 &= 27 + 23 \\ 1\frac{3}{7}x &= 50 \\ \frac{7}{10} \cdot \frac{10}{7}x &= \frac{7}{10} \cdot 50 \\ x &= \mathbf{35} \end{aligned}$$

17.
$$\begin{aligned} 4x + 10 &= 2x - 14 \\ 4x + 10 - 2x &= 2x - 14 - 2x \\ 2x + 10 &= -14 \\ 2x + 10 - 10 &= -14 - 10 \\ 2x &= -24 \\ \frac{2x}{2} &= \frac{-24}{2} \\ x &= \mathbf{-12} \end{aligned}$$

18.
$$\frac{(-7) - (8)(-3) - (-1)^2}{(-1) + (-3)} = \frac{-7 + 24 - 1}{-4} = \frac{16}{-4} = \mathbf{-4}$$

19.
$$\begin{aligned} &100 - \{70 - 2[3 + 2(3^2)]\} \\ &= 100 - [70 - 2(3 + 18)] \\ &= 100 - (70 - 42) = 100 - 28 \\ &= \mathbf{72} \end{aligned}$$

20.
$$\frac{(-6de)(15d^2e)}{-15d^3e} = \frac{-90d^3e^2}{-15d^3e} = \mathbf{6e}$$

TEST 23

1.

	%	$
Original	100	45
Discount	20	d
New	80	n

$$\begin{aligned} \frac{100}{80} &= \frac{45}{n} \\ 100n &= 3600 \\ n &= \mathbf{\$36.00} \end{aligned}$$

2.

	Case 1	Case 2
Amount	20 kg	30 kg
Cost	$31	d

$$\begin{aligned} \frac{20}{31} &= \frac{30}{d} \\ 20d &= 930 \\ d &= \mathbf{\$46.50} \end{aligned}$$

3.

	%	$
Original	100	r
Discount	40	d
New	60	384

$$\begin{aligned} \frac{100}{60} &= \frac{r}{384} \\ 60r &= 38{,}400 \\ r &= \mathbf{\$640} \end{aligned}$$

4. $\dfrac{9 \times 10^7}{3 \times 10^4} = \mathbf{3 \times 10^3}$

5. Mean:

$$\frac{1.5 + 0.4 + 0.8 + 0.85 + 3.4}{5} = 1.39$$

Median: 0.85

Mean − median = 1.39 − 0.85 = **0.54**

6. $P(H, H, H, H) = \frac{1}{2} \cdot \frac{1}{2} \cdot \frac{1}{2} \cdot \frac{1}{2} = \mathbf{\frac{1}{16}}$

7.

$$\begin{array}{rl} \$3500 & \\ \times \quad 1.06 & \\ \hline \$3710 & \text{1st year total} \\ \times \quad 1.06 & \\ \hline \$3932.60 & \text{2nd year total} \end{array}$$

Interest earned: **$432.60**

8.
$$W_P \times 25 = 7.75$$
$$\frac{W_P \times 25}{25} = \frac{7.75}{25}$$
$$W_P = 0.31$$
$$W_P = \mathbf{31\%}$$

9. (a)
$$\text{Volume} = (60 \text{ cm})(30 \text{ cm})(45 \text{ cm})$$
$$= 81{,}000 \text{ cm}^3$$
$$= 81{,}000 \cancel{\text{cm}^3} \cdot \frac{1 \text{ L}}{1000 \cancel{\text{cm}^3}}$$
$$= \mathbf{81 \text{ L}}$$

(b)
$$81 \cancel{\text{L}} \cdot \frac{1 \text{ kg}}{1 \cancel{\text{L}}} = \mathbf{81 \text{ kg}}$$

10.
$$5 \cancel{\text{ft}^2} \cdot \frac{12 \text{ in.}}{1 \cancel{\text{ft}}} \cdot \frac{12 \text{ in.}}{1 \cancel{\text{ft}}} = \mathbf{720 \text{ in.}^2}$$

11. $a = 36$ yd, $b = 48$ yd, $c =$ length of shortcut
$$a^2 + b^2 = c^2$$
$$(36 \text{ yd})^2 + (48 \text{ yd})^2 = c^2$$
$$1296 \text{ yd}^2 + 2304 \text{ yd}^2 = c^2$$
$$3600 \text{ yd}^2 = c^2$$
$$60 \text{ yd} = c$$
Yards saved $= 84 \text{ yd} - 60 \text{ yd} = \mathbf{24 \text{ yd}}$

12. Volume of pyramid
$$= \frac{1}{3} \cdot \text{area of base} \cdot \text{height}$$
$$= \frac{1}{3} \cdot (20 \text{ m})(20 \text{ m}) \cdot 15 \text{ m}$$
$$= \mathbf{2000 \text{ m}^3}$$

13. $y = -x - 2$

x	y
0	–2
1	–3
2	–4

Check student work. Number pairs will vary.

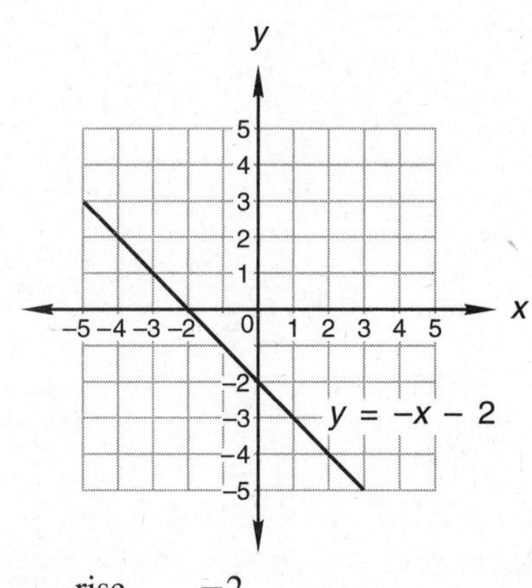

$$\text{Slope} = \frac{\text{rise}}{\text{run}} = \frac{-2}{2} = \mathbf{-1}$$

14.
$$A = \frac{1}{2}bh$$
$$30 = \frac{1}{2}(6)h$$
$$30 = 3h$$
$$\mathbf{10} = h$$

15.

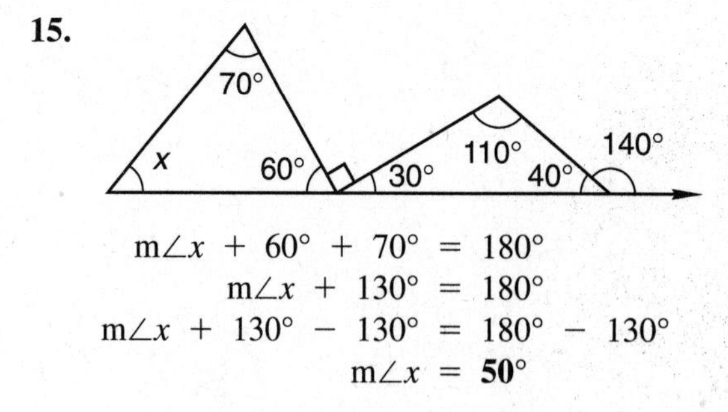

$$m\angle x + 60° + 70° = 180°$$
$$m\angle x + 130° = 180°$$
$$m\angle x + 130° - 130° = 180° - 130°$$
$$m\angle x = \mathbf{50°}$$

16.
$$1\frac{2}{5}w - 18 = 31$$
$$1\frac{2}{5}w - 18 + 18 = 31 + 18$$
$$1\frac{2}{5}w = 49$$
$$\frac{5}{7} \cdot \frac{7}{5}w = \frac{5}{7} \cdot 49$$
$$w = \mathbf{35}$$

17.
$$2x + 4 \leq 6$$
$$2x + 4 - 4 \leq 6 - 4$$
$$2x \leq 2$$
$$\frac{2x}{2} \leq \frac{2}{2}$$
$$\mathbf{x \leq 1}$$

–3 –2 –1 0 1 2 3 4

18.
$$(-2)^2 \cdot 3^{-2} = 4 \cdot \frac{1}{3^2}$$
$$= 4 \cdot \frac{1}{9}$$
$$= \mathbf{\frac{4}{9}}$$

19.
$$\frac{4x \cdot 4x}{4x + 4x} = \frac{16x^2}{8x} = \mathbf{2x}$$

20.
$$\frac{(-4) + (-7) + (3)(-3)}{(-6) - (-2)}$$
$$= \frac{-4 - 7 - 9}{-4} = \frac{-20}{-4} = \mathbf{5}$$